Biodiesel from Flowering Plants

Samuel Paul Raj • Pravin Raj Solomon •
Baskar Thangaraj

Biodiesel from Flowering Plants

Springer

Samuel Paul Raj
School of Energy, Environment and Natural Resources
Madurai Kamaraj University
Madurai, Tamil Nadu, India

Pravin Raj Solomon
Research and Development Centre
Velammal Medical College Hospital and Research Institute
Madurai, Tamil Nadu, India

Baskar Thangaraj
Pilot Plant Development and Training Institute
King Mongkut's University of Technology Thonburi
Bangkok, Thailand

ISBN 978-981-16-4774-1 ISBN 978-981-16-4775-8 (eBook)
https://doi.org/10.1007/978-981-16-4775-8

This Springer imprint is published by the registered company Springer Nature Singapore Pte Ltd.
The registered company address is: 152 Beach Road, #21-01/04 Gateway East, Singapore 189721, Singapore

Preface

Works were initiated during the past three decades on biodiesel being prepared from vegetable oils as raw material so as to combat the pressure built on the petro-diesel which in its quantity is actively dwindling. Ever since such attempts were unfurled, there is a rush on all types of oil including priced edible oils. Thus, the approach competes the edible oil for domestic consumption instead of actively relieving the problems associated with the fast vanishing resource. Meanwhile search commenced predominantly from developing countries to rope in location-specific obscure non-conventional plants in this sector whose oleaginous seeds may provide oil and serve raw material for biodiesel. Some of such species offer bright scope as they can be cultivated in wastelands and can extend job opportunities to the people closer by. The oil obtained from many of such species is esterified to reduce the free fatty acid and is followed with transesterification. This transesterification yields ester which is referred as biodiesel. This biodiesel is often selectively utilized in engines as fuel as the original viscosity and density of the oil gets moderated. In many countries, it has become mandatory to blend the biodiesel with regular petro-diesel at the rate of 5–20% so as to safely engage it in engines without any mechanical modification. Biodiesel by virtue of its rich oxygen in its molecule helps to reduce carbon monoxide and hydrocarbon in the exhaust though NO_x level is invariably high. Research is in progress to circumvent some of the issues such as low temperature performances, instability to oxidation and marginally low heating value apparently due to the stuck in of oxygen in the molecule which stoichiometrically reduced the share of elemental carbon and oxygen.

A total of 96 plants in which 85 obscure (non-conventional) plant species geographically distributed all over the world were studied and reported by different energy researchers focusing on the potentialities of such species in biodiesel production. Among them, information brought out on 12 species is highly hazy and non-comprehensive, and therefore it is observed to be ill defined for favour of consideration in this book. The information on the remaining 84 plant species is presented with suitable illustration in this book. Though the process involved in the biodiesel preparation is fairly common for all the species, the quantity of the reagents involved and the process parameters thereon are specific to each species which necessitates a short account of the process under each chapter. The engine performances reported in this book for each biodiesel are based on the availability of information in the relevant publication.

Thus, there is no omission of species hitherto reported. The 84 plant species dealt in this book are from 77 genera and 41 taxonomical families. Due to the diversities indicated above, the plant species are grouped as families and presented as separate chapters. Each chapter is provided with a very abridged information on the identification features, habit and habitat of the plant along with a note on the oil extraction procedure, biodiesel preparation and physicochemical properties of the oil and biodiesel along with engine performance if studied already by any authors. Besides there were four chapters on Introduction, History of biodiesel, Catalysts for transesterification and Standards for biodiesel. A plethora of catalysts are employed in biodiesel preparation in accordance with the type of plant species and the oil thereon which compels the need for the inclusion of a general chapter on catalyst. As the standards of the biodiesel promulgated among different countries differ, a separate chapter on the above is incorporated in this book. These two general chapters last among the above four are closely associated with the total textual content of this book.

Madurai, Tamil Nadu, India Samuel Paul Raj
Madurai, Tamil Nadu, India Pravin Raj Solomon
Bangkok, Thailand Baskar Thangaraj

Acknowledgement

The limitless encouragement of Dr. V. Annielet Bahmi and the typing assistance received from P. Jayasri and C. Latha of Madurai Kamaraj University are sincerely acknowledged.

Contents

About the Authors

Samuel Paul Raj Graduated in Biology from Madurai Kamaraj University, India, as a rank holder in 1972, and obtained his Master's degree in 1974 from Annamalai University, India, as a university best student and medalist. During his PhD research on Environmental Biology (1983) at Tamil Nadu Agricultural University, India, he received a gold medal for being the best researcher. He became Professor and Head of the School of Energy Environment and Natural Resources of the Madurai Kamaraj University in 1993 and its Senior Professor in 2004. He was made Scientific Adviser to the International Foundation for Science, an organization of 92 countries, in 1985. He received JC Bose Memorial Award in 2010 and became a fellow of the Tamil Nadu Academy of Sciences in 2014. He has authored 11 books, produced 31 PhDs, and 53 publications. He also served in the Editorial Board of two international journals.

Pravin Raj Solomon Obtained his graduate, postgraduate, and doctoral degrees in Biotechnology from Madurai Kamaraj University, India. He was a Postdoctoral Associate of the Yale University, USA. He served as Senior Assistant Professor for 1 year in the Department of Biotechnology at Mepco Schlenk Engineering College in India and for 3 years at MAHSA University, Malaysia. He also served SASTRA—Deemed to be university, Thanjavur, India, for 3 years as an Assistant Professor (Research). He has 26 research publications and authored 2 books.

Baskar Thangaraj Obtained his two Master's degrees in Energy Sciences and Physics from Madurai Kamaraj University, and a Master of Philosophy in Energy Studies from the Gandhigram Rural Institute—Deemed to be University, India. He completed his doctoral (DSc) degree on bionanocatalysts for biodiesel production at the Tsinghua University, Beijing, China. He continued his postdoctoral studies at Jiangsu University, China. Currently, he is at King Mongkut's University of Technology Thonburi, Thailand, as a Postdoctoral Fellow. He has published 23 research papers and authored a chapter in the book *Current and Future Perspectives on Lipid-Based Biofuels*.

Introduction

1

The energy scenario of the world is disquieting with the price of oil increasing at a fast pace as this non-renewable reserve is dwindling rapidly. Intensive use of fossil fuels naturally increases the concentration of CO_2 in the atmosphere and contributes to the climate change. The anthropogenic emission of CO_2 from coal fired thermo-electric plants is partly responsible for the increased level of CO_2 emission. Flue gases from fossil power plants generate rich quantity of CO_2, NOx and SOx depending upon the type of fuel being used on the combustion process. It is felt that CO_2 level if exceeds 450 ppm in the atmosphere will have severe impact on sea levels, global climate patterns and survival of many species. Current technologies for mitigating CO_2 are through storage, utilization and fixation. The most common method of capturing CO_2 is through monoethanolamine scrubbing.

It is also stored in the underground geological cavity at a maximum depth of 800 m as well as in surface sequestration or through deep sea injection (Ohsumi 1995), desiccant adsorption and also through molecular sieve technology (Judkins and Burchell 2001). Unfortunately, these storage methods suffer from major challenges due to the difficulties in the separation and compression of CO_2, inadequate pumping methods, uncertainties in the long term stability and the capital investment involved (Herzog 2001).

Of late two important issues are confronting the minds of energy planners. The foremost between the two is environmental quality maintenance and sustainability. The other is the identification of a suitable path to reach energy security. In most of the situations, both the issues are inseparably linked and it is seldom possible to deal any one in isolation without the interaction of the other. As fossil fuels are depleting fast it has become increasingly necessary to identify alternate sources of energy. Any alternate sources if they do not cause any change in the user structure mechanism with a favourable character of not impinging upon the environmental quality are likely to get good reception. As the cost of petroleum crude has taken a vantage position in governing the economy of the whole world in quick succession, it has a spiraling impact on other associate sectors such as domestic energy and the quality of

S. P. Raj et al., *Biodiesel from Flowering Plants*,
https://doi.org/10.1007/978-981-16-4775-8_1

life. As, per capita energy expenditure has increased over a period, CO_2 emission and other gases connected with the global warming also increased in the recent past. Due to the growing concern on global warming and the elevated level of CO_2 in the atmosphere (Kondili and Kaldellis 2007; Leshkov et al. 2007), the United Nations promoted the Kyoto Protocol (1997) with the objective of reducing the greenhouse gases by 5.2% over the emission level of the year 1990. More than 170 countries have ratified the accord (Gutierrez et al. 2008).

Today supply of liquid fuels world over is almost dependent on petroleum. The demand for petroleum fuel is likely to increase since the number of automobiles on the road increased. As a consequence, the petroleum has become more expensive due to supply and demand principle and has become an inevitable source. Expectedly it has penetrated into political and military conflicts.

Petroleum crudes are continuously extracted from underground starting from the period of Colonel Edwin Drake who drilled the world's first oil-well in 1859. The International Energy Agency has reported that the world's primary energy need is projected to grow by 55% between 2005 and 2030 at an average annual rate of 1.8% per year.

Fossil fuel may continue to remain as the dominant source of primary energy till there is a serious intervention. It is also reported that the world's demand for oil may shoot up by 60% in 2025 while production level will go back to that of the year 1985. Transportation is one of the fastest growing sectors using bulk of the primary energy and thus the fuel need of the transportation sector is needed to be attended (Khan et al. 2009). The development of CO_2 neutral fuels is one of the most urgent challenges facing our society. The synthesis report of IPCC (2007) and the Economics of Climate Change (Stern 2006) made clear the need for a CO_2 neutral fuel. These reports also brought to light the causes and effects of climatic change. The level of CO_2-*e* (*e* is equivalent contribution of all greenhouse gases) has already reached 450 ppm. This threshold level has reached 10 years earlier than that was predicted. Leading countries (e.g. European Union) were expected to impose CO_2 emission restriction targets of 10–20%. Initiations are made by many leading countries to reduce 30% of CO_2 generation if agreed by all other countries. However such reduction targets are not sufficient to stabilize the CO_2 to an accepted level of safe zone, i.e. below 450 ppm CO_2-*e* since at high level several negative consequences are likely. It is also suggested to have a reduction target of 50–85% by 2050 which is considered as an upheaval task and is considered as a global challenge.

Terrestrial vegetation and soil as well as deep sea are the natural sinks of CO_2. The process of increasing the rate of CO_2 sequestration through enhancement of natural sinks includes afforestation, ocean fertigation, rock weathering enhancement, algal culture through photobioreactors and artificial photosynthesis. The CO_2 assimilation rate of trees varies according to the types of tree and geographical location. Although there are no listed adverse effects of energy plantation, locating and dedicating arable land for afforestation programme may bring disagreement among population especially on water availability and shrinkage to food production.

Biomass is the organic material derived from the reaction between carbon dioxide, water, sunlight and other nutrients via photosynthesis. The solar energy absorbed by photosynthesis is converted into chemical bonds of the structural components of biomass. Biomass is one of the organic sources needed to be involved in energy sector as it is renewable and neutral with regard to carbon dioxide emissions. (Cook and Beyea 2000; Coyle 2008). If biomass is processed suitably and converted in to a value added material and used diligently, it can effectively offset the usage of conventional fossil fuels. Biomass has always been a major source of energy for mankind and is presently estimated to contribute 13% of the world's energy supply. The share of biomass energy is relatively high in developing countries (Hall et al. 1991). Energy from biomass if opted is required to contribute a stable energy supply. Biomass to a certain extent can be derived from the cultivation of dedicated energy crops or by optimal harvesting of energy plantations. Garbage from homes and food industries is often used as a raw material for energy extraction through thermo-chemical liquefaction route.

Present trend has invited the planners to exercise conservation measures and to search alternate fuels. It is in this context biodiesel is introduced which is most widely accepted as an alternative fuel for diesel engines due to its technical, environmental and strategic advantages (Bastianoni and Marchettini 1996; Mofijur et al. 2019). Besides, biodiesel does not contribute beyond a level of net carbon dioxide or sulphur to the atmosphere and emits lesser gaseous pollutants than conventional diesel fuel (Lang et al. 2001; Antolin et al. 2002; Vicente et al. 2004). It is also characterized by the rapid biodegradability, reduced toxicity and improved lubricity in comparison with that of the conventional diesel. In addition to this, biodiesel can be straightly used in its pure form or may be blended with the petro-diesel so as to feed it in engines without making any significant modification to the receiving engine as per European Parliament Directive (Benjumea et al. 2008). Global biodiesel production and its utilization in a modest way have taken lead during the last three decades not only to reduce the use of diesel but also to limit the rate of CO_2 emission mostly in the transportation sector. Besides its primary role is to have a true substitute for the conventional fuel.

Any alternatives to diesel should be in a liquid-form and should be compatible to the engines used at present, economically competitive and available in sufficient quantities. Incidentally the engine designed by Rudolf Diesel (1858–1913) ran for the first time in 1893 and two of the fuels then proposed for locomotive were plant-derived oil and pulverized coal (Shay 1993; Altin et al. 2001).

It is in this context research and development has made a headway in producing liquid fuel from the vegetable oils obtained from the flowering plants (Angiosperms). Oils from flowering plants are invariably viscous due to the inherent nature of the triglycerides and certain of the impurities present in it. Therefore, raw oil could not be used directly as an alternative to petro-diesel in *as is where is* form. Many processes were identified to make a change in the consistency of the oil so as to make it amenable for its use either in full or in part (blend) with petro-diesel. Among them transesterification is considered suitable. In transesterification the glyceride of the oil is made to react with an alcohol (typically methanol or ethanol) in the presence of a catalyst, forming fatty acid alkyl ester. Consequently, during the last two decades of the twentieth century, large quantity of edible oil was transesterified to be used in engines. Such massive scale conversion of edible oil for energy-use was eventually viewed as a competitive activity to cooking. Incipient instruction from planners indicates that the biodiesel sourced from edible oils has to be reduced drastically. Thus the production of first generation biodiesel (from edible food crop) gave way for the second generation biodiesel (from non-edible plant sources). Due to the reason cited above the procedure of blending the biodiesel at a maximum of 20% with petro-diesel came into force. This has certainly helped to reduce the usage of petro-diesel.

In continuation of the above, a section of the scientists started investigation on the scope of obtaining oil from obscure non-edible plants hitherto not explored. Consequently, the density of research publications on biodiesel from edible oil dwindled since the start of the first decade of this century. From then onwards, the count of publications on various facets of biodiesel from non-edible oil showed an increase. Such of those new publications mostly generated from the laboratories of the developing countries. These publications brought to light the nature of various obscure plant species whose oils are relevant to biodiesel. Studies on many of such species are cursory, but clearly figure out their usefulness as raw materials to biodiesel as non-conventional sources. In certain of these species harvest of the oil bearing seeds, extraction of the oil, transesterification, characterization of the oil and biodiesel and the engine performance (if available) are reported. On certain species there is minimum information but if works are pursued in a right direction their economical importance on biodiesel may emerge. Short cycle in the production of oil, utilization of non-edible source, emission of low quantity of greenhouse gas, no competition to food resources are the rock-bottom principle behind the selection of certain of the flowering plants (Angiosperms) for this purpose (Ahmad et al. 2011; Hajjari et al. 2017).

In the present book, crisp details on the identification, geographical distribution, vernacular names, oil extraction procedures, physicochemical features of the oil, biodiesel production, the properties of the biodiesel (ester) and the engine performance if any in *the first and second generation* biodiesel species are presented.

1.1 Uniqueness of This Book

Many of the books on biodiesel which came in to being during the last two decades tell us the objective evaluation of biodiesel processes, prospects, environmental benefits, combating carbon foot print, problems in engine performances, benefits of blending with petro-diesel, economic implications, properties of the raw materials and the biodiesel, shelf life and loss of stability against rancidity.

When biodiesel took the centre stage in the recent past substituting the petro-diesel in full or part there is large scale encroachment in the edible oil reserve as raw material which is already in short supply and thus biodiesel started threatening the availability of cooking oil. Therefore, in the next stage there were concerted efforts in identifying new and non-conventional candidate plant species having oleaginous seeds which were hitherto not explored. Many publications dealing with new and obscure plant species were not consolidated and therefore such information did not form in to any contemporary books.

It is in this context the proposed book opened a virtual inventory of all the flowering plants reported to have scope for biodiesel preparation. There were reports on 96 flowering plants (Angiosperms), out of which details obtained on 12 species are highly sketchy and are observed to be not worth recording. The remaining 84 species are documented in this book.

Each chapter is invariably provided with a hand-drawn figure of the plant and a map indicating its geographical distribution. The entire text is treated with a simple language so that student from different strata will be able to cull out information easily since the subject biodiesel is a matter of multiple specialization. It is opined that this book invites demand among readers since such type of book has not yet been brought out by any of the publishers.

The book is prepared by three authors and the senior author has 38 years of research experience in this field and has retired from service recently as a Senior Professor and Chairman of the Dept of Energy, Madurai Kamaraj University, India.

References

Ahmad AL, Yasin NHM, Derek CJC, Lim JK (2011) Microalgae as a sustainable energy source for biodiesel production: a review. Renew Sust Energy Rev 15(1):584–593

Altin R, Cetinkaya S, Yucesu HS (2001) The potential of using vegetable oil fuels as fuel for diesel engines. Energy Convers Manag 42:529–538

Antolin G, Tinaut FV, Briceno Y, Castano V, Perez C, Ramirez AI (2002) Optimization of biodiesel production by sunflower oil transesterification. Bioresour Technol 83(2):111–114

Bastianoni S, Marchettini N (1996) Ethanol production from biomass: analysis of process efficiency and sustainability. Biomass Bioenergy 11:411–418

Benjumea PN, Agudelo JR, Agudelo AF (2008) Basic properties of palm oil biodiesel-diesel blends. Fuel 87:2069–2075

Cook J, Beyea J (2000) Bioenergy in the United States: progress and possibilities. Biomass Bioenergy 18(6):441–455

Coyle W (2008) The future of biofuels: a global perspective. Amber Waves 5(5):24–29

Gutierrez R, Gutierrez-Sanchez R, Nafidi A (2008) Trend analysis using nonhomogeneous stochastic diffusion processes, emission of CO_2: Kyoto protocol in Spain, Stoch. Environ Res Risk Assess 22(1):57–66

Hajjari M, Tabatabaei M, Ghbashlo MA, Ghanavati H (2017) A review on the prospects of sustainable biodiesel production: a global scenario with an emphasis on waste oil biodiesel utilization. Renew Sust Energy Rev 72:445–464

Hall DO, Mynick HE, Williams RH (1991) Cooling the greenhouse with bioenergy. Nature 353 (6339):11

Herzog HJ (2001) What future for carbon capture and sequestration? Environ Sci Technol 35 (7):148–153

Judkins RD, Burchell TD (2001) CO_2 removal from gas streams using a carbon fiber composite molecular sieve. In: Proc. first national conference on carbon sequestration, Washington, DC, 14–17 May

Khan SA, Rashmi, Hussain MZ, Prasad S, Banerjee UC (2009) Prospects of biodiesel production from microalgae in India. Renew Sust Energy Rev 13(9):2361–2372

Kondili EM, Kaldellis JK (2007) Biofuel implementation in East Europe: current status and future prospects. Renew Sust Energy Rev 11(9):2137–2151

Lang X, Dalai AK, Bakhshi NN, Reaney MJ, Hertz PB (2001) Preparation and characterization of bio-diesel from various bio-oils. Bioresour Technol 80(1):53–62
Leshkov RY, Barrett CJ, Liu ZY, Dumesic JA (2007) Production of dimethylfuran for liquid fuels from biomass-derived carbohydrates. Nature 447:982–985
Mofijur M, Rasul MG, Hassan NMS, Nabi MN (2019) Recent development in the production of third generation biodiesel from microalgae. Energy Procedia 156:53–58
Ohsumi T (1995) CO_2 disposal options in the deep sea. Mar Technol Soc J 29(3):58–66
Shay EG (1993) Diesel fuel from vegetable oils: status and opportunities. Biomass Bioenergy 4 (4):227–242
Stern N (2006) The economics of climate change. Cambridge University Press, Cambridge
Vicente G, Martinez M, Aracil J (2004) Integrated biodiesel production: a comparison of different homogeneous catalysts systems. Bioresour Technol 9(3):297–305

History of Biodiesel

2

Detection of a liquid fuel which was later called as diesel in a deep well at Titusville village of Pennsylvania State of USA by Edwin Laurentine Drake (1819–1880) during the year 1859 and the development of a crude internal combustion engine energized with liquid fuel by Rudolf Christian Karl (Carl) Diesel (1858–1913) are contemporary incidents (Figs. 2.1 and 2.2).

In 1893, Rudolf Diesel, developed an internal combustion engine. That was designed to have the whole combustion to take place in a cylinder. Consequently, he published a book on the title *Theorie und Konstruktion eines rationellen Wäremotors* (*Theory and Construction of a Rational Heat Motor*) to replace the steam engine (Springer Verlag, Berlin). The engine he developed in accordance with his theory and design fetched the name diesel engine. Rudolf Diesel obtained patents for his design in Germany and in other countries including the USA. The high compression and thermal efficiency are the special features of his engine which differentiated from hot bulb engine. Of course there was an earlier patent offered to Herbert Akroyd Stuart on his hot bulb compression engine which had a pressure of only 90 PSI with 12% thermal efficiency. Rudolf Diesel in his new patent was able to raise the pressure to 500 PSI with more than 50% thermal efficiency. In the same year Rudolf Diesel's *Internal combustion engine* was granted patent recognition (US Patent number: 542846 and 608845) (Diesel 1895, 1898).

Later in 1894, he improved the above engine and carried out series of trials in it. Most unfortunately, he was seriously injured during the experimentation. That experiment with such dangerous consequence thus proved that diesel could be combusted internally so as to release enormous quantity of energy. Consequent to the recovery from accident he spent 2 years making many improvements in the engine already developed. He understood the need for enhancing the thermo dynamical efficiency of the steam engines which was 10–15%. His continued efforts helped to develop a new version of the above engine. In that engine he could achieve 75% efficiency. However, this model also underwent structural evolution. In 1898,

S. P. Raj et al., *Biodiesel from Flowering Plants*,
https://doi.org/10.1007/978-981-16-4775-8_2

Fig. 2.1 **Edwin Laurentine Drake** (1819–1880). Edwin Laurentine Drake was born on the 29th March, 1819 in Greenville, New York. Drake began drilling in 1858 and stroke oil at a depth of 21M on August 27, 1859. This oil exploration made the communities of Titusville and other north-western Pennsylvania prosperous. In 1873, the State of Pennsylvania granted him an annuity. Drake's work is considered as the foundation of the modern oil industry

Fig. 2.2 **Rudolf Christian Karl (Carl) Diesel** (1858–1913). Rudolf Diesel, the father of diesel engine was born in Paris, France on the 18th March, 1858 as the second of three children to Theodor Diesel and Elise Strobel. Diesel's family migrated to London in 1870. In 1875 he received a merit scholarship from the Royal Bavarian Polytechnic of Munich, and graduated with a highest academic honour. In January 1880, he commenced working with Prof. Carl von Linde. Rudolf Diesel met a tragic end on the 29th September, 1913 in the English Channel. On the fateful day he boarded the SS Dresden on a trip to attend the opening of a factory in Lpswich.. But he never reached the destination. His body was found floating by a coast guard almost 3 days later

he developed a third version which was later used to energize pipeline, power plants, automobiles, marine engines, mines, oil fields and factories.

Subsequently the above engine without any alterations was fueled with peanut oil. In 1912 Rudolf Diesel spelled out that the diesel engine can be fed with vegetable oils and would help the development of agriculture of any country which uses it. The use of vegetable oils for engine fuels may seem insignificant today. But such oils may become in course of time as important as petroleum and the coal tar products of the present time.

Fig. 2.3 Rudolf Diesel's image in a metal badge at Augsburg, Germany

The National Biodiesel Board of USA celebrates the National Biodiesel Day on the 18th of March every year in kind remembrance of Rudolf Diesel's birthday. Similarly, August 10 was declared as the International Biodiesel day since on that day only he exhibited the flywheel type of engine in the world Exhibition Fair in Paris. The German Institute for invention has instituted an award in honour of Rudolf diesel and is being in force since 1953. Rudolf Diesel won the *Grand Prix* for his innovation in that exhibition.

Between 1901 and 1934, the four-stroke trunk type piston engine was introduced. It was the largest diesel engine power plant ever developed at that time. Despite the application of diesel in many type of engines, application of vegetable oil in internal combustion engines became popular in many countries during 1920 to 1930. Such engines were richly used in the Second World War. During those period, many countries such as Belgium, France, Italy, The United Kingdom, Portugal, Germany, Brazil, Argentina, Japan and China started using vegetable oil as fuel. After the Second World War when the desire for modern vehicles increased, the industries once again switched over to petroleum fuel and thus biodiesel became slowly unpopular.

In 1935, the first heavy duty four stroke diesel engine was popularized and in 1952 improvements were made on that. The Queen Elizabeth II had nine such engines generating a combined output of 132,000 bhp.

However, during the operation, users encountered problems such as poor atomization of the fuel resulting into the choking of the injection nozzles, combustion chamber and valve systems. To combat such problems many steps were made which included the raising of the temperature of the oils or blending the vegetable oil either with diesel or with ethanol or passing it through pyrolysis so as to denature the oil to suit the architecture of the engine being in the field. Very often the engine was being first started using diesel and after a few minutes of operation the engine was switched over to vegetable oil (Fig. 2.3).

As the viscosity of the vegetable oil was found to be a major constraint in using the same in internal combustion engines, attempts were made to reduce its viscosity. During 1800 the glycerin content of the oil was distilled out so as to use the same in the manufacture of soap. The other fraction is used as a fuel. Such fuel was derived from many sources such as peanut, hemp, corn and tallow during the past, whereas

Fig. 2.4 Stamps, cover and coin issued in commemoration of Rudolf Diesel in Germany

current sources are from soybean, rapeseed, canola, corn, waste oil from food industry, forest products, sugarcane or by-products of slaughter houses (Fig. 2.4).

Two chemists, E. Duffy and J. Patrick during the year 1853, successfully demonstrated the removal of glycerol from the vegetable oil so as to reduce the viscosity (Feofilova et al. 2010). This process is referred as transesterification. The premier work of those chemists went unnoticed in the history of science.

Though work on transesterification process was initiated as early as 1846 when Rochieder described glycerol preparation through ethanolysis of castor oil (Formo 1979), on the 31st August, 1937 the procedure for the transformation of vegetable oils to fuel came open through a patent (Patent number: 422877) offered to a Belgium scientist G. Chavanne of the university of Brussels (Belgium) (Chavanne 1938). According to this patent, vegetable oil was subjected to alcoholysis so as to separate the fatty acids from glycerol by replacing the glycerol with short linear alcohols. There is no record of the use of mono alkyl esters as a fuel prior to the works of Chavanne in 1937 (Balasubramanian and Steward 2019). Following the above development France, Belgium and the United Kingdom started showing great interest in biodiesel from waste cooking oils collected from their settlements (Fig. 2.5).

Expedito Parente of Brazil developed an industrial process to yield a final product called biodiesel. The biodiesel prepared had qualified to meet the standard. In South Africa active research and development works were in progress to transesterify sunflower oil for large scale application.

A group of researchers investigated the important reaction conditions and the parameters involved in the alcoholysis of fish oil, tallow, soybean, rapeseed, cotton seed, sunflower, safflower, peanut and linseed oils (Fuls and Hugo 1981).

Posorske et al. (1984) used immobilized lipase as catalyst for biodiesel production. The ethanolysis of sunflower oil with lipozyme (immobilized 1,3-specific *Mucor miehei* lipase) on a macroporous anion exchange resin was investigated. Oil to ethanol molar ratio, temperature, added water content, and amount of enzyme were optimized (Selmi and Thomas 1998).

Harrington and Catherine (1985) used sulphuric acid as a catalyst. Freedman et al. (1984) applied alkaline metal alkoxides as catalyst. From 1984 to 1990 many researchers attempted to produce biodiesel using acid and alkaline catalysts. Later non-ionic base catalysts were tried. Schuchardt et al. (1998) evaluated the use of heterogeneous catalysts for biodiesel production. In 1999 for the first time the

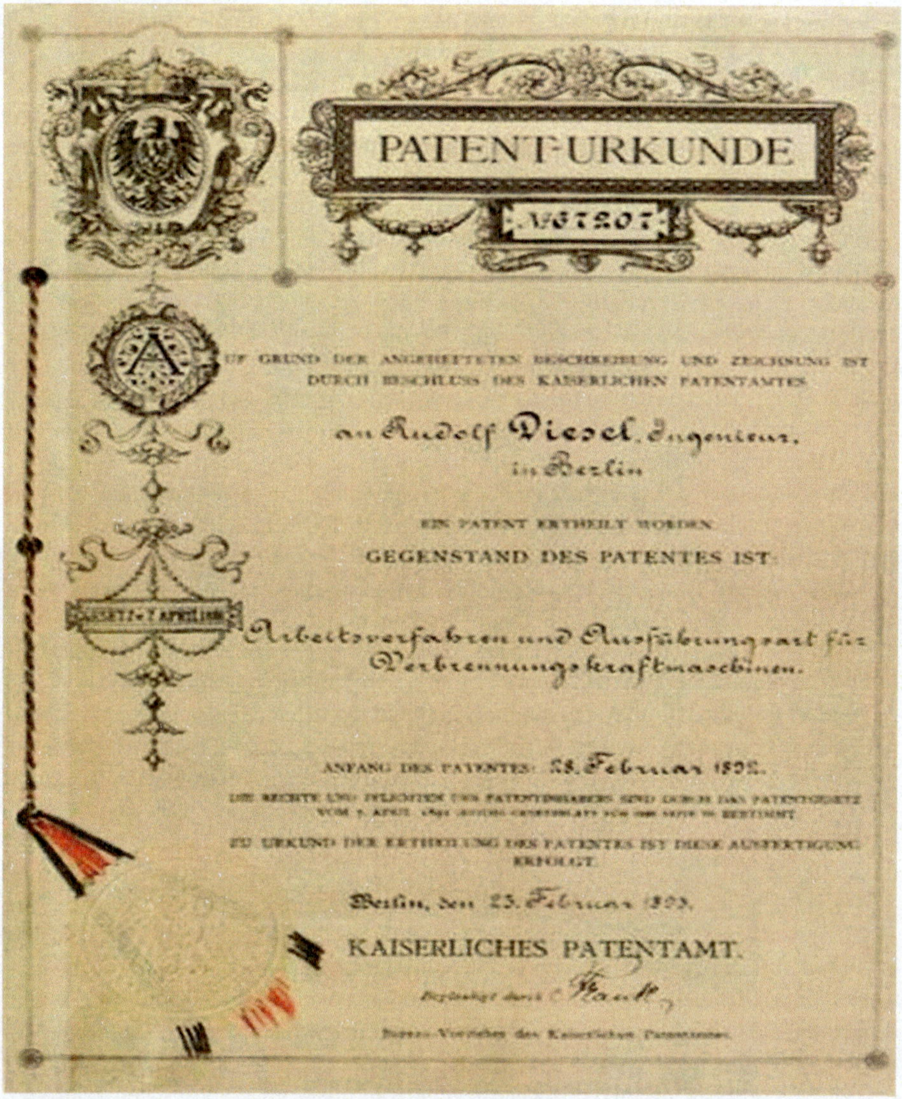

Fig. 2.5 Copy of the Patent document issued to Rudolf Diesel

methodology of supercritical methanol in biodiesel production came in to being for consideration for large scale application.

During the year 1987 an Australian company Gaskohs adopted the technology perfected by the South African Agricultural Engineers. Consequently, a pilot plant was commissioned so as to process a total of 30,000 tonnes of rape seed oil per annum. Subsequently in 1990s processing units came into being in large numbers. Many plants were commissioned in European countries including the Czech

Republic, Germany and Sweden. France started producing biodiesel from rapeseed oil and served it after mixing it at 5% level with diesel.

Isigıgur et al. (1994), Lang et al. (2001), Mittelbach and Gangl (2001) and Goodrum (2002) continued their research to amend the vegetable oil so as to make it suitable to feed in combustion engines. In 1999, Ma and Hanna attempted to produce alternative fuel by direct use of vegetable oils and blends of the oils. But it was observed to be not satisfactory and impractical in direct and indirect diesel engines due to some problems intervened. Chemical and biological catalysts were attempted to activate the biodiesel production. In addition, short chain primary alcohols were preferentially used. Freedman et al. (1986) investigated the transesterification reaction variables affecting the yield and purity of the alkyl esters produced from cottonseed, peanut, soybean and sunflower oils. These variables include molar ratio of alcohol to vegetable oil, types of catalysts (alkaline vs acidic), temperature and degree of refinement of the vegetable oil. Graboski and McCormick (1998) preferred methanol in the transesterification of vegetable oils and fats as methanol is less expensive than ethanol. Alkali catalyst was proposed in 1984 to make large scale production of biodiesel from triglyceride with methanol (Clark et al. 1984).

Many workers attempted lipase immobilization on non-magnetic materials. As a consequence, supermagnetic microspheres came in to being as a carrier for immobilizing Candida *rugosa* lipase and used as a catalyst for biodiesel production. In 2004 magnetic nanoparticles were also employed in enzyme immobilization.

In mass transportation systems biodiesel blend was used even as high as the 30% with diesel. Peugeot as early as 2004 developed engines to use biodiesel at 50% level by blending it with diesel (www.psa-peugeot-citroen.com/fr retrieved in 1st July, 2004). Renault heavy-duty vehicles had been using fuel blends of 30% biodiesel in conventional diesel (van Walwijk 2005). The maximum consumption was in the transport sector. The European Union consumed 13.9 million metric tonnes, during the year 2018, quantity of the biodiesel (eu.support@statista.com).

Many workers now suggest that waste cooking oil discharged from food industry is the best source of raw material to produce biodiesel. However, the quantity of such waste oil available is paradoxically low and hence may not be able to meet a substantial portion of the demand. Of late animal fats which are the by-products of meat industry are being used as a raw material in multi-feedstock biodiesel facilities.

The feasibility of the glycerolysis of vegetable oils with crude glycerol derived from the transesterification of vegetable oils and animal fats was carried out (Muniyappa et al. 1996).

Few Pacific Islands are utilizing coconut oils especially when the ambient temperature does not fall below 17 °C. Mean time Japanese Government offered tax exemption if straight vegetable oil was used in automobiles.

British Train Operating Company has claimed to have run the first biodiesel train with a fuel blend comprising 80% diesel and 20% biodiesel. It was reported to limit 14% of direct emission.

The Royal Train Services commenced working on pure biodiesel on the 15th September 2007. His Royal Highness The prince of Wales and James Hygate the Managing Director of Green Fuels were the first to travel in that train.

The Governor of Massachusetts Deval Patrick ordered that the fuel required for home heating be with biodiesel at the rate of 2% from first July 2010 and 5% from the year 2013.

Of late in 2021 The Volkswagen Group considered the use of B5 and B100 made of rape seed oil in its vehicles as such fuel blends are in accordance with the EN 14214 standard (https://cleantechnica.com/2021/01/09/volkswagen-powers-transport-ships-with-bio-diesel-from-goodfuels).

References

Balasubramanian N, Steward KF (2019) Biodiesel: history of plant based oil usage and modern innovations. Substantia 3(2):57–71

Chavanne CG (1938) Belgian Patent 422,877, Aug 31, 1937; Chem Abs 32:4313

Clark SJ, Wagner LW, Schrock MI, Piennaar EG (1984) Methyl and ethyl soybean esters as renewable fuels for diesel engines. JAOCS 61:1632–1638

Diesel R (1895) Method of and apparatus for converting heat into work. US Patent 542,846. http://www.google.com/patents/US542846

Diesel R (1898) Internal combustion engine. US Patent 607,845. http://www.google.com/patents/US608845

Feofilova EP, Sergeeva YE, Ivashechkin AA (2010) Biodiesel-fuel: content, production, producers, contemporary biotechnology (review). Appl Biochem Microbiol 46(4):369–378

Formo MW (1979) Physical properties of fats and fatty acids. Bailey's industrial oil and fat products, vol 1, 4th edn. Wiley, New York, p 193

Freedman B, Pryde EH, Mounts TL (1984) Variables affecting the yields of fatty esters from transesterified vegetable oils. JAOCS 61:1638–1643

Freedman B, Butterfield RO, Pryde EH (1986) Transesterification kinetics of soybean oil. JAOCS 63:1375–1380

Fuls J, Hugo FJC (1981) On farm preparation of sunflower oil esters for fuel. In: Third international conference on energy use management, pp 1595–1602

Goodrum JW (2002) Volatility and boiling points of biodiesel from vegetable oils and tallow. Biomass Bioenergy 22:205–211

Graboski MS, McCormick RL (1998) Combustion of fat and vegetable oil derived fuels in diesel engines. Prog Energy Combust Sci 24(2):125–164

Harrington KJ, Catherine DV (1985) A comparison of conventional and in situ methods of transesterification of seed oil from a series of sunflower cultivars. JAOCS 62:1009–1013

Isigıgur A, Karaosmonoglu F, Aksoy HA (1994) Methyl ester from safflower seed oil of Turkish origin as a biofuel for Diesel engines. Appl Biochem Biotechnol 45(46):103–112

Lang X, Dalai AK, Bakhshi NN, Reaney MJ, Hertz PB (2001) Preparation and characterization of bio-diesels from various bio-oils. Bioresour Technol 80:53–63

Ma F, Hanna MA (1999) Biodiesel production: a review. Bioresour Technol 70(1):1–15

Mittelbach M, Gangl S (2001) Long storage stability of biodiesel made from rapeseed and used frying oil. JAOCS 78:573–577

Muniyappa PR, Brammer SC, Noureddini H (1996) Improved conversion of plant oils and animal fats into biodiesel and co-product. Bioresour Technol 56:19–24

Posorske LH, Le Febvre GK, Miller CA, Hansen TT, Glenvig BL (1984) Process considerations of continuous fat modification with an immobilized lipase. J Am Oil Chem Soc 65:922–926
Schuchardt U, Sercheli R, Vargas RM (1998) Transesterification of vegetable oils: a review. J Braz Chem Soc 9(1):199–210
Selmi B, Thomas D (1998) Immobilized lipase-catalyzed ethanolysis of sunflower oil in a solvent-free medium. J Am Oil Chem Soc 75(6):691–695
van Walwijk M (2005) Biofuels in France 1990-2005 PREMIA report r PREMIA Number TREN/04/FP6EN/S07.31083/503081

Anacardiaceae

3

The members of the family Anacardiaceae are distinctive in being trees, shrubs, lianas or perennial herbs. This family consists of around 500 species of trees and shrubs, rarely sub-shrubs and lianas. They occur in tropical and subtropical regions and rarely in temperate regions. The genus *Pistacia* (which includes the pistachio and mastic tree) previously placed in Pistaciaceae is now listed in the family Anacardiaceae (Tingshuang et al. 2008). *Pistacia chinensis* is the only species reported to have been linked with biodiesel production.

3.1 Chinese Pistache (*Pistacia chinensis*)

The non-edible oil from the seeds of Chinese pistache *Pistacia chinensis* has become the raw material for biodiesel. *P. chinensis* is a moderate sized tree growing to a maximum height of 20 m and belongs to the family Anacardiaceae. It is a native of Eastern Asia, China, Philippines and Taiwan. It is considered as a sleeper weed in Southern Australia (Fig. 3.1). It prefers to establish at an elevation of 1000–1200 m above means sea level and grows in moist, fertile and well-drained soil. Around ten drooping leaves of 20–25 cm long are available in the rachis. Flowers are produced in panicles of 10–20 cm long. The fruit is a drupe of oval shape with a diameter of 0.7 cm (Fig. 3.2). The fruit is red to orange or pink in colour. This species is locally referred as Pistachio or Chinese pistache.

3.1.1 Oil Extraction

The fruits are collected and dried repeatedly in hot sun, milled and dried again to relieve them from moisture. The dried mass is then extracted with organic solvents (Qin et al. 2010). The pulverized material is leached at room temperature for 48 h using *n*-hexane or it is refluxed at the boiling point for a short duration (2.5 h). The

S. P. Raj et al., *Biodiesel from Flowering Plants*,
https://doi.org/10.1007/978-981-16-4775-8_3

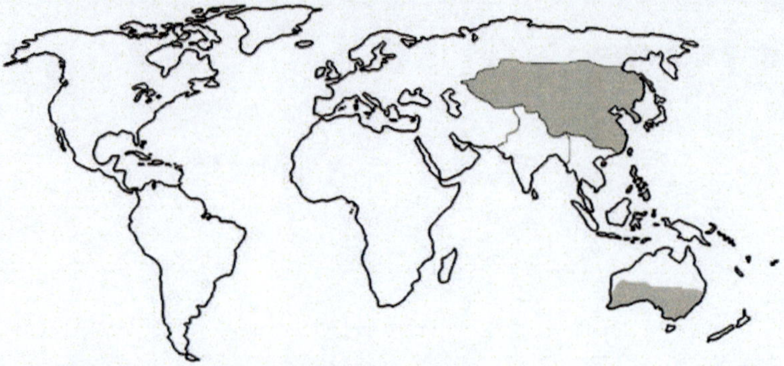

Fig. 3.1 Geographical distribution of *Pistacia chinensis*

Fig. 3.2 The Chinese pistache *P. chinensis*

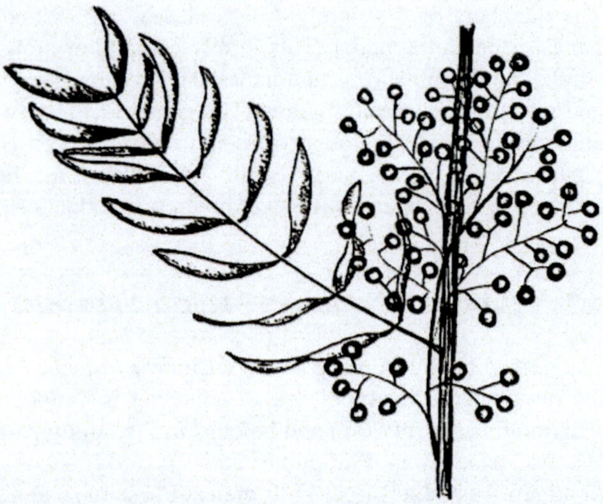

Soxhlet extraction is observed to be very efficient. The micelle is dried over magnesium sulphate, filtered and vacuum distilled. The oil thus obtained is degummed by heating, hydrating, centrifugation and vacuum distillation. It is then deacidified by alkali. Alkali treatment is made using NaOH. The alkali concentration ranges from 0.05 to 0.25%. However, alkali treatment is rarely done since NaOH is likely to react with triglyceride so as to form soap. Esterification also is being carried out using a calculated amount of sulphuric acid. Alcohol also is used in the deacidification process. The removal of free fatty acids due to alcohol treatment depends largely on the volume ratio of oil to ethanol, extraction temperature and the number of repetitive use of alcohol on a single batch of oil. The analytical properties of the oil are given in Table 3.1.

Table 3.1 The physicochemical properties of the oil of *P. chinensis*

Parameters	Li et al. (2012)	Qin et al. (2012)
Density (kg m^{-3})	–	875
Acid value (g KOH kg^{-1})	10.6	33.4
Saponification value (g KOH kg^{-1})	175.7	203.3
Average molecular weight	1019	991
Palmitic acid (C16:0) (%)	16.2	–
Stearic acid (C18:0) (%)	1.7	–
Oleic acid (C18:1) (%)	50.6	–
Linoleic acid (C18:2) (%)	30.4	–

3.1.2 Biodiesel Production

The alkali (NaOH or KOH) is dissolved in a known volume of methanol. The quantity of methanol used should form a methanol to oil molar ratio of 6:1. The alkali concentration is 1% by weight of the oil, the reaction temperature is 60 °C and the reaction duration is 1 h. Soon after the reaction the content is transferred to a separating funnel for the contents to settle thereby to enable phase separation. The upper layer containing the biodiesel is repeatedly washed with warm water and heated to dry. Between KOH and NaOH, KOH shows its superiority in completing the reaction.

Heteropoly acids containing hydrogen and oxygen along with certain metals and non-metals have well defined structure and as catalysts, they do not impose any corrosiveness to the reactors. They normally offer stronger acidity than acidic resins and zeolites. However the heteropoly acids do not have adequate thermal stability. Besides they have low surface area. Often the heteropoly acids are immobilized on SiO_2, ZrO_2 or carbons. But such of those immobilized catalysts function at a high reaction temperature.

In such events Han et al. (2016) suggest the use of cellulose as a catalyst in transesterifying the oil from *P. chinensis*. The prepared cellulose microsphere (CM) is modified using micromolecular amine. Fe_3O_4 powder is dispersed in microporous cellulose to prepare magnetic microporous cellulose (MCM). The MCM is again treated with heteropoly acid $H_3PW_{12}O_{40}$ (HPW) to obtain MCM-HPW catalyst and the same is used in the transesterification of *P. chinensis* oil.

To 5 g cotton, 50 mL sodium hydroxide (19% by weight) is added and kept undisturbed for 2 h. The liquid which remain as soak is then drained and the solid is alkalized for 48 h in room temperature. Five mL of carbon disulphide (CS_2) is added to the alkalized mass and kept waiting for the next 10 h. To this 40 mL of sodium hydroxide (6% by weight) and 8 g of $CaCO_3$ are added. It is then repeatedly but gently stirred till a gel type consistency is obtained. A solution containing 0.1 g potassium oleate and 0.4 g span-60 (sorbitan monostearate) dissolved in transformer oil is first stirred for 30 min. The contents are then transferred to a water bath (90 °C) and maintained for 2.5 h, by then microsphere develops on the catalyst mass. It is

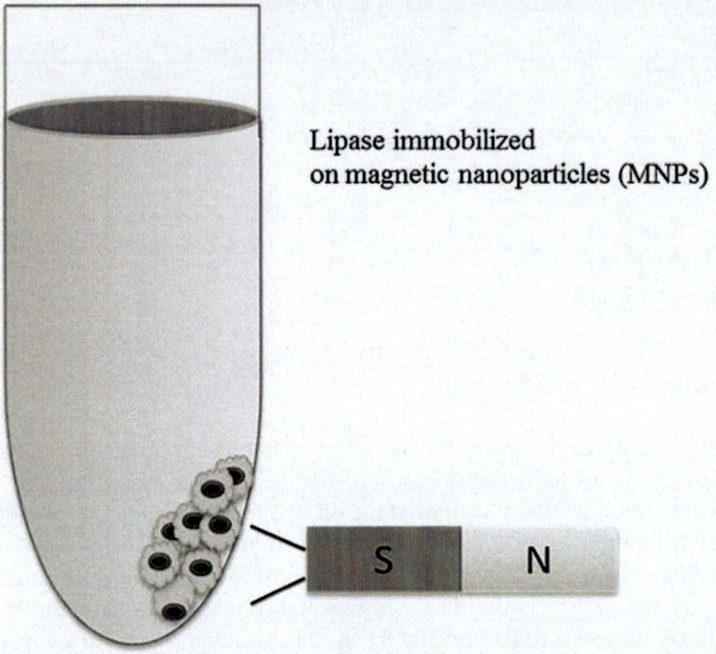

Fig. 3.3 The catalyst is separated with the assistance of a magnet

then washed with dilute HCl (0.1 mol). This creates holes on the surface. To 5 g of the above 5 mL epichlorohydrin and 10 mL of 12% by weight of sodium hydroxide are added and shaken for 8 h. The liquid is subsequently drained and the solid thus obtained is treated with triethylenetetramine (TETA) in the presence of sodium carbonate at 50 °C for 8 h and then repeatedly washed with water. To the above a solution containing 9.75 g of $FeCl_3.H_2O$ and 3.81 g $FeCl_2.H_2O$ in 200 mL deionized water is added. The total content is subsequently loaded in a reactor. The air in the reactor is evacuated to prevent oxidation. The pH of this mixture is slowly raised to 10 using NH_3H_2O and the whole content is heated (60 °C) for 1 h. The MCM thus formed is filtered and washed with distilled water so as to move on to the next step for immobilization. HPW (2 g) is dissolved in 20 mL ethanol and the same is emptied in the washed MCM. The mixture thus formed is kept at 60 °C for 8 h under continuous stirring so as to form MCM-HPW. It is then separated from the mixture by filtration, rinsed with water, dried and used as a catalyst.

This catalyst when used at the rate of 15% on transesterification of *P. chinensis* oil in a methanol to oil ratio of 10:1, at 60 °C in a reaction duration of 80 min gives 93.1% biodiesel (Han et al. 2016). When the reaction comes to an end the catalyst is separated with the assistance of permanent magnet (Fig. 3.3). It is washed first with *n*-hexane then with methanol and dried. The catalyst is separated, cleaned and used in succeeding batches and the conversion is reported to be 85 and 80.7% at third and fourth cycles.

Li et al. (2018) employed a porous microsphere catalyst GO-SO_3H/CM@Fe_3O_4. This type of catalyst has received wide acceptability due to its unique two dimensional effect, reasonably good stability and high surface area (Cheng et al. 2016) of the sulphonated graphene oxide.

Sulphanilic acid (0.5 g) is dissolved in 5 mL of 2% sodium hydroxide. To which 0.2 g $NaNO_2$ is added. Diazonium salt is formed in the above solution due to the addition of 1 mL H_2SO_4 and 10 mL of cold water under stirring for 15 min. The diazonium thus formed is separated and emptied in 50 mL (conc. 10 mg mL^{-1}) of graphene oxide solution and the content is under constant stirring in an ice bath for 2–4 h. The sulphonated graphene oxide (GO-SO_3H) thus formed is repeatedly washed with distilled water till the pH becomes neutral. Separately Fe_3O_4 nanoparticles are prepared by dissolving $FeCl_3.6H_2O$ and $FeCl_2.4H_2O$ at a molar ratio of 2:1 in 150 mL deionized water. The above content is heated to 60 °C and simultaneously the pH is adjusted to 9–10 through drop by drop addition of ammonium hydroxide. The whole mixture is kept on stirring for an hour and at the end, it is centrifuged and washed repeatedly till it becomes neutral. Thus Fe_3O_4 nanoparticles are obtained.

Subsequently cotton is processed by alkalizing with 19% sodium hydroxide to get cellulose as is done by Han et al. (2016). The cellulose thus obtained is mixed with Fe_3O_4 and GO-SO_3H and stirred to get a gel. This gel is taken along with 300 mL of a solution containing 50 mg potassium oleate, 200 mg span-60 dispersed in transformer oil. The whole content is stirred at 700 rpm for 30 min at a water bath (80 °C). After 30 min the stirring speed is reduced to 580 rpm for 2.5 h. In this process, magnetic composite sulphonated graphene oxide cellulose microsphere (GO-SO_3H/CM@Fe_3O_4) is formed. It is washed first with methanol and then with mild acid (0.1 mol HCl) dried and stored. This catalyst at 7% by weight at a methanol to oil molar ratio of 9:1 at 65 °C ambient temperature in a reaction duration of 80 min gives 94% conversion.

In GO-SO_3H/CM@Fe_3O_4, the bond existing between GO and sulphonic acid is proved to be strong (Antunes et al. 2014). In addition to the above, the bond functioning between GO-SO_3H and CM@Fe_3O_4 is strong and stable and is capable of resisting the stirring disturbance caused during the transesterification process in a microwave assisted reactor.

The catalyst separated is reused in the next batches and is recommended to be used up to 5 cycles. The efficiency of such catalyst is observed to be reduced marginally in subsequent cycles which is a general characteristic of all similar sulphonated catalysts (Puna et al. 2014).

Yu et al. (2011) employed CaO-CeO_2 mixed oxides as a catalyst which is prepared by incipient wetness impregnation. The required CaO is prepared by calcinating calcium hydroxide at 873 K for 5 h in a He/O_2 atmosphere (9/1 vv flow). Similarly the needed CeO_2 is prepared from cerium nitrate hydrate at 873 K for 5 h in the same atmosphere. Both are mixed, stirred for 5 h, dried overnight at 393 K and calcined in He/O_2 (9/1 vv flow). The catalyst at a concentration of 9% of the oil is mixed in methanol and kept for 10 min. It is then emptied in the oil, stirred at 1500 rpm at a molar ratio of 30:1 (methanol to oil) at 60 °C for 9 h. On completion

Table 3.2 The operational parameters of the transesterification using the immobilized catalysts AL-ROL and MI-ROL (Li et al. 2012)

Parameters	AL-ROL	MI-ROL
Dose of enzyme (IU g^{-1} of oil)	25	07
Molar ratio of methanol to oil	5:1[a]	5:1[a]
Optimum reaction temperature (°C)	37	37
Reaction duration (h)	60	60
Moisture content of the oil (%)	20	20
Stability in repeated cycles	5	2–3

[a]Methanol is added in 5 split doses at every 12 h

of the reaction, it is centrifuged and the biodiesel which moved to the upper layer is separated. The methanol present in it is distilled out, washed and dehydrated by anhydrous sodium sulphate. The catalyst used in the above transesterification is separated, washed using methanol and dried at 393 K. The catalyst thus made ready can be used 4 times with a marginal change in efficiency (Yu et al. 2011).

Li et al. (2012) employed extracellular enzyme from *Rhizopus oryzae* as a catalyst in the transesterification of *P. chinensis* oil. This enzyme successfully carries out the transesterification even when 4–30% water is present in the oil, but the deterrent is its high cost. Therefore, it is immobilized to enable its recovery and put it back in the next batch. Two types of support materials are employed to immobilize the enzyme (Li et al. 2012). They are Amberlite (IRA-93) and macroporous resin HPD-400. The IRA-93 works on covalent binding principle, whereas HPD-400 works on adsorption method. The respective immobilized catalysts are called then as AL-ROL and MI-ROL and their enzymatic activities are reported to be 95.6 and 57.3 IUg^{-1}, respectively. The procedure for preparing them is as per Wang et al. (2010). The operational parameters of the two immobilized catalysts AL-ROL and MI-ROL are indicated in Table 3.2.

The AL-ROL catalyst requires an optimum concentration of 25 IUg^{-1} oil, whereas it is only 7 in MI-ROL. If the methanol is added as a single dose, there is every likely that methanol coats the surface of the catalyst and poison it. Therefore methanol is recommended to be added in split doses. The properties of biodiesel obtained are given in Table 3.3.

3.1.3 Engine Performance

Xu et al. (2013) studied a few aspects of the performance of the biodiesel from *P. chinensis* oil using a 4 stroke, 4 cylinder DI engine with a compression ratio of 16.7:1. The brake specific fuel consumption (BSFC) decreases slightly when the fuel injection advance angle is increased. Higher fuel injection advance angle makes extended injection delay-duration causing the accumulation of the fuel in the cylinder. This situation may improve the atomization and evaporation thereby promoting ideal combustion leading towards satisfactory engine power. Normally when working on a blend diesel, the BSFC increases as a function of the percentage of blend. Such increase in the BSFC is possibly due to the fact that the engine is

Table 3.3 The physicochemical properties of biodiesel from *P. chinensis* oil

Parameters	Li et al. (2012)	Qin et al. (2012)	Han et al. (2016)	Li et al. (2018)
Kinematic viscosity @ 40 °C (cSt)	4.15	5.24	4.7	–
Density @ 15 °C (kg m^{-3})	880	887	879	–
Flash point (°C)	102	166	–	–
Cetane number	49	52.2	–	–
Pour point (°C)	–	−2	–	–
Cloud point (°C)	–	1.0	–	–
Iodine number (g I_2/100 g)				
Cold filter plugging point (°C)	–	0	–	–
Acid value (g KOH kg^{-1})	–	0.35	0.36	–
Heating value (MJ kg^{-1})	39.8	–	–	–
Methyl palmitate (C16:0) (%)	–	–	18	14.6
Methyl stearate (C18:0) %	–	–	1.5	2.1
Methyl oleate (C18:1) %	–	–	46.7	50.8
Methyl linoleate (C18: 2) %	–	–	33.5	28.5

needed to provide more fuel to discharge the required power output since the heating value is proportionally low. On the other hand, the density is high on a unit volume which naturally makes the mass of the fuel high. At a high load and high injection pressures the temperature of the cylinder increases. The consequent increase in spray causes the effective atomization which tends to increase the combustion thereby the hydrocarbon level in the exhaust reduces. The share of biodiesel in the blend reduces the aromatic hydrocarbon of the mixed fuel. Such reduction shortens the ignition delay. Besides high content of oxygen in the biodiesel marginally increases the oxygen content in the exhaust, along with the increase of fuel injection advance angle. When the piston moves from top dead centre (TDC) to bottom dead centre (BDC) the combustion may retard quickly, during which time the CO generated is unable to get it oxidized to CO_2 and thus CO level slightly increases (Xu et al. 2013). As the combustion progresses, the cylinder temperature is likely to increase which promotes rapid atomization causing complete combustion which reduces the CO level in the exhaust steadily during the next stage of run.

In accordance with the increase in load the NOx emission increases since the cylinder temperature parallelly increases causing rapid oxidation. In addition to the above methyl oleate and linoleate having double bonds present in the biodiesel promote the torsion of the molecule of the fatty acid methyl ester. This releases the free radical which aids in the formation of NOx (Zhihao et al. 2011).

References

Antunes MM, Russo PA, Wiper PV, Veiga JM, Pullinger M, Mafra L, Evtuguin DV, Pinna N, Valante AA (2014) Sulfonated graphene oxide as effective catalyst for conversion of 5-(hydroxymethyl)-2-furfural into biofuels. ChemSusChem 7(3):804–812

Cheng JY, Qiu Y, Huang R, Yang W, Zhou J, Cen K (2016) Biodiesel production from wet microalgae by using graphene oxide as solid acid catalyst. Bioresour Technol 221:344–349

Han Y, Hong L, Wang X, Liu J, Jiao J, Luo M, Fu Y (2016) Biodiesel production from Pistacia chinensis seed oil via transesterification using recyclable magnetic cellulose based catalyst. Ind Crop Prod 89:332–338

Li X, He X, Li Z, Wang Y, Wang C, Shi H, Wang F (2012) Enzymatic production of biodiesel from *Pistacia chinensis* bge seed oil using immobilized lipase. Fuel 92:89–93

Li T, Shen C, Zhang H, Wang X, Jiao J, Wang W, Gai Q, Liu J, Liu T, Fu Y (2018) Transesterification of *Pistacia chinensis* seed oil using a porous cellulose-based magnetic heterogeneous catalyst. Int J Green Energy 16:228–235. https://doi.org/10.1080/15435075.2018.1555759

Puna JF, Gomes JF, Borado JC, Correia MJN, Dias APS (2014) Biodiesel production over lithium modified lime catalysts: activity of deactivation. Appl Catal A Gen 470:451–457

Qin S, Sun Y, Meng X, Zhang SZ (2010) Production and analysis of biodiesel from non-edible seed oil of *Pistacia chinensis,* Energy. Explor Exploit 28(1):37–46

Qin S, Sun Y, Shi C, He L, Meng Y, Ren X (2012) Deacidification of *Pistacia chinensis* oil as a promising non edible feedstock for biodiesel production in China. Energies 5:2759–2770

Tingshuang Y, Wen J, Golan-Goldhirsh A, Parfitt DE (2008) Phylogenetics and reticulate evolution in Pistacia (Anacardiaceae). Am J Bot 95(2):241–251. https://doi.org/10.3732/ajb.95.2.241. PMID 21632348

Wang Y, Shen X, Li Z, Li X, Wang F, Nie X, Jiang J (2010) Immobilized recombinant *Rhizopus oryzae* lipase for the production of biodiesel in solvent free system. J Mol Catal B Enzym 67 (1–2):45–51

Xu B, Luo L, Wu J, Ma Z (2013) The influence of injection timing on emissions characteristics of a D1 diesel engine fueled with *Pistacia chinensis* bunge seed biodiesel. Adv Mater Res 634–638:846–851

Yu X, Wen Z, Hongliang L, Tu S, Yan J (2011) Transesterification of *Pistacia chinensis* oil for biodiesel catalyzed by CaO-CeO2 mixed oxides. Fuel 90:1868–1874

Zhihao M, Xiaoyu Z, Junfa D, Xin W, Bin X, Jian W (2011) Study on emissions of a DI diesel engine fueled with *Pistacia chinensis* bunge seed biodiesel-diesel blends. Procedia Environ Sci 11:1078–1083

4 Apocynaceae

Apocynaceae is commonly known as Oleander family. Most of the members of it have milky and poisonous latex. Updated classification indicates that this family comprises 410 genera, 25 tribes, 49 subtribes and more than 5500 species that include trees, shrubs herbs, stem succulents and lianas. Out of the many species of this family biodiesel works are distinctly available on three non-conventional species, namely Milk weed (*Asclepias syriaca*) of the tribe Asclepiadeae, Milk bush (*Thevetia peruviana) and* Sea mango (*Cerbera odollam*) (both belonging to the tribe Plumerieae) and the same is presented in this chapter.

4.1 Milk Weed (*Asclepias syriaca*)

The milk weed, *Asclepias syriaca* is being investigated as an alternative feedstock for the production of biodiesel. It grows well in Africa, North America, parts of Canada and Eastern USA (Fig. 4.1) where it is seen on roadsides and in undisturbed habitats preferably in sandy soils of rocky terrain.

It is a tall and conspicuous plant having unbranched stem. Normally it grows to the height of 150 cm but can grow exceptionally up to 240 cm. The leaf is oblong, oval, lanceolate and elliptic. The leaves occur in pairs with short petioles. The individual leaf is 10–19 cm long and 5–11 cm wide. The leaf is dark green at its top and dull green at the underside. The fruit is a follicle and is 9 cm long and 4 cm wide. Follicles are covered with hair and soft spikes. This follicle splits open when ripe, releasing silky feathered seeds (Fig. 4.2).

This plant is referred by few vernacular names: milk weed, butterfly flower, silk weed, silky swallow-wort and Virginia silkweed.

S. P. Raj et al., *Biodiesel from Flowering Plants*,
https://doi.org/10.1007/978-981-16-4775-8_4

Fig. 4.1 Geographical distribution of *A. syriaca*

Fig. 4.2 The milk weed, *Asclepias syriaca*

4.1.1 Oil Extraction

The matured fruit is broken to release the seeds and the seeds discharged are mechanically expelled in a roller. The oil thus obtained is centrifuged to remove any solid. It is then refined by mixing 5–10% of activated clay (Harry-O'kuru et al. 2002) and centrifuged it subsequently. The oil thus obtained is stored in a nitrogen atmosphere. The fatty acid profile of the oil is given in Table 4.1.

Table 4.1 Fatty acid profile of the oil from *A. syriaca*

Parameter	Harry-O'kuru et al. (2002)	Holser and Harry-O'kuru (2006)
Palmitic acid (C16:0) (%)	5.7	5.9
Stearic acid (C18:0) (%)	2.5	2.3
Palmitoleic acid (C16:1) (%)	9.6	6.8
Oleic acid (C18:1) (%)	31.0	34.8
Linoleic acid (C18:2) (%)	50.5	48.7

Table 4.2 The Physicochemical properties of biodiesel from *A. syriaca* (Holser and Harry-O'kuru 2006)

Parameters	Methyl ester	Ethyl ester
Kinematic viscosity at 40 °C (mm^2s^{-1})	5.2	5.7
Cloud point (°C)	−0.8	5.1
Pour point (°C)	−6.7	−10.0
Oxidation stability at 100 °C (h)	0.8	1.9

As high as 90% of the fatty acids are in unsaturated form. The bulk of the unsaturated fatty acid is linoleic acid.

4.1.2 Biodiesel Preparation

The transesterification is aided either by methanol or ethanol. The molar ratio between the oil and alcohol is 1:6. The alcohol is initially mixed with alkali (KOH) to form methoxide or ethoxide as the case may be and it is then added to the oil. The conc. of alkali is 1–1.5% of the total content. The mixture is heated to 60 °C and stirred for 1 h. On completion of the reaction the product is taken in a separatory funnel and allowed to have the phase separation by keeping it undisturbed. The ester moves to the top, leaving the dense glycerol at the bottom. First the glycerol is drained. Then the layer comprising the ester (biodiesel) is washed repeatedly with hot water and it is dried in a rotary evaporator. The analytical details of the biodiesel are given in Table 4.2.

The separation of ethyl ester from the glycerol is relatively difficult when compared to that of methyl ester. The cloud point and pour point readings differ between methyl ester and ethyl ester. Such difference is mostly due to the effect of unconverted or partially converted triglycerides. The viscosity values are close to the upper limit of the standard. The viscosity normally shows high value when the number of carbon atoms present in the fatty acid ester are more. This biodiesel is prone to rapid oxidation due to poor oxidation stability and thus causes quality impairment. The rate of oxidation decreases when the degree of unsaturation is low. The oxidation rate of the individual ester is in the order linolenate > linoleate > oleate > stearate. It is reported that the linolenate makes nonlinear response in the oxidative stability tests (Knothe and Dunn 2003). Biodiesels made from different raw materials are unable to meet the oxidative stability requirements since the refining process made on the oil removes the antioxidant compounds. The European standard EN 14214 spells out 6 as a limit for oxidation. The ball-wear scar test run on methyl ester and ethyl ester indicates that the methyl ester shows a low value (100–159 μm) compared to that of ethyl ester (195–218 μm). Similarly, the disc wear scar for the methyl ester is lower (661–721 μm) than that of the ethyl ester (754–771 μm) (Holser and Harry-O'kuru 2006).

4.2 Milk Bush (*Thevetia peruviana*)

The milk bush, *Thevetia peruviana* an evergreen ornamental shrub of no economic value is a native of Central and South America. It grows to a maximum height of 3–5 m. The leaf is 13–15 cm long, linear and is arranged in a spiral. The flowers are tubular, bell shaped and yellow in colour. The fruit is globular with a fleshy mesocarp having a diameter of 4–5 cm (Fig. 4.3). The fruit when young is green which slowly turns to black as it ripens and holds 1–4 seeds. The seed is reported to contain 60–65% oil which is non-edible as it has a toxin cardiac glycoside. Each plant yields around 400–800 seeds in a year. This plant is called by different vernacular names: lucky nut, manjal arali, be-still tree, dioxin, yellow bells, peeli kaner, yellow oleander, nerium oleander and trumpet flower.

This species is distributed in Central and South America, Nigeria, India, Sri Lanka, Kenya, Tanzania, Uganda, Cook Island, Fiji Islands, French Polynesia, Tonga, Marshall Islands, Solomon Islands, Maldives and Bermuda (Fig. 4.4).

Fig. 4.3 The plant, *Thevetia peruviana*

Fig. 4.4 Geographical distribution of *T. peruviana*

4.2.1 Oil Extraction

The fruits are dried and the seeds separated. The seeds are then cracked to remove the kernels. The kernels are sundried and milled. The kernel powder is then subjected to Soxhlet extraction using *n*-hexane as solvent at 60–65 °C. The oil-rich solution thus obtained is heated in a vacuum distillation apparatus so as to distill out the *n*-hexane. The resultant oil is used as a feedstock for biodiesel preparation. The properties of the oil are given in Table 4.3.

This oil is characterized by poor iodine value which is a reference to a low level of unsaturated fatty acids and hence is used as a non-drying oil. The kinematic viscosity is relatively low when compared to that of other oils. Free fatty acid also is low which obviates the acid catalysis in the transesterification process.

4.2.2 Biodiesel Production

Solid NaOH (5 g) is dissolved in 200 mL of methanol and it is mixed in a litre of *T. peruviana* oil and the whole content is heated and maintained at 55 °C for 2 h while it is simultaneously stirred at 600 rpm. On completion of the reaction the mixture is cooled, transferred to a separating funnel and allowed to stand undisturbed overnight. A clear top layer formed of biodiesel is then separated and washed with distilled water repeatedly until the content shows a neutral pH. It is then heated to expel the moisture. The characteristics of the biodiesel are given in Table 4.4.

The kinematic viscosity values of the three works cited in Table 4.4 are within the standard though they are closer to the upper boundary. The density also is within the limit. High density causes delay between the injection and combustion of the fuel in the cylinder and influences the energy release per unit mass. Marginally high density is caused by the presence of high molecular weight of triglycerides. Iodine value

Table 4.3 The physicochemical properties of the seed oil of *T. peruviana* (Balusamy and Marappan 2010; Adebowale et al. 2012; Bora et al. 2014)

Parameters	Balusamy and Marappan (2010)	Adebowale et al. (2012)	Bora et al. (2014)	Arun et al. (2017)	Ighose et al. (2017)
Kinematic viscosity at 40 °C (cSt)	4.8	–	–	–	57.34
Density (kg m^{-3})	920	–	912	–	910
Acid value (g KOH kg^{-1})	–	–	0.66	–	2.86
Iodine value (g I_2/100 g)	–	84.5	71.4	–	–
Free fatty acid (%)	–	3.4	0.35	–	–
Cetane value	42	–	–	–	59.94
Flash point (°C)	128	–	–	–	–
Cloud point (°C)	−4	–	–	–	
Pour point (°C)	−7	–	–	–	
Heating value (MJ kg^{-1})	40.15	–	–	–	–
Palmitic acid (C16:0) (%)	–	8.9	23.28	26.4	16.59
Stearic acid (C18:0) (%)	–	2.4	10.71	3.86	6.09
Arachidic acid (C20:0) (%)	–	–	2.41	1.11	0.68
Oleic acid (C18:1) (%)	–	19.64	43.72	39.4	43.81
Linoleic acid (C18:2) (%)	–	57.13	19.85	27.03	31.10
Linolenic acid (C18:3) (%)	–	10.0	–	–	0.81
Behenic	–	–	–	–	0.50
Palmitoleic	–	–	–	–	23.87

being the reference of unsaturation is relatively low. Iodine value greatly influences the fuel oxidation and a cause for any deposit formed in the injectors.

The cetane number is observed to be satisfactory. The cetane number and iodine value are inversely related. It is in this context an upper limit for cetane number (65) is presented in the standard. This fuel has a good oxidation stability apparently due to the low iodine value.

Esterification and transesterification are also carried out by ultra-sonication method. The mixture (oil + catalyst + methanol) is loaded in a 250 mL glass beaker and exposed to ultrasonic irradiation at a frequency range of 28–20 kHz with an output power of 50 W (Yadav et al. 2016). The comparison of biodiesel prepared by conventional method and ultrasonic method along with conventional diesel is given in Table 4.5.

Table 4.4 The physicochemical characteristics of the biodiesel prepared from *T. peruviana* oil (Balusamy and Marappan 2010; Adebowale et al. 2012; Oladayo and Kemisola 2017)

Parameters	Balusamy and Marappan (2010)	Adebowale et al. (2012)	Oladayo and Kemisola (2017)	Arun et al. (2017)	Ighose et al. (2017)
Kinematic viscosity @ 40 °C (cSt)	4.2	4.5	4.48	–	5.5
Density (kg m^{-3})	872	870	866	–	890
Acid value (g KOH kg^{-1})	–	0.2	–	–	0.43
Flash point (°C)	110	125	135	–	–
Cloud point (°C)	−6	–	–	–	–
Pour point (°C)	−8	–	–	5	–
Cetane number	47	54.2	–	–	125.74
Heating value (MJ kg^{-1})	40.46	–	–	37.56	46.10
Copper strip corrosion test	–	1	–	–	–
Iodine value (g I_2/100g)	–	84.2	–	–	68.82
Oxidative stability (h)	–	30.3	–	–	–

High viscosity causes slow combustion as a consequence of poor spray and uneven mixing with air. High brake thermal efficiency for biodiesel is reported at a static injection timing of 27° bTDC against 23° bTDC for regular diesel. However, further increase is known to cause negative impact (Balusamy and Marappan 2010). In the said injection timing (27° bTDC) CO occurrence in the emission is low. Ideal injection timing reduces HC content in the exhaust. However, there is a feeble increase in NOx. Both injection timing and injector opening pressure (IOP) affect the performance significantly. The brake thermal efficiency improves as the IOP increases.

4.3 Sea Mango (*Cerbera odollam*)

Of late the oil extracted from the seeds of the sea mango, *Cerbera odollam* (syn: *Cerbera lateria*, *C. manghas* Linn) is considered for biodiesel. This species grows on the banks of the backwaters or river mouth, swampy lands, mangrove edges forming savanna and on the banks of estuaries with fluctuating salinity. It is a salt loving coastal tree growing up to 12 m with shiny dark green leaves. The leaves are alternate and ovoid. The fruits are egg shaped with a size ranging from 5 to 10 cm dia. (Fig. 4.5). The tender fruits are green in colour which appears purple red on maturity. The fruit has fibrous shell within which there is almond shaped kernel of 2 × 1.5 cm size. This kernel is white and fleshly which turns to violet first, brown

Table 4.5 Comparison of *Nerium oleander* (NO) biodiesel obtained from conventional and ultrasonic method (Yadav et al. 2016)

Properties	N.O. oil	N. O biodiesel (Con. method)	N. O biodiesel (ultr. method)	Diesel
Moisture content	0.33%	0.55	0.52	–
Free fatty acid	7.5%	0.60	0.56	–
Colour	Greenish	Light greenish	Light greenish	Light brown
Density at 15 °C (g/cc)	0.89	0.88	0.87	0.83
Viscosity at 40 °C (cSt)	37	4.3	4.1	2.9
Flash point(°C)	190	178	175	55
Fire point(°C)	250	186.2	181	63
Pour point(°C)	–2	−7	−8	−15
Cloud point(°C)	2	−3	−5	−10
Calorific.value (MJ kg⁻)	39.8	42.4	43.5	44.49
Acid value (mg KOH g^{-1})	15	0.66	0.64	0.08
Iodine value (gI_2/100 g)	85	76.4	75	4.5
Cetane number	61	71.5	72	49
Oxidation stability (h)	–	> 6.5	> 6.5	–

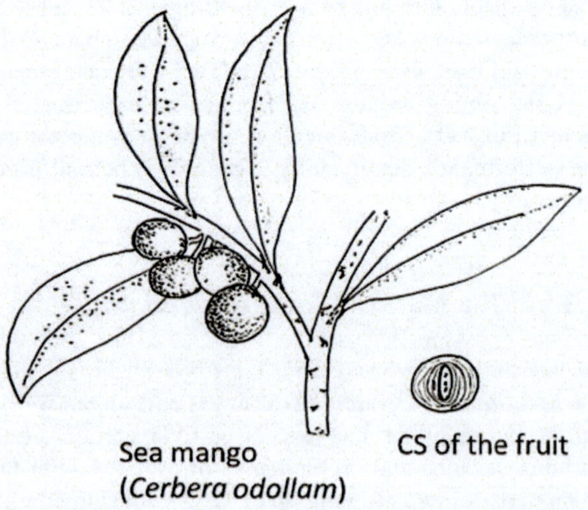

Fig. 4.5 The sea mango plant, *Cerbera odollam*

next and black at last on exposure to air. The seeds are woody and surrounded by hairy tuft. The leaves and fruits contain glycoside cerberin which is a deadly poison. Due to this reason, this plant is called murder tree. Throughout the year it

Fig. 4.6 Geographical distribution of the Sea mango, *Cerbera odollam*

continuously gives forth flowers and the fragrance is just that of jasmine. The oil from sea mango is little known for commercial purpose. Currently, the oil is not in use for any other purposes. Few published works indicate its use exclusively as raw material for biodiesel.

This species is distributed in the coastal creeks of South India and along the river banks of Vietnam, Cambodia, Sri Lanka, Myanmar, Madagascar, Malaysia, French Polynesia and Seychelles (Fig. 4.6) (Khairil et al. 2018). This species is referred by local names in different places: sea mango, Madagascar ordeal bean, odollam tree, pink eyed cerbera, dog-bane, dagor, kattarali, utalam, chattankaya and murder tree.

4.3.1 Oil Extraction

C. odollam kernels are separated manually from the fruits, dried in the hot air oven or in open-sunlight repeatedly to remove the moisture. Moderate temperature is used in drying since elevated temperature would denature the oil (Kansedo and Lee 2013). It is then grinded into fine particles of 5–10 nm size. These powdered materials are then extracted with n-hexane at the rate of 5 L of hexane per kg of nut powder for 24 h (Ang et al. 2015). On completion of extraction, hexane is expelled by rotary evaporator and the oil separated. The kernel has an oil content ranging from 43 to 64% (Khairil et al. 2018).

4.3.2 Properties of the Oil

The physicochemical characteristics of the oil are given in Table 4.6. The kinematic viscosity of the oil is very high. The FFA content also is high which prevents the usual transesterification process. Oleic acid forms the major fatty acid.

Table 4.6 Physicochemical properties of sea mango oil (Kansedo and Lee 2013)

Parameters	Value
Kinematic viscosity @ 40 °C (cSt)	29.57
Density @ 15 °C(kg m^{-3})	919.8
Free fatty acid(%)	6.40
Palmitic acid (C16:0) (%)	24.86
Stearic acid (C18:0) (%)	5.79
Arachidic acid (C20:0) (%)	1.09
Behenic acid (C22.0) (%)	0.37
Lignoceric acid (C24:0) (%)	0.16
Trans (oleic acid) (C18:1) (%)	0.24
Cis (oleic acid) (C 18:1) (%)	52.82
Cis (linoleic acid) (C 18:2) (%)	13.65
n3 (linolenic acid) (C18:3) (%)	0.08
Palmitoleic acid (C16:1) (%)	0.75
Gadoleic acid (C 20:1) (%)	0.19
Average molecular weight (g mol^{-1})	867.10

4.3.3 Preparation of Biodiesel

At the outset the oil is freed from the gum by treating it with dilute phosphoric acid (H_3PO_4) at 70 °C for 30 min. For every litre of oil, 7 mL of H_3PO_4 (20%) is added and stirred. After 30 min the bottom layer containing phosphate compounds is discarded. The refined oil which stays as top layer is yellow in colour and can be readily used as raw material in the preparation of biodiesel.

Since the oil from the sea mango has high FFA, it is recommended to treat it first with acid to reduce the FFA so as to make it available for alkali catalysis. Such dual treatment is fairly cumbersome and the efficiency is low. To overcome such problem supercritical process is followed. In the supercritical process, the oil and alcohol will mix freely and become a single phase (Ang et al. 2015). Methanol and oil are mixed at a molar ratio of 20:1 to 60:1. The mixture is exposed to 320–400 °C for 10–50 min at a pressure of 8–10 MPa. Immediately after this process, the material is air cooled and flashed in rotary evaporator at 80 °C for 20 min to expel the methanol so as to get the biodiesel as a final product.

One of the major issues encountered with the transesterification is the resistance of the oil (non-polar substance) to mix with alcohol (polar substance). The resistance of both for mixing renders the process very difficult to achieve the fullest conversion. Though the usual vigorous mixing by stirrer helps to increase the contact, it is mostly physical in nature and both the reactants fall apart once the stirring is ceased (Kansedo and Lee 2014). To improve the contact between the media, solvents such as tetrahydrofuran (Lam and Lee 2013), propane (Cao et al. 2005), heptane, hexane (Li et al. 2010) and ionic liquids (Fauzi and Amin 2012) are being used. Sometimes mixtures of agents such as acetonitrile with tertiary butanol and isooctane with tertiary butanol are used (Li et al. 2010). Often the molecular structure of oil is altered so as to change its solubility with alcohol. Such molecules when broken

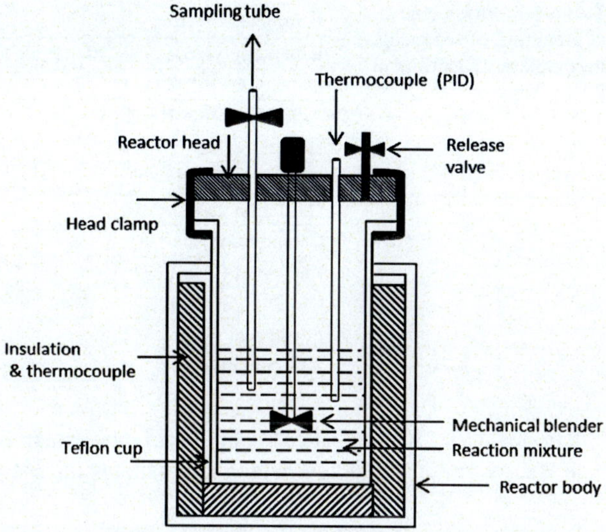

Fig. 4.7 Experimental setup for the transesterification reaction

through lipolysis or hydrolysis, free fatty acids are formed. At the molecular ends of FFA, polar group–COOH develops which shows a partial tendency to mix with alcohol. Hydrolysis is carried out at a higher temperature and pressure. The oil is mixed with water (20 mL water in 100 mL oil) without any catalyst. The catalyst if added normally makes the process costlier and the subsequent disposal of the used catalyst creates another problem. Therefore non-catalytic mode of hydrolysis is greatly recommended.

The oil and de-ionized water are mixed and loaded in the reactor vessel (Fig. 4.7). With the admission of nitrogen gas the internal pressure of the reactor is set at 2.0 MPa and the temperature is raised to 200 °C. The hydrolysis is allowed for a total duration of 60–90 min. On completion of the reaction, the reactor vessel is cooled as swiftly as possible using chill-air. The product from the reactor forms two layers.

1. Oil layer comprising FFA and unreacted oil
2. Aqueous layer comprising water and glycerol

The oil phase containing FFA is treated first with acid (acid catalysts) and then with alkali (alkali catalysts) to produce biodiesel (Table 4.7).

4.3.4 Properties of Biodiesel

The physicochemical properties of sea mango biodiesel are given in Table 4.6. The properties meet the standard (EN14214) of a heating oil (Barabás and Todoruț 2011). The density is higher than that of diesel and the density is dependent on the FFA profile and purity. The viscosity is higher than that of diesel due to the

Table 4.7 Physicochemical properties of sea mango oil biodiesel (Khairil et al. 2018; Pawar and Pawar 2017)

Parameters	Value
Kinematic viscosity @ 40 °C (cSt)	3.157, 5.2
Density @ 15 °C (kg m^{-3})	876, 847.9
Acid value (mg KOH/g)	0.4
Heating value (MJ Kg^{-1})	40.49, 38.5
Flash point (°C)	214.0
Oxidation stability (h)	6.35
Cetane number	50.70
Fire point (°C)	168
Cloud point (°C)	9
Pour point (°C)	3
Ash content (%)	0.05

electronegativity which makes the biodiesel more polar than that of petroleum diesel. The energy value also is low which necessitates the blending of it with diesel.

4.3.5 Engine Performance

Very limited works are available on the engine performance of sea mango oil biodiesel. More quantity of biodiesel enters into the cylinder to pull the load since the calorific value is low. Accordingly, the break thermal efficiency of the biodiesel is roughly 15% lower than that of the regular diesel. Often biodiesel is blended with regular diesel. The blend containing 50% of diesel is reported to perform well (Pawar and Pawar 2017). As indicated already the brake specific fuel consumption also increases. The volumetric efficiency is also low in biodiesel. The quantity of CO released through the exhaust with respect to biodiesel fuelled engine is lower than that of diesel by 50% (Pawar and Pawar 2017). The amount of CO_2 released in the exhaust of biodiesel fuelled engine is also low due to the fact that the biodiesel has low level of elemental carbon to hydrogen ratio than that of diesel. Similarly HC level also is low, due to the active combustion assisted by inbuilt oxygen in the biodiesel molecule.

References

Adebowale KO, Adewuyi A, Ajulo KD (2012) Examination of fuel properties of the methyl esters of *Theveta peruviana* seed oil. Int J Green Energy 9(3):297–307

Ang GT, Ooi SN, Tan KT, Lee KT, Mohamed AR (2015) Optimization and kinetic studies of sea mango (Cerbera odollam) oil for biodiesel production via supercritical reaction. Energy Convers Manag 99:242–251

Arun SB, Suresh R, Avinash E (2017) Optimization of biodiesel production from yellow oleander (*Thevetia Peruviana*) using response surface methodology. Mater Today Proc 4(8):7293–7301

Balusamy T, Marappan R (2010) Effect of injection time and injection pressure on CI engine fuelled with methyl ester of *Thevetia peruviana* seed oil. Int J Green Energy 7(4):397–409

Barabás I, Todoruț IA (2011) Biodiesel quality, standards and properties. In: Montero G, Stoytcheva M (eds) Biodiesel-quality, emissions and by-products. InTech., Rijeka, pp 3–28

Bora MM, Gogoi P, Deka DC, Kakati DK (2014) Synthesis and characterisation of yellow oleander (*Thevetia peruviana*) seed oil-based alkyl resin. Ind Crop Prod 52:721–728

Cao W, Han H, Zhang J (2005) Preparation of biodiesel from soybean oil using supercritical methanol and co-solvent. Fuel 84(4):347–351

Fauzi MAH, Amin NAS (2012) An overview of ionic liquids as solvents in biodiesel synthesis. Renew Sust Energ Rev 16(8):5770–5786

Harry-O'kuru RE, Holser RA, Abbott TP, Weisleder D (2002) Synthesis and characteristics of polyhydroxy triglycerides from milk weed oil. Ind Crop Prod 15(1):51–58

Holser RA, Harry-O'kuru R (2006) Transesterified milk weed (*Asclepias*) seed oil as a biodiesel fuel. Fuel 85:2106–2110

Ighose OB, Adeleke IA, Damos M, Junaid HA, Okpalaeke KE, Betiku E (2017) Optimization of biodiesel production from *Thevetia peruviana* seed oil by adaptive neuro-fuzzy inference system coupled with genetic algorithm and response surface methodology. Energy Conserv Manag 132:231–240

Kansedo J, Lee KT (2013) Process optimization and kinetic study for biodiesel production from non-edible sea mango (*Cerbera odollam*) oil using response surface methodology. Chem Eng J 214:157–164

Kansedo J, Lee KT (2014) Non-catalytic hydrolysis of sea mango (*Cerbera odollam*) oil and various non-edible oils to improve their solubility in alcohol for biodiesel production. Chem Eng J 237:1–7

Khairil R, Aulia R, Iskandar, Jalaludin, Silitonga AS, Masjuki HH, Mahlia TMI (2018) The potential biodiesel production from *Cerbera odollam* oil (Bintaro) in Aceh, IJCAET and ISAMPE, MATEC web of conference Bali, Indonesia, 24–26 Aug 2017

Knothe G, Dunn RO (2003) Dependence of oil stability index of fatty compounds on their structure and concentration and presence of metals. J Am Oil Chem Soc 80(10):1021–1026

Lam MK, Lee KT (2013) Catalytic transesterification of high viscosity crude microalgae lipid to biodiesel: effect of co-solvent. Fuel Process Technol 110:242–248

Li Q, Zheng J, Yan Y (2010) Biodiesel preparation catalyzed by compound-lipase in co-solvent. Fuel Process Technol 91(10):1229–1234

Oladayo RI, Kemisola OO (2017) Assessment of milk bush seed oil as an auspicious feedstock for biodiesel fuel. Int J Environ Chem 2(5):56–62

Pawar AA, Pawar SK (2017) Performance, evaluation and emission testing of sea mango seeds oil biodiesel blends in CI engine. Int J Sci Res 6(6):725–729

Yadav AK, Khan ME, Pal A, Dubey AM (2016) Biodiesel production from Nerium oleander (*Thevetia peruviana*) oil through conventional and ultrasonic irradiation methods. Energy Sources Part A Recovery Utiliz Environ Eff 38:3447–3452

5 Arecaceae

Palms and palm trees form the family Arecaceae. It has 181 genera with around 2600 species. They are known to be geographically restricted to tropical and subtropical climates exhibiting an enormous diversity in physical characteristics and inhabit all habitat within their range from rainforests to deserts. In this Chap. 5 such species are dealt whose oil has high social reception as some of them are used in continental cuisine. These oils are highly saturated with rich source of lauric acid and accordingly known to have high oxidation stability. In few selected pockets of the world such oils are in glut giving promise of being used as supplementary fuel.

5.1 Babassu (*Attalea speciosa*)

The American oil palm, babassu (*Attalea speciosa*) is an important tree in Northern Brazil. It is a monoecious evergreen feather palm growing to a maximum height of 20 m. The trunk is round with a diameter of 30–40 cm. The crown comprises 15–20 large leaves. The tree prefers well-drained loamy soil with adequate moisture and comes to yield in 10 years and flourishes in the region with rainfall of 1200–1700 mm and a temperature range of 25–30 °C. It bears 2–4 inflorescences of pale yellow flowers. The fruits are in bunches of around 1 m long, weighing around 50 kg with 250–500 fruits (Fig. 5.1). The fruit comes to yield twice a year. Each fruit contains 3–6 seeds. The seeds weigh two-thirds of the nut. Each seed is 6 cm long 1–2 cm wide. The oil extracted from the kernel is rich (45%) in lauric acid (C12:0).

A. speciosa tree is known by the following local names: Babassu, American oil palm, motacu, motacuchi, babassou, palma babasu, coco-de-macaco, coco-pindoba and palha-branca. It is distributed in Brazil, Mexico, Suriname, Guyana and Bolivia (Fig. 5.2).

S. P. Raj et al., *Biodiesel from Flowering Plants*,
https://doi.org/10.1007/978-981-16-4775-8_5

Fig. 5.1 The babassu *Attalea speciosa*

Fig. 5.2 Geographical distribution of *A. speciosa*

It is reported that the industrial activity on account of babassu employs 4 million people from an area of 16 million hectares producing as high as 7 million metric tons of oil a year (Silva et al. 2014).

5.1.1 The Oil Extraction

When the fruit ripens, it slips down from the bunch and is being collected from the ground in large numbers. Each nut is cut open by sharp axe and the kernel removed.

Table 5.1 Physicochemical properties of the oil of *A. speciosa*

Parameters	Bouaid et al. (2015)	Ferreira et al. (2012)	Abreu et al. (2004)	Moreira et al. (2018)
Kinematic viscosity@40 °C(cSt)	38.3	–	–	–
Acid value (g KOH kg^{-1})	0.15	–	–	–
Iodine number (g I_2/100 g)	18		–	–
Caprylic acid (C8:0) (%)	5.2	9.2	–	2.2
Capric acid (C10:0) (%)	4.8	9.6	–	3.6
Lauric acid (C12:0) (%)	51.8	54.7	48	46.7
Myristic acid (C14:0) (%)	22.2	11.8	16	17.6
Palmitic acid (C16:0) (%)	3.2	4.8	10	8.2
Stearic acid (C18:0) (%)	3.3	2.0	2	3.1
Oleic acid (C18:1) (%)	7.3	6.5	14	16.4
Linoleic acid (C18:2) (%)	0.3	0.9	5	2.2

The kernels are repeatedly dried in the hot sun and pressed to separate the oil. The residual material is again treated with n-hexane and thus the oil in its entirety is extracted. The physicochemical properties of the oil are given in Table 5.1.

The viscosity is relatively high and the iodine number, acid value and unsaturated fatty acid content are paradoxically low. Low iodine number and low level of unsaturated fatty acids indicate the stability of the oil.

5.1.2 Preparation of Biodiesel

The oil is initially heated to the level of the boiling point of methanol. Methanol is then added so as to form the methanol oil ratio of 6:1. The catalyst, KOH is added at the rate of 0.95% by weight of the total mass. The whole mixture is stirred at 600 rpm for 4 h and transferred to a separating funnel and waited for 8 h to settle. The supernatant phase is the biodiesel which is repeatedly washed to remove the residual catalyst. It is observed that the temperature is a major factor when compared to that of the conc. of catalyst (Bouaid et al. 2015). However, the above two factors when increased beyond optima there is a possibility of soap formation in spite of the fact that the free fatty acid level is low. The soap thus formed may dissolve in glycerol due to the coinciding polarity and consequently, there will be gross interference on the phase separation. On the other hand, it is also possible that low temperature and low conc. of catalyst causes incomplete transesterification, while the high temperature may cause the expulsion of alcohol. Babassu oil is also treated with enzyme in

Table 5.2 Physicochemical characteristics of the *A. speciosa* biodiesel

Parameters	Bouaid et al. (2015)	Khaire and Deshmukh (2017)	Moreira et al. (2018)
Kinematic viscosity @ 40 °C (cSt)	3.0	4.46	3.04
Density @ 20 °C (kg m^{-3})	–	986	854
Flash point (°C)	–	132	–
Pour point (°C)	−6	12	–
Cloud point (°C)	−4	–	–
Cold filter plugging point (°C)	-8	–	–
Oxidation stability at 110 °C (h)	8.3	–	–
Acid value (g KOH kg^{-1})	0.24	–	0.013
Iodine value (g I_2/100 g)	13	–	48.6

microwave to prepare the biodiesel (Da Rós et al. 2014). But the enzymatic process is quite expensive. Often the biodiesel is added with propyl gallate to increase its stability during long storage. The properties of the biodiesel are given in Table 5.2.

The viscosity and the acid value of the *A. speciosa* biodiesel are within the limit prescribed. The pour point, cloud point and cold filter plugging point are quite low. Such low value is caused by the presence of methyl caprylate (C8:0) and methyl caprate (C10:0) with a melting point of −37.3 and −13.1 °C, respectively (Rashid et al. 2008).

In the alkali-catalysed transesterification, there are problems on account of a huge quantity of wash water being released in the environment which demands a separate efficient treatment system, emulsion formation creating complication in phase separation and corrosion of the vessel. To combat the above problems heterogeneous catalysis is in use. The heterogeneous catalysis has the advantage of being recovered from the reaction medium and can be utilized in many cycles to follow (Solis et al. 2015). CaO as a catalyst has strong Lewis base sites and is utilized very effectively. Tin oxide (SnO_2) is also a heterogeneous catalyst, which plays a dual role, as an agent both in esterification and transesterification processes. Tin oxide modified mesoporous silica is also used as a catalyst in biodiesel preparation.

Often, instead of methanol, ethanol is being considered. With respect to the preference of alkali, KOH is favoured as a lead option, as the transesterification is aided by a relatively low quantity of KOH. For example, 1% KOH produces 0.0125 mol of potassium ethoxide, whereas 1% NaOH gives 0.175 mol of sodium ethoxide. With 1% potassium hydroxide, at a molar ratio of 6:1 ethanol to oil in a reaction temperature of 30 °C, at a stirring speed of 400 rpm in a reaction duration of 60 min a recovery of 99% ester has become possible (Paiva et al. 2011).

5.1.3 Engine Performance

Under ideal operation, the three constituents of the biodiesel, namely hydrocarbon, oxygen and nitrogen participate in the combustion process and consequently water, carbon dioxide and nitrogen are released. Normally the biodiesel is blended with diesel and fed in the engine. The brake thermal efficiency decreases with the increase in biodiesel content of the blend. The CO_2 level in the exhaust, which is a reference to the combustion status, increases when the biodiesel level is high in the blend. High level of CO_2 is due to the rich content of oxygen in the biodiesel. The NOx level in the exhaust increases when the biodiesel component increases in the blend. High temperature and high pressure in the combustion chamber increase the NOx in the exhaust (Kale 2017).

When different combinations of biodiesel blend are tried in an internal combustion engine, the B20 blend (i.e. 20% of biodiesel in neat diesel) shows improved brake power at a compression ratio of 17. At the same time, the brake specific fuel combustion lowers in B20 when compared to neat diesel at a compression ratio of 18 (Khaire and Deshmukh 2017). At the same compression ratio, the volumetric efficiency increases and the CO and HC level in the exhaust decreases.

5.2 Coconut (*Cocos nucifera*)

The coconut palm (*Cocos nucifera L.*) also known to taxonomists as *Palma cocos* (Miller) is the only known species of the genus *Cocos* which belongs to the family Arecaceae and subfamily Coccoidea. This species is widely distributed in tropical and subtropical areas and is believed to have been originated from the Malay-Indonesian region. It has been meaningfully referred as tree of heaven and tree of life. The life span of it is around 60 years in which the economically productive period is up to 40 years only. During this period, it reaches an average height of 25 m with a canopy diameter of around 8 m. Such a tree yields around 80 fruits per year on an average (Fig. 5.3).

This tree grows at an elevation of 10–400 m above MSL near the equator having an annual rainfall ranging from 1500 to 2500 mm. It stands on a wide range of soil types. Every year around 65 billion fruits are being harvested globally. The fruits are harvested once in 2 months. In addition to the fruit, every part of this tree is beneficially used and hence there is a lot of interest among peasants and land owners to cultivate this tree, though the cessation period to get beneficial yield is quite high. This tree is called by different names in different countries (Table 5.3).

5.2.1 Distribution

The coastal belts of Malaysia, Indonesia, Melanesia and the Philippines are believed to be the birth places of this tree. It is also believed that the fruits of this tree were drifted by ocean currents to different parts of the world. It grows in tropical Pacific

Fig. 5.3 Coconut palm *Cocos nucifera*

Table 5.3 Vernacular names of the coconut palm in different countries

Names	Country
Coconut, coconut palm	England
Thenkai, thengai	India
Ha'ari	Society Islands
Iru	Palau
Lu	Yap, Kosrae
Ni	Pohnpei, Marshall Islands
Niu	Polynesia, Papua New Guinea, Fiji, Melanesia
Niyog	Guam, The Philippines
Nizok	N. Mariana Islands
Nu	Chuuk, Cook Islands
Te ni	Kiribati
Coco da Bahia, Coco da India, Coqueiro de Bahia	Portugal
Coco, Coco de agua, Cocotero, Palma de coco, Palmera de coco	Spain
Coco, Cocospalm, Klapperboom	The Netherlands
Coco, Cocotier, Cocoyer, Coq au lait, noix de coco	France
Kokospalme	Germany
Kelapa, nyior	Malaysia, Indonesia

Islands Polynesia, India, Sri Lanka, East Africa and many tropical islands, Seychelles, Andaman and Mauritius. It is also established in the coast of Central America, Southeast Asia, Burma, the Philippines, Singapore, Hainan region of South China, Thailand, Vietnam, Bangladesh, Africa, Cameroon, Ivory Coast, Kenya, Madagascar, Mozambique, Nigeria, Tanzania, Caribbean, Brazil, Ecuador, Jamaica, Mexico, Trinidad and Tobago, Venezuela, Fiji, Papua New Guinea, Solomon Islands, Vanuatu, Cook Islands, Hawaii, Kiribati, Line Islands, Nauru, Niue, Samoa, Tonga, Tuamotu Archipelago, Tuvalu Society Islands, Tokelau, Palau,

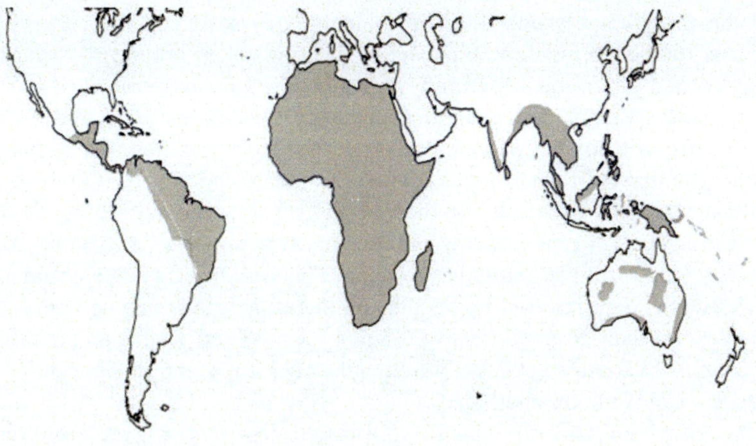

Fig. 5.4 Geographical distribution of the palm *Cocos nucifera*

Table 5.4 Distribution of coconut plantation area in the world in percent (Awaluddin et al. 2010)

Countries	%
Indonesia	31.2
The Philippines	28.5
India	16.0
Sri Lanka	3.7
Thailand	3.1
Other countries	17.5

Chuuk, Guam, Northern Mariana Islands, Pohnpei and Yap (CAB Invasive species compendium) (Fig. 5.4). More than 80% of the coconut plantation exists in five countries as shown in Table 5.4.

5.2.2 Botanical Description

Cocos nucifera L. is a giant palm whose leaf is fused till the completion of the first year (seedlings) and that slowly transforms in to a pinnate shape leaf called fronds as the seedlings grow in to a tree. The fronds have rachis and on either side of it linear lanceolate (200–250 numbers) are arranged. Each frond is 4–5.5 m in length with the petiole making up to a quarter of its length. The leaflet (lanceolate) is 1.5–5.0 cm wide and 50–150 cm long. The base of the petiole is very strong which helps to make firm attachment of the frond with the trunk of the tree. Each tree produces around 12–20 fronds per year. The fronds thus come out will stay in the canopy for a total period of 30 months. Around 30 number of fronds stay at the crown all the time.

Coconut palm is monoecious and in each inflorescence thousands of male flowers are seen arranged along with around 50 female flowers (buttons). Each female flower has six perianth segments in two whorls with a tricarpellate ovary and trifid stigma.

Out of the three carpels one ultimately emerges as seed and the rest will abort consequent to the pollination. First set of fruiting takes around 4–5 years after planting. A total of around 12–15 inflorescence develops in a year.

Based on the height and general appearance, two types (*Tall* and *Dwarf*) of coconut palms are known. *Talls* are by far the most commonly grown palms around the world. The flower of the *tall* type is cross pollinated and thus the *tall* expresses an array of heterogeneity especially on the size and shape of the tree, colour of the fruit and the yield. The *tall* type has two varieties *niu kafa* and *niu vai*. The *niu kafa* is a wild variety whose fruit is triangular in shape. This fruit has a small elongated nut in it. This variety is not preferred by people and therefore it is not being cultivated any longer. West African, Rennell (from Solomon Islands) and Tagnanan are few of the well-known *tall* varieties genetically homogeneous. It is small in size and is known to produce more fruits of small size.

On the other hand, *dwarf* type is self-pollinated. *Dwarf* type is preferred for home gardens, parks and for road side plantations. Dwarf tree comes into bearing mostly in 3 years. *Niu Leka Dwarf* (Fiji) and Samoan Dwarf (Hawaii) are examples for *dwarf* type.

Coconut tree has no tap root and is supported by innumerable (2000–5000) adventitious roots of around 1 cm dia each. The architecture of the roots and the depth of their penetration largely depend on the physical characteristics of the soil and depth of water table. Most of the roots are found within a depth of 1.5 m but they can go up to a depth of 25 m in search of water in drought prone areas.

5.2.3 Propagation

Freshly harvested fruits are kept buried partially (2/3 portion) in nursery bed of course sand with their micropyles facing top. The sand bed is to be kept damp throughout. Within a period of 12 weeks they germinate, the first leaf emerges as a compound leaf which resembles the ear of a rabbit. The seedlings each with two or three leaves are ready for removal in 10 months from the nursery bed for planting in the field. Any seedling with crippled leaf or twisted shoot is discarded.

Field planting is done in pits of the size 1 × 1 × 1 m at a gap of 10 m between pits. Each pit is half filled with course sand and in which the seedling is planted. Irrigation is to be made every day for a total period of 1 year and during the second year weekly irrigation is done.

5.2.4 Pest Management

Yellow leaf disease caused by *Mycoplasm*a is reported to be deadly serious. Bud rot caused by the fungus *Phytophthora palmivora* is a disease of great concern which is also fatal. There is another fungus *P. katsurae* causing coconut heart rot which is also difficult to control. The rhinoceros beetles *Oryctes rhinoceros* and *Scapanes australis* are known to consume the tender leaf primordium of the palm head and the resultant wound is infected by fungus causing the total rot. There are cases of stem rot, also being caused by a fungus.

5.2.5 Fruit

Few of the initial inflorescences of a new tree bear only male flowers through which there may not be any successful setting of fruits. As the age of the tree proceeds, the proportion of female flowers (button) in the inflorescence increases. From the day of pollination, it takes around 12 months for the fruit to mature.

The fruit is a fibrous drupe, which is elongated to spherical in shape, each weighing from 850 to 3500 g on maturity. The fruit has an excellent internal architecture. The outer skin is thin and waxy and is referred as exocarp which helps to waterproof the fruit. As a result, rain water never enters the fruit when the fruit stays on the tree top. The ripe fruit falls down from the tree and moves out taking a watery course for dispersal. Still moisture does not penetrate the exocarp. The next layer is fibrous in nature which is referred as mesocarp or husk. The mesocarp offers a cushioning effect and acts as a shock absorber so as to protect the seed present in it from any damage when the fruit falls down from the top of the tree. Interior to the fibrous mesocarp there exists a rigid shell called endocarp. It is extremely hard to penetrate or break the shell. This shell acts as a protectant to the flesh present inside which is called as endosperm or kernel (Fig. 5.5).

At the centre of the endosperm there exists a space which is filled partially or fully with nutrient water. The very purpose of the nutrient water is to maintain the kernel in healthy and wet condition. The exocarp and mesocarp together form around 40–45% of the fruit mass. The endocarp or shell constitutes around 28–30%, whereas the endosperm or kernel forms around 30% of the total fruit mass calculated without taking into account of the weight of the water which remains locked inside the endosperm. The average quantitative occurrences of various parts of coconut by weight were reported by Colin (2005). The values are given in Table 5.5 which

Fig. 5.5 Section of the nut and seed

Table 5.5 The component of the coconut fruit

Component	%
Husk (exocarp + mesocarp)	33.3
Shell (endocarp)	14.5
Endosperm	31.6
Water present within the endosperm	20.5

however accounts for the water content also. Table 5.5 indicates the component of the coconut fruit calculated from a sample of 1000 fruits (adopted from Colin 2005).

5.2.6 Seed

The seed has a dark brown shell called endocarp which encloses the white endosperm (coco flesh) and within which the nutrient water (coco water) is present. The endosperm yields the oil and hence considered to be an important part of the seed. The endosperm has moisture to the tune of 46%, oil 32% and the resultant cake 22%.

5.2.7 Oil Extraction

The coconut fruit is split open using an axe. The water present inside the endosperm is drained out. The coconut flesh (kernel) is then scooped out using a sharp knife and cut into slices. They were sun dried. Once the material is dried it is collected and stored in gunny bags for crushing.

As the work of scooping the endosperm and slicing are labour intensive the half split of the whole seed containing the endocarp and endosperm is directly dried (without scooping) in hot sun and during which time the kernel (endosperm) shrinks and get detached from the shell (endocarp). The half round dried kernel without being made as chips is further dried and packed. The above material is later made into small pieces by machine just before oil extraction.

Splitting open the fruit, scooping out the endosperm and drying them subsequently are carried out manually as suitable machines with satisfactory performance are not yet developed since there is no uniformity in the size of the fruit.

The kernel is required to be dried quickly so as to free it from moisture. If the moisture content of the dried kernel does not exceed 5–6%, it can be stored without serious loss of quality. If the dried materials contain moisture above the said limit, they will be attacked by microbes as they contain rich quantities of protein and sugar. Such attack by micro-organisms is likely to cause the formation of free fatty acid, aflatoxin and rancidity in the final product, namely oil. Occurrence of free fatty acid and rancidity in oil will adversely affect the processing of the oil for biodiesel production. Similarly, the formation of aflatoxin will affect the food quality of the oil in case the oil is used for cooking. Therefore, depending upon the location, the materials are quickly dried in hot sun or in hot air oven or in smoke.

The dried kernel is partially steamed so as to raise its temperature at a range of 104–110 °C for about 30 min. Exposure to steam for a minimum period of 30 min is recommended so as to have the penetration of the heat in all zones of the dried kernel thereby to keep the oil in a melting condition. The hot kernel is then quickly passed on to the oil extraction machine before the temperature falls below 100 °C. Often screw press is used for the extraction. It contains two type of screws (vertical moving and horizontal moving) along which the hot kernel moves. During the screwing, oil gets out from the mass leaving behind the cake.

Since fresh coconut flesh (endosperm) is used in food industry for the production of food grade milk (coco milk) there is a substantial quantity of resultant residues (solid wastes) which also contain fairly good amount of oil. Often this residue is collected from different food processing industries and pooled. They are then passed through screw press in wet condition. For this purpose, Ram Press (Bielenberg Ram press) is being used. The oil thus obtained is stored in a tank and the water content if there be removed by gravitational separation.

Similarly, fresh coconut kernel is processed through Direct Micro Expelling procedures. Freshly grated kernels are quickly dried so as to get a final moisture level of 9–12%. It is then passed through a direct micro expelling machine at a temperature ranging from 50 to 60 °C. The resultant pulp is then settled in a tank for a week. During the storage period the oil stands as a separate layer which is taken away and stored. The micro expelling machine consists of a press with interchangeable stainless steel cylinders, pistons, electrical grating machine and a dryer. A direct micro expelling machine is designed and fabricated by Kokonut Pacific Pty Ltd., Australia. Roughly 3.5 kg of grated fresh coconut kernel is required to produce around 1 kg of oil. This machine can process 300–600 nuts daily giving an output of 20–50 L of oil. The above machine requires 3–5 workers for routine operation.

Virgin coconut oil is produced by a simple method. Coconut kernel is quickly dried and crushed. Alternatively, fresh kernel without drying also is crushed. In either of the methods the crushing first yields coconut milk. Such coconut milk is then boiled or fermented or refrigerated or treated with enzyme or centrifuged mechanically to get pristine grade oil. Often fresh kernel is ground with the nutrient water present inside the kernel to prepare the coconut milk. This milk is kept undisturbed for half a day. During such period the water settles down to the bottom of the container which is first removed. The solid present in the milk floats at the top which also is removed. A crystal clear layer formed of oil, remains between the solid and water zones is separated, heated marginally to ensure the absence of water and stored. This oil is often referred as virgin coconut oil.

A good quality coconut oil is known by its aroma and also due to low level of free fatty acids. Good quality oil can be obtained if the kernel is processed carefully from a good quality raw material prepared from properly dried kernels. If the kernel becomes black due to extended storage, the oil extracted from it is likely to contain high quantity of free fatty acids, phospholipids or gums with unpleasant odour. In such events the oil is required to be refined. During the process of refining approximately 5–7.5% of the whole material is lost. Such waste oil is discharged along with the other effluents. The process of refining includes physical refining, neutralization, bleaching and deodorization. During the physical refining dilute phosphoric acid is added. The phosphatides and gums which are getting separated during the process are centrifuged, decanted and discarded. The purified oil is then treated with sodium hydroxide so as to remove the free fatty acids. This is known as neutralization. The remaining material is treated with activated charcoal or bentonite to bleach it. The bleaching agents added to it are then filtered out. The finished material is finally heated to a temperature between 150 and 250 °C so as to remove the odour.

5.2.8 Oil

The physicochemical properties of the coconut oil are given in Table 5.6. The heating value of this oil is slightly lesser than that of the diesel. If coconut oil is used as a fuel in a vehicle, it is possible to travel 90% of the distance covered by same quantity of diesel on a volumetric basis. The cetane number which indicates the degree of readiness of the oil to burn when it is under compression is observed to be rather high (58).

Similarly, the viscosity of the oil is several fold higher than that of the diesel. Therefore, there is a major limitation in atomizing the oil in the injector assembly since the oil is less volatile thereby causing poor spraying. The viscosity is observed to be reduced if it is blended with kerosene or diesel or by heating the whole oil. The oil solidifies if the temperature of the environment is low. At 24 °C it starts freezing. As the oil is predominantly formed of saturated fatty acids, it stacks easily due to the linear nature of the bonds resulting into the solidification of oil. Such freezing temperature is referred as solidification point or freezing point. As the coconut oil solidifies in cool climates, blending it with diesel reasonable gives relief to this problem. Iodine value of the coconut oil is relatively low which is an indication of the degree of unsaturation or the presence of bonds among the components of the free fatty acids. It gives an idea of the nature of the oil tending to polymerize. The saponification value is normally measured as the quantity of potassium hydroxide in number of milligrams required to convert 1 g of oil into glycerine or soap. The fatty acid composition is given in Table 5.7. It is to be noted that different fatty acids present in this oil have C8–C18 carbon atoms.

Table 5.6 The physicochemical properties of coconut oil

Parameters	Value
Kinematic viscosity at 40 °C (cSt)	30
Density (kg m^{-3})	0.92
Free fatty acid (%)	2.7
Iodine value (g I_2/100g)	106
Water (%)	0.3
Saponification value (mg KOH g^{-1})	263
Cetane number	58
Heating value (MJ kg^{-1})	38.4
Solidification point (°C)	24

Table 5.7 The fatty acid composition of the coconut oil (Hilditch 1956; Knothe et al. 1997)

Fatty acid	%	Total
Saturated		
Lauric acid C12:0	51.0	
Myristic acid C14:0	18.5	
Caprylic acid C8: 0	9.5	
Palmitic acid C16:0	7.5	
Capric acid C10:0	4.5	
Stearic acid C18:0	3.0	94.0
Unsaturated		
Oleic acid C18:1	5.0	
Linoleic acid C18:2	1.0	6.0

Coconut oil in its pure form is tried as a fuel in many vehicles which are running in remote islands where diesel is a scarce commodity. It is observed that pure coconut oil has a higher lubricating effect than any other fuels and as a result the engine parts are likely to have a long life. As the coconut oil burns slowly it pushes the piston all the way from the bottom of the cylinder instead of rapid explosion at the top of the stroke. As a result, the momentary power release is relatively smooth. The fuel consumption also is reasonably low with less engine wear. The engine also is not getting heated immediately due to the low internal friction and also due to the slow and steady burning rate. Coconut oil when used as a fuel the torque is observed to increase especially on driving the vehicle in steep roads so that a change to the next low gear is not often required as the engine keeps on pulling without failure at a lower rpm since the coconut oil burns slower than diesel. The exhaust of the engine running on coconut oil as fuel has less than 50% particulate matter, less sulphur dioxide and no polyacrylic hydrocarbon compared to the exhaust of the engine running on petro-diesel. It is safe to store and transport the oil. There is a reduced risk of fire and the oil is totally biodegradable. The burning of the coconut oil does not contribute to greenhouse effect since the quantity of CO_2 released to the atmosphere during the burning is very much equal to the amount of CO_2 incorporated from the atmosphere in the same plant through photosynthesis and thus there is no extra load to the environment. This concept is popularly referred as carbon neutral. The CO_2 from the air enters the plant, from plant to oil and from oil to the air. Thus there is no loss or no gain. Besides there is no harmful by-products as no additives are incorporated in the coconut oil. In spite of such advantages coconut oil is not used as a fuel in engines since there is no internationally accepted quality standards pertained to coconut oil as a fuel.

There are few limitations in using coconut oil as a fuel. As the oil normally contains substantial amount of water and free fatty acid, solidification of the oil in the fuel piping is quiet imminent especially when the environmental temperature goes down. To solve this problem blending is being resorted. The oil is often blended with methanol at a maximum of 5% level. But the performance of the above blend is not appreciable as methanol evaporates quickly at a warm condition. Attempts are made to blend the oil with kerosene at 15% level. Such combination is proved to be successful in generators, marine engines and transport vehicles especially in islands where procuring diesel is seldom possible. Alternatively, heat exchangers at fuel lines are being tried to avoid solidification. The water for such a heat exchanger is drawn from the thermostat bypass circuit so as to warm within a minute of the start of the engine. This arrangement eliminates the fuel blockage. On a practical point of view, it is observed to be difficult to avoid the presence of particles in coconut oil. Such particle is reported to chock the fuel filters. In such cases a bypass flow arrangement is normally added with a second filter in position. It is also recommended to spin the oil with ultra-high speed centrifuge or filter the oil with high pressure bag filter.

In many islands crude coconut oil is sold at just 45–55% of the cost of the diesel. The refined coconut oil also costs around 65–70% of the cost of diesel. Similarly, pure kerosene costs around 90% of the cost of diesel. Many engines which are running on 15% kerosene blended with coconut oil spend only around 80% of the cost of the diesel even if the 10% deficiency in the heating value and consequent low distance coverage are taken into account.

5.2.9 Biodiesel

As glycerin is admixed with the fatty acids of the oil, the volatility of the oil is greatly reduced. To improve the volatility, the removal of glycerin is necessary. Such removal parallelly reduces the viscosity and melting point. The removal process is called as transesterification and the resultant product is the ester. Ester having a complex structure inhibits molecular stacking and improves cold flow properties. The melting point of the ester is below zero degree centigrade. Its cetane and iodine numbers are nearly the same as that of the coconut oil.

Coconut oil biodiesel is prepared by treating the oil with methanol in the presence of alkali (KOH). The stoichiometric requirement of KOH is 4.9 g L^{-1}. This value is likely to change on account of the varying quality of the oil. Therefore, it is ideal to analyse the oil to know the actual requirement. Exactly 0.1 g KOH is dissolved in 1 L of dist. water. It is titrated against 1 mL coconut oil dissolved in 10 mL isopropanol with phenolphthalein as indicator. The appearance of pink colour is the end point. The number of mL consumed is added to 4.9 to get the total quantity of KOH to be used as catalyst. After the transesterification reaction, biodiesel (ester) and methanol floats and the lower layer comprising the alkali and glycerin are separated. The methanol present in the biodiesel is then distilled out and stored for the subsequent use. The resultant biodiesel is washed and cleaned repeatedly to free it from alkali. Since methanol is used in the process methyl ester is formed as end product. The methyl ester has a slightly low density, acceptable kinematic viscosity and cetane number. The heating value also has improved from 38.4 to 41.2 MJ kg^{-1} (Table 5.8).

The profile of the various methyl esters presents in a typical biodiesel made from coconut oil is given in Table 5.9. It is known that methyl laurate and methyl myristate form a sizeable portion of the esters.

Table 5.8 Properties of the methyl ester of the coconut oil

Parameters	Value
Density (kg m^{-3})	0.88
Kinematic viscosity at 40 °C (cSt)	4.4
Cetane number	55
Heating value (MJ kg^{-1})	41.2
Flash point (°C)	110
Water and sediment (%)	0
Copper corrosion 3 h at 50 °C	1A
Sulphated ash (% by wt)	0.003
Cloud point (°C)	−6
Acid number (g KOH kg^{-1})	0.2

Table 5.9 Profile of the methyl ester prepared from coconut oil

Ester	%
Methyl laurate	47.5
Methyl myristate	22.0
Methyl caprylate	7.2
Methyl palmitate	8.9
Methyl caprate	6.4
Methyl stearate	2.0
Methyl oleate	5.2
Methyl linoleate	0.4

It is to be noted that the alkali used in the transesterification process is washed away and it goes as an effluent in the environment causing land pollution. To combat such problems heterogeneous catalyst is used which is then separated out easily after the transesterification reaction is completed.

One of the heterogeneous catalysts is calcium oxide (CaO). This is strongly basic and offers a good catalytic activity. At the outset of the transesterification process methanol is mixed with calcium oxide. The methanol is first adsorbed on the surface of the catalyst. Consequently, there is surface diffusion. Accordingly, an unstable ionic compound is formed in which negatively charged oxygen ion of alcohol binds with positively charged calcium. The positively charged hydrogen of alcohol binds with negatively charged oxygen of CaO.

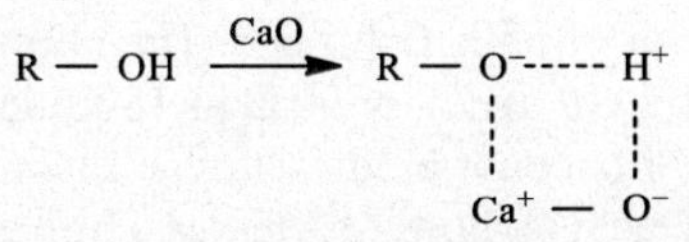

In the second stage RO attacks the carbon atom of the carbonyl group of the triglyceride.

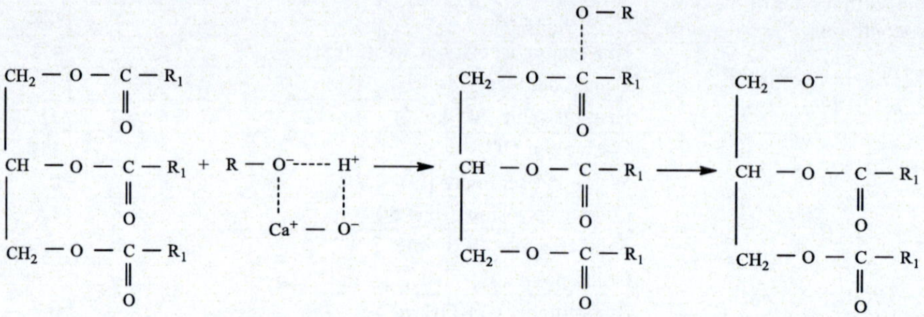

In the final stage, protons (H^+) moves to the diglyceride and binds with the unstable oxygen so as to create a monoester. This mechanism occurs two or more times to produce three alkyl monoester and glycerol. Once again the CaO is formed.

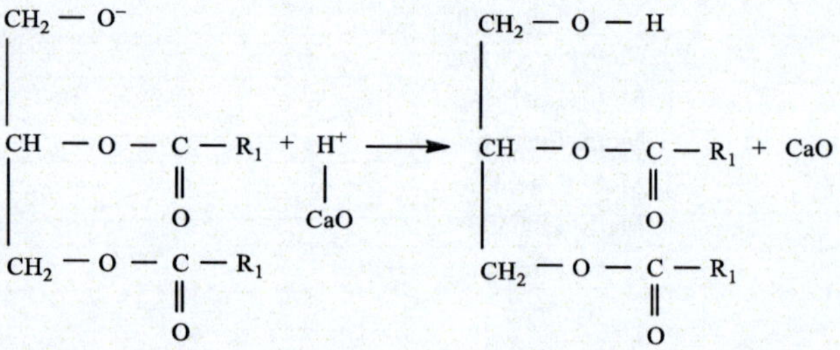

After adding the methanol CaO mixture to coconut oil, the content is heated to 58 °C and uniformly stirred slowly for 2 h. Once the reaction is completed the catalyst is removed by filtration and the resultant filtrate is allowed to settle for 4 h. On settlement two layers are formed. The bottom layer due to glycerin is then removed. The top layer (biodiesel) which stands mixed with methanol is further processed. It is added with warm water at 1:1 ratio, stirred for 5 min and allowed to stand undisturbed for 4 h. The top layer is biodiesel and the cloudy layer at the bottom containing methanol, glycerine and alkali is removed. Gamma alumina (γ-Al_2O_3) with a surface area of 150–300 $m^2\ g^{-1}$ is employed as a catalyst support which enhances the yield. The catalyst is prepared by adding 5 mL of 10% NaOH solution in 5 g of γ-Al_2O_3 and stirred for 30 min. It is then dried overnight in an oven at 100 °C and calcined in air at 550 °C for 3 h. The above catalytic mixture at 3% level gave an oil conversion efficiency of 90% in 180 min (Rasyid et al. 2018).

The distinct advantage of coconut biodiesel over coconut oil is that the biodiesel is formed of esters of fatty acids whose distillation temperature ranges from 125 to 316 °C against the petro-diesel whose distillation temp. ranges from 125 to 371 °C.

At the appropriate distillation temperature, the liquid transforms in to vapour and then burns with oxygen. Around 65% of the ester thus formed is from medium saturated fatty acids which offers better combustion. High level of saturation with medium carbon chain offers stability against oxidation which resists bacterial action. The high cetane number helps to enhance the combustion. The medium carbon fatty acid methyl ester especially lauric acid (C12) has a good lubricity. Therefore, coconut biodiesel may be best used in good old engines.

The carbon deposits formed in the combustion chamber and the clogged fuel nozzles are slowly cleaned by self-cleaning mechanism. The engine also runs smoothly with low vibration and sound. The smoke density of the engine running with coconut oil biodiesel is relatively low. Such low smoke is due to the presence of rich oxygen in esters and also due to the typical molecular structure. The specific fuel consumption of biodiesel is relatively higher than that of the petro-diesel. This is due to the fact that biodiesel has considerable quantity of oxygen thereby other component such as C and H share is lowered.

The hydrocarbon concentration in the exhaust gas is an index of the unburned fuel. The level of hydrocarbon in the exhaust gas of the engine running with pure coconut oil is higher by 20% than that of diesel. But the hydrocarbon release in the exhaust of the engine running with coconut based biodiesel (ester) is lower by 15% than that of the diesel. The high temperature of the cylinder due to the active combustion making use of the rich quantity of oxygen causes the increase in NO_x in the exhaust gas. When pure coconut oil is used as a fuel the temperature of the cylinder is not increased as that of the biodiesel and therefore NO_x is not higher in such cases. The CO_2 emission in the exhaust was around 20 and 10% less for coconut oil and coconut oil biodiesel, respectively, than that of petro-diesel.

5.3 Oil Palm (*Elaeis guineensis*)

The African oil palm *Elaeis guineensis* sparingly referred as *Elaeis melanococca* which is indigenous to the tropical rainforest zone of West and Western Central Africa from Guinea and Northern Angola (11 °N to 10 °S). This species is believed to have its origin from South–Eastern Nigeria and Western Cameroon. It was introduced in Indonesia in 1848 and developed as a major plantation in 1911 which moved to Malaysia in 1917. Commercial plantation of this species came in to being in Africa in 1920. Oil plant plantation and oil extraction activity took off in 1970s in Malaysia, Indonesia, Ecuador, Philippines, Solomon Islands, China, India, Sri Lanka, Brazil, Kenya, Tanzania and Uganda. It also occurs as invasive species in Angola, Benin, Bolivia, Burundi, Cambodia, Central African Republic, Christmas Island, Colombia, Congo, Cook Islands, Dominican Republic, Equatorial Guinea, Fiji, French Polynesia, Gabon, Gambia, Ghana, Guam, Guatemala, Guinea, Guinea Bissau, Honduras, Ivory Coast, Liberia, Madagascar, Malawi, Marshall Islands, Martinique, Mauritius, Mexico, Micronesia, Myanmar, Caledonia, New Caledonia, Nicaragua, Nigeria, Niue Palau, Panama, Paraguay, Peru, Reunion, Sao Tome and Principe, Senegal, Sierra Leone, Sudan, Suriname, Togo, United States, Venezuela, Vietnam and Zimbabwe (Source: Invasive Species Compendium ISC http://www.cabi.org/isc/datasheet/20295; Global Invasive Species Database (GISD) http://www.issg.org/database/species/ecology.asp?si=377) (Fig. 5.6).

The vernacular names of this species being in force in different countries are given in Table 5.10.

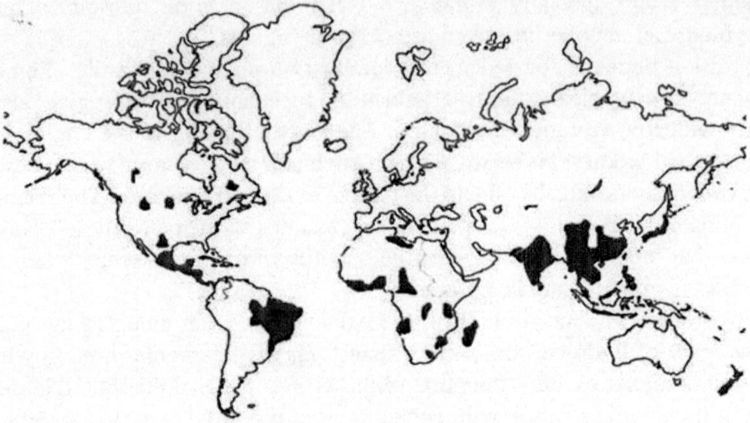

Fig. 5.6 The geographical distribution of oil palm *Elaeis guineensis*

Table 5.10 The vernacular names of oil palm in different countries

Country	Vernacular names
Saudi Arabia	Nakhlet ez Zayt
Brazil	African oil palm, Caiaue
Burma	Sihtan, Si ohn, So Htan
China	You zong,
Danish	Oliepalme
England	Nutamara, African oil palm, Macaw fat tree, Oil palm
Finland	Oljypalmu
France	Palmier a huile, Palmier a huile
Germany	Afrikanische olpalme, Olpalme
Italy	Palm avoira, Palma da olio, Palma oleaginosa Africana
Japan	Palma oleaginosa Africana, Abura yashi
Malay	Kelapa sawit, Kelapa sawil bal
Portugal	Dendenzeiro, Palmera dendem, Palmeira dende
Russia	Maslichnaia palma, Palma maslichnaia
Spain	Palma africana, Palma oleaginosa Africana, Palmera de aceite
Swahili	Mchikichi, Miwesi
Sweden	Oljepalm
Thailand	Paam nam man

5.3.1 Characteristics

The genus *Elaeis* has two species *E. guineensis* and *E. oleifera*. *E. oleifera* has low oil content and hence not well preferred. *E. guineensis* is a tall palm (8–30 m height). The trunk is stout, erect, solitary and has a diameter of 75 cm. Leaves are large and pinnate. The leaf stalk has spines and each tree has 40–50 leaves. These leaves are arranged spirally. The petiole is 1–2 m long.

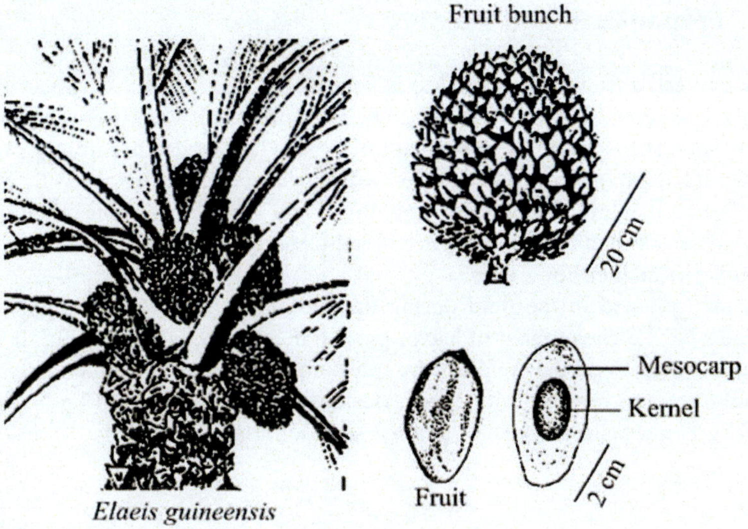

Fig. 5.7 *Elaeis guineensis*

The inflorescence is auxiliary, short and compact. The peduncle is around 35 cm long. The inflorescence is formed as spindle shaped bracts before anthesis. The male inflorescence is oval in shape and 20–35 cm long having around 1000 flowers (Fig. 5.7). The female inflorescence is globose and 25–40 cm long. It has fleshy branches each with 35 cm wide weighing around 15 kg with 2000 fruits. Each fruit is globose or elongated or ovoid weighing around 20 g. The fruit has a woody stigma and a smooth exocarp. The exocarp is often shiny and orange in colour when ripe. The endocarp is brownish in colour with a hard consistency. Its longitudinal fibres are drawn out in to a tuft at its base and the germ pores at the apex encloses a single seed. The seed (kernel) has a deep brown testa. The endosperm is grey white and oily.

5.3.2 Habitat

The oil palm grows well in tropical lowlands and occurs at the edges of the swamps and river banks. It prefers maximum sunlight and unlimited water availability. It requires an annual precipitation of around 2000 mm, high air humidity and 2000 h of sunshine a year. Optimum mean minimum temperature is 29–33 °C. It manages latosols, volcanic soils, alluvial clays and peat soils. It tolerates a wide fluctuation of soil pH (4.2–5.5). It grows in deep soil (>1.5 m) with a soil water availability at a field capacity of 1.0–1.5 mm cm^{-1} soil depth. The soil should drain well with organic carbon >1.5% and cation exchange capacity >100 m mole kg^{-1}.

5.3.3 Propagation and Planting

Freshly harvested fruits are cleaned to isolate seeds. These seeds are dried in shade so as to get a seed moisture of around 20%. They are then stored in low temperature (20 °C). Such seeds are to be sown within a year. The seeds are initially placed in sand flats at a depth of 1 cm and covered with sawdust. These flats are fully exposed to sun. The seeds germinate within 3 months of sowing. The germinated (single leaf staged) seeds are then transplanted in polyethylene bags of the size 40 × 60 cm in which it is maintained for a year.

It is then planted in the field during the pre-monsoon period at a gap of 9 m (140 plants ha^{-1}). The number of leaves present in a tree decides the yield. If the tree has less than 35 leaves, the yield lowers substantially. Hence it is imperative that the tree shall have an average of 40 leaves. Besides organic manure 0.5 kg N, 1.0 kg P and 0.8 kg K per palm per year is applied as basal fertilizer.

5.3.4 Harvesting

Roughly after 3 years of growth the tree puts forth flowers in the form of inflorescence. The first appearing inflorescence is usually ablated so as to activate the plant to produce more inflorescences. The first set of fruit bunches normally have low oil content. Bunches ripen all through the year and harvesting is being done on weekly basis. In the young plants bunch ripening is indicated by the falling down of five fruits per bunch. In older plants when ten fruits fall down the bunch is ready for harvest. Bunches are normally harvested by a sharp knife or chisel.

The fruit bunches at the outset are sterilized in steam under pressure. This helps to loosen the individual fruits from the bunch. Such heating also arrests the action of lipolytic enzyme so as to prevent the formation of free fatty acid in the oil to be extracted. It is then followed by stripping the fruits. These fruits are macerated and repeatedly heated so that the pulp and nuts move apart. The pulp mass is then passed through a screw or hydraulic press. The liquid discharged from the press is centrifuged so as to separate the oil from water and sludge.

In the next stage the nuts are removed, dried, cleaned and cracked in a machine to remove the shell. The shell-free material (kernel) is then subjected to oil extraction. Thus two types of oil are extracted from the same bunch (oil from the mesocarp called palm oil and the oil from kernel called palm kernel oil). Both the oils occur at a volume ratio of approximately 9:1, respectively.

Around 50% of the mesocarp by wt. is formed of oil. Palm oil is used in cooking and for preparing potato chips, pastry, confectionary and ice cream. Besides margarine, vegetable ghee, bakery fats are being manufactured from palm oil. Low quality palm oil is diverted to the manufacture of soap, detergents, candles, resins, lubricating greases, cosmetics, glycerol and fatty acid. It is also utilized in tin plating and sheet-steel manufacturing. It is also employed as a plasticizer and stabilizer in plastic industry. The palm oil is richly utilized as raw material in biodiesel.

Palm kernel oil (made from kernel) is almost equal to coconut oil and is in demand in the manufacture of margarine, edible fats, ice cream and confectionaries. Around 50% of the kernel is formed of oil. The press-cake or kernel meal is being utilized as cattle feed.

5.3.5 Characteristics of Palm Oil

The palm oil is fractionated in to palm olein and palm stearin. The palm olein has two grades, namely standard palm olein and super palm olein. Similarly, palm kernel oil is fractionated in to palm kernel olein and palm kernel stearin.

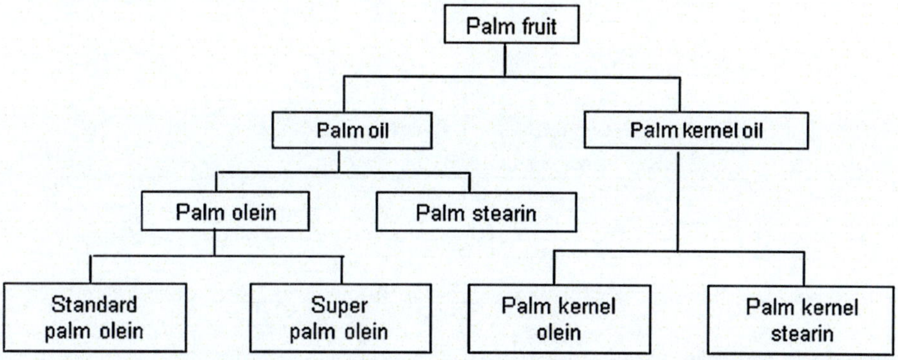

The palm oil is characterized by a higher density, higher viscosity and lower heating value when compared to that of regular diesel. Density and viscosity are important parameters which govern the flow of fuel in the pipeline, nozzles and orifices besides the atomization also is altered. The process of combustion and the constituents of emission largely depend on the properties of the fuel. The palm oil has a balanced saturated and non-saturated fatty acid. The properties of palm oil are given in Table 5.11.

The palmitic acid contributes 32–51% and oleic acid makes 37–52% of the oil. Low level of linoleic acid and linolenic acid causes the acid more stable. The crude palm oil obtained from the mesocarp is cooled to a low temperature and the crystal formed consequently is separated. The liquid thus isolated is referred as palm olein and the crystals are termed as stearin. The palm olein if blended with oils such as soybean oil, corn oil and canola oil gives clear consistency even at 0 °C. The oxidative stability also enhances. The standard palm olein has an iodine value of 56–59 g I 100 g^{-1} and has a cloud point of 10 °C max and it remains as liquid at 20–25 °C, whereas the super palm olein has an iodine value greater than 60 g I 100 g^{-1}. The super palm olein is suited to cooler climate as it has a low cloud point (2–5 °C) and remains in liquid form even at 15 °C. The stearin portion cannot be utilized for biodiesel.

Table 5.11 The properties of palm oil (Salunkhe et al. 1992; Knothe and Dunn 2001; Tyson 2001; de Almeida et al. 2002; Ndayishimiye and Tazerout 2011)

Characteristics	Palm oil
Density at 25 °C (kg m^{-3})	888–915
Specific gravity at 20 °C	0.91
Flash point (°C)	>344
Pour point (°C)	6.0
Cloud point (°C)	14.0
Refractive index—npb—50 °C	1.45
Ash (%)	0.003
Viscosity at 40 °C (cSt)	38.23
Saponification value mg KOH g^{-1}	200
Unsaponifible matter (% wt)	0.3
Iodine value (g I 100 g^{-1})	52.0
Heat of combustion (MJ kg^{-1})	39
Acid value (mg KOH g^{-1})	0.23
Lauric acid C12: 0 (%)	0.35
Myristic acid C14: 0 (%)	0.5–6
Palmitic acid C16: 0 (%)	32–51
Palmitoleic acid C16: 1 (%)	0.30
Stearic acid C18: 0 (%)	1–6.3
Oleic acid C18: 1 (%)	37–52
Linoleic acid C18: 2 (%)	2–12
Linolenic acid C18: 3 (%)	0.5
Arachidic acid C20: 0 (%)	0.5

Table 5.12 Fatty acid profile of palm kernel oil

Fatty acid	(%)
Lauric acid (C12: 0)	48.2
Myristic acid (C14: 0)	16.2
Palmitic acid (C16: 0)	8.4
Capric acid (C10: 0)	3.4
Caprylic acid (C 8: 0)	3.3
Stearic acid (C18: 0)	2.5
Oleic acid (C18: 1)	15.3
Linoleic acid (C18: 2)	2.3

5.3.6 Properties of Palm Kernel Oil

The palm kernel oil is obtained from the kernel of the seed. The characteristics of kernel oil differ substantially from that of the palm oil. The free fatty acid level normally remains lesser than 2%. The palm kernel oil also is utilized as a raw material for biodiesel. The palm kernel oil mainly contains lauric acid (C12: 0) and other saturated fatty acids. Countries which produce palm kernel oil in large quantities are Indonesia, Malaysia, Nigeria, Colombia, Thailand, Zaire and Ecuador (USDA 1998). The fatty acid profile of palm kernel oil is given in Table 5.12.

5.3.7 Biodiesel Production

One of the methods of preparing biodiesel from palm oil is alkali catalysed transesterification. The oil is allowed to react with short chain alcohol (methanol or ethanol). The catalyst used is sodium hydroxide or potassium hydroxide. The catalyst is mixed with alcohol and taken in a closed reaction vessel. With it oil is added and mixed at 60 °C. The reaction time is around 2 h. The biodiesel and glycerin are formed as products. The biodiesel (ester) moves to the top due to its relatively low density and glycerol sinks down which is then separated physically by gravitational segregation. Both the products contain small quantity of alcohol and alkali admixed in them. They are heated and the alcohol is recovered by fractional distillation. They are then washed with hot water till they remain neutral to litmus paper.

The biodiesel is analysed so as to ensure compliance of standard. For global marketing it is required to register the product with the United States Environmental Protection Agency under 40 CFR Part 79 (Anonymous 2009). As high as 98% biodiesel recovery is possible in this method.

Attempts are being made to prepare biodiesel through supercritical methanol mode in which the use of alkali or acid is avoided. The tedious purification and separation steps are not involved. In the supercritical process methanol and oil are mixed and subjected to a high temperature (350 °C) and pressure (8.1 MPa). Under the above reaction conditions methanol enters in to a homogeneous state with oil. The material is loaded in a reactor which can sustain the required temperature and pressure for a given duration. In this method the molar ratio of methanol to oil is relatively high. The effect of molar ratio of methanol to oil in supercritical methanol mode is given in Fig. 5.8. The process flow chart is shown below:

The ester (biodiesel) yield slowly increases when the molar ratio is increased. Beyond the optimum ratio the recovery declines. The optimum methanol to oil ratio is several fold higher compared to normal operation due to the pressure involved. As a result, more ester is formed apparently due to higher contact area between methanol and triglycerides (Kusdiana and Saka 2001). It appears that the reaction

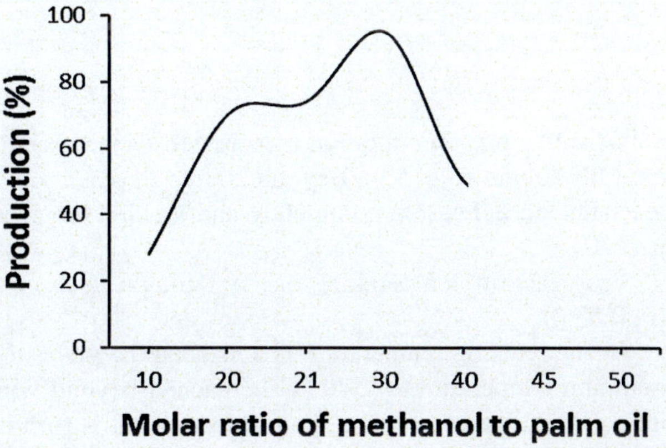

Fig. 5.8 Effect of molar ratio of methanol to palm oil on the ester yield at 350 °C and at a reaction duration of 20 min in supercritical mode

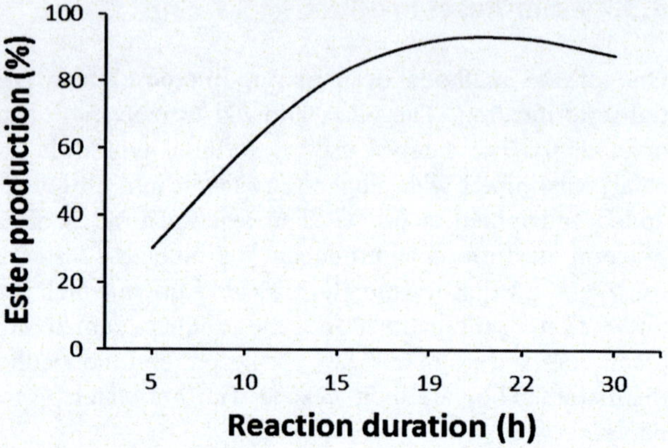

Fig. 5.9 Effect of reaction duration on the yield of ester at 350 °C, 8.1 MPa and methanol to oil molar ratio of 30:1

equilibrium is reached at 30 molar ratios (Song et al. 2008). Similarly, the reaction duration has an effect on the yield.

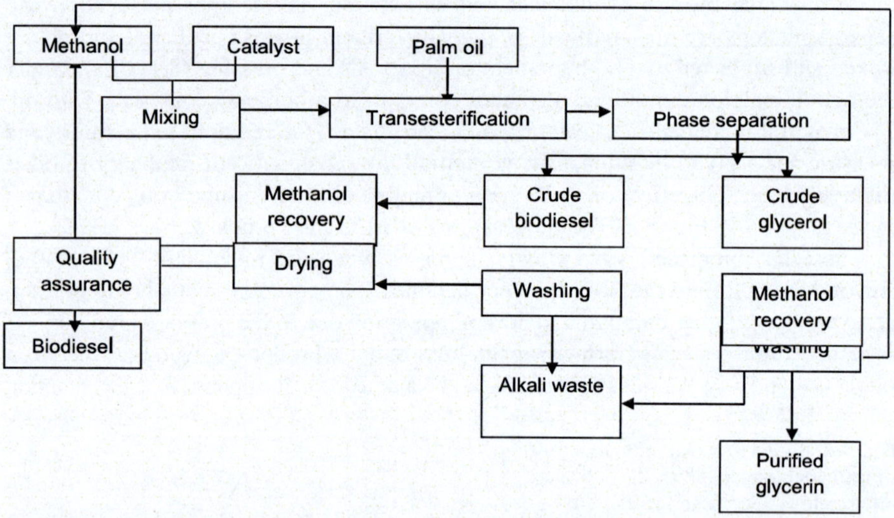

When the reaction duration is increased there is steadfast increase in the yield of ester till 20 min (Fig. 5.9). Beyond 20 min the ester yield is not increasing since the reactants are exhausted completely and yielded the product in full within the said period.

The reaction temperature has a telling effect on the yield of the product (Fig. 5.10).

As the reaction temperature is enhanced the yield of ester also increased till the optimum temperature of 370 °C is reached beyond which the temperature has no influence on the reaction as the product in full is formed in 350 °C itself. Beyond

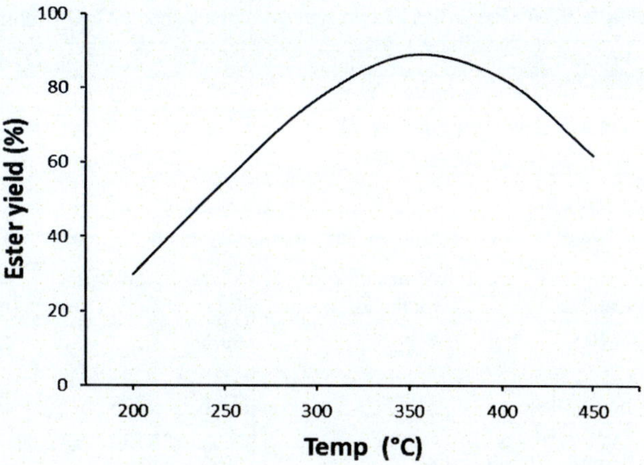

Fig. 5.10 Effect of reaction temperature on the yield of ester at 8.1 MPa, molar ratio of 30 of methanol to oil and duration of 20 min

Table 5.13 Comparison of supercritical process with catalytic transesterification (Tan et al. 2009)

Operational parameters	Supercritical process	Homogeneous catalyst	Heterogeneous catalyst
Temperature (°C)	360	70	150
Reaction duration (min)	20	60	120
Methanol to oil ratio	30:1	6:1	10:1
Catalyst (%)	–	0.1	4.0
Ester yield (%)	72	78	79

350 °C, it appears that there is decomposition of polyunsaturated methyl esters and untreated triglycerides (Imahara et al. 2008).

Tan et al. (2009) made a comparison of the efficiency of non-catalytic supercritical method with that of the homogeneous and heterogeneous catalytic methods (Table 5.13). The supercritical method has a margin over the others especially in lowering the reaction duration. The recalcitrant catalyst is absent in supercritical method which relieved the supercritical procedures from cumbersome waste water treatment. Besides the free fatty acids and the moisture content of the oil do not affect the yield of ester as no unwanted saponification product is formed. The free fatty acids actually act as a raw material for the production of ester and thus the yield increases (Kusdiana and Saka 2004).

Recently cobalt doped MgO as solid oxide catalyst was tested. Application of such catalysts avoids the production of soap through fatty acid neutralization (Albuquerque et al. 2008; Sharma et al. 2008; Li et al. 2009; Sree et al. 2009). Besides there is an increase in the efficiency (Wen et al. 2010).

Cobalt doped MgO heterogeneous catalyst is prepared by co-precipitation and calcination of the precursors. Mg $(NO_3)_2$ (0.1326 M) is mixed with 0.0206 M Co $(NO_3)_2$ in a flask and stirred at 600 rpm. The metal Mg and Co hydroxides are precipitated from their nitrate ion using 4 M potassium hydroxide. It is then filtered

Table 5.14 Properties of two types of cobalt doped MgO (Rahman et al. 2011)

Properties	$Mg_{1.7}Co_{0.3}O_2$	$Mg_{0.3}Co_{1.7}O_2$
BET surface area ($m^2\ g^{-1}$)	5.37	15.82
Total pore volume ($mm^3\ g^{-1}$)	29.59	125.64
Average pore diameter (nm)	22.04	31.76

Table 5.15 Performance of different catalysts on the transesterification of palm oil

Catalyst	Molar ratio of methanol	Yield (%)	Duration (h)	Temperature (°C)	Reference
H_2SO_4	8:1	99.5	1	70	Chongkhong et al. (2007)
KOH/Al_2O_3 KOH/NaY	15:1	91.1	2–3	70	Noiroj et al. (2009)
Montmorillonite	8:1	79.6	3	190	Kansedo et al. (2009)
KSF/KF/Ca-Al	12:1	99.7	3	65	Gao et al. (2010)

and dried at 75 °C for 24 h. Subsequently the above is calcined at 300 °C for 4 h and was used in the transesterification process.

The oil, methanol (methanol to oil molar ratio 9) and the catalysts (5% level) are loaded in the reactor and stirred at 700 rpm at an elevated temperature (150–175 °C). After the reaction (2 h) the catalyst is removed by centrifugation at 2000 g for 15 min. The ester and glycerin are separated gravitationally. The efficiency of ester yield is around 90%. The separated catalyst is repeatedly washed with methanol and dried at 80 °C for 12 h. The resultant product is used again in transesterification. The properties of the catalyst are given in Table 5.14.

The surface area and pore diameter enhance the catalyst activity as the reactant and product pass through the pore easily. The porosity is actually induced by the presence of Mg^{2+} (Olutoye and Hameed 2010) which forms a cluster around the mixed oxides. The function of the catalyst has a limitation if the dose is increased beyond 5%. The reaction rate is largely determined by the surface area and mass transfer rate. Additional amount of catalyst increases the viscosity of the liquid mixture resulting in to resistance to mass transfer.

The KF/ZnO catalyst also is used in the transesterification of the palm oil. KF/ZnO is prepared by simple impregnation of ZnO with KF $2H_2O$. The above is dried at 110 °C for 8 h and then calcinated at 873 K for 5 h in a muffle furnace.

Palm oil and methanol is mixed at a ratio of 1:11 and is joined with the above catalyst at a level of 5.5%. It is then vigorously stirred for 10 h at 65 °C. This process gives a yield of 90% of ester (Hameed et al. 2009). The yield of biodiesel from palm oil using different catalysts is given in Table 5.15.

5.3.8 Microemulsion Based Biodiesel

Microemulsions are colloidal dispersions of homogeneous and thermodynamically stable materials caused by interfacial film of amphiphile occurring due to surfactant and co-surfactant. Microemulsion technique is receiving attention in the fuel production since it helps to lower the viscosity of oil without the involvement of huge quantity of waste water. The mean droplet size in the emulsion is in the range of 2–200 nm. Selecting an appropriate surfactant is considered very important. Non-ionic surfactants are best preferred for this purpose. Structure and nature of the surfactant affect the micelle formation and subsequent aggregation size and thus largely influence the bulk viscosity.

Polyoxyethylene, oleyl alcohol, stearyl alcohol and methyl ester of oleic acid are employed as surfactants. 1-butanol, 1-octanol, 1-decanol and 2-ethyl hexanol are used as co-surfactants (Arpornpong et al. 2014). Ethanol is normally used as polar liquid phase.

A mixture of surfactant and co-surfactant is prepared at a mol ratio of 1:8. Ethanol and palm oil biodiesel are then added to the above mixture and shaken smoothly and then kept undisturbed in a water bath which is set at a pre-determined temperature so as to have the mixture reach an equilibrium. A clear, transparent and homogeneous solution indicates the completion of the process (Arpornpong et al. 2015). The characteristics of the microemulsion based biodiesel are given in Table 5.16.

5.3.9 Winter Grade Palm Oil Biodiesel

Palm oil biodiesel is mixed with equal amount of methanol and chilled at 5 °C for 24 h. The mixture is then filtered using a vacuum suction pump. The methanol present in the filtrate is expelled by nitrogen purging. The finished product is referred as winter grade palm oil biodiesel. The properties of winter grade palm oil biodiesel are given in Table 5.17.

Oil from the kernel of the palm fruit also is used to produce biodiesel (Abigor et al. 2000; Alamu et al. 2007a, b, 2008). The palm kernel oil is transesterified using the conventional method but ethyl alcohol is used instead of methyl alcohol. The ethyl alcohol is mixed with potassium hydroxide at a weight ratio of 10:1 so as to get potassium ethoxide.

The palm kernel oil is heated to a temperature of 60 °C and the potassium ethoxide is added to it at 20% level by weight. The mixture is stirred at 700 rpm

Table 5.16 Characteristics of microemulsion based biodiesel

Properties	Values
Average droplet size (nm)	21.86
Density at 25 °C (kg m^{-3})	0.85
Kinematic viscosity at 40 °C (mm^2 s^{-1})	4.00
Heat of combustion (MJ kg^{-1})	39.20
Flash point (°C)	15.01

Table 5.17 Properties of winter grade palm oil biodiesel (Sadrolhosseini et al. 2011)

Properties	Values
Viscosity at 40 °C ($mm^2\ s^{-1}$)	4.423
Density at 15 °C ($kg\ m^{-3}$)	870
Cloud point (°C)	−18
Pour point (°C)	−21
Cetane number	53
Acid value ($mg\ KOH\ g^{-1}$)	<0.5

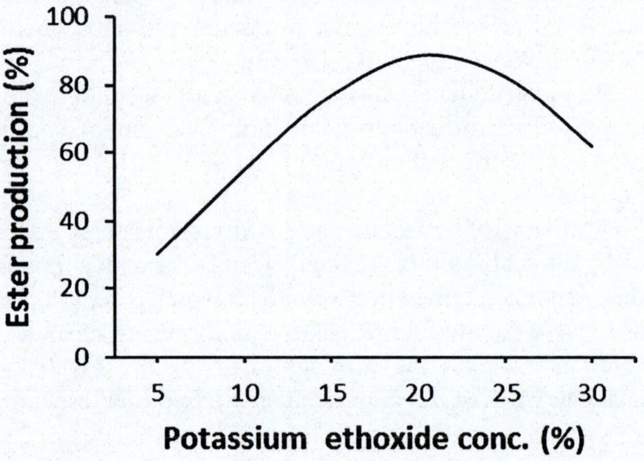

Fig. 5.11 Effect of potassium ethoxide concentration on the ester yield at an ambient temperature of 65 °C and reaction duration of 100 min

for 90–100 min. It is then allowed to settle for a night. The brown glycerol which settles down is removed. The supernatant (biodiesel) is repeatedly washed with hot water (Alamu et al. 2008). It is known that the potassium ethoxide has a decisive role in the yield of ester.

Biodiesel yield increases only up to 20% limit and no further increase in yield is possible beyond the said limit. The highest yield of ester is 96% at a potassium ethoxide level of 20% (Fig. 5.11). When the level of potassium ethoxide is lowered the glycerol formed does not consolidate. When the potassium ethoxide is increased the glycerol quantity tends to increase.

5.3.10 Use of Raw Palm Oil in Electrical Generators

Often raw palm oil is used as fuel in diesel engines. In such events the initial startup operation is carried out by diesel. Once the engine comes to normal state the engine is switched over to raw palm oil. During the time of closure also the engine is once again brought to diesel mode to flush out any palm oil staying in the fuel line, pump and injection assembly so as to avoid any deposit due to active reduction in temperature at the final stage.

In order to manage the problems due to high viscosity the oil is heated before being administered. The preheated palm oil when used in engine emits low quantity of CO, HC and PM but the emission of NO_x increases (Kalam and Masjuki 2004). Crude palm oil causes higher peak pressure, shorter ignition delay and a longer combustion period (Bari et al. 2002a). Normally the combustion chamber suffers carbon deposit, wear of piston rings, uneven spray formation. Besides the cylinder linings are scuffled (Bari et al. 2002b). The carbon deposit in the injectors increases the exhaust temperature if the engine is put on continuous running. The specific fuel consumption biodiesel increases by around 8–10% against regular diesel. The oil of low heating value demands high flow to meet the energy input of the engine. The CO level in the exhaust increases since the oil exhibits low atomization thereby causing poor mixing with oxygen (Altin et al. 2001). Such emission of CO increases along with the load. The hydrocarbon level also increases due to poor ignition. Characteristically the NO_x emission is relatively low apparently due to increase in turbulence intensity in the combustion chamber (Crookes et al. 1997; Altin et al. 2001).

5.3.11 Characteristics of Palm Kernel Oil Biodiesel

The properties of palm kernel oil biodiesel are given in Table 5.18.

Though the viscosity of biodiesel is marginally higher than that of the regular diesel, the pour point and cloud point are phenomenally higher which impose limitation of using this in all occasions.

5.3.12 Engine Performance of Palm Oil Biodiesel

Often palm oil biodiesel is added with 4-nonylphenoxy acetic acid as additive to control the emission of NO_x and CO in diesel engines. A 1% blend increases the brake power and lowers the specific fuel consumption (Kalam and Masjuki 2008). Similarly, the biodiesel is mixed with a limited quantity of ethanol or *n*-butanol or diethyl ether to increase its performance. Among them *n*-butanol is greatly considered as a good additive since it has a relatively high heating value, higher cetane number, low hydrophilic tendency and higher miscibility. Thus addition of *n*-butanol increases the brake power and combustion efficiency due to decrease in density, viscosity and enhanced atomization. The palm oil biodiesel is formed predominantly

Table 5.18 Properties of palm kernel oil biodiesel (Alamu et al. 2008)

Properties	Values
Viscosity at 40 °C ($mm^2\ s^{-1}$)	4.839
Specific gravity at 40 °F	0.883
Cloud point (°C)	6.0
Pour point (°C)	2.0
Boiling point (°C)	320
Flash point (°C)	167
Heat value ($MJ\ kg^{-1}$)	40.56
Sulphated ash (%)	0.018

of saturated molecules due to the presence of limited number of double carbon bonds and therefore it expresses high ignition quality.

The palm oil biodiesel is characterized by a short ignition delay due to high cetane number. Very low quantity of fuel is ignited during premixed combustion. The fuel is injected at the end of piston's compression stroke which is referred as start of injection. Rapidly the fuel disintegrates in to droplets at high temperature. Then it is vaporized, mixed and go in to self-ignition. The time gap between the above phenomena is referred as ignition delay which is very short in palm oil biodiesel. This leads to low pressure and temperature in the cylinder. Naturally such short ignition delay moderates the emission of NO_x and hydrocarbon especially at low temperature and pressure. Normally in diesel engines the emission contains rich quantity of particulate matter which predominantly of soot. Such particles easily move in to the deepest part of human lungs during inhalation. The soot formation is low in the use of biodiesel since the high level of oxygen which exists in the fuel molecule prevents the formation of soot to the best of possibility.

At low temperature, there will be crystal formation in the biodiesel. These crystals are likely to clog the filter. The temperature at which such phenomenon takes place is referred as cold filter plug point. In the case of palm oil biodiesel such plug point develops at 12 °C and therefore to avert any adverse eventualities the fuel is blended with regular diesel.

5.3.13 Prospects of Biodiesel

The demand for vegetable oil is increasing day by day. The world demand is expected to increase to 150 million metric tonnes during the year 2020. It is in this context the role of palm oil assumes prominence. Under best management practice a hectare of land may produce 4–6 times more oil than any other sources. Crop growth model indicates that the palm oil bears a scope of 12–14 t ha^{-1}. The palm oil meets the world vegetable oil demand by 36%. The world production of palm oil against certain prominent oils is given in Table 5.19.

The world production of palm oil over the previous years is given in Fig. 5.12.

In a matter of 20 years (from 2000 to 2020) the palm oil production doubled. Biodiesel as a fuel source in automobile sector is required to comply with the International Standard ASTM D 6751. By and large it is opined that at the first instance palm oil biodiesel may be mixed with regular diesel at 5–10% level. The

Table 5.19 Comparison of the production of certain oils in the world during the year 2015

Oil crop	World oil production (million metric tonnes)	Average oil yield (tonnes ha^{-1} $year^{-1}$)
Oil palm	45.9	3.74
Soy bean	38.9	0.38
Rapeseed	22.3	0.67
Sunflower	12.1	0.48

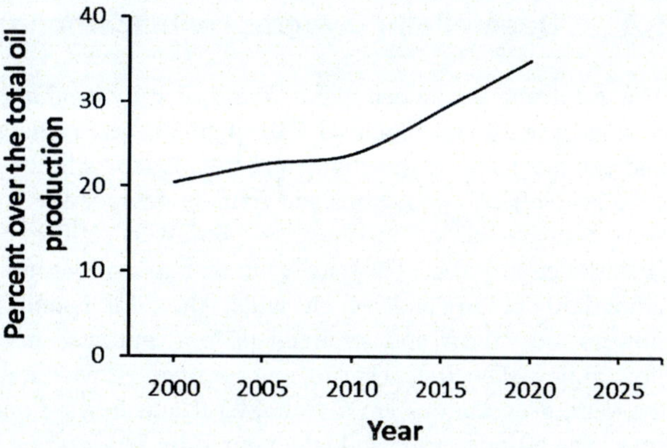

Fig. 5.12 World production of palm oil from 2000 to 2020

diesel engine manufacturers of late are designing the engine to accept the fuel having a minimum of 5% methyl ester meeting the EN 14214 biodiesel standard (Lam et al. 2009). As a result, the consumption of biodiesel all over the world is steadfastly increasing.

Palm oil has relatively more oxidation stability than any other similar oils being used as raw material for biodiesel. But the palm oil biodiesel has more NO_x emission (Lim and Teong 2010) but this can be best controlled even up to 80–90% by installing catalytic converters (Lozada et al. 2010). Similarly, the development of winter grade palm oil has actually overtaken the possibility of using the corresponding biodiesel in countries having low atmospheric temperature (Mekhilef et al. 2011).

Besides, from palm oil extraction mill, sterilizer condensate, hydrocyclone waste and separator sludge are formed. On an average 1–1.5 m^3 of palm oil mill effluent (POME) is generated per tonne of crude palm oil produced. POME is a good source of plant manure as it contains as high as 750 mg L^{-1} of total nitrogen, 2270 mg L^{-1} of potassium and 180 mg L^{-1} of phosphorus and hence be used in agriculture after adequately processing the same (Esther 1997).

The POME has a biochemical oxygen demand of 20,000 mg O_2 L^{-1} which is an excellent source of biogas. Processing of 20 tonnes of palm bunch will generate 100 tonnes of POME which will yield 400 m^3 of biogas (methane) (Gurmit et al. 1999). The mode of generating hydrogen from POME also is identified (Atif et al. 2005; Vijayaraghavan and Ahmad 2006).

Large oil palm vegetation acts as a sequester of carbon dioxide. As per Kyoto Protocol, the carbon sink of the oil palm plantation may be converted to carbon credit which is a promising enterprise (Basiron et al. 2004). The dry matter production from palm throughout its life cycle of 25 years is 44 tonnes ha^{-1} $year^{-1}$ against 26 tonnes by a rain forest. A comparison indicates that the oil palm sequester 29.3 t ha^{-1} $year^{-1}$ of CO_2 releases 21.3 t ha^{-1} against 3.52 t ha^{-1} $year^{-1}$ and 2.56 t ha^{-1}, respectively, from soybean.

5.4 Queen Palm (*Syagrus romanzoffiana*)

The oil from the queen palm *Syagrus romanzoffiana* (Family: Arecaceae) has become a source for biodiesel. This palm grows to a height of 20 m at the maximum and is considered as an ornamental tree. It grows in a variety of diversified habitats such as the banks of streams and rivulets, ravine, hilly coastal lands and heavy clay soils with cracks. It flourishes at a rainfall of 1000 mm and around 80% humidity (Falaska et al. 2012). The leaves are pinnate with around 450 pinnae. Each pinna is around 50 cm long and 3–5 cm wide. The trunk is bold and straight with rings. The flowers are yellow and arranged in long inflorescence guarded with spathe. The flowering will be both at spring and summer. However, it puts forth fruits throughout the year, if conditions are favourable, if not, fruiting occurs twice a year only. The fruit is an elliptic drupe and is devoured by frugivorous arboreal animals. Each fruit measures around 1 cm in diameter. Each seed weighs 2–2.5 g at dry condition. The inflorescence which carries innumerable fruits weigh around 10 kg (Fig. 5.13).

This tree is a native of South America and is seen distributed in Bolivia, Paraguay, Argentina, Brazil and Uruguay (Fig. 5.14).

It is also rarely spotted in Florida, Queensland, Honduras and Mauritius as an ornamental plant, probably brought there by travelers. In Australia this tree is not encouraged to be planted as it is considered as an invasive weed. Common names by which this species is called are pindo, chiriva, jeriva, yba pita, queen palm,

Fig. 5.13 The queen palm *Syagrus romanzoffiana*

Fig. 5.14 The geographical distribution of *S. romanzoffiana*

Table 5.20 Fatty acid composition of *S. romanzoffiana* oil (Moreira et al. 2013)

Fatty acid	%
Caprylic acid (C8:0)	6
Capric acid (C10:0)	6
Lauric acid (C12:0)	37
Myristic acid (C14:0)	11
Palmitic acid (C16:0	8
Stearic acid (C18:0)	3
Oleic acid (C18:1)	24
Linoleic acid (C18:2)	5

keninginpalm, cocos romanzoffiana, cocos plumosa, calapa australis, coquito and guariroba.

5.4.1 Oil Production

The fruits are harvested from the tree or collected from the ground immediately after they fall. They are fermented to loosen the husk and then the husk is removed manually. The seeds are then separated, dried repeatedly in hot sun for more than 10 days. The shell and kernel are separated before extracting the oil. Often not mindful of the shell, the whole seed is ground and the oil extracted using *n*-hexane. But it is felt that it is always better to separate the shell before grinding so as to avoid the soaking up of the oil during extraction. But such efforts need appropriate and specialized machinery. The seeds form around 9% of the whole fruit by volume. It is around 20% by weight of the depulped fruit. (Moreira et al. 2013). The fatty acid composition of the oil is given in Table 5.20.

The saturated fatty acid content is around 70% which indicates the stability of the oil against oxidation. The physicochemical properties of the oil are presented in Table 5.21.

Table 5.21 Physicochemical properties of *S. romanzoffiana* oil (Moreira et al. 2013)

Parameters	Value
Kinematic viscosity at 40 °C ($mm^2 s^{-1}$)	30.12
Acid value (g KOH kg^{-1})	1.8
Density at 20 °C (kg m^{-3})	966
Heating value (MJ kg^{-1})	37.6
Calcium (mg kg^{-1})	39
Phosphorus (mg kg^{-1})	63.2
Magnesium (mg kg^{-1})	13.0
Oxidative stability 110 °C (h)	21.6

Table 5.22 Physicochemical properties of the biodiesel from *S. romanzoffiana* oil (Moreira et al. 2013)

Parameters	Value
Kinematic viscosity at 40 °C ($mm^2 s^{-1}$)	3.16
Density (kg m^{-3})	870
Acid value (g KOH kg^{-1})	0.08
Water content (mg kg^{-1})	453

The oil has a low acid value which permits simple alkali catalysed transesterification and also makes the oil stable against rancidity. Accordingly, the oxidative stability of the oil is observed to be 21.6 h. The phosphorus content of the oil is a consequence of phospholipid. High phospholipid is likely to interfere with the transesterification process. The presence of calcium and magnesium is likely to enhance the particulate matter in the engine exhaust (Lôbo et al. 2009). The properties of the biodiesel obtained by the alkali catalysed transesterification are given in Table 5.22.

The kinematic viscosity, water content and acidity of the biodiesel are within the limit of the standard EN 14214 and ASTM D6751. The water content and acidity are not considered as major factors by their face-value but the ultimate consequences are likely to be high as both are responsible to corrode the tank, fuel conduits and to make rich deposit at the tank bottom.

5.5 Tucuma (Astrocaryum *huaimi*)

The palm tucuma *Astrocaryum huaimi* is one among the 40 species listed under the genus Astrocaryum (Kahn 2008) and is selectively distributed in Bolivia, Brazil and Peru (Fig. 5.15). It is popularly referred as Chontilla, Chonta and huaimi-tococa. It belongs to the family Arecaceae. This palm is spiny and the leaves are pinnately compound. The leaflets emerge on either side of the axis of the leaf. The leaf looks like a feather. The male and female flowers are separate but they occur in a single tree. This tree grows to a maximum height of 15 m. Though it occupies sandy, loamy and clay soil, it prefers to have a well-drained soil. It needs a mean rainfall of 1300–1800 mm. It is not wind tolerant (Fig. 5.16). Each tree bears 2–4 bunches of fruit and in each bunch 200–400 fruits are present. The fruit is around 5 cm long, 4 cm dia. and weighs 60–80 g.

Fig. 5.15 Geographical distribution of *Astrocaryum huaimi*

Fig. 5.16 *Astrocaryum huaimi* plant

The pericarp of the fruit is edible and therefore preferred in confectioneries. The pulp also contains oil, which is separated and used in special food preparations. The seed contains a shell within which there is a white kernel, from which oil is extracted. The kernel as a whole has 30–50% oil.

Table 5.23 The fatty acid profile of the biodiesel (methyl ester) from *A. huaimi seed* oil as per Alexandre et al. (2015)

Fatty acid	Biodiesel (methyl esters)
Lauric (C12:0)	38.3
Myristic (C14:0)	17.9
Palmitic (C16:0)	7.7
Stearic (C18:0)	3.7
Oleic (C18:1)	24.8
Linoleic (C18:2)	5.2

5.5.1 Production of Oil

Fully matured fruits are first depulped and the seeds are taken out. The seeds are cut open and the kernels are separated. The kernels are repeatedly dried and grinded. The oil is extracted from the kernel powder using n-hexane in a Soxhlet apparatus for more than 5 h. The extract is then distilled out and the oil is separated. The oil is then used as a raw material for the transesterification.

The oil content with respect to the whole seed is 21% (Alexandre et al. 2015). The composition of the oil of *A. huaimi* is closer to that of palm, soy and cotton. The acid value is relatively low (1.5 g KOH kg^{-1}) which obviates the twostep process for biodiesel preparation. The iodine value is low which indicates the stability of the oil.

5.5.2 Biodiesel Preparation

The biodiesel is prepared by the conventional alkali catalysed transesterification using potassium hydroxide and methanol (1% KOH and 1:6 oil to methanol molar ratio). The fatty acid profile of the methyl ester is presented in Table 5.23.

More than 80% of the fatty acids of the biodiesel are formed of lauric, oleic and myristic acids. The cold filter plugging point is −7.8 °C (Alexandre et al. 2015) which indicates that the biodiesel from this oil is manageable in colder climates. This biodiesel can be considered as an additive to any other biodiesel with untenable characteristics especially in low temperature conditions (Echim and Greyt 2012).

References

Abigor RD, Uadia PO, Foglia TA, Haas MJ, Jones KC, Okpefa E, Obibuzor JU, Bafor ME (2000) Lipase-catalysed production of biodiesel fuel from some Nigerian lauric oils. Biochem Soc Trans 28(6):979–981

Abreu FR, Lima DG, Hamu EH, Wolf C, Suarez PAZ (2004) Utilization of metal complexes as catalysts in the transesterification of Brazilian vegetable oils with different alcohols. J Mol Catal A Chem 209:29–33

Alamu OJ, Waheed MA, Jekayinfa SO (2007a) Alkali-catalysed laboratory production and testing of biodiesel fuel from Nigerian palm kernel oil. Agricultural Engineering International: the CIGR Ejournal, IX: Manuscript Number EE 07 009, pp 1–11

Alamu OJ, Waheed MA, Jekayinfa SO (2007b) Biodiesel production from Nigerian palm kernel oil: effect of KOH concentration on yield. Energy Sustain Dev 11(3):77–82

Alamu OJ, Waheed MA, Jekayinfa SO (2008) Effect of ethanol–palm kernel oil ratio on alkali-catalyzed biodiesel yield. Fuel 87:1529–1533

Albuquerque MCG, Santamaria-Gonzalez J, Merida-Robles JM, Tost RM, Rodriguez-Castellon E, Jimenez-Lopez A, Azevedo DCS, Cavalcante CL Jr, Mireles-Torres P (2008) MgM (M = Al and Ca) oxides as basic catalysts in transesterification processes. Appl Catal A Gen 347:162–168

Alexandre ECF, Sulveira EV, Castro CFDS, Sales JF, Oliveira LCSD, Viana LH, Barbosa LCA (2015) Synthesis, characterization and study of the thermal behavior of methylic and ethylic biodiesel produced from tucumã (*Astrocaryum huaimi* Mart) seed oil. Fuel 161:233–238

Altin R, Cetinkaya S, Yucesu HS (2001) The potential of using vegetable oil fuels as fuel for diesel engines. Energy Convers Manag 42:529–538

Anonymous (2009) National Biodiesel Board, Myths and Facts about Biodiesel. www.biodiesel.org

Arpornpong N, Attaphong C, Charoensaeng A, Sabatini DA, Khaodhiar S (2014) Ethanol-in-palm oil/diesel microemulsion-based biofuel: phase behavior, viscosity, and droplet size. Fuel 132:101–106

Arpornpong N, Sabatini DA, Khaodhiar S, Charoensaeng A (2015) Life cycle assessment of palm oil microemulsion-based biofuel. Int J Life Cycle Assess 20:913–926

Atif AAY, Fakhrul-Razi A, Ngan MA, Morimoto M, Iyuke SE, Veziroglu NT (2005) Fed batch production of hydrogen from palm oil mill effluent using anaerobic microflora. Int J Hydrog Energy 30(13):1393–1397

Awaluddin A, Padil, Wahyuningsih S (2010) Making biodiesel from coconut oil by reaction metanolisis heterogeneous catalyst. In: PEA—AIT international conference on energy and sustainable development: issues and strategies ESD 2010

Bari S, Lim TH, Yu CW (2002a) Effect of preheating of crude palm oil (CPO) on injection system, performance and emission of a diesel engine. Renew Energy 27:339–351

Bari S, Yu CW, Lim TH (2002b) Performance deterioration and durability issues while running a diesel engine with crude palm oil. J Automobile Eng 216:785–792

Basiron Y, Balu N, Chandramohan D (2004) Palm oil: the driving force of world oils and fats economy. Oil Palm Ind Econ J 4(1):1–10

Bouaid A, Martinez M, Aracil J (2015) Biodiesel production from babassu oil: a statistical approach. J Chem Eng Process Technol 6:232–235

Chongkhong S, Tongurai C, Chetpattananondh P, Bunyakan C (2007) Biodiesel production by esterification of palm fatty acid distillate. Biomass Bioenergy 31(8):563–568

Colin J (2005) Coconut oil as a fuel in the Pacific Islands SOPAC

Crookes RJ, Kiannejad F, Nazha MAA (1997) Systematic assessment of combustion characteristics of biofuels and emulsions with water for use as diesel engine fuels. Energy Convers Manag 38:1785–1795

Da Rós PCM, Silva CW, Grabauskas D, Perez VH, de Castro HF (2014) Biodiesel from babassu oil: characterization of the product obtained by enzymatic route accelerated by microwave irradiation. Ind Crops Prod 52:313–320

de Almeida SCA, Belchior CR, Nascimento MVG, Vieira LDSR, Fleury G (2002) Performance of a diesel generator fuelled with palm oil. Fuel 81:2097–2102

Echim CJM, Greyt WD (2012) Improvement of cold filter plugging point of biodiesel from alternative feedstocks. Fuel 93:642–648

Esther OU (1997) Anaerobic digestion of palm oil mill effluent and its utilization as fertilizer for environmental protection. Renew Energy 10(2):291–294

Falaska SL, Miranda del Frenso C, Ulberich A (2012) Possibilities for growing queen palm (*Syagrus romonzoffiana*) in Argentina as a biodiesel producer under semi-arid climate conditions. Int Hydrogen Energy 37(19):14843–14848

Ferreira BS, Faza LP, Hyaric ML (2012) A comparison of the physicochemical properties and fatty acid composition of India's (*Attalea dubia*) and Babassu (*Orbignya phalerata*) oils. Sci World J 2012:532374, 1–14

Gao L, Teng G, Xiao G, Wei R (2010) Biodiesel from palm oil via loading KF/Ca–Al hydrotalcite catalyst. Biomass Bioenergy 34:1283–1288

Gurmit S, Kim LH, Teo L, David KL (1999) Oil palm and the environment a Malaysian perspective. In: Malaysian oil palm growers council, vol I, p 253

Hameed BH, Lai LF, Chin LH (2009) Production of biodiesel from palm oil (*Elaeis guineensis*) using heterogeneous catalyst: an optimized process. Fuel Process Technol 90:606–610

Hilditch TP (1956) The chemical constitution of natural fats, 3rd rev edn. Chapman & Hall Ltd, London

Imahara H, Minami E, Hari S, Saka S (2008) Thermal stability of biodiesel in supercritical methanol. Fuel 87:1–6

Kahn F (2008) The genus *Astrocaryum* (Arecaceae). Rev Peru Biol 15(Suppl 1):31–48

Kalam MA, Masjuki HH (2004) Emission and deposit characteristics of a small diesel engine when operated on preheated crude palm oil. Biomass Bioenergy 27:289–297

Kalam MA, Masjuki HH (2008) Testing palm biodiesel and NPAA additives to control NO_x and CO while improving efficiency in diesel engines. Biomass Bioenergy 32(12):1116–1122

Kale PT (2017) Experimental investigation of engine parameters influence on performance and emissions of biodiesel fueled CI engine. Int J Eng Res Technol 6(4):11–16

Kansedo J, Lee KT, Bhatia S (2009) Biodiesel production from palm oil *via* heterogeneous transesterification. Biomass Bioenergy 33(2):271–276

Khaire BS, Deshmukh MS (2017) Performance and emission analysis of VCR engine at different compression ratios by using diesel and babassu biodiesel blends. International Engineering Research Journal spl. Edn MECHPGCON Zeal college of Engineering and Research, NARCHE, Pune, 20–21 June 2017

Knothe G, Dunn RO (2001) Biofuels derived from vegetable oils and fats. In: Gunstone FD, Hamilton RJ (eds) Oleochemical manufacture and applications. Sheffield Academic Press CRC, pp 106–150

Knothe G, Robert OD, Bagby MO (1997) The use of vegetable oils and their derivatives as alternative diesel fuels. National Center for Agricultural Utilization Research, Agricultural Research Practices, A.S. Department of Agriculture

Kusdiana D, Saka S (2001) Kinetics of transesterification in rapeseed oil to biodiesel fuel as treated in supercritical methanol. Fuel 80:693–698

Kusdiana D, Saka S (2004) Effects of water on biodiesel fuel production by supercritical methanol treatment. Bioresour Technol 91:289–295

Lam MK, Tan KT, Lee KT, Mohamed AR (2009) Malaysian palm oil: surviving the food versus fuel dispute for a sustainable future. Renew Sust Energ Rev 13:1456–1464

Li E, Xu ZP, Rudolph V (2009) MgCoAl-LDH derived heterogeneous catalysts for the ethanol transesterification of canola oil to biodiesel. Appl Catal B Environ 88:42–49

Lim S, Teong LK (2010) Recent trends, opportunities and challenges of biodiesel in Malaysia: an overview. Renew Sust Energ Rev 14:938–954

Lôbo IP, Ferreira SLC, Cruz RS (2009) Biodiesel: parametros de qualidae e metodos analiticos. Quim Nova 32(6):1596–1608

Lozada I, Islas J, Grande G (2010) Environmental and economic feasibility of palm oil biodiesel in Mexican transportation sector. Renew Sust Energ Rev 14:486–492

Mekhilef S, Siga S, Saidur R (2011) A review on palm oil biodiesel as a source of renewable fuel. Renew Sust Energ Rev 15(4):1937–1949

Moreira MAC, Arrûa MEP, Antunes AC, Fiuza TER, Costa BJ, Neto PHW, Antunes SRM (2013) Characterization of *Syagrus romanzoffiana* oil aiming at biodiesel production. Ind Crop Prod 48:57–60

Moreira FB, Bezerra RCF, Rodrigues FEA, Assunção JCC, Arruda TBMG (2018) In situ transesterification of babassu for production of biodiesel as sustainable energy option for Aratuba community in State of Ceara, Brazil. Electron J Chem 10(2). https://doi.org/10.17807/orbital.vioiz.1088

Ndayishimiye P, Tazerout M (2011) Use of palm oil-based biofuel in the internal combustion engines: performance and emission characteristics. Energy 36:1790–1796

Noiroj K, Intarapong P, Luengnaruemitchai A, Jai-In S (2009) A comparative study of KOH/Al_2O_3 and KOH/NaY catalysts for biodiesel production *via* transesterification from palm oil. Renew Energy 34:1145–1150

Olutoye MA, Hameed BH (2010) Transesterification of palm oil on $K_yMg_{1-x}Zn_{1+x}O_3$ catalyst: effect of Mg-Zn interaction. Fuel Process Technol 91(6):653–659

Paiva EJM, Barboza JCS, da Silva MLCP (2011) Comparative study of biodiesel production from ethanol and babassu oil using mechanical agitation and ultra sounds. In: Proc. international conference on renewable energies and power quality, Las Palmas de Gran Canaria, Spain, 13–15th Apr 2011

Rahman NAA, Olutoye MA, Hameed BH (2011) Synthesis of methyl esters from palm (*Elaeis guineensis*) oil using cobalt doped MgO as solid oxide catalyst. Bioresour Technol 102:9749–9754

Rashid U, Anwar F, Moser BR, Knothe G (2008) *Moringa oleifera* oil: a possible source of biodiesel. Bioresour Technol 99:8175–8179

Rasyid R, Sabara Z, Pratiwi HA, Juradin R, Malik R (2018) The production of biodiesel from a traditional coconut oil using NaOH/γ-Al2O3 heterogeneous catalyst. IOP Conf Ser Earth Environ Sci 175:012025. https://doi.org/10.1088/1755-1315/175/1/012025

Sadrolhosseini AR, Moksin MMM, Nang HLL, Norozi M, Yunus WMM, Zakaria A (2011) Physical properties of normal grade biodiesel and winter grade biodiesel. Int J Mol Sci 12 (4):2100–2111. https://doi.org/10.3390/ijms1204210

Salunkhe DK, Chavan JK, Adsule RN, Kadam SS (1992) World oilseeds chemistry, technology and utilisation. AVI Book, New York, p 545

Sharma YC, Singh B, Upadhyay SN (2008) Advancements in development and characterization of biodiesel: a review. Fuel 87:2355–2373

Silva MCD, daSilva LM, Brandao KS, Souza AG, Cardoso LP, dos Santos AO (2014) Low temperature properties of winterized methyl babassu biodiesel. J Therm Anal Calorim 115:635–640

Solis JL, Alejo L, Kiros Y (2015) Calcium and tin oxides for heterogeneous transesterification of babassu oil (*Attalea speciosa*). J Environ Chem Eng 4:4870–4877. https://doi.org/10.1016/j.jece.2016.04.006

Song ES, Lim JW, Lee HS, Lee YW (2008) Transesterification of RBD palm oil using supercritical methanol. J Supercrit Fluids 44:356–363

Sree R, Babu NS, Prasad PSS, Lingaiah N (2009) Transesterification of edible and non-edible oils over basic solid Mg/Zr catalysts. Fuel Process Technol 90(1):152–157

Tan KT, Lee KT, Mohamed AR (2009) Production of FAME by palm oil transesterification *via* supercritical methanol technology. Biomass Bioenergy 33:1096–1099

Tyson KS (2001) Biodiesel handling and use guidelines. National Renewable Laboratory, US Department of Energy, NREL/TP-580-30004 CO, USA, 17 p

USDA (United States Department of Agriculture) (1998) Production Estimates and Crop Assessment Division, FAS

Vijayaraghavan K, Ahmad D (2006) Biohydrogen generation from palm oil mill effluent using anaerobic contact filter. Int J Hydrog Energy 31(10):1284–1291

Wen Z, Yu X, Tu ST, Yan J, Dahlquist E (2010) Biodiesel production from waste cooking oil catalyzed by TiO_2–MgO mixed oxides. Bioresour Technol 101:9570–9576

6 Asteraceae

The family Asteraceae (Compositae) consists of over 30,000 known annual, biennial or perennial shrub, vine and tree species. They are found in hot desert and cold or hot semi-desert climates and are distributed in all continents from subpolar to tropical regions except Antarctica. Two species known for its use in biodiesel (cardoon *Cynara cardunculus* and sunflower *Helianthus annuus*) are dealt in this chapter. Cardoons are otherwise used in the preparation of enzymes for cheese production and rennet making besides being considered as an ornamental plant. Of late the oils from sunflower seed is being partly diverted to biodiesel production.

6.1 Cardoon (*Cynara cardunculus*)

The cardoon, *Cynara cardunculus* a perennial herb is a native of Central and Western Mediterranean region. It is geographically distributed in Africa, Algeria, Argentina, Asia, Australia, Brazil, Canada, Canary Islands, Chile, China, Croatia, Cyprus, Europe, France, Greece, India, Italy, Libya, Macedonia, Morocco, New Zealand, North America, North Africa, Portugal, Spain, Tasmania, Tunisia, Turkey, Uruguay and USA (Fig. 6.1).

This herb grows well in derelict clay-land having low irrigation with extremes of temperature and paradoxically low rainfall. The flower head is club shaped. The ovate profile of the club measures 3.5–7.5 × 3.0–9.5 cm in its matured state. The flower head contains many tubular florets each 5 cm long. These florets are dark or pale lilac or bluish. The leaf is 80 cm in length and 30 cm in width. They are tomentose in nature with spines of 3–3.5 cm long (Fig. 6.2). The leaves are grey or green in colour. The plant may grow up to 1.5 m height. The seeds are large (around 7 mm dia.) four sided, smooth with longitudinal ridges.

This plant is referred by different local names as shown in Table 6.1.

S. P. Raj et al., *Biodiesel from Flowering Plants*,
https://doi.org/10.1007/978-981-16-4775-8_6

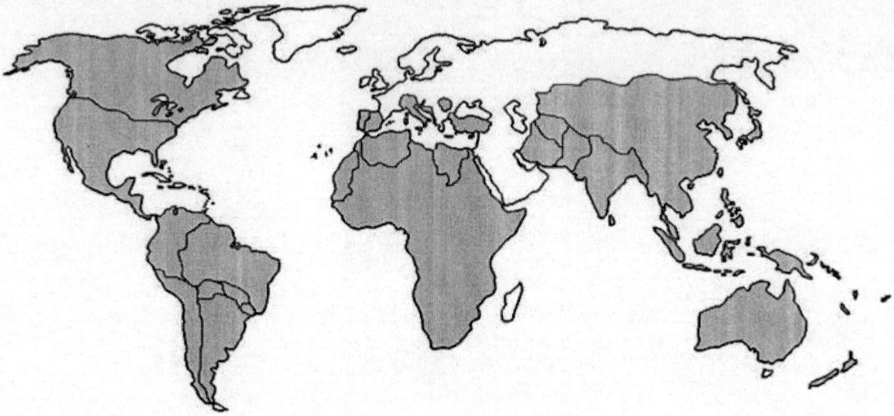

Fig. 6.1 Geographical distribution of *C. cardunculus*

Fig. 6.2 The cardoon Cynara *cardunculus*

Table 6.1 Local names of *C. cardunculus* in different languages

Language	Local names
Arabic	Al harshuff
Dutch	Kardoen
English	March ysgall
Finnish	Isoartisokka
French	Artichaut commun, carde
German	Artishocke, kardone
Italian	Carciofo
Portuguese	Alcachofra, cardo
Russian	Artisok ispanskij
Spanish	Alcachofa, alcaucil, cardo cardo-de-comer
Swedish	Kardon

6.1.1 Oil Extraction

C. cardunculus has a seed potential of 1.5–1.8 tonnes per hectare per year (De Mastro et al. 2011) from 20 to 30 tonnes per hectare of the seed bearing head (Raccuia and Melilli 2007). The seeds are dried and powdered. It is then extracted in *n*-hexane (Wang and Weller 2006). The extract is then distilled to separate the oil. The seeds contain an average of 25% oil with a maximum value of 33% (Curt et al. 2002).

6.1.2 Biodiesel Production

Ethyl alcohol and sodium hydroxide are mixed to prepare ethoxide. The quantity of alcohol is chosen in such a way to form a molar ratio of ethyl alcohol to oil as 9:1. Similarly, the sodium hydroxide concentration is chosen so as to form 1% with that of the oil used. The ambient temperature is 50 °C. The reaction reaches to the final stage in 2 h. After cooling the product, around 25% of glycerin (calculated based on the volume of oil) is added to it. The glycerin now added along with the glycerin being liberated from the reaction product, settles down in the separatory funnel. Along with the glycerin, unutilized alkali, unparticipated alcohol, soap and a feeble quantity of ethyl ester also settle at the bottom and they are drained out. The upper layer containing the ester is distilled at 80 °C to remove the alcohol. The ester thus obtained is washed repeatedly with distilled water. Finally the ester is dried over calcium chloride and then filtered. The physicochemical properties of the biodiesel (ethyl ester) are given in Table 6.2.

Among the various parameters studied viscosity, density, cetane number and the heating values match with that of the diesel. However the cloud point and pour points are higher which are likely to pose difficulties in cold start.

Table 6.2 The physicochemical properties of biodiesel (ethyl ester) from *C. cardunculus* seed oil

Parameters	Encinar et al. (2002)	Carlucci et al. (2015)
Kinematic viscosity @ 40 °C (cSt)	4.17	–
Density @ 25 °C (kg m^{-3})	870	–
Flash point (°C)	185	–
Pour point (°C)	−0.7	–
Cloud point (°C)	−0.4	–
Cetane number	49.2	–
Heating value (MJ kg^{-1})	40.0	–
Palmitic acid (C16:0) (%)	–	14.45
Stearic acid (C18:0) (%)	–	4.90
Oleic acid (C18:1) (%)	–	27.55
Linoleic acid (C18:2) (%)	–	46.64

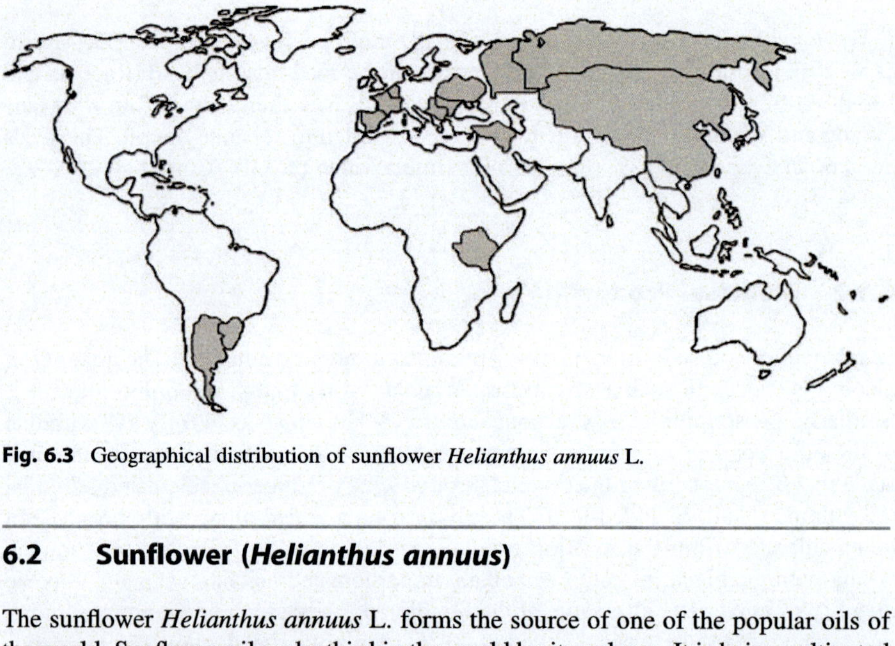

Fig. 6.3 Geographical distribution of sunflower *Helianthus annuus* L.

6.2 Sunflower (*Helianthus annuus*)

The sunflower *Helianthus annuus* L. forms the source of one of the popular oils of the world. Sunflower oil ranks third in the world by its volume. It is being cultivated in an area of 25 million hectares with an annual production of around 32 million tonnes (Ramulu et al. 2011). It is a native of North America. It is now distributed in many countries: Ukraine, The Netherlands, Russia, Uruguay, Argentina, China, Spain, Romania, Turkey, Hungary, Bulgaria, Tanzania and France (Fig. 6.3). This species is referred by different colloquial names in different countries (Table 6.3).

Table 6.3 Colloquial names of sunflower in different countries

Country	Vernacular names
England	Sunflower
French	Farnesol
Portuguese	Girasol

Fig. 6.4 The sunflower *Helianthus annuus*

6.2.1 Characteristics

Sunflower plant grows to a height of 1–3 m tall and is an annual, erect, broad leaf plant with a strong taproot system. The leaves are alternate, strongly phototropic and effectively track the sun. Leaves droop downward from the long petioles. Leaf is as big as 20 cm long and 15 cm broad. The flower is daisy type and consists of numerous florets forming a central disk which is surrounded by 20–40 ray type of florets. Each flower measures 15–25 cm diameter (Fig. 6.4). A single flower may hold as high as 2000 seeds. The seed weight varies from 55 to 75 mg. The total production of seeds ranges from 2 to 2.2 tonnes per hectare.

6.2.2 Cultivation

Sunflower seeds are sown at the rate of 50,000 (3 kg ha^{-1}) to 65,000 (4 kg ha^{-1}) seeds per hectare at a row spacing of 70 cm. Approximately 20 kg N, 20 kg P and 50 kg K are recommended to be applied as basal fertilizer depending on the nutrient status of the soil in which the seeds are sown. A soil pH of 6–7 is ideal. The plant development phase comprises 11 days for seedling emergence, 33 days from emergence to flower budding, 27 days from then to another development, 8 days from first to last another and another 30 days to reach maturity. Thus a plant requires a total of 110–120 days to complete the life cycle and to provide the matured seed for oil extraction.

6.2.3 Oil Extraction

The major producer of sunflower oil is Russia. Two methods of extraction, namely solvent extraction and cold pressing are being followed. In solvent extraction the seeds are crushed along with hexane and the resultant mixture is subjected to fractional distillation to expel the solvent. In cold pressing the seeds are pressed in expeller at cold condition and the crushed out oil is filtered and used. The cold pressed oil is often refined using solvents. The oil is degummed, neutralized and bleached. In such refinement some of the essential nutrients, flavour, colour, free fatty acids, phospholipids, polyphenols and phytosterols are removed. The oil content of the seeds ranges from 22 to 36%.

6.2.4 Oil Quality

Sunflower oil is predominantly formed of less stable polyunsaturated and monosaturated fatty acids and it is amenable to degradation at high temperature, air contact and exposure to light. Storing the sunflower oil at low temperature helps to minimize rancidity and nutrient loss. Unsaturated fatty acids comprise oleic acid (monounsaturated omega-9) and linoleic acid (polyunsaturated omega-6). The saturated fatty acids have a low share. The saturated fatty acids present in it are palmitic acid and stearic acid. Based on the quantum of the above unsaturated fatty acids the sunflower oil is often classified as high linoleic, high oleic and mid oleic. High oleic and mid oleic sunflower oil comprise 80 and 70% oleic acid, respectively. Fatty acid profiles are largely governed by the genetic makeup and the climatic conditions also play a marginal role. The properties of typical sunflower oil are given in Table 6.4.

Table 6.4 Properties of typical sunflower oil

Parameters	Value
Viscosity at 40 °C (mm^2 s^{-1})	5.8
Density at 25 °C (kg m^{-3})	918.8
Smoke point (°C)	107.0
Refractive index at 25 °C	1.4735
Saponification value	188–194
Iodine value	120–145
Unsaponifiable matter (%)	1.5–2.0
Acid value (mg KOH g^{-1})	0.1

Table 6.5 Fatty acid profile and molecular weight of sunflower oil

Fatty acid	Molecular weight	%
Palmitic (16:0)	256	5–6.6
Stearic (18:0)	284	5–6
Oleic (18:1)	282	20–30
Linoleic (18:2)	280	59–69

The fatty acid composition based on GC analysis (Harrington and Arcy-Evans 1985; Georgogianni et al. 2006) is given in Table 6.5.

High content of unsaturated fatty acids in sunflower oil especially of oleic and linoleic acid is reported to be the specific character of the said oil (Turkan and Kalay 2006; Oliveira and Rosa 2006).

6.2.5 Biodiesel Preparation

One of the leading methods of biodiesel preparation is alcoholysis in which the viscous oil is allowed to react with methanol or ethanol or butanol in the presence of a suitable catalyst. This chemical reaction replaces the viscous glycerol with a short chain alcohol thereby the ester (biodiesel) is formed. The optimum level of NaOH needed to maximize transesterification is 2% and the reaction is completed within 15 min at a stirring rate of 600 rpm. Increasing the level of NaOH beyond 2% causes emulsion. In such events the soap molecules and esters are admixed causing the separation of ester extremely difficult. As a result, the quantity of ester ultimately recovered becomes low. Often H_2SO_4 is used as a catalyst especially when there is high free fatty acid and water content. Besides acid catalysis is several times slower than that of the alkali catalysis. Among the various alcohols being employed, methanol is considered favourable due to its physical, chemical and economic advantages. Besides, methanol is a polar short chain alcohol. Notwithstanding some of the advantages of methanol, ethanol also is actively considered as a renewable and environmentally less offensive agent when compared to methanol. Highest yield of ester is obtained at an oil methanol ratio of 1:6 (Fig. 6.5). In the place of NaOH, KOH also is tried but the biodiesel yield is relatively less. It is widely known that the NaOH catalysed in situ transesterification is extremely fast. It requires just 20 min to complete the reaction. Alkali catalysed reaction is around 2500–3500 times faster than that of acid catalysed reaction when same amount of acid is employed. Besides alkali is less corrosive than the acid (Georgogianni et al. 2006). Lipase producing fungus *Rhizopus oryzae* was tested against pure enzyme through immobilization in calcium alginate beads (Bharathiraja et al. 2015). Among the many acyl acceptors (methanol, ethanol, *n*-propanol and *n*-butanol) methanol expressed highest conversion of triglyceride. The recovery was 94% in pure enzyme and with respect to the whole cell (*R. oryzae*), it was 84% only.

The biodiesel from sunflower oil forms 13% of the total world production of biodiesel (Fig. 6.6).

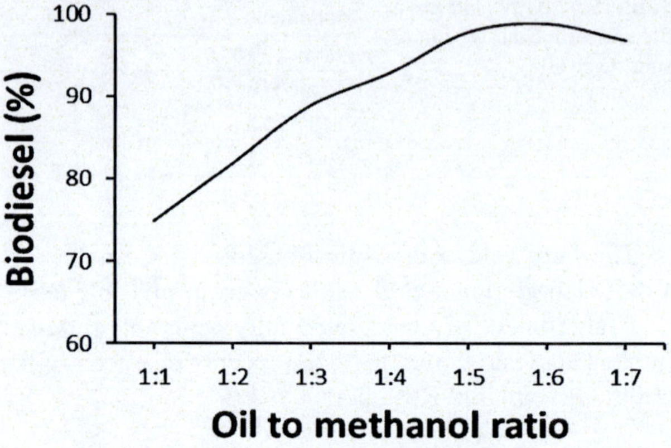

Fig. 6.5 Yield of biodiesel from sunflower oil due to transesterification with different levels of methanol, NaOH 2%, temp. 60 °C, reaction duration 15 min

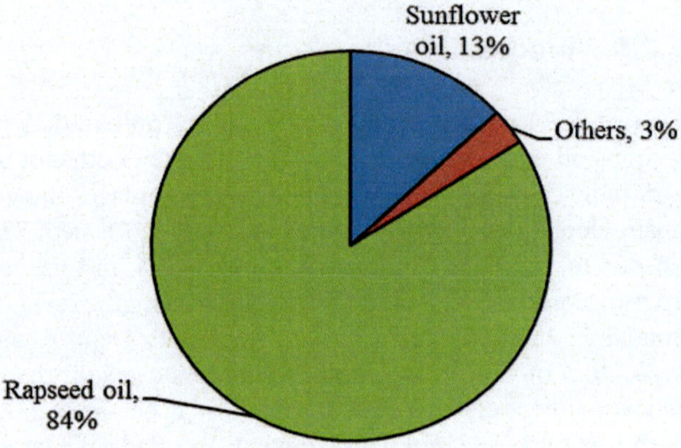

Fig. 6.6 Share of sunflower oil biodiesel among the world production

6.2.6 Quality of the Biodiesel Produced

The properties of sunflower oil biodiesel are shown in Table 6.6.

All the parameters except moisture content comply with the standard.

6.2.7 Fuel Efficiency

The fuel consumption pattern of a typical engine against the speed is given in Fig. 6.7.

The fuel consumption as g kwh^{-1} was plotted against speed (rpm) for biodiesel and regular diesel. The consumption of biodiesel as compared to the regular diesel is higher since the calorific value of this biodiesel is 12–15% lower (Fig. 6.7). The output of power is shown in Fig. 6.8. However due to the high density of biodiesel the total loss may be around 6–7% only (Antolin et al. 2002; Kaplan et al. 2006). The power which being the useful energy at the output shaft when marked against speed indicated identical pattern for both biodiesel and regular diesel (Fig. 6.8).

Table 6.6 Properties of Sunflower oil biodiesel (Antolin et al. 2002; Ghai et al. 2008; Hossain and Boyce 2009; Santos et al. 2013)

Parameters	Value	ASTM—Reference
Viscosity at 40 °C (mm^2s^{-1})	4.3–6.3	ASTM D-445
Density at 15 °C (kg m^{-3})	846–886	ASTM D-4052
Flash point (°C)	>110	ASTM D-93
Moisture (ppm)	615	ASTM D-95
Cold filter plugging point (°C)	−2	ASTM 1P-309
Cloud point (°C)	−1 to 2	ASTM D-2500
Cetane index	46.9	ASTM D-976
Acid value (mg KOH g^{-1})	0.3	ASTM D-664
Calorific value (J kg^{-1})	38–43 × 10^6	ASTM D-240
Ash content (weight %)	0.004	ASTM D-482
Sulphur content (weight %)	0.012	ASTM D-4294
Copper corrosion (3 h at 50 °C)	1A	ASTM D-130
Group I metals (Na + K) (mg kg^{-1})	2.5	
Group II metals (Ca + Mg) (mg kg^{-1})	3.1	

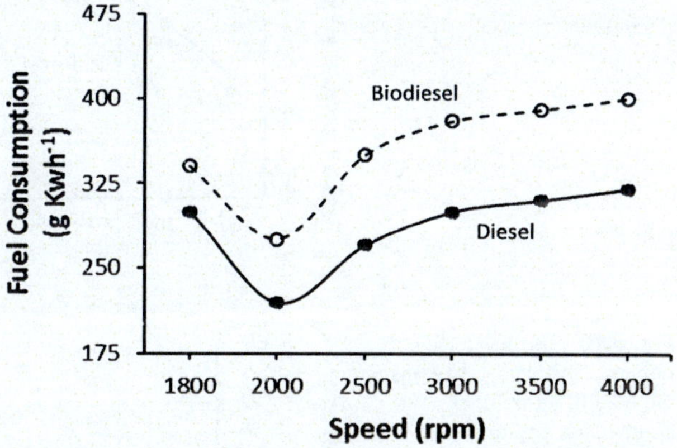

Fig. 6.7 Comparative consumption of biodiesel and regular diesel as a function of speed

The brake thermal efficiency when plotted against the load is seen (Fig. 6.9) that there is a tendency of the regular diesel to show higher efficiency at lower loads. When the load reached the full level the efficiency of the regular diesel and biodiesel is on par due to the fact that at peak loads the biodiesel burns better due to satisfactory combustion caused by the richness of oxygen coupled with the injection of more quantity of fuel.

Since the wall quenching and bulk quenching are low for biodiesel the unburnt hydrocarbon level in the emission is relatively low (Fig. 6.10). At high load the difference between them is substantial.

Contrary to the level of unburnt hydrocarbon, the occurrence of oxides of nitrogen (NO_x) in the exhaust is higher (Fig. 6.11). Such formation is influenced by the prevailing temperature during the combustion of fuel. The biodiesel

Fig. 6.8 The power output at shaft against speed while regular diesel and biodiesel were used as fuel

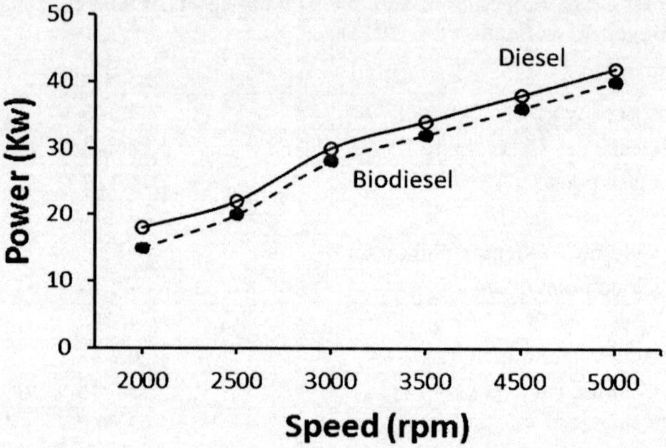

Fig. 6.9 Brake thermal efficiency of diesel and biodiesel when employed as fuel in internal combustion engine in different load conditions

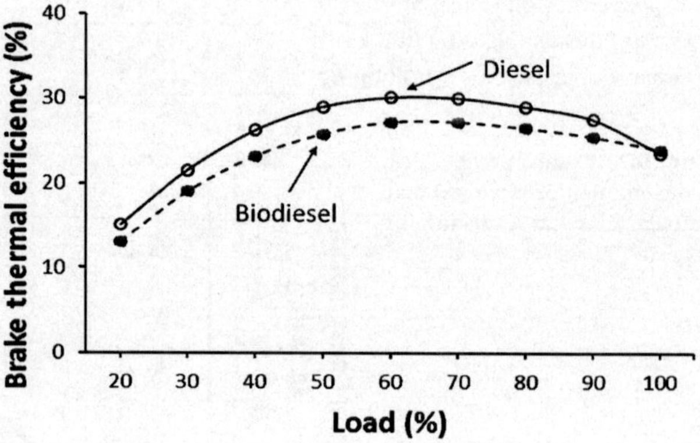

Fig. 6.10 The level of unburnt hydrocarbon against the load of an internal combustion engine

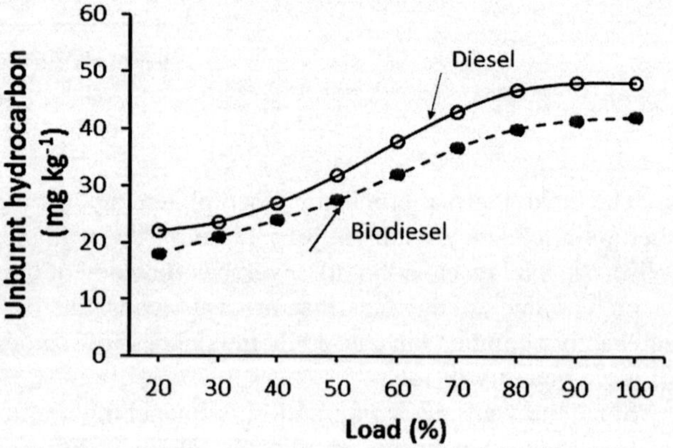

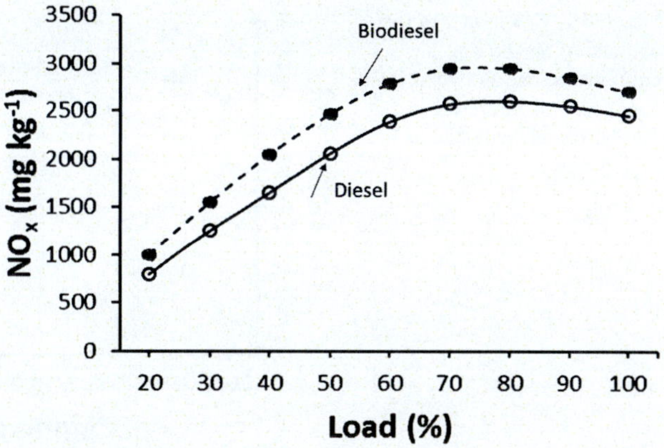

Fig. 6.11 NO_x emission in the exhaust of an internal combustion engine working under different load conditions

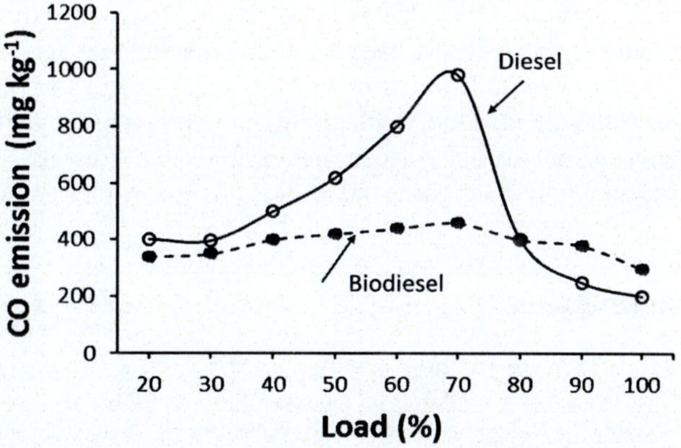

Fig. 6.12 CO level in the exhaust of the internal combustion engine against differential load

molecular structure holds considerable quantity of oxygen which promotes the formation of NO_x.

CO emission analysis indicates that the level of CO in the exhaust is low (Fig. 6.12) and stable in biodiesel fuel when compared to that of diesel indicating a high oxidation capacity of the combustion process being assisted by the rich quantity of molecular oxygen present in the biodiesel.

The CO_2 conc. consequently shows increased value for biodiesel (Fig. 6.13). Higher viscosity and consequent low atomisation of the biodiesel exert change in pressure at the piston of the distribution system causing changes in the injection of fuel.

The fuel characteristics explained above alters the style of injection of the fuel. The spraying characteristics also are influenced due to the change in the size of the droplets of the fuel which in turn corresponds to the density. This phenomenon alters the burning of the fuel. As the biodiesel provides additional oxygen for breaking down the organic molecule the level of oxidized carbon and nitrogen increases in the exhaust. The CO_2 emission for sunflower oil biodiesel is around 14% higher than that of the regular diesel. Similarly, the unburnt hydrocarbon content of the exhaust

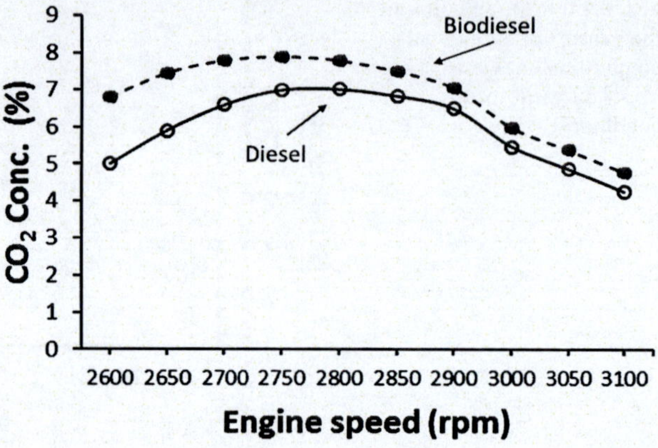

Fig. 6.13 CO_2 conc. in the exhaust of an internal combustion engine when sunflower oil biodiesel and regular diesel are tested as fuel

is as high as 32% (Santos et al. 2013). It is strongly felt that the sunflower oil biodiesel can be best used as supplemental fuel for steady-state non-road diesel engines. As sunflower is one of the major sources of oil, its biodiesel may be best used as a blended fuel (with diesel) in order to take advantage of its lubricity. Besides any attempt to have a comprehension on the use of biodiesel is quite welcome especially in the context of the dependency of fuel on petroleum products.

References

Antolin G, Tinaut FV, Briceno Y, Castano V, Perez C, Ramirez AI (2002) Optimisation of biodiesel production by sunflower oil transesterification. Bioresour Technol 83(2):111–114

Bharathiraja B, Jayamuthunagai J, Praveenkumar R, Jayakumar M, Palani S (2015) The kinetics of interesterification on waste cooking oil (sunflower oil) for the production of fatty acid alkyl esters using a whole cell biocatalyst (*Rhizopus oryzae*) and pure lipase enzyme. Int J Green Energy 12(10):1012–1017

Carlucci AP, Ficarella A, Strafella L, Tricarico A, De Domenico S, Amico LD, Santino A (2015) Behaviour of a compression ignition engine fed with biodiesel derived from *Cynara cardunculus* and coffee grounds. In: Proc. of the 38th meeting of the Italian Section of the Combustion Institute, 20–23 Sept. ISBN 978-88-88104-25-6

Curt MD, Sanchez G, Fernandez J (2002) The potential of *Cynara cardunculus* L. for seed oil production in a perennial cultivation system. Biomass Bioenergy 23(1):33–46

De Mastro G, Grassano N, D'Andrea L, Palumbo AD (2011) GIS based evaluation of cardoon (*Cynara cardunculus* L. var. altilis DC.) suitability in Apulia region. In: Proc. European biomass conference and exhibition, Berlin, Germany, pp 579–584

Encinar JM, Gonzalez JF, Rodriquez JJ, Tejedor A (2002) Biodiesel fuels from vegetable oils: transesterification of *Cynara cardunculus* L. oils with ethanol. Energy Fuels 16:443–450

Georgogianni KG, Kontominas MG, Avlonitis DA, Gergis V (2006) Transesterification of sunflower seed oil for the production of biodiesel: effect of catalyst concentration and ultrasonication. In: Proc. IASME/WSEAS international conference on energy & environmental systems, Chalkida, Greece, 8–10 May, pp 425–429

Ghai S, Das LM, Babu MKG (2008) Emissions and performance study with sunflower methyl ester as diesel engine fuel. ARPN J Eng Appl Sci 3(5):75–80

Harrington KJ, Arcy-Evans CD (1985) Transesterification in situ of sunflower oil. Ind Eng Chem Prod Res Dev 24(2):314–318

Hossain ABMS, Boyce AN (2009) Biodiesel production from waste sunflower cooking oil as an environmental recycling process and renewable energy. Bulgarian J Agric Sci 15(4):312–317

Kaplan C, Arslan R, Surmen A (2006) Performance characteristics of sunflower methyl esters as biodiesel. Energy Sources Part A 28:751–755

Oliveira AC, Rosa MF (2006) Enzymatic transesterification of sunflower oil in an aqueous oil biphasic system. J Am Oil Chem Soc 83:21–25

Raccuia SA, Melilli MG (2007) Biomass and grain oil yields in *Cynara cardunculus* L.genotypes grown in a Mediterranean environment. Field Crops Res 101:187–197

Ramulu N, Krishamurthy HM, Jayadeva MM, Venkatesha, Ravikumar HS (2011) Seed yield and nutrients uptake of sunflower (*Helianthus annuus* L.) as influenced by different levels of nutrients under irrigated condition of eastern dry zone of Karnataka, India. Plant Arch 11 (2):1061–1066

Santos BS, Capareda SC, Capunitan JA (2013) Sunflower methylester as an engine fuel: performance evaluation and emissions analysis. ISRN Renew Energy 2013:352024, 12 p. https://doi.org/10.1155/2013/352024

Turkan A, Kalay S (2006) Monitoring lipase catalysed methanolysis of sunflower oil by reversed–phase high performance liquid chromatography: elucidation of the mechanism of lipases. J Chromatogr A127:34–44

Wang L, Weller CL (2006) Recent advances in extraction of nutraceuticals from plants. Trends Food Sci Technol 17:300–312

Betulaceae

7

Members of the family Betulaceae (birch family) are distributed in northern temperate and mountainous tropical regions. They are either hardwood trees or shrubs with their thin leaves strongly serrated. Many of them are deciduous nut bearing trees with fruits having single seeded nutlet. This family is formed of alders, hazels and hornbeams, all forming around 150 species. The nuts of hazelnut (*Corylus avellana*) are a rich source for oil being used in biodiesel.

7.1 Hazelnut (*Corylus avellana*)

The oil of hazelnut *Corylus avellana* is a raw material for biodiesel production. The name avellana refers to the Avella city of Italy. This plant is often referred as Avellana nux sylvestris (Wild nut of Avella). It is a decidious shrub growing to an average height of 8 m having its leaves 6–12 cm long with a serrated margin. The flowers are monoecious and are wind pollinated. The fruit is a nut, formed as a cluster of four or five (Fig. 7.1). Each nut is protected by a leafy shield involucre which covers more than 60% of the fruit (nut). The nut is oval in shape 15–20 mm across. Eight months after pollination the nut becomes ripe and falls down on the ground (Fig. 7.1). The nut also is called as cobnut. Currently 700,000 tonnes of nuts are produced every year all over the world. The production rate is around 2200 kg ha^{-1} (equivalent to 1000 kg of oil ha^{-1} $year^{-1}$) (Moser 2012). This shrub is cultivated in Europe, Turkey, Iran, Italy and Caucasus (Fig. 7.2).

7.1.1 Oil and Biodiesel Production

The fruits are separated from the involucre and then cracked to remove the kernels. The kernels are sundried and passed through a roller to extract the oil. The oil is centrifuged and refined as per Karabulut et al. (2005). Accordingly, the oil is mixed

S. P. Raj et al., *Biodiesel from Flowering Plants*,
https://doi.org/10.1007/978-981-16-4775-8_7

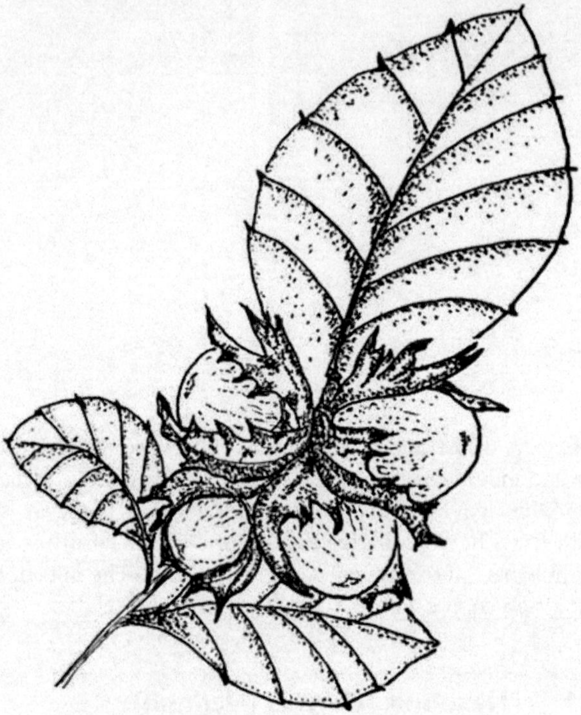

Fig. 7.1 The hazelnut, *Corylus avellana*

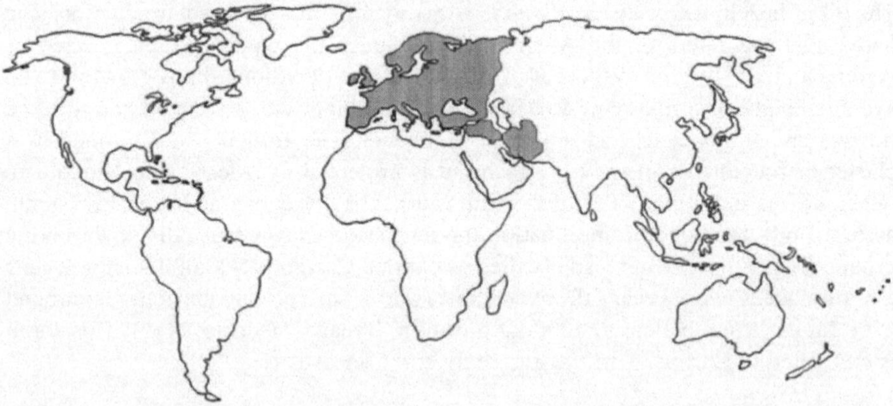

Fig. 7.2 Geographical distribution of *Corylus avellana*

with 6 N sodium hydroxide. The quantity of hydroxide used is around 25% excess by volume of 6 N sodium hydroxide needed to neutralize the oil. The content is heated to 85 °C and washed with water and centrifuged. Following the above, it is heated to 110 °C and stirred well for 30 min with bleaching earth. The resultant oil

Table 7.1 Properties of the refined oil of *C. avellana*

Parameters	Value
Viscosity @ 25 °C (cP)	60.8
Acid value (mg KOH g^{-1})	0.05
Heat of combustion (MJ kg^{-1})	39.95
Palmitic acid(C16:0) (%)	6.6–8.3
Stearic acid (C18:0) (%)	2.8–3.2
Oleic acid (C18:1) (%)	75.7–77.7
Linoleic acid (C18:2) (%)	11.1–13.8

Table 7.2 Properties of *C. avellana* biodiesel (Xu and Hanna 2009; Moser 2012; Saydut et al. 2016)

Parameters	Value
Kinematic viscosity at 40 °C (cSt)	4.67, 4.51, 5.4
Density at 15 °C (kg m^{-3})	891
Flash point (°C)	168, 182.5
Pour point (°C)	−17
Cloud point (°C)	−14.7, −9.3, −11
Cetane number	53.35, 55
Iodine value (g I_2/100 g)	80.5
Heating value (MJ kg^{-1})	40.2, 39.87
Oxidative stability at 110 °C(h)	7.6
Wear scar at 60 °C (μm)	176

is deodorized at 240 °C in 3 mbar vacuum. The properties of the oil thus obtained are furnished in Table 7.1.

Oleic acid is the major fat. Oleic acid which occurs in large quantity gives stability to the oil against spontaneous oxidation. In order to transesterify 1 L of this oil, 220 g of methanol and 4 g of potassium hydroxide are required. The above quantity of methanol with oil forms 6:1 molar ratio. Only 3:1 ratio is sufficient as per the stoichiometry. But double the quantity of methanol is normally added so as to move the reaction steadily forward. At the outset, methoxide is prepared by mixing the alkali in methanol and stirred well for 20 min and this is transferred to the oil and stirred (200 rpm) for 2 h at 50 °C. The content is then cooled and allowed to stand overnight undisturbed. Shortly after, the biodiesel moves to the top and the glycerol settles at the bottom which is removed. The supernatant biodiesel is washed repeatedly with water till the phenolphthalein indicator shows neutral value. The above is then heated to expel the methanol and moisture. Normally the separation of biodiesel takes around 2 h but the biodiesel still looks turbid. If the settlement time is extended beyond (18 h), a clear biodiesel is formed. The properties of the biodiesel prepared are given in Table 7.2.

The properties of the transesterified oil are close to that of the diesel (Saydut et al. 2016). High cetane number is normally preferred for fuels to ensure short ignition delay. Cetane number and iodine values are inversely proportional. Low cetane number indicates the presence of unsaturated fatty acids in rich quantities. Low cetane number is responsible for the long gap between the initiation of combustion and the injection of fuel, whereas high cetane number causes quick auto ignition and in this case NO_x generation is high.

Higher density of biodiesel indicates that the volumetrically operating pumps measure higher mass of biodiesel than that of conventional fuel. The density and NOx are directly proportional. The heating value of biodiesel is around 10% lower than that of the regular diesel since 10% by weight of the biodiesel is formed of oxygen. Less viscosity promotes easy movement of the fuel. The oxygen rich biodiesel causes improvement in burning, reduction in the particulate matter, CO and other pollutants in the exhaust. However NOx level increases in the emission. The brake specific fuel consumption and brake thermal efficiency of biodiesel are higher than that of the diesel. Analysis of other parameters indicates that the biodiesel from *C. avellana* has identical characteristics as that of diesel and no gross deviation is observed (Gumus 2010).

References

Gumus M (2010) A comprehensive experimental investigation of combustion and heat release characteristics of a biodiesel (hazelnut kernel oil methyl ester) fuelled direct injection compression ignition engine. Fuel 89:2802–2814

Karabulut I, Topcu AL, Yorulmaz A, Tekin A, Ozay DS (2005) Effects of the industrial refining process on some properties of hazelnut oil. Eur J Lipid Sci Technol 107:476–480

Moser BR (2012) Preparation of fatty acid methyl esters from hazelnut, high oleic peanut and walnut oils and evaluation as biodiesel. Fuel 92:231–238

Saydut A, Erdogan S, Kafadar AB, Kaya C, Aydin F, Hamamci C (2016) Process optimization for production of biodiesel from hazelnut oil, sunflower oil and their hybrid feedstock. Fuel 183:512–517

Xu YX, Hanna MA (2009) Synthesis and characterization of hazelnut oil-based biodiesel. Ind Crop Prod 29:473–479

8 Brassicaceae

The family Brassicaceae has around 375 genera comprising 3200 species. All the species do not bear oil and not all oil bearing species are cultivated to yield oil for human consumption. The oil from Brassica sp. has rich quantity of erucic acid (>45%) which is a classified toxin to human beings.

8.1 Mustard (*Brassica juncea*)

The mustard *Brassica juncea* (L.) has a universal existence and its oil is actively considered as a raw material for biodiesel. It is an invasive agricultural pest in certain countries. It grows well in orchards, plantation areas, wastelands and derelict areas such as roadsides or near rail tracks. On the cost benefit analysis of biodiesel, it is observed that 70–80% of the total input goes to the cost of raw material, namely the oil. Alternatives to expensive and potentially cost prohibitive oils from peanut, canola, corn, soybean and palm are critical to the economics of biodiesel. In such situation the mustard oil comes handy. *B. juncea* is also known as *B. japonica* (Thunb), *B. integrifolia* (Vahl), *B. willdenowii* (Boiss), *Sinapis juncea* (L.) and *S. timoriana* (Dc). Due to the presence of erucic acid, mustard oil is considered noxious and not considered for human consumption in certain countries. In India mustard oil containing low level of erucic acid is being used for cooking. This oil is subjected to prolonged heating before being used for cooking apparently to expel noxious smell and taste. A variety of this species whose oil contains low level of erucic acid was first discovered in Australia (Kirk and Oram 1981). The mustard oil is much preferred in India for body massage, to improve blood circulation, toning the hair and texturizing the skin. Mustard is reported to tolerate a rainfall ranging from 1000 to 4000 mm, temperature of 5–28 °C and soil pH of 4.3–8.3. It is often grown as a semi-arid crop in the northern and central Africa, northern India and in certain parts of China.

S. P. Raj et al., *Biodiesel from Flowering Plants*,
https://doi.org/10.1007/978-981-16-4775-8_8

8.1.1 Vernacular Names

B. juncea is referred by different vernacular names in different countries as shown in Table 8.1.

8.1.2 Geographical Distribution

B. juncea is grown in North America (North and South Dakota and California), Western China, India, Canada, Pakistan, Poland, Bangladesh, Sweden, West Indies, Burma, Iran, Eurasia, Central Africa, Nepal, Southern Russia and North of Caspian Sea (Fig. 8.1).

Table 8.1 Vernacular names of *B. juncea* being grown in different countries

Country	Names
Europe	Indian mustard, India mustard, Chinese mustard, vegetable mustard, brown mustard
Malaysia	Sawi pahit, Kai choy, Gai choy
Indonesia	Sawi-sawi, sesawi
Philippines	Mustasa, Tagalog
Cambodia	Khat naa
Thailand	Phakkat-khieo, phakkat-khieopli
Vietnam	Cari canh
French	Moutarde indienne, moutarde de chine
India	Rai, kadugu

Fig. 8.1 Distribution pattern of Mustard in the world as seen in the hatched area

8.1.3 Botanical Description

B. juncea is an annual plant which grows up to 1.2 m. The basal leaves of this plant form a rosette. The leaves are alternate with a maximum length of 30 cm and a breadth of 10 cm. It is pinnatifid. The central vein is bold. The leaves are lobed, bluish green, glabrous and glaucous. The stem terminates in a corymbiform raceme of yellow flowers. Each flower has four petals and sepals with a short green pistil, a knobby stigma, several stamens and with yellow anthers. The petals are notched at their edges and have feeble veins running along the entire length. The plants hold the flowers normally for a month. Most of the flowers are self-pollinated. Each pollinated flower transforms in to a seed pod (silique). The silique is 3 times longer than its breadth. It holds 10–20 round, grey black seeds of 1 to 1.5 mm dia. (Fig. 8.2).

8.1.4 Cultivation

In a well ploughed sandy loam the seeds are sown at a gap of 15 cm in rows which are 30–45 cm apart. Around 4–6 kg ha^{-1} of seeds are needed. Around 50–75 kg N, 100–125 kg phosphate and 50–60 kg of potash are applied in split doses. It comes to flowering in 49–52 days and the total life cycle is 90–95 days. The harvesting is being done before the fruits are fully ripe to minimize the quantity of shattering. During the harvest the entire plant is pulled out and bundled or the plants are cut a few centimetres above the ground. The plants are then tied and brought to the

Fig. 8.2 The mustard *Brassica juncea*

threshing field where the seeds are separated. The yield ranges from 1000 to 1250 kg ha^{-1}.

8.1.5 Pests and Diseases

It is considered ideal to cultivate the mustard in the same field after 3–4 years so as to avoid soil borne pests infesting the plant repeatedly. The names of various pest organisms responsible for the infestation are given in Table 8.2.

8.1.6 Mustard Oil

The cold pressed oil is much preferred for human consumption as it retains many of its characteristics since no chemical is added. But, the refined oil does not retain its real characteristics as it is heated to 250 °C and it is also processed using chemicals. The mustard oil has the lowest saturated fatty acids among many edible oils.

The properties of the mustard oil are given in Table 8.3. The oil content of the mustard seed differs according to the nature of cultivation and it ranges from 36 to 47%. These values are considered significant and observed to be higher than certain other oil seeds (e.g. Soybeans: 17–25%). The colour of the oil is relatively dark (The Gardner colour code of this oil is 10.5, code 1 refers the lightest and code 18 refers the darkest). The kinematic viscosity is relatively high which is mostly due to the presence of certain fatty acids having 20 or more carbons. Fatty acids having the long chain length are progressively viscous. The specific gravity and the wear scar (lubricity) are almost identical with other oils. The pour point is very low (−18 °C). Pour point indicates the lowest temperature at which there is active movement of the fuel when the container is tipped. Relatively low level of saturated fatty acid causes the low temperature operability with reference to pour point. This oil is known to have sulphur containing glucosinolates. The glucosinolates are polar in nature. A

Table 8.2 The causative organisms of various diseases in mustard

Pest type	Species
Nematodes	*Heterodera cruciferae, H. schachtii, H. trifolii, Meloidogyne hapla, M. incognita* and *Trichodorus christiei*
Bacteria	*Erwinia carotovora, Xanthomonas campestris*
Virus	Rape mosaic, *Brassica* virus 2, Turnip yellow mosaic, Yellow virus, Cabbage black ringspot, Kukitachina mosaic, Cucumber mosaic, Radish mosaic
Fungi	*Albugo candida, A. macrospore, Alternaria brassicae, A. saccardoi, Ascochyta brassicae-junceae, Cercospora brassicicola, Cercospora brassicae, Cladosporium brassicicola, Colletotrichum higginsianum, Cystopus candidus, Erysiphe polygoni, Ischnochaeta polygoni, Macrophomina phaseolina, Mycosphaerella brassicicola, Ophiolobus graminis, Ovularia indica, Peronospora parasitica, Plasmodiosphora brassicae, Puccinia aristidae, Pythium debaryanum, Rhizoctonia solani, Sclerotinia sclerotiorum* and *Sclerotium rolfsii*

Table 8.3 Physicochemical properties of mustard oil ($n = 3$)

Parameters	Value
Gardner colour	10.5
Kinematic viscosity at 40 °C (cSt)	39.2
Density at 25 °C (g cc^{-1})	0.891
Acid value (mg KOH g^{-1})	0.575
Iodine value (mg g^{-1})	1140
Pour point (°C)	−18.0
Cold filter plugging point (°C)	−3.5
Sulphur (mg L^{-1})	14.5
Phosphorous (mg L^{-1})	2.5
Induction period (h at 110 °C)	5.35
Lubricity wear scar (μm at 60 °C)	153.0
Molar weight (g mol^{-1})	303.6
Free glycerol (%)	0.004
Total glycerol (%)	0.247
Cetane number	62

Table 8.4 Fatty acid composition of the mustard oil (Bannikov 2011)

Fatty acid	Structure	Mol. wt. (g mol^{-1})	Mass fraction (%)
Myristic acid	C14: 0	228.38	0.063
Palmitic acid	C16: 0	256.43	2.377
Heptadecanoic acid	C17: 0	270.46	0.018
Stearic acid	C18: 0	284.48	1.253
Arachidonic acid	C20: 0	312.54	1.338
Heneicosanoic acid	C21: 0	326.57	0.838
Palmitoleic acid	C16: 1	254.41	0.180
Oleic acid	C18: 1	282.47	25.156
Linoleic acid	C18: 2	280.45	14.459
Alpha-linolenic acid	C18: 3	278.44	15.451
Eicosenoic acid	C20: 1	310.52	0.423
Erucic acid	C22: 1	338.58	36.709
Docosadienoic acid	C22: 2	336.56	0.286
Nervonic acid	C24: 1	366.63	1.405

small quantity is likely to enter in to the lipid phase during the separation of the oil from the seed cake thereby leaving a little quantity of sulphur in the oil. The oil also contains a low quantity of phosphorous. The induction period of the oil is relatively lower than that of other oils. Low stability of the oil as indicated by the low induction period is due to the presence of more trienoic fatty acids (fatty acids with three double bonds) as trienoic fatty acids are easily vulnerable to auto oxidation.

The fatty acid composition of the mustard oil is given in Table 8.4. Among the various fatty acids present in the oil erucic acid content is very high. Other fatty acids which are known to occur in large quantities are oleic acid, linoleic acid and alpha-linolenic acid. The saturated fatty acids occur in a limited quantity and the share of unsaturated fatty acid is as high as 94%. Monounsaturated fatty acid forms 63.9%, whereas the polyunsaturated fatty acid content is only 30.2%. The fatty acid composition of the oil indicates that the mustard oil has 0.773 g carbon, 0.120 g hydrogen and 0.107 g oxygen in a molecular mass of 303.5 g mol^{-1}.

8.1.7 The Oil Cake

The oil cake from mustard seed is used in animal feed if it has low glucosinolate and erucic acid. The cake often retains the mustard flavour. The cake if contains more glucosinolate is applied to the field to control weeds and pests.

8.1.8 Preparation of Biodiesel

With 800 mL of methanol (635 g = 19.8 mol) 3 g of KOH is added and mixed well. It is then added with 1100 mL (1000 g) of oil (1000 = 3.3 mol) so as to get an oil alcohol molar ratio of 1:6. This mixture is stirred at 250 rpm in 60 °C for 2 h. Then it is brought to the room temperature and transferred to a separating funnel and allowed to stand undisturbed for 2 h. The glycerol which settles at the bottom is removed. The overlying ester (biodiesel) is transferred to a rotary evaporator to remove the excess methanol at a low temperature. The resultant crude ester is repeatedly washed in warm water till the wash water becomes neutral pH. Then the ester is dried by heating. Residual moisture if any is removed by treating it over magnesium sulphate.

8.1.9 Physicochemical Characteristics of Biodiesel

The colour of the biodiesel made from mustard oil is dark (Gardner colour code is 10). The kinematic viscosity is 5.55 cSt which is a fairly high value within the standard (1.9–6.0 cSt in ASTM D6751). High viscosity of this biodiesel is due to the presence of erucic acid ester in high quantity. The viscosity of the esters of palmitic, oleic and gondoic acids (eicosenoic acid) are 3.67, 4.51 and 5.77 cSt at 40 °C, respectively. The acid value is 0.21 mg KOH g^{-1} which is well within the ASTM standard of <0.5. Iodine value 1120 mg g^{-1} also is within the European standard EN 14214 though ASTM does not specify any standard value (Table 8.5).

Low temperature operability of the biodiesel from mustard oil is largely determined by its cloud point, cold filter plugging point and pour point. Among the three factors cited above, the cloud point is relatively high (4 °C). The esters, 1-monopalmitin (glycerol alpha-monopalmitate), 1-monostearin (glycerol monostearate), 1,3-dipalmitin (glyceryl dipalmitate) and 1–3 distearin (glyceryl distearate) are characterized by high melting points. As the chain length of saturated fatty acid methyl ester increase, there is increase in melting point. The melting point of palmitic (C16: 0), stearic (C18: 0), arachidonic (C20: 0) and behenic (C22: 0) fatty acids are 31, 39, 46, 54 °C, respectively. Presence of certain phytosterols is

Table 8.5 Physicochemical characteristics of biodiesel from mustard oil adopted from Jham et al. (2009) and Bannikov (2011)

Parameters	Value
Gardner colour	10
Kinematic viscosity at 40 °C (cSt)	5.55
Density at 25 °C (g cc^{-1})	0.877
Acid value (mg KOH g^{-1})	0.21
Iodine value (mg I_2g^{-1})	1120
Cloud point (°C)	4.0
Pour point (°C)	−16.5
Cold filter plugging point (°C)	−3.0
Induction period (h at 110 °C)	4.8
Lubricity wear scar (μm at 60 °C)	151
Molecular weight (g mol^{-1})	327.9
Free glycerol (%)	0.04
Total glycerol (%)	0.197
Cetane number	61.1
Sulphur (mg kg^{-1})	8.5
Phosphorous (mg kg^{-1})	2.0
Flash point (°C)	158
Heating value (KJ kg^{-1})	38,000

responsible for the high cloud point. These phytosterols exist as steryl glucosides in vegetable oil which are not soluble in biodiesel and have a high melting point (240 °C). The oxidative stability in terms of induction period at 110 °C is 4.8 h against >6 h (as per European standard EN 14214 it should be >6 and as per American Standard ASTM D 6751 it should be >3).

The lubricity value (wear scar reading made using a high frequency reciprocating rig HFRR) is 151 μm at 60 °C. The lubricity analysis made as per the ASTM D 6079 is not monitored by the standard in ASTM D 6751 or EN 14214. But, standards are prescribed for petro-diesel in ASTM D 975 or EN 590. The maximum prescribed wear scar is 520 and 460 μm for ASTM and EN standards, respectively. The low value for the wear scar in the biodiesel from mustard oil indicates that it possesses an inherent lubricity. The sulphur content is less than the limit (15 mg kg^{-1}) as per ASTM D 6751 and 10 mg kg^{-1} as per EN 14214. The sulphur content of the ester is the mere reflection of its presence in the oil. Attempts to regulate the sulphur content include refining, bleaching and deodorization of the raw material. Similarly, the phosphorous level in the biodiesel (2.0 mg kg^{-1}) is also within the standard (<15 mg kg^{-1} as per ASTM D 6751 and <10 mg kg^{-1} as per EN 14214).

The cetane number and heating value of the biodiesel increase with the increase in the chain length and decrease with the increase in unsaturation. The increase in the unsaturation also causes the increase in NO_x in the emission to the tune of 10% over that of the regular diesel.

8.1.10 Fuel Efficiency

Employing the biodiesel from the mustard oil in the place of regular diesel may not alter the mechanical efficiency and exhaust temperature but the brake specific fuel efficiency is likely to increase by 10–15% and the brake fuel conversion efficiency may be reduced by 3–5%. The bulk modulus, viscosity and density of the biodiesel are on the higher side which cause the increase in injection pressure. The increase in consumption increases the injection duration. The ignition delay period of this biodiesel is shorter than that of the regular diesel.

8.2 Canola (*Brassica napus* and *B. rapa*)

The word canola is an abbreviation of **Can**adian **O**il **L**ow **A**cid which came in to a popular reference during the year 1978. Of late the same word refers to an oil extracted from the seeds of two notified species of the genus *Brassica,* namely *Brassica napus* L. and *B. rapa* (*B. campestris* L.) belonging to the order Brassicales, family *Brassicaceae* (*Cruciferae*), tribe *Brassiceae* and sub-tribe *Brassicinae*. Canola is the third leading vegetable oil in the world in terms of its quantity after that of soybean and oil palm. Though the oil from *Brassica* sp. generally has rich quantity of erucic acid (>45%), the erucic acid content of the modified B. *napus* and *B. rapa* is lower than 2% and such low level is considered to be within the toxic limit. The above two species have common set of chromosomes signaling a great chance for interspecific crossing.

8.2.1 Different Names

B. napus is referred popularly as rapeseed, rape, oilseed rape, rapa, rappi, swede, oil turnip, rutabaga, rape kale, oil colza, neep and rappi. *B. rapa* is often called as rapa, rapum, turnip rape, field mustard, turnip mustard, polish rapeseed.

8.2.2 Distribution

The two species *B. napus* and *B. rapa,* jointly referred as canola are distributed in a vast stretch of geographic domain (Fig. 8.3). They are being cultivated in Australia, Bangladesh, Canada, China, Czech Republic, Denmark, Europe, France, Germany, Greenland, Hungary, India, Pakistan, Poland, Russia, Sweden and USA (North Dakota).

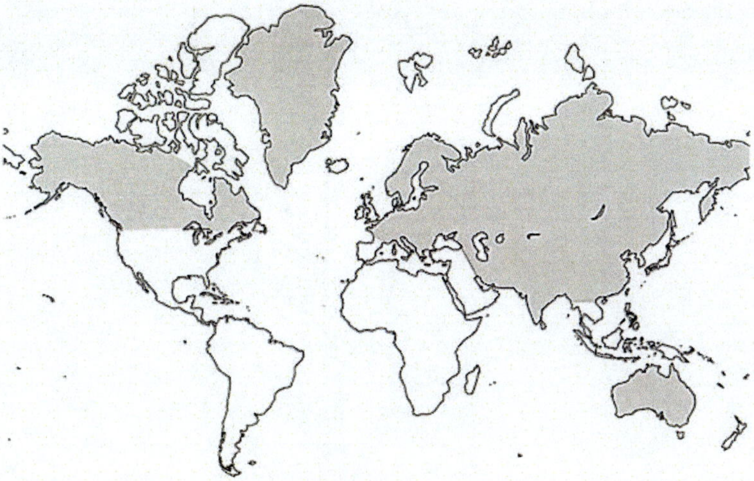

Fig. 8.3 Canola cultivation zone of the world shown as hatched

8.2.3 Botanical Features

B. napus is a multistemmed plant with a slender and tuberous taproot. The foliage is bluish green and glaucous. The lower leaves are pinnatifid or lobed with petioles of 10–30 cm long. The leaves at the upper stem are lanceolate, sessile and clasping type. The stem has multiple branches and each ends in an inflorescence. Each inflorescence is an elongated raceme with yellow flowers. Inflorescence is much branched (Fig. 8.4). Each inflorescence is around 0.75 m height and the silique is around 10 cm long and 4 mm wide. The flower is around 1.5 cm long.

B. rapa is an annual herb. Each lower leaf has 3–5 lobes. The leaves are light to medium green, lyrate-pinnatifid with clasping base. Flowers are bright yellow. Each flower has four green sepals and four petals. Each fruit is 3–7 cm long with an extended tapering beak. The seeds are blackish or reddish brown of 1.5–2 mm dia.

Canola in general is characterized by entomophilous flowers which are amenable to cross or self-pollination. However, fertilization mostly takes place through self-pollination since each flower has abundant pollen grains which out-number the extra pollen grains reaching the flowers from elsewhere. There are abundant chances of cross pollination also depending on the situation. The propagation is through seed.

8.2.4 Cultivation

There is a growing hope that canola oil can act as a trustful raw material for biodiesel production and therefore many countries are gearing up the cultivation of these species. The plant (both the species) performs well in medium textured loam. Sloped land with a character of easy drainage is ideal. Neutral soil with marginally low pH is

Fig. 8.4 The canola *Brassica napus*

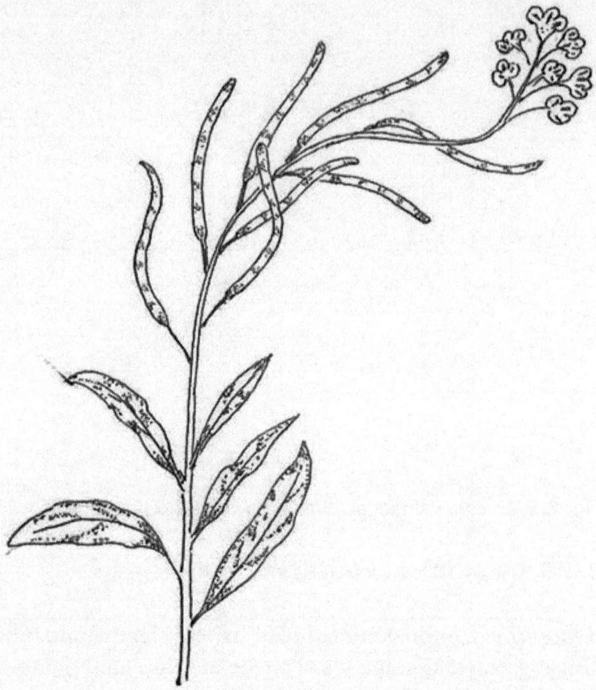

considered good. It also manages medium saline soils. The land is initially tilled to a depth of around 10 cm. Around 10 kg of P_2O_5 and 10 kg of K_2O and 8 kg of urea are applied as basal fertilizers before sowing. Seeds are sown by broadcasting at the rate of 4–6 kg ha^{-1}. The above rate of seeding may provide a density of 50–70 plants m^{-2}. Season specific varieties of B. *rapa* are available to suit the fall or spring.

Around 30 kg of urea may be applied during the growth phase in 3–4 split applications. Canola is a sulphur demanding crop and hence sulphur supplementation to the soil may be required. Around 40–50 cm of water is required to irrigate the crop during its full lifespan.

The plant emerging from the seed is vulnerable to pest attack. Soon after emerging, the plant increases its leaf area and tries to resist the impending pests. When 30–60% of the leaf area is reached it may put forth buds. In 15–20 days flowers open and around 50% of the flowers end in pod formation. Shortly after that, the petals start falling from the flowers. Pod filling is completed in 35–45 days after flower initiation. Initially the seeds are green in colour. As days pass by, the colour of the seed turns to brown. When 30–50% of the seeds turn brown it is harvested.

8.2.5 Pests and Diseases

Often the canola plant is damaged by a host of fungi. The fungus *Leptosphaeria maculans* causes blackleg disease. This disease which transfers through the infected stubble destroys the seedlings and reduces the yield of seeds in mature plants. The fungus *Phytophthora megasperma* causes root rot. Other diseases caused by the fungi are the downy mildew disease by *Peronospora parasitica*, leaf spot disease by *Alternaria brassicae*, white rust disease by *Albugo candida* and the stem rot disease by *Sclerotina* sp.

Beet western yellows virus (*Luteoviridae*) and mosaic virus are known to spread by aphids. The insect pests redlegged earth mite *Halotydeus destructor*, blue oat mite *Penthaleus major*, aphids *Brevicoryne brassicae* and *Lipaphis erysimi*, diamond-back moth *Plutella xylostella* and the Rutherglen bug *Nysius vinitor* cause extensive seasonal damages. The cutworm *Agrotis infusa*, the caterpillars of *Helicoverpa punctigera* and *H. armigera* and the larvae of the bronzed field beetle *Adelium brevicorne* also are known to destroy the crop.

8.2.6 Fruit and Seed

The fruiting body is called silique or pod. Around 15–25 seeds are produced per pod. Each seed measures 1–2 mm in diameter and 2.5–3.0 lakh seeds weigh a kg.

Timely harvest is one of the most important parameters. Delay in harvesting the crop would cause the seed to shatter. The crop has to be closely monitored for the purpose of swathing once the pod begins to show yellow colour. A reliable parameter of the maturity of the seed is its colour. If 50–70% of the pod becomes yellowish brown it is considered ready for swathing.

8.2.7 Oil Extraction

During the earlier days edible oil was extracted from canola only for domestic consumption. Canola seeds normally contain 40% oil. Oil is extracted by two sequential processes, namely pressing and solvent extraction, the former has to be followed by the latter.

The seeds are first cleaned from foreign matter such as debris, pebbles, metal pieces of the threshing field and other plant matters. Such removal may be effected by sieves or magnetic separators. The seeds are then dried to reduce the moisture content. Seeds with 6–12% moisture are considered ideal. The seeds are subsequently conditioned by heating them to have a temperature just by 20 °C above the ambient temperature.

The warmed up seeds are quickly pressed through rollers. This would cause the rupture of the seed wall and as a result the oil moves out of the roller. The residual seed mass takes the shape of flakes. The roller and the rate of flow of the warm seeds are adjusted in such a way to have the resultant flakes around 0.2–0.4 mm thickness.

Too low a thickness spoils the consistency and too high a thickness reduces the efficiency of the subsequent process.

The flake thus obtained is passed along with steam so as to raise its temperature to 80–90 °C. The rise in temperature shall be rapid so as to inactivate the enzymes present in the flake. Inactivation of the enzymes inherently controls the breakdown products which would otherwise unnecessarily destroy the quality of the oil and also the cake. The heating duration shall not extend beyond 20 min as extended heating or higher temperature may volatilize certain essential compounds which ultimately causes unnecessary odour to the oil besides the protein content of the meal also is likely to be denatured.

The heating of the flake causes the rupture of the oil cells present in the flake. Once the oil cell breaks, the oil squeezes out. These oils gradually coalesce and form droplets. The coalescence and formation of the droplets are ably aided by lowering in viscosity caused by the enhanced temperature.

The hot flakes are then passed under pressure in screw press or through expeller. The rotating screw of the press propels the flake towards an adjustable choke which partially constricts the discharge of the cake at the end of the screw press. Thus the oil present in the flake is removed to the maximum. However, there may be residual oil in the cake. The cake thus being discharged from the screw press is therefore extracted using a solvent. The cake is broken in to small pieces and loaded in an extractor. Hexane is then passed over it repeatedly by spraying so that the residual oil present in the cake dissolves in the hexane. The hexane-oil mixture is called as miscella. The miscella percolates down through the mass in the extractor and drains out. The same miscella is repeatedly sprayed over the cake so that maximum removal of oil is ensured. The miscella is then distilled out to separate the hexane from the oil.

The meal is again passed with steam to expel the residual hexane. Care is taken not to expose the cake in steam for more than 20 min. It is then dried at 103 to 107 °C. The cake moving out at this stage may have less than 1% oil and 8–10% moisture. This is then granulated or pelleted before it is sent as cattle feed.

8.2.8 Degumming of Oil

The oil is inadvertently seen mixed with an organophosphorus compound, namely phospholipids. These phospholipids may remain as colloid or in dissolved form. The phospholipid is referred as phosphatide or gum. If the phosphatide is not removed in time, the oil may become dark. Therefore, the oil is treated with dilute phosphoric acid at a warm temperature (70 °C). Alternatively, citric acid also is used. The hydrated phosphatides are removed by centrifugation. It is then neutralized with dilute sodium hydroxide. The soap thus formed is washed away and the wash water is then separated. Any residual water present in the oil is removed by drying in vacuum.

8.2.9 The Physicochemical Characteristics of Oil

The physicochemical properties of the degummed canola oil are given in Table 8.6. The analytical details are not exclusively for *B. napus* or *B. rapa*. The data are considered common for both the species. The viscosity of the canola oil is relatively high (35–40 cSt) caused by the unique combination of certain fatty acids with glycerol. The viscosity may change according to the temperature (Fig. 8.5).

The smoke point is above the recommended minimum limit of 200 °C. The smoke point refers to a temperature at which the oil produces a wisp of smoke on heating. The evaluation of smoke point is highly subjective and visual and therefore there can be error in fixing the point causing a maximum variation of 20% in the value. The flash point being the ignition point of canola oil is well within the standard limit of 275–330 °C.

Certain grade of canola oil has chlorophyll content as one of the impurities. When immature seeds are processed for oil the resulting oil may contain chlorophyll. The canola plant if caught in frost, the seed development is hampered and the harvested seeds ultimately may carry chlorophyll. The best quality canola has chlorophyll level lesser than 25 mg kg^{-1}. Medium grade canola seed has 26–45 mg kg^{-1} chlorophyll.

Table 8.6 Physicochemical properties of degummed canola oil

Parameter	Value
Viscosity at 40 °C (cSt)	35–40
Density at 30 °C (g cc^{-1})	0.912–0.918
Smoke point (°C)	215–240
Flash point(open cup) (°C)	280–290
Hydrogen (%)	10.5
Carbon (%)	80.0
Oxygen (%)	9.5
Acid value (mg KOH g^{-1})	0.4–0.6
Heating value (KJ kg^{-1})	39,720
Free fatty acid (%)	0.33
Chlorophyll (mg L^{-1})	65.0

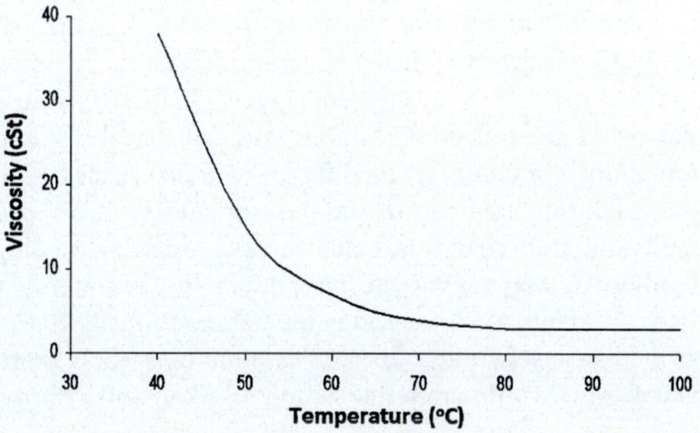

Fig. 8.5 Changes in the viscosity of canola oil according to temperature

Table 8.7 The fatty acid composition of canola oil

Fatty acid	Percent
Myristic acid (C14: 0)	0.2
Palmitic acid (C16: 0)	4.2
Stearic acid (C18: 0)	1.9
Arachidic acid (C20: 0)	0.7
Behenic acid (C22: 0)	0.3
Palmitoleic acid (C16: 1)	0.3
Oleic acid (C18: 1)	58.3
Linoleic acid (C18: 2)	21.5
Linolenic acid (C18: 3)	11.5
Erucic acid (C22: 1)	1.1

Lower grade canola has 46–100 mg kg^{-1} chlorophyll. The chlorophyll content causes adverse effect on the stability of the oil. The chlorophyll pigment degrades to pheophytin if the chelating metal (Mg) is removed. The enzyme chlorophyllase often assists in converting the chlorophyll into chlorophyllide. In order to remove the chlorophyll pigments from the oil certain adsorbents are used. They comprise certain special types of clay (montmorillonite K10, montmorillonite KSF and Attapulgus) and activated carbon. Among them montmorillonite K10 is reported to be efficient. Around 5–10 g of montmorillonite K10 is added to every 100 g of oil and stirred vigorously for 1 h. In this process around 30–40% of the chlorophyll is removed. The fatty acid composition of the oil is given Table 8.7.

The unsaturated fatty acids (oleic acid and linoleic acid) form the bulk of the fatty acids (>75%). Due to the above reason the stability of this oil is low. Certain saturated fatty acids such as myristic acid (C14: 0), arachidic acid (C20: 0) and behenic acid (C22: 0) occur in insignificant quantities.

8.2.10 Biodiesel Production

The oil is treated with alcohol such as methanol or ethanol at a molar ratio of 6:1 (alcohol: oil) along with 1% KOH or NaOH. It is then stirred at 600 rpm for 90 min. at 60 °C. The transesterified material is taken in a separating funnel and is allowed to settle. After 1–2 h the glycerol layer which settles at the bottom is removed. The unreacted alcohol present along with ester at the top layer is distilled out and the remaining content is washed thrice with dist. water. The washed biodiesel is finally passed through anhydrous sodium sulphate. In rare occasions ethanol or a mixture of methanol and ethanol at equal molar ratio is also being used. Thus three types of biodiesel (ester) are known. They are methyl ester, ethyl ester and methyl-ethyl ester. In case ethanol alone is used in the transesterification process there may be a problem in the phase separation. In such eventualities dilute tannic acid is mixed in the wash water. This would arrest the action of alkali and improve the phase separation.

8.2.11 Characteristics of Biodiesel

The characteristics of biodiesel such as methyl ester, ethyl ester and methyl-ethyl ester are furnished in Table 8.8.

The viscosity of all the esters is within the standard. Relatively high viscosity of ethyl ester is mostly due to the low conversion of triglyceride. As a result, such ethyl esters contain more glyceride which is a known source for higher viscosity. Therefore, the heating value of the ethyl ester also is higher. Low iodine value of the ethyl ester is apparently due to the low degree of unsaturation compared to that of the methyl esters. Assuming that methyl and ethyl esters have identical configuration with respect to the number of double bonds per molecule, the ethyl alcohol which has a higher molecular weight (46.07 g mol^{-1}) than that of methyl alcohol (32.04 g mol^{-1}) has a lower concentration of double bond leading to a relatively low level of unsaturation as contended by Knothe and Dunn (2003) and thus the iodine content shows a relatively low value. The fatty acid composition of the methyl and ethyl esters is given in Fig. 8.6.

In the methyl ester the saturated component forms only 7.3% and the unsaturated portion is 92.7%. Similarly, the ethyl ester also has a limited quantity of saturated component (6.45%), whereas the unsaturated component is very high (91.4%). Because of the very limited extent of saturation the stability of canola biodiesel also is low.

The stability of the biodiesel is studied by applying the Rancimat method. The Rancimat method also is referred as *Automated Swift* test or A*ccelerated oxidation* test. In this method, under hot condition a stream of air is passed through the sample of the biodiesel in a sealed reaction vessel. From a short while from then the biodiesel sample get oxidized. As an initial oxidation product, peroxide is generated which is followed by the complete destruction of the fatty acid resulting in to the formation of organic acids and volatile organic compounds. These end products are transferred by the same stream of air in to the distilled water present in the next vessel where the conductivity is being constantly monitored and the concentration of the organic acid released in the distilled water is known by the increase in conductivity. The time taken for the formation of the ultimate reaction product as known

Table 8.8 The characteristics of canola biodiesel (methyl ester, ethyl ester and methyl-ethyl ester) (Kulkarni et al. 2007; Issariyakul and Dalai 2010)

Parameters	Unit	Methyl ester	Ethyl ester	Methyl-ethyl ester
Viscosity at 40 °C	(cSt)	4.8	5.3	5.0
Density at 30 °C	(g cc^{-1})	0.875	0.869	0.883
Acid value	(mg KOH g^{-1})	0.48	0.65	0.7
Heating value	(KJ kg^{-1})	40,070	40,290	40,110
Cloud point	(°C)	1.0	−1.0	0
Pour point	(°C)	−15.0	−12.0	−15.0
Iodine value	(mg I_2 g^{-1})	109.5	103.0	103.5
Sulphur	(mg L^{-1})	<1.0	<1.0	<1.0

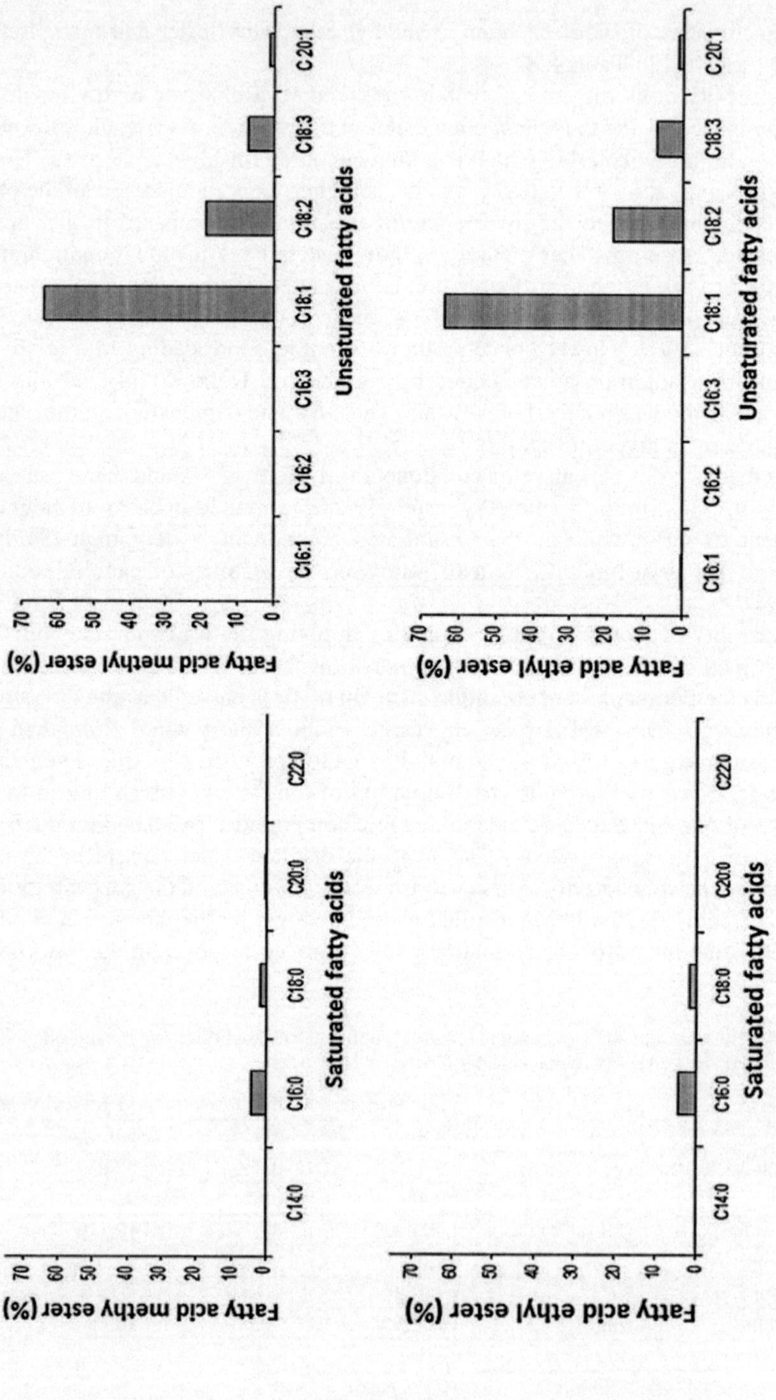

Fig. 8.6 Fatty acid composition of methyl and ethyl esters of canola

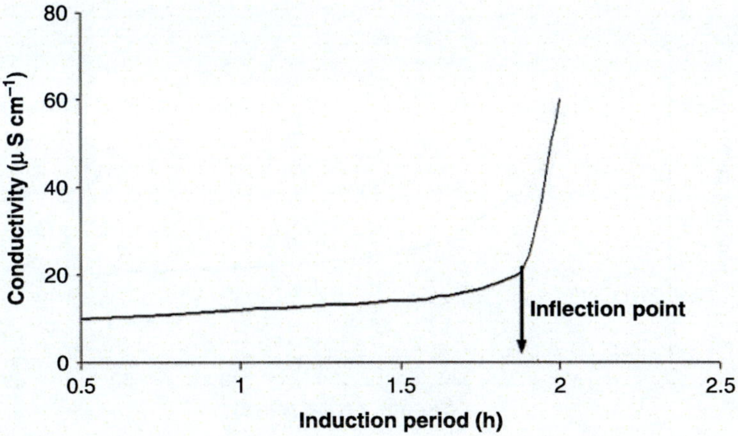

Fig. 8.7 The Induction period of biodiesel from canola oil stored for 60 days. The arrow indicates the inflection point

from the inflection point of the curve (Fig. 8.7) is termed as Induction Period or Oil Stability Index. As per the European Standard for biodiesel (DIN EN 14214) the induction period is 6 h at 110 °C. The ASTM (ASTM D 6751–07) specification is 3 h. It is to be observed that biodiesel from canola oil does not meet this standard.

Due to the presence of more quantity of unsaturated fatty acid esters, there is every likely that the biodiesel from canola oil loses its stability badly if it is stored for 60 days (Fig. 8.5). However, the rate of degradation depends largely on many factors such as storage temperature, iodine number, etc.

Many reports indicate that the biodiesel (from canola oil) of less than 5 days old has a good induction period (>6 h). As days pass by, the induction period slowly decreases (<6 h) indicating that the biodiesel is unfit for sale as per the standard prevailing in force (Fig. 8.8). The ethyl ester of the canola has a relatively longer induction time than that of methyl ester since ethyl ester has a lower molar concentration of double bond.

Another factor responsible for the loss of stability is the presence of chlorophyll pigments in the oil especially when such oil is extracted from the seeds of the frost affected crop. The chlorophyll pigments present in the oil are removed to a greater extent as water soluble chlorophyll in salt during the alkali catalysed transesterification. The chlorophyll pigment present in the oil is required to be removed using montmorillonite K10 before the transesterification process is commenced. On the other hand, if montmorillonite K10 is incorporated along with the alkali and alcohol during the transesterification process the catalytic function of the alkali (KOH) is hampered as potassium cation is inadvertently trapped in the interlayer of the montmorillonite K10 thereby reducing the concentration of the alkali. As a result, the ester production is seriously affected. Besides, when potassium as cation stays between the layers of montmorillonite K10, the adjacent layers are greatly polarized. Thus the adsorption characteristics of the montmorillonite K10

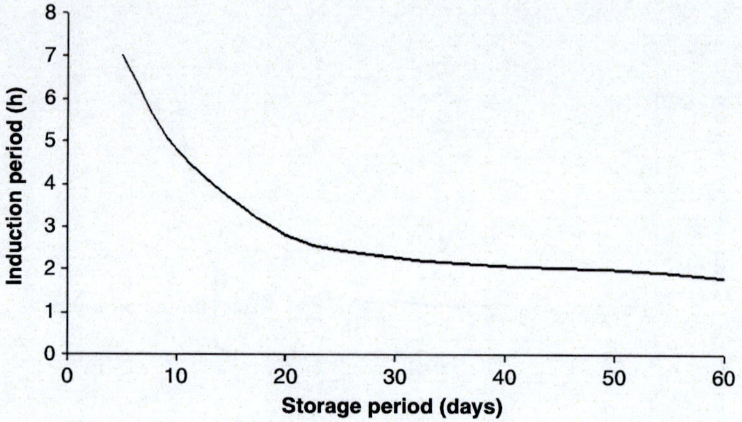

Fig. 8.8 Loss of stability of biodiesel from canola oil as inferred from the induction period

Fig. 8.9 The ester yield from canola oil treated with montmorillonite K10. (Single stage = Transesetrification and chlorophyll removal made simultaneously, Double stage = Transesterification made on chlorophyll removed oil)

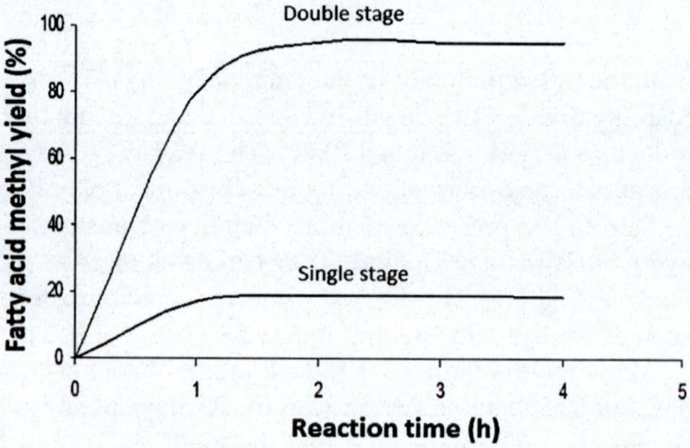

are altered and the adsorbent is ultimately decapacitated with reference to its job of removing chlorophyll pigment from the oil.

The pattern of methyl ester yield from canola oil containing chlorophyll at a conc. of 90 mg kg^{-1} is given in Fig. 8.9.

It is recommended to pretreat the oil with an adsorbent before the same is taken for transesterification. In such case the efficiency of ester formation is more than 95%. This is referred as double stage process. When the adsorption of chlorophyll by montmorillonite K10 and transesterification process are carried out as a single step the ester yield is relatively low (as low as 15–20% in a reaction time of 4 h) indicating that the function of the catalyst in the formation of biodiesel is seriously affected by the montmorillonite K10. Therefore, if montmorillonite K10 is used along with the catalyst and alcohol in the process of transesterification (single stage), chlorophyll pigment continues to remain in the biodiesel (Fig. 8.10) which greatly affects the quality during long storage.

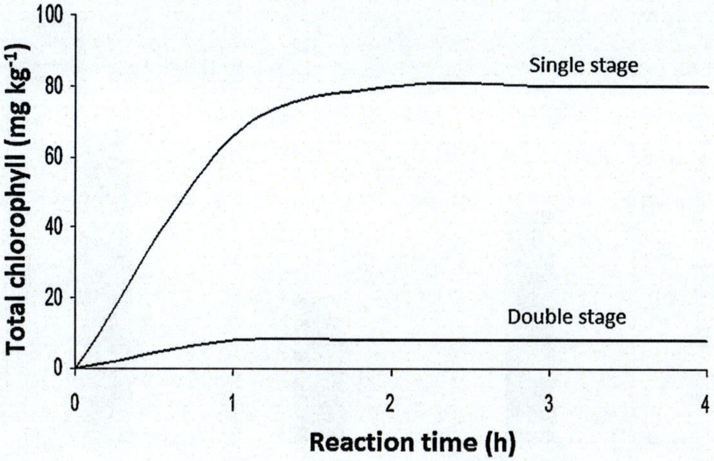

Fig. 8.10 Chlorophyll content of the canola ester prepared from the canola oil containing chlorophyll (Single stage = Ester prepared by treating the oil jointly with montmorillonite K10, methanol and KOH in the same reactor, Double stage = Ester prepared from chlorophyll removed oil)

The ester prepared from the oil (having chlorophyll at a conc. of 90 mg kg^{-1}) pretreated in isolation with montmorillonite K10 has a chlorophyll content of around 5 mg kg^{-1}, whereas the ester prepared using alcohol and alkali along with montmorillonite K10 in the same reactor has the chlorophyll content to the tune of 80 mg kg^{-1}. Thus, it is known that the process of chlorophyll removal has to be done as a separate step.

There is an overall awareness on the need for using low sulphur diesel in automobiles. Many countries therefore have gone in to framing suitable standards and regulations. As a result, the regular diesel is hydroprocessed by petroleum industries to remove sulphur and polar and aromatic compounds which are otherwise known to give lubricity to the petroleum diesel. Unfortunately, such rectified fuels are known to cause engine problems due to low lubricity. Hence, safe additives are needed to be incorporated in such petroleum diesel to improve the lubricity. The ASTM suggests *High Frequency Reciprocating Rig* (HFRR) *Test* (ASTM D 6079–99) as a standard method to evaluate the lubricity. Parallel test known as Scuffing Load Ball on cylinder Lubricity Evaluator *tests* (STM D 6078–04, 2010) also is recommended. However, no accurate interconversion factors between the above two tests are available. Such measurement of the lubrication and the study of the mechanism of lubrication are called tribology. The lubricity is measured in terms of wear scar diameter. The ASTM standard for wear scar diameter is 0.45 mm for an additive conc. of 1% by volume.

It is reported that ester made from vegetable oil can act as a lubricating agent in regular diesel so as to meet ASTM standard (VanGerpen et al. 1999). According to Lang and Dalai (2001), incorporation of canola ester in petro-diesel at 1% level increases the lubricity of the diesel substantially. It is interesting to note that the

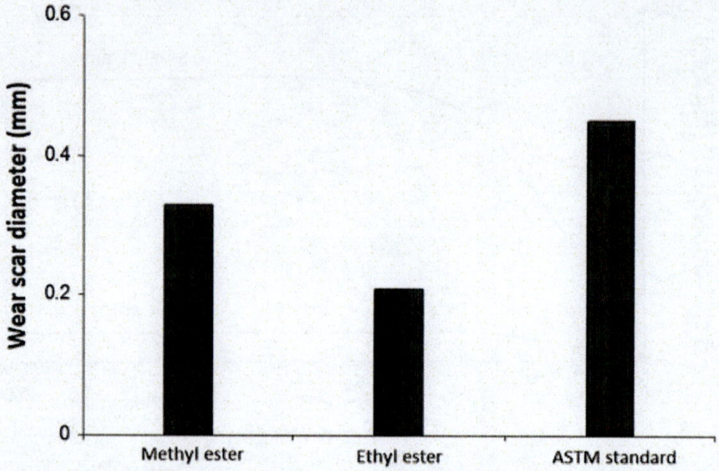

Fig. 8.11 The wear scar diameter shown by methyl ester and ethyl ester. The ASTM standard also is shown for comparison

lubricity offered by the methyl ester developed from the commercial canola oil is higher than the lubricity of the methyl ester developed from the analytical grade individual fatty acid (Geller and Goodrum 2004). Ethyl ester has good lubricity when compared to methyl esters. The process of preparing ethyl esters is environmentally benign since the ethanol required is manufactured from renewable sources. Besides, ethanol is relatively more soluble in oil than methanol. However, the separation of ethyl ester is slightly cumbersome (Meher et al. 2006). On the other hand, the separation of methyl ester is quite simple as distinct zonation develops (Boocock et al. 1996). It is also to be noted that the reaction rate with respect to methyl ester formation is relatively slow. The rate of formation of methyl ester is 15 fold lower than that of butyl ester (Boocock et al. 1996). The ASTM standard and the wear scar diameter showed by methyl and ethyl ester is compared in Fig. 8.11. It is reported that lubricity and the chain length of alcohol used in the transesterification are directly proportional (Anastopoulos et al. 2001; Drown et al. 2001).

The performance of canola biodiesel in engines was reported (Lapuerta et al. 2008). When canola biodiesel is used in engines there is a marginal reduction in torque and power which is naturally a cumulative effect of low heating value and high density. The torque increases along with the speed but such effects always remain lower with respect to biodiesel by 5–9% when compared to that of regular diesel (Fig. 8.12). However, such decrease depends upon the engine rating and the source of the biodiesel.

Similarly, the engine power also increases as the speed is increased. When the power generated by the biodiesel is compared to that of regular diesel it is known that it falls short by 2–11% depending on the speed (Fig. 8.13).

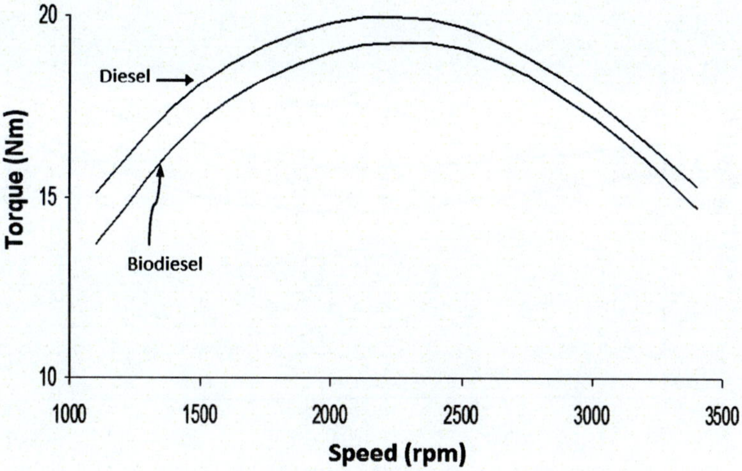

Fig. 8.12 The relationship between engine speed and torque when biodiesel is used as a fuel

Fig. 8.13 The relationship between speed and power in an engine run with biodiesel

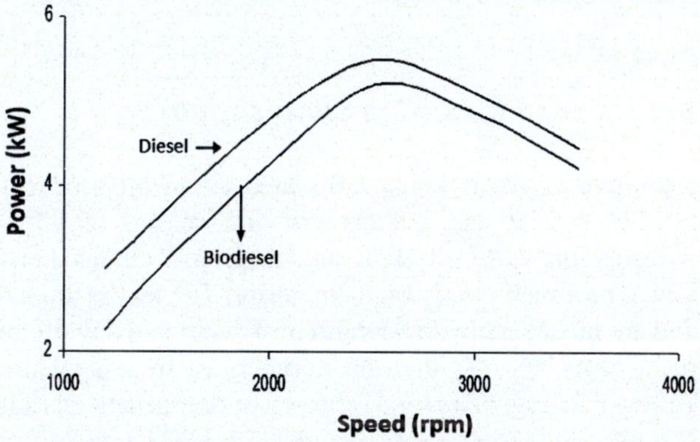

The brake specific fuel consumption (BSFC) of biodiesel which refers to the rate of fuel consumption against the power produced may also increase in accordance with the increase in speed (Fig. 8.14).

Many experimental data suggest that the BSFC of biodiesel expressed as kg kW. h^{-1} is always higher than that of the petroleum diesel. The difference may be in the range of 5–20% according to the speed, type of biodiesel and the nature of engine. Such difference is largely due to the lower heating value and higher viscosity of the biodiesel.

The CO emission in the exhaust is relatively low due to the increased combustion of the biodiesel making use of the oxygen offered by the air jet and also additionally due to the presence of more oxygen atom in the biodiesel. Therefore, the rate of discharge of unburned hydrocarbon also is low. Higher quantity of oxygen present in the biodiesel offers high local peak temperature causing the discharge of higher quantity of NO_x.

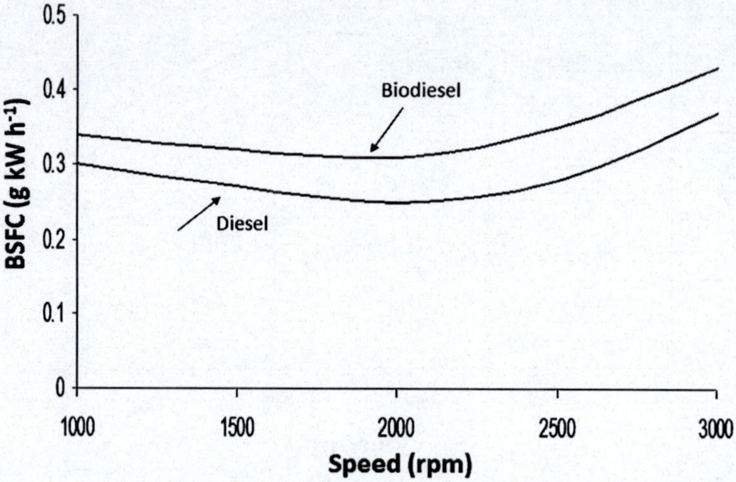

Fig. 8.14 The relationship between engine speed and brake specific fuel consumption when biodiesel is used as a fuel

8.3 Wild Flax (*Camelina sativa*)

Camelina sativa is being cultivated as an oil seed crop. It grows to a height of 30–120 cm. It has a short growing season of 85–100 days. It is capable of withstanding cold, drought and semi-arid climates. Its stem has many branches which becomes woody on its maturity. The leaves are fixed on the stem alternatively and are lanceolate with a length of 2–8 cm and a width of 0.2–1 cm (Fig. 8.15). This plant bears not less than 50 numbers of four petal flowers. The flowers are pale yellow in colour. The seed is brown or orange and has a length of 2–3 mm. The seeds are tiny and 1000 seeds weigh 1–2 g. The yield of seed ranges from 336 to 2240 kg ha^{-1} (Moser 2010).

C. sativa is distributed in Asia, China, India, Kazakhstan, Korea, Mongolia, Pakistan, Russia, Tajikistan, Africa, North America, Canada and USA (Fig. 8.16).

This species is vernacularly called as camelina, gold-of-pleasure, dutch flax, false flax, wild flax, linseed dodder, German sesame and Siberian oilseed.

8.3.1 Oil Extraction

Dried seeds are powdered and extracted with *n*-hexane in a Soxhlet apparatus. The hexane present in the extract is then removed by a rotary evaporator under low pressure (10 mbar at 30 °C) and temperature. Alternatively the oil is extracted by passing it through a press. Solids present if any in the crude oil is removed by settling the oil for an extended period (1 week). It is then refined by adding diluted sodium

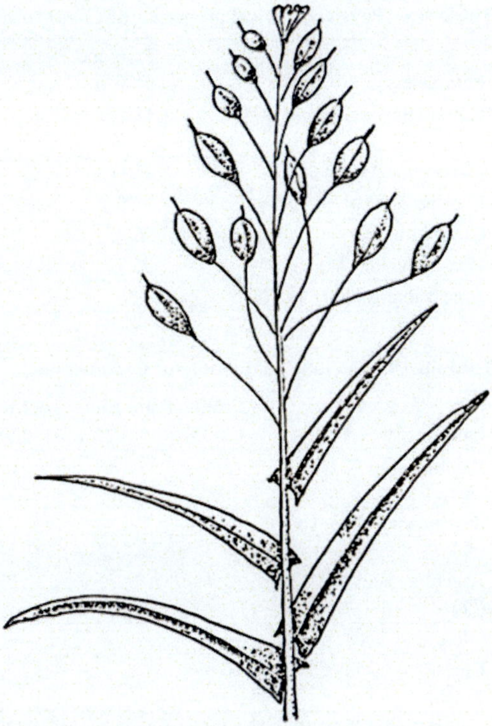

Fig. 8.15 The wild flax *Camelina sativa*

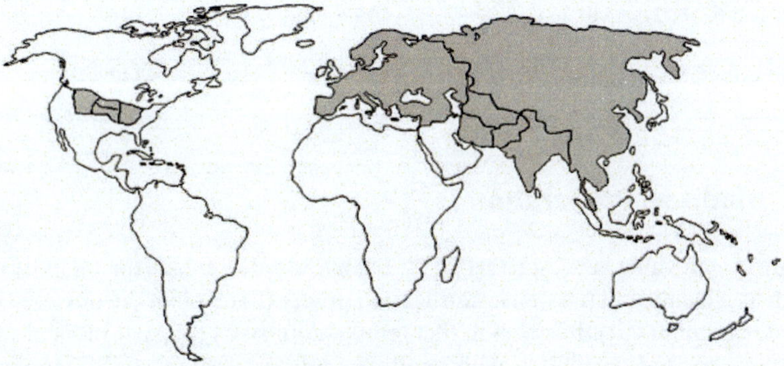

Fig. 8.16 The geographical distribution of *C. sativa*

hydroxide just in excess needed to neutralize the free fatty acids. The precipitated solids are then separated. The physicochemical properties of this oil are given in Table 8.9 and the amino acid profile is given in Table 8.10.

The viscosity is relatively high, whereas the cetane number is low.

Table 8.9 Physicochemical properties of *C. sativa* oil

Parameters	Patil and Deng (2009)	Patil et al. (2009)	Hangun-Balkir (2016)
Kinematic viscosity at 40 °C (mm^2s^{-1})	15.12	14.03	14.6
Flash point(°C)	260	–	–
Pour point(°C)	−20	−23	−23
Cetane number	36.25	35.16	35.8
Heating value (MJ kg^{-1})	46.5	44.5	44.5
Acid value(g KOH kg^{-1})	–	3.16	3.2

Table 8.10 The fatty acid profile of *C. sativa* oil

Fatty acid	Fröhlich and Rice (2005)	Moser and Vaughn (2010)	Ciubota-Rosie et al. (2013)
Palmitic acid (C16: 0) (%)	5.4	6.8	5.16
Stearic acid (C18: 0) (%)	2.6	2.7	2.68
Arachidic acid (C20: 0) (%)	1.4	1.5	1.44
Oleic acid (C18: 1) (%)	14.3	18.6	15.21
Eicosenoic acid (C20: 1) (%)	16.8	12.4	15.4
Erucic acid (C22: 1) (%)	2.9	2.3	2.57
Linoleic acid (C18: 2) (%)	14.3	19.6	17.9
Linolenic acid (C18: 3) (%)	38.4	32.6	34.64

8.3.2 Biodiesel Preparation

The highest acid value is 3.2 g KOH kg^{-1}. Such low value entails the oil to be treated by alkali as a catalyst in the transesterification process. The oil at the outset is heated to 100 °C and maintained for 1 h in that temperature so as to expel moisture if any.

In a three necked flask fitted with a reflux condenser 200 mL (0.202 mole) oil is taken. In a separate flask sodium methoxide is prepared by mixing sodium hydroxide with methanol so as to have 1% sodium hydroxide in the combined mixture. This methoxide is then poured into the oil present in the three necked flask and stirred well for 1.5 h at 60 °C. The resultant material is transferred to a separatory funnel. The glycerol present in the mixture sinks down at the bottom of the funnel and it is separated gravitationally and discarded. The upper zone containing the biodiesel is taken out distilled at low temperature in a vacuum distillation unit so as to remove methanol if any. The purified biodiesel is then added with Magnesol® at the rate of 2.5 g per 100 mL biodiesel. The biodiesel and Magnesol® mixture is stirred at

Table 8.11 Conditions of transesterification on the use of heterogeneous catalysts in *C. sativa* oil (Patil et al. 2009)

Catalyst	Conc.of catalyst (%)	Molar ratio of methanol to oil	Temp, (°C)	Duration (min)	Biodiesel yield (%)
BaO	1.0	9:1	100	180	84
SrO	0.5	12:1	60	120	82
CaO	0.5	15:1	100	90	33
MgO	1.0	15:1	100	90	30

500 rpm at 65 °C. Finally the Magnesol® is filtered out and thus the biodiesel is prepared.

Ethanol also is used in the place of methanol, as ethanol is ecofriendly and sourced from biomass. To 200 mL of oil, 110 mL of ethanol is added and heated to 70 °C. Following the above KOH is added so as to make a conc. of 1% by weight and stirred for 1.5 h. The contents obtained finally are distilled in a rotary evaporator set at 10 mbar and at 35 °C. The whole material is then transferred to a separatory funnel and the same is allowed to settle till phase separation is completed. The glycerin which settles at the bottom of the funnel is discarded. The upper phase being the ethyl ester is taken in a beaker, repeatedly sprinkled with warm water and washed. After Magnesol® treatment at the rate of 2.5% the pure ester is obtained by filtration.

Instead of NaOH or KOH, heterogeneous metal oxide catalysts BaO, CaO, MgO and SrO are tried. These catalysts are first dissolved in methanol so as to form an ultimate concentration of 0.5–1% of catalyst and 9:1 to 15:1 methanol to oil ratio. The above solution is added to the purified oil and stirred for 90–180 min at 60 °C (Patil and Deng 2009).

The content is then transferred to a separatory funnel for phase separation. Consequent to the separation the lower phase having the catalyst and the glycerol is drained and discarded. The upper phase containing the biodiesel is gravitationally drained into a beaker and repeatedly washed with hot dist. water so as to free it from trace of glycerin and any other entrained impurities. The biodiesel is then passed through sodium sulphate so as to free it from moisture. The conditions of operation with respect to the use of heterogeneous catalyst are given in Table 8.11.

Among the four heterogeneous catalysts tried, high efficiency is recorded in BaO. CaO prepared by the calcinations (900 °C for 3 h) of seashells and eggshells are low-cost catalyst employed in the transesterification. (Hangun-Balkir 2016). The shells are rinsed in water to remove any adhering organic materials. They are then dried at 110 °C for 2 h. This material is grinded and calcined at 900 °C in air raising the temperature at the rate of 10 °C per minute and maintaining it at 900 °C for 3 h. The calcined shell is first mixed with methanol and stirred vigorously for 1 h at 65 °C. It is then added to the oil and stirred incessantly at 65 °C so as to complete the transesterification. The molar ratio of methanol to oil is 12:1. The properties of *C. sativa* biodiesel are given in Table 8.12.

Table 8.12 Physicochemical properties of *C. sativa* biodiesel

Parameters	Patil and Deng (2009)	Moser and Vaughn (2010)	Ciubota-Rosie et al. (2013)	Hangun-Balkir (2016)
Kinematic viscosity at 40 °C (mm^2s^{-1})	3.15	4.15	4.3	3.1
Density at 15 °C (g kg^{-1})	–	–	888	–
Acid value (g KOH kg^{-1})	–	0.31	0.15	–
Flash point (°C)	140	–	152	–
Pour point (°C)	−8	−4	–	−10
Cloud point (°C	–	3	0	–
Iodine value (gI_2/100 g)	–	151	152	–
Cetane number	51.17	52.8	42.76	49
Oxidation stability 110 °C (h)	–	2.5	1.3	–
Cold filter plugging point(°C)	–	−3	−4	–
Heating value (MJ kg^{-1})	46.15	–	–	46
Copper strip corrosion	–	–	1A	–

The high content of linolenic acid (C18: 3) in the crude oil which ranges from 32.6 to 38.4% has a telling effect on the iodine value, cetane number and oxidation stability of the biodiesel. As the unsaturated fatty acid is phenomenally high in the oil, the iodine value is high in biodiesel (>150). The iodine value, when it is high the fuel tends to polymerize and form deposits on the fuel nozzles, piston rings and grooves. The cetane number in the present case is low as it depends on the chain length of the fatty acid. The cetane number is known to be low when the number of double bonds increases. If the cetane number is low there will be engine knocking and incomplete combustion, notwithstanding the fact that it contains rich oxygen. This incomplete combustion causes high emission of gaseous and particulate pollutants in the exhaust. A low oxidation stability is the reflection of the increased unsaturation of fatty acid esters. The methylene groups adjacent to the double bond are susceptible to denaturation. The fatty acid esters having three or more double bonds even in a low quantity influence the oxidative stability due to the bis-allylic position in the ester molecule attracting oxidation. The oxidation products may generate aldehydes, ketones and short-chain carboxylic acids which ultimately cause the corrosion of the injection system. In such events, antioxidant additives are added to minimize the rapid auto oxidation of the fuel. This situation implies that the biodiesel obtained from *C. sativa* cannot be used as a fuel in its pure form, rather it warrants blending with petro-diesel. The wear scar generated in *C. sativa* biodiesel is lower (122 μm) than that of certain other biodiesel, namely canola (169 μm), palm (172 μm) and soybean (135 μm). (Moser and Vaughn 2010). With reference to the

Table 8.13 Effect of *Lubrizol* 7670 on the low temp properties of *C. sativa* biodiesel when added at the rate of 2000 ppm (Fröhlich and Rice 2005)

Properties	Without additive	With additive *Lubrizol 7670*
Cloud point (°C)	4	3
Pour point (°C)	−8	−16
Cold filter plugging point (°C)	−4	−5

Table 8.14 Cold temp. properties of the blends of *C. sativa* biodiesel with petro-diesel (Fröhlich and Rice 2005)

C. sativa biodiesel (%)	Petro-diesel (%)	Cloud point	Cold filter plugging point (°C)	Pour point (°C)
100	0	3	−3	−4
60	40	3	−9	−9
20	80	3	−13	−18

Table 8.15 Fuel consumption studies in two engines (Fröhlich and Rice 2005)

Fuel	Isuzu Trooper (km L^{-1})	Peugeot 306 (km L^{-1})
C. sativa biodiesel	10.92	17.61
Petro-diesel	11.18	18.07
Difference	0.26	0.46

low temperature properties, attempts are made (Fröhlich and Rice 2005) to improve the properties. The effect of a fuel additive, namely Lubrizol 7670 on the biodiesel made from *C. sativa* oil in improving the low temperature properties is given in Table 8.13.

Attempts are also made to blend this biodiesel with petro-diesel at different levels (Fröhlich and Rice 2005). The low temperature properties of the blend are given in Table 8.14.

Table 8.14 indicates that the blend did not alter the cloud point but the cold filter plugging point and the pour point improved considerably.

With reference to the volumetric efficiency of the biodiesel from *C. sativa* and petro-diesel, two trials were made, one in Isuzu Trooper engine and the other in Peugeot 306 engine. The values are reported in Table 8.15.

The consumption of biodiesel is marginally high and is reflected on the rate of consumption per km. The life cycle greenhouse gas emissions for biodiesel and jet fuels produced from *C. sativa* oil are 80 and 75% lower than that of the petro-diesel.

8.4 Desert Mustard (*Lesquerella fendleri*)

The desert mustard (Bladder pod) *Lesquerella fendleri* is a tufted annual plant and has a strong tap root system with many branched secondary roots. It grows to a maximum height of 0.4 m. resisting environmental vagaries.. It is a semi-arid plant

Fig. 8.17 The desert mustard, *Lesquerella fendleri*

Fig. 8.18 Geographical distribution of *Lesquerella fendleri*

and occupies an elevation between 600 and 2000 m above MSL. It has the ability to colonize disturbed lands and tolerates extreme cold and drought. It manages a rainfall from 250 to 400 mm and an atmospheric temperature of around 15 °C. The leaves are densely pubescent and silvery. The plant comes to yield in 200 days. The pod carries 10–20 seeds and each seed weighs 0.54–2.3 mg (Salywon et al. 2005) (Fig. 8.17). This plant is reported to grow in the deserts of Southwest United States and Northern Mexico (Fig. 8.18). The seed yield is around 1.1 ton per hectare.

8.4.1 Oil and Biodiesel

The seeds are partially cooked for 20 min and then passed through a screw press to release the oil. The oil is filtered and bleached using clay and activated carbon. Lesquerella seed contains around 25% oil, in which hydroxy fatty acids form around 65%. Refining the oil reduces the free fatty acid to the extent of 90% (Knothe et al. 2012). The chosen properties of the oil are indicated in Table 8.16.

The oil has lesquerolic acid (14-hydroxy-11 (Z)–eicosenoic acid) content to the tune of 57.6% by weight which is a 20-carbon homolog of ricinoleic acid. It also contains auricolic acid (14-hydroxy-11(Z), 17(Z)-eicosadienoic) and densipolic acid (12-hydroxy-9 (Z), 15 (Z)-octadecenoic). The estolides content (not shown in the table) allows the oil to flow freely than petroleum at low temperature (Isbell et al. 2006). The cloud point is 28 °C but the pour point is −7 °C. The oxidative stability is very low and the viscosity is high (Knothe et al. 2012).

The low acid value of oil does not require any pretreatment for alkali based transesterification. The oil is treated at 65 °C for 1 h with 1% sodium hydroxide. The sodium hydroxide is dissolved in a known quantity of methanol so as to have a molar ratio of 6:1 (methanol to oil). Consequent to the reaction the content is allowed to settle for 8–12 h and the bottom lying layer containing glycerol is separated. The top layer having the biodiesel is gently washed with distilled water and dried. Certain properties of the biodiesel from *L. fendleri* is given in Table 8.17.

The kinematic viscosity is not within the limit. When the content of methyl linoleate (4.51 cSt) and methyl oleate (3.65 cSt) is low the tendency of the concerned biodiesel is to show high viscosity (Knothe et al. 2012). High viscosity of the biodiesel naturally gives an option of blending it with regular diesel. The density of the biodiesel is close to the upper limit (860–900 kg m^{-3}) of the standard EN14214. The oxidative stability of the biodiesel is relatively high when compared to that of the oil though it does not satisfy the standard.

Table 8.16 Properties of *L. fendleri* oil (Knothe et al. 2012)

Properties	Value
Kinematic viscosity at 40 °C (cSt)	138
Oxidative stability at 110 °C (h)	0.05
Acid value (mg KOH g^{-1})	0.34
Myristic acid (C14: 0)(%)	0.3
Palmitic acid (C16: 0) (%)	1.0
Stearic acid (C18: 0) (%)	1.8
Eicosanoic acid (C20: 0) (%)	0.8
Palmitoleic acid (C16: 1) (%)	0.4
Oleic acid (C18: 1) (%)	13.6
Linoleic acid (C18: 2) (%)	6.7
Linolenic acid (C18: 3) (%)	10.2
Ricinoleic acid (C18: 1–OH) (%)	0.6
Lesquerolic acid (C20: 1–OH) (%)	57.6
Auricolic acid (C20: 2–OH) (%)	3.6

Table 8.17 Properties of *L. fendleri* biodiesel (Knothe et al. 2012)

Parameter	Value
Kinematic viscosity at 40 °C (cSt)	11.22
Density at 40 °C (kg m^{-3})	895
Cetane number	45.6
Pour point (°C)	−20
Acid value (mg KOH kg^{-1})	0.026
Oxidative stability at 110 °C (h)	1.9

Fig. 8.19 The turnip, *Raphanus* sativus

8.5 Turnip (*Raphanus sativus*)

The seeds of the annual or perennial herb turnip *Raphanus sativus* pose scope for biodiesel production although the quantity of seed available world over for commercial exploitation is not clearly specified in the literature. It is an erect plant growing up to a height of 60 cm. The leaves have a length of around 30 cm. The petals of the flower are blue to purple and are radially symmetrical (Fig. 8.19). The fruit is endospermic with silicle seeds. Each seed weighs around 10 mg.

Though *R. sativus* occurs universally as a vegetable crop it is being richly cultivated in Asia, Southern USA, Africa, West China and India (Fig. 8.20).

The seeds are harvested from the plants, dehulled and are passed through a mechanical pressing unit. Pressing the seed yields around 25% oil. To enhance the oil recovery the cake obtained is again refluxed with n-hexane (Shah et al. 2013). The micelle thus obtained is dried by passing through $MgSO_4$, filtered and distilled to expel the solvent. The total recovery of the oil is around 40–54% (Domingos et al. 2008). The oil characteristics are given in Table 8.18.

Fig. 8.20 Geographical distribution of *R. sativus*

Table 8.18 The physicochemical characteristics of the oil of *R.sativus*

Parameters	Valle et al. (2009)	Chammoun et al. (2013)	Shah et al. (2013)
Kinematic viscosity @ 40 °C (cSt)	–	–	22.8
Iodine number (g I_2/100 g)	–	97	111
Cold filter plugging point (°C)	–	6.0	–
Oxidation stability at 110 °C (h)	–	–	8
Acid number (g KOH kg^{-1})	–	2.16	2.56
Heating value (MJ kg^{-1})	–	–	38.5
Palmitic acid (C16: 0) (%)	7.0	–	–
Stearic acid (C18: 0) (%)	3.6	–	–
Oleic acid (C18: 1) (%)	27.9	23.9	–
Linoleic acid (C18: 2) (%)	7.6	13.5	–
Linolenic acid (C18: 3) (%)	4.6	10.3	–
Gadoleic acid (C20: 1) (%)	11.2	–	–
Erucic acid (C22: 1) (%)	33.3	31.8	–
Molecular weight (g mol^{-1})	–	–	916.8

8.5.1 Biodiesel Preparation

Shah et al. (2013) ran a two-step process as the acid value of the oil was relatively high (2.56 g KOH kg^{-1}). Methanol and a calculated amount of sulphuric acid are added to the oil and the mixture is refluxed gently for 2 h and then it is transferred to a separating funnel. A phase separation takes place after 2 h. The ester thus obtained is washed repeatedly with sodium bicarbonate solution (5%) until it becomes neutral. The oil phase is dissolved in hexane and passed through magnesium sulphate and filtered. The filtered oil is vacuum distilled to clear the solvent. The resultant esterified oil is treated with sodium methoxide at a low stirring speed for 2 h. The above material is transferred to a separating funnel and waited till the phase separation is completed. The ester phase is removed and the excess methanol present in it is expelled by a rotary evaporator. It is washed first with 5% phosphoric acid and then with sodium chloride solution. This is again dissolved in hexane, passed over magnesium sulphate and filtered. The hexane is distilled out through vacuum distillation and the biodiesel thus obtained is reported to be of 99% purity.

Chammoun et al. (2013) followed a different procedure which is as follows. Potassium hydroxide 9 g is dissolved in 200 mL methanol so as to prepare potassium methoxide. Around 85% by volume of the above potassium methoxide solution is added to 1 L of oil as a first instalment, heated to 55 °C and stirred at a low speed (270 rpm) for 1 h. It is then taken to a separating funnel and allowed to settle for another 1 h. The glycerin and waste material staying at the bottom are removed and the supernatant ester is washed in a flask using a minimum quantity of (100 mL) distilled water while the whole content is kept on continuous motion (170 rpm) for 1.5 min. The content is allowed to stand for an hour and the ester phase is serially washed thrice as per the above procedure. The remaining 15% portion of potassium methoxide kept reserved already is then added as the second instalment to the above-finished material and the process of transesterification and washing (3 times) as explained above continued. Finally, the ester obtained is heated to 100 °C to expel the moisture so as to get the biodiesel. The biodiesel thus obtained is once again passed through the transesterification process using newly prepared potassium methoxide (9 g KOH in 200 mL methanol). This time the entire potassium methoxide is mixed as a single instalment and all the steps continued as before and the finished biodiesel is obtained. The physicochemical properties of the biodiesel thus obtained are given in Table 8.19.

If the viscosity and the specific gravity are favourable, the fuel flows through the conduits and orifices very smoothly without any undue pressure and thus ensuring appropriate atomization. This causes ideal combustion with low carbon deposit. The copper corrosion value indicates that the metallic parts are comfortable. The cold filter plugging point is within the prescribed limit. The molecular weight of the oil (917 g mol^{-1}) calculated from the fatty acid profile is reduced due to the transesterification (306 g mol^{-1}) which also helps in the smooth flow of biodiesel (Shah et al. 2013).

Domingos et al. (2008) carried out the transesterification of *R. sativus* oil using ethyl alcohol and sodium hydroxide. The molar ratio of ethanol to oil is 8:1, NaOH concentration is 0.6% and the temperature is 55 °C. The sodium hydroxide is dissolved in ethyl alcohol and the resultant ethoxide is added to the oil and stirred for 60 min. Excess alcohol if any is evaporated under reduced pressure at a temperature less than 60 °C. The content is phase separated and the ester is treated with Magnesol (2%) at 65 °C for 20 min under continuous stirring. Magnesol is a product comprising magnesium silicate (Cooke et al. 2003). Magnesol is then filtered out. Often second stage ethanolysis (repetition) is performed. In that case, the ester formed as above just before Magnesol treatment is once again treated with ethoxide. The molar ratio is 6:1, temperature 55 °C and the conc. of sodium hydroxide $<0.1\%$ by weight. After stirring the content for 2 h, the mixture is allowed for phase separation and the ester is treated with Magnesol and filtered. The

Table 8.19 The physicochemical properties of *R. sativus* biodiesel

Properties	Valle et al. (2009)	Shah et al. (2013)
Kinematic viscosity @ 40 °C (cSt)	5.5	5.0
Density @ 20 °C (kg m^{-3})	872	–
Flash point (°C)	157	–
Iodine number (g I_2/100 g)	96	111
Cold filter plugging point (°C)	5	−2
Oxidation stability at 110 °C (h)	–	1.6
Acid number (g KOH kg^{-1})	0.21	–
Heating value (MJ kg^{-1})	–	38.8
Pour point (°C)	–	−4
Cloud point (°C)	–	10
Copper strip corrosion	–	1a
Palmitic acid (C16: 0) (%)	–	8.5
Stearic acid (C18: 0) (%)	–	2.4
Oleic acid (C18: 1) (%)	–	28.4
Linoleic acid (C18: 2) (%)	–	20.4
Linolenic acid (C18: 3) (%)	–	11.5
Erucic acid (C22: 1) (%)	–	15.5
Gondoic acid (C20: 1) (%)	–	7.4
Molecular weight (g mol^{-1})	–	306

characteristics of methyl ester and ethyl ester for which the procedures are given above do not differ much in quality except on the oxidation stability. The performance of the methyl ester when compared to that of diesel does not change seriously. However, the consumption of biodiesel as a fuel in the internal combustion is relatively higher.

An economic appraisal made by Chammoun et al. (2013) on *R. sativus* biodiesel indicates that out of the total cost of biodiesel production, the seed cost alone works out to 36%. Thus the turnip seeds which are hitherto not utilized for any benefit are best suited for biodiesel enterprise yielding a marginal gain.

References

Anastopoulos G, Lois E, Zannikos F, Kalligeros S, Teas C (2001) Influence of aceto acetic esters and di-carboxylic acid esters on diesel fuel lubricity. Tribol Int 34(11):749–755

Bannikov MG (2011) Combustion and emission characteristics of mustard biodiesel. In: Poc. 6th international advanced technologies symposium IATS'11, 16–18 May 2011, Elazig, Turkey

Boocock DGB, Konar KSK, Mao V, Sidi H (1996) Fast one phase oil rich process for the preparation of vegetable oil methyl esters. Biomass Bioenergy 11(1):43–50

Chammoun N, Geller DP, Das KC (2013) Fuel properties, performance testing and economic feasibility of *Raphanus sativus* (oilseed radish) biodiesel. Ind Crop Prod 45:155–159

Ciubota-Rosie CJ, Ruiz R, Ramos MJ, Perez A (2013) Biodiesel from *Camelina sativa*: a comprehensive characterization. Fuel 105:572–577

Cooke BS, Abrams C, Betram B (2003) Purification of biodiesel with adsorbent materials, US Patent 0509959 P

Domingos AK, Saad EB, Wilhelm HM, Ramos LP (2008) Optimization of the ethanolysis of Raphanus sativus (L.Var) crude oil applying the response surface methodology. Bioresour Technol 99(6):1837–1845

Drown DC, Harper K, Frame E (2001) Screening vegetable oil alcohol esters as fuel lubricity enhancers. JOCS 78(6):579–584

Fröhlich A, Rice B (2005) Evaluation of *Camelina sativa* oil as a feedstock for biodiesel production. Ind Crop Prod 21:25–31

Geller DP, Goodrum JW (2004) Effects of specific fatty acid methyl esters on diesel fuel lubricity. Fuel 83:2351–2356

Hangun-Balkir Y (2016) Green biodiesel synthesis using waste shells as sustainable catalysts with *Camelina sativa* oil. J Chem 106715232, 10 p

Isbell TA, Lowery BA, Dekeyser SS, Winchell ML, Cermak SC (2006) Physical properties of triglyceride estolides from Lesquerella and castor oils. Ind Crops Prod 23(3):256–263

Issariyakul T, Dalai AK (2010) Bioseed production from greenseed canola oil. Energy Fuel 24:4652–4658

Jham GN, Moser BR, Shah SN, Holser RA, Dhingra OD, Vaughn SF, Berhow MA, Winkler-Moser JK, Isbell TA, Holloway RK, Walter EL, Natalino R, Anderson JC, Stelly DM (2009) Wild Brazilian mustard (*Brassica juncea*, L.) seed oil methyl esters as biodiesel fuel. J Am Oil Chem Soc 86:917–926

Kirk JTO, Oram RN (1981) Isolation of erucic acid free lines of *Brassica juncea*. Indian mustard now a potential oil seed crop in Australia. J Int Agric Sci 47:51–52

Knothe G, Dunn R (2003) Dependence of oil stability index of fatty compounds on their structure and concentration and presence of metals. J Am Oil Chem Soc 80:1021–1026

Knothe G, Cermak SC, Evangelista RL (2012) Methyl esters from vegetable oils with hydroxy fatty acids: comparison of Lesquerella and castor methyl esters. Fuel 96:535–540

Kulkarni GM, Dalai AK, Bakhshi NN (2007) Transesterification of canola oil in mixed methanol/ethanol system and use of esters as lubricity additive. Bioresour Technol 98:2027–2033

Lang X, Dalai AK (2001) Biodiesel esters as lubricity additives effects of process variables and evaluation of low temperature properties. Fuels Int 1–3

Lapuerta M, Armas O, Fernandez JR (2008) Effect of biodiesel fuels on engine emissions. Prog Energy Combust Sci 34:198–223

Meher LC, Vidyasagar D, Naik SN (2006) Technical aspects of biodiesel production by transesterification—a review. Renew Sust Energ Rev 10:248–268

Moser BR (2010) Camelina (*Camelina sativa* L.) oil as a biofuel feedstock: golden opportunity or false hope? Lipid Technol 22(12):270–273. https://doi.org/10.1002/lite.201000068

Moser BR, Vaughn SF (2010) Evaluation of alkyl esters from *Camelina sativa* oil as biodiesel and as blend components in ultra low-sulfur diesel fuel. Bioresour Technol 101:646–653

Patil PD, Deng S (2009) Transesterification of *Camelina sativa* oil using heterogeneous metal oxide catalysts. Energy Fuel 23:4619–4624

Patil PD, Gude VG, Deng S (2009) Biodiesel production from *Jatropha curcas*, waste cooking and *Camelina sativa* oils. Ind Eng Chem Res 48:10850–10856

Salywon AM, Dierig DA, Rebman JP, de Rodriguez DJ (2005) Evaluation of new *Lesquerella* and *Physaria* (Brassicaceae) oil seed germplasm. Am J Bot 92(1):53–62

Shah SN, Iha OK, Alves FCSC, Sharma BK, Erhan SZ, Suarez PAZ (2013) Potential application of turnip oil (Raphanus sativus L.) for biodiesel production: physical chemical properties of neat oil, biofuels and their blends with ultra low sulphur diesel (ULSD). Bioenergy Res 6:841–850

Valle PWPA, Rezende TF, Souza RA, Fortes ICP, Pasa VMD (2009) Combination of fractional, factorial and Doehlert experimental design in biodiesel production: ethanolysis of Raphanus sativus L. var. oleiferous stokes oil catalysed by sodium ethoxide. Energy Fuel 23 (10):5219–5227

VanGerpen JH, Soylu S, Mustafa ET (1999) Evaluation of the lubricity of soybean oil-based additive in diesel fuel. ASAE Paper No. 996134

9 Caryocaraceae

Caryocaraceae is a small family of flowering plants comprising around 25 species in two genera in which the genus Caryocar has 16 species with several medicinal uses. They are native to tropical regions of Central and South America and West Indies. Member species are evergreen trees to shrubs. Fruits and seeds of many species are over-exploited due to their wide range of uses. It is reported that pequi (*Caryocar brasiliense*) yields oil and the same is used in biodiesel production.

9.1 Pequi (*Caryocar brasiliense*)

Caryocar brasiliense (Syn: Acantacaryx pinguis, Caryocar intermedium) is one of the 16 species of the genus caryocar of the family Caryocaraceae. It is popularly referred as pequi or souari nut. It is a semideciduous tree of a maximum height of 10 m with a trunk diameter of 30 cm (Fig. 9.1). It grows at an annual rainfall of 1000–1500 mm with a dry season extending from 3–5 months having a low humidity. It manages to survive in untenable soils having low nutrients, rich clay with high content of iron and aluminium. This tree bears 1000–2000 fruits each weighing around 100 g and the seed (stone) weighs around 20 g. The fleshy pericarp of the fruit forms 75% of the total weight (Quirino et al. 2009). These trees grow at South America, Northern Argentina, Brazil, Paraguay and Bolivia (Fig. 9.2).

9.1.1 Oil and Biodiesel Preparation

The pericarp (the fleshy portion of the fruit) being used as a raw material for biodiesel is removed by a knife, made into small pieces and dried at a low temperature, preferably at vacuum. Similarly the stone containing the kernel also is cut open and the kernel dried. The dried pericarps and the dried kernels are made into granules and extracted using a suitable solvent (Aquino et al. 2011). Then the solvent is

S. P. Raj et al., *Biodiesel from Flowering Plants*,
https://doi.org/10.1007/978-981-16-4775-8_9

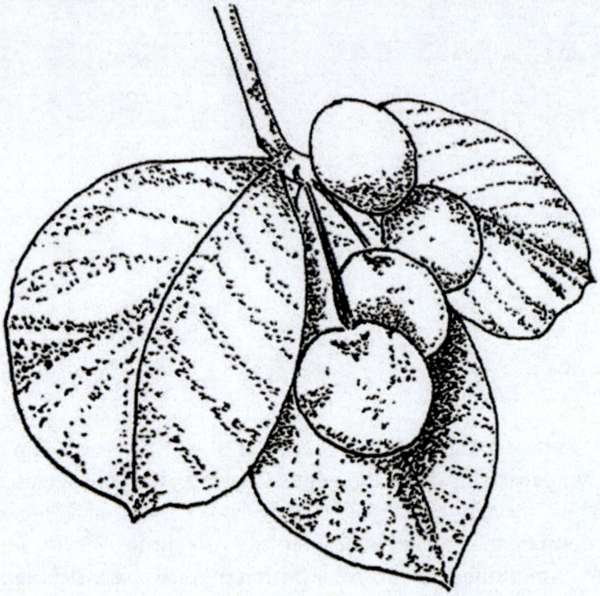

Fig. 9.1 Pequi plant, *Caryocar brasiliense*

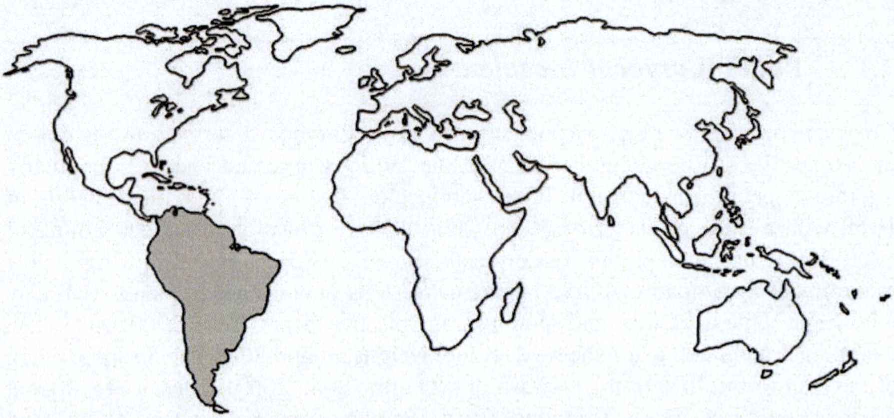

Fig. 9.2 Geographical distribution of Caryocar *brasiliense*

distilled out and the oil is collected. Among the various solvents (hexane, acetone, ethanol and methanol), hexane is observed to be effective in extracting the oil (Table 9.1).

This oil has a relatively low acidity which gives a lead to the course of action on the nature of transesterification process to be followed. In such low acidity, base catalyst transesterification is observed to be ideal. The density and the kinematic viscosity of the oil are observed to be high. However, the oxidative index is quite promising (Table 9.2).

Table 9.1 The recovery of oil from the pericarp of *C. brasiliense* using different solvents (Borges et al. 2012)

Solvent	Recovery (%)
Hexane	68
Acetone	54
Ethanol	42
Methanol	22

Table 9.2 Properties of *C. brasiliense* oil (Silva et al. 2014)

Parameters	Value
Kinematic viscosity at 40 °C (cSt)	35
Density at 20 °C (kg m^{-3})	906
Acidity (mg KOH g^{-1})	2.2
Oxidative stability 110 °C (h)	14.4
Palmitic acid (C16:0) (%)	31.4
Stearic acid (C18:0) (%)	2.2
Oleic acid (C18:1) (%)	47.3
Linoleic acid (C18:2) (%)	15.6

Table 9.3 The properties of *C. brasiliense* biodiesel (Borges et al. 2012; Silva et al. 2014)

Parameters	Value
Density at 20 °C (kg m^{-3})	866, 870
Acidity (mg KOH g^{-1})	0.8, 0.47
Kinematic viscosity at 40 °C (cSt)	4.3
Oxidative stability at 110 °C (h)	5.8, 4.5

Sodium sulphate is added to the oil at the rate of 5 g L^{-1} and allowed to stay overnight and the moisture free clear oil is collected which stands at the top of the sodium sulphate residue. In a separate beaker 4 g of KOH is added in 30 mL of methanol and stirred so as to form methoxide. After 20 min of stirring the methoxide formed is transferred to a flask containing 100 mL of moisture free oil and stirred for an hour. Then the whole content is transferred to a separating funnel. The glycerol which settled at the bottom is removed and the overlying biodiesel is collected and washed repeatedly with warm water so as to remove the residual soap, alkali and methanol. It is then dried by adding sodium sulphate. The property of the biodiesel thus prepared is presented in Table 9.3.

The oil is richly used as an emollient and therefore there is less chance to be used as a raw material for biodiesel. Besides this tree is not distributed universally. Not much work is available on the biodiesel made from this oil. Large scale cultivation of this tree is needed for commercialization of this biodiesel. Thus at present the scope of biodiesel from *C. brasiliense* is quite limited.

References

Aquino LP, Borges SV, Queiroz F, Antoniassi R, Cirillo MA (2011) Extraction of oil from pequi fruit (*Caryocar brasiliense* Camb.) using several solvents and their mixtures. Grasas Aceites 62 (3):245–252

Borges KA, Batista CF, Rodrigues HDS, Terrones MH, Vieira AT, Oliveira MFD (2012) Production of methyl and ethyl biodiesel fuel from pequi oil. Chem Technol Fuels Oils 48(2):83–89

Quirino GDS, Leite GDO, Rebelo LM, Campos ADR (2009) Healing potential of pequi (*Caryocar coriaceum* Wittm.) fruit pulp oil. Phytochem Lett 2:179–183

Silva TA, Assunção RMND, Vieira AT, Oliveira MFD, Batista ACF (2014) Methylic acid and ethylic biodiesels from pequi oil (*Caryocar brasiliense* Camp.): production and thermogravimetric studies. Fuel 136:10–16

10 Chrysobalanaceae

Chrysobalanaceae is a pan-tropical family of trees, shrubs and lianas, consisting of more than 500 species in 18 genera and is among the most representative families of the Amazon in terms of number of species. They are discontinuously distributed in USA, Mexico, West Indies, Central America, South America, South Asia, Africa, Pacific Islands and Australia. These trees are deciduous to evergreen, fruits are drupes with one seed per fruit. One of the species oiticica (*Licania rigida*) is a source for oil for making biodiesel.

10.1 Oiticica (*Licania rigida*)

The evergreen tree *Licania rigida* of the family Chrysobalanaceae is a short tree growing to a maximum height of 12 m. It grows in sandy, loamy and clay soil of a wide range of pH. It grows in semi-shade, moist soil of the banks of the temporary streams. However, it tolerates drought. The bole is often forked and short. Leaves measure 12 cm long and 6 cm wide (Fig. 10.1).

The flower is yellow and 3 mm in dia. The fruit is a drupe and is oval in shape. Each tree produces around 100 kg of nuts per season. This species has a limited global distribution (Brazil, Puerto Rico, Mexico, Guatemala, Colombia and Peru) (Fig. 10.2). The seeds of this tree are considered for biodiesel production since it has as high (60% oil) oil content (Macedo et al. 2011). The additional advantage is that it is not a part of human food chain and is a preferential species of caatinga region.

10.1.1 Oil Extraction

The seeds of oiticica are sundried for many days and are pressed in an oil mill. The oil is collected and stored in a dark room for 24–48 h. During this period the solid

S. P. Raj et al., *Biodiesel from Flowering Plants*,
https://doi.org/10.1007/978-981-16-4775-8_10

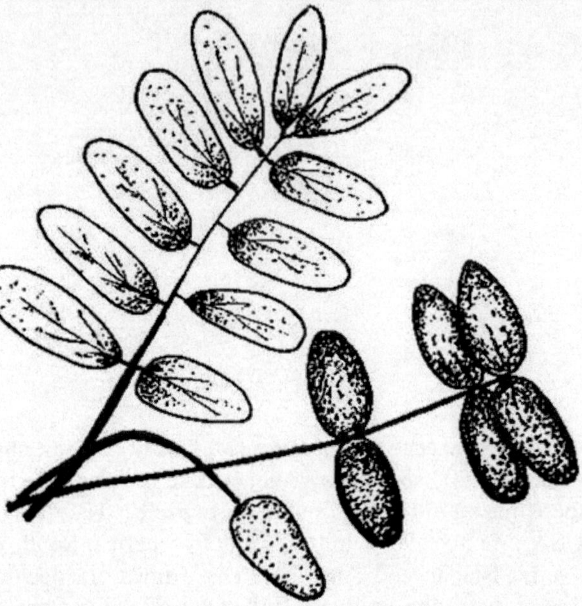

Fig. 10.1 The oiticica *Licania rigida*

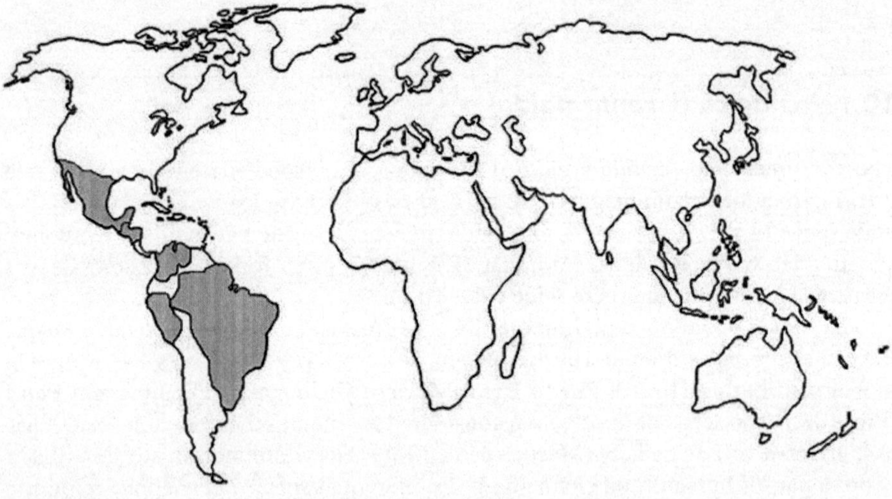

Fig. 10.2 Geographical distribution of *L. rigida*

present in it settles at the bottom of the container. The oil is then filtered and preserved in a light-free room to safeguard it from photo-oxidation.

10.1.2 Properties of the Oil

Chosen physicochemical properties are given in Table 10.1.

This oil has a high kinematic viscosity, density and acid value. Among the unsaturated fatty acids the bulk (43%) is formed of licanic acid (trienoic fatty acid $C_{18}H_{28}O_3$). It is susceptible to oxidation and polymerization. During polymerization C–H and C=O bonds are broken and consequently a molecular rearrangement takes place.

The yield of oil from the seed mass depends on the storage duration of the seed after it is harvested. Ageing due to storage causes the seeds attacked by pests and microorganisms. Therefore, it is recommended to have the pressing of the seed, as immediately as possible after the harvest or immediately after the collection of the seeds from the ground. The oil recovery in percent from the mechanical extraction of seed as a function of storage duration is given in Table 10.2.

10.1.3 Preparation of Biodiesel

Esterification process is carried out before the transesterification to bring down the acid value, if not there may be a problem of soap formation. For esterification, sulphuric acid and ethanol are mixed at the first instance. The above mixture is poured in hot oil and stirred continuously. The product obtained is taken in a separatory funnel for phase separation. The oil which is seen separated is drained

Table 10.1 Physicochemical properties of the *L. rigida* oil (Macedo et al. 2011; Guimarães et al. 2018)

Parameters	Macedo et al. (2011)	Guimarães et al. (2018)
Kinematic viscosity at 40 °C ($mm^2 s^{-1}$)	133.9	138.6
Density at 25 °C (kg m^{-3})	962.6	966.7
Acid value(g KOH kg^{-1})	13.0	1.52
Iodine value (gI_2/100 g)	–	95.4
Palmitic acid (C16:0) (%)	–	5.0
Stearic acid (C18:0) (%)	–	8.3
Oleic acid (C18:1) (%)	–	10.6
Linoleic acid (C18:2) (%)	–	3.0
Linolenic acid (C18:3) (%)	–	16.9
Licanic acid (C18:3) (%)	–	43.2

Table 10.2 The quantity of the oil extracted from the seed in accordance with the storage period (values in percent) (Guimarães et al. 2018)

Storage of oiticica seeds (months)	Oil content (%)
3	44.2
4	36.3
6	21.0
9	11.7
12	6.5

Table 10.3 The acid value of the oil of *L. rigida* before and after esterification process (Macedo et al. 2011)

Molar ratio of oil to ethanol	Catalyst (H_2SO_4) conc. (%)	Acid value (g KOH kg^{-1})	
		Before esterification	After esterification
1:12	2.0	13	7.2
1:12	0.5	13	7.0
1:9	2.0	13	4.4
1:9	0.5	13	8.1

Table 10.4 Chosen properties of *L. rigida* biodiesel obtained by using 1.5% catalyst (KOH) at 50 °C using methanol (Guimarães et al. 2018)

Parameters	Value
Kinematic viscosity at 40 °C (mm^2s^{-1})	9.61
Density at 20 °C (kg m^{-3})	927
Acid value (g KOH kg^{-1})	0.55

from the funnel, washed with warm distilled water until the acid is completely washed out as known from neutral pH. Certain trials made by Macedo et al. (2011) on the molar ratio of oil to ethanol and the percentage of catalyst (H_2SO_4) used in the esterification process and the reduction in the acid value of the oil are given in Table 10.3. The oil to ethanol molar ratio of 1:9 and a catalyst (H_2SO_4) concentration of 2% are able to reduce the acid value from 13 to 4.4 g KOH kg^{-1}.

The esterification procedure is repeated so as to bring down the acid value to an acceptable level.

The esterified oil is then subjected to alkali (KOH) catalysed transesterification at a molar ratio of 1:9 methanol to oil and 1.5% catalyst (KOH) at a reaction temperature of 50 °C in a closed system for 60 min (Guimarães et al. 2018). At the end, the resultant product containing ester (biodiesel), glycerine, catalyst and alcohol is loaded in a separatory funnel and kept undisturbed for 12 h so as to complete the phase separation. The upper layer containing the ester is removed and repeatedly washed with warm distilled water until it reaches a neutral pH. It is then dried by heating and subsequently cooled in a desiccator. The chosen properties of the biodiesel thus obtained are presented in Table 10.4.

The high viscosity of biodiesel is caused by many factors such as the long chain length of carbon in the fatty acid, the degree of saturation and the presence of free fatty acid (Melo et al. 2014; Yaakob et al. 2014). The *cis* double bond configuration causes low viscosity and the *trans* double bond creates high viscosity. The position of a double bond and the branch in the chain of the ester has no effect on the viscosity. According to Zhou et al. (2017) high viscosity has a tendency of rapid oxidation which is caused by the isomerization of the double bond (*cis* to *trans*) and with a high molecular weight. Polymeric secondary oxidation generates soluble gum which later transforms into sediment and thus increases the viscosity (Yaakob et al. 2014).

References

Guimarães AKV, Jesus AA, Oliveira HNM, Neto ELB, Gandim AD, Chiavone-filho O (2018) Biodiesel from oiticica oil (*Licania rigida*, Benth): production, thermogravimetric and oxidative stability studies. Braz J Pet Gas 12(2):107–122

Macedo FL, Candeia RA, Sales LLM, Dantas MB, Souza AG, Conceicão MM (2011) Thermal characterization of oil and biodiesel from oiticica (*Licania rigida* Benth). J Therm Anal Calorim 106:531–534

Melo MAMF, de Melo MAR, Pontes ASGC, Farias AFF, Dantas MB, Calixto CD, de Sougza AG, de Filho JRC (2014) Non conventional oils for biodiesel production: a study of thermal and oxidative stability. J Therm Anal Calorim 117(2):845–849

Yaakob Z, Narayanan BN, Padikkaparambil S, Unni KS, Akbar PM (2014) A review on the oxidation stability of biodiesel. Renew Sust Energ Rev 35:136–153

Zhou J, Xiong Y, Liu X (2017) Evaluation of the oxidation stability of biodiesel stabilized with antioxidants using the Rancimat and PDSC methods. Fuel 188:61–68

11 Clusiaceae

The family Clusiaceae comprises more than 1200 species of tropical trees and shrubs. They are economically important as several of them are known for their fruits, resins, dyes, incense, essential oils, timbers and ornamentals. Among the plants of this family exhaustive reports are available for two species, namely Kamani (*Calophyllum inophyllum*) and kokum (*Garcinia indica*) for biodiesel and hence they are dealt in this chapter. According to certain taxonomists *Calophyllum inophyllum* is distinctly placed in a special family called Calophyllaceae. Some taxonomists continue to list this species under Clusiaceae).

11.1 Kamani (*Calophyllum inophyllum*)

The Kamani tree *Calophyllum inophyllum* L. also known as *Balsamaria inophyllum* (L) grows widely in the tropical regions of the world, including many Pacific Islands. The genus *Calophyllum* has around 190 species. Due to the attractive leaves, fragrant flowers and spreading crown, the members of this genus are considered as ornamental plants. *C. inophyllum* is an evergreen tree growing to a height of 8–20 m with its crown broad spreading formed of irregular branches. This tree supports a dense canopy of glossy, elliptical leaves with fragrant white flowers and large round nuts. It grows along the coastal areas, adjacent to low lying lands although it occasionally occurs at inland. It does not prefer high elevations, chill areas or very dry lands, but prefers a mean annual rainfall of 1000–5000 mm and a mean annual temperature of 18–33 °C. It performs well in sandy well-drained soils of coastal lands. However, it tolerates clay, calcareous soil, rocky soil, sandy loam and sandy clayey loam. It is occasionally found growing in soils suffering from impeded drainage or seasonal water logging. It grows at a soil pH of 5–7.5 and manages untoward circumstances such as wind, salt spray and drought. It does not tolerate much shade.

S. P. Raj et al., *Biodiesel from Flowering Plants*,
https://doi.org/10.1007/978-981-16-4775-8_11

11.1.1 Geographical Distribution

C. inophyllum is a native of East Africa, India and Southeast Asia. It richly occurs in the coastal stretches of India (Tamil Nadu and Kerala states), Sri Lanka, Maldives, Hawaii, Pacific Islands, Caribbean, Australia, South Pacific, the Philippines, the Marianas, Melanesia, Polynesia, Aruba, Cambodia, Cook Islands, Fiji, French Polynesia, Indonesia, Japan, Kiribati, Laos, Madagascar, Malaysia, Marshall Islands, Myanmar, New Caledonia, Norfolk Island, Papua New Guinea, Reunion, Samoa, Solomon Islands, Taiwan, China, Thailand, Tonga, Vanuatu, Vietnam, Djibouti, Eritrea, Ethiopia, Kenya, Nigeria, Somalia, Tanzania, Uganda and the USA (Fig. 11.1).

11.1.2 Botanical Description

The generic name *Calophyllum* came from two Greek words *kalos* and *phyllon* (kalos = beautiful, phyllon = leaf). The leaves are opposite, dark green, shiny and hairless with broadly elliptical blades of 10–20 cm long and 6–9 cm wide. Both the tip and base of the leaves are round. Leaf veins run parallel to each other and perpendicular to the midrib.

Sap is milky white. Young trees begin to flower after 7–8 years. The flower is white, fragrant, sweetly scented, 25 mm wide and 8–14 mm long on sturdy stalk in leaf axil, and occurs in racemose or paniculate inflorescences comprising 4–15 flowers each with 4–8 oblong petals in several whorls. The stamens are numerous, yellow which occur in groups of four bundles. The colour of the anthers changes

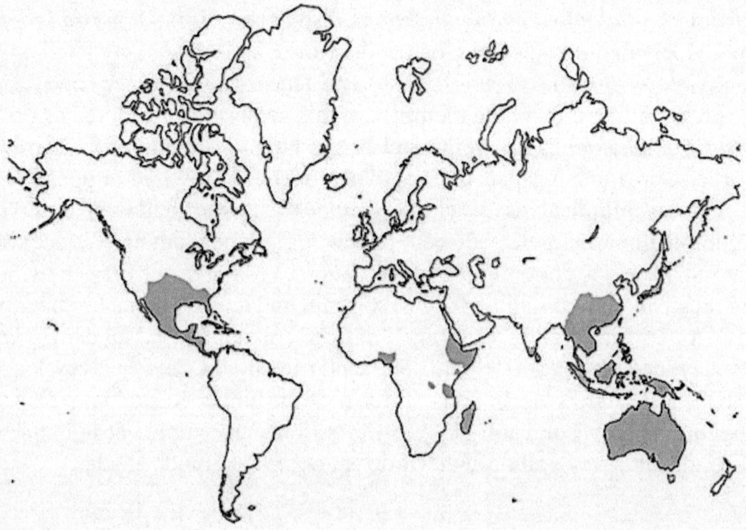

Fig. 11.1 Distribution pattern of Kamani in the world as seen in the hatched area

Fig. 11.2 The fruit of *C. inophyllum*

from deep yellow to brown. Flowering occurs all through the year. However two distinct flowering seasons are observed in a year. In the northern hemisphere the tree gives forth flowers in the late spring or early summer and late fall. Flowers are usually bisexual but rarely they are unisexual. Only the hermaphroditic flower has an ovary looking like a bright pink ball. Ovary can be seen at the tip of the stem after the petals drop (Fig. 11.2).

The fruit is round, green or greyish-green drupe enclosing a single large seed. When ripe, the fruit appears wrinkled and its colour varies from yellow to brownish red. It has a thin smooth compact outer layer called exocarp or skin. The fruit holds a seed at the centre and this seed has a hard shell which encloses a spongy layer within which a ball like kernel is embedded.

11.1.3 Propagation

The tree is not invasive in nature since the fruits are not dispersed by wind or birds and they are also not likely to be carried gregariously to inland areas. The buoyant fruits are carried away through moving water and therefore these trees mostly stand on the banks of the rivers. It is also easy to propagate this tree by seed. The fruits are often soaked overnight to remove the skin. Just prior to sowing the fruit is hammered softly to crack the shell (endocarp) of the seed to enable easy germination. Seeds are then sown at a depth of 2.5 cm from the surface in pits of the size 60 × 60 × 60 cm filled with a mixture of farmyard manure and native soil. Seeds of current year are to be sown immediately after collection. The aged seeds normally lose their viability quickly since they contain rich load of oil. Seeds with their shell fully removed germinate in 3 weeks. Seeds with their shells partially broken germinate in 4 weeks. Seeds with their shells intact will take around 8 weeks to germinate. Wildlings collected from the vicinity of the mother tree also are normally being transplanted.

Initial growth of the seedling is relatively slow. No additional treatments are required to enhance the growth. Germination rates of fresh seeds are more than 90%. Seedlings may be established in small containers. Partial shade is needed to the germinating plant during the first few weeks especially in hot areas. The seedlings are then grown in full sun after 1–2 months. These seedlings are to be made ready in containers at least 3 months ahead of their actual plantation in the field. It has a non-aggressive root. It tolerates shade marginally and therefore may not come up under dense forest canopies. As its growth is very slow it is often overwhelmed by the weeds especially during the young stage. The seedlings are normally ready for planting in the field 20–24 weeks after the germination. Seedlings are needed to be exposed to full sun for a few days before taken to field for planting as a matter of acclimatization. Young trees are benefited by the application of fertilizers containing 1–3 kg N per 100 m^2 of canopy per year. Fertilizers containing N and K are best applied in several split applications over the course of the year rather than full at once. Fertilizers are placed in holes dug around the drip-line of the canopy of the trees. In addition to the chemical fertilizers, well composted manures or other organic fertilizers can also be supplied to the plant.

11.1.4 Fruit

The fruit of this tree is a round drupe 3–5 cm in diameter. The single large kernel is surrounded by a shell (endocarp) which is turn is embedded in a thin 3–5 mm layer of pulp. The ball-shaped fruits occur in clusters twice a year totaling around 100 kg $tree^{-1}$ which on ripening fall down and are mostly collected from the ground under the canopy. The fruit appears pinkish green at first turning bright green later and when ripe it turns in to dark brown. The fruits come to ripe after 77 days of anthesis.

11.1.5 Seed

One large seed of 2–4 cm dia. is present in each fruit. The kernel is taken out of the shell of the seed after the skin and husk are removed. Around 150 seeds weigh a kg.

11.1.6 Oil Yield

Yield of oil from the fruit largely depends on the age, size and the location of the tree. Annual yield of fruits ranges from 20 to 100 kg $tree^{-1}$ (Venkanna and Reddy 2009). The kernel contains 50–70% oil (Azam et al. 2005; Adeyeye 1991; Hemavathy and Prabhakar 1990) and a mature tree may produce 1–10 kg oil $year^{-1}$ depending on the productivity of the tree and the efficiency of the extraction process. Azam et al. (2005) reported a production of 4680 kg oil $ha^{-1}year^{-1}$ under the assumption that 400 trees standing in a hectare could yield oil at the rate of 11.7 kg $tree^{-1}$. This is one of the highest values reported.

11.1.7 Pests and Diseases

Tender parts of this tree are being attacked by insects and thrips. The fungus *Fomes dochmius* causes cuboidal rot in the plant. Another fungus *Trichocoma* sp. often causes the death of this tree. The tree is the host for the fruit fly *Bactrocera facialis* in Tonga. It is also a host of the weevil *Trigonops inusitata*, a psyllid *Leptinoptera sulfurea* a plant hopper *Lamenia caliginea* and palm scale *Hemiberlesia palmae*. But this tree mostly manages such pests.

11.1.8 Vernacular Names

C. inophyllum is called by different names in different countries as given in Table 11.1.

In India this tree is called by different names in many local languages being spoken in different states (Table 11.2).

Table 11.1 The local names of *Calophyllum inophyllum* in different countries

Country	Names
England	Alexandrian laurel, beauty leaf, dilo oil tree, oil nut tree
Papua New Guinea	Beach calophyllum
Yap	Biyuch
Palau	Btaches
Guam, N. Marianas	Daog, daok
Fiji	Dilo
Kosrae	Eet
Tonga	Fetau
Pohnpei	Isou
Hawaii	Kamani, kamanu
Marshalls	Lueg
Chuuk	Rakich
Cook Islands, Society islands, Marquesas	Tamanu
Kiribati	Te itai
Burma	Phong, ponnyet
Malaysia	Bentagor bunga, pegana laut, penaga pudek
Sri Lanka	Domba
Swahili	Mtomondo, mtondoo
Thailand	Krathing, naowakan, saraphee neem
Vietnam	C[aa]y, m[uf]u
The Philippines	Bitaog, palo maria

Table 11.2 The local names of *C. inophyllum* in vernacular languages being in force in certain states in India

Language	Names
Bengali	Kath champa, punnang sultanachampa
Hindi	Sultanachampa, surpan, surpanka, undi
Sanskrit	Nagachampa, punnaga
Tamil	Pinnay, punnagam, punnai

Table 11.3 The physico-chemical characteristics of the oil of *C. inophyllum.* Adopted from Sahoo et al. (2007), Venkanna and Reddy (2009) and Atabani et al. (2011)

Parameter	Value
Kinematic viscosity at 40 °C (cSt)	32–72
Density (g cc^{-1})	0.896–0.940
Acid value (mg KOH g^{-1})	5–60
Free fatty acid (%)	2.5–30
Iodine value (mg I_2 g^{-1})	82–98
Saponification value (mg KOH g^{-1})	191–202
Triacylglycerol (%)	77
Diacylglycerol (%)	7
Saturated fatty acid (%)	29.2
Unsaturated fatty acid (%)	70.8
Flash point (%)	221–224
Heating value (KJ kg^{-1})	39,250

11.1.9 Extraction of Oil

The fruits collected are dehulled to separate the kernels and the kernels are then sliced. These slices are sun dried for a total of 3–5 days. The slices at dry stage appear deep brown. These slices with 8–12% moisture are kept stored in gunny bags till the oil is extracted. They are then passed through hydraulic press or through screw extruder. The hydraulic press may not be able to expel the entire oil. Besides, hydraulic press is often used to handle a small quantity of raw material, whereas screw extruder is used if the quantity of raw material is more. Though the kernel has a high quantity of oil (50–70%) around 20–30% alone is extracted by the above machines. As a result, the weight of the cake will be around 70–80% of the mass of the kernel.

The physicochemical characteristics of the oil are presented in Table 11.3.

The oil of *C. inophyllum* is greenish yellow in colour with a high viscosity and pungent odour. The viscosity is around 10 times higher than that of the petro-diesel and is dominated by triacylglycerol.

The raw oil has many impurities and one such impurity is gum. The gum comprises phosphate, protein, carbohydrate and resin. These impurities are likely to interfere with the process of making biodiesel besides such oil may suffer oxidation stability and therefore it is necessary to degum the oil.

As a first step in degumming, the oil is freed from the sediment either by filtration or by decanting. The oil is then mixed with water at 1:1 ratio and stirred well. It is then allowed to settle. The lower layer formed of water is discarded. Thus any water soluble impurities are simultaneously removed. The resultant overlying oil is heated

Table 11.4 Physicochemical properties of *C. inophyllum* oil before and after degumming. Adopted from Atabani et al. (2011)

Properties	Before degumming	After degumming	Change (%)
Water content (%)	0.25	0.41	64.0
Density at 20 °C (g cc^{-1})	0.94	0.94	0
Kinematic viscosity at 40 °C (cSt)	56.7	53.4	5.8
Acid value (mg KOH g^{-1})	59.94	54.18	9.6
Free fatty acid (%)	29.53	27.21	7.9
Iodine value (mg I_2 g^{-1})	86.42	85.04	1.6
Saponification value (mg KOH g^{-1})	198.1	194.7	1.7
Colour	Dark green	Reddish yellow	

Table 11.5 The average oil content of *C. inophyllum* kernels on dry basis during winter and autumn. Adopted from Hathurusingha et al. (2011). Winter = July–August, Autumn = April–May values are the average of three provenances

Season	Oil (%)
Winter (2008)	33.81
Autumn (2009)	44.25

to 80 °C in vacuum so as to remove the moisture. Consequently it is added with phosphoric acid at the rate of 0.5 g L^{-1} and was stirred. While stirring, the clean oil moves to the upper zone which is removed and washed with hot water. The characteristics of this degummed oil are presented in Table 11.4.

The estimation of the oil content made by different authors indicates that there is no uniformity among the values. There is variation on account of provenance and season, out of which variations caused due to season is much pronounced. The oil content of the kernels collected from the Queensland provenance in two fruiting seasons, namely winter, 2008 and autumn, 2009 is compared in Table 11.5. The values are the average of 4–8 trees in three northern provenances (Cardwell, Townsville and Yeppoon) of Queensland.

It is known that the season characterized by higher temperature during fruit development induces lipid generation with more saturated fatty acids.

The synthesis of triglyceride does not take place during the early stage of the seed development (Fig. 11.3). The oil content of the fruit of *C. inophyllum* increases day by day as it slowly ripens from anthesis. The fatty acid composition of the oil of the *C. inophyllum* seed also varies during the various stages of the seed development. Palmitic acid (16.0) and linoleic acid (C18:2) are the two fatty acids formed abundantly during the early stages.

The levels of palmitic acid (C16:0) in seed slowly declines as the fruits are progressing towards maturity. On the other hand, the level of oleic acid (C18:1) steadfastly increases. The total saturated fatty acids marginally decrease and in turn the unsaturated fatty acids increase as the development towards maturity proceeds (Fig. 11.4). The fruits normally come to harvest around 75–77 days after anthesis.

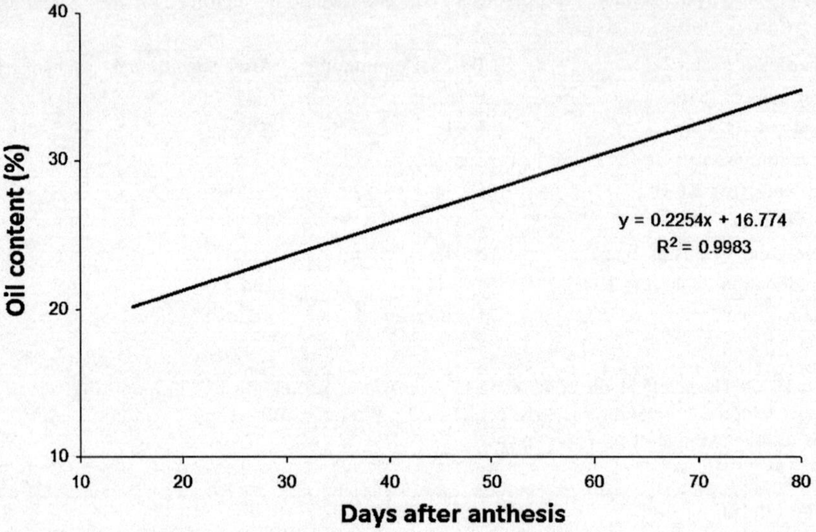

Fig. 11.3 Oil content of the seed of *C. inophyllum* since anthesis

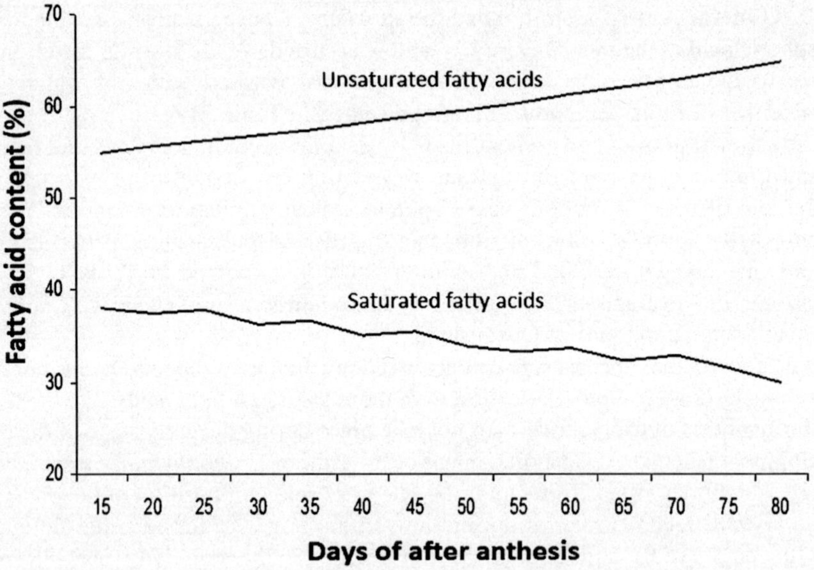

Fig. 11.4 Saturated fatty acid levels in the oil of *C. inophyllum* since anthesis

Esters of saturated fatty acid in rich quantity are known to increase the cloud point and cold filter plugging point of the biodiesel. At the cloud point temperature, the liquid fuel solidifies and creates problem to engines. Similarly, at the cold filter plugging point the wax starts settle on the filters and fuel lines. It is known that the oil

Table 11.6 The fatty acid composition of the oil of *C. inophyllum* (Ajayi et al. 2008; Pinzi et al. 2009; Sahoo et al. 2007; Atabani et al. 2011; Hathurusingha 2011)

Fatty acid	%
Myristic acid (C14: 0)	0.09
Palmitic acid (C16: 0)	12.01–18.5
Stearic acid (C18: 0)	6.10–19.96
Arachidic acid (C20: 0)	0.60–0.80
Behenic acid (C22: 0)	0.20–0.40
Oleic acid (C18: 1)	33.40–53.10
Linoleic acid (C18: 2)	13.70–38.26
Linolenic acid (C18: 3)	0.30–0.50
Erucic acid (C22: 0)	3.3

extracted from the fully matured seed (harvested on the 77th day from anthesis) is relatively of poor quality with respect to the ideal characteristics of the biodiesel when compared to that of the oil of the seed harvested around 45th day (Hathurusingha et al. 2011). However as the quantitative recovery of oil increases beyond 45 days of anthesis, fruits harvested on a fully riped state (i.e. 77th day) are accepted and recommended as the best.

The fatty acid composition of the oil of *C. inophyllum* is given in Table 11.6.

One of the most suitable unsaturated fatty acids, namely oleic acid (C18:1) is available in plenty in this oil (33–53%). However the other unsaturated fatty acid, namely palmitoleic acid (C16:1) is totally missing in this oil. The linolenic acid (C18:3) is present in low quantities. By and large the share of unsaturated fatty acids is reported to be more than 70% (Sahoo et al. 2007; Hathurusingha 2011; Kumar and Sharma 2011; Tong et al. 2011). The free fatty acid content of this oil is more than 25%. The acid value also is around 60 mg KOH g^{-1}.

11.1.10 Biodiesel Preparation

It is necessary to know certain intrinsic details of the oil such as its moisture and free fatty acid content before the transesterification process is initiated. The oil is heated to raise its temperature to 110 °C for 30 min. to expel the residual moisture. Since the oil also has higher level of free fatty acid it is necessary to reduce the free fatty acid level. Such attempt is referred as pretreatment. Once the moisture content and free fatty acid are reduced the next step, namely the alkali transesterification is to be taken up. Finally a post-treatment refinement also is carried out. All the above aspects, namely moisture removal, free fatty acid reduction, transesterification and the refinement are carried out in a vessel which has many ports for individually inserting the temperature probe, funnel with a valve arrangement for adding the required reagents, drain pipe with a control valve, a condenser and a safety valve. Such a reactor is to be made of a material which has to resist rust and corrosion. After expelling the moisture, the next in the pretreatment process is carried out to reduce the free fatty acid. It is totally an esterification reaction using H_2SO_4 as an acid catalyst. To 1 L of moisture free oil, 3–5 mL of conc. H_2SO_4 and a measured

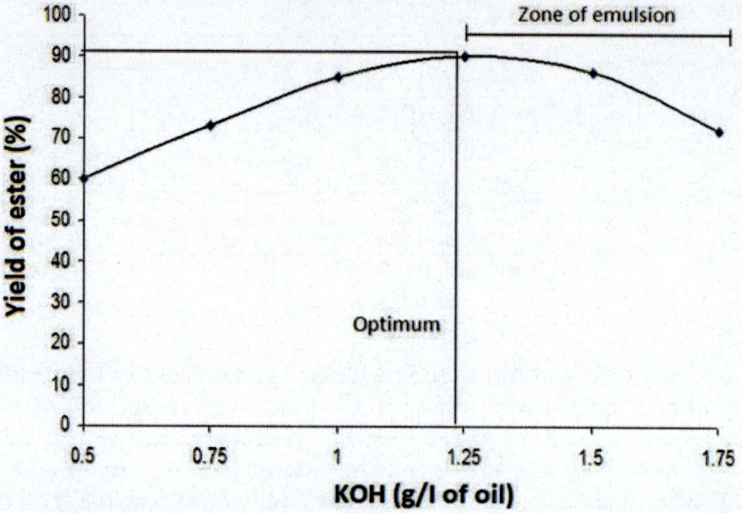

Fig. 11.5 The effect of KOH on the ester formation in the oil of *C. inophyllum* at a methanol oil ratio of 7:1

quantity of methyl alcohol so as to have a molar ratio of 4:1 with oil are added. The reaction is carried out at a temperature of 60 °C for 2 h. The content is then transferred to a separating funnel and kept undisturbed. After 2 h the top layer containing excess methanol and impurities is removed. The oil which moves down to the bottom containing very less quantity of free fatty acid is then transferred for transesterification.

The product thus obtained in the above process may contain acid, which is then washed away using water and the complete removal is ensured with litmus. Once again the moisture content is expelled and subsequently mixed with potassium hydroxide-methanol solution. This solution is prepared by adding a known quantity of KOH in methyl alcohol. The quantity of methyl alcohol thus taken is required to establish a molar ratio of 6–8: 1 with oil. If the quantity of KOH taken is lower than that of the actual quantity needed, the reaction may not progress well due to low pH. On the contrary if the quantity is increased more than the need, the yield of ester is hampered due to excessive gelling.

The formation of such emulsion may lead to difficulty in separating the finished product from glycerol. The ester recovery as a function of KOH is given in Fig. 11.5.

Much like that of KOH the methanol to oil molar ratio also plays a crucial role in the effective completion of transesterification. It is observed that the yield of ester increases as the molar ratio is increased as more triglycerides are prone to react with methanol molecules. But beyond a certain level the methanol makes the glycerol to remain unsettled and shift the reaction equilibrium to go reverse. When glycerol stands as a solution with methanol the separation of glycerol also is very difficult. In extremely high ratio, instead of the usual two layers (ester and glycerol) three layers are likely to be formed. They are the glycerol as bottom layer, ester as middle layer

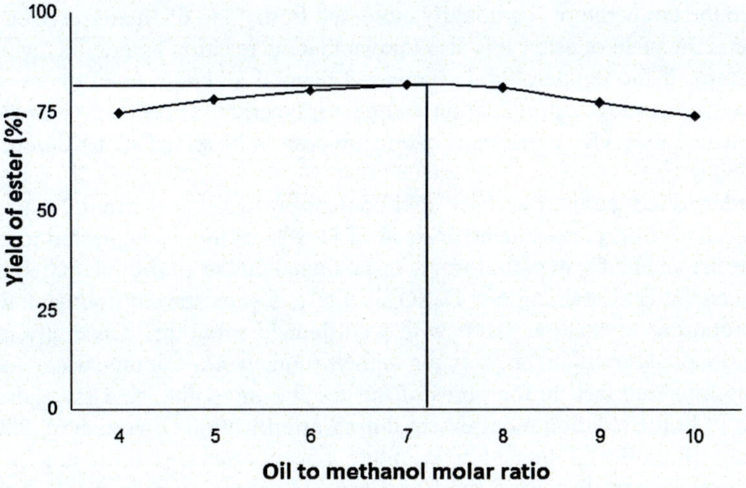

Fig. 11.6 The effect of molar ratio of oil to methanol on ester production in *C. inophyllum* when 1.25% of KOH is used at 60 °C at a reaction duration of 120 min

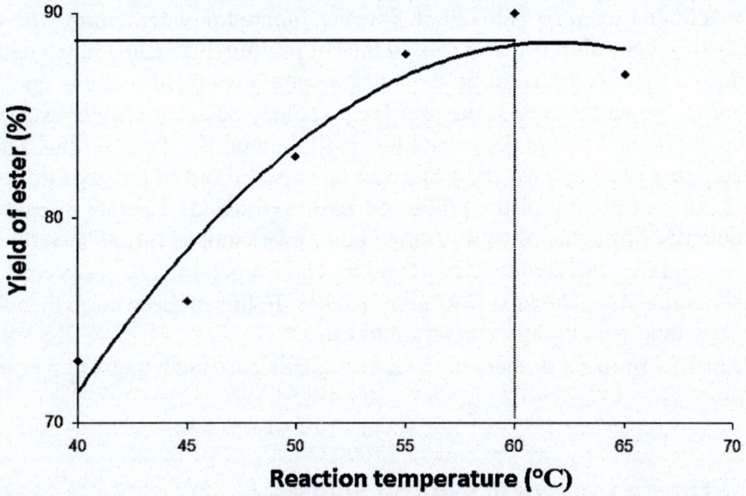

Fig. 11.7 The effect of temperature on ester production in C. inophyllum oil when 1.25% of KOH and methanol to oil ratio of 7:1 at a reaction duration of 120 min

and the unreacted excess methanol as the top layer. A molar ratio of 1: 6–8 (oil: methanol) is found to be optimum (Fig. 11.6).

The temperature of the transesterification process is normally kept closer to the boiling point of methanol. The yield of biodiesel with a KOH level percent and a molar ratio to oil: 1 for different temperature is given in Fig. 11.7.

When the temperature is gradually increased from 40 to 65 there is a phenomenal increase in the yield of ester. It is also known that the reaction kinetics is favoured by temperature. If the temperature is increased beyond a certain level, there is every likelihood of setting saponification of the triglyceride before the completion of transesterification. The transesterification process is observed to be completed in 90–120 min.

In order to reduce the free fatty acid content of the oil at the initial stage of the process (i.e. during pretreatment) instead of H_2SO_4 zeolite is suggested to be used since the use of H_2SO_4 in plain form is discouraged due to environmental reasons as the washed water containing raw H_2SO_4 is likely to enter the environment. Zeolites are microporous crystalline solids with well-defined structures. Generally, zeolites contain silicon, aluminium and oxygen in their frame work. Cations, water and other molecules are stationed in the pores of the zeolite. In zeolite, acid strength can be adjusted to fit the reaction requirement during esterification (Lotero et al. 2005). At the outset β zeolite is treated with dilute phosphoric acid for using it in the pretreatment process to reduce the free fatty acid content in the oil. The acid sites thus developed in the pore absorb the water being released during the treatment.

Around a kg of β zeolite is mixed with 10 L of 5% phosphoric acid. The above content is vigorously stirred at room temperature for about 12 h. It is then dried at 120 °C in a hot air oven so as to obtain a white coloured powder. During the normal acid catalysis when sulphuric acid is used the equilibrium is reached much before the completion of the esterification process as the water generated remains open in the reaction phase. Alternatively, if the processed zeolite is used, the water as and when generated gets stored in the active site located in the pores of the zeolite. Once the reaction process is completed the water can be expelled out of the zeolite.

To 1 L of the oil, 10 g of the processed zeolite is added. It is then mixed with a known quantity of methanol so as to maintain a methanol to oil molar ratio of 9:1. After 1 h of mixing the zeolite is removed by centrifugation. The recovered zeolite can also be utilized for the next few more batches. In this process more than 90% of the free fatty acid is straightly converted to ester.

The finished product is then subjected to alkali catalysed transesterification as usual.

11.1.11 The Characteristics of the Biodiesel

The density, kinematic viscosity, flash point and iodine value of *C. inophyllum* biodiesel are higher than the prescribed standard (Table 11.7) and these parameters may be brought within the standard limit by blending the biodiesel with regular diesel at a rate not more than 20%.

High iodine value is a good index of the presence of mono, di or tri unsaturated fatty acids. In general high iodine value indicates a low oxidation stability causing unwanted deposits in the engine nozzles and piston rings. The influence of unsaturated fatty acids on oxidation is actively induced by the exposure of oil to air, light and certain metals. Mono unsaturated fatty acid methyl ester (C18: 1) is considered

Table 11.7 The physico-chemical characteristics of ester (biodiesel) of *C. inophyllum* (Sahoo et al. 2007; Sahoo and Das 2009; Atabani et al. 2011; Sathyaselvabala et al. 2011)

Parameters	Value
Density (g cc^{-1})	0.869–0.889
Kinematic viscosity at 40 °C (cSt)	3.99–7.7
Cetane number	52–57
Flash point (°C)	130–151
Copper strip corrosion	1 b
Conradson carbon residue (%)	0.43
Water and sediment (%)	0
Dist. temperature (°C)	340–356
Sulphated ash (%)	0.03
Sulphur (mg kg^{-1})	16
Phosphorus (mg kg^{-1})	0.223
Acid value (mg KOH g^{-1})	0.76
Total glycerol (%)	0.232
Iodine value (mgI_2g^{-1})	850
Heating value (KJ kg^{-1})	38,660–41,397
Cloud point (°C)	13.2
Pour point (°C)	4.3

to be better than polyunsaturated fatty acid methyl esters such as linoleic and linolenic fatty acid esters (C18: 2 and C18: 3) in terms of the oxidation stability.

The oxidation of the esters of linoleate and linoleates is 41 and 98 fold higher than that of the oleate methyl esters (Knothe 2007).

Relatively higher viscosity noted in this case leads to a higher drag in the injection pump, causing higher pressures and injection volumes especially at low operating temperatures of the engine. As a result, the timing for fuel injection and ignition is likely to be advanced marginally which may cause the increase in NOx emission due to higher maximum combustion temperatures. Among the various factors the presence of glyceride is highly responsible for the viscosity. With reference to the saturated fatty acid esters the viscosity is observed to increase along with the carbon number in a curvilinear pattern. Regarding the unsaturated fatty acids a reverse trend is observed. The viscosity decreases while the number of bond increases.

11.1.12 Engine Performance

The brake thermal efficiency of the biodiesel is satisfactory especially in lower loads. This is by and large considered to be due to the better combustion and satisfactory lubricity. It is reported that the biodiesel from the oil *C. inophyllum* has 8.5% higher peak pressure than that of the regular diesel. Few works are available on the emission characteristics of automobiles fed with biodiesel made from *C. inophyllum* oil. The discharge of smoke is reported to be low during the use of biodiesel blend B60, B80 and B100 (i.e. the mixture containing 60, 80 and 100% of biodiesel in regular diesel). Such advantage is due to the presence of oxygen atom at high quantity in the biodiesel molecule. When the engine loading is increased, the NOx content of the smoke also increases. When the load of the engine is increased, the temperature of the combustion chamber tends to increase. Naturally when the temperature increases the NOx emission also increases. Similarly the level of hydrocarbon in the exhaust of the automobile reduces if biodiesel is used. The emission of CO_2 also is reported to be reduced by 40%.

11.2 Kokum (*Garcinia indica*)

The kokum, *Garcinia indica* grows richly in the Western Ghats of India. It is a slender tree of the tropical rainforest and grows to a maximum height of 15 m. It is popularly referred as wild mangosteen or goa butter tree. The leaves are ovate and lanceolate. The fruit is globose, succulent and purple in colour when ripe, having a size of 4 cm dia with 5–8 seeds embedded in it (Fig. 11.8) (Baliga et al. 2011).

G. indica is a threatened species according to IUCN. Besides the Western Ghats, it occurs at the forest of Assam, Maharashtra, Goa, Karnataka and Kerala states of India (Fig. 11.9). It is distributed below the altitude of 800 m. It is reported to yield 8.5 tonnes of fruit per ha (Ananthakrishnan and Rameshkumar 2017). It is a source of hard and brittle fat and is used in soap and creams (Padhye et al. 2009).

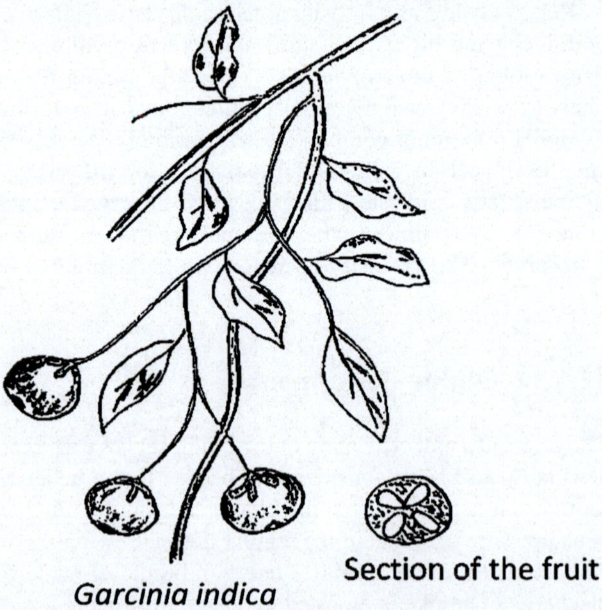

Fig. 11.8 Kokum plant *Garcinia indica*

Fig. 11.9 Geographical distribution of *Garcinia indica*

Table 11.8 Properties of *G. indica* oil biodiesel (Hosamani et al. 2009; Attal et al. 2014; Attal and Mahalle 2015)

Parameter	Value
Viscosity (cSt)	5.8
Density (kg m^{-3})	862
Flash point (°C)	135
Cetane number	52.8
Heating value (MJ kg^{-1})	39.12
Pour point (°C)	9.5
Cloud point (°C)	12
Oleic acid (C18:1) (%)	50.2
Linoleic acid (C18:2) (%)	26.1
Palmitic acid (C16:0) (%)	8.0
Stearic acid (C18:0) (%)	5.5

11.2.1 Oil Extraction and Biodiesel Preparation

The seeds are isolated from the riped fruits, air dried, powdered and extracted in petroleum ether (boiling point 40–60 °C). The oil thus obtained is treated with 0.8 N potassium hydroxide in ethanol and stored overnight at 27 °C. The non saponifiable fraction is then removed. The resultant content is fractioned into oxygenated and non-oxygenated portion (Bharucha and Gunstone 1956). The non-oxygenated portion is refluxed with methanol sulphuric acid mixture for 1 h. It is then extracted with ether solvent and washed with distilled water. The solvent is then removed by distilling and the resultant content is freed from moisture by passing it over sodium sulphate (Hosamani et al. 2009). The properties of the biodiesel thus obtained are given in Table 11.8.

The viscosity, density, cloud point and pour point of it are unfavourable to be considered as a biodiesel. However, attempts are made to use it by blending with diesel. The cetane number is observed to be a favourable parameter. Higher the value of CN, better the ignition quality. Biodiesel standards ASTM D6751, DIN 51606, EN14214 prescribe CN values as 47, 49 and 51, respectively.

11.2.2 Engine Performance

The brake thermal efficiency is low since the calorific value is low. High viscosity results into poor atomization and combustion. The air fuel ratio is also low. This is due to the increased consumption of biodiesel to produce the same level of power output though the air consumption remains the same. The smoke opacity of exhaust is less due to the satisfactory burning of the soot as a consequence of rich oxygen. The CO and HC levels in the exhaust are low probably because of shorter ignition delay. Low CO_2 and high O_2 levels in the exhaust are reported (Attal et al. 2014). Due to the high combustion temperature NOx emission is high. Additional oxygen content of the biodiesel along with the oxygen of the aspirated air readily reacts at an elevated temperature and generates oxides of nitrogen.

References

Adeyeye A (1991) Studies on seed oils of *Garcinia kola* and *Calophyllum inophyllum*. J Sci Food Agric 57:441–442

Ajayi IA, Orderinde RA, Taivo VO, Agbedana EO (2008) Short term toxicological evaluation of *Terminalia catappa, Pentaclethra macrophylla* and *Calophyllum inophyllum* seed oils in rats. Food Chem 106:458–465

Ananthakrishnan R, Rameshkumar KB (2017) Phytochemicals and bioactivities of Garcinia indica (Thouars) choisy—a review. In: Koranappallil BR (ed) Diversity of Garcina species in the Western Ghats: phytochemical perspective. Jawaharlal Nehru Tropical Botanic Garden and Research Institute, Palode, Thiruvananthapuram, pp 142–149. ISBN 978-81-924674-5-0

Atabani AE, Silionga AS, Mahila TMI, Masjuki HH, Badruddin IA (2011) *Calophyllum inophyllum* as a future feedstock for biodiesel production. In: Proc. 6th international green energy conference, 5–9 June 2011, Anadolu University, Turkey, pp 281–291

Attal MH, Mahalle AM (2015) The emission analysis of diesel engine using Garcinia indica and rice bran oil based methyl esters as fuels with an ANN approach. Int J Innov Emerg Res Eng 2 (1):15–19

Attal M, Dave C, Mahalle AM (2014) Experimental investigation on performance of diesel engine using Garcinia indica (kokam) and rice bran oil based biodiesel as fuels. J Mech Civil Eng 11 (5):01–07

Azam MM, Waris A, Nahar NM (2005) Prospects and potential of fatty acid methyl esters of some non traditional seed oils for use as biodiesel in India. Biomass Bioenergy 29:293–302

Baliga MS, Bhat HP, Pai RJ, Baloor R, Princy LP (2011) The chemistry and medicinal uses of the underutilized Indian fruit tree Garcinia indica choisy (kokum): a review. Food Res Int 44:1790–1799

Bharucha KE, Gunstone FD (1956) Vegetable oils-IV a new method of determining the component acids of oils containing epoxy and or hydroxyl acids. J Sci Food Agric 6:373–380

Hathurusingha S (2011) Periodic variation in kernel oil content and fatty acid profiles of *Calophyllum inophyllum* L.: a medicinal plant in northern Australia. Ind Crop Prod 33:775–778

Hathurusingha S, Ashwath N, Subedi P (2011) Variation in oil content and fatty acid profile of *Calophyllum inophyllum* L. with fruit maturity and its implications on resultant biodiesel quality. Ind Crop Prod 33:629–632

Hemavathy J, Prabhakar JV (1990) Lipid composition of *Calophyllum inophyllum* kernel. J Am Oil Chem Soc 67(12):955–957

Hosamani KM, Hiremath VB, Keri RS (2009) Renewable energy sources from Michelia champaca and Garcinia indica seed oils: a rich source of oil. Biomass Bioenergy 33:267–270

Knothe G (2007) Some aspects of biodiesel oxidative stability. Fuel Process Technol 88:669–677

Kumar A, Sharma S (2011) Potential non edible oil resources as biodiesel feed stock—an Indian perspective. Renew Sust Energ Rev 15:1791–1800

Lotero E, Liu Y, Lopez DE, Suwannakaran K, Bruce DA, Goodwin JC Jr (2005) Synthesis of biodiesel *via.* acid catalysis. Ind Eng Chem Res 44:5353–5363

Padhye S, Ahmad A, Oswal N, Sarkar FH (2009) Emerging role of garcinol, the antioxidant chalcone from Garcinia indica choisy and its synthetic analogs. J Hematol Oncol 38(2):1–13

Pinzi S, Garcia IL, Lopez-Gimenez FJ, Luque de Castro MD, Dorado G, Dorado MP (2009) The ideal vegetable oil based biodiesel composition: a review of social economical and technical implications. Energy Fuel 23:2325–2341

Sahoo PK, Das LM (2009) Combustion analysis of jatropha, karanja and polanga based biodiesel as fuel in a diesel engine. Fuel 88:994–999

Sahoo PK, Das LM, Babu MKG, Naik SN (2007) Biodiesel development from high acid value Polanga seed oil and performance evaluation in a C1 engine. Fuel 86:448–454

Sathyaselvabala V, Selvaraj DK, Kalimuthu J, Periyaraman PM, Subramanian S (2011) Two step biodiesel production from *Calophyllum inophyllum* oil: optimization of modified β–zeolite catalysed pre-treatment. Bioresour Technol 102:1066–1072

Tong D, Hu C, Jiang K, Li Y (2011) Cetane number prediction of biodiesel from the composition of the fatty acid methyl esters. J Am Oil Chem Soc 88:415–423

Venkanna BK, Reddy CV (2009) Biodiesel production and optimization from *Calophyllum inophyllum* linn oil (hone oil)—a three stage method. Bioresour Technol 100:5122–5125

Combretaceae 12

The family Combretaceae, often called as white mangrove family comprises around 500 trees, shrubs, lianas and sub-scadents. Members of this family are known for their use in medicine, gum, tannin and timber. Terminalia (Terminalia *catappa and T. bellerica*) is dealt in this chapter due to its reported connection in biodiesel preparation.

12.1 Terminalia (Terminalia *catappa* and *T. bellerica*)

Terminalia is the largest genus of the family Combretaceae with more than 200 species, in which around one fourth of the members are the native of Africa. Terminalia *bellerica*, *T. catappa* and *T. chebula* are the three popular species in which *T. catappa and T. bellerica* received importance in biodiesel. These species are grown as shade giving trees for the public. They are medium to large deciduous trees, growing to the height of 30–40 m with wide spreading branches and having disk shaped crown. The fruit is considered as a king of medicines. The seed oil is a potential source for biodiesel.

T. bellerica is distributed in India especially in the moist valleys of Tripura and Assam States. The bark is bluish and ash grey with longitudinal cracks. It grows in all types of soil but prefers loam and red soil. A tree of 6–8 years age yields around 500 kg of fruits annually. Fruits are globular with ridges containing hard thick shelled stone.

T. catappa is also a native of India but seldom grows in south East Asia, and in Taiwan. This tree is well adopted to the coastal climate and resists strong winds, salt breeze and moderately high saline soils. A tree of 3 years old starts yielding fruits of ellipsoidal shape with a characteristic purple colour (Fig. 12.1).

The genus Terminalia is characterized by spiral-stalked, papery, thin-leathery leaves. Leaves at the top of the canopy are green in colour, in the middle layer of the canopy they are yellowish green and are red at the falling stage. The leaf is egg

S. P. Raj et al., *Biodiesel from Flowering Plants*,
https://doi.org/10.1007/978-981-16-4775-8_12

Fig. 12.1 Geographical distribution of *T. bellerica and T. catappa*

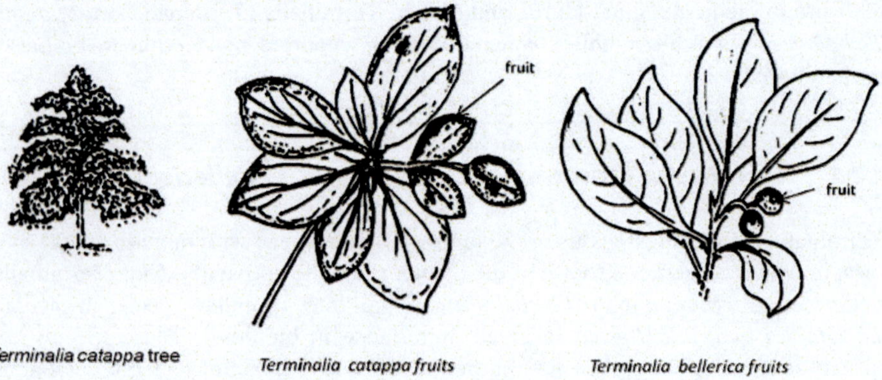

Fig. 12.2 Terminalia *catappa* and *T. bellerica*

Table 12.1 Vernacular names of *T. catappa and T. bellerica*

Language	*T. catappa*	*T. bellerica*
English	Country almond	Belleric myrobalan
French	Amandier	Myrobalan belleric
Malay	Mentalun	Jelawai
Cambodia	–	–
Hindi	Badambo	Bahe
Tamil	Saraparuppu	Akkam

shaped 10–35 cm long and around 10 cm wide. The fruits are green when tender, yellow or red when ripe. Each fruit has a single seed (Fig. 12.2).

The two species of Terminalia are referred by different location specific vernacular names as in Table 12.1.

12.1.1 Oil Extraction

The seeds are collected, dried and the kernels isolated by cracking the nut. The kernels are dried to remove the moisture by exposing them in hot sun for a few days. The dried kernels are then powdered in hammer mill and the oil extracted using *n*-hexane solvent. The oil recovery is around 50% (Adewuyi et al. 2011).

12.1.2 Composition of the Oil

The analytical details of the oil of *T. catappa* and *T. bellerica* are given in Tables 12.2 and 12.3. The oil of *T. catappa* and *T. bellirica* are almost identical in their viscosity, density and iodine value.

Table 12.2 The physicochemical characteristics of the oil of *T. catappa* and T. *bellerica* (dos Santos et al. 2008; Sarin et al. 2010; Adewuyi et al. 2011; Anjaneyulu et al. 2014; Yatish et al. 2016)

Parameters	*T. catappa*	*T. bellerica*
Viscosity at 40 °C (mm^2/s)	39.8	34.1, 39.8, 35.6
Density (kg/m^3)	910	915, 929, 894
Saponification value (mg KOH/g)	206.5	186
Iodine value (g I_2/100g)	67.6, 83.9	85.0, 71.2
Free fatty acid (%)	7.0	0.45
Acid value (mgKOH/g)	0.5	1.3, 3.36
Oxidation stability (h)	–	8.68, 23.8
Copper corrosion (50 °C, 3 h)	–	1a
Calorific value (MJ/kg)	–	38.7

Table 12.3 Fatty acid composition of *T. catappa* and *T. bellerica* seed oil (dos Santos et al. 2008; Sarin et al. 2010; Adewuyi et al. 2011; Anjaneyulu et al. 2014; Yatish et al. 2016)

Fatty acid	*T. catalpa*	*T. bellerica*
Lauric acid (12:0)	–	0.08
Myristic acid (14:0)	–	–
Palmitic acid (16:0)	35, 35.8	11.6, 28.3, 21.8
Stearic acid (18:0)	5, 6.8	3.9, 10.6, 8.2
Eicosanoic acid (20:0)	0.8, 2.3	0.8, 0.2
Behenic acid (22:0)	–	0.2, 10.3
Lignoceric acid (24:0)	–	1.75
Palmitoleic acid (16:1)	0.18	0.5
Oleic acid (18:1)	32, 29	61.5, 43.1, 31.5
Linoleic acid (18:2)	28, 23.3	18.5, 17, 24.7
Linolenic acid (18:3)	–	0.1, 0.38
Paullinic acid (20:1)	0.48	–
Eicosadienoic acid (20:2)	1.3	–
Mead acid (20:3)	0.75	–
Arachidonic acid (20:4)	–	1.17

12.1.3 Biodiesel Preparation

Transesterification of the oil is carried out by mixing the oil with methyl alcohol at 1:6 molar ratio followed by slow addition of sodium methoxide at the rate of 0.6% of the weight of the oil. The whole content is continuously stirred for about 1 h, by that time the biodiesel gets separated due to difference in gradient. The underlying glycerine is separated and the supernatant biodiesel is repeatedly washed with warm water so as to remove the trace of glycerine, unreacted catalyst and soap which are formed during the reaction. The resultant biodiesel is of 98% purity. The recovery is around 95–98%.

12.1.4 Properties of *T. catappa* Biodiesel

The density and viscosity though slightly higher, they are within the acceptable limit of the standard EN 14214 and ASTM D 6751. The heat value is lower than that of the petroleum diesel (46.65 MJ/kg) and the values are comparable to that of the soybean and palm oil. The fatty acid composition revealed that the share of unsaturated fatty acids is low. The Rancimat study of the biodiesel indicated a high oxidative stability which is due to a low iodine value. The flash point is observed to be lower than the limit. The copper corrosion test also is in agreement with the standard EN 14214 (Table 12.4).

12.1.5 Properties of *T. bellerica* Biodiesel

The kinematic viscosity of the biodiesel of *T. bellerica* is higher than that of the *T. catappa*. The kinematic viscosity is known to increase according to the chain

Table 12.4 Physicochemical properties of biodiesel from *T. catappa* and *T. bellerica* (dos Santos et al. 2008; Sarin et al. 2010; Adewuyi et al. 2011; Yatish et al. 2016)

Parameters	*T. catappa*	*T. bellerica*
Kinematic viscosity at 40 °C ($mm^2\ s^{-1}$)	4.3, 4.1	4.7, 4.8
Density at 20 °C (kg m^{-3})	873, 840	880
Iodine value (g I/100 g)	83.2, 65.3	–
Heating value (MJ kg^{-1})	37.0	40
Flash point (°C)	128	160, 162
Acid value (mg KOH g^{-1})	0.1	0.2, 0.01
Copper corrosion test	1a	1a
Oxidation stability (h)	32.5	4.4, 2.1
Cloud point (°C)	–	6.0, 5.0
Pour point (°C)	–	3.0, 0
Cetane number	–	53.4, 54.0

length and high degree of saturation of the fatty acid or fatty ester (Demirbas 2008). The flash point is observed to be higher than the limit prescribed.

The density is higher than the conventional diesel (815 kg m^{-3}) which is apparently due to the presence of residual triglyceride molecule having higher molecular weight and hence it is difficult to use the biodiesel as such in engines. The corrosion is not a major issue in this biodiesel since it has no sulphur. The cetane number also is higher than that of the conventional diesel. Higher cetane number makes the ignition to be faster resulting in to incomplete combustion causing increased smoke in the exhaust. Again the cetane number depends on the molecular structure which decreases with the increase in the number of double bonds and increases with the increase in the carbon atoms. The cloud point and pour point are higher than the conventional diesel but they are within the prescribed standards. The cloud point refers to the operability of the biodiesel in extremely low temp. The pour point is 3 °C for *T. bellerica* which means the biodiesel will quickly agglomerates and form crystals when compared to that of the regular diesel (−4 °C). The acid value is quite satisfactory in this biodiesel. Acid value is often called as acid number or neutralization number which refers to the quantity by milligram required to neutralize the acidic constituents of 1 g of the biodiesel.

The heating value is 40 MJ kg^{-1} against 44 MJ kg^{-1} of conventional diesel. The value is bound to decrease with the degree of unsaturation and increases with molecular chain length. The oxidation stability being the index of long storage is an important parameter. The unsaturated fatty acid negatively influences the stability. Long storage usually causes the increase in viscosity.

12.1.6 Engine Performance

The specific fuel consumption is largely governed by heating value, density, viscosity and surface tension. (Usta et al. 2005). Specific fuel consumption increases when the heating value is low. The amount of fuel entering in to the combustion chamber is greatly restricted if the viscosity is high. High viscosity increases the fuel line pressure and consequently more fuel injection is required due to the loss in delivery. Fuel having high surface tension reduces the atomization and combustion thereby the specific fuel consumption shows a tendency to increase.

In order to manage the application, the biodiesel is blended with conventional diesel. The biodiesel blend causes the raise in pressure instantaneously due to early premixed combustion. The peak pressure thus generated may cause severe engine knock affecting the durability of the engine. The peak pressure being formed inside the cylinder at the premixed combustion stage depends on the amount of fuel actually taking part in the process. The peak pressure variation is prominent at lower loads. There will be short ignition delay with respect to the use of blend.

References

Adewuyi A, Oderinde RA, Ojo DFK (2011) Production of biodiesel from *Terminaliacatappa* seed oil with high free fatty acids using an acid catalyzed pretreatment step. Int J Sust Energy 30 (Suppl 1):S59–S65. https://doi.org/10.1080/1478646X.2011.587518

Anjaneyulu B, Ravinder T, Kanjilal S, Chakrabarti PP (2014) Lipid classification and characterization of *Terminalia belerica* seed oil from Tripura. J Lipid Sci Technol 40(40):145–149

Demirbas A (2008) Relationship derived from physical properties of vegetable oil and biodiesel fuels. Fuel 87:1743–1748

dos Santos IC, de Carvalho SH, Solleti JI, de la Salles WF, da Silva de La Salles K, Meneghetti SM (2008) Studies of Terminalia catappa. L. oil: characterization and biodiesel production. Bioresour Technol 99(14):6545–6549

Sarin R, Sharma M, Khan AA (2010) *Terminalia belerica* Roxb, seed oil: a potential biodiesel resource. Bioresour Technol 101:1380–1384

Usta N, Ozturk E, Can O, Conkur ES, Nas S, Con AH, Can AC, Topcu M (2005) Combustion of biodiesel fuel produced from hazelnut soap stock/waste sunflower oil mixture in a diesel engine. Energy Convers Manag 46:741–755

Yatish KV, Lalithamba HS, Suresh R, Omkaresh BR (2016) Synthesis of biodiesel from *Garcinia gummi-gutta, Terminalia belerica* and *Aegle marmelos* seed oil and investigation of fuel properties. Biofuels 9(1):121–128. https://doi.org/10.1080/17597269.2016.1259524

Compositae

13

Few publications came into being on safflower (*Carthamus tinctorius*) and niger (*Guizotia abyssinica*) as their oils are being utilized as raw material for biodiesel. These plants are from daisy group and also the members of the family Compositae. Though many taxonomists group these plants in family Compositae, few taxonomists consider them under Asteraceae since they are homologous families and their characteristics are synonymous. These plants are normally cultivated either as a horticultural crop or as an oil seed crop. Some are weeds and uninvited partners of garden.

13.1 Safflower (*Carthamus tinctorius*)

Safflower (*Carthamus tinctorius*), an oilseed agro-crop is an annual herb which grows to a height of 1.5 m. The stem is glabrous, cylindrical and woody. The leaf is sessile and is 10–20 cm long and 5 cm wide. The leaves bear sharp spines at their edges. The stem ends up with an inflorescence which carries a maximum of 70 orange coloured flowers. Each flower bears a fruit. These fruits are achenes and are called as seeds. These seeds are enclosed in a fibrous hull. The seeds are smooth, shiny and with dark stripes (Fig. 13.1). The hull constitutes 50% of the seeds and the rest is formed of the kernel. Around 1 million tonnes of seeds are produced globally during the year 2016 and it is 0.6 million tons during 2019 (FAOSTAT 2019).

Safflower has originated from Eastern Europe, Central Asia and is being cultivated in India, Ethiopia, Kenya, USA, Mexico, Argentina and Kazakhstan (Fig. 13.2) and in countries located between 20°S and 40°N with a rainfall ranging from 800 to 1000 mm and at an elevation of 900–1800 M above mean sea level. It gives a yield of around 1500 kg of seeds per ha in rainfed condition.

Safflower is being called by various names in different countries (Table 13.1).

S. P. Raj et al., *Biodiesel from Flowering Plants*,
https://doi.org/10.1007/978-981-16-4775-8_13

Fig. 13.1 Safflower, *Carthamus tinctorius*

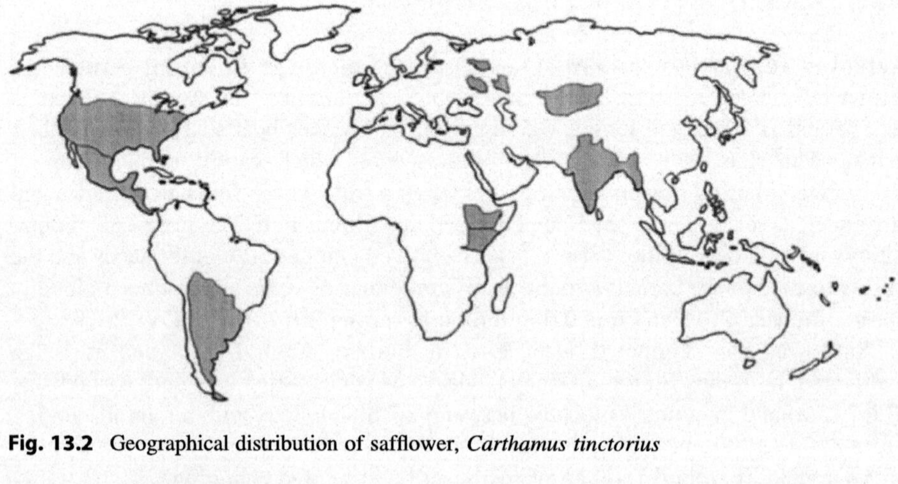

Fig. 13.2 Geographical distribution of safflower, *Carthamus tinctorius*

13.1.1 Oil Extraction

The seeds are sun-dried prior to oil extraction. The oil is extracted either through mechanical pressing or through solvent extraction. In mechanical expellers or

Table 13.1 Common names of safflower in different languages

Languages	Names
English	False saffron, thistle oil
French	Carthame des teinturiers, safran des teinturiers, safran batard
Portuguese	Acafrão-bastardo
Dutch	Saffloer
German	Färberdistel, saflor
Indonesian	Kesumba
Italian	Zafferanone
Turkish	Aspir

Table 13.2 Physicochemical properties of *Carthamus tinctorius* oil (Ilkiliç et al. 2011; Anonymous 2014)

Parameters	Value
Kinematic viscosity @ 40 °C ($mm^2 s^{-1}$)	67
Density @ 15 °C (kg m^{-3})	950
Iodine value (g I_2/100 g)	129.5, 141.52
Acid value (mg KOH g^{-1})	0.07, 0.35
Heat value (MJ kg^{-1})	39
Peroxide value (meq O_2 kg^{-1})	0.30
Flash point (°C)	225
Cloud point (°C)	−2
Average molecular weight (g mol^{-1})	885.47
Palmitic acid (C16:0) %	6.34
Stearic acid (C18:0) %	2.95
Oleic acid (C18:1) %	36.11
Linoleic acid (C18:2) %	53.31
Alpha linolenic acid (C18:3) %	0.11

presses, the whole seed or kernel or both are used as raw materials. In case solvent extraction is resorted, kernel alone is used as raw material. In case of solvent extraction, the kernel is to be dried repeatedly and pulverized and then extracted using n-hexane at 1:3 volume ratio of kernel powder and solvent. The analytical details of safflower oil are given in Table 13.2.

13.1.2 Biodiesel Preparation

Transesterification of oil is carried out using sodium hydroxide or potassium hydroxide or sodium methoxide or potassium methoxide as catalyst. The alcohol being used is methanol. The methoxide of sodium or potassium is observed to be very effective when compared to the performance of their corresponding hydroxides (Fig. 13.3). The other conditions of transesterification are temperature 60 °C, oil to methanol molar ratio 1:6, reaction duration 120 min and the stirring rate is 600 rpm. The optimum level of catalyst is 1%. If the concentration of catalyst is increased, the viscosity of these reactants goes high.

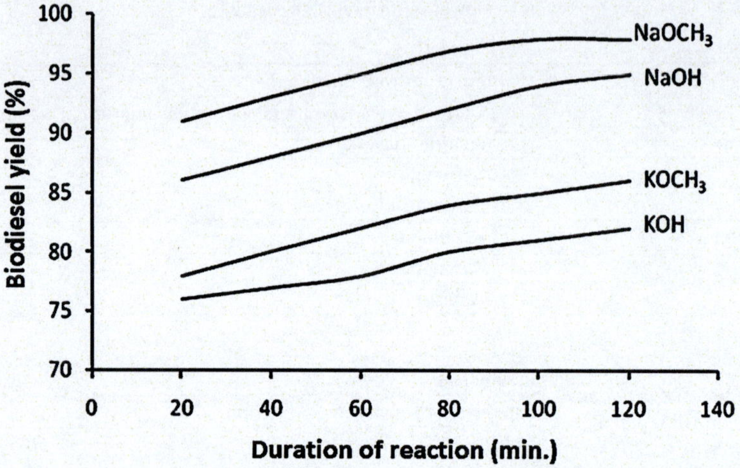

Fig. 13.3 Performance of various catalysts in the preparation of safflower biodiesel

Table 13.3 Properties of safflower biodiesel (Rashid and Anwar 2008)

Parameters	Value
Kinematic viscosity @ 40 °C (mm^2s^{-1})	4.29
Density @ 15 °C (kg m^{-3})	874
Acid value (mg KOH g^{-1})	0.28
Cetane number	52.3
Copper strip corrosion (50 °C, 3 h)	1a
Flash point (°C)	176
Cloud point (°C)	−2.0
Pour point (°C)	−8.0
Cold filter plugging point (°C)	−6.0

The oil methanol molar ratio is 1:6 and is observed to be ideal. If the ratio is increased beyond the optimum, separation of glycerol is difficult. The excess methanol interferes with the phase separation and as a result, the quantity of ester to be recovered in pure form is reduced, since a portion of the glycerol is likely to remain in the ester phase (Usta 2005).

Higher mixing normally promotes homogenization and increased reaction. However, it is observed that at 600 rpm maximum recovery is obtained.

13.1.3 Quality of Biodiesel

The physicochemical properties of the biodiesel are given in Table 13.3. The density (874 kg m^{-3}) is marginally higher than that of the diesel (825–835 kg m^{-3}). As fuel injection is being done by volume and not by weight, higher density fuel when injected increases the mass of fuel. The kinematic viscosity though marginally higher may create operational inconveniences by interfering with the process of injection. The higher heating value of it being the index of the energy produced when the fuel is burnt is almost on par with that of the diesel. The cetane number which actually indicates the tendency of the fuel to self-ignite at the given temperature and pressure in the cylinder when the biodiesel is injected is relatively higher when compared to that of the diesel.

The flash point is higher than that of the diesel which expresses that the biodiesel poses less risk with reference to the catch of fire. The cloud, pour and cold filter plugging point is higher than that of the diesel and these are negative factors. The engine torque, brake power, brake thermal efficiency, brake specific fuel consumption, brake specific energy consumption are being studied so as to qualify the biodiesel for use in combustion engines where chemical energy of the fuel is converted into mechanical energy. The engine torque for the biodiesel is marginally lower since the calorific value is lower than that of the diesel and another reason may be the higher viscosity. The engine power also is low due to low calorific value. The brake specific fuel consumption is relatively higher. The exhaust gas temperature also is relatively higher in biodiesel which is attributed to the complete combustion mainly due to the inherent oxygen present in the biodiesel.

The particulate matter in the exhaust is low with respect to the biodiesel. Similarly, smoke opacity also is reduced. It is mainly due to the oxygen present in the molecule of the ester which may even break down the soot particle. The oxidation stability of the soot from the biodiesel is reported to be 6 times higher than that of the diesel (Ilkiliç et al. 2011). The safflower biodiesel has low sulphur content and consequently the sulphate as particulate matter in the exhaust is considerably low in its occurrence. The effective combustion due to rich oxygen yields more NOx in the exhaust. Besides, the high cetane number of biodiesel causes rapid combustion through the shortening of the ignition delay. The iodine number is high in biodiesel which is also known to induce the NOx emission (Ilkiliç et al. 2011). Interestingly, the hydrocarbon level in the exhaust increases in certain biodiesel (Sensöz and Angın 2008). CO emission level is reduced. The advanced injection time of biodiesel helps to reduce the CO emission.

13.2 Niger (*Guizotia abyssinica*)

Guizotia abyssinica is a crop on cultivation in India and Ethiopia and it belongs to the family Compositae. *G. abyssinica* is an erect, stout herb, grown for its seed rich in oil. The genus *Guizotia* comprises six species, out of which *G. abyssinica* alone is being cultivated. This plant grows to a maximum height of 1.5 m and comes to yield within 120 days of sowing. The leaves are arranged opposite. The leaf is 10–20 cm long and 3–5 cm wide. The stem is hollow and fragile. The flower is yellow or greenish yellow. The flower head is a daisy of 1.5–5.0 cm diameter with 5–20 mm long florets (Fig. 13.4). Each flower head harbours 20–30 mature seeds. Each seed is 3.5 mm long and 1.5 mm wide. The seeds are sown by broadcasting at the rate of 10 kg ha^{-1}. Rarely they are sown in rows 40–50 cm apart at the rate of 4–5 kg ha^{-1}.

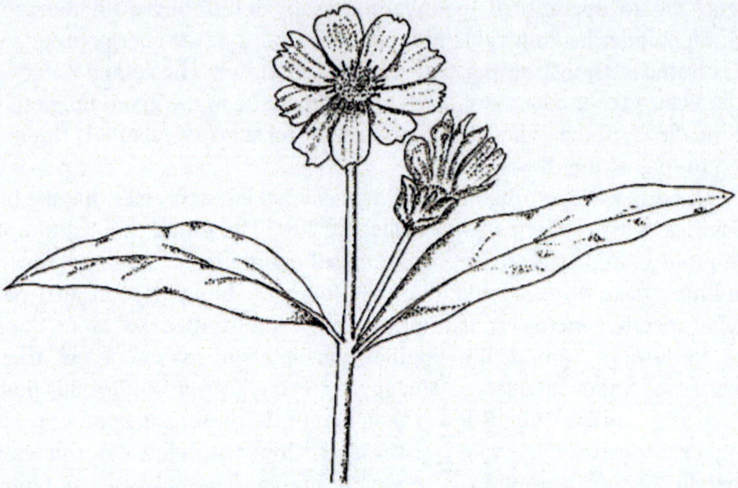

Fig. 13.4 The niger *Guizotia abyssinica*

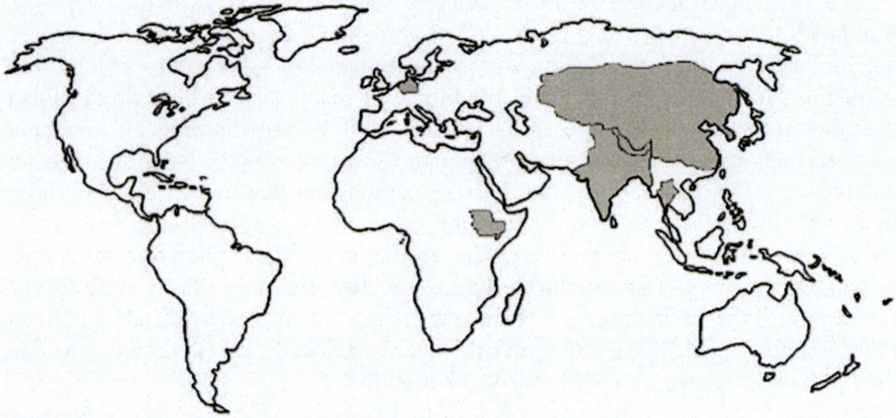

Fig. 13.5 The geographical distribution of *G. abyssinica*

The ripened flower heads start drying in 120 days. These heads are harvested and threshed so as to separate the seeds. The seed heap is cleaned of earth particles and dusts by winnowing and sieving. Each seed weigh around 4–5 mg. This plant is reported to grow in an altitude of 1600–2200 m above mean sea level with a rainfall of 1000–1300 mm (Burnette 2010). It is grown in India, West Indies, Germany, China, Ethiopia, Nepal and Thailand (Fig. 13.5). It is called by different location specific names: niger, nigerseed, noug, guizotia oleîfêre, ramtil, nigersaat, verbesina da India, abisin, negrillo and ramtilla, noog, nug, nyjer, inga seed and black seed.

13.2.1 Oil Preparation

The seeds are dried repeatedly in hot sun for several days. The dried seeds are crushed, powdered and extracted using *n*-hexane as solvent. The refluxing is done at 60 °C with 6–8 cycles in an hour for a total of 16–18 h. The seed solvent ratio is 1:15. After complete extraction the hexane is distilled off leaving a pale yellow oil. The oil is degummed using phosphoric acid (Jaikumar et al. 2018). With every litre of oil 1 mL H_3PO_4 and 0.5 mL water are added and the content warmed up to 60 °C for 30 min. The content is then allowed to settle and the gum leaving the oil is discarded. To the degummed oil, 20 g kg^{-1} of bleaching powder is added and the content heated to 110 °C for 2 min. It is then filtered to get rid of the impurities. The cleaned oil is deodorized by raising the temperature to 250 °C at a vacuum pressure of 760 torr (15 psi) using nitrogen as a carrier gas. The physicochemical properties of the oil thus prepared are given in Table 13.4.

Normally the viscosity is the function of monosaturated fatty acid and negatively correlates with the amount of polyunsaturated fatty acids. The iodine value is high which indicates the occurrence of unsaturated fatty acid in rich quantity. The low oxidation stability is also a consequent of the high level of unsaturated fatty acids.

13.2.2 Biodiesel Preparation

To a known volume of methanol which is as equivalent to the molar ratio 8:1 to oil, sodium hydroxide as catalyst at a concentration of 6 g L^{-1} of oil is added and dissolved by stirring at 500 rpm for 15 min. The above solution (sodium methoxide) is emptied in oil at 65 °C and stirred at 500 rpm for 2 h by then the biodiesel is formed. The content is then transferred to a separating funnel and waited till the phase separation is completed. The upper zone is the biodiesel. The ester conversion is reported to be more than 98% (Sarin et al. 2009).

The use of base catalysts, namely NaOH or KOH is often creating corrosion to the reactors and NaOH is a likely to generate emulsion which affects the smooth phase

Table 13.4 Physicochemical properties of *G. abyssinica* oil (Sarin et al. 2009)

Parameters	Value
Kinematic viscosity at 40 °C (cSt)	28.2
Density (kg m^{-3})	901
Acid value (g KOH kg^{-1})	6.03
Iodine value (g I2//100)	132
Cloud point (°C)	−2
Pour point (°C)	−9
Oxidation stability at 110 °C (h)	1.4
Palmitic acid (C16:0) %	9.2
Stearic acid (C18:0) %	10
Oleic acid (C18:1) %	9.0
Linoleic acid (C18:2) %	71.7

separation. To counter the above problems Yerranguntia et al. (2013) recommended the use of $MnCO_3$ as a heterogeneous catalyst. The shape of the manganese carbonate and the images obtained through Scanning Electron Microscopy (SEM) shows distinct spatial variations and crystallographic structures. The crystal structure of the surface shows uniform undulations due to the particle of the size range 5–10 μm which offers maximum catalytic exposure.

Manganese carbonate at a concentration of 1% by weight at a molar ratio of 5:1 (methanol to oil) in 45 min at 60 °C gave a conversion of 95%. On completion of the transesterification, the manganese carbonate is recovered by filtration and washed with distilled water 4–5 times, dried at 50 °C for 35 h and stored for further use. The manganese carbonate is reported to be environmentally safe, does not require any neutralization and can be recovered from the medium by simple filtration (Yerranguntia et al. 2013).

Similar to manganese carbonate, ZnO also is used as a catalyst (Kumari et al. 2014). To 1 molar solution of NaOH, zinc nitrate (0.5 m) solution is added drop wise and the content is kept under constant stirring at 40 °C for 2 h. A white deposit (ZnO) formed at the bottom is collected, washed repeatedly with ethanol and finally rinsed with distilled water. Then it is dried and calcined at 400 °C for 4 h. The reaction steps are given under:

When zinc nitrate is added to NaOH, zinc hydroxide is formed

$$Zn(NO_3)_2.6H_2O + 2\,NaOH \rightarrow Zn(OH)_2gel + 2\,NaNO_3 + 6H_2O$$

The $Zn(OH)_2$ gel slowly transforms to a super saturation state

$$Zn(OH)_2 + 2\,H_2O \rightarrow Zn(OH)_4{}^{2-} + 2\,H^+$$

The $Zn(OH)_4{}^{2-}$ dissociates and form ZnO + H_2O + $2OH^-$

The ZnO catalyst is added to the oil along with methanol for the transesterification.

The performance of ZnO is considered better than that of $MnCO_3$ (Kumari et al. 2014). The properties of the biodiesel are given in Table 13.5.

The kinematic viscosity and the acid value are within the standard limit. The flash point also is higher than the recommended minimum of 130 °C. The cloud and pour point are slightly higher which is a limitation in this fuel. Similarly the oxidation stability is poor due to the presence of high level of unsaturated linoleate. To improve the stability, phenolic and aminic antioxidants are used. The biodiesel also is blended with palm methyl ester to improve the stability (Sarin et al. 2007).

13.2.3 Engine Performance

Manjunatha et al. (2015) evaluated the performance of this biodiesel in an engine (Cylinder bore D 80 mm, cylinder stroke L 110 mm, brake horse power 10 hp, rated speed 1500 rpm) and evaluated the brake power, specific fuel consumption, brake

Table 13.5 Physicochemical properties of *G. abyssinica* biodiesel

Parameters	Sarin et al. (2009)	Yerranguntia et al. (2013)	Jaikumar et al. (2018)
Kinematic viscosity at 40 °C (cSt)	4.3	4.1	4.1
Density (kg m^{-3})	–	890	889
Acid value (g KOH kg^{-1})	0.15	–	0.3
Flash point (°C)	157	128	169
Cloud point (°C)	4	3	3
Pour point (°C)	3		
Pour point (°C)	–	2.5	–
Copper corrosion	1		1a
Cetane number	57		60
Oxidation stability at 110 °C (h)	1	–	–
Heating value (MJ kg^{-1})	–	–	41.1
Methyl palmitate %		7.9	
Methyl stearate %		8.1	
Methyl oleate %		8.25	
Methyl linoleate %		69	

thermal efficiency, brake mean effective pressure and came to a conclusion that blending this biodiesel with neat biodiesel to the extent of B10-B20 will give performance as equal to that of neat diesel.

References

Anonymous (2014) The soap kitchen, Safflower oil analysis. www.thesoapkitchen.co.uk

Burnette R (2010) The recent introduction of niger seed (*Guizotia abyssinica*) production in Northern Thailand. ECHO Asia Notes 6

Ilkiliç C, Aydın S, Behcet R, Aydin H (2011) Biodiesel from safflower oil and its application in a diesel engine. Fuel Process Technol 92(3):356–362

Jaikumar S, Bhatti SK, Srinivas V (2018) Effect of catalyst on transesterification of niger seed oil methyl ester. Int J Mech Prod Eng Res Dev 8(3):309–316

Kumari MA, Patnaik KSKR, Rao YR (2014) Production of biodiesel from Pongamia pinnata and *Guizotia abyssinica* seed oil using crystalline manganese carbonate ($MnCO_3$) and nano zinc oxide (ZnO) a green catalysts. J Environ Stud 3(1):17–22

Manjunatha MV, Gowda VTK, Murthy GP (2015) Performance analysis of CI engine using Niger biodiesel blends. Int J Mech Eng 3(10):1–11

Rashid U, Anwar F (2008) Production of biodiesel through base-catalyzed transesterification of safflower oil using an optimized protocol. Energy Fuel 22(2):1306–1312

Sarin R, Sharma M, Sinharay S, Malhothra RK (2007) Jatropha-palm biodiesel blends: an optimum mix for Asia. Fuel 86(10–11):1365–1371

Sarin R, Sharma M, Khan AA (2009) Studies on *Guizotia abyssinica* L. oil: biodiesel synthesis and process optimization. Bioresour Technol 100(18):4187–4192

Sensöz S, Angın D (2008) Pyrolysis of safflower (*Charthamus tinctorius* L.) seed press cake in a fixed-bed reactor: part 2. Structural characterization of pyrolysis bio-oils. Bioresour Technol 99 (13):5498–5504

Usta N (2005) Use of tobacco seed oil methyl ester in a turbocharged indirect injection diesel engine. Biomass Bioenergy 28(1):77–86

Yerranguntia RR, Zubaidha PK, Jakku NR, Kondhare D, Deshmukh S, Saiprakash SP (2013) Production of biodiesel from *Guizotia abyssinica* seed oil using crystalline manganese carbonate $MnCO_3$ a green catalyst. Catal Sust Energy. https://doi.org/10.2478/cse-2012-0003

14 Cornaceae

Cornaceae is a cosmopolitan family of plants in the order Cornales. It contains less than 150 species in two genera. They are mostly trees and shrubs, which may be deciduous or evergreen, although a few species are perennial herbs and rhizomatous half shrubs. The cornelian cherry *Swida wilsoniana* of this family occurs in China and contributes richly for biodiesel production.

14.1 Cornelian Cherry (*Swida wilsoniana*)

The cornelian cherry *Swida wilsoniana* is a tree belonging to the family Cornaceae and occurs exclusively at the landscape of China (Fig. 14.1).

It flourishes at an elevation of 100–1000 m above sea level. This tree prefers a temperature of 18–25 °C with a rainfall of 1000–1570 mm. It continues to bear fruits till its age of 50. Its life span is 200 years. Each tree bears around 50 kg of fruits per year yielding around 15–18 kg of oil. This tree is popularly referred as walteri, walters dogwood, cornus, European comel, Cornelian cherry, dog berries, hound berries and whipple tree. This tree is deciduous, leaves opposite and pinnately veined. It has small flowers surrounded by an involucre of large, white petal like bracts. The fruits are drupes with one or two seeds (Fig. 14.2).

14.1.1 Oil Extraction

The drupe is dried in hot sun for several days and then pressed to release the oil. The crude oil is degummed by hydrated method, discoloured by clay treatment and neutralized using alkali. The moisture content is removed by heating. Chosen properties of the oil are given in Table 14.1.

S. P. Raj et al., *Biodiesel from Flowering Plants*,
https://doi.org/10.1007/978-981-16-4775-8_14

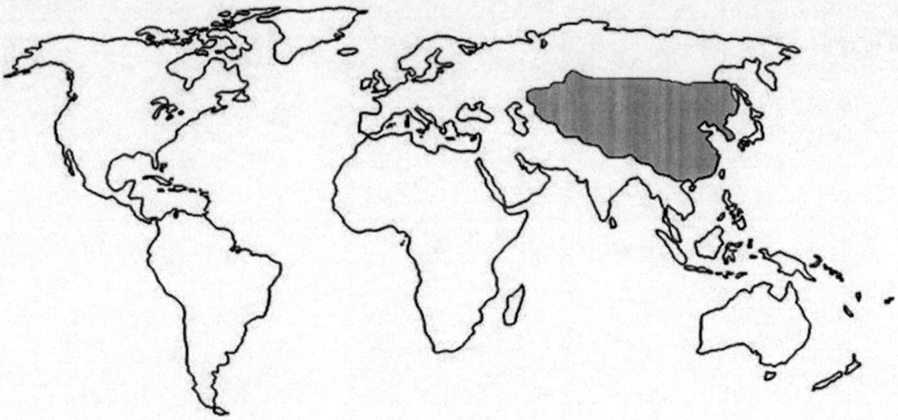

Fig. 14.1 Geographical distribution of *Swida wilsoniana*

Fig. 14.2 The Cornelian cherry *Swida wilsoniana*

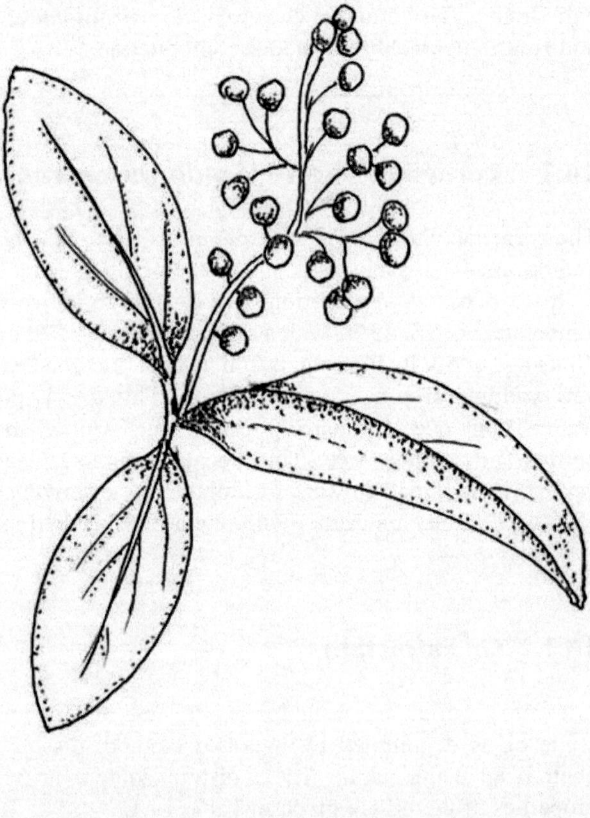

Table 14.1 Properties of *S. wilsoniana* oil (Ding et al. 2010; Zhang and Li 2012)

Parameters	Value
Acid value(g KOH kg^{-1})	0.97
Saponification value(g kg^{-1})	195
Molecular weight	867
Oleic acid (C18:1) (%)	38.3
Linoleic acid (C18:2) (%)	38.85

Table 14.2 Physicochemical characteristics of *Swida wilsoniana* (Li et al. 2015) biodiesel obtained by methoxide process

Parameters	Value
Kinematic viscosity ($mm^2.s^{-1}$)	5.6
Density (kg m^{-3})	891
Flash point (°C)	>105
Cetane number	38–41

14.1.2 Biodiesel Production

The biodiesel is prepared through the routine methoxide process. The catalyst (KOH) is first dissolved in methanol. This is then added to the oil at room temperature and is stirred at 600 rpm for 60 min. The methanol to oil molar ratio is 5:1, the treatment temp. is between 40 and 60 °C and the catalyst conc. is 10 g kg^{-1} of oil. The range comparison of the operation parameters according to the nature of importance is methanol to oil molar ratio > reaction duration > reaction temp. > catalyst dosage. The physicochemical parameters of the biodiesel thus obtained are given in Table 14.2.

The transesterification is also carried out using Lipozyme TLIM in $MgCl_2$ saturated solution (Ding et al. 2010). The enzyme is immobilized on a macroporous ion-exchange resin. The ratio of $MgCl_2$ saturated solution to methanol is 0.3 by volume. The enzyme conc. is 20% of the oil. Molar ratio of methanol to oil is 5:1. The stirring rate is 150 rpm and the reaction duration is 8 h. The yield of the ester is 86.5%. Zhang and Li (2012) used 1-(4-sulphonic acid) butyl-pyridinium hydrosulphate as a catalyst at the rate of 0.06 g L^{-1}. The transesterification yield is between 77 and 94% when the molar to oil ratio is between 12 and 18, respectively. The reaction temp is 160 °C in a high pressure chamber as methanol used to move to a gaseous state at that temp. The reaction duration is 6 h. The catalyst can be used in 6 cycles without much loss in efficiency. The physicochemical characteristics of the biodiesel obtained through 1-(4 sulphonic acid) butyl-pyridinium hydrosulphate are given in Table 14.3.

The viscosity, density and cetane number of the biodiesel obtained using 1-(4 sulphonic acid) butyl-pyridinium hydrosulphate are well compared to that of methoxide method. Besides the oxidation stability and the value for copper corrosion analysis are good.

Table 14.3 Physicochemical characteristics of biodiesel obtained through 1-(4 sulphonic acid) butyl-pyridinium hydrosulphate (Zhang and Li 2012)

Parameters	Value
Kinematic viscosity at 40 °C ($mm^2.s^{-1}$)	3.24
Density at 20 °C (kg m^{-3})	856
Acid value (g KOH kg^{-1})	0.15
Cetane number	49
Copper corrosion test	1a
Oxidation stability 110 °C (h)	7.9

References

Ding R, Zhong S, Li N, Yang J (2010) Transesterification of *Swida wilsoniana* oil with methanol to biodiesel catalyzed by Lipozyme T.L.IM in MgCl2 saturated solution. J Fuel Chem Technol 38 (3):287–291

Li C, Zhang L, Xiao Z, Li P, Liu R, Chen J, He Z, Fu J (2015) Production of biodiesel using a vegetable oil from *Swida wilsoniana* fruits. Period Polytech Chem Eng 59(4):283–287

Zhang A, Li C (2012) *Swida wilsoniana* oil based biodiesel catalyzed by 1-(4 sulfonic acid) butyl pyridinium hydrosulfate. Human Academy of Forestry, Hunan, 410004, China, pp 503–507

Cucurbitaceae

15

Family Cucurbitaceae has 965 species in 95 genera. Most of the plants in this family are annual vines. They are native to temperate and tropical areas. The most recent classification of Cucurbitaceae delineates 15 tribes. The seeds are exalbuminous, flattened, oleaginous with leafy cotyledons and are hitherto rejected for want of definite economical applications. Of late works were initiated for biodiesel production in four species belonging to two tribes, namely Citrullus *colocynthis* and *Cucumis melo* of Benincaseae and *Hodgsonia macrocarpa* and *Luffa cylindrica* of Sicyoeae.

15.1 Egusi (*Citrullus colocynthis*)

The egusi, *Citrullus colocynthis* grows in the coastal sand dunes, brackish zones and hyper arid deserts with less than 50 mm annual rainfall (Qasim et al. 2011). It also grows in inland derelict lands suffering due to paucity of rain. The oil extracted from it is one of the candidates of biodiesel. This species is referred by different names: thumba, bitter apple, handhal, colocynth, bitter cucumber, desert gourd, wild gourd and vine of sodom.

C. colocynthis is a creeper of 1–1.5 m long and its seed sprouts when there is moisture and bears flowers during winter and gives forth fruits during hot summer. The flower is bright yellow and the fruit looks like a watermelon and is the size of an orange (Fig. 15.1).

The fruit is greenish yellow. The leaf has 3–7 lobes in which the centre lobe is of 5–10 cm long. It yields 2–36 fruits and 15–350 g seeds per plant (Menon et al. 2016). The seed holds around 53% oil (Giwa et al. 2014). The seeds are around 6 mm long, 3 mm wide, smooth, compressed and ovoid in shape. The kernel has 52% oil, 28.4% protein, 2.7% fibre, 3.6% ash and 8.2% carbohydrate (Ahmed and Mohamed 2020). This plant is a native of Turkey (Chavan et al. 2014). It is geographically distributed

S. P. Raj et al., *Biodiesel from Flowering Plants*,
https://doi.org/10.1007/978-981-16-4775-8_15

Fig. 15.1 The egusi *Citrullus colocynthis*

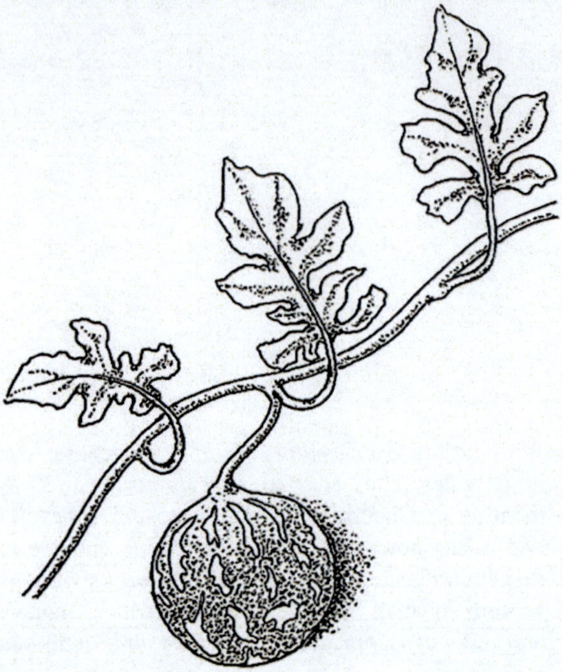

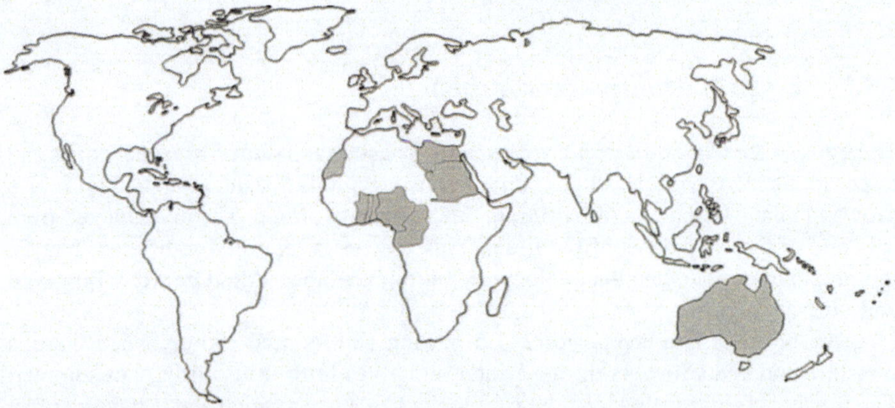

Fig. 15.2 Geographical distribution of *C. colocynthis*

in Nigeria, Ghana, Togo, Benin, Cameroon, Sudan, Egypt, Sahara, Cyprus and Australia (Fig. 15.2).

15.1.1 Extraction of Oil

The seeds are removed from the fruits and dried till it is brittle. It is then powdered and extracted in *n*-hexane for 8 h. The extract obtained is distilled to remove hexane. The residual oil is then heated at 105 °C for 2–3 h to expel the moisture and to ensure a moisture level lower than 0.06%. The oil is filtered subsequently. The physicochemical properties of *C. colocynthis* oil are presented in Table 15.1.

Low acid value indicates that the oil can be transesterified straight using alkali as a catalyst. The kinematic viscosity is high and on par with many other oils. The cetane number is relatively low.

15.1.2 Preparation of Biodiesel

The alkali catalysed transesterification is carried out to prepare biodiesel. The oil and methanol are in the molar ratio of 1:7. The catalyst sodium hydroxide is at 1.2% (w/w) of the oil. The reaction temperature is 65 °C, the stirring rate is 600 rpm and the duration is 60–90 min. At the outset sodium hydroxide is dissolved in methanol to prepare sodium methoxide. The sodium methoxide is then added to the oil present in the reactor and is stirred. On completion of the reaction, the heating and stirring are stopped. The resultant mixture is then transferred to a separating funnel and made to stand overnight, by then two layers are formed. The top layer is the biodiesel which is isolated, and the residual methanol present in it is distilled out at 80 °C using a rotary vacuum evaporator. The resultant material is washed with warm distilled water repeatedly to make it devoid of any alkali. It is then passed over anhydrous sodium sulphate and finally filtered. The properties of *C. colocynthis* oil biodiesel are presented in Table 15.2.

Table 15.1 Physicochemical properties of *C. colocynthis* oil (Chavan et al. 2014; Elsheikh and Akhtar 2014; Giwa et al. 2014)

Parameters	Value
Kinematic viscosity @ 40 °C (cSt)	40.2, 31.5, 35.2
Density (kg m^{-3})	927, 905, 926
Acid value (g KOH kg^{-1})	2.3, 0.98
Flash point (°C)	225, 236
Pour point (°C)	6, 12
Cloud point (°C)	3.5, 6
Calorific value (MJ kg^{-1})	39.6, 39.4
Iodine value (g I_2/100 g)	114.5, 107.4
Free fatty acid (%)	0.5
Molecular weight (g mol^{-1})	874
Cetane number	37.4
Palmitic acid (C16:0) (%)	10, 10.3, 10.5
Stearic acid (C18:0) (%)	7.9, 9.8, 9.6
Oleic acid (C18:1) (%)	18.2, 15.9, 14
Linoleic acid (C18:2)	56.9, 64.0, 64.7

Table 15.2 Properties of *C. colocynthis* oil biodiesel (Chavan et al. 2014; Giwa et al. 2010, 2014)

Parameters	Value
Kinematic viscosity @ 40 °C(cSt)	4.78, 3.83, 3.9
Density (kg m^{-3})	870, 883, 884
Acid value (g KOH kg^{-1})	0.42, 0.2
Iodine value (g I_2/100 g)	113.8
Flash point (°C)	164, 142
Cloud point (°C)	2, 0.5
Cold filter plugging point (°C)	−8, −4
Calorific value (MJ kg^{-1})	40.0, 42.0
Cetane number	41.7, 53.7
Oxidative stability at 110 °C (h)	1.32
Palmitic acid (C16:0) (%)	10.5
Stearic acid (C18:0) (%)	9.7
Oleic acid (C18:1) (%)	18.0
Linoleic acid (C18:2) (%)	61.4

The transesterification reduced the kinematic viscosity and density by 88 and 4%, respectively. Viscosity is considered very important since it governs the fluidity of the fuel. The viscosity of this biodiesel satisfies the standard. The calorific value (40–42 MJ kg^{-1}) is relatively high compared to that of certain other biodiesels, low compared to that of gasoline (46 MJ kg^{-1}) and petro-diesel (43 MJ kg^{-1}) and high compared to that of coal (32–37 MJ kg^{-1}). The acid value is observed to be satisfactory and the low acid value helps to prevent polymerization. The iodine value is 113.8 g I_2/100 g which is within the limit (120 g I_2/100 g) prescribed. The oxidative stability is low due to the presence of high level of polyunsaturated fatty acid (Linoleic acid).

Biodiesel produced from *C. colocynthis* oil by alkali transesterification process with NaOH 0.7%, oil to methanol molar ratio 1:6, reaction temperature 60 °C and stirring rate 650 rpm is analysed by GC MS and presented in Table 15.3.

The *C. colocynthis* oil biodiesel is known to give better performance in internal combustion engines when it is blended with diesel at 1:1 ratio. The blend containing 25% biodiesel and 75% diesel (B 25) expresses better brake thermal efficiency when compared to that of the diesel (Sase et al. 2016). Variation in the torque for different blends of biodiesel (B20, B50 and B80), pure biodiesel and pure diesel is investigated by feeding in an IC engine under various loads. The torque was maximum (44.44 N m) in B20 at 60 kg load in 1100 rpm. This value is higher than that of the pure diesel. Between 1000 and 1500 rpm the torque remains almost constant with speed and beyond which the torque decreases sharply, indicating that further upsurge in speed leads to a reduction in torque.. (Al-Hwaiti et al. 2020).

Table 15.3 GC MS analysis of *Citrullus colocynthis* biodiesel (Ahmed and Mohamed 2020)

S. no.	Chemical compound	Retention time (min)	Area	Area %
1.	Hexadecanoic acid methyl ester	15.298	800,917	1.39
2.	Hexadecanoic acid ethyl ester	15.957	2,297,573	4.00
3.	9, 12-Octadecadienoic acid (Z, Z)-methyl ester	16.945	1,040,495	1.81
4.	9-Octadecenoic acid (Z)-methyl ester	16.990	621,088	1.08
5.	Methyl stearate	17.209	651,528	1.13
6.	Linoleic acid ethyl ester	17.551	9,656,002	16.79
7.	Ethyl oleate	17.590	2,219,349	3.86
8.	Octadecanoic acid ethyl ester	17.807	2,022,783	3.52
9.	Octadecanoic acid 2-hydroxy-1, 3-propanediyl ester	18.744	1,204,736	2.10
10.	9, 12-Octadecadienoic acid (Z, Z)	19.127	203,776	0.35
11.	Eicosanoic acid ethyl ester	19.509	69,241	0.12
12.	Trilinolein	19.817	1,352,118	2.35
13.	12-Methyl-E, E-2, 13-octadecadien-1-ol	19.901	196,759	0.34
14.	9-Octadecenoic acid, 1, 2, 3-propanetriyl ester	20.032	488,253	0.85
15.	9, 12-Octadecadienoic acid methyl ester	20.206	12,975,094	22.56
16.	Glycidol stearate	20.408	1,314,627	2.29
17.	Cyclopropane, 1, 1-dichloro-2, 2, 3, 3-tetramethyl	21.568	14,549,166	25.31
18.	Glycine, *N*-butoxycarbonyl-, isohexyl ester	21.746	1,535,268	2.67
19.	9, 12-Octadecadienoic acid (Z, Z)-2-hydroxy-1-ethyl ester	22.489	2,796,769	4.86
20.	Squalene	22.817	1,507,312	2.62

15.2 Musk Melon (*Cucumis melo*)

The trailing creeper musk melon, *Cucumis melo* is being considered as a source for biodiesel as the oil from the seeds is being tested as a raw material. The musk melon grows at an elevation up to 2200 m and occupies derelict marshy lands, creeks, fallow fields, abandoned home sites, roadsides, medians and trash heaps. The stem of this plant has unbranched tendrils emerging from the base of the petioles. The leaves are simple, alternate and with 3–7 shallow palmate lobes. The flower is solitary, yellow, with five lobed campanulate calyx. The flowers remain open for a single day only. The fruit is a berry and round to ellipsoid. The fruit has its body covered with hairs which later withers so that the mature fruit is relatively smoother. The fruits vary in colour showing shades of yellow, green, orange and rarely white. The exterior of the fruit is either mottled or carries stripes (Fig. 15.3). The seeds are smooth, elliptic and flattened. They germinate epigealy. Melon seeds are rich in

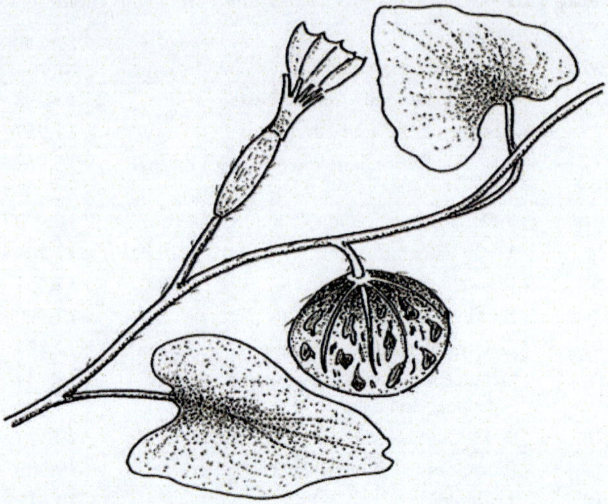

Fig. 15.3 The musk melon *Cucumis melo*

Fig. 15.4 Geographical distribution of *C. melo*

proteins (14.9–27.4%), lipids (25.7–30.8%), fibre (19.0–25.3), carbohydrates (20.8–24.8%) and ashes (3.2–4.8%) (Rabadán et al. 2020).

This plant is distributed in Brazil, Uzbekistan, the USA, Spain and France (Fig. 15.4).

Its common names are melon, musk melon, cantaloupe, honey dew, sweet melon, Queen Anne's melon, Armenian cucumber, wild melon, orange melon, round melon, casaba, winter melon, metao, kharbuz, kharbuza and tian gua.

15.2.1 Oil Production

The seeds are removed from the fruits and cleaned. They are cracked and combed to remove the shell. It is then roasted and pressed (Aktaş et al. 2016). According to Petkova and Antova (2015), the seeds are first air dried for 72 h. after removing them from the fruit. It is powdered and extracted using *n*-hexane in a Soxhlet apparatus for 8 h. The extract is then taken in a rotary vacuum evaporator and the solvent removed. If any solvent exists as residue, it is removed under a stream of nitrogen. The above oil is then filtered using a cheese cloth. The moisture content is removed by heating the oil at 90 °C for 1 h.

15.2.2 Biodiesel Production

At the outset sodium methoxide solution is prepared in a flask by dissolving NaOH in methanol. This mixture is poured into the hot oil (60 °C) and stirred continuously. It is then allowed to settle in a separatory funnel. Once the phase separation is completed the bottom layer containing the glycerin is drained out and the top layer having the biodiesel is washed by bubbling hot water (Aktaş et al. 2016). It is then dried by heating. Methanol to oil ratio of 5.8:1, catalyst (NaOH) conc. of 0.79%, reaction temp. of 55 °C and a reaction duration of 72.5 min at a stirring speed of 720 rpm are found to be optimum. The physicochemical properties of the biodiesel are given in Table 15.4.

The FTIR spectra (Rashid et al. 2011) show bands at 3000–2853 cm^{-1} which correspond to CH_3 stretching vibration (–CO–O–CH_3), 1499 cm^{-1} for CH_3 deformation, 1169 cm^{-1} for CH_3 rocking vibration and 1244–1015 for C–O ester groups indicating the satisfactory conversion of triglyceride to methyl esters. The viscosity is an essential parameter responsible for the smooth ignition and for the formation of engine deposits. The cetane number also agrees with the standard prescribed. The oxidative stability is paradoxically low which needs to be improved by the addition of a suitable antioxidant. The loss in stability is suspected to be due to the nature of the transesterification process. Other parameters such as density, heating value and copper strip corrosion values are within the standard limit.

15.2.3 Engine Performance

The engine performance of the biodiesel is tested by Aktaş et al. (2016). They used a four stroke, air cooled diesel engine with the following specification (Table 15.5).

It is reported that the engine torque in the biodiesel run engine is reduced by 1–6% from that of the diesel when it is run at a speed between 1800 and 3400 rpm. This reduction is due to the relatively low heating value of the biodiesel. Specific fuel consumption increases by 13 kw^{-1} h^{-1} (267 g kw^{-1} h^{-1} in biodiesel against 254 in diesel). Such increase is a compensatory mechanism of the engine in the context of low heating value so as to contribute the same energy needed to manage the load.

Table 15.4 Physicochemical properties of the biodiesel prepared from the oil of *C. melo*

Parameters	Rashid et al. (2011)	Aktaş et al. (2016)	Ameen et al. (2017)
Kinematic viscosity at 40 °C (mm^2s^{-1})	4.75	4.8	5.35
Density at 25 °C (kg m^{-3})	892	887.1[a]	873[a]
Acid value (mg KOH kg^{-1})	0.45	0.07	0.24
Flash point (°C)	148	149	91
Pour point (°C)	−3	–	−13
Cloud point (°C)	1.0	–	−10
Cold filter plugging point (°C)	1.0	−5	–
Cetane number	58.0	55.1	–
Oxidative stability 110 °C (h)	2.65	–	–
Copper strip corrosion (50 °C 3 h)	1a	1	–
Lubricity (HFRR; μm)	136.5	–	–
Heating value (MJ kg^{-1})	45.02[b]	36.96	–
Palmitic acid methyl ester (C16:0) (%)	17.68	–	–
Stearic acid methyl ester (C18:0) (%)	10.84	–	–
Oleic acid methyl ester (C18:1) (%)	21.12	–	–
Linoleic acid methyl ester (C18:2) (%)	50.34	–	–

[a]At 15 °C
[b]Higher heating value

Table 15.5 Specification of the engine used in testing the biodiesel from *C. melo* (Aktaş et al. 2016)

Specification	Value
Number of cylinder	1
Bore	70 mm
Cylinder volume	210 cm^3
Engine power	3.46 kw (3600 rpm)
Compression ratio	18:1
Injection pressure	19 megapascal
Stroke	55 mm

Low heating value is the reason for the low exhaust gas temperature. Low conc. of CO and HC is reported in the exhaust. Consequent to each injection, few drops of fuel is retained in the nozzle which cannot be removed in full. This small amount gets charged in the cylinder during the expansion regime which then burn with inadequate oxygen thereby developing low quantity of CO and HC in the exhaust. Besides, the fuel which gets locked up in the piston ring gap also expresses improper ignition (Heywood 1988). However, the level of CO and HC is relatively lower. The rich oxygen in the fuel causes the generation of NO and NO_2 (NO_x). The generation of NO_x is related to the cylinder-temp., pressure, oxygen level and the combustion duration (Kegl 2006). At high engine speed the level of NO_x also reduces mainly because of the low residence time of the gas in the cylinder.

15.3 Lard Seed (*Hodgsonia macrocarpa*)

The lard seed *Hodgsonia macrocarpa* is a fast growing woody creeper and grows enjoying the support of other plants. This plant grows to a maximum length of 30 m. It is a dioecious plant and hence male and female plants are to be grown parallelly in an area if seeds are to be collected. A single plant bears 40–100 fruits (Meng et al. 2007). The plant comes to bear fruit in 3 years. The fruit (12–25 cm dia.) is a drupe, hard skinned and a depressed globose (Fig. 15.5). Each plant may yield around 15 kg of non-drying oil per year. The leaves are three lobed. The flower comes to bloom at night and the fruit setting there upon strongly depends on the climatic conditions. The life span of the plant is 70 years. Each fruit has 6–12 seeds and each seed weighs 50–70 g.

H. macrocarpa is called by few local names: pig fruit, ribbed melon, lard, kadam seed, kapayang and kadamzaad.

This plant establishes itself at river-sides, low lands and foot hills of 50–1000 m altitude. They are geographically distributed in Thailand, Malaysia, Sumatra, Java, part of China, Borneo and Myanmar (Fig. 15.6).

Fig. 15.5 The lard seed *Hodgsonia macrocarpa*

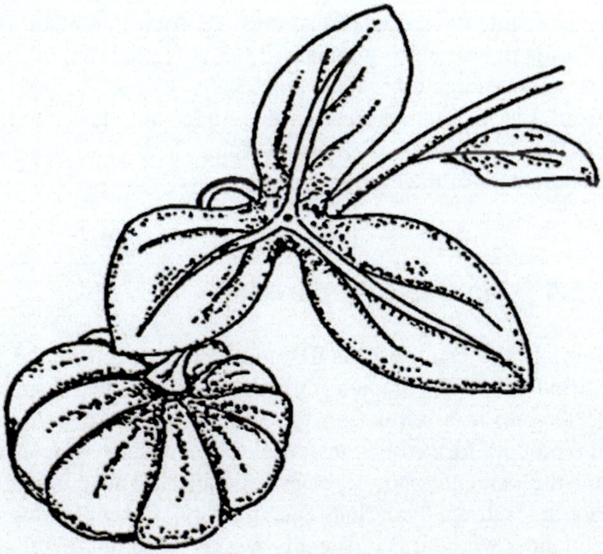

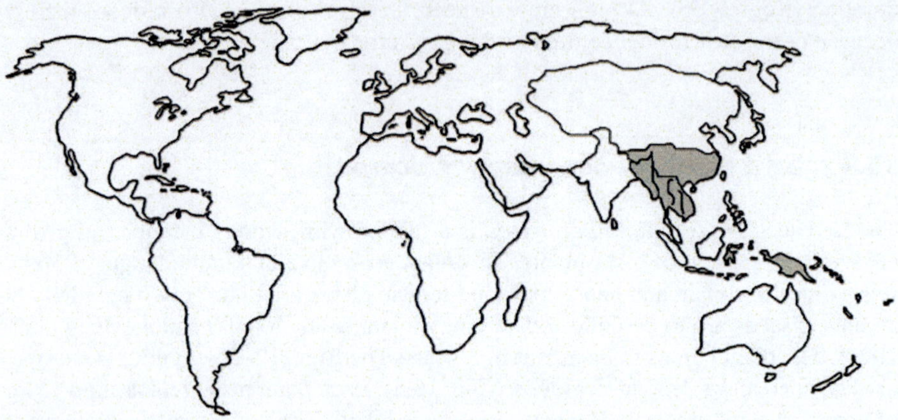

Fig. 15.6 The geographical distribution of *H. macrocarpa*

Table 15.6 The fatty acid profile of the crude oil of *H. macrocarpa*

Fatty acids (%)	Meng et al. (2007)	Cao and Zhang (2015)
Palmitic acid (C16:0)	33.5	28.45
Stearic acid (C18:0)	7.1	7.87
Oleic acid (C18:1)	19.2	23.69
Linoleic acid (C18:2)	43.9	39.36
Linolenic acid (C18:3	<0.5	0.41

15.3.1 Oil Extraction

At the outset, the seeds from the drupe are plugged out. These seeds are then stripped to extract the kernels. The kernels are dried in hot sun repeatedly. The dried kernel mass is powdered to granules of 0.3 mm size. The oil from the granule is extracted using petroleum ether (Cao and Zhang 2015). The petroleum ether is then expelled from the extract to retain the crude oil. The fatty acid composition of the *H. macrocarpa* oil is given in Table 15.6. The unsaturated fatty acids form the major share of the total fatty acids.

15.4 Biodiesel Preparation

Initially the catalyst (NaOH) and methanol are mixed separately in a beaker and stirred till it is completely dissolved. It is then poured over the oil and stirred (500 rpm) in a reflux unit for 120 min at 60 °C. When the reaction is completed the product formed is transferred to a separatory funnel where it is allowed to stand till the contents show a clear separation. The glycerol which configurates at the bottom is drained out first. Then the top layer containing the biodiesel is drained in a container where it is repeatedly washed with hot distilled water till the pH of the oil

Table 15.7 The physicochemical properties of *H. macrocarpa* biodiesel (Cao and Zhang 2015)

Parameters	
Kinematic viscosity @ 40 °C (mm^2s^{-1})	4.1
Density @ 15 °C (kg m^{-3})	883
Acid value (g KOH kg^{-1})	0.35
Flash point (°C)	153
Pour point (°C)	−9
Cold filter plugging point (°C)	−7
Cetane number	51.2
Oxidative stability 110 °C (h)	5.4
Copper strip corrosion 50 °C, 3 h	1

Fig. 15.7 Geographical distribution of *Luffa cylindrica*

becomes neutral. It is then centrifuged and dried with anhydrous sodium sulphate. In this method it has become possible to get 95% biodiesel at an alcohol to oil molar ratio of 8:1 and at a catalyst conc. of 1%. The physicochemical properties of the biodiesel obtained from the oil of *H. macrocarpa* are given in Table 15.7.

It is to be seen that the values for viscosity, density, acid value, flash point, copper strip corrosion and cetane number are within ASTM D6751 and EN 14214 standards. Besides it has good cold flow properties.

With reference to the oxidative stability, it is known that it needs to be improved marginally to reach the standard of EN 14214. Such limitation in the stability is due to the high content of linoleic acid in the oil. The oxidative stability is improved by the addition of (400 ppm) tertiary butyl hydroquinone (Moser and Vaughn 2010).

15.5 Loofah (*Luffa cylindrica*)

Loofah, *Luffa cylindrica* is a herbaceous plant of tropical area and is considered to be the native of India which later naturalized to Eastern Africa (Fig. 15.7).

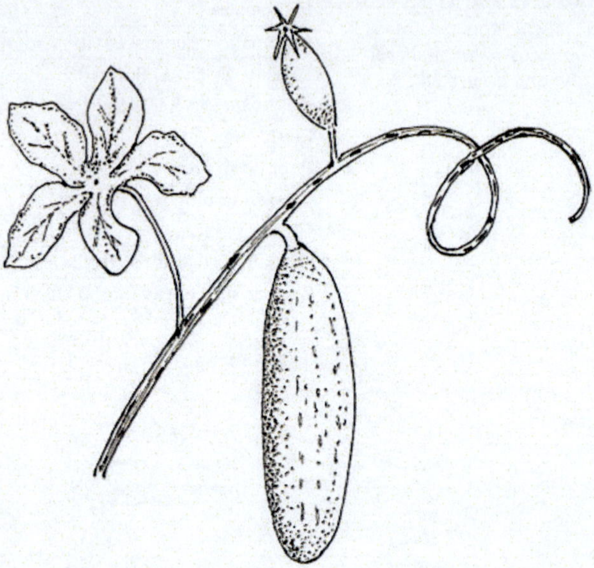

Fig. 15.8 The loofah Luffa *cylindrica*

It is an annual fast growing monoecious creeper which grows to a maximum length of 10 m and climbs with a support. The leaf is 7–20 cm dia., lobed and fringed. It has a tendency to invade virgin areas. It also occupies abandoned desolate buildings, and grows on fences. The flower is bright yellow. The fruit is cylindrical in shape and makes up to a length of 60 cm. It is pale green and becomes brown when mature. Often the fruits are harvested at tender stage and are used as vegetables (Fig. 15.8).

The ripe fruit contains around 30 oleaginous seeds. These seeds are invariably discarded as there is no tangible use hitherto. As these seeds are reported to have around 32% oil (Ndukwe et al. 2001) it is largely opined that the oil from these seeds are eligible to be converted to biodiesel (Ajiwe et al. 2005; Yusuf and Sirajo 2009; Bamgboye and Oniya 2012; Oniya and Bamgboye 2012). This plant is popularly referred as loofah, smooth luffa, sponge luffa, climbing okra, dishcloth gourd and Chinese okra, wash sponge, gourd towel and galka.

15.5.1 Oil Extraction

The hull of the seeds is removed easily as the seeds undergo dehydration. Therefore, seeds are initially sun dried for a week until the seed coat splits and the kernel exposed. The seed coat is then separated by winnowing. The kernels are again dried and crushed to have them as small particles. These crushed mass is then extracted using *n*-hexane in a Soxhlet apparatus for 5–10 h (Adewuyi et al. 2012; Ibeto et al. 2012). The extractant (*n*-hexane) of the mixture obtained from the Soxhlet apparatus is expelled by heating in a rotary evaporator. The physicochemical properties of the

Table 15.8 The physicochemical properties of *L. cylindrica* oil

Parameters	Adewuyi et al. (2012)	Ibeto et al. (2012)	Udo et al. (2020)
Kinematic viscosity (mm^2s^{-1})	–	0.92	4.9
Specific gravity	–	0.93	0.92
Acid value (g KOH kg^{-1})	–	2.47	20.62
Flash point (°C)	–	338	134
Iodine value (gI_2/100 g)	108	66.37	130
Free fatty acid (%)	5.1	1.23	–
Palmitic acid (C16:0) (%)	17.3	–	–
Stearic acid (C18:0) (%)	9.0	–	–
Oleic acid (C18:1) (%)	26.0	–	–
Linoleic acid (C18:2) (%)	46.8	–	–

above oil are given in Table 15.8. Iodine value is an index of stability of the oil. Besides the iodine value gives inference to the drying feature of the oil. Any value above 125 indicates that the oil is prone to dry easily. If the value is between 110 and 140 that oil is indicated as semidrying and when the oil is less than 110 that is in the non-drying category. The oil susceptible to drying is prone to be rancid and such oil cannot be stored long. Characteristically this oil has more unsaturated fatty acids. The flash point of this oil is adequately high ensuring high safety.

15.5.2 Biodiesel Preparation

The free fatty acid content of the oil is quite high which implies that this oil is to be passed through double stage process. In the first stage the level of free fatty acid is reduced through esterification process. In esterification, sulphuric acid (2% of the oil) methanol (98 mL) mixture is prepared, added to the hot oil and stirred for 2 h. On completion of the esterification, the whole content is cooled and phase separated in a separatory funnel. The esterified oil is washed repeatedly using distilled water to free it from acid and then dried by gentle heating. Subsequently in the second stage the transesterification process is carried out. Accordingly the esterified oil is stirred (600 rpm) with alkali methanol mixture (1% KOH in methanol by wt.) for 2 h at 70 °C. It is then transferred to a separatory funnel and kept waiting for the phase separation to complete. The overlying material (biodiesel) is separated and washed repeatedly with warm water and dried over sodium sulphate. The physicochemical properties of the biodiesel obtained are presented in Table 15.9.

The oxidative stability of this biodiesel is paradoxically low due to the presence of rich quantity of polyunsaturated esters. A minimum of 6 h Rancimat induction period is the standard norm. The auto-oxidation which takes place in such oils largely depends on the number and position of the double bond in the fatty acid. The limit of 6 h of oxidative stability refers to the degradation of methyl ester at 110 °C in a stable air stream as known by the formation of volatile acids.

Table 15.9 Physicochemical properties of *L. cylindrica* biodiesel (Adewuyi et al. 2012)

Parameters	Adewuyi et al. (2012)	Udo et al. (2020)
Kinematic viscosity at 40 °C mm^2s^{-1})	4.1	3.237
Density at 15 °C (kg m^{-3})	0.89	0.875
Acid value (gKOH/kg^{-1})	0.06	1.48
Iodine value (g I_2/100 g)	108.2	106
Pour point (°C)	0.4	–
Copper corrosion test	1a	–
Oxidative stability 110 °C (h)	0.58	–
Heating value (MJ kg^{-1})	39.0	–

Table 15.10 GC MS analysis of *L. cylindrical* oil

Peak	Retention time	Chemical compound	Formula	Area (%)	Weight (g mol^{-1})
1	30.2893	Acridine,9-anilino-	$C_{19}H_{14}N_2$	12.6811	270.3
2	33.899	11Octadecenoicacid,methyl ester	$C_{19}H_{36}O_2$	63.8748	296.48
3	34.2944	Methyl stearate	$C_{19}H_{38}O_2$	23.4441	298.5

Table 15.11 GC MS analysis of *L. cylindrical* oil biodiesel

Peak	Retention time	Chemical compound	Formula	Area (%)	Weight (g mol^{-1})
1	30.4064	Benz(a)anthracene-6,7,12-trimethyl	$C_{21}H_{18}$	16.5618	270.4
2	46.026	15-Octadecenoic acid, methyl ester	$C_{19}H_{36}O_2$	34.0161	296.5
3	34.4774	Methyl stearate	$C_{19}H_{38}O_2$	34.1791	296.4879
4	37.8894	Eicosanoic acid, methyl ester	$C_{21}H_{42}O_2$	1.56	326.56
5	41.6602	Serine methyl ester-*N*-[2-oxo4-phenylbutyryl]-	$C_4H_9NO_3$	1.6732	119.12

Udo et al. (2020) made GC MS analysis of oil and biodiesel and the data obtained are presented in Tables 15.10 and 15.11, respectively.

References

Adewuyi A, Oderinde RA, Rao BVSK, Prasad RBN, Anjaneyulu B (2012) *Blighia unijugata* and *Luffa cylindrical* seed oils: renewable sources of energy for sustainable development in rural Africa. Bioenergy Res 5:713–718

Ahmed FBM, Mohamed SFEA (2020) Production of biodiesel from *Citrullus colocynthis* seeds oil. J Energy Environ Chem Eng 5(2):14–20

Ajiwe VIE, Ndukwe GI, Anyadiegwu IE (2005) Vegetable diesel fuels from *Luffa cylindrica* oil, its methylester and ester-diesel blends. Chemclass J 2:1–4

Aktaş A, Sekmen Y, Sekmen P (2016) Biodiesel production from waste melon seeds and using it as alternative fuel in direct injection diesel engine. J Energy Inst 83(2):69–74

Al-Hwaiti MS, Alsbou EM, Al Haddad RM, Osman AI, Jrai AA, Al-Muhtaseb AH, Hasan AO, Morgan K, El-Sayed E-SM, Al-Fatesh AS, Fakeeha AH, Rooney DW, Al-Rawashdeh HA (2020) Spatio-temporal analyses of extracted *Citrullus colocynthis* seeds (Handal seed oil) as biofuel in internal combustion engine. Renew Energy 166:234–244

Ameen M, Zafar M, Ahmad M, Shaheen A, Yaseen G (2017) Wild melon: a novel non-edible feedstock for bioenergy. Pet Sci 15:405–411. https://doi.org/10.1007/s12182-018-0227-0

Bamgboye IA, Oniya OO (2012) Fuel properties of loofah (*Luffa cylindrica* L.) biofuel blended with diesel. Afr J Environ Sci Technol 6(9):346–352

Cao L, Zhang S (2015) Production and characterization of biodiesel derived from *Hodgsonia macrocarpa* seed oil. Appl Energy 146:135–140

Chavan SB, Kumbhar RR, Sharma YC (2014) Transesterification of *Citrullus colocynthsis* (Thumba) oil: optimization for biodiesel production. Adv Appl Sci Res 5(3):10–20

Elsheikh YA, Akhtar FH (2014) Biodiesel from *Citrullus colocynthis* oil: sulfonic-ionic liquid-catalysed esterification of a two step process. Sci World J 2014:1–11

Giwa S, Abdullah LC, Adam NM (2010) Investigating egusi (*Citrullus colocynthis* L.) seed oil as potential biodiesel feed stock. Energies 3:607–618

Giwa S, Chuah LA, Adam NM (2014) Fuel properties and rheological behaviour of biodiesel from egusi (*Colocynthis citrullus* L.) seed kernel oil. Fuel Process Technol 122:42–48

Heywood JB (1988) Internal combustion engine fundamentals. McGraw-Hill Book Company, pp 620–648

Ibeto CN, Okoye COB, Ofoefule AU (2012) Comparative study of the physicochemical characterization of some oils as potential feedstock for biodiesel production. Int Sch Res Netw (Renew Energy) 2012:621518. https://doi.org/10.5402/2012/621518

Kegl B (2006) Experimental Investigation of Optimal Timing of the diesel engine injection pump using biodiesel fuel. Energy Fuels 20:1460–1470

Meng B, Long CL, Cheng ZY, Liu Y (2007) Development strategy for a biodiesel plant, large fruited Hodgsonia. North Hortic 10:51–54

Menon K, Sood N, Rao NK (2016) Study of morpho-agronomic diversity and oil content in desert gourd (*Citrullus colocynthis* (L.) Schrad). Aust J Crop Sci 10(7):1000–1006

Moser BR, Vaughn SF (2010) Evaluation of alkyl esters from *Camelia sativa* oil as biodiesel as a blend components in ultra-low sulfur diesel fuel. Bioresour Technol 101:646–653

Ndukwe GI, Amupitan JO, Badmus NK (2001) Preliminary investigation on the antibiotic and fungicidal activities of the extracts from the seeds of *Luffa cylindrica*. J Eng Technol 8 (2):3202–3207

Oniya OO, Bamgboye IA (2012) Loofah (*Luffa cylindrica,* l.) biodiesel as fuel for diesel engine. J Eng Appl Sci 4:8–18

Petkova Z, Antova G (2015) Proximate composition of seeds and seed oils from melon (*Cucumis melo* L.) cultivated in Bulgaria. Cogent Food Agric 1:1018779

Qasim M, Khan MA, Gulzar S (2011) Halophytes as medicinal plants. In: Ozturk M, Mermut AR, Celik A (eds) Urbanization, land use, land degradation and environment. Daya Publishing House, Turkey

Rabadán A, Nunes MA, Bessada SMF, Pardo JE, Beatriz M, Oliveira PP, Alvarez-Ortí M (2020) From by-product to the food chain: melon (*Cucumis melo* L.) seeds as potential source for oils. Foods 9:1–11

Rashid U, Rahman HA, Hussain I, Ibrahim M, Haider MS (2011) Musk melon (*Cucumis melo*) seed oil: a potential non-food oil source for biodiesel production. Energy 36:5632–5939

Sase A, Ardhapurkar PM, Bobade SN (2016) Biodiesel from *Citrullus colocynthis* oil: performance testing and combustion characteristics in IC engine. Int Res J Eng Technol 3(12):518–524

Udo GJ, Etesin UM, Awaka-Ama JJ, Nyong AE, Uwanta EJ (2020) GCMS and FTIR spectroscopy characterization of *Luffa Cylindrica* seed oil and biodiesel produced from the oil. Commun Phys Sci 5(3):378–390

Yusuf N, Sirajo M (2009) An experimental study of biodiesel synthesis from groundnut oil. Part I: synthesis of biodiesel from groundnut oil under varying operating conditions. Aust J Basic Appl Sci 3(3):1623–1629. INSI Net Publication

Cyperaceae

16

The plants of the sedge family *Cyperaceae* are mostly perennial herbs, rarely annual known to perennate the intermittent seasons through creeping rhizomes or tubers. Around 4000 plant species constitute this family. They inhabit damp places. Their underground rhizomes, tubers or stolons and aerial shoots are smooth, solid, glaucous or glabrous, without distinction into nodes and internodes. These plants are seen distributed throughout the world but most abundantly occur in temperate zones. The tubers yield 25–30% oil which is considered as a raw material for biodiesel.

16.1 Tiger Nut (*Cyperus esculentus*)

The tiger nut *Cyperus esculentus* plant of the family *Cyperaceae* contributes raw material for biodiesel production. Its tuber contains reasonably good quantity of oil. It is a slender annual or perennial plant and grows to a maximum height of 90 cm with a solitary stem standing straight over the underground tuber. This plant propagates by its seeds or rhizomes or tubers. Once the plant establishes in a location, it spreads tenaciously. The stem is triangular (in section), and the leaves are slender and narrow (3–10 mm width). The fibrous roots bear small hard spherical tubers. Each plant produces hundreds and thousands of tubers of 3–20 mm dia. each (Fig. 16.1). It requires mild climate. The tubers lay position at 30 cm depth. It prefers sandy loam at pH 5–7.5. These plants grow densely at wetlands. The tuber yield is around 10–20 t ha^{-1} per year.

This plant is distributed in California, Guatemala, India, Mexico, Chile, Brazil, Lebanon, Syria, Jordan, Saudi Arabia, Oman, Iran, Iraq, Pakistan, Yemen, Morocco, Ivory coast, Sudan, Gambia, Guinea, Ghana, Nigeria and Cameroon (Fig. 16.2).

It is locally called as chufa sedge, nut grass, yellow nutsedge, tiger nutsedge, edible galingale, water grass, earth almond, earth nut, rush nut and zulu nut.

S. P. Raj et al., *Biodiesel from Flowering Plants*,
https://doi.org/10.1007/978-981-16-4775-8_16

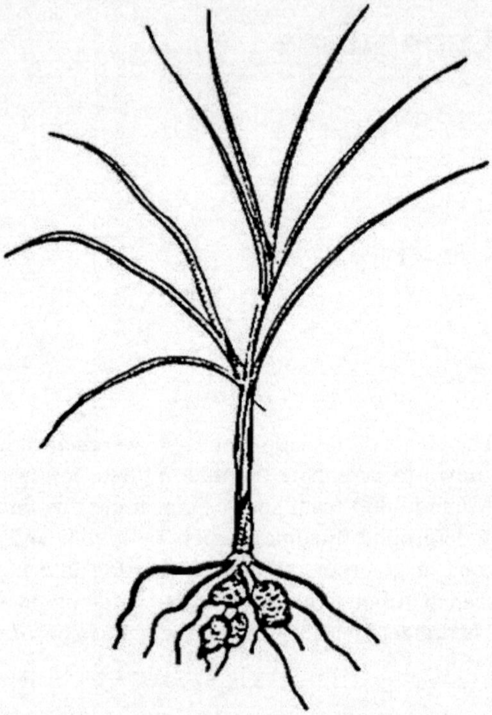

Fig. 16.1 The tiger nut
Cyperus esculentus

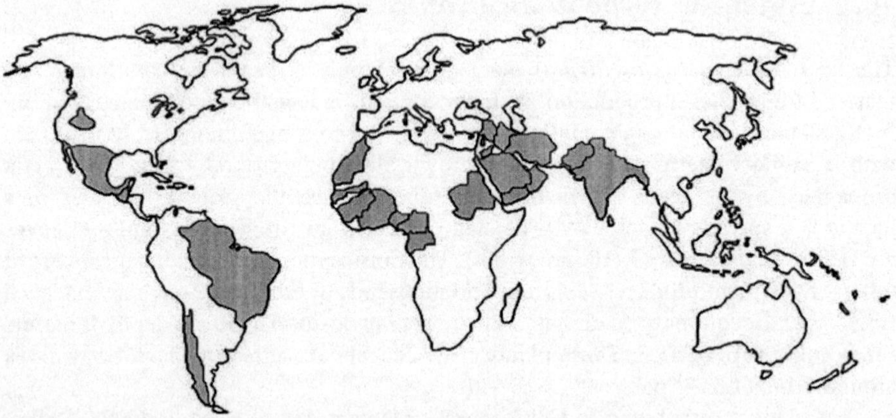

Fig. 16.2 Geographical distribution of *C. esculentus*

16.1.1 Harvest and Processing of Tubers

When the leaves are scorched it is time to remove the tubers. The plant is pulled out along with the tuber. The tuber is isolated washed and sliced. The slices are spread in hot sun for drying. It takes 2–3 months for a complete drying so as to have less than

12% moisture content. While drying the slices are to be turned twice a day to ensure uniform drying and to prevent bacterial infestation. This dried tuber contains around 20% oil. These slices are ground to powder and sieved through a 600-μm mesh.

16.1.2 Oil Extraction

Tiger nut powder is extracted using *n*-hexane (Aremu et al. 2016). The extract containing oil and other impurities is treated with sodium sulphate so as to remove the moisture. It is subsequently distilled to expel the *n*-hexane. Instead of *n*-hexane, petroleum ether (40–60 °C) also is used to extract (12 h) the oil from the dried powder (Barminas et al. 2001; Ofoefule et al. 2013; Ezeh et al. 2014).

The recovery of oil ranges from 15.9% to 41.2% (Lasekan and Abdulkarim 2012; Yeboah et al. 2012; Ofoefule et al. 2013). Supercritical carbon dioxide was also tried which gave 26.3% oil (Ezeh et al. 2014).

16.1.3 Physicochemical Properties of the Oil

The physicochemical properties of the crude oil are given in Table 16.1. The viscosity of the oil is higher several fold than petro-diesel. The energy content is

Table 16.1 Physicochemical properties of *C. esculentus* oil

Parameters	Zhang et al. (1996)	Barminas et al. (2001)	Aremu et al. (2016)	Ofoefule et al. (2013)
Kinematic viscosity at 40 °C ($mm^2\ s^{-1}$)	43.4	–	–	–
Density at 15 °C ($kg\ m^{-3}$)	–	896	–	–
Flash point (°C)	–	228	282	–
Free fatty acid (%)	–	0.85	–	4.49
Iodine value (g I_2/100 g)	113.3	88.46	57.3	143.4
Acid value (g KOH kg^{-1})	–	–	9.12	8.97
Heating value (MJ kg^{-1})	39.2	40.7	–	–
Cetane number	41.2	–	–	–
Cloud point (°C)	−0.3	–	–	–
Pour point (°C)	−9.0	–	–	–
Palmitic acid (C16:0) (%)	14.99	–	12.96	10.21
Stearic acid (C18:0) (%)	2.56	–	4.35	1.47
Arachidic acid (C20:0) (%)	–	–	2.34	0.32
Behenic acid (22:0) (%)	–	–	3.68	–
Oleic acid (C18:1) (%)	69.32	–	50.84	75.72
Linoleic acid (C18:2)	13.11	–	24.08	11.64

lower than diesel by 10–13%. Cloud point and pour point make deviation from the standard. The iodine value is within the recommended range.

16.1.4 Production of Biodiesel

Potassium hydroxide (23 g) is dissolved in a litre of methanol so as to form potassium methoxide. This makes the ultimate concentration of 0.65% in the reaction mixture (Ofoefule et al. 2013). The whole methoxide is emptied in 4 L of oil and heated to 60 °C at constant stirring and the reaction is allowed to progress for 1 h. At the end, the product is transferred to a separatory funnel and kept undisturbed for a period of 24 h. The layer at the top (biodiesel) is separated, washed five times in warm distilled water and heated to dry.

16.1.5 The Physicochemical Properties of Biodiesel

The physicochemical properties of the biodiesel are furnished in Table 16.2.

The viscosity is higher than the ASTM standard which warrants blending with petro-diesel. The iodine value is within the standard. The flash point is observed to be favourable. The ash content is well above the ASTM standard. High ash content is an index of the presence of mineral which is likely to create excessive sulphur and nitrogen in the exhaust.

Table 16.2 Physicochemical properties of *C. esculentus* biodiesel (Ofoefule et al. 2013)

Parameters	Value
Kinetic viscosity ($mm^2 s^{-1}$)	8.08
Acid value (g KOH kg^{-1})	1.122
Iodine value (g I_2/100 g)	98.4
Free fatty acid (%)	0.56
Flash point (°C)	178
Cloud point (°C)	13
Pour point (°C)	−3
Ash content (%)	1.13

References

Aremu MO, Ibrahim H, Aremu SO (2016) Lipid composition of black variety of raw and boiled tiger nut (*Cyperus esculentus* L.) grown in North-East Nigeria. Pak J Nutr 15(5):427–438

Barminas JT, Maina HM, Tahir S, Kubmarawa D, Tsware K (2001) A preliminary investigation into the biofuel characteristics of tiger nut (*Cyperus esculentus*) oil. Bioresour Technol 79:87–89

Ezeh O, Gordon MH, Niranjan K (2014) Tiger nut oil (*Cyperus esculentus* L): a review of its composition and physicochemical properties. Eur J Lipid Sci Technol 116(7):783–794

Lasekan O, Abdulkarim SM (2012) Extraction of oil from tiger nut (*Cyperus esculentus* L) with supercritical carbon dioxide (SC-CO_2). LWT Food Sci Technol 47:287–292

Ofoefule AU, Libeto CN, Okoro UC, Onukwuli OD (2013) Biodiesel production from tiger nut (*Cyperus esculentus*) oil and characterization of its blend with petro-diesel. Phys Rev Res Int 3 (2):145–153

Yeboah SO, Mitei YC, Ngila JC, Wessjohann L, Schmidt J (2012) Compositional and structural studies of the oils from two edible seeds: tiger nut, *Cyperus esculentus* and asiato *Pachira insignis* from Ghana. Food Res Int 47:259–266

Zhang HY, Hanna MH, Ali Y, Nan L (1996) Yellow nut-sedge(*Cyperus esculentus* L.) tuber oil as a fuel. Ind Crops Prod 5:177–181

Dipterocarpaceae

17

The dipterocarpaceae is a family of around 700 known species. They are mostly trees of giant size (rarely shrubs). They occur mainly in tropical low land and in rainforest. They also discontinuously form lushy patches of forest in Africa, Ceylon, Western and Eastern Indian Peninsula, Asia, Malaysia, and South America. Few publications are hitherto made on sal tree *Shorea robusta* whose seeds yield oil for use in biodiesel preparation.

17.1 Sal Tree (*Shorea robusta*)

The sal tree *Shorea robusta* is a species exclusive to India, Myanmar and Nepal and covers approximately 5% of the forest area (Vedaraman et al. 2012). It grows to a maximum height of 35 m and a diameter of 2.5 m. It is a deciduous tree belonging to the family Dipterocarpaceae and is seen distributed at a latitude from 20 to 32°N and longitude from 75 to 95°E. The bole is straight and round. The leaf is broad and oval and is 10–25 long. The leaves, when formed, are red in colour which slowly turn to green as they mature. Flowers are yellowish white, arranged in terminal or axillary racemose panicles. The fruit is 1.3–1.5 cm long and around 1 cm in diameter. The fruits appear once in 2 or 3 years (Fig. 17.1). The geographical distribution of this tree is delineated in Fig. 17.2. This species is referred by different vernacular names: sal, sal tree, sakhua, shala tree (English), sāla, dammar de Ílnde (French), salboom (Dutch), salbaum (in German), sala (Vietnamese), can (Russian), sal, sala agnivallabha, karimaruthu, kungiliam, maramaram, sajara, guggilu, attam, asina, asu and hal (Indian).

S. P. Raj et al., *Biodiesel from Flowering Plants*,
https://doi.org/10.1007/978-981-16-4775-8_17

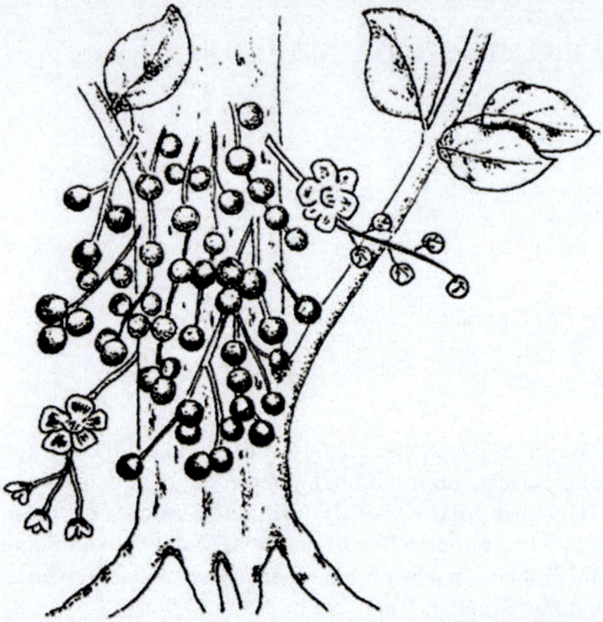

Fig. 17.1 The plant *Shorea robusta*

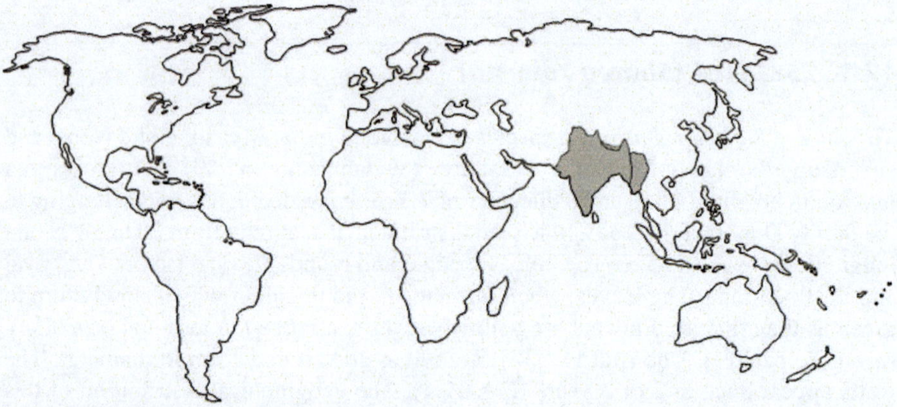

Fig. 17.2 Geographical distribution of *S. robusta*

17.1.1 Oil Preparation

The seed has around 19–20% oil (butter) which is stored in the cotyledons and the cotyledons constitute around 70% by weight of the seeds. It is reported that India has a potential of around 180,000 tons of seeds in a year (Pali and Kumar 2014). As the seed is embedded in winged fruit it is dried and dewinged. The dewinging is done by beating the seeds with a wooden stick. The brittle wings are broken and the round seeds are separated. The round seeds thus obtained are passed through a roller.

Table 17.1 Physicochemical properties of *S. robusta* oil

Parameters	Vedaraman et al. (2012)	Pali and Kumar (2014)	Hajra et al. (2015a)
Kinematic viscosity at 40 °C (cSt)	–	157	45
Density at 15 °C (kg m^{-3})	879	899	879
Acid value (g KOH kg^{-1})	0.8	–	1.2
Flash point (°C)	–	–	238
Pour point (°C)	–	–	38
Iodine number (g I_2/100 g)	39.6	–	49
Heating value (MJ kg^{-1})	–	39.8	–
Palmitic acid (C16:0) (%)	4.8	–	–
Stearic acid (C18:0) (%)	44.2	–	–
Arachidic acid (C20:0) (%)	6.1	–	–
Oleic acid (C18:1) (%)	42.4	–	–
Linoleic acid (C18:2) (%)	2.5	–	–

Under pressure, the seeds break and the kernels come out. The oil is extracted by water rendering or through expeller or by solvent extraction (*n*-hexane at 60 °C) the properties are given in Table 17.1.

The kinematic viscosity and flash point are relatively high. The refined sal oil is directly transesterified by alkali catalysts since the acid value and the corresponding free fatty acid are low. Before transesterification, the removal of moisture is ensured by passing the oil repeatedly through Na_2SO_4 (Hajra et al. 2015b). To 1 L of oil 1500 mL of methanol and 2.25 g anhydrous sodium hydroxide are needed. At the outset, sodium methoxide solution is prepared by dissolving sodium hydroxide in methanol. The methoxide thus prepared is then mixed with oil and stirred at 600 rpm for 90 min at 65 °C. The contents, on completion of the reaction are transferred to a separating funnel and allowed to stand undistributed overnight, by then the phase separation completes. A viscous glycerol layer moves to the bottom and it is discarded. The top layer containing the biodiesel is collected and distilled (at 70 °C in 1500 mmHg vacuum) to remove the methanol. A trace amount of alkali if present is washed with dilute acetic acid and at the end, the pH of the product is checked for neutrality. It is then washed with sodium chloride (10%) so as to remove glycerin or unreacted triglycerides. Finally, it is rinsed with distilled water and then heated to remove moisture if any. Hajra et al. (2015a) used acidic ion-exchange resin, namely INDION 225H which easily manages high temperature, high pressure and high free fatty acid. According to them the role played by various factors in the transesterification process is in the order: reaction temperature > reaction duration > methanol to oil molar ratio > catalyst concentration. The share of activity of the above parameters is 53%, 32%, 11% and 4%, respectively (Hajra et al. 2015a) (Fig. 17.3). The contribution of the interaction of factors, as reported by Hajra et al. (2015a) indicates that the interaction between the reaction duration and temperature plays a major role (58.2%). The interaction of catalyst with reaction duration

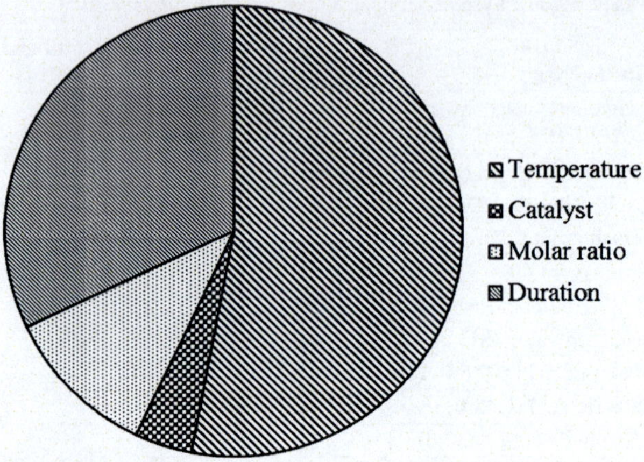

Fig. 17.3 The role of individual factors in the transesterification process

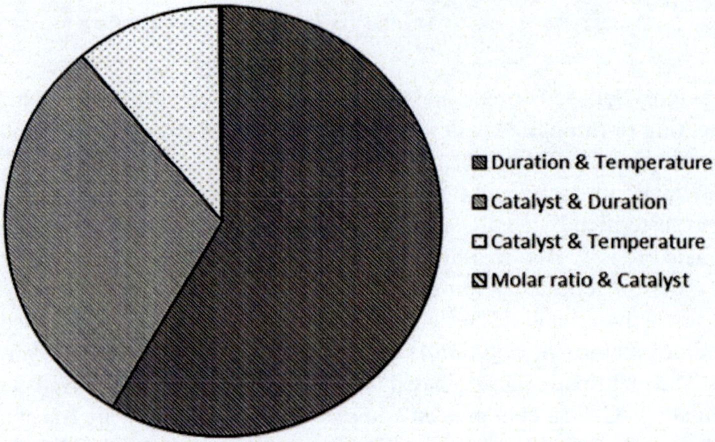

Fig. 17.4 The role of interacting factors on the transesterification

(30.3%) and that of the catalyst with temperature (10.7%) plays at least role (Fig. 17.4). Thus the three major interactors are duration and temperature, catalyst and duration and catalyst and temperature. The physicochemical properties of biodiesel are given in Table 17.2.

The viscosity of the above biodiesel is higher than that of the neat diesel although the values obtained are within the standard. Similarly, the density also is slightly higher than the diesel, but within the acceptable limit. The heating value of biodiesel is low by 5 MJ kg^{-1} which is around 11% lower than that of the neat diesel. This is due to the presence of oxygen in the biodiesel molecule to the tune of 11% which apparently does not contribute any energy. This biodiesel has a rich quantity of saturated fatty acid ester which naturally impairs the cold filter plugging point and pour point. However, saturation has the benefit of enhancing oxidation stability.

Table 17.2 Physicochemical properties of *S. robusta* biodiesel

Parameters	Vedaraman et al. (2012)	Pali and Kumar (2014)	Hajra et al. (2015a)	Pali et al. (2015)
Kinematic viscosity at 40 °C (cSt)	5.0	5.9	4.9	5.9
Density at 15 °C (kg m^{-3})	875	884	874	877
Acid value (g KOH kg^{-1})	–	–	0.2	–
Flash point (°C)	160	–	160	127
Pour point (°C)	18	–	18	–
Cetane number	52	–	53	–
Heating value (MJ kg^{-1})	40	40.3	39.9	39.7
Oxidative stability 110 °C (h)	–	>6	–	<6
Iodine number (g I_2/100 g)	–	–	52	–
Cold filter plugging point (°C)	–	–	–	12
Palmitic acid (C16:0) (%)	–	–	–	3.7
Stearic acid (C18:0) (%)				47.0
Arachidic acid (C20:0) (%)	–	–	–	3.5
Oleic acid (C18:1) (%)	–	–	–	44
Linoleic acid (C18:2) (%)	–	–	–	1.2

17.1.2 Engine Performance

The brake specific fuel consumption is high on two counts, namely high density and low heating value. The average density of biodiesel in the present case is 877.5 kg m^{-3} against the neat diesel (824 kg m^{-3}). The difference is 53.5 kg m^{-3}. On volume basis, in every litre therein there is 53.5 g increase in weight. This increase in density naturally reflects in the brake specific fuel consumption. In addition to that, due to the low heating value, there is a possibility of more biodiesel drawn to the cylinder in order to match the energy need reflecting in the increase of brake specific fuel consumption. The specific energy consumption is same both in biodiesel and in neat diesel in the entire range of brake power tested .This is a consequence of the combined effect of low heating value and high density. Since the high density and viscosity affect the atomization, the brake thermal efficiency also reduces with the increase in brake power (Vedaraman et al. 2012). It is known that CO and HC reduce in the exhaust as the brake power is increased. The CO_2 level is seen increased at all level of the brake power. As cetane number of this biodiesel is higher than that of the diesel there is a short ignition delay period promoting increased combustion. The level of NOx is directly proportional to the brake power apparently due to the proportionate combustion according to the load. The smoke discharge is low in the case of biodiesel which is attributed to high oxygen content, low carbon-hydrogen ratio and absence of aromatic compounds in the fuel.

References

Hajra B, Sultana N, Pathak AK, Guria C (2015a) Response surface method and genetic algorithm assisted the optimal synthesis of biodiesel from high free fatty acid sal oil (*Shorea robusta*) using ion exchange resin at high temperature. J Environ Chem Eng 3:2378–2392

Hajra B, Pathak AK, Guria C (2015b) Optimal synthesis of methyl ester of sal oil (*Shorea robusta*) using ion exchange resin catalyst. Int J Ind Chem 5(3–4):95–106

Pali HS, Kumar N (2014) Biodiesel production from sal (*Shorea robusta*) seed oil. NIET J Eng Technol 5:24–29

Pali HS, Kumar N, Alhassan Y (2015) Performance and emission characteristics of an agricultural diesel engine fueled with blends of sal methyl esters and diesel. Energy Convers Manag 90:146–153

Vedaraman N, Puhan S, Nagarajan G, Ramabrahmam BV, Velappan KC (2012) Methyl ester of sal oil (*Shorea robusta*) as a substitute to diesel fuel - a study on its preparation, performance and emissions in a direct diesel engine. Ind Crop Prod 36:282–288

18 Euphorbiaceae

Euphorbiaceae, the spurge *family* has 6745 species in 218 genera. Though they are cosmopolitan in occurrence, they are distributed primarily in the tropics and are xerophytes. Many of these species carry milky latex, and some are useful as a source of oil or wax. Out of the huge number of species 8 of them are observed to be useful as resource in biodiesel production and hence dealt in this chapter. Out of the above, 5 (*Aleurites moluccanus*, *Croton megalocarpus*, *Hevea brasiliensis*, *Jatropha curcas* and *Vernicia montana*) belong to a single subfamily Crotonoideae. Other 3 species belong to Euphorbioideae (*Euphorbia lathyris* and *Triadica sebifera*) and Acalyphoideae (*Ricinus communis*).

18.1 Candle Nut (*Aleurites moluccanus*)

The candle nut tree, *Aleurites moluccanus* whose seed oil is considered as a raw material for biodiesel has assumed important as a non-conventional species. It grows to a height of 15–25 m with a large and spreading canopy comprising pendulous branches. It prefers a soil with a pH 5–8 and flourishes at an elevation ranging from 0 to 1200 m, temp 18–28 °C and rainfall 650–430 mm (Elevitch and Manner 2006; Kibazohi and Sangwan 2011). The leaf is simple, ovate, trilobed and 10–20 cm long. The flower is greenish-white and monoecious (both male and female flowers occur in the same tree). Many small male flowers surround a single female flower and are seen as a cluster. Each flower has five petals (Fig. 18.1).

The fruit is an indehiscent drupe of 5–6 cm long and 5–7 cm wide. This species is distributed in India, Indonesia, Malaysia, Japan, Australia and part of Asia and Africa (Fig. 18.2) and is referred by different vernacular names (Table 18.1).

S. P. Raj et al., *Biodiesel from Flowering Plants*,
https://doi.org/10.1007/978-981-16-4775-8_18

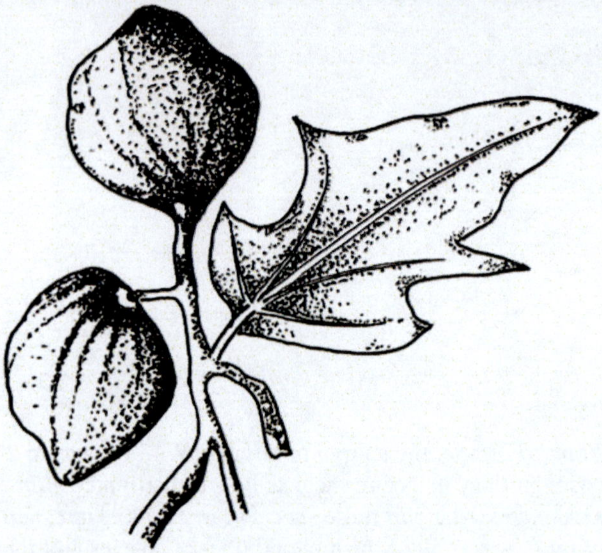

Fig. 18.1 The candle nut *Aleurites moluccanus*

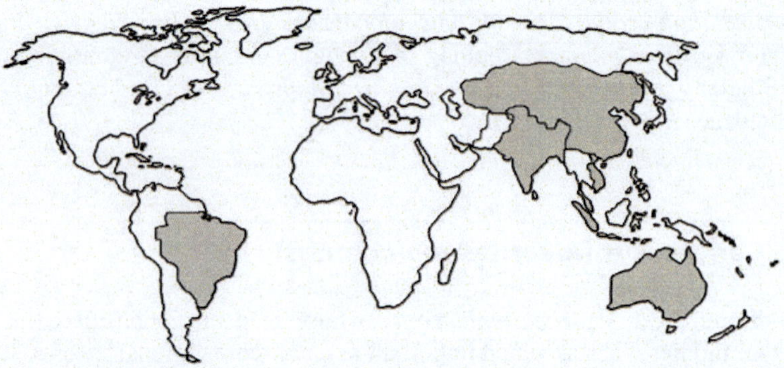

Fig. 18.2 Geographical distribution of *A. moluccanus*

18.1.1 Oil Extraction

A. moluccanus kernel is dried repeatedly and crushed to small granules. This is then extracted for oil using *n*-hexane as solvent at 50 °C in a Soxhlet apparatus until the oil is completely extracted. This oil is then degummed using hot water (Cabral et al. 2016). To every litre of oil 30 mL of hot water is added and vigorously stirred for 30 min at 70 °C. The content is then allowed to settle and the residue is removed. This oil is filtered and freed from moisture using anhydrous sodium sulphate. The analytical details of the oil are given in Table 18.2.

The kinematic viscosity, free fatty acid and iodine number are observed to be high. The unsaturated fatty acids are more than 89% which is also reflected in the iodine number. This also causes high pour point (Pham et al. 2018).

Table 18.1 Vernacular names of *A. moluccanus* in vogue in different countries (Krishnawati et al. 2011)

Country	Vernacular names
England	Candle nut, candle berry, varnish tree, Indian or Belgaum walnut
Fiji	Lauci, nggerenggere, sikeci, sikeli, sikethi, toto, tuitui, waiwai
France	Bancoulier, noyer de bancoul, noyer de moluques, aleurites, noisette, noix, noyer des indes
Germany	Kerzennussbaum, lichtnussbaum
Indonesia	Buwa kare, kembiri, kemili, kemiling, kerch, madang ijo, tanoan
Java	Kamere, kemiri, komere, midi, miri
Portugal	Calumban, noz da India
Spain	Arbol iloran, avellano, avellano criollo nogal de la India, nuez

Table 18.2 Physicochemical properties of *A. moluccanus* oil (Pham et al. 2018)

Parameters	Value
Kinematic viscosity ($mm^2\ s^{-1}$)	24.89
Density at 20 °C ($kg\ m^{-3}$)	914
Free fatty acid (%)	6.9
Iodine number (g I_2/100 g)	138
Palmitic acid (C16:0) (%)	6.3
Stearic acid (C18:0) (%)	3.2
Oleic acid (C18:1) (%)	21.8
Linoleic acid (C18:2) (%)	41.1
Linolenic acid (C18:3) (%)	26.8

18.1.2 Biodiesel Preparation

Transesterification with alkali as catalyst is normally being followed. For the purpose of transesterification, the free fatty acid has to be reduced, otherwise there will be soap formation. In order to reduce the free fatty acid the oil is first esterified. To 1 L of oil, 300 g of acetonitrile is added. To this 10 g sulphuric acid and 230 g of methanol are also added and stirred vigorously at 65 °C for 1 h. In such esterification process the free fatty acid level is reduced to less than 1%. In esterification a small quantity of biodiesel (ester) also is formed. The whole content is then taken in a separatory funnel and allowed to settle. The lower layer is discarded and the upper layer comprising the oil and a little ester is washed repeatedly until the wash water shows neutral pH. It is then dried under low pressure.

The esterified oil is then transesterified using alkali as a catalyst. With 1000 mL of the esterified oil, 200 g acetone is added in a round bottom flask kept in a water bath maintained at 40 °C. In a separate flask sodium methoxide is prepared by dissolving sodium hydroxide in methanol so as to have an ultimate molar ratio of 1:5 (oil to methanol). The sodium methoxide is then poured in the round bottom flask containing the oil and stirred for 1 h. The content is then taken in a separatory funnel and waited until the phase separation is complete. The lower layer containing the glycerol is removed. The upper layer formed of biodiesel admixed with methanol is

Table 18.3 The fatty acid profiles of biodiesel obtained from *A. moluccanus* using methanol and ethanol (Cabral et al. 2016)

Esters of fatty acid	Methyl (%)	Ethyl (%)
Palmitic	7.0	7.2
Stearic	2.7	2.9
Oleic	27.1	25.5
Linoleic	36.6	34.4
Linolenic	26.3	29.4

Table 18.4 Physicochemical properties of *A. moluccanus* biodiesel

Parameters	Imdadul et al. (2016)	Pham et al. (2018)
Kinematic viscosity at 40 °C ($mm^2\ s^{-1}$)	4.8	4.24
Density at 20 °C ($kg\ m^{-3}$)	885.7	887
Acid value ($g\ KOH\ kg^{-1}$)	0.4	0.21
Flash point (°C)	161.0	–
Cloud point (°C)	6.0	5.0
Pour point (°C)	6.84	6.3
Oxidation stability at 110 °C (h)	5.9	–
Heating value ($MJ\ kg^{-1}$)	40.3	–
Methyl palmitate (C16:0)	6.5	–
Methyl stearate (C18:0)	3.2	–
Methyl oleate (C18:1)	24.3	–
Methyl linoleate (C18:2)	39.6	–
Methyl linolenate (C18:3)	25.2	–

distilled to recover the methanol. The remaining biodiesel is neutralized using 5% aqueous phosphoric acid and then washed with warm dist. Water repeatedly until the pH of the wash water becomes 7. Then it is heated (85 °C) under reduced pressure for 40 min so as to expel the moisture. Instead of sodium hydroxide, potassium hydroxide also is used. These alkalis while joining with methanol react as follows and form methoxides.

$$NaOH + CH_3OH \rightleftharpoons CH_3ONa + H_2O$$

$$KOH + CH_3OH \rightleftharpoons CH_3OK + H_2O$$

Potassium hydroxide is more soluble in methanol than sodium hydroxide. In the place of methanol, other alcohols such as ethanol and 1-butanol are also tried. The yield is in the order methanol > ethanol > 1-butanol. It is known that the reaction rate depends on the bulkiness of the alkoxy anion produced from the alcohol during the catalytic reaction. Among the three alcohols, methanol easily attacks the carboxyl groups of the glycerides and moves the reaction rapidly forward. The profiles of fatty acid esters thus produced making use of methanol and ethanol are presented in Table 18.3.

The properties of the biodiesel obtained from *A. moluccanus* oil are given in Table 18.4.

In FTIR analysis (Szybist et al. 2005; Habibullah et al. 2015; Imdadul et al. 2016) the fatty acid methyl ester formation is confirmed by the wave number 1170.6 cm^{-1} (ester C-O) and 1740 cm^{-1} (ester C=O). The absence of band in 3500–3200 cm^{-1} indicates that neither alcohol nor water molecules are available in the biodiesel.

Though the oxidative stability is stable (5.9 h), there is a chance of forming polymerized compounds automatically (Rashed et al. 2016). Besides insoluble sediments may develop which may generate clogs in the fuel pipeline. There is high brake specific fuel consumption which is caused by the low heating value of the biodiesel. There is increase in NO_x and reduction in the HC and CO of the exhaust. Presence of marginally high levels of CO_2 is the characteristic features of the biodiesel (Ashraful et al. 2014; Can 2014). Increase in NO_x is a consequence of high heat release due to high cetane number and also due to the presence of oxygen molecule in the biodiesel.

18.2 Croton (*Croton megalocarpus*)

The oil extracted from the dry nuts of croton, *Croton megalocarpus* is being considered as a raw material in the production of biodiesel. It grows in East Africa at an altitude ranging from 1200 to 2000 m and has a maximum height of 35 m. The pole is cylindrical to a height of 18–25 m. It starts bearing nuts after 3 years of planting. This tree is being maintained in fields to offer shades for shade-loving agro-crops. It prefers a mean rainfall of 800–1600 mm and an annual temp. of 11–26 °C. It has a flat crown. The leaf is long (12 cm), oval with a pointed tip (Fig. 18.3).

The upper surface of the leaf appears dull green and the underside is pale and silvery. The flowers are conspicuous and short lived. It is yellow or white in colour. Many flowers join together to form a cluster. Female flowers appear first and stay at the bottom and the male flowers occupy the rest of the raceme. The fruit setting takes 5 months and the mature fruits are collected from the ground. The fruits are normally green in colour which turns brown as they mature. The endocarp is hard and woody.

Fig. 18.3 The croton *Croton megalocarpus*

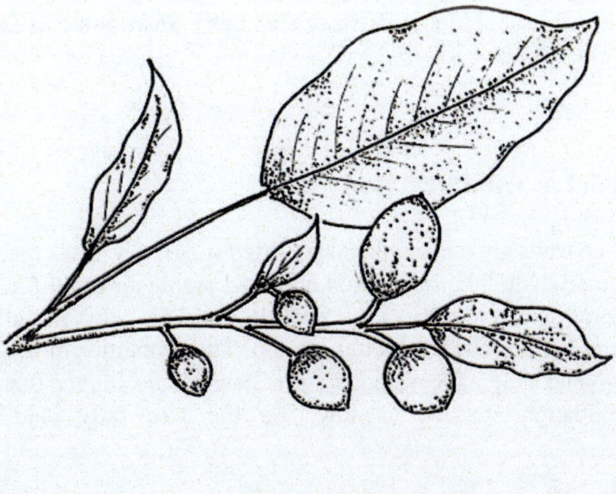

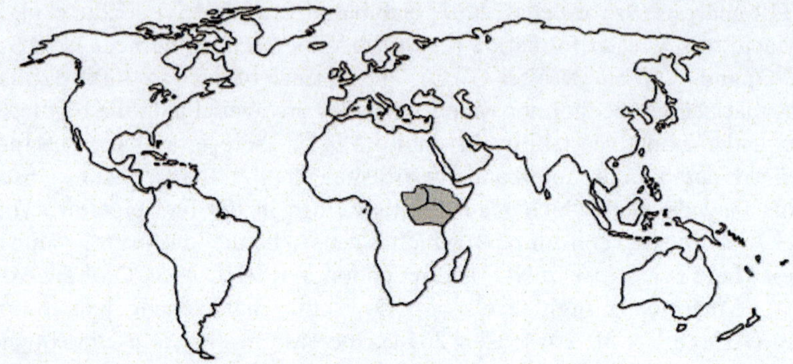

Fig. 18.4 Geographical distribution of *C. megalocarpus*

Table 18.5 The physicochemical properties of the *C. megalocarpus* oil

Parameters	Kafuku and Mbarawa (2010)	Kivevele and Mbarawa (2010)	Kafuku et al. (2011)	Aziz et al. (2017)
Kinematic viscosity at 40 °C ($mm^2\ s^{-1}$)	64	52.5	49.4	30.85
Specific gravity	0.918	0.921	–	–
Density ($kg\ m^{-3}$)	–	–	916.8	910
Free fatty acid (%)	1.68	1.73	2.4	5.04
Oxidation stability 110 °C (h)	–	3.12		–
Palmitic acid (C16:0)	–	7.2	6.5	–
Stearic acid (C18:0)	–	3.7	3.8	–
Oleic acid (C18:1)	–	13.7	11.6	–
Linoleic acid (C18:2)	–	69.0	72.7	–
Linolenic acid (C18:3)	–	4.6	3.5	–

Each fruit holds three ovoid seeds of 2–2.5 cm long and 1.2–1.4 cm wide. It is frequently being referred by the local names: croton, Kenya croton, nkulumire, mbula, msenefu and musine. This tree is distributed in Tanzania, Kenya and Uganda (Fig. 18.4.)

18.2.1 Oil Extraction

The fruits are collected and sundried for 10 days. It is then decorticated to remove the outer shell. The seeds thus obtained are again dried for 2 days and are then passed through a screw press to expel the oil. This oil is initially filtered using a filter bag and later through a vacuum pump. The properties of the oil are given in Table 18.5.

The kinematic viscosity and density are higher than that of the standard. The oxidation stability is low. As the free fatty acid is high, alkali catalysed transesterification is not considered ideal since there is a possibility of the formation of emulsion during transesterification.

18.2.2 Biodiesel Preparation

The oil at the outset is esterified using sulphuric acid (1 g 100 g^{-1} oil) as a catalyst along with methanol (20 mL of methanol in 100 g of oil) at a stirring speed of 350 rpm for 1 h. The lowest free fatty acid value obtained in the esterified oil is reported to be 0.3 g KOH kg^{-1}. The esterified oil is then transesterified using KOH as a catalyst. The KOH is dissolved into methanol so as to form potassium methoxide. This methoxide is poured in hot esterified oil and stirred. The resultant product is settled in a separatory funnel. The upper phase having the biodiesel is removed and washed. This is treated with fused silica gel for 24 h so as to remove the moisture. The recommended level of KOH is 10 g l^{-1} of oil and the molar ratio of oil to methanol is 1:6.

Tungsten trioxide (WO_3) supported on silica mesoporous-macroparticles is also used as a catalyst in the transesterification process. The above catalyst is synthesized as per Zhang et al. (2010). Cetyltrimethyl ammonium bromide ($C_{19}H_{42}BrN$), acetone (C_3H_6O), ammonium hydroxide (NH_4OH) and water are mixed at a molar ratio of 14:1.4:150:20, respectively. The contents are stirred vigorously for 20 min at a temp. 298 K. It is then added with 2.8 mL of tetraethyl orthosilicate and the stirring continued for another 2 h. The product obtained is dried at 333 K overnight and then calcined at 823 K for 6 h. Thus the silica mesoporous-macroparticles (SMP) are prepared. Aqueous solution of ammonium meta tungstate $(NH_4)_6[H_2W_{12}O_{40}]\cdot nH_2O$ is impregnated onto SMP at a temp. of 333 K, dried overnight at 383 K and calcined at 823 K for 3 h (Aziz et al. 2017). The catalyst is then mixed with methanol. To this, oil is added. The oil to methanol molar ratio is 1:7 and the catalyst conc. is 4% by weight with oil. The above mixture is taken in a flask fitted with a reflux condenser and stirred for 30 min in room temp. The whole content is then transferred to a separatory funnel and waited for a full day for phase separation. The reaction is based on the Lewis and Bronsted acid site of the catalyst. The incorporation of tungsten trioxide onto SMP has activated the Lewis and Bronsted acid sites. The first step in the reaction pathway is the adsorption of methanol and oil onto the surface of WO_3/SMP. This adsorption is aided by the rapid stirring. Being very precise the oil is adsorbed at the W^{6+} surface site which acts as Lewis acid sites, whereas the methanol is adsorbed at the lattice oxygen atom of the Lewis basic sites at the surface of the catalyst ultimately forming oxygen anion. The protonation of the carbonyl group of the ester causes the carbocation which is unstable. The nucleophilic attack of alcohol over the carbocation leads to the formation of tetrahedral intermediate which eliminates the glycerol.

Mirie et al. (2012) employed *Candida antarctica* B lipase as catalyst in the transesterification of *C. megalocarpus* oil. In a flask the oil and the enzyme at 4% by weight of the oil are taken and the flask is then kept in an agitator under hot condition. The methanol quantity (1:4 oil to methanol ratio) is split into three equal portions and the first portion is added to the flask at the start of the experiment. The second part is added after 10 h and the third part is added after 24 h. The experiment is continued till 56 h (Mirie et al. 2012). On completion of the reaction the flask is rinsed with chilled acetone, the content is filtered and the catalyst separated. The filtrate is then heated to expel acetone and methanol. Consequently the same is loaded in a separatory funnel and waited until the phase separation is completed. The glycerin which sinks down to the bottom of the separatory funnel is discarded. Though the transesterification procedure is eco-friendly, the biodiesel obtained has a high free fatty acid. But it is recommended to add caustic soda so as to form soap and the excess free acid can thus be contained. However, the longer reaction time is one of the major limitations in the above transesterification process.

A non-catalytic mode of producing biodiesel is reported (Kafuku et al. 2011) wherein a yield of 74.91% is obtained through supercritical procedures. The *C. megalocarpus* oil and methanol are mixed at a molar ratio of 1:50. The above content in a reactor tube is placed in a furnace at 330 °C for 20 min. It is then transferred to a bath to arrest the reaction. The yield of biodiesel is just 74.91% since this oil is not able to withstand high temperature and the polyunsaturated methyl linoleate, a dominant component of the biodiesel partially denatures.

Kafuku et al. (2010) used sulphated tin oxide enhanced with SiO_2 as catalyst. The catalyst $SO_4^{2-}|SnO_2\text{-}SiO_2$ with a weight ratio of $SnO_2:SiO_2 = 3:1$ is prepared by slowly mixing 100 g SnO_2, 33.3 g SiO_2 in 1000 mL of 2.0 M sulphuric acid. It is put on continuous stirring at 600 rpm for 6 h at room temp. The precipitate formed in it is separated through filtration and the same is calcined at 300 °C by raising the temp slowly at the rate of 10 °C min^{-1}. The calcinations process continued for 2 h. Making use of the calcined catalyst the transesterification is carried out. To 1000 mL of oil 23 g of sulphated tin oxide catalyst (2.3% by wt) and methanol to form a molar ratio of 15:1 to oil are added. The contents are stirred at 350 rpm for 2 h at 180 °C. During the reaction the vessel is kept in nitrogen atmosphere under pressure (10 bars) in order to ensure that the contents continue to remain in liquid state. On completion of the process the reactor vessel is cooled to reach the room temperature. It is then filtered to remove the residual catalyst. The filtrate is taken in a separatory funnel and kept undisturbed until phase separation is completed. The lower layer containing the glycerin is drained retaining the upper layer which is the biodiesel. In low temp. with high methanol to oil molar ratio the yield of biodiesel is higher than that of low temp. with low methanol to oil molar ratio. But in higher temp. the yield is high even when the methanol to oil ratio is low. It appears that the methanol tends to enhance the solubility of glycerin causing the glycerolysis. This is not favoured since the ester reacts with the glycerol to form monoglyceride (Lin et al. 2009). Apparently there is no appreciable relationship between the reaction duration and reaction temp. It is also known that the ester yield is promoted by longer reaction duration as sufficient period is necessary to push the reaction forward to reach the equilibrium. High temp. (150–180 °C) enhances the mass transfer among the three immiscible phases, namely oil, methanol and solid catalyst. The physicochemical properties of the biodiesel obtained are given in Table 18.6.

Table 18.6 The physicochemical properties of the biodiesel prepared from *C. megalocarpus* oil

Parameters	Kafuku and Mbarawa (2010)	Kivevele and Mbarawa (2010)	Osawa et al. (2014)
Kinematic viscosity at 40 °C ($mm^2\ s^{-1}$)	4.56	4.78	4.51
Density at 15 °C ($kg\ m^{-3}$)	889.9	883	885.8[a]
Acid value ($g\ KOH\ kg^{-1}$)	0.16	0.20	0.34
Flash point (°C)	189	192	>200
Pour point (°C)	−9	–	−6.5
Cloud point (°C)	−4	−6	−1.5
Cold filter plugging point (°C)	–	−11	–
Oxidative stability 110 °C (h)	–	2.88	–
Cetane number	–	47.52	–
Heating value ($MJ\ kg^{-1}$)	–	37.24	39.2

[a]At 40 °C

The viscosity, density, acid value, flash point and cold flow properties are observed to be favourable and conform to the standard prescribed. The cold flow and pour point indicate that the biodiesel from this species is suitable as a fuel in cold climates. This shift in properties from that of the oil is due to the gross conversion of big and branched molecules into light and straight chain methyl ester molecules. Poor oxidation stability is causing concern. The biodiesel when oxidizes forms alcohols, aldehydes peroxides, gums and sediments. Such phenomena are reported to plug the filters, foul the injector and form deposit in the combustion chamber. The stability and the iodine value are inversely proportional. Presence of rich quantity of polyunsaturated fatty acid is the sole reason for such instability (Kivevele and Zhongjic 2015). In such context, certain synthetic antioxidants are added to the biodiesel. Pyrogallol or propyl gallate at the rate of 200 ppm each enhances the oxidation stability. These antioxidants are characterized by the presence of three hydroxyl (–OH) groups in their aromatic rings. The –OH group present in these chemicals donates hydrogen atom after it is being abstracted from it. Consequently it moves to the oxidized free radicals of the biodiesel and thus inhibits the oxidation. Butylated hydroxyanisole also is being used but it has only one –OH group and hence its efficiency as an antioxidant is low.

18.3 Paper Spurge (*Euphorbia lathyris*)

The oil from the seeds of paper spurge *Euphorbia lathyris* is now being considered as a raw material for biodiesel. This plant which looks erect takes 2 years (biennial) to come to yield. It grows to a maximum height of 1.5 m with glaucous blue green stem. The leaves are lanceolate and arranged decussate. Each leaf is 5–15 cm long and 1–2.5 cm broad. The leaves are of waxy texture and pale green with white midrib

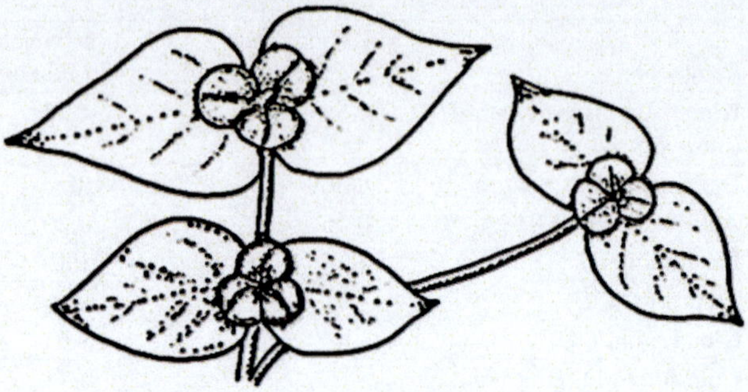

Fig. 18.5 The paper spurge, *Euphorbia lathyris*

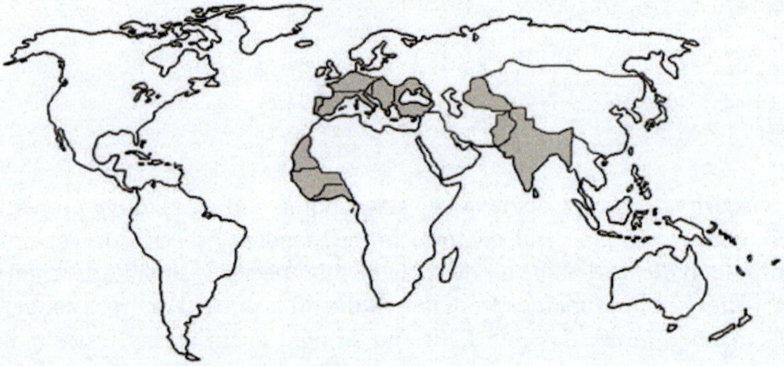

Fig. 18.6 Geographical distribution of *E. lathyris*

and vein (Fig. 18.5). The flowers are green to yellow, 4 mm dia. with no petal. The seeds are green in colour and change to brown or grey shade when ripe. The seed yield is around 3 t ha^{-1}.

They grow in a region with an annual temp. of 10–15 °C and a rainfall of 500–1500 mm. This species is a native of Mediterranean region and distributed in Southern Europe, North Western Africa and Western China (Wang et al. 2011) (Fig. 18.6). This plant is referred by certain local names: caper spurge, paper spurge, gopher spurge, gopher plant, mole plant and tartago.

The oil from the seeds of this plant is considered useful as a raw material for biodiesel since it has a rich content of oleic acid (Zhang et al. 2018). Oil containing myristic, palmitic, stearic fatty acids is a good source for biodiesel with a high cetane number and high cloud point. Biodiesel prepared from any oil with a high level of polyunsaturated fatty acids such as linoleic acid and linolenic acid may denature easily due to low oxidative stability.

18.3.1 Oil Extraction

The seeds are dried in hot sun for 3–4 days and then passed through a screw press. The oil obtained is allowed to settle and the resultant clear oil is decanted. The properties of this oil are presented in Table 18.7.

The acid value of this oil is considerably high, demanding a two stage esterification process for making biodiesel. The level of unsaturated fatty acid is high which is considered very ideal in oil as a raw material for biodiesel (Wang et al. 2011).

18.3.2 Biodiesel Production

As the oil contains considerable quantity of free fatty acid, the oil is esterified prior to the transesterification. In esterification, the oil is treated with H_2SO_4 at an optimum conc. of 0.8% at an oil to methanol molar ratio of 1:10 with the reaction duration of 45 min (Wang et al. 2011). The product obtained is taken in a separatory funnel and allowed to stand overnight. The separated oil is then washed repeatedly and dried by anhydrous sodium sulphate.

Mesoporous Al–Mo oxide also is used as an effective catalyst in the esterification process (Zhang et al. 2017). Aluminium tri propanolate ($C_9H_{21}AlO_3$) and molybdenum pentachloride ($MoCl_5$) are mixed with stearic acid (The molar ratio of $C_9H_{21}AlO_3$ to stearic acid is 1:3 and the molar ratio of $MoCl_3$ to stearic acid is 1:5). This mixture is kept stirred for 1 h at room temperature. It is heated and maintained at 100 °C for 30 min and then at 170 °C for 1 h. Following the above it is calcined at 600 °C for 5 h at a heating rate of 4 °C min^{-1}.

The catalyst thus prepared is mixed at the rate of 3% by weight of the methanol–oil mixture and kept at 180 °C for 2 h at a pressure of 1.0–1.2 MPa. The oil to methanol molar ratio is 1:20. On completion of the reaction, the content is cooled and the catalyst separated by filtration. The esterified oil is then purified. The catalyst can be reused for another 3–5 runs without any significant loss.

The above esterified oil containing low quantity of free fatty acid is transesterified using alkali (0.8%) as catalyst. The alkali (KOH) and methanol are mixed to form

Table 18.7 Physicochemical properties of *E. lathyris* oil

Parameters	Wang et al. (2011)	Zapata et al. (2012)	Zhang et al. (2018)
Density (kg m^{-3})	–	920	–
Iodine value (g I_2/100 g)	83.37	–	83.59
Acid value(g KOH kg^{-1})	25.17	–	24.59
Molecular weight (g mol^{-1})	1184.7	–	1182.6
Palmitic acid (C16:0) (%)	6.8	5.53	6.89
Stearic acid (C18:0) (%)	1.98	–	2.08
Oleic acid (C18:1) (%)	81.46	78.2	82.05
Linoleic acid (C18:2) (%)	3.71	4.62	4.13
Linolenic acid (C18:3) (%)	2.78	7.85	2.34

Table 18.8 Physicochemical properties of biodiesel from *E. lathyris*

Parameters	Wang et al. (2011)	Zapata et al. (2012)
Kinematic viscosity at 40 °C ($mm^2\ s^{-1}$)	4.6	4.9
Density at 20 °C ($kg\ m^{-3}$)	876.1	880
Flash point (°C)	181	241.7
Acid value(g KOH kg^{-1})	0.19	–
Oxidation stability 110 °C (h)	10.4	–
Cetane number	59.6	52.46
Copper strip corrosion	1a	–
Cold filter plugging point (°C)	−11	–
Iodine value (g I_2 100 g^{-1})	–	87.5
Heating value (MJ kg^{-1})	–	38.17

pot.methoxide which is then poured in hot (60 °C) oil and stirred (600 rpm) for 30 min. The molar ratio of oil to methanol is 1:6. The resultant product is taken in a separatory funnel and allowed to stay undisturbed. The top layer containing the biodiesel is separated, washed with distilled water and dried using anhydrous sodium sulphate. The properties of biodiesel are presented in Table 18.8.

The biodiesel yield due to transesterification is observed to be as high as 91.1% (Zapata et al. 2012) and 8.81% of the oil is converted to glycerin. The iodine value is within the limit and is observed to be reasonable. High iodine value is likely to make rapid oxidation and form gel or polymerization during long storage (Ma and Hanna 1999). According to Pinto et al. (2005) any oil containing rich quantity of monounsaturated fatty acids is a good source for an ideal biodiesel. Low cetane number in a biodiesel may cause poor ignition and as a result there will be high pressure due to the high lag time at the onset of combustion (Benaides et al. 2007). The heating value of this biodiesel is observed to be lower than that of the diesel. Therefore, the engine ultimately demands more volume of biodiesel to support the required torque resulting into increased specific fuel consumption.

Of late, magnetic-acidic poly ionic liquid catalysts with varying hydrophobicity and acidity are being used as a catalyst in biodiesel preparation. Zhang et al. (2018) used Fnm S-PIL(1a,C8) as core shell structure catalyst. It has the BET surface area of 128.1 $m^2\ g^{-1}$, mesoporous pore dia. of 4.2 nm, magnetic effect of 12.4 $emug^{-1}$ and an acid site of 2.14 m mol g^{-1}. This catalyst [FnmS-PIL (1a,C8)] can be isolated very easily from the finished product by applying external magnet. It is known to be employed in five successive cycles. The catalyst is first mixed with methanol and the resultant mixture is poured in hot oil. It is then stirred at 650 rpm. The conc. of catalyst is 5% by weight of the oil–methanol mixture. The molar ratio of oil to methanol is 1:18 and the duration of the reaction is 6 h at an ambient temp. of 120 °C. The esterification and transesterification reactions are often carried out in a single jar concomitantly. The esterification and transesterification flow chart is presented in Fig. 18.7.

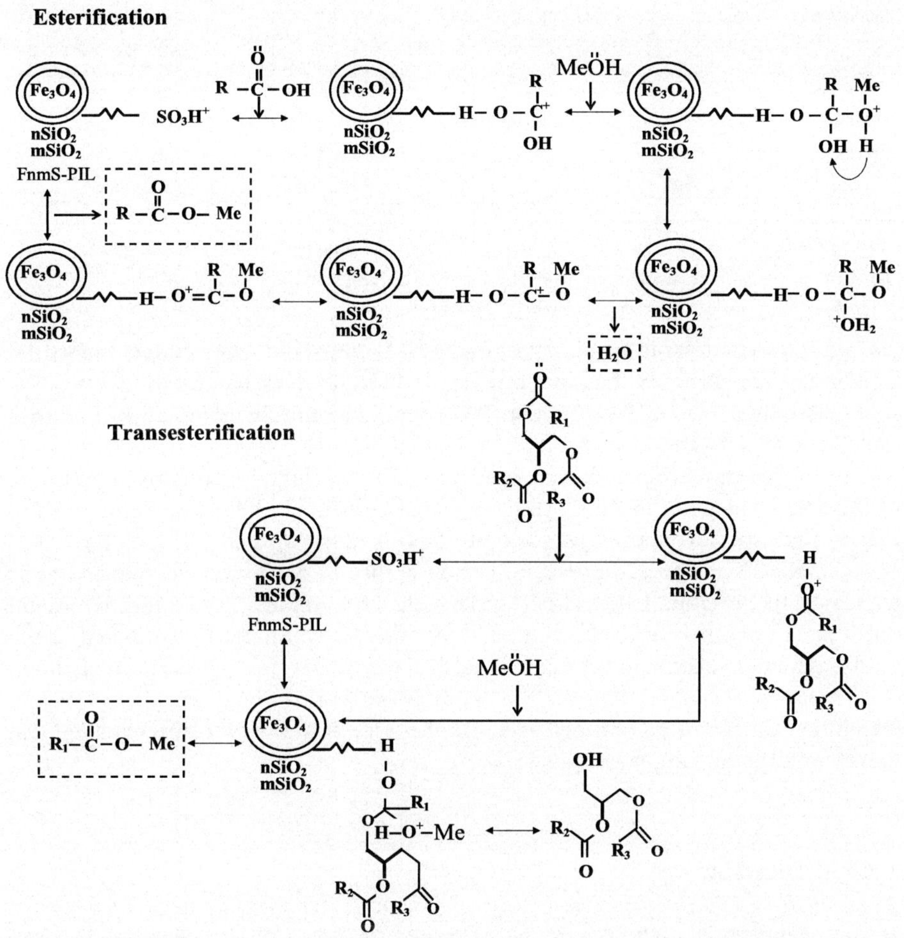

Fig. 18.7 The scheme illustrating the mechanism of esterification and transesterification catalysed by FnmS-PIL(1a,C8) catalyst. The catalyst FnmS-PIL(1a,C8) due to its strong repulsion to water shows high acid strength, stability, steady activity in few cycles to follow and offers easy separation by magnet

18.4 Rubber Tree (*Hevea brasiliensis*)

The rubber tree *Hevea brasiliensis* Muell.Arg (*Siphonia brasiliensis*) is the native of Amazon rain forest and is very popularly referred as para rubber tree (heve = rubber). The name of the tree is derived from *Para*, which is the second largest Brazilian state. Its capital lies on the banks of the river Amazon in the northern part of Brazil. Its trunk is smooth and straight. This tree is unbranched till a height of 5 m from the ground and then it branches profusely to form a leafy canopy. The crown is conical in shape. The bark is greyish. Though ten species are recognized under the genus

Table 18.9 Vernacular names of the rubber tree

Language	Name	Language	Name
Sino-Tibetan	Jaang	Amharic	Yegoma zaf
Malay	Kayu getah, Pokok getah para	Burmese	Kyetpaung
Spanish	Caucho	English	Rubber
Swahili	Mpira	French	Caoutchouc
Thai	Katoh, Yang phara	German	Parakautschukbaun
Vietnamese	Cao sau	Indonesian	Kayu getah
Khmer	Kausuu		

Hevea, *H. brasiliensis* is the only species being largely cultivated all over the world. This species is currently distributed in ten million hectares around the world. The rubber tree lives around 100 years but its economic return dwindles when it reaches the age of 25–35 years.

The rubber trees efficiently sequester carbon in day light. The photosynthetic rate of this tree is 11 μmol m^{-2} s^{-1} and this value is relatively high when compared to that of other species. The values for other species range from 5 to 13 μmol m^{-2} s^{-1} (Chapman 2007). The carbon sequestration ability of a 30-year-old rubber tree is estimated to be around 270 t ha^{-1}. The value for tropical rain forest is around 230 t ha^{-1} only. It is around 150 t ha^{-1} by other trees. Similarly the extent of the total leaf area in relation to the total biomass of the rubber tree also is relatively high. It is reported that *Hevea* plantation around the world would fix approximately 90 million tonnes of carbon per year. Rubber tree is known by different regional names in different languages (Table 18.9).

18.4.1 Distribution

H. brasiliensis was a wild tree in the Amazon river basin till last century. It started spreading as a plantation crop to the eastern tropics and central Africa in the early twentieth century. Cultivation of it in Uganda commenced during the year 1901 using the seedlings received from Sri Lanka (Ceylon). Thus Uganda had a total of 15,000 acres of land planted with this species in 1912. Firestone Company also planted this species in Liberia during the year 1924. Similarly the Goodyear Company undertook the planting of this tree in the Philippines in 1928. This species is now successfully cultivated in humid tropics extending roughly between 15° N and 15° S (i.e. between 1126 km north and 1126 km south of the equator). It is now being distributed in Bolivia, Brazil, Colombia, Peru, Venezuela, Brunei, Cambodia, China, Ethiopia, India, Indonesia, Laos, Liberia, Malaysia, Myanmar, Philippines, Singapore, Sri Lanka, Thailand, Uganda, Vietnam, Nigeria, Guatemala, Mexico, French Guyana, Trinidad, Papua New Guinea and Ghana (Fig. 18.8)

It grows well in hilly slopes with a good rainfall and drainage. It prefers an altitude ranging from 300 to 500 m and a mean annual temperature of 23–35 °C. The preferred rainfall is 2000–3000 mm. It tolerates a soil pH ranging from 4 to 8. The

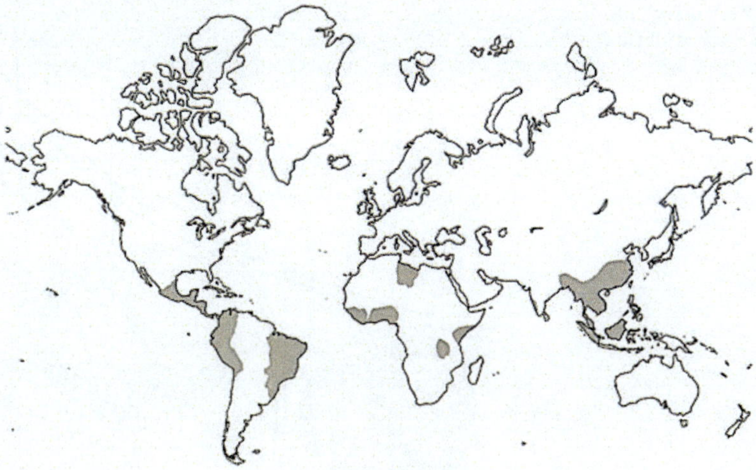

Fig. 18.8 Distribution pattern of *H. brasiliensis* in the world as seen in the hatched area

growth is observed to be low in poorly drained terrain. Peaty soil also is not preferred. The performance is observed to be good especially when the soil depth of the plantation is satisfactory with a good loam. The distribution pattern of the rubber tree is largely governed by many factors such as

1. Annual rainfall ranging from 2000 to 3000 mm distributed in 125–150 rainy days without any major dry spell.
2. Highest atmospheric temperature of 34 °C and the lowest temperature of 20 °C with a monthly mean of 25–30 °C.
3. Monthly average of 80% atmospheric humidity throughout the year.
4. Sunshine of around 700 kW h^{-1} for a total of 2000 h $year^{-1}$ and 6 h day^{-1} throughout the year.
5. Absence of strong wind.

18.4.2 Botanical Features

The tree grows up to a height of 40 m with a maximum trunk diameter of 50 cm (at 1 m above the ground). It is a dicotyledonous flowering tree with palmately compound leaves. The leaves are alternate, each with three leaflets (Fig. 18.9).

Leaflets are petiolated, glabrous with pinnate venation and pungent odour. The female flowers are larger than the male ones. The gynoecium of the female flower comprises a three celled ovary with single ovule in each cell. Thus the fruit is a tripartite capsule containing three mottled seeds, one in each compartment. The seeds are ovoid and slightly compressed. The testa is grey or pale brown with irregular dark brown dots, lines and blotches. In young seeds the endosperm is white in colour while it is pale yellow in matured seeds.

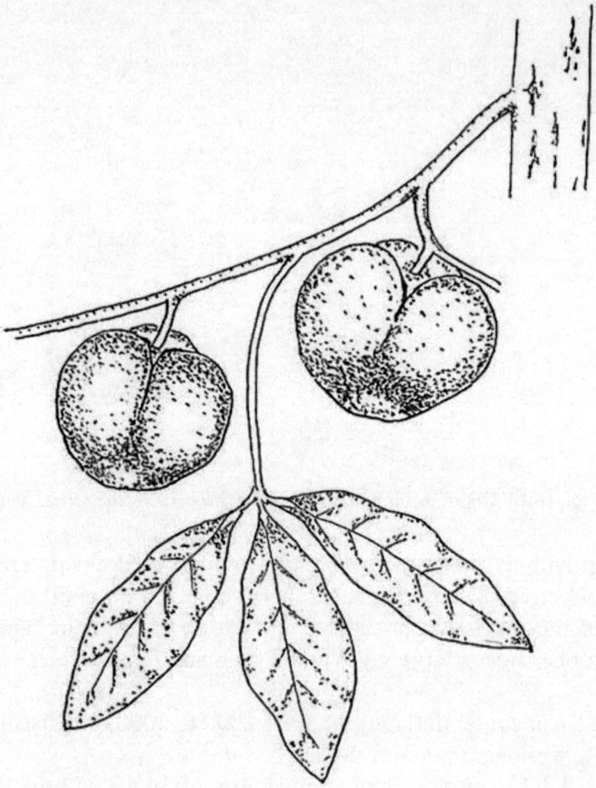

Fig. 18.9 A branch of *H. brasiliensis* with fruit

Male and female flowers are produced in the same inflorescence at a ratio of 1:60–80. Flowers normally remain intact in the tree for 2 weeks. They do not secrete nectar. But the leaf is characterized by an extra floral nectarine. Sweet secretions appear at the tip of the young leaf petioles and shoots. The pollination is aided by honey bees and midges. Ripe fruits appear in 6–7 months after pollination and they dehisce violently at the air so as to scatter the seeds far away from the mother tree. The trees yield matured seeds only after 7–8 years of their planting. The seed setting takes place once a year.

18.4.3 Propagation

The propagation of this species is carried out through any one of the following means:

1. Seed at stake
2. Stump planting

3. Polybag planting
4. Budded stump planting

In seed at stake, the seeds which are collected within 7–10 days of falling from the mother tree are sown straight in a new field kept ready for planting. Such seeds have a good rate of germination. In case, the seeds are to be preserved without loss of viability, it has to be stored by mixing it in charcoal powder or in saw dust with 15–20% moisture in wooden containers with aeration. Such seeds are viable for a maximum period of 4–6 weeks. However, sowing the seed straight in field pit is not encouraged as the emerging young plant may not be able to face the vagaries of the field.

In stump planting, seedlings raised in nurseries are transplanted after pruning the stem at a height of 45–60 cm from the collar. For the purpose of polybag planting, seedlings are raised in polyethylene bags of the size 18 × 45 cm. These bags are first filled with a mixture of native soil and farmyard manure. After a week the seed is sown in it. Once the collar of the seedlings reaches a thickness of 2–2.5 cm, the seedlings are budded with buds collected from the bud-wood trees or from clones. After the establishment of the bud the baby plants are transplanted to the field. In certain adverse conditions the budded stumps are transferred to big polybags in nurseries and transplanted in the field after they are well established in such polybags. During the transplantation the lateral roots of the budded stumps are trimmed at a distance of 7.5–10 cm from the tap root. At any cause the roots shall not be trimmed too close to the tap root since close trimming may adversely affect the root formation. Planting is carried out during favourable climate.

The recommended planting distance between trees is 7 m thereby the total number of standing trees in a hectare is around 200. The size of the pit is 50 × 50 × 30 cm. Well composted cattle manure (10–15 kg) along with 150 g of rock phosphate is mixed with the surface soil and filled in the newly created pit. In virgin land rock phosphate alone is applied as a chemical fertilizer. Coffee (*Coffea arabica*) or Cocoa (*Theobroma cacao*) is grown as intercrop in young plantation. Crops, such as *Calopogonium*, *Centrosema*, *Flemingia*, *Psophocarpus*, *Pueraria* are cultivated as a soil cover in the rubber plantation, so as to increase the fertility and to minimize the infestation of the weed. Fodder crop such as *Cajanus* also is grown in the inter space which is parallelly used for lac production. If the cover crop stands for many years, the dose of nitrogenous fertilizer to rubber tree may be minimized. The press cake also is applied as a manure to the tree. In the usual course the fertilizers such as phosphate, magnesium and potassium are needed to be applied regularly.

18.4.4 Pests and Control

The tree is often affected by trunk disease, root disease and leaf disease. Black stripe, mouldy rot and panel necrosis are caused by fungi and are being effectively controlled by the application of fungicides. Trees with small stems are attacked by fungi *Pellicularia salmonicolor* and *Gloeosporium heveae* causing pink disease, canker and dieback.

White, red and brown rot diseases caused by the fungus *Fomes lamaensis* are prevented by applying prophylactic coatings.

Leaves are also often affected by *Gloeosporium* leaf disease, powdery mild dew disease caused by the fungus *Oidium heveae* and birds eye spot disease. They are controlled by spraying copper oxychloride, wettable sulphur, bordeaux mixture and sulphur dust. The fruit rot disease is caused by the fungus *Phytophthora palmivora.*

There are parasitic plants such as *Loranthus* sp. which suck the tree sap. The trees are also commonly infested with nematodes *Helicotylenchus caveness*, *H. dihystera*, *H. erythrinae* and *Meloidogyne incognita acrita* and white ants. The tender green leaves of the seedlings are often affected by leaf fall disease (*Phytophthora* sp.) which is controlled by spraying bordeaux mixture at the rate of 1%. Since the leaf emergence is relatively rapid in seedlings, repeated spraying in short intervals are being carried out.

18.4.5 Seeds

A grown up tree bears around 150–200 fruits with three seeds in each. The seed is ellipsoidal in shape. The surface of the seed is woody (Fig. 18.10).

Each seed is 2.5–3 cm long weighing 2.1–4.7 g (average 4.3 g). The fruits dehisce while they are falling down from the tree and their seeds are scattered due to the violent explosion consequent to the crack taking place in the exocarp. Each tree yields around 400–500 seeds. From 200 standing trees in a hectare the total yield of seeds per year is around 500 kg. The estimated availability of the seed across the world is around 0.7 million tonnes a year which has the potentiality of yielding

Fig. 18.10 A section of the fruit showing a seed

Table 18.10 The physical properties of the rubber seed (Haque et al. 2009)

Parameter		Value
Average length	(mm)	21.5
Average weight	(g)	4.3
Thickness	(mm)	17.3
Slender ratio (length/thickness)		1.1
Volume	(cc)	14.3
Bulk density (weight/volume)	g cc^{-1}	0.3

around 0.11 million tonnes of raw oil (Ramadhas et al. 2005). It is reported that the seeds if happen to stay on the ground for more than 7 days they deteriorate. During such period, the moisture content of the seeds increases as the seeds are hygroscopic. Consequently such seeds suffer from microbial infection which parallelly enhances the endogenous lipase and free fatty acids (Bringi 1987). Therefore, it is imperative to collect the seeds from the field soon after they fall on the ground and process them without any loss of time.

The rubber seed is more or less rectangular in shape with a round tip. The body is slightly flattened with a slender ratio of 1.1 (Table 18.10). The density of the whole seed is low (0.3) which is well compared to that of jatropha (0.3) and far lower than that of the castor (0.8). The presence of hollow space between the endosperm and the outer hard shell is the reason for the low density. Besides, the endosperm is relatively spongy.

The free fatty acid content and acid value of the whole seed are high (Table 18.12) indicating that complication may arise in the esterification process of the oil extracted from it and large quantity of polyols may be required to complete the reaction. The high iodine value indicates the presence of huge quantity of unsaturated fatty acids. Though the protein content of the kernel is favourable for using it as an animal feed it contains hydrogen cyanide. Concentration of this toxin in the kernel may go up to 2000 mg kg^{-1}. But such high values get reduced gradually as the seeds are stored for a long time extending up to 2 months. As a matter of fact, processed kernels are reported to be consumed by certain people in the Amazon valley.

Drying the seed mostly halts the activity of the fat splitting enzymes and reduces the lipase activity. In the oil extracted from properly dried seeds, the content of free fatty acid (FFA) is observed to be low. Drying also greatly reduces the cyanide level. The drying of the seed helps to detach the kernel easily from the husk and thus decortications are made easy. For long term storage the moisture content of the seed shall be kept lower than 5% (UNIDO 1987). The scope of air drying the seed is limited during rainy season. In such eventualities the seeds may be dried in hot air at 60–70 °C. During the drying process, the FFA is likely to be reduced causing the rupture of the cell wall. The viscosity of the oil also is reduced. However, any temperature above 60–70 °C may darken the colour of the oil.

The rubber seed with 4.0% moisture has a high hardness value of 8.6 MPa. The high hardness is due to the extremely rigid outer shell. The corresponding values for jatropha and castor are 2.7 and 1.8 MPa, respectively. The crushing strength of the rubber seeds is observed to be 121.1 kg cm^{-2} which is 38.1 kg cm^{-2} for jatropha and

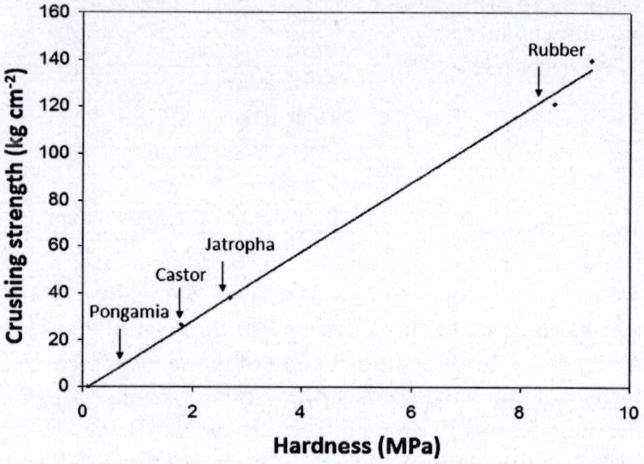

Fig. 18.11 The relationship between the hardness and crushing strength of the oil seeds. The relationship is established from values obtained from the seeds of pongamia, castor, jatropha and rubber

26.5 kg cm^{-2} for castor. Thus the rubber seed offers an extremely tough resistance to crushing. It indicates that the hardness is directly proportional to the crushing strength. The crushing strength is in the order: rubber seed > jatropha seed > castor seed > pongamia seed (Fig. 18.11).

18.4.6 Oil Extraction

The kernel forms 50–55% of the seed by weight. But invariably the whole seeds without decortications are used for the extraction of oil. The extraction is carried out either through screw press or through rollers. The screw press comprises a horizontally moving screw which rotates and simultaneously directs the seed mass into an orifice of a sleeve where the seed is chocked. During the chocking, the oil squeezes out through the perforations of the sleeve. Parallelly the meal (cake) is ejected out. In such process a maximum of 30 kg of oil is recovered from 100 kg of seeds.

When the oil extraction is carried out through rollers the operation is made either at an ambient temperature or at a hot condition. At ambient temperature a maximum of 35 kg of oil may be extracted from 100 kg of seeds and thus the recovery remains incomplete. If the extraction is carried out in hot condition, 40–45 kg of oil may be obtained from every 100 kg of seeds. But in such hot condition the oil invariably becomes dark and partially denatured. Drying or heating the seeds at a higher temperature tends to polymerize the oil and as a result the original pale colour of the oil often turns dark brown (Njoku et al. 1996).

Solvent extractions also are being tried so as to increase the recovery percentage. Petroleum ether or hexane or chloroform is frequently used as a solvent and around 50–55 kg of oil is recovered from every 100 kg of seeds (Christie 1982).

18.4.7 Characteristics of the Oil

The oil present in the kernel remains as a semi-dried substance with a pasty consistency (Aigbodion and Pillai 2000) and it is a rich source of polyunsaturated fatty acids. It is also known that the seeds of the rubber tree contain a lipolytic enzyme, which makes the refining of the oil fairly difficult. The level of FFA increases if the seed is damaged during dehiscence or the seed happens to over stay on the surface of the moist soil.

The oil is golden yellow in colour at normal temperature and has properties similar to that of soybean and linseed. The colour of the oil is a fair index of the level of FFA content in it. Pale yellow colour of the oil indicates that it has a low level of FFA. If the oil is brown in colour, it indicates that the FFA content of that oil is high. Brownish colour of the oil also is due to the presence of high moisture and endogenous lipase which ultimately increases the FFA (Bringi 1987). Quick collection of the seeds from the field, immediate decortication, rapid drying of the seeds in the hot sun and storing them in a warm dry place will help to have an oil low in free fatty acid.

Proximate analysis of the rubber seed oil is presented in Table 18.11. The FFA content of the rubber seed oil is found to be high (25.2 and 35.6 wt%) (Devi 2010; Le et al. 2018a, b) and the acid value is around 50 mg KOH g^{-1}. The specific gravity of the oil is closer to that of groundnut oil (0.91 g cc^{-1}) and less dense than neem seed oil (0.94 g cc^{-1}) (Peterson and Scarrah 1984; Akpan 1999). Usually the saponification value falls within the range of 185–195 mg KOH g^{-1} for most of the vegetable oils (Pearson 1976). Relatively lower saponification value of the rubber seed oil indicates that the rubber seed oil has a higher molecular weight than that of many common oils.

Peroxide value is a quick indicator of the quality of the oil. The standard for peroxide value is 10 meq kg^{-1}. Values lesser than 10 meq kg^{-1} normally indicate

Table 18.11 Characteristics of the rubber seed oil (Devi 2010)

Parameters		Value
Free fatty acids	(%)	25.2 ± 0.85
Acid value	(mg KOH g^{-1})	50.8 ± 1.90
Specific gravity at 15 °C	(g cc^{-1})	0.9 ± 0.01
Kinematic viscosity at 40 °C	(cSt)	41.1 ± 0.66
Iodine value	(mg I_2 g^{-1})	822.0 ± 35.0
Saponification value	(mg KOH g^{-1})	168.7 ± 0.33
Peroxide value	(meq kg^{-1})	2.8 ± 0.22
Carbon	(%)	77.8 ± 0.04
Hydrogen	(%)	12.9 ± 0.03
Nitrogen	(%)	0.2 ± 0.02
Sulphur	(%)	0.2 ± 0.01
Heating value	(MJ kg^{-1})	37.7 ± 0.11

the freshness of the oil. The low value for peroxide indicates a low rate of oxidation of the oil. Any values between 20 and 40 meq kg^{-1} indicate a state of rancidity.

The heating value of the rubber seed oil lies around 38 MJ kg^{-1} which is marginally lesser than that of the diesel whose heating value is 42.5 MJ kg^{-1}. The low heating value is apparently due to high unsaturation of hydrogen in the oil molecule. The carbon, hydrogen, nitrogen and sulphur content of the biodiesel are around 78%, 13%, 0.2% and 0.2%, respectively. The flash point of the rubber seed oil is relatively higher than that of the diesel which is considered to be favourable as for as the safe storage is concerned. The viscosity also is far higher than that of the diesel. Diesel comprises saturated non-branched hydrocarbon molecules with the number of carbon atoms ranging from 12 to 18, whereas the vegetable oil has triglycerides which are predominantly with non-branched chains having different lengths and different degrees of saturation.

The saturated fatty acids, namely palmitic acid and stearic acid which do not contain any double bond constitute 17–20% of the total fatty acid. The unsaturated fatty acid is about four times higher than the quantity of the saturated fatty acids (Table 18.12). The unsaturated fatty acids having one or more double bonds are formed of oleic acid, linoleic acid and linolenic acid. Rubber seed oil is not used for any edible purpose at present. However, it has been utilized as a partial replacement of linseed oil especially in some of the non-edible applications. The most abundant fatty acids present in the oil are linoleic, oleic and linolenic acid. In contrast to many vegetable oils it has a fairly rich quantity of linolenic acid (an omega 3 fatty acid). Each of these fatty acids has 14–22 carbon atoms with 0–3 double bonds.

18.4.8 Press Cake

The press cake is a good source of protein (Table 18.13). It has more than 30% protein. However due to the presence of toxic constituents, it is seldom used as a feed for lower animals (Ensminger and Olextine Jr 1978). However, it is recommended to use the cake at a rate lesser than 20% of the total meal.

Many efforts are being made to detoxify the cake so as to incorporate it in animal feeds. Since complete recovery of oil is not possible, the oil cakes normally contain oil to the tune of 5–10%.

Table 18.12 Fatty acid composition of Malaysian rubber seed oil (Aigbodion and Bakare 2005)

Fatty acids	Percent	Total
Saturated fatty acids		
Palmitic acid	8.56 ± 0.07	
Stearic acid	10.56 ± 0.02	19.12
Unsaturated fatty acids		
Oleic acid	22.95 ± 0.15	
Linoleic acid	37.28 ± 0.10	
Linolenic acid	19.22 ± 0.021	79.45

Table 18.13 Constituents of rubber seed press cake (UNIDO 1987)

Constituent	%
Moisture	9.1
Residual oil	6.2
Protein	30.0
Ash	6.5
Total carbohydrate	19.9

18.4.9 Biodiesel Production

The usual procedure followed in the transesterification utilizing alkaline catalyst cannot be employed straight away in the case of rubber seed oil as it contains exceedingly high quantity of FFA.

The course of treatment of the oil for biodiesel production and the choice of a catalyst to be employed is largely based on the quantity of FFA present in it. The FFAs are also highly susceptible to oxidation resulting into oxidative rancidity. If the FFA content of the oil is around 3%, the usual base catalyst reaction can be employed without any technical impediment. In case the FFA content is more, the base catalyst such as NaOH or KOH cannot be employed since a portion of the alkali is likely to be spent on non-target activity such as the saponification of the free fatty acid which automatically results into the formation of soap causing the wastage of catalyst.

$$RCOOH + NaOH \rightleftarrows R'COONa + H_2O$$

$$RCOOH + KOH \rightleftarrows R'COOK + H_2O$$

In addition to the saponification, mucilaginous substances, phospholipids and pigments present in the oil also react with alkali and form complex substances. When the FFA level is exceedingly high the contents present in the whole reaction chamber are likely to be neutralized so that hardly any forward reaction is possible. In case any ester is formed during this process it is likely to be reverted with the help of H_2O to fatty acids through base catalysis.

$$\underset{\text{Ester}}{R'COOR} + H_2O \underset{\text{base catalyst}}{\rightarrow} \underset{\text{FFA}}{RCOOH} + R'OH$$

This newly generated FFA once again may enter into the formation of more soap and thus this vicious activity continues till the catalyst is completely consumed and deactivated. Even if ester is formed to a limited extent at the final stage, its isolation from such mixture is nearly impossible since the phase separation boundary is not visible clearly. Besides, there may be considerable loss of oil through hydrolysis engineered by the alkali. In few cases the quantity of soap is reported to be as high as 50% of the weight of the oil which ultimately reduces the yield of ester. As a result additional process to contain the frothing and consequent removal of water may also be needed. Huge quantity of fresh water may also be needed to wash away the impurities present in the ester phase under such circumstances. The waste water thus generated may also pose additional problems and the treatment of it increases the cost of production, since the polluting stream is to be treated to meet the statutory standards. Therefore, rubber seed oil is normally subjected to pretreatment.

When the FFA content of the rubber seed oil is 3–15%, its level is required to be reduced to 3 or less than 3% so as to make such oil amenable for the usual base catalysed transesterification. This dual process involving the reduction of FFA as a preliminary step and the consequent transesterification as the second step is referred as double stage transesterification.

18.4.10 Double Stage Transesterification

18.4.10.1 First Stage

In rubber seed oil, if the FFA level is observed to be between 3% and 15% there is no scope of considering comprehensive alkali catalysed transesterification. Therefore, such oil is first subjected to acid catalysed reaction using any one of the following catalysts prior to the actual alkali catalysed transesterification.

1. Sulphuric acid (Homogeneous)
2. Hydrochloric acid (Homogeneous)
3. Orthophosphoric acid (Homogeneous)
4. Ferric sulphate (Heterogeneous)

The oil is treated with a mixture comprising methanol and any one of the above catalysts. The reaction causes the reduction of FFA in the oil with the simultaneous conversion of it to ester. Such reaction is very slow due to the low solubility of the reactants. At the initial stage of the reaction the rate of mass transfer is feeble. As the reaction proceeds further, a kinetically favoured change takes place with a rapid formation of the product causing the simultaneous reduction of FFA.

The performance of three homogeneous acid catalysts and a heterogeneous catalyst is given in Table 18.14. The acid catalysts normally cause the reduction of FFA in 60–70 min. Sulphuric acid is the preferred catalyst and the optimum dose of it is around 5 mL L^{-1} or 0.5%. (Ramadhas et al. 2005; Devi 2010). Sulphuric acid

Table 18.14 Comparative performance of various catalysts in reducing the free fatty acid in rubber seed oil treated with methanol at a molar ratio of 9:1 with oil (Devi 2010)

Acid catalysts	Conc. per litre	Maximum reduction of FFA (%)	Optimum reaction (min)
Sulphuric acid (mL)	4.9	85.4	66.0
Hydrochloric acid (mL)	4.3	75.0	69.0
Orthophosphoric acid (mL)	3.4	64.6	68.7
Ferric sulphate (g)	2.5	89.3	135.0

offers the highest catalytic activity resulting into the reduction of FFA to reach a level of around 2–4% from an initial level of even 15%. In sulphuric acid treatment more H^+ is released to protonate the carboxylic moiety of the fatty acid. The proton migration is followed with the breakdown of the intermediate products (Kocsisova et al. 2005; Lotero et al. 2005; Aranda et al. 2008). Le et al. (2018a, b) used sulphuric acid as catalyst at 1.5% level in esterification process. Acetonitrile was added as solvent. As a result, the FFA came down to $2.1 \pm 0.1\%$. It is also known that the acid dissociation index (p*K*a) is not the only criterion to identify an acid as an effective catalyst in such reactions (Dean 1985; Ting et al. 2008). Sulphuric acid has another advantage of being converted into an insoluble salt so that it can be recovered easily.

In the event of using ferric sulphate as a catalyst it can be recovered easily after the process is completed. The recovered ferric sulphate can be reused with a reasonable success. In the acid catalytic reaction, on using any one of the above catalysts two layers are formed; the upper being the ester and unreacted triglyceride. The lower layer is water which is removed by separation. This process catalysed by acid is often referred as pretreatment. The upper phase is then collected and analysed for the reduction of FFA. Once the FFA is reduced to the accepted level it becomes eligible to pass on to the second stage. The reaction conditions for the first stage (pretreatment stage) with acid catalysts are as follows:

1. Higher molar ratio of alcohol as high as 45:1 (Freedman et al. 1986; Liu 1994; Crabbe et al. 2001).
2. Higher reaction temperature (60–120 °C) (Freedman et al. 1986; Liu 1994; Demirbas 2003).
3. Use of long chained alcohols (Nye et al. 1983)

The reduction in the FFA level is monitored as a function of the efficiency of the first stage process. Though the stoichiometric requirement of alcohol is only 3:1 on molecular basis, the rubber seed oil demands more alcohol so as to maintain the steadiness of the forward reaction. The suggested molar ratio to reduce the FFA through acid catalysis is around 9–10:1 (alcohol to oil) (Fig. 18.12).

The esterification reaction catalysed by acid requires high temperature. In an open system a maximum temperature of 55 °C may be maintained. In case the temperature is raised beyond, alcohol tends to evaporate. Therefore, any increase in the reaction temperature beyond 55 °C has to be made in a closed chamber where the reaction may take place under pressure.

According to Devi (2010), every raise in a degree centigrade the FFA content of the oil is reduced by 0.67% in 65 min when the rubber seed oil is treated with 3.9 mL sulphuric acid per litre in a methanol to oil ratio of 9:1.

In the acid catalysed pretreatment, i.e. in the first stage the type of alcohol chosen for use has its own importance. Long chain alcohol (*n*-butanol) has the capacity to yield higher quantity of ester when it is mixed with the oil containing significant quantities of FFA as long chain alcohol is relatively more soluble in oil.

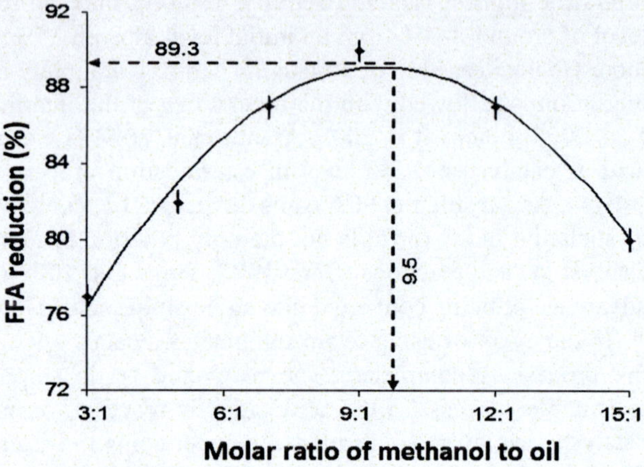

Fig. 18.12 Effect of methanol to oil ratio on the reduction of FFA in the rubber seed oil during the first stage (pretreatment) process (Devi 2010)

Table 18.15 Comparison of the efficiency of the catalyst as a function of the yield of methyl ester when acid treated rubber seed oil is transesterified at 60 °C (Devi 2010)

Catalyst	Dose (g L^{-1})	Yield (%)	Duration (min)
Sodium hydroxide	3.7	83.4	65
Potassium hydroxide	4.1	78.5	68
Sodium methoxide	3.0	77.8	65
Calcium oxide	2.2	79.5	132

18.4.10.2 Second Stage

Once the FFA level of oil is reduced to around 3% after the first stage acid catalysed reaction (pretreatment) the regular alkali catalysed reaction is taken up as a second stage operation. Before entering into the second stage it is necessary to remove the acid present in the contents of the reactor by washing the same with warm water till the litmus paper present neutrality to the wash water. The washed oil thus obtained is then treated with sodium hydroxide at the rate of 2–6 g L^{-1} and methanol 200–250 mL L^{-1} at 50–55 °C and stirred gently for 60–75 min. Care is taken not to stir the mixture vigorously so as to prevent foaming and soap formation. When the reaction is completed the contents are either centrifuged or allowed to stay undisturbed for a long time so as to have the separation of the biodiesel. The biodiesel being lighter moves up and the glycerol settles down as a bottom layer. The biodiesel is then separated and heated to 80 °C so as to expel and distill out the unreacted excess methanol. It is subsequently washed with warm water till the wash water is neutral. The purified biodiesel is then dried either by heat treatment or by passing it through a suitable dehydrant.

Homogeneous base catalysts such as sodium hydroxide, potassium hydroxide, sodium methoxide and heterogeneous base catalysts such as calcium oxide are used as base catalysts. It is known that sodium hydroxide has proved to be a best catalyst among the other base catalysts such as potassium hydroxide, sodium methoxide and calcium oxide (Table 18.15). However if sodium hydroxide is added beyond the

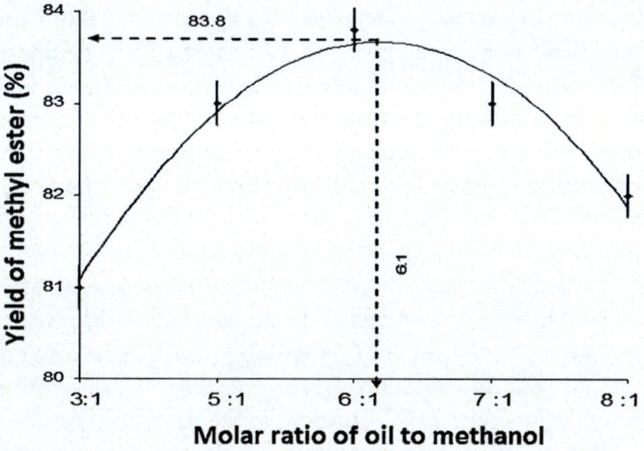

Fig. 18.13 Effect of methanol to oil molar ratio on the yield of ester at 60 °C in 60 min using sodium hydroxide as catalyst in the second stage of the double stage transesterification (Devi 2010)

optimum requirement, emulsion is likely to be formed which may cause the increase in viscosity and formation of gel. Similarly the esterification process remains incomplete if sodium hydroxide quantity is less than the optimum. Thus the catalyst plays a crucial role in the process.

Sodium hydroxide at a concentration of around 3.7 g L^{-1} effectively carries out the second stage reaction. The yield will be high when the molar ratio of methanol to oil is around 6:1 (Fig. 18.13). When the ratio is increased beyond 6:1, the yield of ester decreases slowly and steadily.

The formation of ester is observed to be very high at 60 °C but any temperature between 45 and 60 °C is considered ideal. Temperature greater than 60 °C has adverse effects by accelerating the saponification during the process of alcoholysis. It is known that the duration of the transesterification process of this oil ranges from 65 to 70 min when sodium hydroxide or potassium hydroxide or sodium methoxide is used, whereas the duration time doubles (120–140 min) when calcium oxide is used as a catalyst since it is a heterogeneous catalyst.

18.4.11 Deacidification of Oil

Deacidification is another process for reducing the FFA. Many methods of deacidification are being practiced. They are physical deacidification, miscella deacidification, biological deacidification, chemical re-esterification, solvent extraction, supercritical fluid extraction, membrane technology, etc.

In physical deacidification, the FFA is stripped away by steam which parallelly removes any unsaponifiable substances and foul odour. In this method the soap formation is averted and as a result the loss of oil is low. The refining of crude oil prior to solvent stripping in an extraction plant is termed as miscella refining. Oil and hexane are mixed with sodium hydroxide solution and the soap produced is then removed by centrifugation.

Microorganisms (*Pseudomonas* sp.) isolated from soil is employed to assimilate fatty acids containing less than 12 carbon atoms. Besides, oil containing high level of FFA is treated with fungal (e.g. *Mucor miehei*) lipase and glycerol to bring down the FFA level through the formation of triglyceride. This process greatly relies on certain essential variables such as enzyme concentration, reaction temperature, duration, glycerol concentration, moisture level of the contents and pressure in the reaction vessel. The specific advantage of the enzymatic deacidification is the increase in triglyceride content parallel to the reduction of FFA.

Chemical re-esterification is another process in which the FFA content of the vegetable oil is converted to neutral glyceride with the free hydroxyl group remaining in the oil. In this method the FFA is converted to glycerides. However this process may not convert the entire FFA. Therefore residual FFA is likely to be present. Residual FFA if present in the finished product, it is treated by other modes.

Supercritical fluid extraction has emerged as a favourable method to reduce the FFA content with many advantages. If extraction is made using a suitable solvent at temperature and pressure in its critical state, it is referred as supercritical fluid extraction. At such critical point the system functions without any distinction between liquid and gas. Carbon dioxide is often used as a solvent at the reaction pressure of 20 MPa and at a temperature of around 60 °C. Other solvents used in the process are ethylene, propane, nitrogen, nitrous oxide and monochlorofluromethane.

Membrane technology, reverse osmosis, ultra-filtration and microfiltration methodologies are tried with limited success in the reduction of FFA. The chemical, physical and miscella deacidification methods used in the reduction of FFA are confronted with many drawbacks. Novel methods such as biological mode of action and solvent extraction are not been followed in industrial scale operation.

18.4.12 Deacidification at Laboratory Scale

In a 2 L beaker 1 kg of rubber seed oil, 50 g of glycerol and 2.5 g of zinc chloride as catalyst are taken. The contents of the beaker are then heated and maintained at 180 °C for 2 h with mild stirring. The FFA reacts with the hydroxyl groups of the glycerol causing 60–70% reduction of the FFA. Parallelly, the triglyceride content also increases due to the conversion of monoglyceride and diglyceride present in the oil.

18.4.13 Esterification (Direct)

In the esterification process the FFA present in the oil is directly converted to alkyl ester through acid catalysed reaction. Such process is fairly slow and requires a higher temperature. Often sulphuric acid is used as the catalyst. While the FFA present in it is converted to alkyl ester, the triglycerides are also parallelly transesterified to esters. This reaction is fairly unstable and reversible. According

Table 18.16 GC-MS analysis of rubber seed oil biodiesel

Retention time (min)	Chemical compound	Molecular formula	Molecular weight	Percentage of conversion
7.353	Decanoic acid, methyl ester	$C_{11}H_{22}O_2$	186	5.62
10.226	Dodecanoic acid, methyl ester	$C_{13}H_{26}O_2$	214	19.27
10.501	Nonanedioic acid, dimethyl ester	$C_{11}H_{22}O_4$	216	1.49
12.691	Tetradecanoic acid, methyl ester	$C_{15}H_{30}O_2$	242	19.59
14.860	9-Octadecenoic acid (Z)-, methyl ester	$C_{19}H_{36}O_2$	296	4.17
14.923	9-Hexadecenoic acid, methyl ester, (Z)-	$C_{17}H_{32}O_2$	268	19.48
15.046	9-Hexadecanoic acid methyl ester, (Z)-CAS	$C_{17}H_{32}O_2$	268	2.08
15.195	Hexadecanoic acid, methyl ester	$C_{17}H_{34}O_2$	270	28.30

to Lotero et al. (2005) the whole reaction takes place in four successive steps. They are as follows:

1. Protonation of the carbonyl group.
2. Protonation of the alcohol group.
3. Actual migration of the proton.
4. Liberation of the alkyl ester.

During the protonation, charge develops at the carbonyl group and at alcohol due to the addition of proton. In the next stage, proton movement takes place so as to liberate the esters at the terminal stage. In recent years solid state catalysts are being employed in the place of liquid catalysts, since liquid catalysts are mostly corrosive. Besides, solid state catalysts help to reduce the operation cost with a high turnover coupled with ideal absorption characteristics, repeated usability and stability with high porosity so as to offer increased sites for action. Such materials include zeolites, silica, molecular sieves doped with metals such as aluminium or titanium and sulphonated zirconia.

Calcined eggshells containing CaO are impregnated with Al_2O_3 for use as heterogeneous catalyst for biodiesel production from rubber seed oil. The catalyst yielded a maximum conversion of 98.9% at optimum reaction conditions of methanol to oil molar ratio (12:1), catalyst 3% by weight and reaction time 4 h. The biodiesel obtained was analysed by GC-MS (Lakshmi et al. 2020) and the data are given in Table 18.16. It is known that the biodiesel (methyl ester) is predominantly formed of hexadecanoic acid, tetradecanoic acid, 9-hexadecenoic acid, dodecanoic acid.

18.4.14 Properties of Biodiesel

The fuel properties of biodiesel prepared from rubber seed oil are given in Table 18.17. The standard for acid value of biodiesel is 0.5 mg KOH g^{-1} (ASTM D 6751). Any value higher than 0.5 mg KOH g^{-1} may damage the injector assembly and adversely affect the pump and filters. Similarly high level of free fatty acid may also affect the fuel properties such as low temperature performance, oxidative stability, kinematic viscosity and lubricity. It is also known that free fatty acid level in biodiesel is considered as a pro-oxidant which has a negative influence on the oxidative stability. When glycerol is removed from the oil during transesterification, the specific gravity of the oil tends to go down. The recommended specific gravity according to ASTM D 6751 is 0.87–0.91 g cc^{-1}. The conventional diesel has a specific gravity of 0.85 g cc^{-1}. Higher specific gravity causes the delivery of greater mass of fuel.

Kinematic viscosity is another factor which indicates the quality of the biodiesel. According to ASTM D 6751 the viscosity of the biodiesel should be 3–6 cSt at 40 °C. Higher viscosity affects the injection of the fuel which naturally leads to

Table 18.17 Fuel properties of biodiesel prepared from rubber seed oil (Devi 2010)

Parameters		Value
pH		6.9–7.1
Acid value	(mg KOH g^{-1})	0.1–0.12
Free fatty acids	(%)	0.05–0.06
Specific gravity at 30 °C	(g cc^{-3})	0.87–0.90
Viscosity at 40 °C	cSt	5.5–6.5
Iodine value	(mg I_2 g^{-1})	1200–1500
Saponification value	(mg KOH g^{-1})	180–200
Peroxide value	(meq kg^{-1})	0.8–2.0
Heating value	(MJ kg^{-1})	35.5–36.5
Sulphated ash	(%)	0.01–0.02
Carbon	(%)	76.5–76.7
Hydrogen	(%)	12.2–12.3
Oxygen	(%)	11.0–12.0
Sulphur	(%)	0.02–0.03
Water and sediments	(%)	0.01–0.02
Copper corrosion(3 h at 50 °C)		1a
Vacuum distillation at 90 °C	(%)	345–350
Carbon residue	(%)	0.02–0.03
Flash point (closed cup)	(°C)	135–235
Pour point	(°C)	−8.0 to (−9)
Cloud point	(°C)	0.4–4.8
Cetane number		44–49
Total glycerol	(%)	0.21–0.22
Free glycerol	(%)	0.01–0.02

Fig. 18.14 Percentage distillate of biodiesel with respect to temperature

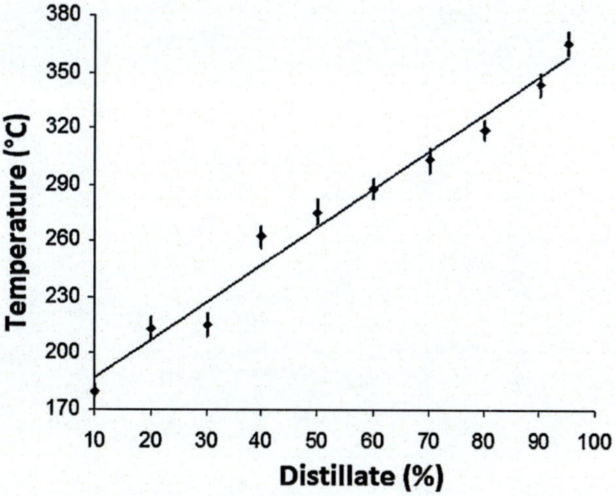

larger droplet size, inadequate atomization and unsatisfactory spraying of the fuel in the combustion cylinder. These conditions are likely to cause deposit of soot in the fuel conduits of the engine. The viscosity of the biodiesel changes if the temperature is altered. Such change is rapid in the case of biodiesel, whereas in the conventional diesel such change is not conspicuous. In a distillation trial it is observed that the percentage of distillate increases when the temperature is increased (Fig. 18.14).

High viscosity occurring in biodiesel is often due to incomplete processing of the oil or inefficient purification of the final product leaving the conjugated or free glycerol as a contamination constituent in the methyl ester phase. The viscosity increases with the increase in the chain length and decreases with the increase in the number of double bonds.

The iodine value of the biodiesel from rubber seed oil is around 1250 mg I_2 g^{-1}. It indicates the level of unsaturated fatty acids present in it. The iodine value of the biodiesel is largely governed by the type of oil used as the raw material. The auto oxidation, polymerization and hydrolysis which are taking place during the period of stay of seed on the ground normally enhance the iodine value of the oil which in turn reflects on the biodiesel prepared out of it. High iodine value though not welcome, too low a value also is not good. Unsaturated fatty acids as known from the high iodine value cause deposits in the fuel line or cause deterioration of the lubricating quality of the oil. Such effects are likely to increase as the number of double bonds of the fatty acid present in the biodiesel increase. Therefore it is ideal to have a biodiesel with a low level of certain unsaturated fatty acids such as linolenic acid than to limit the overall level of unsaturated fatty acids. The presence of a limited level of unsaturated fatty acids is welcome as it restricts the methyl ester from solidification. In the rubber seed oil, the linolenic acid content is lower than the other fatty acids such as oleic acid and linoleic acid (Table 18.12). The ASTM D 6751 is silent on the limit of the saponification value in biodiesel. The peroxide level in biodiesel is an index of rancidity and therefore it is appreciated if its level is low.

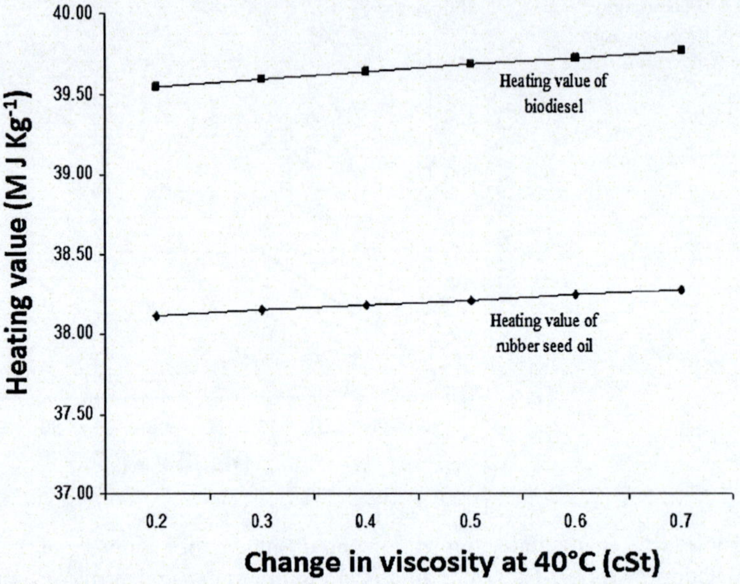

Fig. 18.15 The relationship between the heating value and the change in viscosity of rubber seed oil and biodiesel produced out of it

The heating value of the biodiesel is approximately 5–8% lower than that of the diesel. This indicates that around 8–10% volume of biodiesel would be additionally required to make a travel of the same distance normally covered through diesel. It is due to the presence of oxygen as an extra component in the biodiesel causing lower calorific value. However the extra oxygen present in it helps to enhance the combustion of the fuel. Besides it is to be known that the heating value of a biodiesel is normally lower than that of the corresponding oil from which the biodiesel is prepared due to the removal of various factors contributing to the viscosity of the oil. It is reported that there is a feeble positive correlation between viscosity and heating value of both oil and biodiesel. The heating value of the fuel increases with the increase in the carbon number of the fuel. The heating value also is directly related to the ratio of carbon and hydrogen to oxygen and nitrogen. The change in heating value in accordance with the change in viscosity of the rubber seed oil and biodiesel is presented in Fig. 18.15. When the viscosity is altered by 0.2–0.7 cSt from its original value, there is feeble change in the heating value of the oil and biodiesel.

The limit of sulphated ash in biodiesel is 0.02% (ASTM D 874). Sulphated ash indicates the presence of metallic contaminants. Insoluble metallic soap, residual catalyst and other solids are the possible sources of sulphated ash. If the catalysts are not washed away in full, there is a possibility of having high ash content.

The copper strip corrosion value is an index of the relative degree of corrosivity of petroleum product due to certain corrosive chemicals present in the fuel. A polished copper strip is immersed in 30 mL of the fuel sample at an elevated temperature. After a fixed time the strip is removed and examined for any sign of corrosion. A scale 1–4 is assigned as per ASTM 130 standards. Low value for copper strip corrosion is preferred.

Distillation is a known process to purify the biodiesel. The distillation temperature provides a rough idea of the boiling point of the fuel. It is also used to characterize the fuel in terms of the boiling temperatures of the various components. The distillation characteristics of the biodiesel prepared from rubber seed oil are observed to be far different from that of the regular diesel. Biodiesel does not contain any highly volatile compounds and therefore it normally evaporates at a high temperature which is around 345 °C at 90% distillate. The distillate of the biodiesel increases with the increase in temperature.

When the FFA is high, the volatility of the biodiesel is observed to be low. High distillation temperatures shorten the ignition delay of the biodiesel and decrease the occurrence of knocking of the engine. The atmospheric boiling point of biodiesel generally ranges from 330 to 357 °C. The flash point of the biodiesel is higher than 130 °C and hence biodiesel is nonhazardous especially during storage. The limit specified for carbon residue according to ASTM D 6751 is 0.05%. The carbon residue increases exponentially when the biodiesel is blended with the conventional diesel. This is a major setback being faced in biodiesel blend.

The pour point of the rubber seed oil is around −9 °C. In such low temperature the biodiesel solidifies and forms crystals. Such crystals later agglomerate and clog the fuel line causing the breakdown of the engine. If the biodiesel has considerable amount of saturated fatty acids, the crystallization takes place at a higher temperature itself. Cloud and haze of crystals appear in the biodiesel made from rubber seed oil when the temperature is around 5 °C. Since the saturated methyl esters are the first to precipitate, the quantity of these esters, namely methyl palmitate and methyl stearate determines the cloud point.

The cetane number of the biodiesel produced from rubber seed oil ranges from 45 to 50. The cetane number is mainly decided by the length of the carbon chain and level of saturation of the fatty acid. Higher cetane number enhances the combustion efficiency which improves the cold start and lowers the exhaust gases. The cetane number of biodiesel is higher than that of the regular diesel. Consequently, the ignition delay becomes shorter.

The presence of glycerol in any form is not good for the best performance of the biodiesel. As glycerol has higher density than oil it needs to be removed to the extent possible. Presence of glycerol indirectly indicates the incomplete transesterification. Glycerol being denser than biodiesel settles down on the floor of the fuel tank and it is the first to enter the fuel outlet immediately after the parked vehicle commences the drive. Besides, free glycerol is a source of carbon deposit in the engine. The limit of the glycerol in biodiesel is 0.02% as per ASTM D 6751.

18.4.15 Shelf Life of Biodiesel

The degradation of biodiesel is relatively faster. If the oil from which the biodiesel is prepared has more saturated fatty acids, the stability of the biodiesel is higher than that of a biodiesel prepared from oil containing more unsaturated fatty acids. Biodiesel prepared from the oil containing polyunsaturated fatty acid is more susceptible to auto oxidation and polymerization under high temperature, moisture, ultraviolet radiation and long storage (Wu et al. 1998; Ferrari et al. 2005). The stability of the biodiesel largely depends on the temperature, exposure to air and the length of the chains of the fatty acid present in it. The first stage of chemical change is the oxidation which produces peroxides. This peroxide is unstable and consequently it forms short chain fatty acids, aldehydes, ketones, hydrocarbons and alcohols. Such change is often catalysed by the light. Biodiesel kept stored in day light has more acid value than that of the one stored in darkness. The acid value increases when the storage duration is increased (Fig. 18.16).

Though the peroxide is likely to decompose gradually a situation may develop in which the rate of production of peroxide is higher than the rate of its decomposition. Consequently such increase in peroxide value is known to enhance the cetane number which tends to alter the ignition delay time (Dunn 2008). Higher peroxide content in the biodiesel is also due to the presence of a high level of oxygen present in the vegetable oil (Vicente et al. 2005). Figure 18.16 gives an indication that the peroxide value increases in accordance with the length of storage. Hydrogen peroxide is the primary product of the oxidative degradation. This degradation is activated by the elevated storage temperature. However peroxide value is never considered as a tool to monitor the shelf life of the biodiesel as its concentration usually does not change in accordance with the age of the biodiesel. Similarly, when biodiesel is hydrolysed due to its exposure to air, alcohol is formed. Such generation of alcohol

Fig. 18.16 Increase in the acid value in biodiesel during the storage for a period of 1 year (Devi 2010)

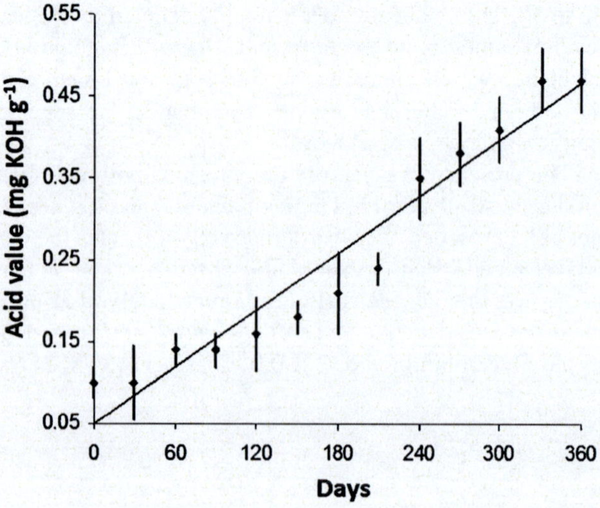

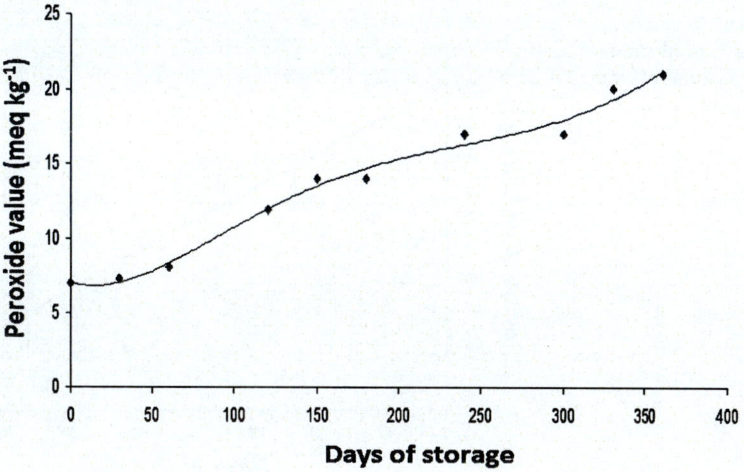

Fig. 18.17 Increase in peroxide value of biodiesel during the storage for a period of 1 year (Devi 2010)

causes the reduction in the freezing point. Parallelly the total acid number increases. Thus the biodiesel loses its natural character. Though the raw oil containing more saturated fatty acid is relatively stable, the cold flow properties are adversely affected. The product of oxidation of the biodiesel interferes with the normal functioning of the engine. The rate of oxidation of the 18 carbon esters are linolenic > linoleic > oleic (Cosgrove et al. 1987). Such rating is due to the fact that the di- and tri-unsaturated fatty acids have more reactive sites for the initiation of the auto oxidation chain reaction sequence (McCormick et al. 2007). The auto oxidation is not exclusively due to the presence of the double bond but also governed by the position of the double bond. The acid value slowly increases as the period of storage increases. The increase is very feeble during the first few days and it tends to increase rapidly at the later stage (Fig. 18.17).

The iodine value unlike other parameters is inversely proportional to the storage time. The decrease in iodine value is an indication of lipid oxidation since there is a decline in unsaturation during oxidation. In few occasions, continued storage of biodiesel for a long time at a temperature higher than 180 °C also causes the iodine value to increase due to the elimination of the oxidized triglyceride molecules.

If biodiesel is prepared using a base catalyst, the FFA level will be very low due to the neutralization effect. However upon degradation of the biodiesel due to exposure to water and air the FFA reappears. At high temperature the FFA reacts with the engine parts made of metals such as zinc, lead, manganese, cobalt and tin causing serious damage.

The kinematic viscosity also is known to increase during the long storage. Besides the generation of peroxide parallelly develops polar and oxidized polymeric compounds. These compounds lead to the formation of microbial mass which causes the increase in viscosity. Saturated fatty acid having a long carbon chain with a high

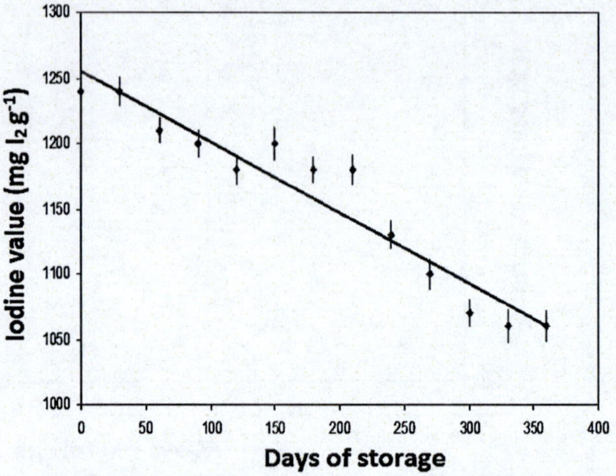

Fig. 18.18 Decrease in iodine value of biodiesel during the period of storage (Devi 2010)

molecular weight and strong molecular bonding causes the biodiesel to be viscous due to oxidation.

Short duration of storage may not cause any change in the specific gravity. Minor deviation in specific gravity takes place when the storage duration is increased. Even such minor change depends on the saturation level of the FFA present in the biodiesel. Higher specific gravity of the biodiesel is directly correlated to the emission of particulate matter and nitrous oxide. Water and sediment also increase with the increase in storage duration (Fig. 18.18).

The water content of the biodiesel should be less than 0.5%. Presence of water also opens the chance for microbial growth besides the water may settle at the bottom of the fuel tank of the automobile due to its relatively high density. The water thus settled is likely to corrode the tank. It may also stand on the fuel gate and reduce the flow of the fuel from the tank to the combustion cylinder. The water may also develop emulsion with oil along with the hydrolysis of triglyceride. Thus the fuel injector's efficiency is affected. It is known that the increase in water content and the quantity of sediment are a function of storage duration (Fig. 18.19)

18.5 Jatropha (*Jatropha curcas*)

Jatropha, *Jatropha curcas* is a small tree with glabrous branches offering great excitement to scientists working in the field of bioenergy. This tree is favoured due to its short gestation period, low cost cultivation, endurance to adversities and easy establishment in unfavourable agro-climatic conditions. It can tolerate extremes of temperature but cannot withstand frost. It is a deciduous and succulent tree and is known to shed its leaves to tide over drought conditions or winter season. It is not sensitive to the change in length of the day. Under very favourable agro-climatic conditions this tree grows up to a height of 10 m. This tree is known to live for

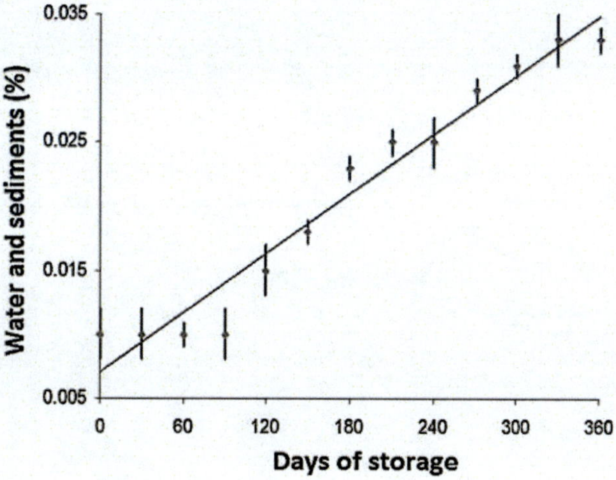

Fig. 18.19 Increase in water and sediments of biodiesel during the storage (Devi 2010)

40–50 years and produces seeds thrice a year. The genus Jatropha has approximately 170 known species and its name is derived from the Greek words *Jatros* which means a doctor and *trophe* means food.

It is a native of tropical America, which naturalized steadily throughout the tropical and sub-tropical parts of Asia and Africa. This tree is believed to have been spread by Portuguese seafarers via the Cape Verde Islands and Guinea to other countries in Africa and Asia (Heller 1996). It is now distributed in Argentina, Australia, Benin, Jamaica, Bolivia, Brazil, China, Egypt, El Salvador, Ethiopia, Fiji, Ghana, Guinea, Honduras, India, Madagascar, Mali, Mexico, Mozambique, Panama, Paraguay, Peru, Puerto Rico, Namibia, Senegal, South Africa, Sudan, Tanzania, Uganda, Zambia and Zimbabwe (Henning 1997; USDA 2000). However, of late it has a pan tropical distribution and has now been listed as a weed in Australia, South Africa, India, Brazil, Fiji, Honduras, Panama, El Salvador, Jamaica, Puerto Rico and other parts of Caribbean (Fig. 18.20). The *J. curcas* is being called by different vernacular names in various parts of the world (Table 18.18).

18.5.1 Habitat

Three distinct varieties of *J. curcas* are known. They are

1. Cape Verde variety which has slowly spread all over the world
2. Mexican variety with non-toxic fruits
3. Nicaraguan variety with larger fruits

J. curcas is well adapted to arid and semi-arid conditions. It can withstand wide range of rainfall. It is also reported to establish well in a place with an annual precipitation level ranging from 450 to 2300 mm (mean 1440, $n = 58$) and at an

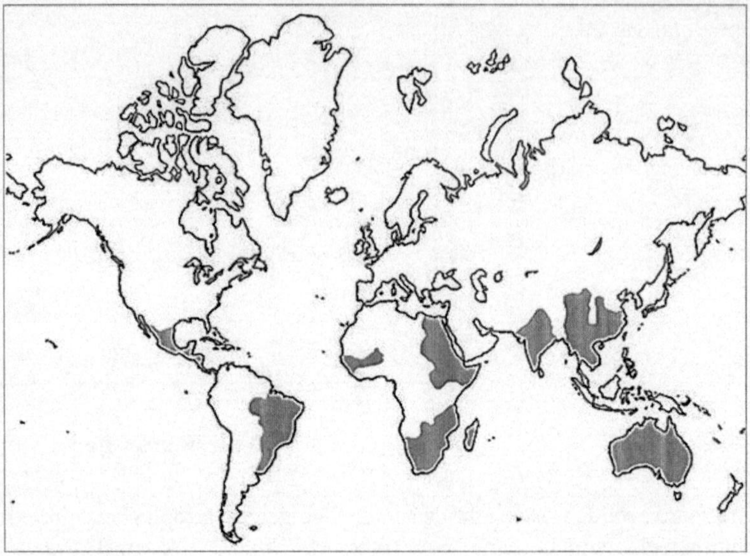

Fig. 18.20 Distribution pattern of *J. curcas* in the world as seen in the hatched area

annual temperature ranging from 18 to 28.5 °C (mean 25.2 °C, $n = 45$). Barring water logged lands it can establish in any worst terrain and soil conditions such as eroded and gravelly lands, sandy saline soils, sand dunes, crevices of hard rocks and in poor rainfall conditions. It can survive in long drought of 7–8 months depending on the air humidity. It can rarely withstand mild frost. Even in an untenable barren land it can grow. During prolonged drought it tends to control its evapotranspiration rate by shedding the leaves. The leaves, which fall on the ground, are known to enhance the humus level of the land. Such increase in humus often helps to establish the soil biota. The fumic acid and fulvic acid which develop in such situation assist in increasing the fertility of the land. The *J. curcas* cultivation thus helps to reclaim the soil besides it is grown as a live fence to exclude farm animals.

The leaves and stems are relatively succulent which will be able to guard its water even in an exceedingly high desiccating environment. The stem contains extracellular mucilage which helps to bind water in the plant body. The roots are capable of penetrating deep into the soil in search of ground water for its survival. Besides, a large gap remains between the rows of trees are observed to be netted with bundles of absorbing roots of this tenacious tree. Such tree which struggles to exist during the drought period will be relieved from the state of idle growth and swiftly enter into the growth phase within a week of rainfall. This tree is not browsed by goat or cattle due to the presence of toxic substances such as *phorbol* ester and *curcin* a highly toxic protein.

Table 18.18 Vernacular names of *J. curcas* (Invasive Species Compendium CABI)

Country	Vernacular names
Brazil	Figo do inferno Mandubiguasu, munduyguasu, pihao de purge, pinhao do paaguay, pinhao manso
Cambodia	Lohong khvangsu
Cameroon	Botije, botuje, botuje-ubo, lobotuje, olobontuje
Cape Verde	Pulguiera
Caribbean	Feved'enfer, herbe du bon dieu, herbe du diable mancenillier benit, medicinier benit
Cook Islands	Fiki, pakarani, piki, tuitui pakarangi, tuke
Cuba	Piñón criollo, piñón lechero, piñón vómico
Egypt	Habbel-meluk
Ethiopia	Ehanduejot, erundi, jangli-yarandi
Fiji	Banidakai, fiki, manggele, maqele, mbanindakai, ndrala, uto ni vavalangi, wiriwiri, vavalangi
French Guiana	Barane, medeicinier
India	Kadalmanakku, kaitta, kananeranda, kattamanakku, kattavanakku, kattukkottai
Indonesia/ Java	Dijark
Iran	Dandebarri dandenahri
Italy	Fava purgatrice, giatrofa catarcita, ricino maggiore
Mali	Baga-ni, iridingue
Mauritius	Pignon d'Inde
Mexico	Avellanes purgantes, pinon Mexicano, sangregaod
Mozambique	Sassi
Myanmar	Kesugi, thinbankyekku, thinbaukyeksu, thinbawkyetsu
Nepal	Kadam
Netherlands	Purgeernoot
Philippines	Bolongcauit, casta, cator, kator, taatava, tuba
Saudi Arabia	Pignon d'Inde
Sri Lanka	Kaddamanakku
USA/Hawaii	Kuikui Pake, kuku'

18.5.2 Botanical Features

It has large, green to pale green, 3–7 lobed leaves arranged alternate to sub-opposite pattern with spiral phyllotaxis. The tender green leaves are often used as a feed to tasar silkworms. Dark blue dye and latex obtained from the leaves have many pharmaceutical applications. The bark is rich in tannin.

These trees normally come to flowering immediately after first few showers of rain. Flowering has two peaks, one at summer and the other at autumn. However there will be continuous flowering throughout the year especially in humid areas. The inflorescence is auxiliary paniculate with polychasial cymes. The plant is monoecious with staminate (male) and pistillate (female) flowers arranged in the

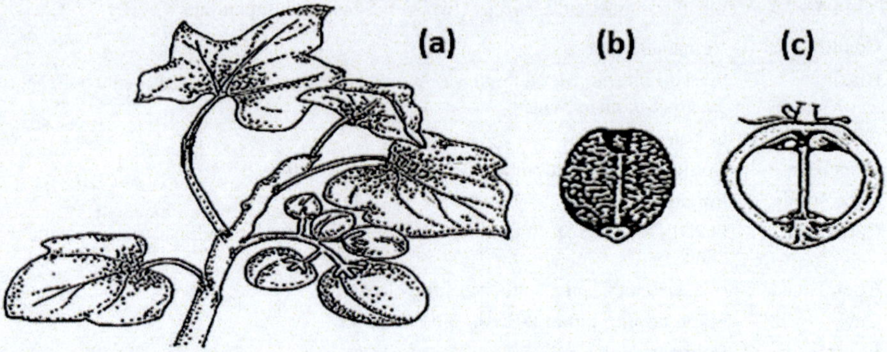

Fig. 18.21 (**a**) A branch of *J. curcas* (**b**) full seed (**c**) longitudinal section of the seed

same inflorescence. In the paniculate inflorescence, the female flowers which are less in number (10–20%) are seen arranged at the apices of the main stem. The male flowers are more in number (80–90%) and occupy subordinate position in the inflorescence. The flowering stage is considered to be crucial since the unaborted flowers solely decide the total yield. The total number of female flowers present in the tree and the rate of their fertilization determine the number of fruits being developed. Male flowers remain open for 8–10 days, whereas female flowers are in open condition only for 2–4 days. Most often, in a single branch all stages of floral development are represented. Mature fruits, known by their yellow colour normally stay at the base of the branch, whereas green fruits which are partially ripe are seen in the middle of the branch. The fresh immature flowers are seen located at the top of the branch. Good soil, adequate rainfall, nutrient availability and humidity decide the success of the fruit formation. Around 50% of the flowers are normally aborted during the first stage of the fruit formation. A branch of *J. curcas*, full seed and a longitudinal section of the seed may be seen in Fig. 18.21. A bunch of fruits is shown in Fig. 18.22.

The fruit is ellipsoid in shape and 2–3.5 cm long. Each fruit has three seeds. The seeds roughly resemble that of the castor in their shape. The seed has a maximum oil content of 44%. The oil content is observed to be high only in fully matured seeds. In tender seed the free fatty acid level is high which is replaced slowly with triacylglycerol as the seed reaches maturity.

The oil extracted from the seed is viscous and a large portion of it is currently being used in the manufacture of candles, soaps, cosmetics and also utilized for biodiesel. The first commercial scale application of jatropha oil was reported from Lisbon where the oil imported from Cape Verde was already in use for the production of soap and for burning lamps.

Fig. 18.22 A bunch of *J. curcas* fruit

18.5.3 Propagation

J. curcas is not capable of self-propagation. It is propagated through seeds and also through stem cuttings. Seeds of the current year only will give the maximum rate of germination. Aged seeds normally have poor viability. When seeds are dropped in a measuring cylinder containing water, good seeds rapidly sink down to the bottom. Such seeds are to be chosen and sown either straight in pits dug in the field or they are to be sown first in containers and then transferred as seedlings to the field. For maximum survival, planting through seedlings is recommended. For the best of convenience, seedlings may be raised in polyethylene containers. Black polyethylene bags containing sandy loam and farmyard manure mixed at a ratio of 3:1 are taken. Seeds soaked in dilute cow dung slurry for 12 h for maximizing the rate of germination are sown in the above polyethylene bags at a depth of 2 cm. The seeds germinate within 9 days. After 30–40 days the seedlings may be transferred to the field. It is better that seeds are to be sown in bags roughly 2 months ahead of the monsoon so that at the onset of monsoon such seedlings may be transferred for planting in the field. In such arrangements the tender plants may escape drought. Normally five roots are formed; one tap root (central) and four peripheral roots. These roots fix the plant firmly in the soil and therefore the plant can withstand any type of wind. Besides, the penetrating roots help to take the water from deep underground. In the field the tree grows to a height of 4 m in 2–3 years. It comes to the first batch of yield in 9–12 months. However effective economic yield as indicated in many survey reports is possible only after 24 months that too after the second rainy season. The dry seed yield during different growth periods is indicated in Fig. 18.23. Trees derived from seeds may express heterogeneity.

If the tree is to be propagated through stem cutting, selection of stem is considered very important. Stem should be of the length 25–30 cm and it has to be obtained from 1 year old branch. Planting may be carried out during the rainy season. If the mother plant from which the stem is taken is derived from the seed, aged stem may

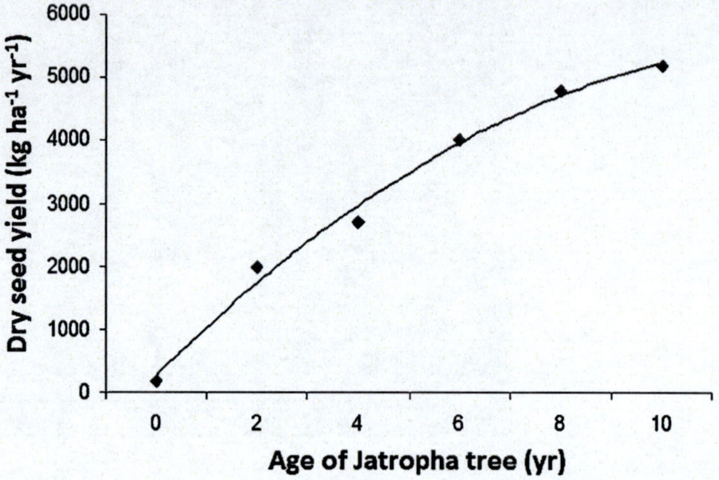

Fig. 18.23 The dry seed yield of *J. curcas* according to its age ($n = 11$)

be avoided as it normally has poor sprouting ability. Cuttings made from tender stem have high sprouting ability (Hartmann and Kester 1983). Trees raised from stem cutting will be devoid of tap root. Alternatively such trees produce pseudo and shallow tap roots which may not be able to support the tree effectively during strong winds. Such trees have relatively low longevity and have low resistance to agro-climatic conditions. Trees raised from stem cuttings are known to yield fruits earlier than that of the trees raised from seeds.

18.5.4 Cultivation

Mass plantation of this tree is the first step in the production of biodiesel. There is a wide spread apprehension that the large scale cultivation of this species is likely to take away the agricultural lands thereby competing with food production rather than complementing each other. Therefore it is all the more necessary to have a resolve to cultivate this tree in waste lands. Waste lands are not useless lands. They may be ideal lands in all respects but not put to plant husbandry effectively due to one reason or the other.

J. curcas is also being cultivated in irrigated and rainfed conditions. The land which is chosen for this purpose has to be necessarily with ideal slope to enable quick drainage. Even nutritionally impoverished hilly slopes can also be considered. If a land is known to be water logged for a long time after a good rainfall, it is not suitable to plant this tree. Prolonged wetness causes stress to this tree since water saturated soils do not provide sufficient aeration to its roots. The roots of jatropha cannot tolerate exceedingly high wet conditions for more than 2–3 days as pores of water logged soil are filled totally with water. In such conditions, deep furrows may be formed at definite gaps so as to promote fast drain of rainwater (Fig. 18.24).

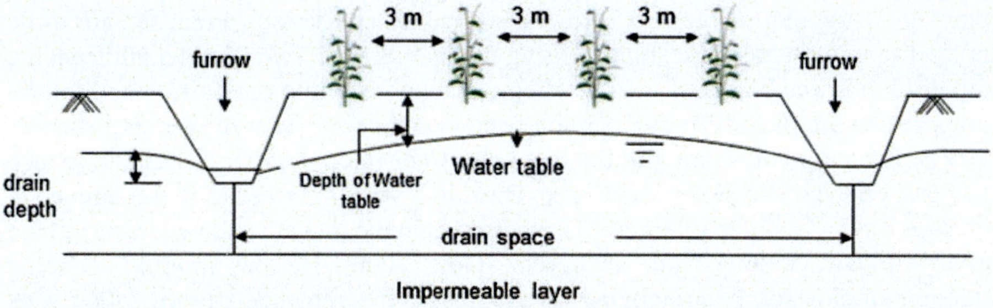

Fig. 18.24 Section of the land indicating the drainage to be created in a water logged derelict area

Table 18.19 The spacing and the number of jatropha trees standing in a hectare

Spacing (M)	Trees ha^{-1}
2.0 × 2.0	2500
2.5 × 2.5	1600
3.0 × 3.0	1100

There is no need to remove minor bushes and jungles during the initial stage of the plantation. These bushes in the field are known to protect the newly developing saplings against hot sun or wind. Once the saplings are established, slowly the land may be cleared of the bushes, weeds and minor jungles. The terrain need not be leveled since, in most of the cases irrigation is not resorted except during the young stage. Soil pH should be between 6 and 8. Wherever necessary minor civil works may be taken up as it deems fit so as to limit the soil erosion thereby to control the leaching of the native nutrients. Such efforts will help to build the soil nutrients caused by the decaying of old leaves which fall from the same trees. Soil depth should be at least 45 cm. Land slope shall not exceed 10°.

Different spacing in planting is being followed. In nutritionally impoverished soils closer spacing is recommended as jatropha may not grow as a large tree. Periodical trimming of the branches will help to establish the tree in the limited space. Wherever the soil conditions and water availability are favourable distant spacing can be considered. The spacing and the corresponding number of trees per hectare are given in Table 18.19.

The recommended spacing in hedge grows for soil conservation is 15–25 cm within and in between in case of double row fencing.

For planting in field, pits of the size 30–45 cm (*l*) × 30–45 cm (*b*) × 30–45 cm (*d*) are taken and they are filled three fourth with the local soil. For faster establishment the pits are to be filled with a mixture of local soil, sand and farmyard manure. Regular weeding, pruning and canopy management are recommended. Pruning helps to develop more number of branches which in turn promotes more inflorescence so as to have increased number of fruits due to high incidence of sunlight. It is good to clip the terminal shoots once in 6 months so as to induce lateral branching (Gour 2006). Pruning is to be carried out at 30–45 cm down from the tip of the branch. At the end of the first year the secondary and tertiary branches are to be pruned so as to induce branching. During the second year, two-third length from the top portion of the branches is to be chopped away retaining one-third length of the branches on the plant. Pruning may be done at any period after the tree has shed its leaves. Such type of shaping of the tree helps to develop more branches and reduce any pest on the down storey of the tree. Once in 5 years the tree has to be trimmed at its base leaving a stump of 50 cm. This would help the tree to re-grow afresh and the tree will start yielding again in a year (Gour 2006). Besides pruning, periodic thinning of plantation by removing certain trees is recommended. From 1600 plants per hectare, the stand density may be reduced to 1000 trees per hectare at the final stage.

Basal application of super phosphate or NPK is reported to increase the yield of fruits. The tree demands high dose of nitrogen and phosphorus. The required level of inorganic fertilizer depends largely on the age and size of the tree (Patolia et al. 2007). In degraded land, application of organic manure is greatly helpful. A plantation supposed to yield 1 t ha^{-1} $year^{-1}$ has to be fertilized with 25–35 kg N, 5–7 kg P and 25–30 kg K ha^{-1} $year^{-1}$ either through organic or inorganic means.

18.5.5 Pests and Control

The jatropha tree harbours insect and fungal pests but does not offer permanent niche for them. As the parts of this tree are toxic to pests, very limited instances of pest damages are reported (Grainge and Ahmed 1988; Sauerwein et al. 1993; Solsoloy 1993). Bugs are reported to damage the fruits (Heller 1996; Grimm and Guharay 1997). During the seed setting stage, *Pachycoris klugii* and *Leptoglossus zonatus* attack the plants and cause economic damage. Stem borer *Lagocheirus undatus*, grass hoppers, leaf browsing beetles and caterpillars also damage the plant and bring economic damage (Meshram and Joshi 1994; Grimm and Maes 1997).

However, jatropha tree needs pest management if yield has to be enhanced. The trees are often infested with the Scutellaridae bug *Scutellaria nobilis* and the inflorescence cum capsule borer *Pempelia morosalis*. Other major pests include *Pachycoris klugii*, *Leptoglossus zonatus*, the blister miner *Stomphastis thraustica*, the semi-looper *Achaea janata* and the flower beetle *Oxycetonia versicolor*. Regular irrigation and fertilizer application actively invite these pests.

18.5.6 Fruits and Seeds

A total of three seeds are present in each fruit. The seeds mature in 3–4 months after flowering. The yield of fruits largely depends on the soil type, spacing, crown characteristic, canopy management, germplasm, age of the tree, propagation methods, climatic conditions, irrigation pattern, water quality, fertilizer regimes and rainfall. The ripe fruit is known by the change of colour from green to lemon yellow and then to deep brown. Ripe fruits are hand plucked. All the fruits available

Table 18.20 Dry weight of fruit, seeds, kernels and husk of *J. curcas*

Fruit part	Weight (g)
Full fruit	1.67 ± 0.43
Seeds (3 no.)	1.08 ± 0.26
Kernels (3 no.)	0.67 ± 0.18
Husk (shell)	0.40 ± 0.16

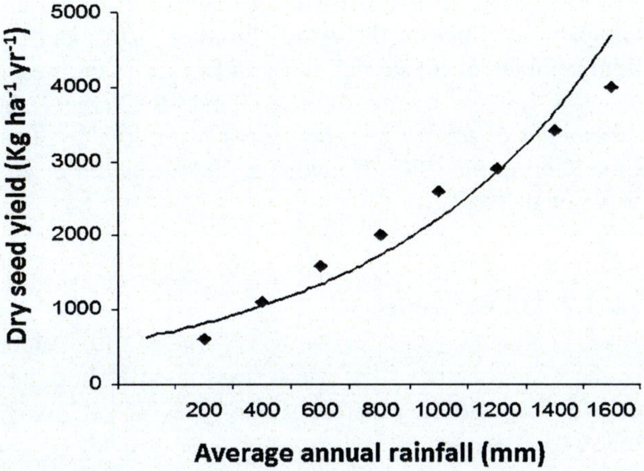

Fig. 18.25 The dry seed yield of *J. curcas* with respect to the rainfall

in a tree do not mature at a time. Therefore the matured fruits are to be harvested at weekly intervals or as and when they are ripe. The duration of the harvest in relation to the time of ripening of the fruit varies according to the climatic conditions, geographic location of the field and the variety of the tree. In semi-arid conditions the harvest duration is approximately 2 months. In humid areas the harvest is made throughout the year. The exocarp (peel) is normally recycled in the same field as an organic manure to enhance the fertility of the soil.

The fruits are sundried immediately after the harvest and peeled to separate the seed. A ripe fruit on an average weighs 1.67 g on dry weight basis. A healthy seed may weigh around 0.35 g and accordingly a set of three seeds have the total weight ranging from 1 to 1.1 g (Table 18.20). The seed is diploid with 2n = 22 chromosomes. The annual seed production from a tree ranges from 0.2 to 2 kg. The achievable production of seed from a semi-arid area or from a cultivable waste land, in a 4–5-year-old plantation is 3–5 t ha^{-1} $year^{-1}$.

If the field is ideal in all respects together with a rainfall ranging from 900 to 1200 mm, the seed yield may go up to 5 t ha^{-1} $year^{-1}$. The dry seed yield as a function of rainfall is shown in Fig. 18.25. The hard seed coat is normally referred as shell or hull or husk. The husk forms around 37% of the seed by weight. The husk has a heat value of 19–20 MJ kg^{-1}, whereas the internal kernel has the heat value of 28–32 MJ kg^{-1}.

The husk may be used for direct combustion in gasifiers as it has a reasonably high heat value. In an open core down draft gasifier, maximum efficiency of 68.3% is

Table 18.21 Protein, lipid and ash contents of the kernel, husk and exocarp (peel) of *J. curcas*

Constituent (%)	Kernel	Husk	Exocarp
Protein	26.3	4.8	6.1
Lipid	58.0	0.8	1.0
Ash	4.1	5.8	15.1

observed at a gas flow rate of 5.5 $m^3\ h^{-1}$ and with specific gasification rate of 270 kg $h^{-1}\ m^{-2}$. The evolved gas has an energy value of 4.6 MJ m^{-3} which is comparable to that of the woody biomass (Vyas and Singh 2007). The protein and lipid contents of the kernel, husk and exocarp are presented in Table 18.21.

More than one quarter of the weight of the kernel (26.3%) is made of protein. The oil content of the pure kernel varies from 40% to 60% (Miinch and Kiefer 1986; Liberalino et al. 1988; Neuwinger 1994; Gandhi et al. 1995; Sharma et al. 1997; Wink et al. 1997).

18.5.7 Oil Extraction

Highest oil content of the whole seed is 44%. In practice such a high level of oil recovery is seldom possible. Four types of oil extraction procedures are being followed. They are

1. Solvent extraction (Extraction by hexane)
2. Pressing the kernels mechanically
3. Aqueous oil extraction
4. Three phase partitioning

Each methodology has its own merits and deficiencies. The efficiency of the four methodologies is presented in Table 18.22.

18.5.8 Solvent Extraction

Through solvent extraction, oil from oil-bearing materials is effectively removed leaving only less than 1% of the total oil un-extracted in the raw material. This method is observed to be efficient if the raw materials contain low level of oil. It is also suitable to extract the residual oil from pre-pressed oil cakes obtained from oil rich raw materials. Due to the high efficiency in oil recovery, solvent extraction has become the most popular method. In this method, crushed kernel is repeatedly refluxed with hexane. The liquid obtained from the refluxing system is distilled to recover the oil. The hexane evaporated during the distillation is condensed back and collected. The recovered hexane is reused in the next batch of extraction. The low boiling point of hexane (68.7 °C) and the high solubility of oil make it a popular and favourable organic solvent throughout the world. Easy availability and the low cost add merits to it. However, hexane is categorized as a hazardous air pollutant (HAP)

Table 18.22 Methodologies of oil extraction in *J. curcas* seed and the corresponding yield of oil

Method		Oil yield (%)
Solvent extraction		44.0
Press	1. Hand press	22.5
	2. Animal driven stone press	23.0
	3. Motor (light duty) press	23.0
	4. Motor (heavy duty) press	27.0
Aqueous oil extraction	1. Simple aqueous oil extraction	17.0
	2. Aqueous enzymatic oil extraction	28.0
	3. Aqueous enzymatic oil extraction cum sonication	32.5
Three phase partitioning (TPP)	1. Basic three phase partitioning	36.0
	2. Enzyme assisted three phase partitioning	40.5
	3. Enzyme assisted three phase partitioning cum sonication	42.7

by the US Environmental Protection Agency. It is extremely difficult to prevent the inadvertent leakage of hexane in the environment during the oil extraction. In addition, some residual hexane may remain in the oil and cake.

18.5.9 Press

If the farm is small with a maximum yield of around 15 t of seed per year, the oil extraction can be made conveniently using a hand press. Mostly oil is extracted by pressing the whole seeds. The kernels are not separated from the seeds before pressing. The seeds are sun dried for a period of 15–20 days so as to expel the moisture. If the pressing is carried out in hot condition, oil separation is favourable as the pasty fat melts down and moves out freely. The hand press may yield oil up to half (22%) of the maximum content. In certain villages oil pressing is still being carried out by animal driven stone press which gives a maximum yield of around 23% only.

Mechanized mode of oil press with the help of a motor is in vogue if the seed yield of a farm is more than 15 t a year. By all means oil recovery in motorized operation is slightly higher than that of the hand operated press. Motorized operation may yield around 250 L per tonne of seeds. Very often decorticator is employed to remove the husk (hull) of the seed. In the decorticator the seed is split open and a current of air blows out the low density woody materials including the husk, twigs and leaves. Such rejected materials do not contribute any oil. These waste materials roughly constitute 30–35% of the seed mass. In the decorticator, the seed at the first instance enters the dehuller. The removal of the husk (shell) is essential otherwise it may give additional unnecessary load to the processing machine in the next stage. Then the clean kernels are passed through a crushing unit (expeller). The expeller is the main equipment for the extraction of the oil. It consists of a pressing box. At the feeding end of the expeller there is a gate, through which the seed received from the decorticator is fed. At the exit point there is a cone which practically restricts the passage. The rotation of the horizontal shaft causes the seed to push forward by its screw action thereby increasing the internal pressure resulting into the squeezing out of the oil. The internal pressure of the expeller is regulated by the adjustment of the cone. The oil thus extracted flows through a perforated box while the cake passes out through the opening around the cone. The oil which flows out of the expeller contains suspended particles. Therefore the oil is pumped into a filter press formed of fabric. The oil when passes through the filter gets filtered and the resultant solid matter is retained.

In many places the use of Bielenberg ram press is in practice. This type of press is known to give an oil yield of around 25%. Diesel energized expellers are in operation for this purpose. The popular Sayari oil expeller (Strainer press) energized with diesel, originally developed in Nepal is now being in use in Tanzania and Zimbabwe. The raw material is repeatedly passed through the expeller so as to enhance the oil recovery. The screw type Komet expeller gives a higher yield (30%) than that of Ram press which gives a recovery of around 27% only.

18.5.10 Aqueous Oil Extraction

In aqueous oil extraction process, it is necessary to separate the kernel from the shell. Aqueous oil extraction is of three types. They are

1. Simple aqueous oil extraction (SAOE)
2. Aqueous enzymatic oil extraction (AEOE)
3. Aqueous enzymatic oil extraction cum sonication (AEOES)

In SAOE the kernel is first made as a paste. It is then suspended in sixfold quantity of water and kept at 40 °C for 2 days under constant agitation. The contents are then centrifuged at 10,000 × *g* for 20 min and the upper layer having the oil is separated. Oil recovery in this process is observed to be relatively poor.

AEOE is a modification of SAOE in the sense that enzymes are added to the above broth so as to aid the rapid separation of the oil from the paste. The enzymes normally used are proteases developed from the fungus Aspergillus flavus and xylanase, pectinase and cellulase from *Aspergillus* niger. Around 250 mg of enzyme is needed for every 5 g of jatropha kernel. The oil which remains trapped in the paste formed by proteins, cellulose and hemicellulose is released by the enzymes. It is observed that the separation of oil in this method is around 350 L for every tonne of seed. However there are no takers for this methodology since it is not cost effective. If the above process is carried out along with sonicator the oil recovery is known to increase further by 4–4.5%.

18.5.11 Three Phase Partitioning

In three phase partitioning the pH is first adjusted to 7. With every 100 mL of the paste 30 g ammonium sulphate and 100 mL *t*-butanol are added and the content is then stirred gently. The whole content is incubated for an hour at 25 °C. Consequently three clear phases are formed on centrifugation at 2000 × *g* for 10 min. The upper, less dense layer comprising the oil is collected and evaporated at a low pressure and at 50 °C for 5 min. This methodology is considered to be efficient. Around 250–400 L of oil can be recovered from every tonne of seed. Approximately 5 t of *t*-butanol is needed to process 1 t of seed. Thus, this methodology also is not cost effective.

18.5.12 Cake

During the extraction, the entire oil may not be removed from the seed mass. The cake may contain considerable amount of residual oil. In practice the seed cake after the oil extraction may contain residual oil to the tune of 9–12%. As this cake contain toxic materials, it cannot be utilized as a cattle feed. Nevertheless, it is a good organic manure as it is a good source of N (4–7%), P (2–3%), K (0.8–1.8%), Ca (0.5–0.8%) and Mg (1.2–1.4%). The increase in the yield of various crops due to the application of *J. curcas* cake as a manure is given in Table 18.23.

The phorbol ester present in the cake acts as a biopesticide which is capable of controlling the root borne pests of various crops. It is reported that the phorbol ester is degraded in a week's time under ideal field conditions. However it is feared that this toxin may get into the edible parts of the crop, through systemic path. The cake is also considered as a feed stock for biogas. Around 0.4–0.5 m^3 kg^{-1} of biogas having 70% methane may be produced from the cake. Therefore it is prudent to extract the gas first from the cake and then to use the resultant slurry as a manure. If a technology is identified to remove the toxins from the cake, the resultant cake can be best used as a component in the supplementary feed of the poultry, cattle and for the farm grown fish.

Table 18.23 The yield response of few crops due to *J. curcas* cake application as manure

Crop	Dose (t ha^{-1})	Yield over zero input (%)
Pearl millet (*Pennisetum glaucum*)	5	46
Cabbage (*Brassica oleracea*)	2.5–10	40–113
Rice (*Oryza sativa*)	10	11
Jatropha (*J. curcas*)	0.75–3	13–120

18.5.13 Oil

The oil which is often referred as curcas oil is a good substitute for kerosene in rural areas as a fuel for cooking. The quality and quantity of this oil greatly rely on the genetics, age of the tree, cultivation mode, seed size, agro-climatic conditions, soil characteristics and the extraction methodologies. The unsaturated fatty acids content of this oil is observed to be high. The unsaturated fatty acids comprise palmitoleic acid (C16:1), oleic acid (C18:1) linoleic acid (C18:2) and linolenic acid (C18:3). The saturated fatty acids are formed of myristic acid (C14:0) palmitic acid (C16:0), margaric acid (C17:0) stearic acid (C18:0) and arachidic acid (C20:0). The fatty acid composition of the oil of J. *curcas* is given in Table 18.24.

The triglycerides form the major portion (63.7%) of the oil. The free fatty acid content also is observed to be high (8.7%). It is known that if free fatty acid is higher than 3.5%, double stage transesterification is recommended to produce biodiesel. The lipid classes of *J. curcas* oil are given in Table 18.25.

The oil is relatively viscous (15–35 cSt at 30 °C). The unsaturated fatty acids present in high quantities (>70%) in this oil are likely to be denatured by atmo-

Table 18.24 Fatty acid composition of the *J. curcas* oil (Akbar et al. 2009)

Fatty acid	%
Unsaturated fatty acids	
Palmitoleic acid (C16:1)	0.7
Oleic acid (C18:1)	44.7
Linoleic acid (C18:2)	32.8
Linolenic acid (C18:3)	0.2
Total	78.4
Saturated fatty acids	
Palmitic acid (C16:0)	14.2
Myristic acid (C14:0)	0.1
Margaric acid (C17:0)	0.1
Stearic acid (C18:0)	7.0
Arachidic acid (C20:0)	0.2
Total	21.6

Table 18.25 The lipid classes of *J. curcas* oil (Adebowale and Adedire 2006)

Lipid classes	%
Hydrocarbons	4.8
Triacylglycerols	88.2
Diacylglycerols	2.5
Monoacylglycerols	1.7
Unsaponifiable matter	3.8
Free fatty acids	3.4
Sterols	2.2
Polar lipids	2.0

spheric oxygen thereby peroxides are formed. This may cause the oil to get thickened further due to polymerization. The polymerization process is likely to be intensified when the temperature increases. Therefore the degree of unsaturation of the oil is to be known clearly for which iodine value is a tool. The iodine value refers to the quantity of iodine in milligrams absorbed in a gram of the oil. In the oil of *J. curcas*, the iodine value is around 100 mg g^{-1} and therefore this oil is likely to solidify during long storage. Any oil with less than 25 mg g^{-1} iodine value is considered good especially if it is to be stored for a long time before being used in engines. Oil with high iodine value can be chemically processed through hydrogenation in which the double bonds are broken so as to convert the oil more to a saturated form. This may help to reduce the tendency of the oil to solidify. However, such step is likely to increase the melting point of the oil. If the oil has the iodine value ranging from 25 to 50 mg g^{-1}, the engine performance may not be seriously affected. But the engines are required to be actively maintained by changing the lubricating oil frequently and also by cleaning the exhaust system.

Similarly, the triglyceride level of the jatropha oil is around 60%. If the triglyceride level is more than 50%, it is not considered good. Certain specific characteristics of the oil are given in Table 18.26.

The high viscosity of the oil due to its large molecular mass and chemical structure causes problems in fuel pumping, atomization and combustion. Long term usage results in the gumming of the fuel injection system. Marginal scale technical incompatibility of biodiesel with the diesel engines is also being reported. Therefore, the viscosity has to be reduced through transesterification to make the oil more suitable as an alternate fuel. The viscosity of the oil is tackled in several other ways such as preheating the oil, blending or dilution with other fuels, thermal cracking and through pyrolysis.

Table 18.26 Certain essential characteristics of the jatropha oil (Akintayo 2004; Becker and Makkar 2008; Achten et al. 2008)

Parameter		Value	Parameter	Unit	Value
Crude fat	(%)	47.25	Flash point	(°C)	240
Colour		Light yellow	Refractive index	At 25 °C	1.47
Acid value	(mg KOH g^{-1})	3.5	Viscosity	(cSt at 30 °C)	32
Saponification value	(mg KOH g^{-1})	193.6	Heating value	(MJ kg^{-1})	37.8
Iodine value	(mg g^{-1})	105.2	Water content	(%)	0.07
Unsaponifiable matter	(%)	0.8	Calcium	(mg kg^{-1})	56
Density at 15 °C	(g cc^{-1})	0.92	Iron	(mg kg^{-1})	2.4
Phosphorus	(mg kg^{-1})	290	Flash point	(°C)	240
Magnesium	(mg kg^{-1})	103			

18.5.14 Biodiesel

The oil is converted into biodiesel or ester through a process called transesterification in which the viscous glycerol is separated from the oil through alcoholysis in the presence of an alkali or acid as catalyst so as to reduce the viscosity of the oil. The constituents of the oil have a major say in the chemically catalysed esterification process. If the free fatty acid content of the oil is more than 3.5%, soap may be formed in high quantities during the process of transesterification and therefore free fatty acid has to be kept under control. If the free fatty acid is lesser than 3.5%, transesterification can be straightly proceeded using alkali as the catalyst. In such case the chemicals needed are alcohol and alkali. Either methanol or ethanol is popularly employed. Similarly NaOH or KOH is used as a catalyst. Based on the alcohol used, methyl ester or ethyl ester is formed.

In case, methyl ester is to be prepared through chemical process, 30 g KOH (0.53 mol) is first dissolved in 331 g (10.34 mol) of methanol (CH_3OH). This solution is known as methoxide. Two-third quantity of the above methoxide is first mixed with 2000 mL (2.3 mol) of crude oil and stirred slowly for 30 min maintaining it at 50 °C. It is then transferred to a separating funnel so as to enable the separation of the glycerol from ester. After 5 h the glycerol which remains as a lower layer is removed. The upper layer is then mixed with the remaining quantity (one part) of the methoxide, stirred for 30 min and kept undisturbed for another 5 h. The glycerol which settles down at the bottom is then removed. The upper layer is the methyl ester which is washed thrice each with 500 mL of warm (50 °C) distilled water so as to completely remove the traces of soap and glycerol. The ester is then dried by passing it over anhydrous sodium sulphate (Na_2SO_4). A scheme is given in Fig. 18.26.

For preparing ethyl ester through chemical process 30 g KOH (0.53 mol) is first dissolved in 317.7 g (6.9 mol) of ethanol (C_2H_5OH). The same is mixed with 1000 mL of jatropha oil and stirred for 90 min at 60 °C. It is then transferred to a

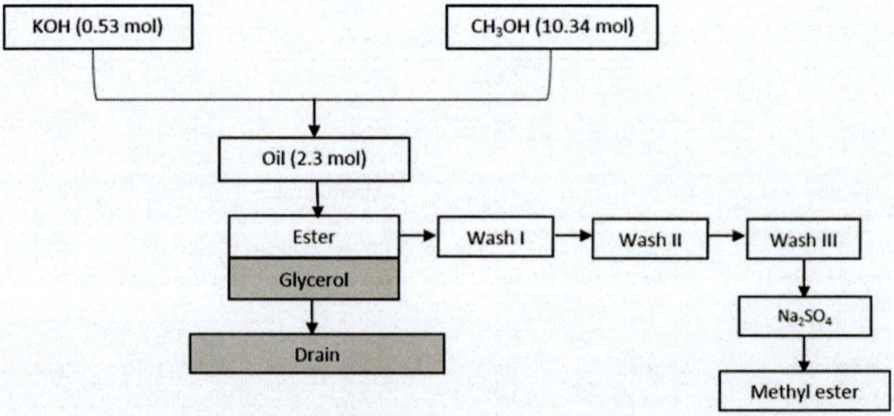

Fig. 18.26 Scheme for the preparation of methyl ester

separating funnel. After 5 h two layers appear. In order to have a rapid separation, the glycerol obtained from a previous trial is added as seeding material to the content of the separating funnel which actively assists the separation. The upper layer is the ethyl ester and the lower layer is the glycerol which remains mixed with the alkali. The glycerol is then separated, evaporated to remove alcohol if any, warmed up to 70 °C and acidified with 50% sulphuric acid so as to reach pH 5. The residue formed if any is removed. The overlying liquid phase has two clear layers. The bright red layer is glycerol which is separated. Part of the glycerol thus formed is utilized to enhance the separation of glycerol in the subsequent transesterification process. Another layer of glycerol admixed with the free fatty acid is separated and treated with 6 g conc. H_2SO_4 and 70 g ethanol. This mixture is stirred at 60 °C for 6 h. After cooling, the lower layer having glycerol is separated. The top lying liquid layer is the ester. This is then pooled with the ester originally formed and washed with 500 mL of 1 N HCl. Subsequently it is washed thrice with 500 mL each of warm (50 °C) distilled water. The resultant ester is then treated with 500 mL of 1 N Na_2CO_3 to remove any fatty acid residues. The final product is dried over Na_2SO_4 to obtain the ethyl ester.

During the chemical mode of transesterification large quantity of wash water is discharged which contains alkali or acid and soap. On account of environmental safety this wash water is needed to be treated. Besides, in such chemical processes it is extremely difficult to separate the glycerol from ester since part of the glycerol always remains as emulsion. Therefore attempts are being made to transesterify the oil of *J. curcas* using lipase enzyme as a catalyst. In such process there will not be any discharge of waste water. Besides, glycerol also will not be formed. The lipase enzyme, a product of microorganisms is used at the rate of 4–7 g per 100 g of oil. Two microorganisms capable of producing the concerned lipase are *Candida rugosa* and *C. antarctica*. One of the major problems in this method is the high cost of the enzyme. Moreover the enzyme used in its original form cannot be recovered at the end of the process. Therefore to offset the above problem the practice of incorporating the enzyme in certain porous particles known as vehicles is being followed. The porous particle also is often referred as support. The process of incorporating the enzyme in the support is called as immobilization. After the transesterification reaction the particles laden with the enzyme can be removed easily from the process system. It is also known that the enzyme thus recovered along with the support can be used repeatedly up to five times or more in the subsequent processes. In recent times the microorganisms capable of producing the lipase are straightly employed which release the required quantity of enzymes directly in the reaction mixture. One such organism is *Rhizopus oryzae*. The alcohol used mostly is methanol. The efficiency of methanol is higher than that of the ethanol. The relatively low molecular weight and high polarity of methanol execute the transesterification process efficiently.

18.5.15 Quality of Biodiesel

Depending on the type of alcohol (methanol or ethanol) being used in the transesterification process the corresponding ester (methyl ester or ethyl ester) is produced. Certain standards are prescribed by various agencies functioning in different countries so as to qualify a good ester. They are EN 14214: 2003 (European), DIN V 51606 (Deutsches Institut for Normung = German Institute for Standardisation) and ASTM D6751 (American Society for Testing and Materials (ASTM) which is an international standards organization that develops and publishes voluntary consensus technical standards for a wide range of materials.). The EN 14214:2003 is now superseded by a new version EN 14214:2008 (Nov. 2008). This standard is specific to fatty acid methyl ester (FAME). However, the biodiesel fatty acid ethyl ester is not covered in EN 14214.

A comparative account of the quality of the methyl ester, ethyl ester and the standard prescribed is given in Table 18.27. The density, flash point and cetane number of the methyl ester and ethyl ester produced from the oil of *J. curcas* are within the standard prescribed. The Conradson carbon residue is determined from the original sample and then calculated for the residue, which is assumed to be 10%.

18.5.16 Engine Performance

Experiments carried out indicate that the performance of biodiesel in the internal combustion engine is quite satisfactory (Takeda 1982). A 50 h test carried out in a

Table 18.27 Chosen characteristics of methyl ester and ethyl ester produced from the oil of *J. curcas* and the standards

Parameter	Unit	Methyl ester	Ethyl ester	EN 14214:2008	DIN V 51606	ASTM D6751
Density	(g cm^{-3})	0.88	0.89	0.86–0.90	0.87–0.90	–
Heating value	(MJ kg^{-1})	39.65	–	–	–	
Flash point	(°C)	186.0	190.0	Min 120	Min 110	Min 130
Cetane number		52.0	59.0	Min 51	Min 49	Min 47
Viscosity	(cSt at 30 °C)	5.1	5.5	–	–	1.9–6.0
Conradson carbon residue[a]	(% by wt)	0.02	0.05	–	–	Max 0.05
Acid number	(mg KOH g^{-1})	0.27	0.08	Max 0.5	Max 0.5	Max 0.5
Phosphorus	(mg l^{-1})	18.0	16.0		Max 10.0	Max 0.001
Ester	(% by wt)	99.6	99.3	Min 96.5		
Water	(% by wt)	0.1	0.16	Max 0.5	Max 0.3	Max 0.5

[a]Conradson carbon residue is calculated for 10% of the residue

small pre-combustion chamber type diesel engine using biodiesel blend as fuel proved to be promising (Ishii and Takeuchi 1987). In Mali diesel engines of Indian origin with pre-combustion chamber are in function with pure jatropha oil as fuel which is reported to give a reasonably good result at a maximum load conditions. Fuel conversion efficiency, specific consumption rate as well as exhaust gas emissions are compared by feeding jatropha oil in direct and indirect injection fuel engines. Jatropha oil gives the lowest emission when compared to other oils. As high as 1000 h test gives good performance. For industrial power output, specific fuel consumption is observed to be lower and the efficiency is higher than that of the engine fed with regular diesel. In order to use this oil in automobile, it is needed to be transesterified using methanol or ethanol.

Direct injection diesel engine operating on pure oil of *J. curcas* gives higher performance and satisfactory emission level. The measured emission of hydrocarbon and oxides of nitrogen as NO from biodiesel run engine is around 20–30% lower than that of the engines running with regular diesel (in both the parameters). The smoke level also is lower by 20–25%. The brake thermal efficiency (BTE) with the use of this oil as fuel is however lower (by 25–28%) than that of the regular diesel. Since the BTE is observed to be lower for this oil, esterification is the solution. Alternatively such esters can be used straight or in combination with petro-diesel. But it is widely felt that in tropical countries, where *J. curcas* can grow easily, its oil even without transesterification has a great potential to be used in non-automobile applications. It may also be considered in certain low speed diesel engines especially on use in water lifting purposes or for power generations.

18.5.17 Economic Appraisal

Economic analysis is the bottom line of any enterprise. The tree *J. curcas* which ramified as a weed in many tropical countries has now reached the centre stage as a cash crop offering tremendous hope as an energy source. This tree has confined its habitat in tropical countries whose geographical extent is around 15% of that of the world. These tropical countries have varied economic profiles. While some of them are in the process of development, few have already developed. Some of them are still underdeveloped. Therefore, in such situations no common scale for economic understanding may be adopted. Instead of applying currency values, the material turnover is considered as a best mode of analysis. Again economics of *J. curcas* biodiesel greatly relies on the dimension of the operation. A scan of literature indicates that the production level of seeds ranges from 0.4 to 7.8 t ha^{-1} $year^{-1}$. This range appears to be very broad indicating the instability in the quantity of the production. However this is apparently due to a differential dimension during the different stages of operation from planting to biodiesel production. Therefore it is necessary to have a closer look at the entire cycle of the operation starting from the plantation to the esterification of the oil.

From plantation to the stage of producing biodiesel, every step is considered important as the energy turnover is low if the biodiesel alone is taken as the end product. The quantity of oil production largely depends on ever so many combination of factors such as ideal soil, favourable agro-climatic conditions, a good genetic stock, optimum labour management, efficient oil extraction and suitable transesterification process. Besides, there should be efficient utility of the fleshy exocarp, husk and the leafy biomass obtained due to canopy and bush management. It is seldom possible to have all the above aspects working to the favourable end. All the fruits may not come to harvest uniformly at a time and therefore mechanization is not possible. In such situation manual operation schedule has to be schemed properly so as to have it well commensurate with the fruit stock available for ready harvest at a given point of time. The climate for drying the fruit should be favourable during the period of harvest. As per the material recovery is concerned, from every 100 kg of fruit 30–35 kg of exocarp (dry fruit peel) is available which is to be composted properly in case it has to be utilized as a manure or it has to be dried to the extent of making it as a feed stock for thermal application as it has a heat value of 11.0 MJ kg^{-1}. In large scale practice the seeds are not dehulled as it demands huge labour involvement resulting into negative economics. The *J. curcas* plantations may be conveniently grouped into two categories depending upon the magnitude of operation. They are:

1. A low input plantation on a derelict area having poor soil quality, limited water facilities for irrigation and with low management. The yield of fruit from such a system may be 500 kg ha^{-1} $year^{-1}$ in a 3 year old crop. The estimated woody biomass stand of this plantation is around 8000 kg ha^{-1} $year^{-1}$ in which 10% can be clipped off either through canopy management or through natural process.
2. A plantation with a high input operation in an ideal soil, satisfactory agro-climatic conditions together with optimum irrigation, manure, intensive pest and canopy management and with required labour force, giving a fruit yield of around 7000 kg ha^{-1} $year^{-1}$ from a standing biomass of 48,000 kg.

In many countries the cultivation of *J. curcas* is being made in waste lands only. Waste land is normally any area which remains fallow and unoccupied due to one reason or the other. Waste land is not always considered as a useless piece of land. Due to constraints in operational variables, intensive crop management with high technical input is not possible in such waste lands. A 3-year-old low input system with 1600 trees per hectare standing at a space of 2.5 m × 2.5 m yielding 500 kg of fruits at the third year is taken as an example of a low input system for analysing the return. If the individual plant has an average dry biomass of 5 kg, the total standing biomass is calculated to be 8000 kg (1600 × 5). A portion of such biomass comprising woody material and leaf is clipped off during the canopy management. The quantity of such biomass (800 kg) is worked out to be around 10% of the total standing biomass. Parallelly another 3-year-old high input system with a fruit yield of 7000 kg ha^{-1} $year^{-1}$ is considered for analysis. The standing biomass in this case is 48,000 kg ha^{-1} $year^{-1}$ if the average weight of the plant is 30 kg. The biomass to be clipped off is 10%, i.e. 4800 kg ha^{-1} $year^{-1}$. The details considered for analysis of a low input and a high input system are furnished in Table 18.28.

Table 18.28 The parameters for analysis in a low input and a high input system

No	Parameters	Unit	Low input	High input
Main product				
1	Fruit yield	(kg ha^{-1} $year^{-1}$)	500	7000
2	Seed yield at 70% of the weight of fruit	(kg ha^{-1} $year^{-1}$)	350	4900
3	The oil yield at 30% of the weight of seed	(kg ha^{-1} $year^{-1}$)	105	1470
4	Heating value of the oil	(MJ kg^{-1})	38	38
5	Heating value of the oil totally produced	(MJ)	3990 (105 × 38)	55,860 (1470 × 38)
6	Ester recovery from oil	(%)	90	90
7	Total quantity of the ester produced	(kg)	94.5 (105 × 0.9)	1323 (1470 × 0.9)
8	Heating value of the ester	(MJ kg^{-1})	39.65	39.65
9	Total heating value of the ester prod.	(MJ)	3747 (94.5 × 39.65)	52,457 (1323 × 39.65)
By-product				
10	Glycerol generated during the esterification	(%)	7	7
11	Total quantity of glycerol thus generated	(kg)	7.4 (105 × 0.07)	103 (1470 × 0.07)
12	Heating value of glycerol	(MJ kg^{-1})	19	19
13	Total heating value of the glycerol produced	(MJ)	141	1957
14	Biomass (wood + leaf) clipped of	(kg $year^{-1}$)	800	4800
15	Heating value of the biomass (wood + leaf)	(MJ kg^{-1})	15.5	15.5
16	Total heating value of all the biomass	(MJ)	12,400 (800 × 15.5)	74,400 (4800 × 15.5)
17	Exocarp (peel) generated (30% of the weight of fruit) (Column 1–2)	(kg)	150	2100
18	Heating value of the exocarp	(MJ kg^{-1})	11	11
19	Total heating value of the exocarp generated	(MJ)	1650 (150 × 11)	23,100 (2100 × 11)
20	Cake generated	(kg)	245 (350–105)	3430 (4900–1470)
21	Heating value of cake	(MJ kg^{-1})	25	25
22	Total heating value of the cake generated	(MJ)	6165 (245 × 25)	85,750 (3430 × 25)
24	Total heating value of all the by-products (column 13 + 16 + 19 + 22)	(MJ)	20,356	185,207

Table 18.29 Item wise expenditure of energy in the *J. curcas* life cycle system for a calculated recovery of 100 MJ biodiesel (adopted from the data given in Achten et al. 2008)

Item	Energy expenditure			
	Low input (MJ) %		High input (MJ) %	
Cultivation expenses: The agricultural input such as labour for planting, harvesting, seed removal, canopy management, transportation, etc.	2.7	16.9	44.2	49.9
Extraction of the oil	1.3	8.1	9.1	10.3
Transesterification process	12.0	75.0	35.3	39.8
Total	16.0	100.0	88.6	100.0

Three important inputs are normally taken into consideration. They are crop management, oil extraction and transesterification processes. Achten et al. (2008) who studied the life cycle assessment of input contended that the low input system has a total of 160 MJ as expenditure (input) for a total output of 1000 MJ as ester (biodiesel) yield when biodiesel alone is considered as the final product. The corresponding input in the higher input (high tech operation) system is 886 MJ. This means that the benefit from a low input system is 840 MJ (1000–160). For a 100 MJ equivalent of biodiesel production the gain is 84 MJ, whereas the corresponding value for a high tech system with high input is 114 (1000–886) for a 1000 MJ biodiesel output system which works out to be 11.4% which is paradoxically low. Thus the economic benefit of the high input system is tending towards zero. The item wise energy expenditure is listed in Table 18.29 which is adopted from the value furnished in Achten et al. (2008). They have also considered the earlier works in this field.

It is to be noted that the high input system has a poor turnover

The energy incurred as input in the low input system is calculated as follows:

For every 100 MJ output as biodiesel the expenditure as per Table 18.28 = 16 MJ

∴ For 3747 MJ output as per Table 18.27 (3747 × 16/100) the total expenditure is = 600 MJ

Similarly the total energy spent in the high input system is calculated as follows:

For every 100 MJ recovery as biodiesel the expenditure as per (Table 18.28) = 88.6 MJ

∴ For 52,457 MJ the total expenditure is (52,457 × 88.6/100) = 46,477 MJ

The item wise total expenditure of energy in both the systems is furnished in Table 18.30.

The Table 18.30 indicates that the cultivation expenses cover around 50% of the total input in high input operations. When the biodiesel production alone is considered the total input and output ratio in the high input system is 1:1.3 (52,457/46,477) against 1:6.2 (3747/600) of a low input system. This indicates that the economic return of the low input system is around five times higher than that of the high input system.

Table 18.30 Item wise expenditure of energy in low input and high input systems of *J. curcas* biodiesel production

Items	Low input (MJ)	High input (MJ)
Cultivation	101 (2.7 × 600/16)	23,186 (44.2 × 46,477/88.6)
Extraction of oil	49 (1.3 × 600/16)	4774 (9.1 × 46,477/88.6)
Transesterification process	450 (12 × 600/16)	18,517 (35.3 × 46,477/88.6)
Total	600	46,477

In order to improve the economic benefits, the corporate sectors involved in *J. curcas* based biodiesel system using high tech mechanism and by incurring high input have to hatch out comprehensive plan of action so as to encash the by-products such as cake, glycerol and wood biomass. If such outputs are taken into consideration, the ratio will move to 1:5, i.e. $\frac{52,457+185,207}{46,477}$ for high input system, whereas it is 1:40 $\frac{3747+20,356}{600}$ for the low input system. This indicates that the energy spent on irrigation, fertilizer and other farm expenditures do not completely pay off in an extra energy production. Thus it is known that the degree of economic benefit largely depends on the type of input and the parallel utilization of various by-products. The energy balance efficiency will go to the beneficial scale if improvements are made on the cultivation and transesterification steps.

Out of the many methods being followed for oil extraction, mechanical oil extraction is considered favourable as it covers only 8–10% of the total life cycle energy. The solvent extraction methodology is energy intensive and it can be considered in large scale production systems. As transesterification takes the lion share (40–75%) of the energy expenditure research focus is found necessary to make use of the raw oil as a fuel source with necessary engine modification so as to avert huge expenditure on transesterification.

The seed cake which is a by-product of the oil extraction can be considered as an organic manure and also can be best utilized as a feedstock for biogas generation before being utilized as manure. The seed cake also is suitable for animal feed (cattle, poultry and for aquaculture) provided technological advancement is made to detoxify it from plant toxins. The glycerol is another by-product which can be considered as a fuel in certain selected industries besides having it used sufficiently in cosmetic industries. The woody biomass obtained from annual pruning and from coppicing can be considered as a feedstock in gasification process.

18.5.18 Global Warming Abatement Potential

J. curcas is a C_3 plant and therefore involved in direct fixation of CO_2. The CO_2 combines with ribulose bisphosphate to produce two molecules of three carbon compound, namely 3-phosphoglycerate with the active assistance of the enzyme

Table 18.31 The pattern of greenhouse gas emitted from a high input biodiesel production system (expressed for 1000 MJ biodiesel production)

Item of expenditure	Kg CO_2 *eq.*	%
Land preparation	2.7	4.7
Cultivation	0.1	0.2
Irrigation	14.8	26.1
Fertilizer	17.1	30.3
Cracking (labour)	1.7	3.0
Oil processing	6.5	11.4
Transesterification	13.8	24.3
Total	56.7	100

rubisco. Jatropha being grown in untenable rough climate it tends to close the stomata to reduce the evapotranspiration. In such situation rubisco reacts with O_2 instead of CO_2 and results in photorespiration which naturally causes the wastage of CO_2.

Prueksakorn and Gheewala (2006) indicated that 90% of the total life cycle greenhouse gas (GHG) emission in a biodiesel system is released during the burning process of the finished product, namely biodiesel at the engine. They calculated that the global warming potential due to the production and utilization of the biodiesel is just 23% of the total global warming potential of the petro-diesel. Tobin and Fulford (2005) and Prueksakorn and Gheewala (2006) worked on the low input and high input system and their role in greenhouse gas emissions. It is known that during the production of 1000 MJ biodiesel 56.7 kg CO_2 *eq.* greenhouse gas is generated in a high input system. In a low input system the corresponding value is 16.5 kg CO_2 *eq.* only. The greenhouse gas generated during the production process of a high input system is given in Table 18.31.

In the high input system, the greenhouse gas takes up a complex dimension due to the generation of NO_2 from the fertilizer being used in the jatropha plantation.

N_2O is a by-product of nitrogen application in agriculture having a global warming potential of 296 times (Prather et al. 2001) higher than an equal mass of CO_2 and therefore high input system using nitrogenous fertilizers practically derail the benefit of global warming abatement on this account. The biomass produced in the cultivation system, namely wood, leaf, exocarp and oil seed cake is assumed to contain carbon at 50% basis of the wet biomass whose equivalence in terms of CO_2 is 183.5 g as each gram carbon is accumulated from 3.67 g of CO_2 (44/12). It is also necessary to calculate the amount of CO_2 being generated due to the biodiesel and glycerol being produced. For the purpose of computation, it is assumed that both the above products are burnt completely. In a 1000 MJ equivalent biodiesel system the actual quantity of biodiesel being produced is 25.22 kg (1000/39.65) since the heat value of biodiesel is 39.65. The above quantity of biodiesel would have been produced from 27.12 kg of raw oil (25.22/0.93) as the biodiesel recovery from oil is 93% and the rest (7%) being glycerol. Thus the quantity of oil (27.12 kg) now calculated would account for the glycerol also. The oil has 57 C, 107 H and 6 O_2. Therefore each gram mole of oil weighs 887 g [$(57 \times 12) + (107 \times 1) + (6 \times 16)$] in which there is 684 g C. This means that in each gram of oil there is 0.77 g C. Thus the

Table 18.32 The CO_2 sequestration of *J. curcas* biodiesel production cycle

Plant parts	Low input (kg CO_2 ha^{-1} $year^{-1}$)	High input (kg CO_2 ha^{-1} $year^{-1}$)
Biomass (wood + leaf)	1468 (800 × 0.5 × 3.67)	8808 (4800 × 0.5 × 3.67)
Husk	275 (150 × 0.5 × 3.67)	3854 (2100 × 0.5 × 3.67)
Cake	450 (245 × 0.5 × 3.67)	6294 (3430 × 0.5 × 3.67)
Oil	297 (105 × 0.771 × 3.67)	4160 (1470 × 0.771 × 3.67)
Total	2490	23,116
Less CO_2 generated in the process including the end product use	304	4082
Balance	2186	19,034
Net CO_2 sequestrated (%)	87.8	82.3

CO_2 generated from the end product, namely biodiesel of a 1000 MJ biodiesel system (including glycerol) is 76.64 kg (27.12 × 0.77 × 3.67). From this the total CO_2 generated from the low input system is calculated to be 349 kg $\frac{76.64}{1000} \times 3747 + \frac{16.5 \times 3747}{1000}$ in which 16.5 is the CO_2 released from the low input system except that of the biodiesel. Similarly the CO_2 generated from the high input system is calculated to be 6995 kg CO_2 $\frac{76.64}{1000} \times 52,457 + \frac{56.7 \times 52,457}{1000}$ in which 56.7 is the CO_2 released from a high input system. The CO_2 sequestration potential is thus worked out for the full cycle and presented in Table 18.32.

18.6 Castor (*Ricinus communis*)

Oil from the castor seeds of *Ricinus communis* has tremendous industrial applications. *Ricinus* is a Latin word for tick (acarid blood sucking arachnids). The seed of castor looks like a typical tick having a bump at its end. The seed has a toxic protein ricin, alkaloid ricinine and lectins. Castor oil is a less expensive raw material for biodiesel which can be best utilized in cold winters due to its low cloud and pour points. As *R. communis* is distributed all over the world it is being known by different names according to the different vernacular languages as given in Table 18.33.

R. communis is a native of Arica and Asia. It has slowly invaded Australia, USA, Mexico, South America, New Zealand and many Pacific Islands (Samoa, Fiji, Hawaii and Solomon Islands), India, Pakistan, Nepal, Bangladesh, China, Brazil, Ethiopia, Algeria, Egypt, Greece, France, Argentina, Thailand and Philippines (Fig. 18.27).

Table 18.33 The vernacular names of castor in different countries

Country	Vernacular names
India	Castor, krapata, eragach, arandi, aavannaku, amandam, Amandam, Chittamanakku, erandam, Gandharvahastakam, Kottamaram, Vilakkennai, Kottaimuthu, amanakku
Finland	Risiini
France	Risin commun
Germany	Christus palme, hundsbaum, Kreuzbaum, Lausebaum, Palmachristi, Rizinus, Romische bohne, Wunderbaum, Wunderstrauch
Portuguese	Ricino
Spain	Agaliya, ricino, aciete de ricino, higuerilla
Fiji	Benenivavalagi, mbele ni vavalagi utouto
Mariana Islands	Agaliya
Hawaii	Kaapeha, Kamakou
England	Castor bean, maple weed
Poland	Wonderboom
Thailand	La hoong
Sumatra	Doelang bajora
Java	Djarak kaliki
Philippines	Katana, lansina, taca taca

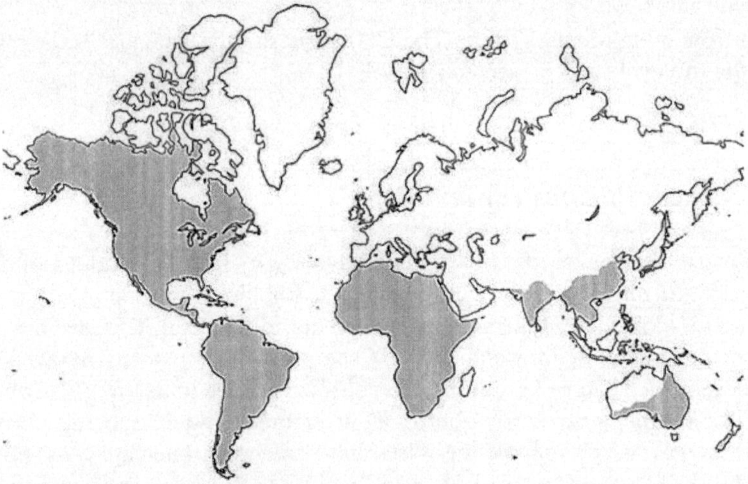

Fig. 18.27 Geographical distribution of Castor *Ricinus communis*

18.6.1 Habitat

The castor plant is often found grown on stream sides. In fertile soil the plant grows well and gives forth lushy leaves with few flowers. It is a perennial plant in fertile soils and annual in frosty areas. In Australia this plant is placed in Class B noxious weed (it is desirable to control the growth and spread of the plant) and Class C noxious weed (it is desirable to control the introduction of the plant) of the Noxious Weeds Act of the Weeds Management Act 2001, No. 2 of the Northern Territory of Australia.

18.6.2 Distinguishing Features

The leaves are palmately divided which discharges awful smell when squeezed. The leaf has a width of 30–45 cm with a long stalk and 6–12 lobes with coarse teeth. Roots are thick and fibroid. The flowers are arranged in panicles emerging from auxiliary positions. The flowers do not have petals and are monoecious. The male flowers are located at the base of the spikes and the female flowers occupy the upper part of the spike (Fig. 18.28).

The plant is at the mercy of the wind for pollination. The fruit is greenish and spiny and carries a seed in its deeply grooved capsule. The fruit coat is leathery outside and brittle and polished inside. The seed is egg shaped, 2 cm long and mottled grey. It has a fleshy projection at one of the tips. These plants propagate through seed. They grow to a height of 1–8 m and come to flowering in 6 months. They continue to yield throughout the year without any interruption. Depending on the size and height, a 2 year plant may yield 1500–15,000 seeds. The details of seed production from various countries are given in Table 18.34.

18.6.3 Cultivation

The distribution of castor plants is restricted to the geographic area lying between 45° N and 50° S. It can grow up to a maximum altitude of 2500 m. In colder regions (average temp. lesser than 10 °C) its growth is restricted to a maximum altitude of

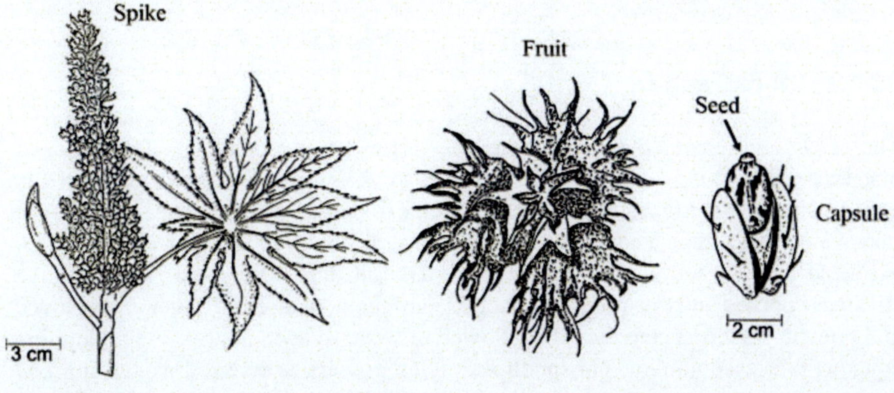

Fig. 18.28 Castor *Ricinus communis*

Table 18.34 Production of castor seeds from various countries[a]

Country	t $year^{-1}$
India	830,000
China	210,000
Brazil	92,000
Ethiopia	15,000
Paraguay	12,000
Thailand	11,000
Vietnam	5000
South Africa	5000
Philippines	5000
Angola	4000

[a]As per the year 2011

500 m. Castor can withstand agro-climatic extremes such as drought and torrential rain. They do not perform well in water logged area.

A geographic area having a frost-free period of 150–200 days, average ambient atmospheric temperature ranging from 20 to 25 °C with an annual rainfall of 400–600 mm is ideal for the cultivation of castor. Castor is cultivated in a wide range of soil having a good network of water drainage. Slightly acidic (around pH 6) and moderately fertile sandy loam is well preferred.

The land is repeatedly ploughed to free it from clumps and clods. Chemical fertilizer to the tune of 20N-20P-20K kg ha^{-1} is given as basal application. Seeds treated with pesticide are sown at a gap of 50 cm between plants and 100 cm between rows. Around 10–12 kg of seeds may be required to sow a hectare. After 45 days, another dose of toping fertilizer as equal to that of basal application is given. Three to four irrigations in a period of 2 months are preferred. The fruits come to ripe in 200 days and they are harvested in batches as and when they are ready. The seeds are then separated from the capsules by beating them with sticks or by treading them under the feet of bullocks. The yield of seeds per hectare per year is around 250, 400 and 600 kg in mixed crop, rain-fed crop and irrigated crop, respectively.

18.6.4 Oil Extraction

The seeds are separated from the capsules, cleaned and dehulled. The dehulled materials are then cooked and partially dried so as to coagulate the protein present in it. It is then passed through a high pressure expeller so as to separate the oil. The oil cake discharged as a solid by-product having residual oil (around 9%) is further extracted using a suitable organic solvent. The oil thus discharged from both the streams is pooled and the impurities such as particulate matter and water are removed and refined. As a first step the oil is allowed to settle so as to remove the underlying aqueous phase containing phospholipids. This process is referred as settling and degumming. It is then neutralized on account of the free fatty acid content. This is then followed with a bleaching process so as to get rid of any colouring materials, phospholipids and oxidation products. As a last step the oil is deodorized. Normally the recovery of oil will be around 40% by weight of the seed mass.

18.6.5 Oil Quality

Castor oil is rich in hydroxy fatty acid, ricinoleic acid, $C_{18}H_{14}O_3$ (*cis*-12-hydroxy octadec-9-enoic acid, 18-carbon hydroxylated fatty acid having 1 double bond) and has a molecular weight of 298.5. Its density is 0.94 g mL^{-1}. The quality of the castor oil changes according to the location from where the seeds are collected, the genotype of the castor, environmental conditions, cultural packages and the time of harvesting. The characteristics of the castor oil are given in Table 18.35.

Table 18.35 The characteristics of the castor oil

Parameters	Value	References
Lipid content (%)	43, 48	Salimon et al. (2010), Abitogun et al. (2008)
Moisture content (%)	0.2, 0.3	Salimon et al. (2010), Abitogun et al. (2008), Nangbes et al. (2013)
Viscosity (cps)	0.42	Nangbes et al. (2013)
Refractive index at 25 °C	1.47, 1.8, 1.8	Salimon et al. (2010), Abitogun et al. (2008), Nangbes et al. (2013)
Iodine value (mg g^{-1})	84.5, 90.0, 58.0	Salimon et al. (2010), Sreenivas et al. (2011), Abitogun et al. (2008)
Acid value (mg g^{-1})	4.9, 14.8	Salimon et al. (2010), Abitogun et al. (2008)
Free fatty acid (%)	3.4, 7.4	Salimon et al. (2010), Abitogun et al. (2008), Nangbes et al. (2013)
Peroxide value (meq kg^{-1})	10.2, 158	Salimon et al. (2010), Nangbes et al. (2013)
Saponification value (mg g^{-1})	182.9, 178.0, 180	Salimon et al. (2010), Abitogun et al. (2008), Nangbes et al. (2013)
Unsaponifiable matt. (mg g^{-1})	3.4	Salimon et al. (2010)
Molecular weight	937.7	Salimon et al. (2010)
Density (kg m^{-3})	925, 948	Sreenivas et al. (2011), Nangbes et al. (2013)
Flash point (°C)	200, 225	Sreenivas et al. (2011), Abitogun et al. (2008), Nangbes et al. (2013)
Cetane number	42	Sreenivas et al. (2011)
Net calorific value (mJ kg^{-1})	37.5	Sreenivas et al. (2011)
Specific gravity	0.948	Abitogun et al. (2008)
Fire point (°C)	256	Abitogun et al. (2008), Nangbes et al. (2013)
Smokepoint (°C)	215	Abitogun et al. (2008), Nangbes et al. (2013)
pH	5.8	Abitogun et al. (2008)
Congealing temp (°C)	−18.0	Abitogun et al. (2008)

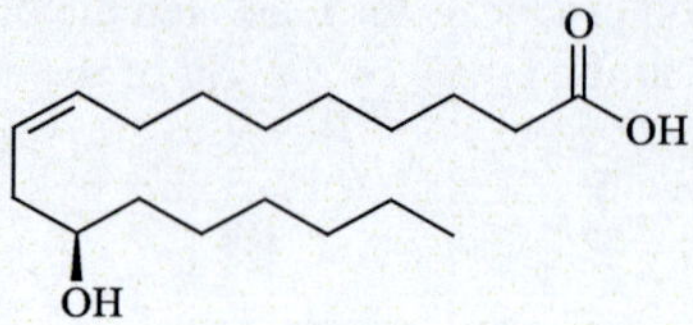

Ricinoleic acid $C_{18}H_{14}O_3$

Low moisture content of the oil guarantees adequate shelf life. Acidic pH is an expression of the high content of free fatty acids. The congealing temp. (−18 °C) and high iodine value are references to the presence of high level of unsaturated fatty acid and therefore this oil is considered as a non-drying liquid-oil enabling its application in the manufacture of lubricants hydraulic fluids, paints, cosmetics, softener for tanning, solar cells, textiles, small components of computers, mobile phones and the like. High peroxide content also indicates its amenability to rapid rancidity. The fatty acid composition of the castor oil is given in Table 18.36.

The unsaturated fatty acids comprising ricinoleic acid and oleic acid form a bulk share. The saturated fatty acids constitute a low percentage.

18.6.6 Production of Biodiesel

18.6.6.1 Acid Catalysis

To 1 L of castor oil, 350 mL of methanol and 10 mL of conc. sulphuric acid are added. It is then uniformly stirred for 6 h at 60 °C and made to settle for 8 h. The biodiesel moves to the upper zone and the lower layer containing the glycerol and gum is drained out. The biodiesel is recovered and heated to 100 °C so as to distil out the methanol through a condenser. The resultant biodiesel is then washed with 200 mL of hot water (40 °C) and the whole content is allowed to settle for 8 h. The wash water remaining at the bottom of the vessel is drained. This process is

Table 18.36 Fatty acid composition of the castor oil

Fatty acid	Percent	References
Ricinoleic (C18:1)	81.9, 84, 94	Abitogun et al. (2008), Nangbes et al. (2013), Salimon et al. (2010)
Oleic (C18:1)	2.3, 5.5	Abitogun et al. (2008), Nangbes et al. (2013), Salimon et al. (2010)
Linoleic (C18:2)	0.6, 7.3	Abitogun et al. (2008), Nangbes et al. (2013), Salimon et al. (2010)
Linolenic (C18:3)	0.3, 0.5	Abitogun et al. (2008), Nangbes et al. (2013), Salimon et al. (2010)
Stearic (C18:0)	0.5, 1.2	Abitogun et al. (2008), Nangbes et al. (2013), Salimon et al. (2010)
Palmitic (C16:0)	0.46, 1.3	Abitogun et al. (2008), Nangbes et al. (2013), Salimon et al. (2010)

repeated thrice or more till the litmus paper indicates neutrality. The final product is heated and stirred to remove the moisture if any and cooled.

18.6.6.2 Solid Acid Catalysis

Potassium bisulphate ($KHSO_4$) was considered as a solid acid catalyst in the production process. One of the prime characteristics of potassium bisulphate is that it does not facilitate any methanol olefin esterification (Goodain et al. 2002) despite it has olefinic bonds. As a matter of immobilization micro-porous silica is soaked in a saturated solution of potassium bisulphate. It was then dried in hot air oven at 150 °C for a day and preserved in a vacuum desiccators till its use. In the above process the potassium bisulphate get dispersed evenly over micro-porous silica so as to form Bronsted acid site which is responsible for the increased activity. It is known that in such immobilization treatments, potassium bisulphate occupies as high as 80% of the surface area thereby offering large Bronsted sites (Furuta et al. 2004). Around 80–90% of the spaces of silica are occupied by potassium bisulphate. Thus the immobilization support offers uniform dispersion to potassium bisulphate thereby offering high plugging effect. The silica often offers wide micro-pores to accommodate the giant molecules of triglyceride so as to have an increased exposure of the active sites. The mechanism of the catalytic activity of $KHSO_4/SiO_2$ is illustrated below.

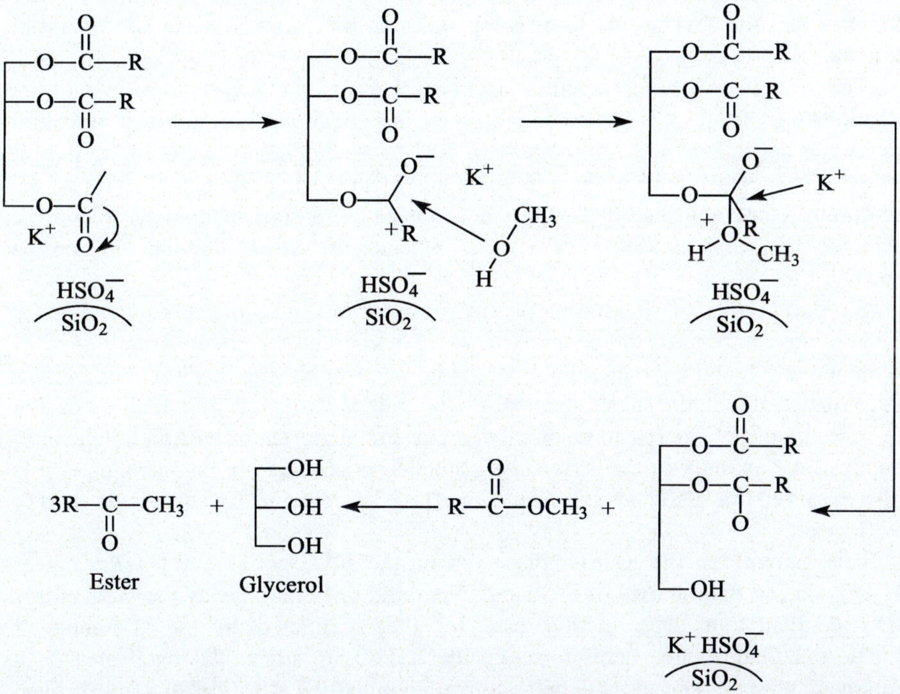

The changes in triglyceride molecules take place in three stages leading to the formation of monoglyceride and glycerol at the terminal stage. In the initial step the

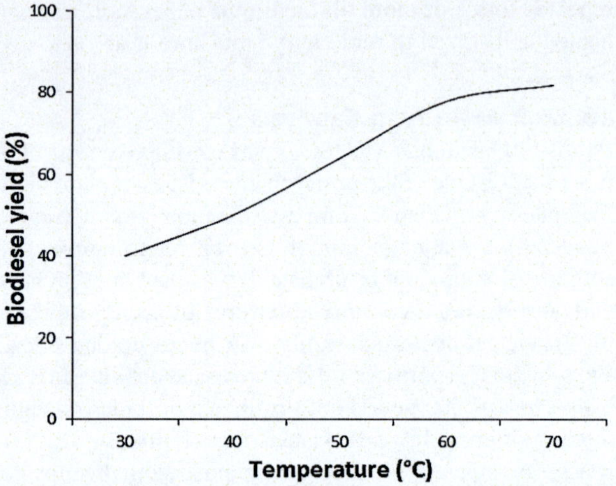

Fig. 18.29 The effect of temperature on the esterification of castor oil with 5% $KHSO_4/SiO_2$

reaction is characterized by a low miscibility of the catalyst with the reactants. As a result, the non-polar oil remains separated from the polar phase of the alcohol.

In the second phase the product thus formed actively plays the role of an emulsifier. In the terminal stage the equilibrium is reached. The forward reaction in the transesterification process in the presence of alcohol is a pseudo first order kinetics and the reverse reaction taking place in the process is the second order kinetics.

The transesterification reaction accomplished in the presence of potassium bisulphate as catalyst is governed by many optimizations. The transesterification process is carried out at a volume ratio of 40:1 of alcohol and oil with 5% by weight of catalyst (potassium bisulphate in silica gel). The stirring rate is 600 rpm and the optimum temperature is 60 °C. On the optimum conditions mentioned above, the recovery is as high as 95%. Figure 18.29 indicates the effect of temperature on the recovery.

Similarly the effect of duration on the transesterification process is studied at 5% level of the catalyst at a temperature of 60 °C. It is noted that the maximum yield is obtained at 5 h and beyond which there is no positive change in yield (Fig. 18.30). As a matter of fact, the substrate is optimally utilized in the reaction duration of 5 h.

The amount of catalyst to be employed plays an important role. At a low level of catalyst, the quantity of the active site generated is observed to be inadequate. It is also observed that 5% catalyst is adequate to take the reaction to a successful yield of 95% ester.

It is known that the $KHSO_4$ dispersed on the silica gel slowly leaches out as H_2SO_4 during the removal of methanol. To avoid this contingency the methanol is first distilled out. The methyl ester is then extracted in the presence of dichloromethane which helps to retain the $KHSO_4$ in silica. The catalyst is then removed, washed repeatedly in petroleum ether and dried at 150 °C in a hot air oven for a maximum of 12 h. The sustainable reusability can be gauged from Fig. 18.31.

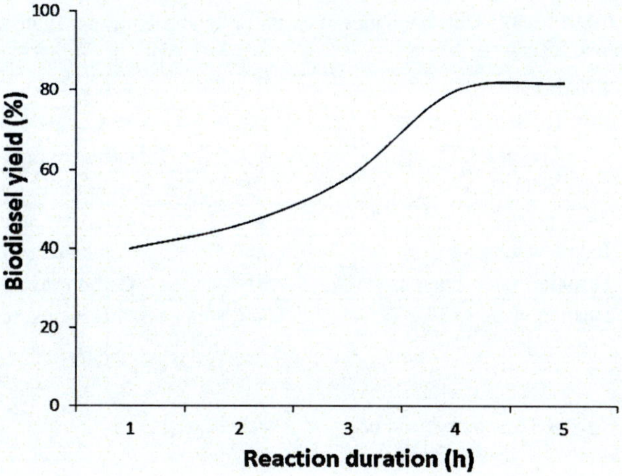

Fig. 18.30 Effect of reaction duration on the esterification of castor oil using 5% $KHSO_4/SiO_2$ at 60 °C

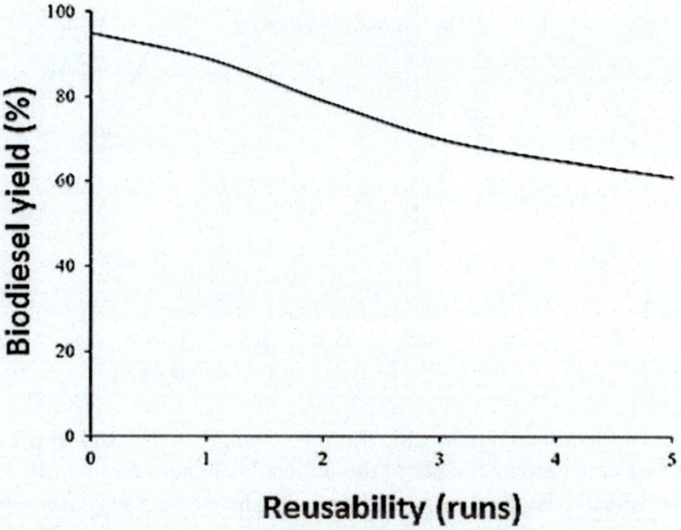

Fig. 18.31 Ester yield as a function of reusability of the $KHSO_4/SiO_2$ catalyst at 5% level

It is known that till five runs, the same catalyst can be effectively reused without much loss of potency. The properties of the castor oil biodiesel are given in Table 18.37. High flashpoint of castor oil biodiesel guarantees safety at the place of storage.

The viscosity of castor oil biodiesel is high. High viscosity is caused by the presence of a small amount of triglyceride in it (Conceicao et al. 2005). High viscosity alters the atomization characteristics. As a result the biodiesel is normally blended with regular diesel. Blends ranging from 5% to 25% (v/v) express rheopexy.

Table 18.37 Characteristics of castor oil biodiesel (as per Bello and Makanju 2011 and Sreenivas et al. 2011)

Parameters		Parameters	
Relative density (35 °C)	0.886	Kinematic viscosity ($mm^2\ s^{-1}$)	10.4
Cloud point (°C)	4	Cetane number	53
Pour point (°C)	−15	Acid value mg KOH g^{-1}	0.01
Flash point (°C)	447	Free fatty acid (%)	0.30
Iodine value (mg g^{-1})	0.8	Sulphated ash (%)	0.006
Peroxide value (meq kg^{-1})	50	Carbon residue (%)	0.037
Heating value (mJ kg^{-1})	38.0	Phosphorus mg kg^{-1}	3.02

Fig. 18.32 Comparison of the torque against the speed of a diesel engine

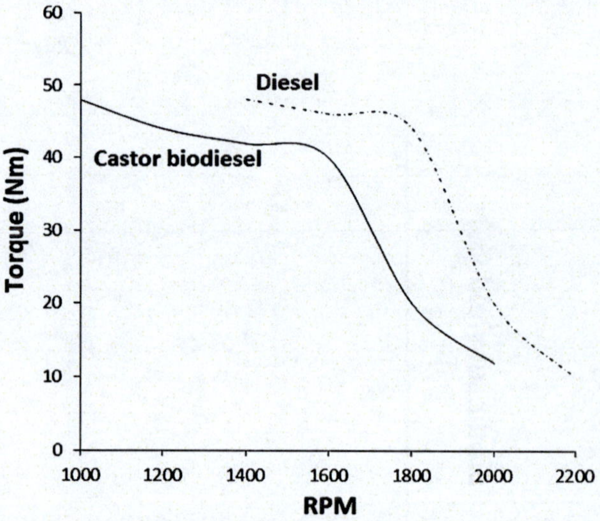

Rheopexy has a functional relationship with the change in viscosity. It is known that the three dimensional arrangements of the molecule changes temporarily during the application of shear. Rheopexy reverses when the shear of the engine ceases. In high temperature operational conditions, high molecular weight components are likely to be formed in the biodiesel due to oxidative degradation and polymerization (Conceicao et al. 2005).

The torque and power output of a stationary engine running on biodiesel is observed to be around 9–10% lower (Fig. 18.32) than that of the regular diesel. Since the biodiesel has high oxygen content guaranteeing complete combustion, the load carrying capacity increases by 15–20%.

Specific consumption of fuel is low by 8–10% in biodiesel run engine (Fig. 18.33) which by and large is a function of the heating value (Bello and Makanju 2011) of the fuel.

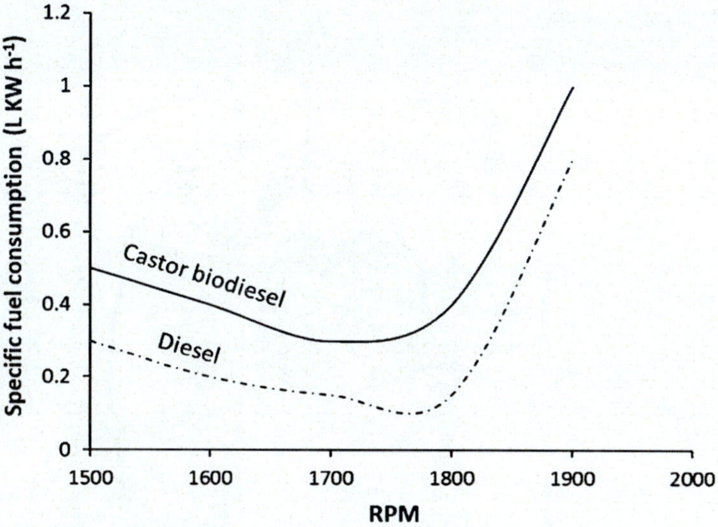

Fig. 18.33 Specific fuel consumption against the speed of the engine

Fig. 18.34 Geographical distribution of *Triadica sebifera*

18.7 Chinese Tallow (*Triadica sebifera*)

The Chinese tallow tree *Triadica sebifera* (syn: *Sapium sebiferum*) is a native of Eastern China, Taiwan and Japan. The seeds of this tree bear wax and hence the tree is called by the popular name Chinese tallow. Chinese tallow oil dedicated to biodiesel production ranks third by volume at world level. This tree is geographically represented in South Eastern United States, Puerto Rico, Australia, Africa, Cuba, Bangladesh, Pakistan, Madagascar, Myanmar, Vietnam, Costa Rica, Taiwan, India, China, Japan, Martinique, Zimbabwe, Sudan and Southern France (Fig. 18.34). It grows in an altitude of 0–2800 m above MSL. A single tree can produce 100,000

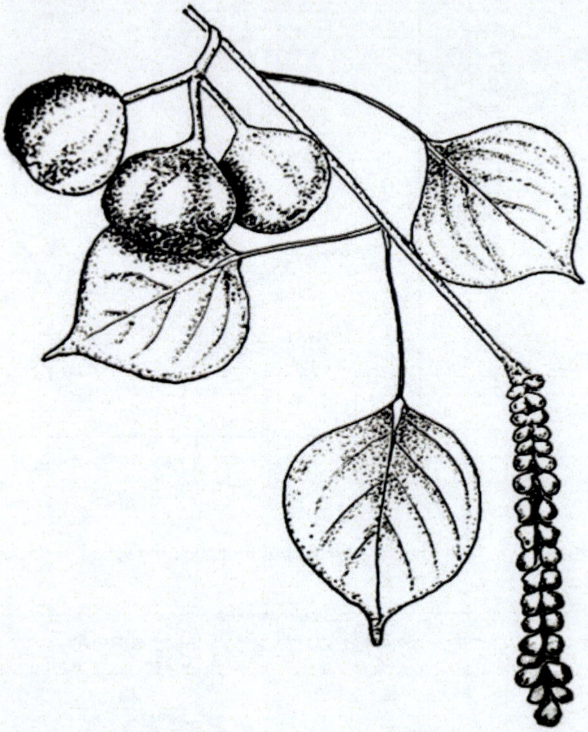

Fig. 18.35 *Triadica sebifera* plant

seeds a year which by weight is 4500 kg seeds $ha^{-1}year^{-1}$. It is popularly referred as Chinese tallow tree, Florida aspen, Chicken tree, gray popcorn tree, payaung and candleberry tree.

T. sebifera is a quick growing deciduous tree growing up to 11 m in height and 1 m in diameter. The leaves are alternate, simple, oval, 3–8 cm long and 3–7 cm wide. The flowers occur in a terminal spike of 15–20 cm long with a maximum of 15 staminate flowers along the spike (Fig. 18.35). The fruit is a capsule of 10–15 mm diameter. Each capsule has three seeds of 5–10 mm long and 5–6 mm wide. It is estimated that this tree yields around 2000 L of oil per acre per year (Picou and Boldor 2012).

18.7.1 Oil Production

The lipid is distributed at the surface of the seed and also in the kernel. Both the oils are suitable for biodiesel production. Though the oil extraction is made through many processes, mechanical pressing and organic solvent extraction are currently in practice.

The seeds are dried, cleaned, powdered and the oil extracted using n-hexane solvent in a Soxhlet extractor. Alternatively microwave assisted extraction is being

Table 18.38 Physicochemical properties of the whole seed oil obtained through microwave assisted extraction and the stillingia oil obtained through *n*-hexane extraction

Parameters	Whole seed oil	Stillingia oil
Acid value (mg KOH g^{-1})	–	2.11
Iodine value (g I_2/100 g)	–	127
Density (kg m^{-3})	–	930
Myristic acid (C14:0) (%)	0.13	–
Palmitic acid (C16:0) (%)	17.03	7.72
Stearic acid (C18:0) (%)	–	1.75
Eicosanoic acid (C21:0) (%)	10.92	–
Palmitoleic acid (C16:1) (%)	0.85	–
Oleic acid (C18:1) (%)	13.22	15.93
Linoleic acid (C18:2) (%)	13.30	29.05
Linolenic acid (C18:3) (%)	42.65	39.54

carried out. In microwave treatment, the contents are heated through dipolar rotation and ionic conduction. The heat is generated through internal friction between polar molecules and surrounding media. Thus the reaction is smooth and efficient. The ground seeds are mixed with three times by weight of absolute ethanol and kept in a bath maintained at 60 °C for 20 min. The material is then loaded in a Soxhlet apparatus to extract the oil. The extract is then filtered. The solvent present in the oil-solvent mixture is evaporated in vacuum. Through the conventional Soxhlet extraction, the oil recovery is observed to be low (around 25%), whereas in microwave assisted extraction the recovery is as high as 35% (Bordor et al. 2010). The seeds contain two types of oil. First type is extracted from the waxy coating of the seed referred as Chinese vegetable tallow fat which will be around 12–35% of the total weight of the seed. The second type is extracted from the kernel which is called as stillingia oil and is around 13–32% of the kernel mass (Liu et al. 2009). For biodiesel production, such a discrimination is not done and the necessary oil is extracted from the whole seed. The oil content varies according to the place where the seeds are harvested. The physicochemical properties of the whole seed oil obtained through microwave assisted extraction and the stillingia oil obtained from the kernel through *n*-hexane extraction are given in Table 18.38.

The iodine value of the stillingia oil is high, since the percentage of unsaturated fatty acids is high (84.5%). It has relatively low content of bis-allylic methylene carbons on fatty acid chains which brings in low resistance to oxidation thereby shorten the shelf life. Heptadecanoic acid (C17:0) being rarely found in other vegetable oil is detected in the stillingia oil.

18.7.2 Biodiesel Production

Biodiesel is prepared by a short cut method wherein instead of oil being used as a raw material, the seed itself is directly used as a raw material. The seeds are dried repeatedly to remove the moisture and then powdered. The powder is mixed with hexane and sodium methoxide and is heated in a microwave. The mixture thus

Table 18.39 Physicochemical properties of the biodiesel obtained through in-situ transesterification (Barekati-Goudarzi et al. 2016)

Property	Value
Kinematic viscosity at 40 °C (cSt)	2.02
Density at 40 °C (kg m^{-3})	888
Acid value (g KOH kg^{-1})	0.34
Cetane number	62.7
Cloud point (°C)	8–9
Pour point (°C)	2.0

obtained is then centrifuged (3000 rpm) so as to recover the liquid. This liquid is then allowed to settle overnight and the supernatant phase is the biodiesel. This is washed with water to remove the chemical impurities. The hexane is distilled out and the FAME thus obtained is freed from any moisture by heating. The physicochemical characteristics of the ester thus obtained are given in Table 18.39.

Often lipase catalysed transesterification also is carried out using Novozyme 435, Lipozyme TLIM and Lipozyme RMIM. These enzymes are added along with methanol. Tertiary butanol also is added to that mixture so as to enhance the efficiency of the process (Liu et al. 2009). The efficiency of the enzymes is in the order Novozyme 435 > Lipozyme TLIM > Lipozyme RMIM. Normally the methanol and glycerol which emerge as by-product are likely to retard the function of the enzymes. It is at this point the tertiary butanol which is added to the mixture ably solubilizes the methanol and glycerol and thus prevent the negative effect on the enzyme. The oil to methanol molar ratio is 1:5 and the conc. of enzyme is 15% by weight.

KF/CaO and KF/CaO-Fe_3O_4 nanocatalysts are also used as catalysts. The KF/CaO is prepared as follows: CaO is powdered and soaked in KF for 1 h and baked at 105 °C for 4–6 h. Further it is calcined in a muffle furnace at 600 °C. It is then used as a catalyst in transesterification (Wen et al. 2010) subjecting the process with all optimum conditions.

Similarly, nano-magnetic catalyst KF/CaO-Fe_3O_4 is also tried in the transesterification of this oil (Hu et al. 2011). $FeSO_4 \cdot 7H_2O$ and $Fe_2(SO_4)_3$ are dissolved in deionized water. To this $NH_3 \cdot H_2O$ is added in drops while the solution is maintained at 65 °C and the pH at 12.0. After 60 min the black solid is separated by magnet, the pH moderated to 7.0, dried at 60 °C and pulverized. The above material along with MgO, CaO and SrO is soaked in KF and subsequently dried at 105 °C and calcined at 600 °C in a muffle furnace. This material is used as a catalyst. The properties of the kernel oil (stillingia oil) biodiesel are given in Table 18.40.

Table 18.40 gives the properties of biodiesel from stillingia oil of *T. sebifera* catalysed by the catalase enzyme Novozyme 435. The cold flow properties (Cloud point and cold filter plugging point) are considered good and the cetane number also is satisfactory. The unsaturated fatty acid of the oil has a bearing on the cold flow properties of the biodiesel. The rich level of linoleic acid and linolenic acid (Table 18.1) of the oil is responsible for the poor oxidation stability. Therefore, such biodiesel requires the addition of antioxidants to maintain the shelf life.

Table 18.40 The physicochemical properties of the *T. sebifera* kernel oil (stillingia oil) biodiesel (Liu et al. 2009)

Parameters	Value
Kinematic viscosity at 40 °C (cSt)	4.81
Density at 15 °C (kg m^{-3})	900
Flash point (°C)	137
Acid value (g KOH kg^{-1})	0.007
Cetane number	50
Cloud point (°C)	−13
Cold filter plugging point (°C)	−10
Oxidation stability 110 °C (h)	0.6

Though the oil obtained from *T. serbifera* has a scope of being considered as a raw material for biodiesel, no tangible experimental data is available on the engine performance apparently due to the limitations of the total volume of oil available in the market.

18.8 Tung Tree (*Vernicia montana and V. fordii*)

The tung oil is extracted from the seeds of the species *Vernicia montana* or *V. fordii*. The oil of *V. montana* is called abrasin oil or Chinese wood oil. This tree attains a maximum height of 20 m and grows at an elevation of 800–2000 m above MSL and with a rainfall of 800–2000 mm a year. The flower is white in colour and the leaves are with three distinct lobes. The fruit is a drupe 5–7.5 cm dia. with three seeds. The tree comes to yield in 5–6 years. The surface of the fruit bears distinct wrinkles (Fig. 18.36). The fruits when ripe fall on the ground and are picked up for use. Alternatively they are harvested straight from the tree. The normal yield of fruit is around 3.5 t ha^{-1} year^{-1}. The kernel yield is around 1 t ha^{-1} year^{-1}. The seeds contain 40–60% oil. This tree is also referred as abrasin, cantonese wood oil tree, Chinese tung oil, mu oil tree, tung tree and tung nut tree.

V. montana is distributed in Myanmar, Thailand, Vietnam, Taiwan, Southern China, Kenya, Tanzania, Malawi, Zambia, Zimbabwe, Mozambique Madagascar and Cambodia (Oyen 2007) (Fig. 18.37).

Vernicia fordii (formerly *Aleurites fordii*) is a small tree of 15–20 m height. It bleeds latex if the bark is injured. It has a red gland at the base of the leaves. The leaves are heart shaped with three shallow lobes. The petiole is 5.5–26 cm long. The flower is pale pink to purple. The fruit is hard and peer shaped, 4–6 cm long, 3.5 cm dia. with 4–5 seeds per fruit (Fig. 18.38).

The oil content of the seeds of this tree is lower (<20%) than that of *V. montana.* This tree is popularly referred as airy shaw, youtong and oil tong. It is distributed in Alabama, Australia, Argentina, California, Florida, South and Central China, East Himalayas, Hainan, Korea, Leeward Is, Mozambique, Myanmar, New South Wales, Paraguay, Papua New Guinea, Puerto Rico, Taiwan, Thailand, Trinidad-Tobago, Vietnam and Zimbabwe (PFAF Plant data base).

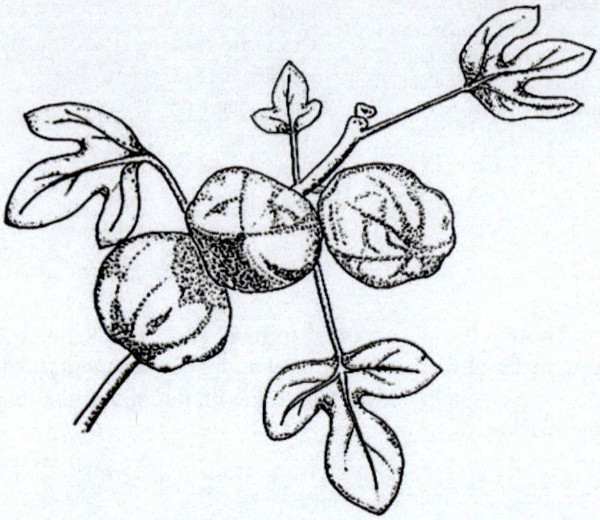

Fig. 18.36 *Vernicia montana* plant

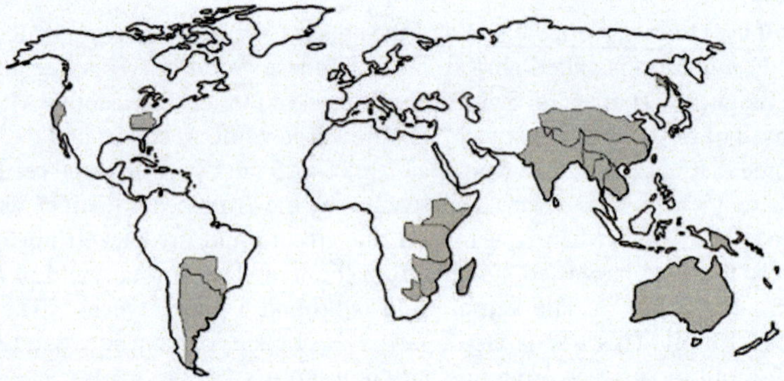

Fig. 18.37 Geographical distribution of *V. montana* and *V. fordii*

18.8.1 Oil Production

The seeds (*V. montana* or *V. fordii*) are decorticated and the kernels are sundried repeatedly. The kernels are then powdered and the oil extracted using *n*-hexane. The solvent kernel ratio during the extraction is 10:1. The extraction duration is an hour. The extract thus obtained is allowed to settle. The upper layer contains oil and the lower layer has the residue. This residue is separated and repeatedly treated (3–5 times or till the oil is completely removed) with fresh *n*-hexane. The combined oil-hexane mixture is pooled, filtered and the hexane removed by vacuum distillation (40 °C). The physicochemical properties of *V. montana* and *V. fordii* oils are given in Table 18.41. The percentage of oil extracted from the seed, kinematic viscosity and

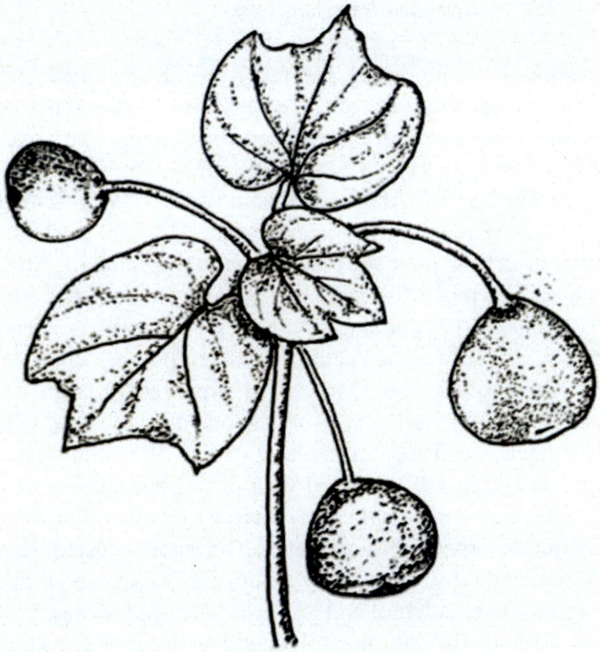

Fig. 18.38 *Vernicia fordii* plant

Table 18.41 Physicochemical properties of the oil from *Vernicia* sp. (Chen et al. 2010; Chung 2010; Le et al. 2018a, b)

Parameters	*V. montana*	*V. fordii*
Oil content of the seeds (max wt%)	33.7	24.8
Kinematic viscosity at 40 °C (cSt)	103	75
Density at 25 °C (kg m^{-3})	912, 941	892
Acid value (g KOH kg^{-1})	0.2, 1.45	1.85
Iodine value (g I_2/100 g)	118, 161	–
Oxidative stability (110 °C) (h)	0.8	–
Palmitic acid (C16:0) (%)	2.2	5.3
Stearic acid (C18:0) (%)	1.9	9.1
Oleic acid (C18:1) (%)	6.5	57.2
Linoleic acid (C18:2) (%)	8.5	15.6
Linolenic acid (C18:3) (%)	–	12.8
α-Eleostearic acid (C18:3) (%)	80.3	–

density are high with respect to *V. montana*. Oleic acid content is high in *V. fordii*. A major portion of the fatty acids in *V. montana* is by 9*Z*, 11*E*, 13*E*-α eleostearic acid. This acid contains three conjugated carbon–carbon double bonds. The low acid value obviates the two stage transesterification process.

18.8.2 Biodiesel Production

Biodiesel is produced by alkali catalytic mode of transesterification. In the transesterification of oil, the methoxide reagent is prepared at the outset by dissolving potassium hydroxide in methanol. The conc. of potassium hydroxide is 1% of the wt. of oil to be used. Similarly the methanol quantity is decided so as to have a molar ratio of 6:1 (methanol to oil). The methoxide is slowly added to the hot oil and the reaction progressed at 60 °C. The contents are stirred at 800 rpm for 90 min so as to have the completion of the reaction. After 90 min the heat source and the stirring motor are switched off and the contents are allowed to settle overnight. The upper layer containing the diesel is washed repeatedly with distilled water till the biodiesel becomes neutral and then it is heated to expel the moisture.

Alternatively the biodiesel is prepared through a co-solvent method. In the co-solvent method, 20 g of acetone is added to 100 g oil and the mixture is homogenized using an electric stirrer. This homogenate is kept in a water bath maintained at 40 °C, added with potassium methoxide and maintained for 60 min. The whole content is then transferred to a separating funnel and kept undisturbed till the phase separation completes. The upper phase is a mixture containing acetone, fatty acid methyl ester, triglyceride, diglyceride and monoglyceride. The lower phase contains the methanol and glycerol. The upper phase is collected and heated to 60 °C for 30 min. By then the solvents dispel and the residual content (ester) is washed repeatedly with water so as to remove any trace of alkali. The physicochemical characteristics of the biodiesel thus obtained from the oil of *V. montana* and *V. fordii* are given in Table 18.42. The kinematic viscosity is slightly high and therefore the biodiesel needs to be blended to bring down the viscosity to the acceptable level. The cold filter plugging point is very low suggesting the acceptability of the biodiesel in cold conditions. The oxidative stability is poor due to the presence of more unsaturated fatty acids. The eleostearic acid is highly unstable and its level reduces from 82.2% to 11% when the biodiesel is aerated continuously for 7 days. The oxidative

Table 18.42 Physicochemical characteristics of the biodiesel from the oils of *Vernicia* sp. (Chen et al. 2010; Chung 2010; Le et al. 2018a, b)

Parameters	*V. montana*	*V. fordii*
Kinematic viscosity at 40 °C (cSt)	7.84, 7.7	2.5
Density (kg m^{-3})	903, 897	864
Acid value (g KOH kg^{-1})	0.12, 0.3	0.19
Cetane number	39	53
Cold filter plugging point (°C)	−11	
Iodine value (g I_2/100 g)	161, 158.5	–
Oxidative stability (110 °C) (h)	0.3	–
Palmitic acid (C16:0) (%)	2.3	–
Stearic acid (C18:0) (%)	2.4	–
Oleic acid (C18:1) (%)	5.6	–
Linoleic acid (C18:2) (%)	6.3	–
α-Eleostearic acid (C18:3) (%)	82.2	–

degradation is in the order: eleostearates (C18:3) > linoleates (C18:2) > oleates (C18:1) (Le et al. 2018a, b).

Normally α-eleostearic acid has a high melting point and good oxidation stability than its corresponding *cis*-isomers (Knothe 2007). However it is responsible for the cold filter plugging point of the biodiesel due to its high share among the fatty acid contents of the biodiesel.

References

Abitogun A, Alademeyin O, Oloye D (2008) Extraction and characterization of castor seed oil. Int J Nutr Wellness 8(2):7

Achten WMJ, Verchot L, Franken YJ, Mathijis E, Singh VP, Aerts R, Muys B (2008) Jatropha biodiesel production and use. Biomass Bioenergy. https://doi.org/10.1016/j.biombioe.2008.03.003

Adebowale KO, Adedire CO (2006) Chemical composition and insecticidal properties of the underutilized *Jatropha curcas* seed oil. Afr J Biotechnol 5(10):901–906

Aigbodion AI, Bakare IO (2005) Rubber seed oil quality assessment and authentication in Nigeria. J Am Oil Chem Soc 82:465–469

Aigbodion AI, Pillai CKS (2000) Preparation, analysis and applications of rubber seed oil and its derivatives in surface coatings. Prog Org Coat 38:465–469

Akbar E, Yaakob Z, Kamarudin SK, Ismail M, Salimon J (2009) Characteristic and composition of *Jatropha curcas* oil seed from Malaysia and its potential as biodiesel feedstock. Eur J Sci Res 29 (3):396–403

Akintayo ET (2004) Characteristics and composition of *Parkia biglobosa* and *Jatropha curcas* oil and cakes. Bioresour Technol 92:307–310

Akpan UG (1999) Extraction and characterization of neem seed oil. In: Eyo AA, Aloku PO, Garba SA, Ali UD, Lamai SL, Olufeagba SO (eds) Biotechnology and sustainable development in Nigeria. Proceedings of the 12th annual conference of the Biotechnology Society of Nigeria, pp 63–66

Aranda DAG, Santos RTP, Tapanes NCO, Ramos ALD, Antunes OAC (2008) Acid-catalysed homogeneous esterification reaction for biodiesel production from fatty acids. Catal Lett 120:20–25

Ashraful AM, Masjuki HH, Kalam MA, Fattah IMR, Imtenan S, Shahir SA, Mobarak HM (2014) Production and comparison of fuel properties, engine performance and emission characteristics of biodiesel from various non-edible vegetable oils: a review. Energy Convers Manag 80:202–228

Aziz MAA, Puad K, Triwahyono S, Jalil AA, Khayoon MS, Atabani AE, Ramli Z, Majid ZA, Prasetyoko D, Hartanto D (2017) Transesterification of *Croton megalocarpus* oil to biodiesel over WO_3 supported on silica mesoporous-macroparticles catalyst. Chem Eng J 316:882–892

Barekati-Goudarzi M, Boldor D, Nde DB (2016) In-situ transesterification of seeds of invasive tallow trees (*Triadica sebifera* L.) in a microwave batch system ($GREEN^3$) using hexane as co-solvent: biodiesel production and process optimization. Bioresour Technol 201:97–104

Becker K, Makkar HPS (2008) *Jatropha curcas*: a potential source for tomorrows oil and biodiesel. Lipid Technol 20:104–107

Bello EI, Makanju A (2011) Production characterization and evaluation of castor oil biodiesel as alternative fuel for diesel engines. J Emerg Trends Eng Appl Sci 2(3):525–530

Benaides A, Benjumea P, Pashova V (2007) El biodiesel de aceite de higuerilla como combustible alternative para motores diesel. Dyna 74:141–150

Bordor D, Karitkar A, Terigar BG, Leonardi C, Lima M, Breitenbeck GA (2010) Microwave assisted extraction of biodiesel feedstock from the seeds of invasive Chinese tallow tree. Environ Sci Technol 44:4019–4025

Bringi NV (1987) The non-traditional oilseeds and oils in India. Oxford & IBH Publishing Co. Pvt. Ltd., New Delhi

Cabral RPM, dos Santos ALS, Stropa JM, da Silva RCR, Cardoso ALC, de Oliveira LCS, Scharf DR, Simionatto EL, Santiago EF, Simionatto E (2016) Chemical composition and thermal properties of methyl and ethyl esters prepared from *Aleurites moluccanus* (L.) Willd (Euphorbiaceae) nut oil. Ind Crops Prod 85:109–116

Can Ô (2014) Combustion characteristics, performance and exhaust emissions of a diesel engine fueled with a waste cooking oil biodiesel mixture. Energy Convers Manag 87:676–686

Chapman AV (2007) Natural rubber and NR based polymers: renewable materials with unique properties. In: Proc. 24th International H.F. Mark symposium on Advances in the field of elastomers and thermoplastic elastomers, Vienna, 15–16 Nov 2007

Chen Y, Chen J, Chang C, Chang C (2010) Biodiesel production from tung (*Vernicia montana*) oil and its blending properties in different fatty acid compositions. Bioresour Technol 101:9521–9526

Christie WW (1982) Lipid analysis, isolation, separation, identification and structural analysis of lipids, 2nd edn. Pergamon, Oxford

Chung K (2010) Transesterification of *Camellia japonica* and *Vernicia fordii* seed oils on alkali catalysts for biodiesel production. J Ind Eng Chem 16:506–509

Conceicao MM, Candeia RA, Dantas HJ, Soledade LEB, Fernandes VJ, Souza AG (2005) Rheological behaviour of castor oil biodiesel. Energy Fuels 19:2185–2188

Cosgrove JP, Church DF, Pryor WA (1987) The kinetics of the auto-oxidation of polyunsaturated fatty acids. J Lipid Res 22(5):299–304

Crabbe E, Nolasco-Hipollito C, Kobayashi G, Sonomoto K, Ishizaki A (2001) Biodiesel production from crude palm oil and evaluation of butanol extraction and fuel properties. Process Biochem 37:65–71

Dean JA (1985) Lange's handbook of chemistry. McGraw-Hill, New York

Demirbas A (2003) Biodiesel fuels from vegetable oils *via* catalytic and non-catalytic supercritical alcohol transesterifications and other methods: a survey. Energy Convers Manag 44:2093–2109

Devi RM (2010) Studies on biodiesel production and quality analysis. PhD thesis, Madurai Kamaraj University, Madurai, India

Dunn RO (2008) Antioxidants for improving storage stability of biodiesel. Biofuels Bioprod Biorefin 2(4):304–318

Elevitch CR, Manner HI (2006) *Aleurites moluccana* (kukui) traditional trees of pacific Islands: their culture, environment and use. Permanent Agriculture Resource, Holualoa, pp 41–56

Ensminger ME, Olextine CG Jr (1978) Feeding sheep. In: Feeds and nutrition. The Ensminger Publishing Company, Clovis, CA, pp 743–786

Ferrari RA, Oliveira VS, Scabio A (2005) Oxidative stability of biodiesel from soybean oil fatty acid ethyl esters. Sci Agric 62(3):291–295

Freedman B, Butterfield RO, Pryde EH (1986) Transesterification kinetics of soybean oil. JAOCS 63:1357–1380

Furuta S, Matsuhashi H, Arata K (2004) Catalytic action of sulphated tin oxide for etherification and esterification in comparison with sulphated zirconia. Appl Catal A Gen 269:187

Gandhi VM, Cherian KM, Mulky MJ (1995) Toxicological studies on ratanjyot oil. Food Chem Toxicol 33:39–42

Goodain JG, Natesalehawat S, Nikolopolous AA, Kim SY (2002) Etherification on zeolites: MTBE synthesis. Catal Rev Sci Eng 44:287–320

Gour VK (2006) Production practices including post-harvest management of *Jatropha curcas*. In: Singh B, Swaminathan R, Ponraj V (eds) Proc. of the biodiesel conference towards energy independence-focus of Jatropha, Hyderabad, India, June 9–10 Rashtrapati Bhavan, New Delhi, pp 223–251

Grainge M, Ahmed S (1988) Handbook of plants with pest control properties. Wiley, New York

Grimm C, Guharay F (1997) Potential of entomopathogenous fungi for the biological control of true bugs in *J. curcas* L. plantations in Nicaragua. In: Giibitz GM, Mittelbach M, Trabi M (eds) Biofuels and industrial products from *Jatropha curcas*. DBV, Graz, pp 40–46

Grimm C, Maes (1997) Arthropod fauna associated with *Jatropha curcas* L. in Nicaragua: a synopsis of species; their biology and pest status. In: Giibitz GM, Mittelbach M, Trabi M (eds) Biofuels and industrial products from *Jatropha curcas*. DBV, Graz, pp 40–46

Habibullah M, Futtah IMR, Masjuki HH, Kalam MA (2015) Effects of palm-coconut biodiesel blends on the performance and emission of a single cylinder diesel engine. Energy Fuel 29:734–743

Haque MA, Islam MP, Hussain MD, Khan F (2009) Physical, mechanical properties and oil content of selected indigenous seeds available for biodiesel production in Bangladesh. Agric Eng Int 1419

Hartmann HT, Kester DE (1983) Plant propagation. Principles and practices, 4th edn. Prentice-Hall, Englewood Cliffs, NJ

Heller J (1996) Physic nut. *Jatropha curcas* L. Promoting the conservation and use of underutilized and neglected crops. International Plant Genetic Resources Institute, Rome

Henning K (1997) Fuel production improves food production: the Jatropha project in Mali. In: Giibitz GM, Mittelbach M, Trabi M (eds) Biofuels and industrial products form *Jatropha curcas*. DBV, Graz, pp 92–97

Hu S, Guan Y, Wang Y, Han H (2011) Nano-magnetic catalyst KF/CaO-Fe_3O_4 for biodiesel production. Appl Energy 88:2685–2690

Imdadul HK, Zulkifli NWM, Masjuki HH, Kalam MA, Kamruzaman M, Rashed MM, Rashedul HK, Alwi A (2016) Experimental assessment of non-edible candle nut biodiesel and its blend characteristics as diesel engine fuel. Environ Sci Pollut Res. https://doi.org/10.1007/s11356-016-7487-y

Ishii Y, Takeuchi R (1987) Transesterified curcas oil blends for farm diesel engines. Trans Am Soc Agric Eng 30:605–609

Iyayi AF, Akpaka PO, Ukpeoyibo U, USDA (2000) ANGRPNGRLBM *Jatropha curcas* information from Germplasm Resources Information Network (GRIN) [on line database]. http://www.ars.grim.gov/cgi.bin/npgs/html/taxon.pl?20692. Accessed 9 Nov 2007 last updated on 16 Apr 2010

Kafuku G, Mbarawa M (2010) Biodiesel production from *Croton megalocarpus* oil and its process optimization. Fuel 89:2556–2560

Kafuku G, Lam MK, Kansedo J, Lee KT, Mbarawa M (2010) *Croton megalocarpus* oil: a feasible non-edible oil source for biodiesel production. Bioresour Technol 101:7000–70004

Kafuku G, Tan KT, Lee KT, Mbarawa M (2011) Non-catalytic biodiesel fuel production from *Croton megalocarpus* oil. Chem Eng Technol 34(11):1827–1834

Kibazohi O, Sangwan RS (2011) Vegetable oil production potential from *Jatropha curcas*, *Croton megalocarpus*, *Aleurites moluccana*, *Moringa oleifera* and *Pachira glabra*: assessment of renewable energy resources for bio-energy production in Africa. Biomass Bioenergy 35 (3):1352–1356

Kivevele TT, Mbarawa MM (2010) Comprehensive analysis of fuel properties of biodiesel from *Croton megalocarpus* oil. Energy Fuel 24:6151–6155

Kivevele T, Zhongjic H (2015) Review of the stability of biodiesel produced from less common vegetable oils of African origin. S Afr J Sci 111:9–10

Knothe G (2007) Some aspects of biodiesel oxidative stability. Fuel Process Technol 88:669–677

Kocsisova T, Cvengro J, Lutisan J (2005) High temperature esterification of fatty acids with methanol at ambient pressure. Eur J Lipid Sci Technol 107:87–92

Krishnawati H, Kallio M, Kanninen M (2011) *Aleurites moluccana* (L.) Wild: ecology, silviculture and productivity. CIFOR, Bogor, Indonesia

Lakshmi SBAVS, Subramania Pillai N, Mohamed MSBK, Narayanan A (2020) Biodiesel production from rubber seed oil using calcined eggshells impregnated with Al_2O_3 as heterogeneous catalyst: a comparative study of RSM and ANN optimization. Braz J Chem Eng 37:351–368

Le HNT, Imamura K, Watanabe N, Furuta M, Takenaka N, Boi LV, Maeda Y (2018a) Biodiesel production from rubber seed oil by transesterification using a co-solvent of fatty acid methyl esters. Chem Eng Technol 41(5):1013–1018

Le HNT, Imanura K, Furuta M, Boi LV, Maeda Y (2018b) Production of biodiesel from *Vernicia montana* Lour oil using a co-solvent method and the subsequent evaluation of its stability during storage. Green Process Synth 7:170–179

Liberalino A, Bambirra EA, Moraes-Santos T, Vieira EC (1988) *Jatropha curcas* L. seeds: chemical análysis and toxicity. Arq Biol Technol 31:539–550

Lin L, Ying D, Chaitep S, Vittayapadung S (2009) Biodiesel production from crude rice bran oil and properties as fuel. Appl Energy 86:681–688

Liu KS (1994) Preparation of fatty acid methyl esters for gas chromatographic analysis of lipids in biological materials. JAOCS 71:1179–1187

Liu Y, Xin H, Yan Y (2009) Physicochemical properties of stillingia oil: feasibility for biodiesel production by enzyme transesterification. Ind Crop Prod 30:431–436

Lotero E, Liu Y, Lopez DE, Suwannakarn K, Bruce DA, Goodwin JG (2005) Synthesis of biodiesel *via* acid catalysis. Ind Eng Chem Res 44:5353–5363

Ma F, Hanna MA (1999) Biodiesel production: a review. Bioresour Technol 70:1–15

McCormick RL, Ratcliff M, Moens L, Lawrence R (2007) Several factors affecting the stability of biodiesel in standard accelerated tests. Fuel Process Technol 88:651–655

Meshram PB, Joshi KC (1994) A new report of *Spodoptera litura.* (Fab.) Boursin (Lepidoptera: Noctuidae) as a pest of *Jatropha curcas* L. Indain Forest 120:273–274

Miinch E, Kiefer J (1986) Die purgiernuB (*Jatropha curcas*). M.Sc. thesis, Universitat Hohenheim

Mirie SN, Kioni PN, Thiong'o GT, Kariutci PN (2012) Immobilized *Candida antarctica* lipase catalyzed transesterification of *Croton megalocarpus* seed oil for biodiesel production. J Energy Technol Policy 2(5):20–24

Nangbes JG, Nvau JB, Babu WM, Zukdimma AN (2013) Extraction and characterization of castor *Ricinus communis* seed oil. Int J Eng Sci 2(9):105–109

Neuwinger HD (1994) Afrikanische Arzneipflanzen and Jagdgifte. W.V. GesmH, Vienna, Austria, pp 450–457

Njoku OU, Ononogbu IC, Owusu JY (1996) An investigation of oil of rubber (*Hevea brasiliensis*). J Rubber Res Inst Sri Lanka 78:52–59

Nye MJ, Williamson TW, Deshpande S, Sehrader JH, Snively WH, Yuskewich TP, French CR (1983) Conversion of used frying oil to diesel fuel by transesterification: preliminary tests. J Am Oil Chem Soc 60:1598–1601

Osawa WO, Onyari JM, Sahoo PK, Mulaa FJ (2014) Process optimization for production of biodiesel from croton oil using two-stage process. IOSR J Environ Sci Toxicol Food Technol 8(11):49–54

Oyen LPA (2007) In: van der Vossen HAM, Mkamilo GS (eds) *Vernicia montana* Lour. PROTA (Plant Resources of Tropical Africa/Ressources végétales de l'Afrique tropicale), Wageningen, Netherlands

Patolia JS, Chikara J, Chaudhary DR, Parmar DR, Bhuva HM (2007) Response of *Jatropha curcas* grown on wasteland to N and P fertilization. In: Proc. of the FACT seminar on *Jatropha curcas L.* agronomy and genetics, Wageningen, the Netherlands, 26–28 March, Article No. 34. FACT Foundation, Wageningen

Pearson D (1976) The chemical analysis of foods, 7th edn. Churchill Living stone, London

Peterson GR, Scarrah WP (1984) Rapeseed oil transesterification by heterogeneous catalysis. JAOCS 61:1593–1597

Pham LN, Luu BV, Phuoc HD, Le HNT, Truong HT, Luu PD, Furuta M, Imamura K, Maeda Y (2018) Production of biodiesel from candle nut oil using a two-step co-solvent method and evaluation of its gaseous emissions. J Oleo Sci 67(5):617–626

Picou L, Boldor D (2012) Thermophysical characterization of the seeds of invasive Chinese tallow tree: importance for biofuel production. Environ Sci Technol 46:11435–11442

Pinto AC, Guarieiro LL, Rezende MJ, Ribeiro NM, Torres EA, Lopes WA, Pereira PA, Andrade JB (2005) Biodiesel: an overview. J Brazil Chem Soc 16:1313–1330
Prather M, Ehhalt D, Dentener F, Derwent R, Dlugokencky E, Holland E, Isaksen I, Katima J, Kirchhoff V, Matson P, Midgley P, Wang M (2001) Atmospheric chemistry and greenhouse gases. In: Contribution of working Group I to the Third Assessment Report of the International Panel on Climate Change. Cambridge University Press, Cambridge, UK, pp 239–287
Prueksakorn K, Gheewala SH (2006) Energy and greenhouse gas implications of biodiesel production from *Jatropha curcas*. In: Proc. of the Second Joint International conference on sustainable energy and environments (SEE, 2006), Bangkok, Thailand, 21–23 November
Ramadhas AS, Jeyaraj S, Muraleedharan C (2005) Biodiesel production from high FFA rubber seed oil. Fuel 84:335–340
Rashed MM, Masjuki HH, Kalam MA, Alabdulkarem A, Rahman MM, Imdadul HK, Rashedul HK (2016) Study of the oxidation stability and exhaust emission analysis of *Moringa oleifera* biodiesel in a multi-cylinder diesel engine with aromatic amine antioxidants. Renew Energy 94:294–303
Salimon J, Noor DAM, Nazrizawati AT, Firdaus MYM, Noraishah A (2010) Fatty acid composition and physicochemical properties of Malaysian castor bean *Ricinus communis* L seed oil. Sains Malaysiana 39(5):761–764
Sauerwein M, Sporer F, Wink M (1993) Insect toxicity of phorbol esters from *Jatropha curcas* seed oil. Plant Med Suppl Iss 59:686
Sharma GD, Gupta SN, Khabiruddin M (1997) Cultivation of *Jatropha curcas* as a future source of hydrocarbon and other industrial products. In: Giibitz GM, Mittelbach M, Trabi M (eds) Biofuels and industrial products from *Jatropha curcas*. DBV, Graz, pp 19–21
Solsoloy AD (1993) Insecticidal action of the formulated product and aqueous extract from physic nut, *Jatropha curcas* L on cotton insect pests. Cotton Res J 6:24–35
Sreenivas P, Mamilla VR, Sekhar KC (2011) Development of biodiesel from castor oil. Int J Energy Sci 1(3):192–197
Szybist JP, Boehman AL, Taylor JD, McCormick RL (2005) Evaluation of formulation strategies to eliminate the biodiesel NO_x effect. Fuel Process Technol 86:1109–1126
Takeda Y (1982) Development study of *Jatropha curcas* (Sabu Dum) oil as a substitute for diesel engine oil in Thailand. J Agric Assoc China 120:1–8
Ting WJ, Huang CM, Giridhar N, Wu WT (2008) An enzymatic acid – catalysed hybrid process for biodiesel production from soybean oil. J Chin Inst Chem Eng 39:203–210
Tobin J, Fulford DJ (2005) Life-cycle assessment of the production of biodiesel from Jatropha. MSc Dissertation, The University of Reading
UNIDO (1987) Rubber seed processing for the production of vegetable oil and animal feed. In: Rubber seed processing for value-added latex production in Nigeria. Rubber Research Institute of Nigeria, Nigeria
Vicente G, Martinez M, Aracil J (2005) Optimization of *Brassica carinata* oil methanolysis for biodiesel production. J Am Oil Chem Soc 82(12):899–904
Vyas DK, Singh RN (2007) Feasibility study of Jatropha seed husk as an open core gasifier feedstock. Renew Energy 32:512–517
Wang R, Hanna MA, Zhou W, Bhadury PS, Chen Q, Song B, Yang S (2011) Production and selected fuel properties of biodiesel from promising non-edible oils: *Euphorbia lathyris*, L. *Sapium sebiferum* L. and *Jatropha curcas* L. Bioresour Technol 102:1194–1199
Wen L, Wang Y, Lu D, Hu S, Han H (2010) Preparation of KF/CaO nanocatalyst and its application in biodiesel production from Chinese tallow seed oil. Fuel 89:2267–2271
Wink M, Koschmieder C, Sauerwein M, Sporer F (1997) Phorbol esters of J. curcas-biological activities and potential applications. In: Giibitz GM, Mittelbach M, Trabi M (eds) Biofuels and industrial products form *Jatropha curcas*. DBV, Graz, pp 160–166
Wu WH, Foglia TA, Marmer WN, Dunn RO, Goering CE, Briggs TE (1998) Low-temperature property and engine performance evaluation of ethyl and isopropyl esters of tallow and grease. J AOCS 75:1173–1178

Zapata N, Vargas M, Reyes JF, Belmar G (2012) Quality of biodiesel and press cake obtained from *Euphorbia lathyris*, *Brassica napus* and *Ricinus communis*. Ind Crops Prod 38:1–5

Zhang J, Liu M, Zhang A, Lin K, Song C, Guo X (2010) Facile synthesis of mesoporous silica nanoparticles with controlled morphologies using water acetone media. Solidstate Sci 12:267–273

Zhang H, Pan H, Yang S (2017) Upgrading of cellulose to biofuels and chemicals with acidic nanocatalysts. Curr Nanosci 13:513–527

Zhang H, Li H, Pan H, Wang A, Souzanchi S, Xu C (2018) Magnetically recyclable acidic polymeric ionic liquids decorated with hydrophobic regulators as highly efficient and stable catalysts for biodiesel production. Appl Energy 223:416–429

Fabaceae

19

Fabaceae is the third largest family in the plant kingdom which comprises around 19,500 species forming 7% of the total flowering plants. Around 8% of them are oil bearing, on which the world seriously relies on its vegetable oil need. This chapter deals with 10 oil bearing species in which 2 are edible and 8 are non-edible. Among them 7 are mostly obscure with reference to their role on biodiesel production. Such species are also taken into active consideration as biodiesel resource as they are emerging as location-specific complementary energy alternatives. Biodiesel prepared from a portion of the edible oil from groundnut (*Arachis hypogaea*) and soybean (*Glycine max*) is being in use world over. Among the non-edible sources pongam (*Pongamia pinnata*) is richly used in biodiesel preparation. Active investigations are pursued during the last one decade on another seven species: African oak (*Afzelia africana*), babul (*Acacia nilotica*), diesel tree (*Copaifera langsdorffii*), mesquite (*Prosopis juliflora*), shikakai (*Acacia concinna*), shittim (*Acacia raddiana*) and brebra (*Millettia ferruginea*) though large scale economic exploitation is still on progress.

19.1 Groundnut (*Arachis hypogaea*)

The groundnut often referred as peanut is an annual herbaceous plant and grows up to 50 cm in height. It is reported to have been first grown in Paraguay 7000 years ago. Now, it is being cultivated in many countries throughout the tropical and warm temperate zones of the world between 40°S and 40°N latitude (Fig. 19.1).

It is grown in 22 million ha worldwide making a production of 46 million metric tons of nut during the year 2019–2020 (USDA 2021). The oil production was 6.25 million metric tons. As high as 70% of the world production comes from China, India and USA. The plant comes to ripe within 150 days of sowing. It is a frost sensitive crop. It grows well in well-drained sandy loams of pH 6–8. It tolerates an ambient temperature ranging from 15 to 28 °C. Groundnut is distributed in

S. P. Raj et al., *Biodiesel from Flowering Plants*,
https://doi.org/10.1007/978-981-16-4775-8_19

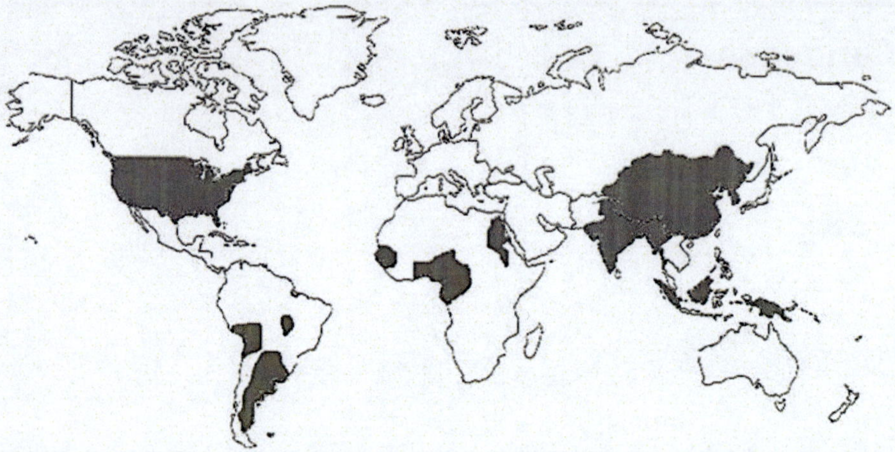

Fig. 19.1 Map showing the distribution of the ground nut *Arachis hypogaea* (dark shade refers to the cultivation area)

Table 19.1 The vernacular names of groundnut *Arachis hypogaea* in different countries

Country	Vernacular names
England	Groundnut, monkeynut, peanut, goober, pinder
China	Lotua sheng
Finland	Maapaehkinae
France	Arachide souterraine
Germany	Erdnuss
Paraguay	Manduvi
India	Groundnut, monkeynut, earthnut, pignut, nilakadalai, moongphali, shengdhana
Japan	Nankin-mame
Malaysia	Katjang tanah
Portugal	Amendoim, mendubi, mudubim
Bolivia	Inchik, inchi
Russia	Arachis, zemljanoj orech
Spain	Cacahuete, mani
Sweden	Jordnot

Argentina, Asia, Africa, Burkina Faso, Chad, China, Congo, Ghana, India, Indonesia, Myanmar, Nigeria, Senegal, Sudan, North America and Viet Nam (Anonymous 2004). It is grown over an area of 26.4 million ha with a total production of 36 million tonnes of nut (Fukuda et al. 2001). The groundnut is being referred by different vernacular names in different countries (Table 19.1).

Table 19.2 Countries produced groundnut during the year 2019

Country	(Million tonnes)	Country	(Million tonnes)
China	17.6	Myanmar	1.6
India	7.3	Tanzania	1.0
USA	2.7	Argentina	1.0
Nigeria	2.8		

19.1.1 Habitat

Groundnut is called as the king of oil seeds. It tends to grow in relatively average to poor soil and prefers slightly acidic soil. It manages well in stiff soil with clay like property. The plant prefers lime soil and faces drought condition easily. This plant also fixes atmospheric nitrogen in the soil with the help of nitrogen fixing bacteria. Some of the countries contributed the world production of groundnut are given in Table 19.2. China and India are the major producers.

19.1.2 Distinguishing Features

The leaves of the groundnut plant are compound, opposite, pinnate with four leaflets having no terminal leaflet. Each leaflet is 1–7 cm long and 1–3 cm wide. The leaf does not have a teeth or lobe. The flower is bilaterally symmetrical and 2–4 cm long. There are four petals and sepals with 10 stamens. The flower stalk elongates after pollination till it touches the ground. The stalk continues its growth and pushes the ovary underground where the ovary transforms in to a pod. These pods are about 3 cm long encasing 1–4 seeds. The pods are positioned several inches deep into the ground (Fig. 19.2).

19.1.3 Oil Extraction

Groundnut is sown in a prepared soil at the rate of 300,000 seed ha^{-1}. The seed is placed 5–6 cm deep in to the soil and covered. The spacing is 30 × 10 cm. The pod matures in 120–150 days. Yellowing of the foliage is the indication of maturity. Then the plant bunch is hand pulled and the pod separated. The separated pods are sun dried and stored. The total yield largely depends on the soil health and irrigation. Average yield of nut is 3500 kg ha^{-1} $year^{-1}$. The nuts form around 50–65% of the pod weight depending on the moisture level and the oil forms around 40–50 of the nut weight (Davis et al. 2008).

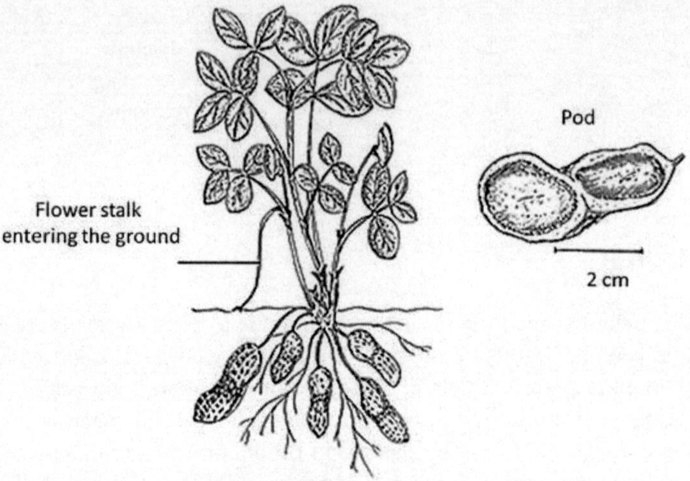

Fig. 19.2 The groundnut plant *Arachis hypogaea*

The pods are first cleaned from dust, leaves, stems and stones. It is then decorticated and the kernel is separated from the hulls. The nut is then mixed with hot water and ground to make a paste. It is subsequently kneaded by hand till the oil separates in to an emulsion. This emulsion is mixed with common salt to coagulate the protein and to activate the isolation of the oil. Then the above mass is pressed through a roller to release the oil. Trace of water if any is removed by heating. The cake also is extracted with food grade hexane. The oil obtained is degummed with water to remove the phospholipids and metals. A small amount of phosphoric or citric acid is added to remove the residual phospholipids. It is then neutralized with sodium hydroxide and washed with water repeatedly. The resultant material is deodorized to free it from trace amount of aldehydes and ketones. The deodorization is carried out through steam distillation at high temperature so as to get purified oil. The process flow chart is shown below:

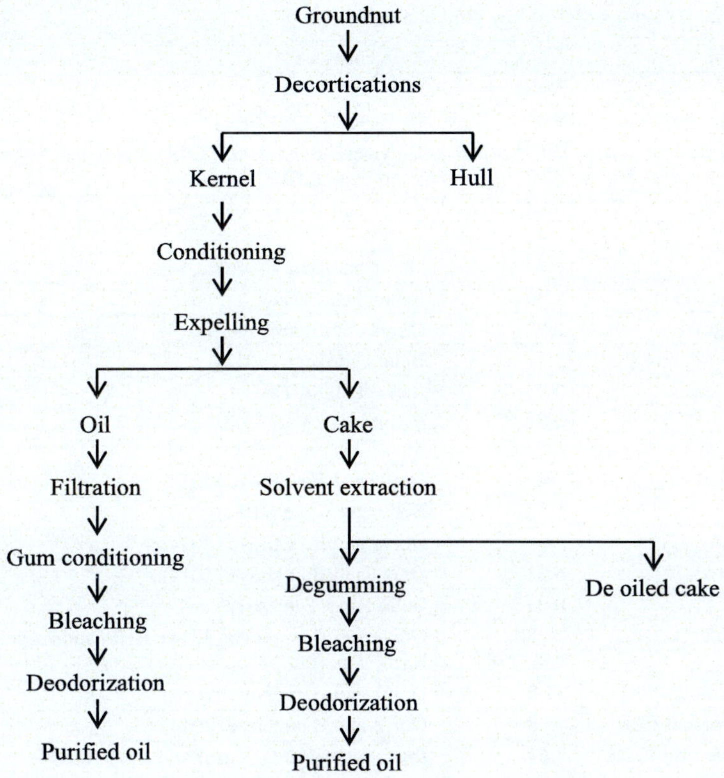

The characteristics of raw groundnut oil are given in Table 19.3. It is known that the viscosity was high which is needed to be reduced. The specific gravity is low which indicates a good ignition property. The pH value of the oil also is low which is due to the presence of sulphur (Oniya and Bamgboye 2014). The ash content is low. Since the iodine value is less than 110 it is considered as a non-drying oil and its keeping quality is fairly good. Acid value 2.16 mg KOH g^{-1} is higher than the standard (0.8 mg KOH g^{-1}) which refers to the risk of corrosion due to the free fatty acid and oxidation products. High free fatty acid interferes with the transesterification process and affects the separation of glycerol from the ester. This often warrants two stage process, namely esterification and transesterification. The saponification value is more than 100 mg g^{-1} which reconfirms the risk of soap formation during biodiesel formation and therefore there is every chance of poor recovery of biodiesel. High peroxide value indicates lipolytic hydrolysis and oxidative deterioration causing rancidity. Therefore these oils are to be utilized without long storage. High flash point indicates that there is no risk of fire during storage. The refractive index is fairly high. Such high refractive index is the function of molecular weight, fatty acid chain length, level of unsaturation and conjugation.

The groundnut oil is a rich source of unsaturated fatty acid comprising oleic acid and linoleic acid. The fatty acid profile is given in Table 19.4.

Table 19.3 The properties of raw groundnut oil

Parameters	Value	References
Colour	Yellow	Ibeto et al. (2012)
Density (kg m^{-3})	956	Bello and Agge (2012)
Viscosity (Cst at 40 °C)	39.2, 24.4, 32.7	Oniya and Bamgboye (2014), Bello and Agge (2012), Ibeto et al. (2012)
Heating value (MJ L^{-1})	37.4, 39.3	Oniya and Bamgboye (2014), Bello and Agge (2012)
pH	3.4	Oniya and Bamgboye (2014)
Refractive index	1.46	Ibeto et al. (2012)
Specific gravity (at 15 °C)	0.9	Oniya and Bamgboye (2014)
	1.1	Bello and Agge (2012)
	0.93	Ibeto et al. (2012)
	0.91 (25 °C)	Yusuf and Sirajo (2009)
Cloud point (°C)	−8.0	Oniya and Bamgboye (2014)
	−3.0	Bello and Agge (2012)
Pour point (°C)	3.0	Oniya and Bamgboye (2014)
Ash content (%)	0.01	Oniya and Bamgboye (2014)
	0.05	Bello and Agge (2012)
Flash point (°C)	>280, 458, 178	Oniya and Bamgboye (2014), Bello and Agge (2012), Ibeto et al. (2012)
Sulphur content (%)	12.8	Oniya and Bamgboye (2014)
Carbon content (%)	24.8	Oniya and Bamgboye (2014)
Acid value (mg KOH g^{-1})	2.61	Ibeto et al. (2012), Yusuf and Sirajo (2009)
Iodine value (mg g^{-1})	89.46, 84.0	Ibeto et al. (2012)
Peroxide value (meq kg^{-1})	18.0	Bello and Agge (2012)
Saponification value (mg g^{-1})	148.7, 188.0	Ibeto et al. (2012), Yusuf and Sirajo (2009)
Free fatty acid (%)	11.0	Oniya and Bamgboye (2014)
	1.3	Ibeto et al. (2012)
Cetane number	26.0	Bello and Agge (2012)
Phosphorus (mg kg^{-1})	2.13	Bello and Agge (2012)

19.1.4 Biodiesel Production

The methodology of producing groundnut biodiesel was reported by many authors (Kaya et al. 2009; Perez et al. 2010; Davis et al. 2009). If the level of free fatty acid is high, there is problem in alkaline catalysed transesterification as high free fatty acid is known to react with catalyst and form soap. The transesterification process of the oil is seriously hampered if the free fatty acid content is more than 3% (Anggraini and Wiederwertung 1999; Dorado et al. 2002). To avoid such contingency the oil is pretreated with 1% hydrochloric acid as catalyst. The acid containing oil is treated

Table 19.4 The fatty acid profile of groundnut oil

Fatty acid	Percent	References
Palmitic acid (C16:0, $C_{16}H_{32}O_2$)	11.69, 7.50, 8.0	Oniya and Bamgboye (2014), Oghome (2012), Bello and Agge (2012)
Stearic acid (C18:0, $C_{18}H_{36}O_2$)	3.84, 4.5, 2.9	Oniya and Bamgboye (2014), Oghome (2012), Bello and Agge (2012)
Oleic acid (C18:1, $C_{18}H_{34}O_2$)	41.64, 52.0, 44.9	Oniya and Bamgboye (2014), Oghome (2012), Bello and Agge (2012)
Linoleic acid (C18:2, $C_{18}H_{32}O_2$)	32.04, 27.0, 32.8	Oniya and Bamgboye (2014), Oghome (2012), Bello and Agge (2012)
Arachidic acid (C20:0, $C_{20}H_{40}O_2$)	1.84, 3.0, 5.6	Oniya and Bamgboye (2014), Oghome (2012), Bello and Agge (2012)
Arachidonic acid (C20:1, $C_{20}H_{38}O_2$)	0.89	Oniya and Bamgboye (2014), Oghome (2012), Bello and Agge (2012)
Behenic acid (C22:0, $C_{22}H_{44}O_2$)	4.05, 1.50, 3.8	Oniya and Bamgboye (2014), Oghome (2012), Bello and Agge (2012)
Lignoceric acid (C24:0, $C_{24}H_{48}O_2$)	1.53, 1.50, 1.0	Oniya and Bamgboye (2014), Oghome (2012), Bello and Agge (2012)
Erucic acid (C22:1, $C_{22}H_{32}O_2$)	1.0	Bello and Agge (2012)

with methanol at a molar ratio of 9:1 and the whole mixture is stirred for 1 h at 60 °C at 400 rpm. The content is then poured in a separating funnel and allowed to stay undisturbed for more than 3 h. The methanol moves to the top and the underlying oil is first removed and washed repeatedly till the content reaches neutral. The oil is tested for free fatty acid. If the free fatty acid content is less than 1.5% alkaline transesterification is carried out. Sodium hydroxide (3.5 g) is dissolved in 100 mL of methanol to get the sodium methoxide needed for alkali catalysed transesterification. The sodium methoxide is mixed with oil so as to get 6:1 molar ratio of methanol to oil. Methanol is often preferred due to its low cost and fast action on account of its short chain. If potassium hydroxide is used the ester production marginally increases since the solubility of KOH is greater in alcohol and thus the reaction rate is bolstered. The mixture is stirred for 1 h and the whole content is allowed to stand undisturbed for 12 h. The glycerol content which stayed at the bottom is first discarded and the supernatant phase is separated. The methanol content of the supernatant layer is isolated by distillation and the resultant material is repeatedly washed and dried by increasing the temperature. Alternatively, the moisture present in the finished product is removed by passing it through anhydrous sodium sulphate (Na_2SO_4).

Often ethanol is employed in the transesterification process as it is renewable and derived locally from agricultural products. Similarly KOH also is employed as a substitute to NaOH. The process flow sheet is given below:

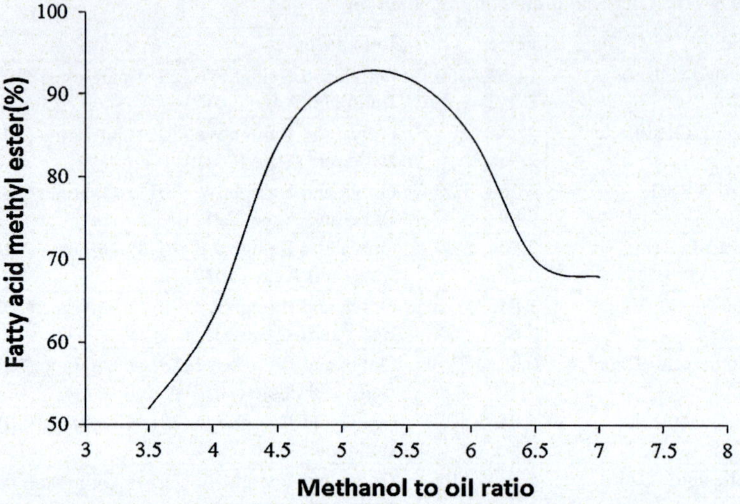

Fig. 19.3 Effect of molar ratio of methanol with oil on the ester yield of groundnut oil. Experimental para-meters are temperature 60 °C, 0.5 g KOH L^{-1} and the duration of reaction is 60–90 min

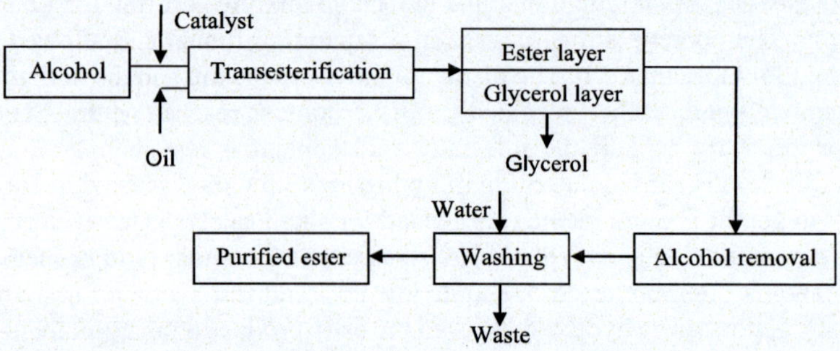

More than 85% conversion of oil in to biodiesel is possible in this process. Such conversion occurs at a reaction temperature of 60 °C (Ahmad et al. 2009). Stoichiometrically the methanolysis requires only 3 mol of methanol. As the reaction is largely reversible excess methanol is added to drive the reaction forward towards the formation of ester. The maximum yield is obtained at a molar oil ratio of 5:1 (Fig. 19.3). The calculated value of the molecular weight of the groundnut oil is 891 g mol^{-1}. Ratio of alcohol with oil if goes lower than the optimum the transesterification will not be complete. If the ratio is higher than optimum, the separation of ester layer is greatly affected since methanol and hydroxyl group may cause emulsification (Leung and Guo 2006).

If the reaction temperature is increased beyond 60 °C, saponification is accelerated by the alkali glyceride interaction before the actual completion of alcoholysis (Karaosmanoglu et al. 1996). Much like the methanol concentration, the quantity of alkali catalyst (NaOH) also exerts influence on the recovery of

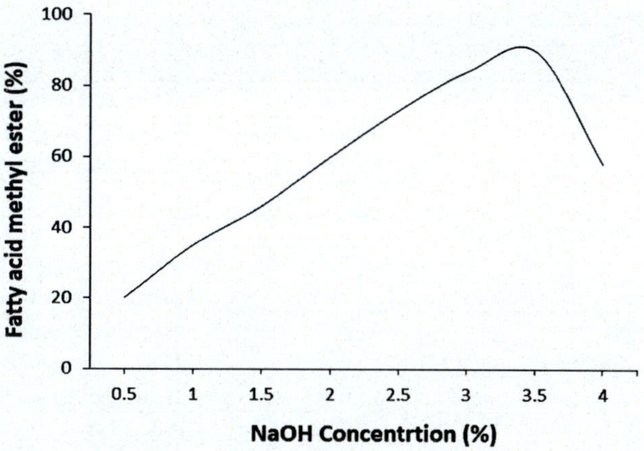

Fig. 19.4 Effect of NaOH concentration in percent w/w with groundnut oil on the ester yield

biodiesel. Figure 19.4 indicates the effect of alkali concentration in percent over the weight of oil used after acid catalysis. The reaction is carried out at 65 °C and the methanol to oil molar ratio is 5:1.

When the raw oil is transesterified the yield is paradoxically low since the NaOH concentration cannot be raised to more than 1 g kg^{-1} of oil due to the excessive soap formation. When the alkali transesterification is carried out consequent to acid catalysis the NaOH administration goes around 3.5 g kg^{-1} of the oil since the soap formation is under control enabling rich yield of ester. In order to overcome the problems associated with alkali catalyst, heterogeneous catalyst. Cesium tungstophosphoric acid (0.03%) is employed by Anitha and Dawn (2010) at a methanol to oil ratio of 7.8:1.

In order to assess the actual quantity of ester produced, chromatography or FTIR method is suggested. FTIR spectroscopy is considered to be fast and accurate (Gelbard et al. 1995) though this technique is fairly old. The spectra were obtained at 500–4000 cm^{-1} region. The height of absorbance at wave number 1741 cm^{-1} is chosen to assess the quantity of the ester. The height of the band is calibrated using analytical grade methyl ester. From the calibration curve, actual quantity of ester produced in g mL^{-1} is known. The total yield is then calculated as follows:

$$\text{Yield (\%)} = \frac{\text{Volume of biodiesel layer (mL)} \times \text{Concentration of ester} \left(\text{g mL}^{-1}\right)}{\text{Volume of oil used (mL)}}$$

19.1.5 Properties of Groundnut Biodiesel

The properties of biodiesel prepared from groundnut oil are given in Table 19.5.

The viscosity of biodiesel is reduced considerably when compared to that of the raw oil. However such reduction is not satisfactory to the extent needed. The viscosity reduction helps to increase the fluidity in the diesel engine. As a matter

Table 19.5 Fuel properties of groundnut oil biodiesel

Parameters	Value	References
Viscosity at 40 °C (Cst)	7.6	Oniya and Bamgboye (2014)
	6.6 $mm^2 s^{-1}$	Bello and Agge (2012), Ahmad et al. (2009)
	5.9 (ny)	Santos et al. (2013)
	7.0, 4.8	Oghome (2012)
Heating value (MJ L^{-1})	30.1	Oniya and Bamgboye (2014)
	39.6 MJ kg^{-1}	Bello and Agge (2012)
	39.2	Santos et al. (2013)
pH	2.9	Oniya and Bamgboye (2014)
Specific gravity at 15 °C	0.85	Oniya and Bamgboye (2014)
	0.96	Bello and Agge (2012)
	0.92	Ahmad et al. (2009)
	0.84	Oghome (2012)
Cloud point (°C)	7.0	Oniya and Bamgboye (2014)
	10.0	Bello and Agge (2012)
	6.0, 15.0, 8.0	Ahmad et al. (2009), Santos et al. (2013), Oghome (2012)
Pour point (°C)	4.0	Oniya and Bamgboye (2014)
	−6.0	Bello and Agge (2012)
	3.0	Ahmad et al. (2009)
	−3.0	Oghome (2012)
Ash content (%)	0.01	Oniya and Bamgboye (2014)
	0.01	Bello and Agge (2012)
Flash point (°C)	200	Oniya and Bamgboye (2014)
	182, 190	Bello and Agge (2012), Ahmad et al. (2009)
	192, 167	Santos et al. (2013), Oghome (2012)
Sulphur content (%)	9.73	Oniya and Bamgboye (2014)
Carbon content (%)	10.4	Oniya and Bamgboye (2014)
Iodine value	0.34	Oniya and Bamgboye (2014)
Peroxide value (meq KOH^{-1})	0.13	Oniya and Bamgboye (2014)
Sap value (mg KOH^{-1} g^{-1})	0.17	Oniya and Bamgboye (2014)
Free fatty acid (g kg^{-1})	70.0	Oniya and Bamgboye (2014)
Cetane number	51	Bello and Agge (2012)
Refractive index	1.366	Bello and Agge (2012)

of fact the difference in viscosity between the oil and the ester is used to monitor the production of biodiesel (Filippis et al. 1995). Viscosity affects the lubrication of the injector and fuel atomization. On the other hand, fuel with very low viscosity may not meet the needed lubrication, which causes leakage and increased wear in the fuel injection assembly. High viscosity may form larger droplets on injection resulting in to poor combustion, increased exhaust smoke and emissions (Encinar et al. 2002).

Therefore, it is recommended to blend the ester with conventional diesel so as to meet the appropriate viscosity.

The specific gravity of the biodiesel matches well with that of the mineral diesel. The flash point of the biodiesel is higher than the regular diesel which determines the safety of the biodiesel during handling and storage. During the transesterification process there is drop in the heating value mainly due to the reduction of carbon content during the transesterification process. Higher cloud and pour points may create complications in the engine during cold weather. The ash content which being the measure of metals present in the fuel is considered as an important parameter. Low ash content conveys that there will be reduction in the plugging of the injector system. Iodine value is reduced in the process of transesterification. Iodine value is a measure of the stability of the biodiesel during storage. Low iodine value makes the fuel more stable. Higher iodine value of the biodiesel causes shorter ignition delay and longer combustion duration. As a result particulate emission is reduced. The cloud and pour point of the groundnut oil was not very satisfactory to meet out contingencies in winter season. Perez et al. (2010) gave the methodology of winterization to improve the cold flow properties (Gunstone and Hamilton 2001).

19.1.6 Engine Performance

One of the parameters in the engine performance is the net brake power. According to the Society of Automotive Engineers (SAE J: 1349 2008) the net brake power is a measure of engines horsepower delivered directly to the crank shaft of the engine without any loss in power caused by gearbox, alternator, coolant pump and muffle exhaust system. When the speed of the engine fed with groundnut biodiesel is increased gradually slight increase in net brake power is discerned. Further increase in engine speed decreases the net brake power due to increased friction at high speed (Fig. 19.5).

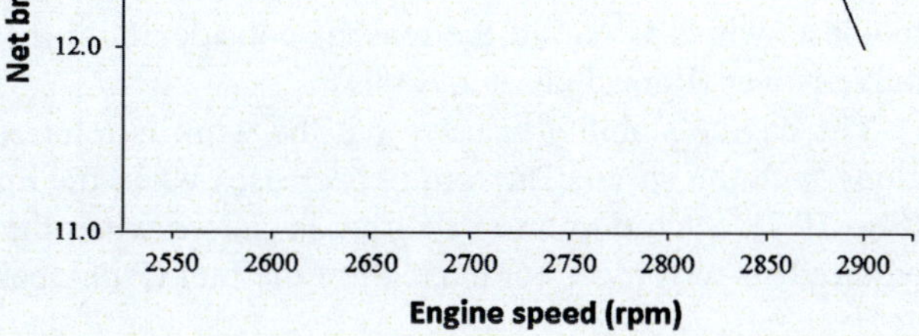

Fig. 19.5 Effect of engine speed on the net brake power (kW) when the engine is fed with groundnut biodiesel

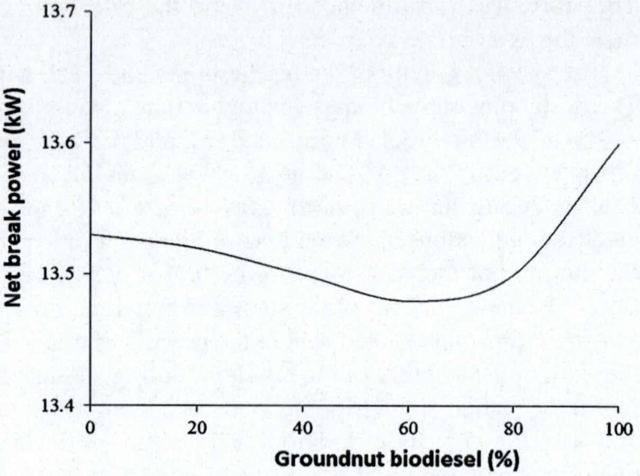

Fig. 19.6 Effect of groundnut biodiesel (ester) blend in percent with petro-diesel on the net brake power of the engine

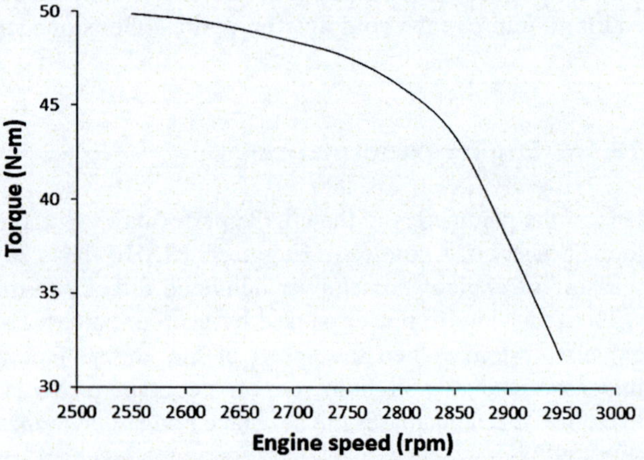

Fig. 19.7 The relationship between engine speed and torque in an internal combustion engine when groundnut biodiesel is used as fuel

If the fuel fed in to the engine is a mixture of groundnut biodiesel and regular diesel, serious change in the net brake power may take place (Fig. 19.6). As a matter of fact, there can be slight increase in the net brake power when pure biodiesel is used (Al-Widyan et al. 2002).

A slight increase in net brake power in biodiesel is attributed to its high viscosity which enhances the fuel spray penetration thereby causing increased air-fuel mixing (Lin et al. 2009). The frictional loss also is reduced due to the lubricity of the biodiesel which is one of the reasons considered on the slight improvement in the brake power (Ramadhas et al. 2005).

The engine's ability to carry out the work is referred as torque which changes along with the speed. The torque decreases when the maximum speed is increased (Fig. 19.7). Such decrease in torque is caused when the engine is unable to inhale required air for proper combustion of the fuel (Pulkrabek 2004).

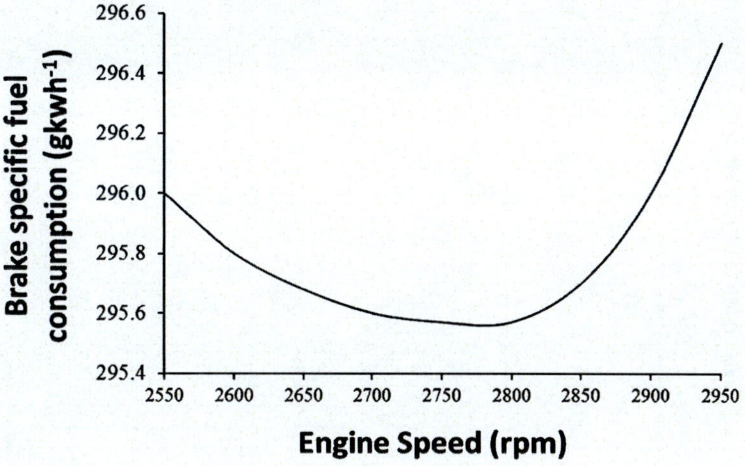

Fig. 19.8 Brake specific fuel consumption in relation to the change in engine speed in an internal combustion engine when groundnut biodiesel is used as fuel

Fig. 19.9 The relationship between the groundnut biodiesel blend and brake specific fuel consumption

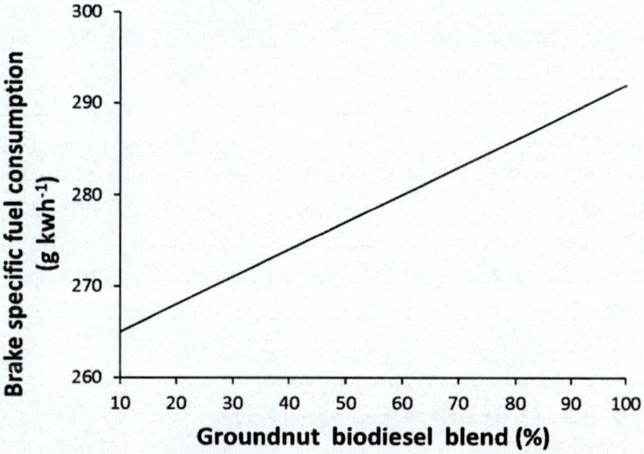

Break specific fuel consumption is a parameter indicating the fuel efficiency in the crank shaft of the internal combustion engine. Brake specific fuel consumption is calculated by dividing the rate of fuel consumption by the net brake power. When the rpm of an engine is slowly increased from 2600 to 2900 the brake specific fuel consumption gradually reduces. Further increase in rpm enhances the consumption (Fig. 19.8).

At low speed the duration taken for completing a cycle is higher which causes loss through heat demanding the need for more fuel. Similarly the frictional loss at higher speed also demands more fuel. Therefore speed optimization is essential for maximizing the output and minimizing the consumption of fuel. The brake specific fuel consumption increases steadily when the blend percent of biodiesel increases (Fig. 19.9)

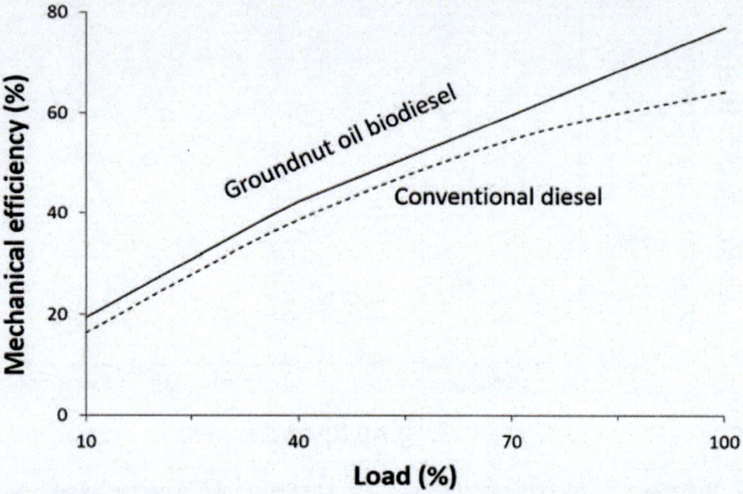

Fig. 19.10 The relationship between load and mechanical efficiency

At a high torque the brake specific fuel consumption increases when the level of biodiesel blend is increased. Such increase in fuel consumption is due to low heating value, high density, lubricity and high viscosity of the biodiesel compared to that of the regular diesel (Xue et al. 2011). The rise in mass flow of the biodiesel is yet another reason. Because of the low energy value of the biodiesel more flow is needed to meet out the total demand of energy. Mechanical efficiency of the biodiesel prepared from waste groundnut oil is observed to be better (77.1%) than conventional diesel (64.3) at 90% load (Fig. 19.10) as biodiesel has better lubricity which reduces the frictional losses (Anitha and Dawn 2010).

19.1.7 Emission Characteristics

High content of oxygen in biodiesel is one of the reasons for low particulate matter in the emission. The sulphur generated during the combustion is adsorbed in the aerosol drawn from sulphuric acid formed during the combustion of biodiesel. The NO_x level of the exhaust from an engine fed with groundnut biodiesel is far higher compared to that of the pure diesel. The fuel spray is found to be modified due to increase in the size of droplets of the fuel thus altering the combustion. More oxygen content of the biodiesel molecule provides additional oxygen for the formation of NO_x.

The CO_2 level in the emission from the engine fed with biodiesel is characteristically high. High level of hydrocarbon emission due to biodiesel was observed at high speed and load (Munoz et al. 2004). High density and viscosity alter the size of the droplets and penetration of the fuel in the injector assembly. Any quantity retained in the interior portion of the nozzle may be a cause for the high hydrocarbon release since the biodiesel does not reach the combustion chamber in full at the ignition time. The CO concentration is very low in the exhaust of the engine run by biodiesel as higher oxygen content of biodiesel promotes active combustion thereby the level of CO in the exhaust is low but it causes the CO_2 to go up.

19.2 Soybean (*Glycine max*)

Soybean (soyabean) oil is quite popular, being extracted from the seeds of a leguminous plant *Glycine max* belonging to the family: Fabaceae. It is a herbaceous annual bush and stands upright. It grows to a maximum height of 1.5 m and is often referred as golden bean. The leaves are compound and trifoliate. The leaves, pods and stems are clothed with soft hairs. The flowers are white, yellow or purple and the floral alignment shrinks from base to tip. The younger flowers occupy the tip of the inflorescence. The flower has a tubular calyx having 5 sepals, corolla with 5 petals, 1 pistil and with a fused stamen. The fertilized flower puts forth pod having 1–4 seeds enclosed in it (Fig. 19.11). The seeds are either round or elliptical. A single plant produces as many as 400 pods with up to 20 pods in a single node. Each seed weigh 125–250 mg. The pod is either straight or partially curved. The length of pods varies from 2 to 7 cm. The worldwide seed production during the year 2019–2020 was 336.46 million metric tons with an oil yield of 57.87 million metric tons (Table 19.6).

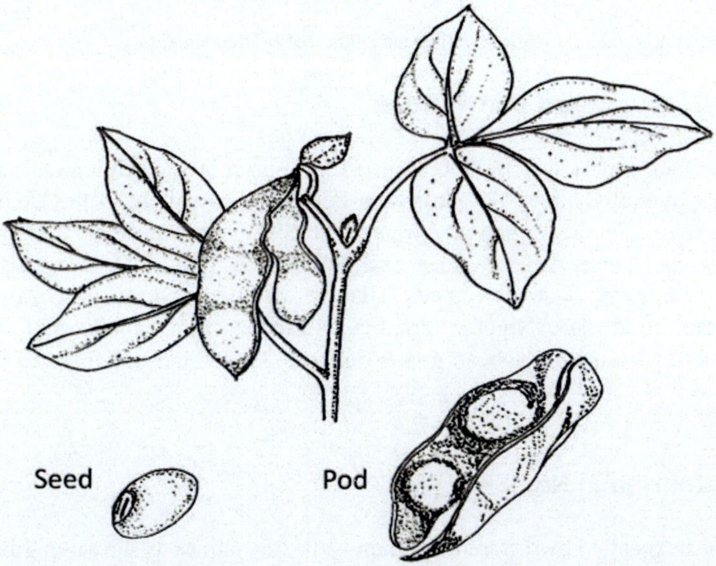

Fig. 19.11 The soybean *Glycine max*

Table 19.6 World production of soybean seed and oil (million metric tons) during 2019–2020 (USDA 2021)

Country	Seed	Oil
Brazil	126.0	8.5
USA	96.7	11.3
Argentina	48.8	7.7
China	18.1	16.4
India	9.3	1.5
Paraguay	9.9	–
Canada	6.1	–
Mexico	–	1.1
Others	21.5	11.4

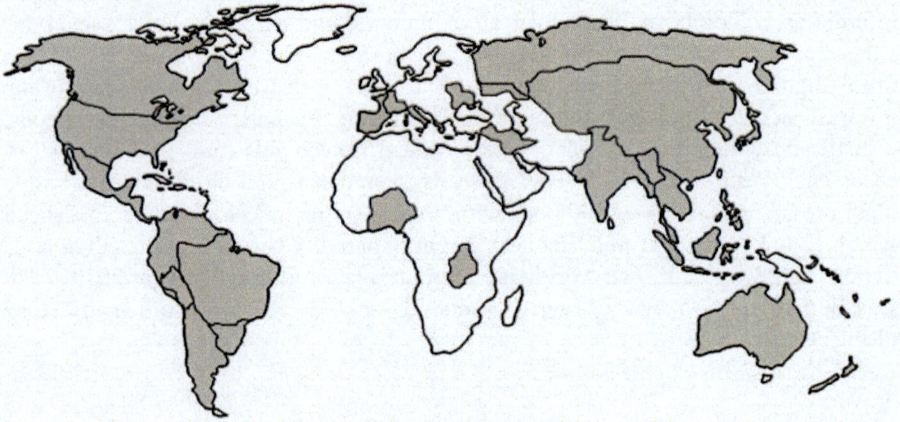

Fig. 19.12 Worldwide cultivation of soybean (indicated by dark shade)

19.2.1 Geographical Distribution

Soybeans were reported to have first grown as a crop in East-China during seventeenth to eleventh Century BC. It is now being grown widely in the USA, South America, Brazil, India, Argentina, Africa, Australia, Caribbean, West Indies, Central Asia, Mexico, Central America, East Asia, South Asia, France, Greece, Italy, Spain, Portugal, Paraguay, Canada, Japan, Ukraine, Bolivia, Uruguay, North Korea, Kazakhstan, Indonesia, Nigeria, Zambia, Romania, Iran, Thailand, Cambodia, Vietnam and Myanmar. Soybean grows richly in 35–45° latitude (Fig. 19.12).

19.2.2 Colloquial Names

The soybean is called by different location—specific names as shown in Table 19.7.

Table 19.7 Colloquial names of soybean in different countries

Country	Names
Europe	Soyabean, sojabean, Ybean, sojaboon, soya soy, Frijol-de-soya, soja söjabona
Finland	Soijapapu
France	Soja
Germany	Sojabohne
India	Bhat, Bhatwar, soya chikkudu, suntha kadalai
Japan	Daizu
Korea	Kong
Russia	Soja
Vietnam	Däu tuog, däu nänh

19.2.3 Cultivation and Dehulling

It is a preferred crop to cultivate in hot summer at a temp. range of 20–30 °C. The growth retards in temp. below 20 °C or above 40 °C and its flowering also is affected in adverse temperatures. They are being cultivated in different types of soil. However they prefer alluvial soil having good organic content with deep, well-drained soil of good water holding capacity. Typically, soybean requires 60–70 kg P_2O_5, 300 kg K_2O and 8 kg ha^{-1} of manganese sulphate. It can manage low pH, but any pH lower than 5 is not preferred since it interferes in the nitrogen fixation. The seeds are sown at a depth of 3–5 cm. The seedlings emerge within 5–7 days. The spacing between the rows is 40–90 cm and between plants it is 15 cm. The overall plant population is 250,000–400,000 ha^{-1}. The crop comes in harvest within 120 days of sowing. High yield of 2.5 metric tonnes per hectare is expected in an annual rainfall of 500–900 mm. As the plant comes to harvest, the seeds get hardened and the leaves become dry and fall. If the harvest is delayed beyond, there will be serious loss due to seed shattering. The pods from the dead plants are normally handpicked and stacked till the seeds are isolated from the pods. The recommended moisture of the seed for storage is around 12%.

The ripe pod is collected and dehulled. Dehulling is carried out by any one of the three methods. They are conventional dehulling, hot dehulling and Escher Wyss dehulling. In the conventional dehulling the pods are tempered followed by cracking, screening and aspiration. In hot dehulling the pods are passed over a fluidized bed to expel the moisture before cracking. In Escher Wyss dehulling system, the pods on harvest are cracked and passed over the fluidized bed. This would provide a uniform drying. Through cracking, the pod is made to pieces so as to get an ideal sized dry material to enable easy separation of hulls. Separation of hulls from the seed is achieved by aspiration. The hulls form around 8% of the seed weight.

19.2.4 Oil Extraction

The seed normally have 17–22% oil. Good quality beans are essential for a satisfactory recovery of oil. The quality of the beans is decided by the test weight which refers to the weight of grain against unit volume. Foreign materials such as sand, pods, leaves and stems of other plants are removed first. In order to get rid of the above materials the soybean grains are normally sieved and the grains which stay on the mesh are used for oil extraction. The split beans present due to mechanical damage or over-heat during drying will affect the yield of oil. Besides, it is known to increase the free fatty acid. Beans with high free fatty acid produce dark oil. Loss of quality of the beans due to frost, insect bites, mould or sprouting is not desired due to the presence of undesired pigments in the resultant oil.

Soybeans with low moisture content (less than 12%) stay for long in storage. When the moisture content is low, the beans split in to two half pieces which affects the oil recovery. Similarly if the moisture content is high, mould formation takes place which also affects the oil due to the formation of aflatoxin. After cleaning, the soybean seeds are processed to extract the oil by any one of the following three methods: (a) solvent extraction, (b) continuous pressing and (c) batch pressing.

The beans at the first instance are heated to 74 °C so as to condition the seed mass. Using roller, the beans are then made as flakes. The above treatment causes the cell wall and pseudo membranes around oil bodies of the seed to rupture. Through the ruptured region the solvent makes contact with the oil contained in the seed. High recovery of oil is possible if the flaked materials are of 0.25 mm size. Often the flakes are either sent to the solvent extractor or to an expander where the cell distortion is intensified so as to form small porous pellets. Such small porous pellets help to increase the extraction of oil. Normally the solvents are allowed to percolate through the pellets rather than flooding the mass. The solvent moves down through the crushed beans. The extraction process takes around an hour. The solvent normally being used is hexane or any other suitable solvents with a boiling point of 65–71 °C. Hexane is mostly employed in this extraction due to its low vaporization temperature, high stability, low corrosiveness and low greasy residual effects (Seth et al. 2007) (Fig. 19.13). Isopropyl alcohol also is being used.

Oil from heated beans is also extracted through continuous screw pressing. The squeezed out oil is then settled. The solids thus retained at the bottom of the hopper are once again taken to the screw press. The combined oil is then filtered (Fig. 19.14).

Supercritical carbon dioxide also is used to extract oil from soybean (Mendes et al. 2002). High temp. and pressure in supercritical process help to increase the extraction rate (Salgin 2007). Supercritical carbon dioxide extraction proves to be advantageous over solvent extraction (Kao et al. 2008). Ultrasound-assisted extraction gives an oil yield of 93.3% which is observed to be the highest among many reports (Luthria et al. 2007). Microwave assisted extraction is also attempted as it is known to improve the oil quality and quantity (Uquiche et al. 2008).

Microwave process is often used as a pretreatment to mechanical process so as to improve the quality and quantity of the oil. The ultrasound and microwave technique when used in combination helps to reduce the extraction timing and enhance solvent quantity (Cravotto et al. 2008).

Enzymatic hydrolysis is also applied after pretreatment. It is reported to be very efficient and cost intensive but the recovery is relatively low (Kashyap et al. 2007).

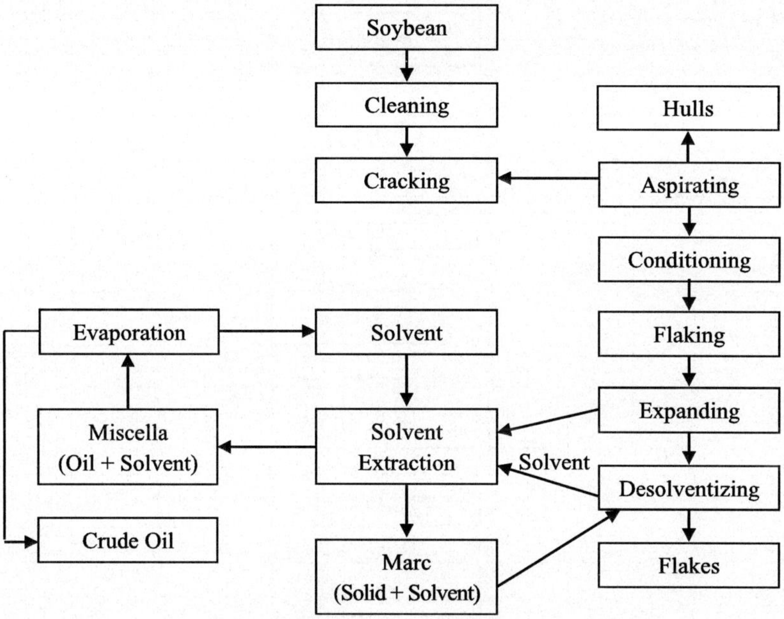

Fig. 19.13 Process flow chart for direct solvent extraction of soybean

19.2.5 Purification of Oil

The soybean oil is predominantly formed of triacylglycerols. The other components are phosphatides, free fatty acids, oxidation materials, unsaponifiable matter which includes tocopherols, sterols and hydrocarbons. The levels of such components in the crude oil vary according to the source of oil, season of the harvest, geographical origin of the plant and the extraction process being employed.

The crude oil is initially degummed to get the oil freed from hydratable phosphatides. The hydratable phosphatides are removed by washing with water. A known quantity of water is added and mixed thoroughly for a fixed time and then the whole content is centrifuged to isolate the gum. During the degumming process care is taken to avoid the formation of any emulsion so as to achieve high quality product.

Following the degumming, the oil is neutralized to remove free fatty acids, phosphatides, protein fractions, glycerol, carbohydrates, resins and metals. Caustic soda (sodium hydroxide) in pre-estimated conc. is allowed to react with the oil. The free fatty acid and the residual phosphatides join with alkali and generate soap and hydratable gums. The soap and other impurities emerging out of the neutralization are separated by centrifugation. There shall be minimum reaction time between oil and caustic soda since continued contact may cause some portion of the triglycerides becoming soap. Expertise is needed to separate the three major phases, the light phase being the oil, the heavy phase being the soap and the third middle phase being the emulsion.

Fig. 19.14 Process flow chart for screw pressing of soybean

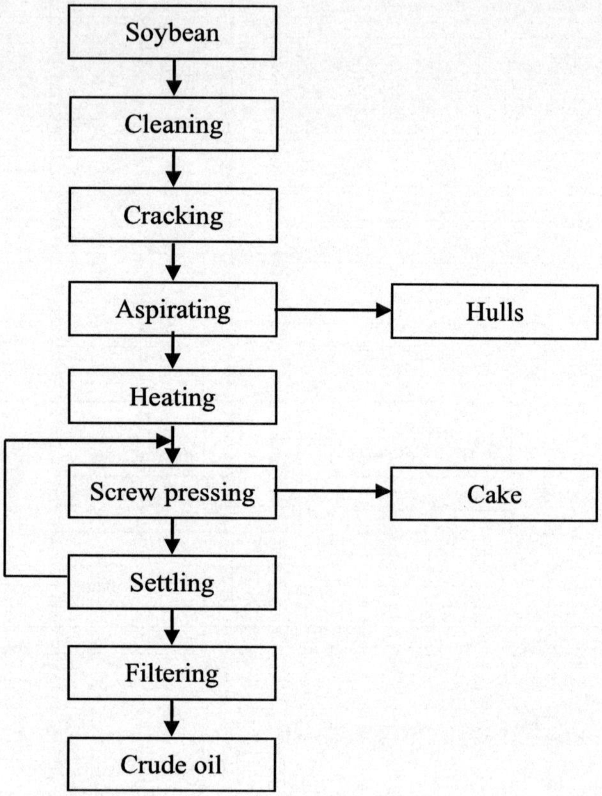

The oil thus purified is treated with acid activated clay. A calculated quantity of the clay is added in to the neutralized oil and the clay is allowed to adsorb coloured pigments for 20–25 min at a temperature ranging from 104 to 106 °C and at a vacuum of 50 mmHg. Subsequently the clay is separated. The function of the clay is often hampered by the phosphatides, soap and polymerized oil as they block the adsorption site of the clay. Thus the bleaching efficiency of the clay particles depends on the adsorption sites present on the surface. Smaller the particles, higher will be the presence of adsorption sites.

Often the oil is passed through winterization process. In winterization the saturated triglycerides or wax are removed by temperature adjustment. The oil is cooled slowly and maintained at a low temperature (5–6 °C) for a long time so as to enable the crystallization of saturated triglycerides and waxes. On completion of the crystallization the temperature is slowly increased to reach 15 °C. The oil is then filtered.

The oil thus obtained is deodorized and heated to a maximum of 260 °C under vacuum (2–10 mmHg). The non-triglyceride components are thus converted to a vapour state and removed. The deodorization process is largely governed by a variety of variables such as temperature, pressure, height of the oil column, duration, steam stripping, molecular weight of the component to be removed and the quality of

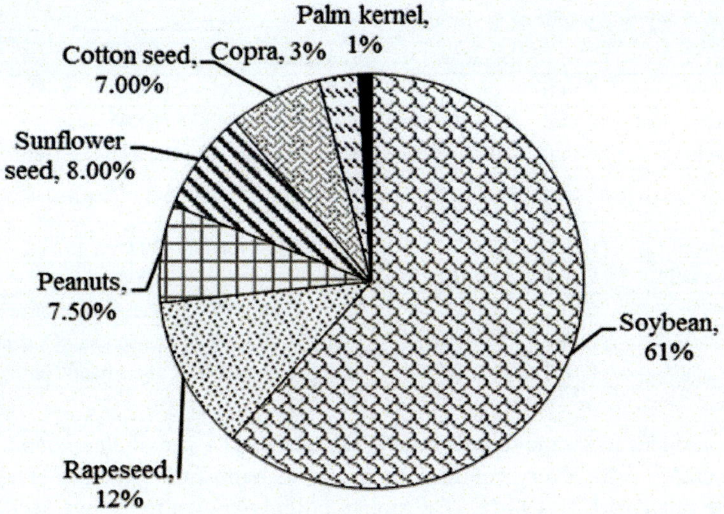

Fig. 19.15 The share of soybean among the oil seeds of the world

Table 19.8 Physical properties of soybean oil

Parameters	Value	Reference
Density at 20 °C (g mL^{-1})	0.9165–0.9261	Wesolowski (1993)
Viscosity at 20 °C (cP)	58.5	Hammond et al. (2005)
Surface tension at 30 °C (dyne cm^{-1})	27.6	de Alvarado (1995)
Melting point (°C)	0.6	Graef et al. (1985).
Cloud point (°C)	−9.0	Ali et al. (1995).
Pour point (°C)	−12 to −16	Ali et al. (1995)
Heat of combustion (Cal g^{-1})	9135–9450	Koseki et al. (2001)
Smoke point (°C)	245	Hammond et al. (2005)
Flash point (°C)	324	Hammond et al. (2005)
Fire point (°C)	360	Hammond et al. (2005)
Diesel index	16.6	Goswami and Usmani (2014)

the raw oil. Triglyceride with short fatty acids and low molecular weight is easy to deodorize. Properly deodorized oil will have low fatty acid (<0.05%) with no oxidation product.

The oil if needs to be stored for long, it has to be preserved in nitrogen blanket. The oil has to be agitated occasionally to avoid the separation of triglyceride components. Among the various sources of oil, soybean occupies the central stage (61% of the world production) (Fig. 19.15). The properties of soybean oil are given in Table 19.8.

The physical properties are reported to vary in accordance with the chain length of the fatty acids, the degree of unsaturation, different constituents contained in it and the temperature. The kinematic viscosity, one of the leading parameters for

Table 19.9 Fatty acid profile of soybean oil

Parameters		% w/w	References
Myristic acid	C14:0	<0.10	Canakci and Gerpen (2003)
Pentadecyclic acid	C15:0	<0.10	Canakci and Gerpen (2003)
Palmitic acid	C16:0	10.3, 10.58	Wang et al. (2010), Canakci and Gerpen (2003)
Margaric acid	C17:0	0.11	Canakci and Gerpen (2003)
Stearic acid	C18:0	4.5, 4.76	Wang et al. (2010)
Arachidic acid	C20:0	0.36	Canakci and Gerpen (2003)
Palmitoleic acid	C16:1	<0.10	Canakci and Gerpen (2003)
Oleic acid	C18:1	23.7, 22.52	Wang et al. (2010), Canakci and Gerpen (2003)
Linoleic acid	C18:2	51.5, 52.34	Wang et al. (2010), Canakci and Gerpen (2003)
Linolenic acid	C18:3	7.2, 8.19	Wang et al. (2010), Canakci and Gerpen (2003)

biodiesel reduces when the temperature is increased. The rate of change in viscosity is 0.027 cm^2 s^{-1} for every centigrade change in temperature (Miller et al. 1994) within the range of 170–190 °C. The smoke point is known to change according to the purity of the oil, presence of free fatty acid and according to the process of oil extraction (Hammond et al. 2005). The smoke point increases if the phenolic impurities are high. The flash point which refers to the temperature at which the vapour emanating from the oil catches fire is reported to be between 304 and 331 °C (Hammond et al. 2005). The fatty acid composition of soybean oil indicates that the linoleic acid forms more than 50% of the fatty acids in the oil (Table 19.9).

19.2.6 Biodiesel Production

Biodiesel is produced by the transesterification of oil in the presence of a suitable catalyst. The reactant is alcohol. Various alcohols such as methanol, ethanol, butanol, propanol and amyl alcohol are being used. The guiding principle in selecting an alcohol is the efficiency, cost consideration, easiness in purifying the finished product and the recovery of unreacted alcohol. Methanol has an advantage over other alcohols. Methanol is low cost and its chemical configuration brings out higher efficiency. According to the stoichiometric principle 3 mol of alcohol reacts with 1 mol of triglyceride to yield 3 mol of fatty acid ester and 1 mol of glycerin. As the transesterification process is reversible, more reactants, namely alcohol is added to ensure that the reaction is positive and reverse process is hampered. Normally the methanol-oil ratio may go up to 12:1. With reference to many of the catalysts used, alkali is preferred due to its fast reaction but its use is often limited especially when the oil has more free fatty acids and water which aid soap formation as a result the resultant phase separation is not easy. If acid catalysts are used, the above-mentioned problem can be solved reasonably. The acid catalysts act tough on the free fatty acids and prevent the soap formation. Among the various acids sulphuric acid is preferred. Other catalysts are alkoxides, carbonates, non-ionic bases such as amines, amidines, guanidines and triamino phosphoranes (Mittelbach 1990; Schuchardt et al. 1998;

Freedman et al. 1986; Noureddini and Zhu 1997). Enzymes are used but it has the limitation of cost and the action duration is fairly long. Ultrasound-assisted transesterification is yet another process which has a high prospect due to the swift reaction and the product obtained is relatively clean. The oil is treated with alcohol and methanol under high power ultrasounds (Chand et al. 2010; Koc and McKenzie 2010; Yu et al. 2010). When ultrasound is used at the oil extraction stage the free fatty acid content is considerably reduced (Li et al. 2004).

Often the oil extraction and the transesterification are coupled so that both the processes take place simultaneously. Thus the lengthy process of the oil extraction, followed by alkyl ester production is shortened. In doing so both the processes are completed within 20 min (Georgogianni et al. 2008).

In conventional method, the quantity of alkali required for the transesterification is calculated based on the titration value of alkali with a known quantity of oil. The alcohol as quantified by the mol. weight is mixed with the known quantity of alkali and heated to 50 °C. The above is then mixed with the heated oil and stirred at 700 rpm for 5 min. The content is then allowed to stand for 18–24 h to enable the separation of the biodiesel from glycerin. The biodiesel thus obtained may contain certain quantity of alcohol which is separated by distillation. The biodiesel left over is mixed thoroughly with equal quantity of warm water at 50 °C and the whole content is settled for 24 h. This washing process is continued till the wash water becomes neutral to phenolphthalein. The washed biodiesel is heated for 6 h at 104 °C so as to expel water if any present.

Alternatively, when the oil extraction and transesterification are to be coupled, fresh soybean is ground to a fine consistency. A sample of the pasty material is treated against alkali of known strength so as to calculate the total requirement of alkali pellet. Calculated amount of alkali pellet is then dissolved in methanol of required quantity so that the methanol and the oil present in the smash have to establish a molar ratio of 6:1. The alkali-methanol mixture is then added to the paste and ultrasonified for 30 min. After ultrasonification the mixture is kept settled for 2 h and the liquid phase containing methanol and the ester is separated by centrifugation at 1000 rpm for 10 min. The residue is again mixed with two portion of methanol and was allowed to settle for 2 h. The supernatant is then separated and centrifuged at 1000 rpm for 10 min. The overlying solution comprising the ester and methanol is added with hexane at 1:1 ratio. To this, water is added at a volumetric ratio of 3:1. The whole content is heated to 50 °C and maintained in that temp. for 20 min under constant stirring. The whole content is then settled for 24 h. The upper phase containing hexane and biodiesel is isolated and the methanol removed. It is then washed with copious amount of hot water repeatedly till the biodiesel becomes neutral. It is then dried using sodium sulphate. The hexane is then distilled out at 70 °C so as to recover biodiesel. The spent residue is dried for 24 h at 104 °C and was then treated with hexane so as to recover any oil if present.

Dimethyl carbonate is often employed in the production of soy biodiesel since dimethyl carbonate is considered eco-friendly. The dimethyl carbonate is made to react with soybean oil so as to get a product called fatty acid glycerol carbonate which is considered equivalent to biodiesel. The reaction is given below:

Table 19.10 Conversion efficiency of different catalysts in the production of dimethyl carbonate biodiesel from soybean oil (Fabbri et al. 2007)

Catalyst	Conversion (%)
Sodium methoxide	>99.5
Na_2 polyethylene glycol (300)	>98.00
1,5,7, Triazabicyclo (4,4,0) dec-5-ene (TBD)	>99.5

Table 19.11 Properties of dimethyl carbonate biodiesel (Fabbri et al. 2007)

Parameter	Value
Density at 15 °C (kg m^{-3})	0.887
Viscosity at 40 °C (cSt)	5.6
Cetane number	50.1
Heat value (MJ kg^{-1})	38.77
Flash point (°C)	160
Cold filter plugging point (°C)	0
Pour point (°C)	−2
Acid number (mg KOH g^{-1})	0.3

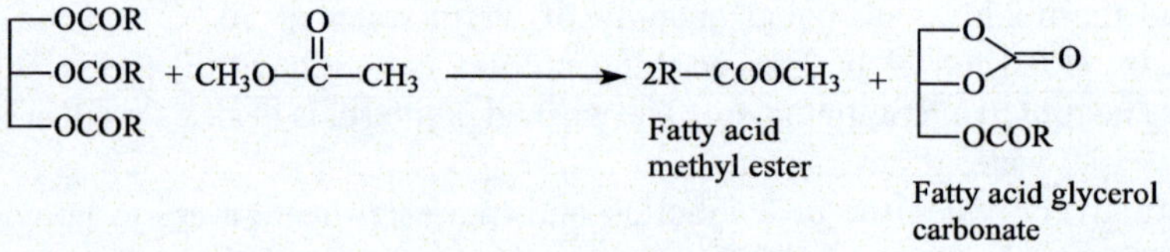

The soybean oil (2.5 L) is mixed with 800 g of dimethyl carbonate and is distilled azeotropically in atmospheric pressure so as to enable the expulsion of water present in the mixture. The above mixture is then made to react with sodium methoxide at 30% level on weight basis. The above content is stirred for 6 h at 90 °C. Consequent to the reaction the content is cooled and neutralized with phosphoric acid. The unreacted dimethyl carbonate is subsequently removed by vacuum distillation. As the reaction is proceeded, precipitate is formed due to sodium phosphate, glycerol carbonate and dicarbonate which is isolated by filtration. The resultant oil is mixed with methyl-*tert*-butyl ether and washed repeatedly with dist. water. The washed organic layer is separated and vacuum distilled to expel the methyl-*tert*-butyl ether. The final product is dimethyl carbonate biodiesel. Other catalysts were also tested and their performance is compared in Table 19.10.

The properties of dimethyl carbonate biodiesel are shown in Table 19.11.

Hydrotalcite ($Mg_6Al_2(OH)_{16}CO_3{\cdot}4H_2O$), a clay mineral having a carbonate anion intercalated among lamellar double hydroxide, magnesium and aluminium is used as a catalyst (Martins et al. 2013). The hydrotalcite is synthesized by mixing NaOH (0.84 mol) with Na_2CO_3 (0.25 mol) at room temperature under vigorous stirring at 60 °C for 12 h. The content is filtered and washed with dist. water till the pH is neutral. The hydrotalcite is then dried at 60 °C for 12 h followed by calcinations at 450 °C for 6 h by raising the temp. by 2 °C min^{-1}. The hydrotalcite is added at 5% level in the methanol-oil (20:1 molar ratio) mixture and stirred for 10 h at 64 °C.

After the reaction the hydrotalcite is isolated by centrifugation at 8000 rpm for 10 min and the resultant solution is then distilled to separate excess methanol. The content is phase separated from glycerol by allowing the solution to stay undisturbed for 24 h. Around 95% recovery is possible through this method.

Since methanol is popularly employed in the transesterification process there is large scale criticism that the process is not fully non-conventional as methanol is obtained from fossil sources. Therefore, in the place of methanol, ethanol is used since it is obtained from sugarcane. When soybean oil is allowed to react with a mixture containing ethanol and sodium hydroxide two products are formed. They are biodiesel and saponification product such as soap. In the soap formation, part of sodium hydroxide is spent leaving a minimum portion of sodium hydroxide to act on oil in the formation of biodiesel. The temperature plays a major role in the above reaction. Higher temperature though favours transesterification reaction, in reality it promotes saponification. Hence low temperature and higher molar ratio of ethanol is preferred. High production of ethyl esters is favoured by high molar ratio of alcohol and oil since the ratio is an effective variable in the transesterification. However, when ethanol level is increased the glycerol separation has become difficult, since the solubility of glycerol increases. As the glycerol's solubility in ethanol is high it tends to remain in solution thereby lowers the yield of ester. Higher yield of ester is possible under high molar ratio and with high concentration of catalyst. Excess catalyst also favours emulsion. The optimum values for the various parameters to enable the increase in the yield of ester are ethanol oil ratio 9:1, sodium hydroxide 1.3%, temp. 40 °C in a total reaction duration of 80 min (Silva et al. 2010).

Nano sized SiO_2/ZrO_2 catalyst also is being used in the soybean oil transesterification process (Faria et al. 2009). Tetraethyl orthosilicate (66.6 g) is added to 300 mL of ethanol and 2 mL HNO_3 and stirred vigorously for 30 min. A 50 mL aqueous solution containing 15 g of $ZrOCl_2$ is added to the above mixture drop by drop keeping the whole content to remain at 50 °C to produce $SiO_2/ZerO_2$ at a mole ratio of 8:2. The whole content is continuously stirred for 120 min and the gel thus obtained is placed in an autoclave and heated at 150 °C for 4 h. The resultant material is dried and calcined at 550 °C for 4 h to produce SiO_2/ZrO_2 catalyst. This has a surface area higher than 100 $m^2\ g^{-1}$. The transesterification is carried out by mixing 5 g of the catalyst in 1000 g of oil and 3500 g of methanol in a reaction duration of 5 h at a temp. of 65 °C. The process reaction does not promote soap or emulsion. There is a bond formation between carbonyl oxygen of the oil and that of zirconium. The carbon present in it is subsequently attacked by methanol thereby forming a C–O bond which rearranges and releases the fatty acid methyl ester (Fig. 19.16). The catalyst settles down to the bottom and makes it amenable to be separated. The over lying layer is formed of biodiesel and glycerin. The glycerin is then gravitationally separated and the biodiesel is recovered. The catalyst can be continuously used up to 6 cycles without any deterioration in its quality. This process has a biodiesel recovery of over 95%.

One of the important facts concerned with the use of solid catalyst is the spontaneous removal of water and carbon dioxide from the surface of the catalyst. Another issue concerned with the solid catalyst is that during transportation it

Fig. 19.16 SEM image of the SiO_2/ZrO_2 catalyst (Faria et al. 2009). (Source: https://doi.org/10.1590/S0103-50532009000900023)

undergoes sorption of the atmospheric carbon dioxide and water so as to form carbonates and hydroxyls. Calcium oxide is a best example to this which undergoes carbonation and hydration in few minutes on contact with air. The CaO on contact with biodiesel leads to form a paste. The CaO reacts with 15 times of its weight of biodiesel. It also combines with glycerol and forms calcium glyceroxide. Of late, lithium carbonate (Li_2CO_3) is used as a catalyst. Unlike CaO, lithium carbonate is not affected on its exposure to air. The soybean oil is initially heated to 65 °C on constant stirring at 300 rpm. The methanol oil ratio is 32:1. The catalyst quantity is 8% of the total mixture. The reaction temp. is 65 °C and the reaction duration is 2 h. On completion of the reaction, distilled water is added and the glycerol is isolated by settlement since the glycerol has a high density (1.126 g cm^{-3}) compared to that of the biodiesel (0.86 g cm^{-3}). The overlying biodiesel is separated and the excess methanol is evaporated. In this process a maximum conversion of 94% is reported (Wang et al. 2012).

The conversion efficiency is actively influenced by the level of free fatty acid content. Increased level of free fatty acid negatively impacts the recovery. Reusability of the catalyst is considered as an economic option. Consequent to the transesterification, the catalyst is isolated by decantation, washed with methanol and dried at 95 °C for 1 h. The Li_2CO_3 catalyst is repeatedly used in many transesterification batches to follow. Repeated use marginally lowers the conversion efficiency. The loss in weight of the catalyst due to repeated applications and the coating of the glycerol on the surface area of the catalyst are the two major reasons for the reduction in efficiency. However, the crystalline character of the Li_2CO_3 does not change even in repeated application indicating its stability.

Cesium fluoride (CsF) impregnated CaO is also tried as a solid based catalyst in the transesterification of soybean oil. The morphological characteristic of the synthesized catalyst of $CsCaF_3$ has cubic crystalline structure with the concentration of 0.5 mmol CsF/CaO (Fig. 19.17b, c) than pure CaO (Fig. 19.17a). The catalyst 0.9 mmol CsF/CaO shows higher basicity of 28.4 mmol g^{-1} than 0.1 mmol CsF/CaO (25.6 mmol g^{-1}) and 0.5 mmol CsF/CaO (28.0 mmol g^{-1}). In addition,

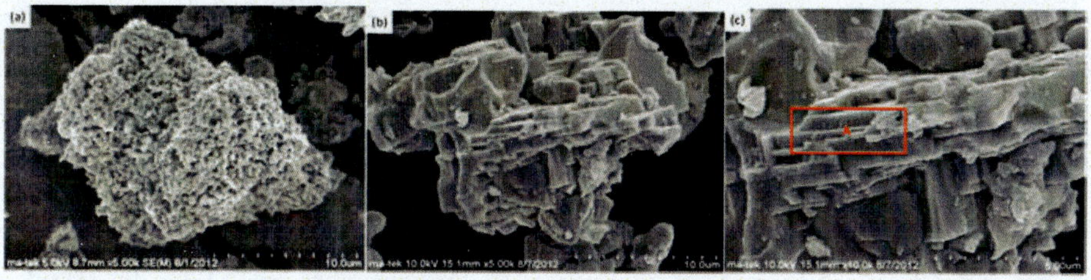

Fig. 19.17 Scanning electron microscopy images (SEM images) of the catalyst: (**a**) pure CaO, (**b**) 0.5 mmol CsF/CaO, 5000×, (**c**) 0.5 mmol CsF/CaO, 10,000×. (Reprinted (adapted) from Liu et al. 2012. Reproduced with permission from the American Chemical Society)

the biodiesel yield is also higher (96.3%) in 0.9 mmol CsF/CaO than that of 0.5 mmol CsF/CaO (95.6%) and 0.1 mmol CsF/CaO (90%) (Liu et al. 2012).

19.2.7 Biodiesel Property

The properties of the soybean oil biodiesel are presented in Table 19.12.

Certain essential properties of the soybean biodiesel are needed to be improved. They are the formation of the oxidation products and the issue connected with the cold filter plugging point. The oxidation of biodiesel causes problems at the fuel infection point of the engine. During the oxidation of biodiesel insoluble gums, organic acids and aldehydes are formed. Therefore as a standard, the oxidation time is maintained to a minimum of 6 h (Liang et al. 2006). The oxidation stability of a biodiesel largely depends on the fatty acid composition and antioxidants content. Besides the biodiesel has the characteristics of developing crystals at a colder temperature. At colder temperatures the biodiesel tends to plug the fuel-filter due to the formation of crystals or gelatin (Knothe 2005). The oxidation stability at colder climates largely depends on the fatty acid composition and therefore the soybean biodiesel often does not fulfill the 6 h oxidation stability criterion. The soybean biodiesel mainly comprises linoleic and oleic acids (Table 19.13).

The unsaturated fatty acid has a telling effect on the oxidation stability. The relative rates of oxidation are 1 for oleate, 41 for linoleate and 98 for linolenate. The highly oxidative fatty acid linoleic and linolenic acid when put together showed 60.04% and thus the oxidative stability is observed to be poor. When the content of the unsaturated fatty acid is high the flow of the biodiesel through the engine orifice is not easy. The total unsaturated fatty acid content is 83.16%. The oleic acid content is low (23.12%) and such low level is not desirable with reference to the cold filter plugging point property. On both account the soybean biodiesel is in a disadvantage position. To improve this, attempts were made to mix the soybean oil biodiesel with palm and rapeseed biodiesel. The improvement observed in the oxidation stability of the soybean biodiesel when it is blended with palm biodiesel and rapeseed biodiesel is shown in Fig. 19.18.

Table 19.12 Properties of soybean oil biodiesel

Parameters	Value	References
Density (g mL^{-1})	0.86, 0.84, 0.88, 0.85	Koc et al. (2011), Goswami and Usmani (2014), Wang et al. (2010), Canakci and Gerpen (2003)
Viscosity (40 °C $mm^2\ s^{-1}$)	4.66, 4.06, 4.27, 4.08, 6.22, 4.2	Koc et al. (2011), Schumacher et al. (1996), Canakci and Gerpen (2003), Azad et al. (2015), Wang et al. (2012)
Cloud point (°C)	−5.56, −1.0, −1.0 to 4.0	Koc et al. (2011), Schumacher et al. (1996), Wang et al. (2010), Azad et al. (2015)
Flash point (°C)	150, 92, 179, 69, 165	Koc et al. (2011), Goswami and Usmani (2014), Schumacher et al. (1996), Azad et al. (2015), Wang et al. (2012)
Pour point (°C)	−5.56, −7.0, −3.0	Koc et al. (2011), Schumacher et al. (1996), Azad et al. (2015)
Sulphur (mg L^{-1})	0.9, 0.5, 2.1	Koc et al. (2011), Canakci and Gerpen (2003), Wang et al. (2012)
Acid value (mg KOH g^{-1})	0.23, 0.10	Koc et al. (2011), Wang et al. (2012)
Cetane number	18	Goswami and Usmani (2014)
Diesel index (min)	25.5	Goswami and Usmani (2014)
Cold filter plugging point (°C)	3.0	Park et al. (2008)
Carbon (%)	76.5, 77.2, 77.1	Schumacher et al. (1996), Wang et al. (2010), Canakci and Gerpen (2003)
Hydrogen (%)	12.5, 11.9, 11.81	Schumacher et al. (1996), Wang et al. (2010) Canakci and Gerpen (2003)
Oxygen (%)	11.0, 11.3, 10.97	Schumacher et al. (1996), Wang et al. (2010), Canakci and Gerpen (2003)
Heating value (MJ kg^{-1})	37.44, 42.64, 39.76	Wang et al. (2010), Canakci and Gerpen (2003), Azad et al. (2015)
Aniline point (°C)	25.0	Goswami and Usmani (2014)
Copper corrosion	1A	Wang et al. (2012)
Typical formula	$C_{18.74}H_{34.43}O_2$	Canakci and Gerpen (2003)
Molecular weight	291.6	Canakci and Gerpen (2003)
Oxidation stability (h)	3.87	Park et al. (2008)
Iodine number g I_2/100 g	118	Wang et al. (2012)

It is clear that when the content of soybean oil biodiesel is reduced in the blend the oxidation stability increases. The optimum oxidation stability (6 h) is reached when the blend contained 38% of soybean oil biodiesel with palm oil biodiesel and 15% of soybean oil biodiesel with rapeseed oil biodiesel. Regarding the cold filter plugging

Table 19.13 Fatty acid composition of soybean oil biodiesel (Park et al. 2008)

Fatty acid	(%)
Myristic acid (C14:0)	0.07
Palmitic acid (C16:0)	10.95
Stearic acid (C18:0)	4.31
Arachidic acid (C20:0)	0.27
Oleic acid (C18:1)	23.12
Linoleic acid (C18:2)	53.27
Linolenic acid (C18:3)	6.77
Total saturated fatty acid	15.60
Total unsaturated fatty acid	83.16

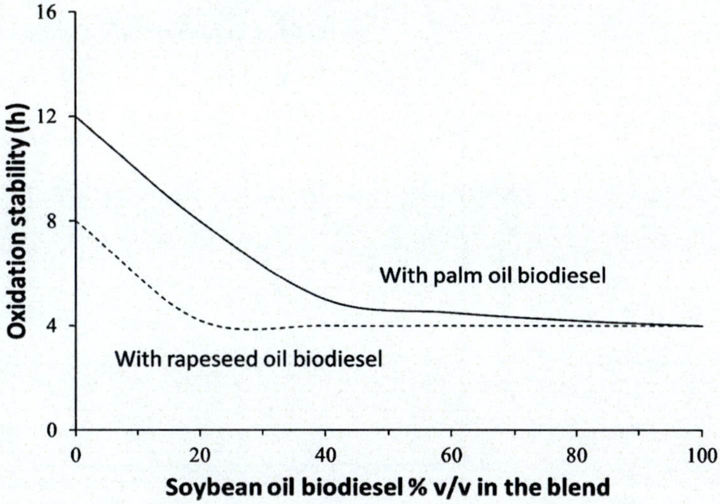

Fig. 19.18 The oxidation stability of the blend of soybean oil biodiesel with palm oil biodiesel and with rapeseed oil biodiesel

point, when the percentage of soybean oil biodiesel is increased in the blend with palm oil biodiesel the cold filter plugging point tends to be low and when the palm oil biodiesel level is high in the blend the cold filter plugging point increases and thus expresses a negative relationship with the actual percentage of the soybean biodiesel in the said blend. But when the share of the soybean oil biodiesel is increased in the blend with rapeseed oil biodiesel the cold filter plugging point value tends to increase and when the level of the soybean biodiesel is reduced the cold filter plugging point also decreases and the relationship is a straight line function (Fig. 19.19).

When methanol is used in the transesterification process, the resultant fatty acid methyl ester has a poor low-temperature performance with a reduced lubricity. But the kinematic viscosity of ethyl ester is higher than that of the methyl ester. It has an enhanced cold filter plugging property. Hence as a best management option, a mixture of methanol and ethanol was employed. Methyl and ethyl esters in the

Fig. 19.19 The cold filter plugging point of the blend of soybean oil biodiesel with palm oil biodiesel and rapeseed oil biodiesel

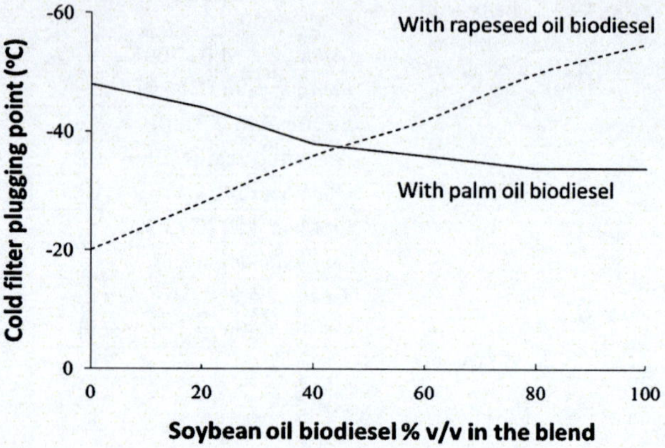

Fig. 19.20 Soybean oil biodiesel blend and the kinematic viscosity

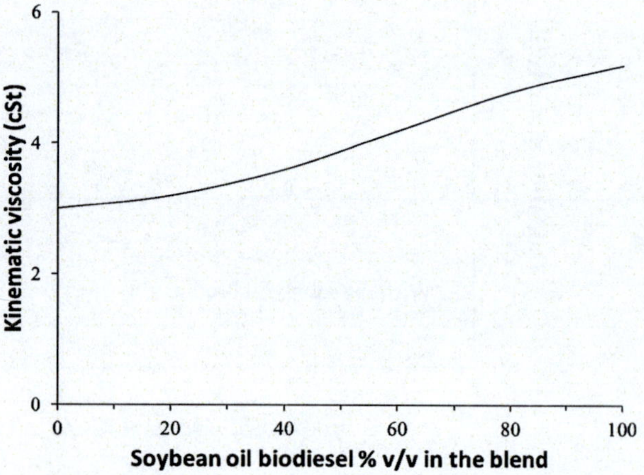

final product form a ratio of 1.7:1 indicating that methanolysis precedes ethanolysis. Mixtures of both esters exhibited improved lubricity and oxidation stability. Thus it is known that the properties of the soybean oil biodiesel may be improved by substituting a portion of the methanol with ethanol (Joshi et al. 2010).

The kinematic viscosity of the soybean biodiesel decreased when it is blended with diesel. The viscosity value of the biodiesel blend is reduced by 40% (3.3 cSt) of that of the biodiesel when the biodiesel was mixed with regular diesel at a maximum of 30% (Fig. 19.20). Thus the blend helps to favour proper ignition due to the reduction in viscosity.

As the diesel index of soybean oil biodiesel (25.5 min) is fairly low when compared to that of the diesel (57 min) it expresses poor ignition quality. The diesel index is known to increase to a standard level when the biodiesel is blended with diesel at a minimum of 20% by volume. As the percentage of blend increases the diesel index naturally reduces (Fig. 19.21).

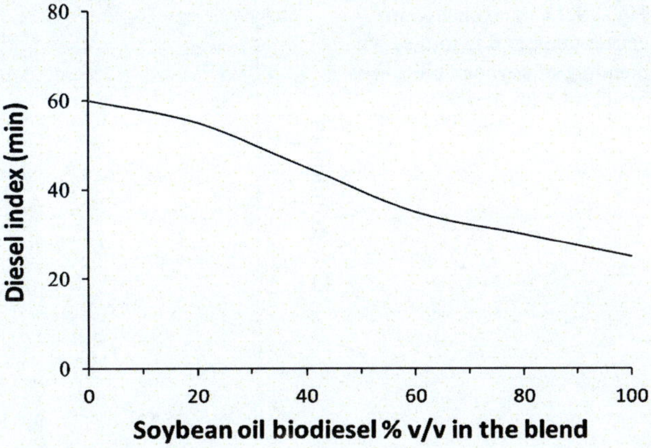

Fig. 19.21 The change in diesel index when soybean oil biodiesel is blended at different levels with diesel

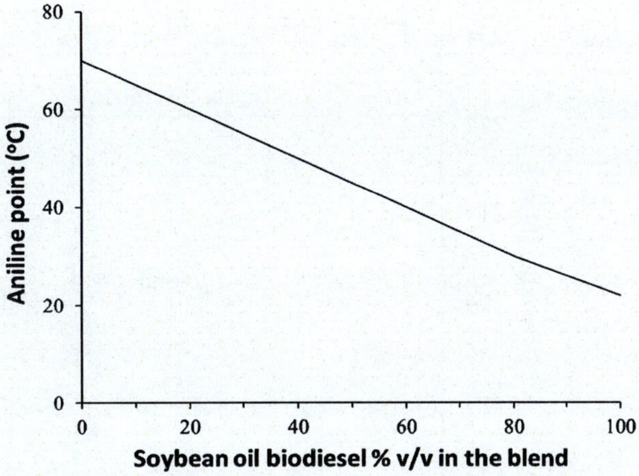

Fig. 19.22 The change in aniline point due to the blending of soybean oil biodiesel with diesel

The aniline point is an additional index of the ignition property of the fuel. The commercial diesel has an aniline point of 66 °C and the biodiesel has the value of 25 °C. Blending of the biodiesel with normal diesel will help to improve the aniline point (Fig. 19.22).

The cetane number of the soybean oil biodiesel is 18 against 50 of the diesel. Thus the ignition characteristics of the biodiesel are needed to be improved by blending with diesel (Fig. 19.23). A mixing level of 20% of the biodiesel (v/v) with diesel has a cetane number of 43 (Goswami and Usmani 2014).

19.2.8 Engine Performance

It is known that pure soybean oil biodiesel cannot be fed in the engines due to the limitations of viscosity and consequent poor energy conversion. Therefore it has become the practice of blending the biodiesel with diesel. If 5 mL of biodiesel is

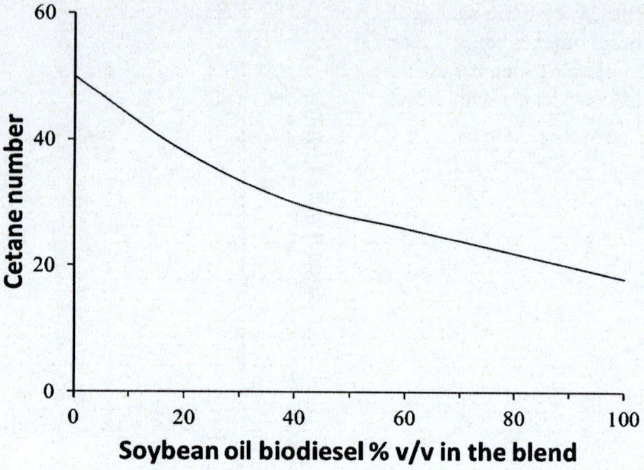

Fig. 19.23 The change in cetane number due to the blending of soybean oil biodiesel with diesel

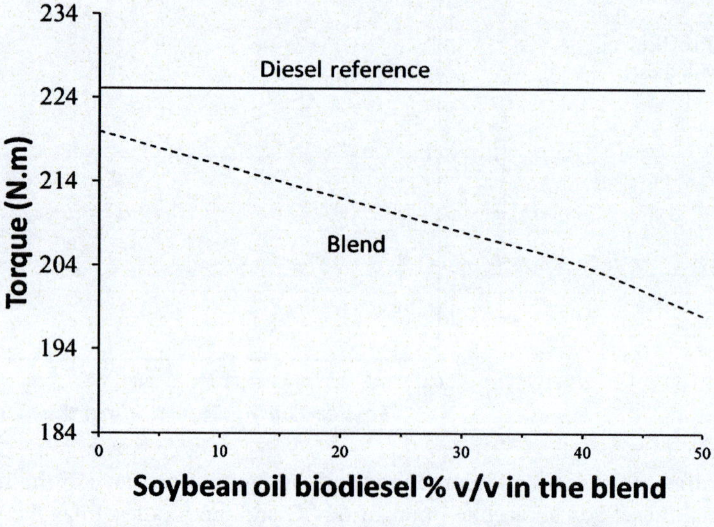

Fig. 19.24 The change in torque when soybean oil biodiesel is blended with diesel

mixed with 95 mL of diesel, it is referred as 5% blend or B5. In such blending the level of biodiesel is kept minimum (i.e. B5 to B20) since the soybean oil biodiesel has a low energy. If the biodiesel content in the blend is more, there will be low brake power and torque since the torque and power are directly related. As the share of biodiesel in the blend is increased the combustion pattern is altered. One of the principal reasons is the reduction in the spray of the fuel due to the presence of linoleic acid which by its nature is viscous. In soybean oil biodiesel the torque gradually reduced to 219, 216, 213, 211 and 198 as the blending increased to B5, B10, B15, B20 and B50 blends (Fig. 19.24). The brake power also follows the same pattern with brake power (Fig. 19.25).

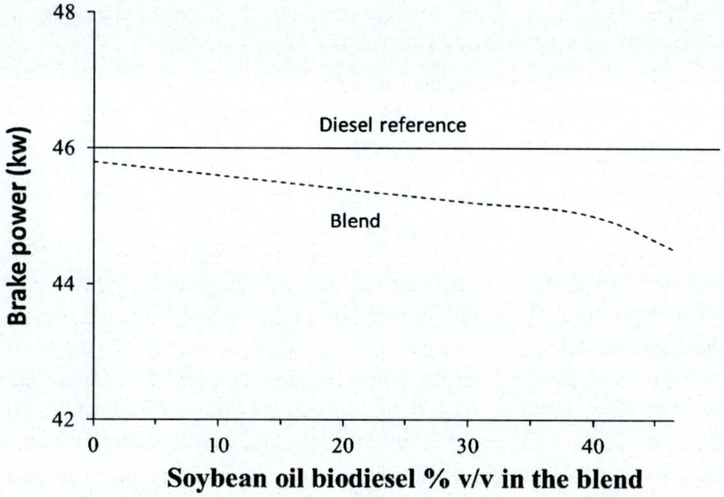

Fig. 19.25 The change in brake power when soybean oil biodiesel is blended with diesel

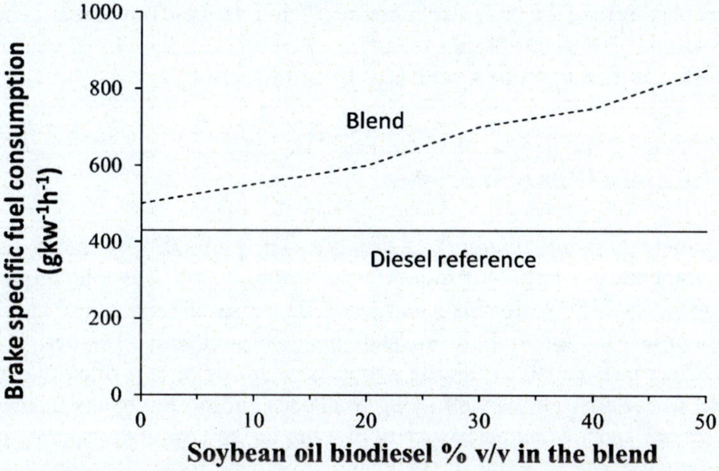

Fig. 19.26 The brake specific fuel consumption of soybean oil biodiesel when blended with diesel

The biodiesel shows higher brake specific fuel consumption due to its low energy content. Higher specific fuel consumption is an offsetting mechanism to transfer energy in the context of low energy fuel (Fig. 19.26).

The CO_2 emission in the exhaust shows marginally low values for the blend when compared to that of the diesel (Azad et al. 2015). The CO_2 reduction reaches the peak if more biodiesel is incorporated in the blend. CO emission in the exhaust is governed by ever so many factors such as engine speed, injection pressure, fuel characteristics and the composition of the fuel. By and large the level of CO is lower

Table 19.14 The change in the conc. of pollutants in the exhaust of internal combustion engine fed with blends of soybean oil biodiesel with diesel (Azad et al. 2015)

	% Change in the exhaust				
Parameters	B5	B10	B15	B20	B50
CO_2 reduction (%)	0.36	0.56	0.83	1.10	1.23
CO reduction (%)	28.45	36.03	40.96	45.89	52.32
NO_x increase (%)	13.57	19.60	–	35.18	50.75

in the biodiesel blend when compared to that of the diesel (Table 19.14). Such reduction is due to the rich oxygen content in the biodiesel molecule which helps to convert the CO in to CO_2.

Hydrocarbon is an another component in the exhaust which offers a reference to the quality of the fuel. Hydrocarbon conc. increases in the exhaust when the laminar flames enter the region of intense turbulence without adequate flow of air at a high temperature. Normally the conc. of hydrocarbon in the exhaust gas is around 20 ppm (0.002%). In pure soybean oil biodiesel and also in its blend with diesel having higher share of biodiesel the hydrocarbon in exhaust tends to reduce conspicuously. Oxides of nitrogen are high in the exhaust when biodiesel is used as a fuel. When the blend level is increased the NO_x also increases (Table 19.14). The potential influence of NO_x is around 300 times higher than that of the CO_2 in the global warming and therefore NO_x in its minimum level has a major impact.

19.3 Pongam (*Pongamia pinnata*)

The pongam (*Pongamia pinnata*), a fast growing glabrous deciduous tree with drooping branches is a native of India (Osman et al. 2009). It is a leguminous tree growing up to 15–25 m with a lushy canopy. The leaves are soft, shiny burgundy in early summer and mature to a glossy deep green subsequently. The bark is smooth and grey. This tree is capable of fixing atmospheric nitrogen. It is often planted as an ornamental tree which grows well along roadsides and on the banks of the canals. This species has special characteristics of binding the soil through its dense network of roots. Because of the extensively lushy growth of roots, *P. pinnata* is not a preferred species for agro-forestry. It belongs to the order: Fabales and family: Fabaceae. This species is also named as *P. glabra* (Vent.), *Derris indica* (Lam.) and *Millettia pinnata* (Lewis) and is popularly referred as Indian beech tree.

The Indian born *P. pinnata* had its first prevalence in Burma. Next it spread to the Asian sub-continent. Now it is distributed in many parts of the world.

19.3.1 Geographic Distribution

The natural distribution of pongam is along the sea coasts and river banks of India and Burma. This species were later introduced to humid tropical lowlands of the

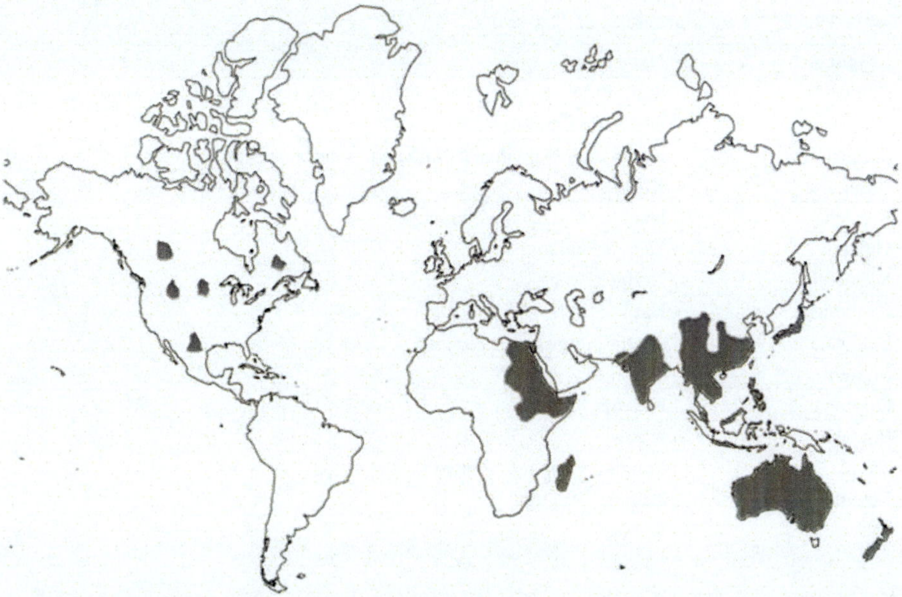

Fig. 19.27 Distribution pattern of *P. pinnata* in the world (indicated as dark shade)

Philippines, Malaysia, Australia, Seychelles, United States of America and Indonesia. It is also distributed in Bangladesh, Nepal, Thailand, China, Egypt, Fiji, Japan, Mauritius, New Zealand, Pakistan, Papua New Guinea, Samoa, Solomon Islands, Sri Lanka, Sudan and Tonga (Fig. 19.27).

This species is called by different names as given in Table 19.15.

P. pinnata prefers an annual rainfall ranging from 500 to 2500 mm. It tolerates a maximum temperature range of 27–38 °C. The minimum tolerant temp is 10–16 °C. It grows at a maximum elevation of 1200 m in moist soil whose type ranges from stony to sandy to clayey including verticals. However it does not perform well in dry sands. It can tolerate moderate salinity and therefore can exist in sea shores. Good growth is recorded in well-drained soil with adequate moisture.

P. pinnata is a medium sized tree with a long and thick tap root and numerous well developed lateral roots. The leaves are compound pinnate having 5 or 7 leaflets which are arranged in 2 or 3 pairs with single terminal leaflets. The leaflets are 5–10 cm long, 4– 6 cm wide and are pointed at the tip. The tree comes to flowering after 3–4 years of planting. The flowers borne on racemes are pink to light purple or white in colour. A fully grown tree may yield around 100 kg of fresh seeds per annum and the tree continues to yield fruits up to 70 years. The tree bears flowers throughout the year which are arranged in pairs along reaches in auxiliary pendent, long racemes or panicles, calyx companulate or cup shaped, truncate, short dentate, lower most lobe longer and suborbicular, broad with inflexed basal ears appearing silky with the wings oblique, long adherent to the obtuse keel with the keel petals

Table 19.15 Vernacular names of *P. pinnata*

Language	Names
English	Smooth leaved pongamia, Indian beech tree, Karum, seashore mempari
Hindi	Dithouri, Karuain
Sanskrit	Karanjaka, Naktamala, Naktahva, Ghritakaranja
Assamese	Korach
Bengali	Nata karanja, Dahara karanja
Gujarati	Kanojo, karanja
Kannada	Honge, Hulagilu
Kashmiri	Kath
Malayalam	Avittal, ungu, unu, Pungu
Marathi	Karanja
Oriya	Karanja
Punjabi	Karanj
Tamil	Pungan, Pongana, Pungai
Telugu	Lamiga, Kanuga
Urdu	Karanj

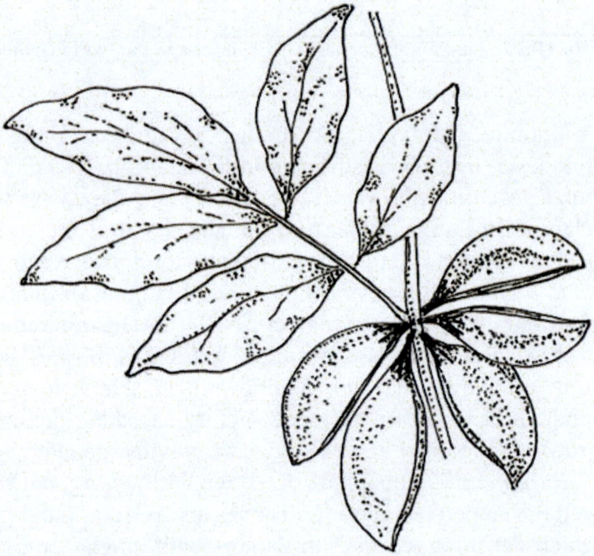

Fig. 19.28 The pongam *Pongamia pinnata*

remaining coherent at apex, ovary appearing as subsessile to short stalked, 2–3 ovules, stigma small and terminal (Fig. 19.28).

This tree can grow well in soil having the pH 7.5 and express deficiency syndrome beyond the above pH. The propagation is mostly through seeds which germinate in a week to a month. Quality seed normally does not require any treatment and may remain viable for a year when stored in air tight container. Seeds are sown in pits of the size 45 × 45 × 45 cm at a spacing of 5 × 5 m.

Germination is hypogeal and the radical develops before the plumule emerges. Alternatively, seedlings can be raised in polyethylene bags till it reaches a height of 25–30 cm. It may then be transplanted in the field. It becomes an adult tree in 4–5 years and starts bearing fruits. A yield of 3500–5000 kg seeds can be obtained from a hectare in a year. Each tree may yield 10–50 kg seeds. These trees are resistant to high winds and drought but cannot withstand chill climate. A growth of around 1.5 m in height and 4 cm in diameter can be attained in a year. Pruning and sucker management also are recommended.

The leaves of pongamia often have whitish streaks and galls caused by the larvae of various insects: *Parnara mathias*, *Indarbela quadrinotata*, *Myllocerus curvicornis* and *Acrocercops* spp. Adults and nymphs of the bug *Aspongopus brunneus* suck the sap and cause extensive damage even up to 30% of the nursery plants. The fungus *Ganoderma lucidum* causes the rotting of the root of the seedlings. Similarly, the fungus *Fomes merilli* damages the tender shoots of the seedlings.

19.3.2 Fruits and Seeds

The flowers mature into elliptical, 3–6 cm long, 2–3 cm wide thick walled pods each containing 1–2 seed. From fifth year onwards the trees start yielding fruits. However economically advantageous production of seed starts from the tenth year. The pod is woody, flat, smooth, oblong, indehiscent, 3–6 cm long, 2–3 cm wide and thick walled. Half of its weight is constituted by the kernel.

The composition of the air dried kernels is given in Table 19.16. It has a protein content of 18%. The oil content is 35%.

The seeds are oblong and light brown in colour. They are compressed ovoid, bean shaped with a dimension of 1.5–2.5 $l \times$ 1.2–2.0 $b \times$ 0.8 cm t. The average weight of the seed is 1.2 g. The seed coat is brittle and oily.

19.3.3 Oil Extraction

The oil extracted from this seed is being used for lighting the lamps, making soaps besides being used as a lubricant. The oil content of this seed is around 30–35%. If the seeds are expelled through crushers, the oil recovery is in the range of 25–28%

Table 19.16 Composition of the dried kernels of *P. pinnata*

Components	%
Moisture	25
Oil	35
Protein	18
Carbohydrate	9
Crude fibre	8
Ash	4

Table 19.17 Oil recovery through different means of extraction of *P. pinnata* seeds

Extraction method	Yield (%)
Mechanical expeller	24
Cold percolation technique	22
Soxhlet extraction using hexane	34

Table 19.18 The physicochemical characteristics of the oil of *P. pinnata* (partial data from Meher et al. 2004)

Parameters	Value
Colour (Lovibond, 2.54 cm cell)	10Y + 0.7R
Acid value (mg KOH g^{-1})	5.06
Saponification value (mg KOH g^{-1})	187.00
Unsaponifiable matter (% w/w)	2.60
Iodine value (mg I_2 g^{-1})	865
Peroxide value (meq kg^{-1})	11.00
Specific gravity	0.94

Table 19.19 Fatty acid composition of pongam oil

Fatty acid		%	
Saturated	Palmitic	C16:0	11.8
	Stearic	C18:0	7.7
	Arachidic	C20:0	1.8
	Behenic	C22:0	4.2
	Lignoceric	C24:0	1.3
Mono unsaturated	Oleic	C18:1	51.2
	Eicosenoic	C20:1	1.3
	Erucic	C22:1	2.0
Poly unsaturated	Linoleic	C18:2	16.0
	Linolenic	C18:3	2.7

only. The oil is thick yellowish orange in colour, non-edible and has a bitter taste with disagreeable aroma. Presence of erucic acid, toxic flavonoids karanjin, pongapin and pongaglabrin renders the oil non-edible. This oil has a specific gravity of 0.937 at 15 °C. Certain physical and chemical properties of this oil resemble that of the diesel. However, the Conradson carbon residue and viscosity are relatively higher in this oil.

The pongam seeds are extracted for oil both through mechanical and chemical means. Mechanical expellers are used in large scale operation but the yield is relatively low. For small scale operation Soxhlet extraction is considered to be good. Highest recovery of oil is possible if hexane is used as a solvent. The yield of oil from pongam seeds by different methods is given in Table 19.17.

The oil is dark in colour containing saturated, mono unsaturated and poly unsaturated fatty acids. The physicochemical characteristic of the oil is given in Table 19.18. The oil has a high acid value and saponification value.

The fatty acid composition of the oil is given in Table 19.19 and Fig. 19.29 which indicates the percentage occurrence of saturated, monounsaturated and

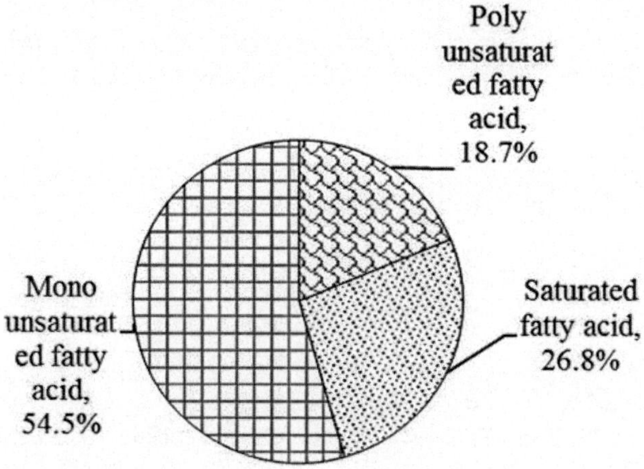

Fig. 19.29 The percentage occurrence of polyunsaturated, monounsaturated and saturated fatty acid in pongam oil

polyunsaturated fatty acids. It is known that monounsaturated fatty acids occur in a high percentage (>50%).

19.3.4 Biodiesel Production

The pongam oil is transesterified to produce ester by which the viscosity of it is reduced substantially. In transesterification, the alcohol is displaced from an ester by another in a process identical to that of hydrolysis except the fact that alcohol is employed instead of water. Most often methanol is used in the process. The whole reaction is steered up in the presence of an acid or alkali as a catalyst. The reaction is summarized and presented below:

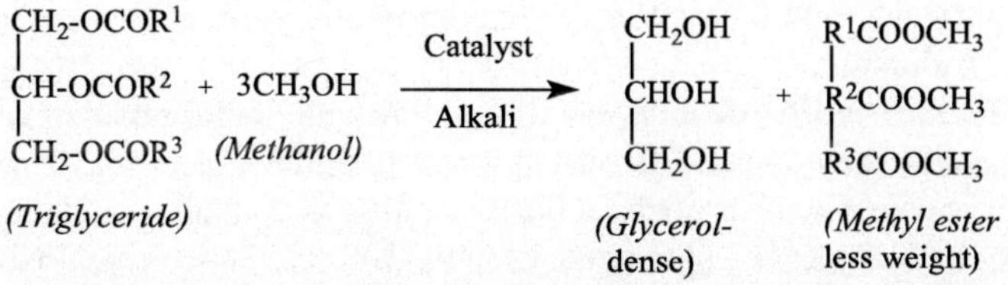

Potassium hydroxide is first dissolved in moisture free methanol. This solution is then mixed with oil and allowed to react at a fairly high temperature and at a constant stirring. On completion of the reaction two major products, namely glycerin and biodiesel (ester) are formed. The glycerin being dense settles down and is drained out from the bottom. The biodiesel which stays as an upper layer is separated after a day of settlement. The excess methanol which remains in the biodiesel is recovered by heating and condensation. The biodiesel is then washed by warm water and dried. Normally the quantity of alkali has a decisive role in the transesterification. When the quantity of alkali is low the reaction may not proceed in the forward direction. When

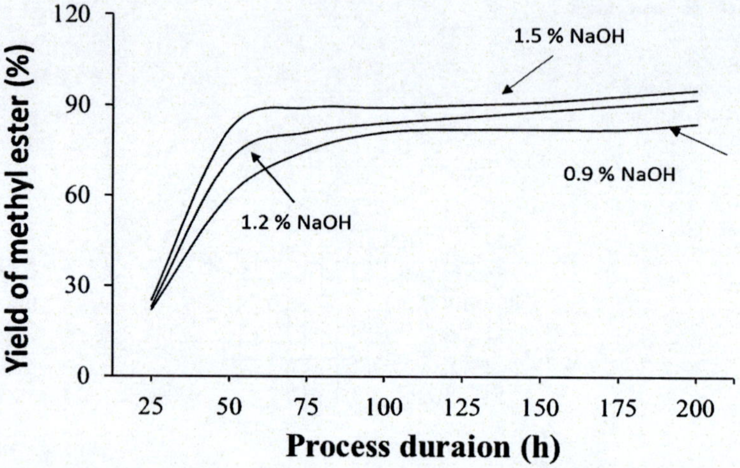

Fig. 19.30 The effect of different levels of NaOH as a catalyst in pongam biodiesel recovery during transesterification at a methanol to oil ratio of 6:1 and at 60 °C

the concentration of alkali is optimum the reaction is slow and steady which has the highest recovery of biodiesel. When the conc. of alkali is increased the yield is fast in the initial stage but becomes static subsequently. On the best interest of the environment which receives the wash water containing alkali, it is good to use low level of alkali and to wait for the reaction to complete. The amount of alkali (NaOH or KOH) required is to be decided based on the quantity of free fatty acid present in the oil.

Figure 19.30 indicates that NaOH when applied in the required quantity (0.9%), the yield of biodiesel increased steadily to the tune of 95% in 200 min. When the conc. of NaOH is raised to 1.2% the recovery is 96%. In that concentration during the initial stage the recovery is rapid. When the NaOH level is further increased (1.5%) the ultimate yield is low (86%) apparently due to over dose causing cloudiness and gel formation.

Besides sodium hydroxide and potassium hydroxide, other types of catalysts are being employed. They include Bronsted acids (sulphuric acid and sulfonic acid, etc.), heterogeneous catalysts such as basic zeolites and alkaline metal compounds montmorillonite K10, ZnO and enzymes such as lipases. Supercritical methodology without the application of catalyst also is perfected. In supercritical process the conversion takes place in 4 min at an ambient temperature of 240 °C and at a pressure of 8 MPa. As high as 98% conversion is achieved. In this process there is complete transformation of free fatty acid. Certain basic metal oxides such as CaO and ZnO are also employed. In certain cases, CaO loaded with alkali metal is used which often increases the conversion to even 100% in a short duration.

Alkali metals such as Li or Na or K is impregnated in CaO and used in the transesterification of pongamia oil containing high levels (>5%) of free fatty acid. The nitrate salt of Li or Na or K is impregnated in CaO and stirred for 2 h in wet condition. It is then dried and heated to 100 °C for 24 h. The impregnated CaO is

Fig. 19.31 Biodiesel (ester) production from pongam oil when treated with different levels of Li/CaO as a catalyst at a methanol to oil ratio of 6:1 and at 60 °C

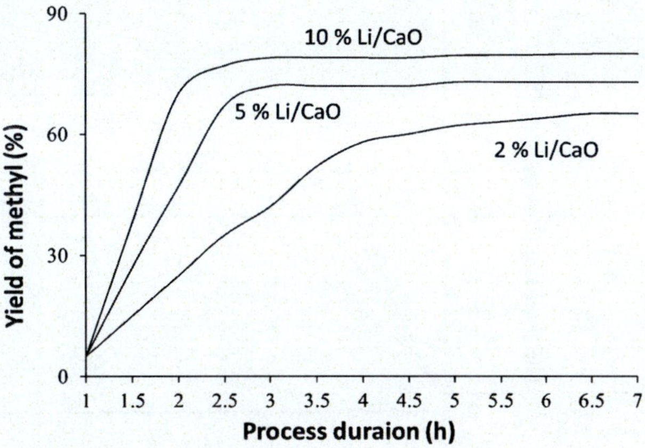

Fig. 19.32 Biodiesel (ester) production from pongam oil treated with 10% Li/CaO catalyst at a methanol to oil ratio of 6:1 at four different temperatures

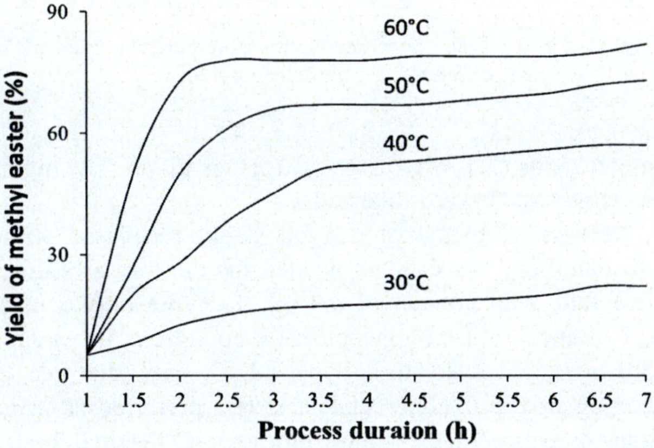

used along with methanol to prepare biodiesel. The Li/CaO is used at the rate of 2–10% of the whole mixture at a reaction temperature of 60 °C in a closed chamber for 8 h. The effect of three different doses of Li/CaO on pongam oil at a methanol to oil molar ratio of 6:1 and at 60 °C is given in Fig. 19.31. In 2% level of Li/CaO the recovery is low and in 10% level the recovery is 86%.

The whole reaction is temperature dependent (Fig. 19.32). At a reaction temperature of 30 °C the recovery is around 28%. When the temperature is raised to 60 °C the recovery of ester rises to 86% at a catalyst (Li/CaO) concentration of 10% in an experimental duration of 8 h.

Methanol to oil ratio is considered as one of the major factors deciding the rate of reaction. The stoichiometry indicates that 3:1 is the required ratio. As the whole reaction is reversible it is needed to flood the mixture with alcohol so as to maintain a decisively forward reaction. Double the ratio (6:1) is normally recommended to meet such contingencies. The heterogeneous catalysts are usually less efficient and therefore needs a high molar ratio of alcohol to oil to speed up the reaction and also to increase the recovery rate. In Fig. 19.33 the yield of methyl ester in four distinct

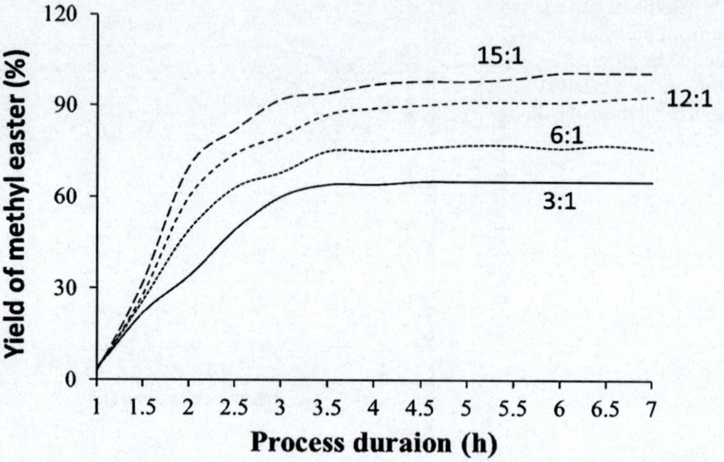

Fig. 19.33 Biodiesel (ester) production from pongam oil treated with 10% Li/CaO as a catalyst at 60 °C with four different ratio of methanol to oil

molar ratios (3:1, 6:1, 12:1 and 15:1) is given. The higher molar ratio definitely has an edge over lower molar ratio.

Excess of methanol can be safely recovered without loss. The quantity of methanol and the catalyst needed for the transesterification largely depend on the free fatty acid content of the oil. The free fatty acid content of the oil varies in accordance with the agro-climatic conditions in which the tree grows, cultivation practices and seed-storage period. In case the free fatty acid content is high, homogeneous alkaline catalyst is less preferred as it induces saponification rather than assisting in the transesterification. In such high levels of free fatty acids heterogeneous catalysts prove to be efficient. The outcome of transesterification using Li/CaO and KOH as catalyst in different concentrations of free fatty acid in pongamia oil is illustrated in Fig. 19.34.

In Li/CaO catalyst the ester production is marginally affected when the free fatty acid level is high. If NaOH or KOH is used, the yield reduces due to saponification and associated isolation problems. Anjana et al. (2016) used CaO impregnated potassium iodide as a catalyst at 4% by weight which gave a maximum conversion of 95.7% at a methanol to oil molar ratio of 12:1 in a process duration of 2 h. Glycerol derivative of SO_3-H carbon catalyst at 20% by weight gave >99% conversion at an oil to methanol molar ratio of 1:45 at 160 °C in 4 h. This catalyst performed well up to five cycles without any deactivation under optimized reaction conditions (Manneganti et al. 2014). Single step supercritical methyl acetate process achieved 96.6% conversion in 45 min at 300 °C at 20 MPa, with 45:1 molar ratio of methyl acetate to oil. This method produces glycerol-free biodiesel (Goembira and Saka 2015). But, it requires sophisticated reactor system which consume high energy thereby the cost of production increases. Microwave was also used to reduce the free fatty acid in oil. At an irradiation duration of 150 s, at 180 W, with methanol to oil

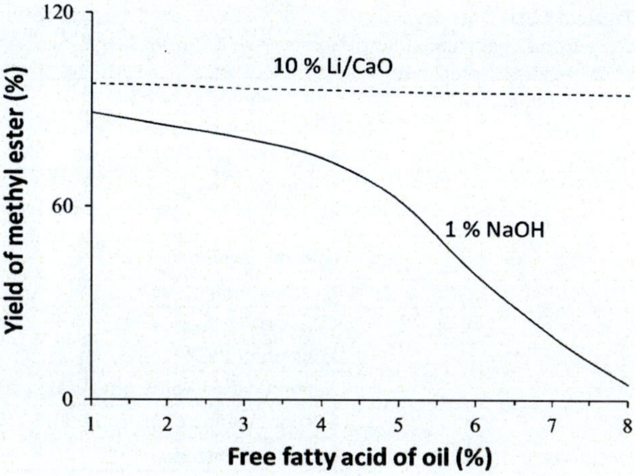

Fig. 19.34 Effect of free fatty acid concentration on the production of methyl ester when Li/CaO at 10% and NaOH at 1% level are used as catalyst at a methanol to oil molar ratio of 6:1 at 60 °C

ratio 34:1 (w/w) and 3.5 mL L^{-1} of sulphuric acid as catalyst the free fatty acid content reduces in the first stage. In the second stage at a methanol to oil ratio of 34:1 (w/w), 1.5 g L^{-1} of NaOH and at an irradiation time of 150 s complete transesterification is carried out.

19.3.5 Biodiesel Property

The analytical characteristics of the biodiesel prepared from pongam oil are given in Table 19.20.

The characteristics of the biodiesel are largely based on the source and age of the raw oil used. Iodine value which is the index of the degree of unsaturation is found to be relatively low. Similarly, higher cetane number indicates a better ignition of the fuel.

The viscosity actively controls the fuel properties especially at the injection assembly of the engine. The viscosity of the ester from pongam oil is lower than that of the oil. Flash point refers to a temperature at which it ignites when the material is exposed to open flame or spark. The flash point of the pongam biodiesel is observed to be relatively higher. The pour point of the biodiesel which is otherwise known as cold filter plugging point refers to the performance of a fuel under cold conditions. Under low temperature the fuel is likely to thicken and may not flow freely in the fuel conduits of the engine. Pongam biodiesel has poor cold flow properties. There will be formation of gum and crystallization of particles .This can be improved by winterization, blending and by the addition of cold flow improvers. Pongam biodiesel and conventional diesel are blended at various weight ratios 80:20, 60:40 and 20:80. The cold flow properties (cloud point and pour point) improve significantly in 20:80 ratio. Ethanol is used as cold flow improver with biodiesel at different blending ratios (98:02, 92:08, 90:10, 85:15 and 80:20). The ratio 80:20 shows conspicuous improvement (Dwivedi and Mahendra 2015).

Table 19.20 The physicochemical characteristics of the biodiesel prepared from *P. pinnata* oil

Properties	Value
Colour	Yellow to orange
Acid value (mg KOH g^{-1})	5.2
Kinematic viscosity (at 40 °C)	5.8
Density at 40 °C (cSt)	0.865
Flash point (°C)	164
Cloud point (°C)	22
Specific gravity	0.925
Sulphur content (%)	0.11
Centane number	53
Pour point (°C)	15.8
Saponification value (mg KOH g^{-1})	128.8
Carbon residue (%)	0.64
Iodine value (mg I_2 g^{-1})	965
Carbon chain	C16–C28
Bulk modulus at 20 MPa	1810
Oxygen content (%)	11.0
Heating value (J g^{-1})	38,450
Sulphur content (%)	0
Dist. temperature (°C)	360
Molecular weight	281
Water and sediment (% vol)	0.005
Total glycerin (% w)	0.18

Oxidation resistance is an essential parameter and instability leads to the formation of aldehydes, alcohols, short chain carboxylic acids, gums and insoluble sediments in biodiesel. Thermal instability is also based on the rate of oxidation at high temperature which creates gums and insoluble substances thereby the density alters. Storage stability is a favourable property which resists changes in physical and chemical characteristics such as sedimentation and colour changes. The above problems are overcome by the addition of natural and synthetic additives.

The cake obtained after oil extraction is recommended to be used in biogas plant since it has a higher biodegradability than that of the jatropha cake due to the high content of volatile solid (7.5%) and low content of non-volatile solid (5.5%). A first order model based on a single linear regression was used to study the kinetics of anaerobic (batch) digestion of the oil cakes of jatropha and pongam in pure form and also in combination with cattle dung (Gupta et al. 2009). The cake was made as slurry with 4 parts of water and used as a feed stock. It was then mixed with cow dung at the ratio of 1:0.5 and aerobically digested. This gave a high yield of biogas in 30 days when compared to that of jatropha cake.

The oil cake is reported to inhibit the nitrification process so as to improve the N use efficiency in fertilizers (Osman et al. 2009). Analysis of the oil cake is given in Table 19.21.

Table 19.21 Chemical composition of *P. pinnata* oil cake (Osman et al. 2009)

Nutrients	%	Nutrients	mg kg^{-1}
Total nitrogen	4.28	Zinc	59
Phosphorus	0.40	Iron	1000
Potassium	0.74	Copper	22
Calcium	0.25	Manganese	74
Magnesium	0.17	Boron	19
		Sulphur	1894

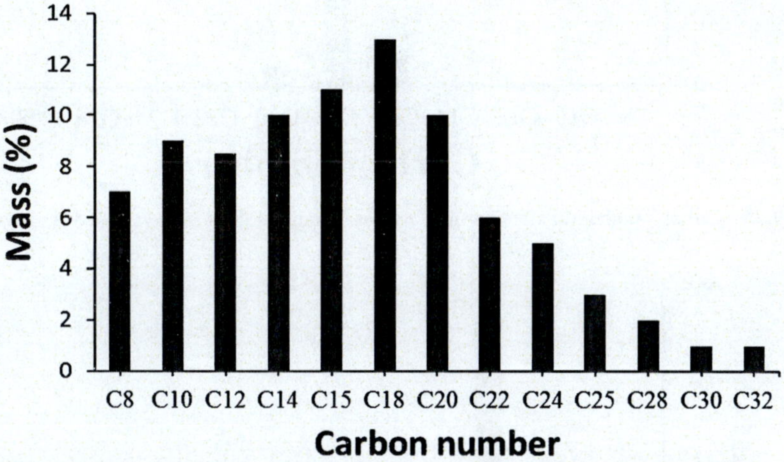

Fig. 19.35 Carbon number distribution of petro-diesel (Satyanarayana and Rao 2009)

19.3.6 Pongam Biodiesel and Petroleum Diesel

In the regular petro-diesel the distribution of carbon atoms in a hydrocarbon molecule ranges from 8 to 32 (C8–C32) which is grouped into two types of carbon chains such as short chain 8–12 (C8–C12) and medium chain 14–32 (C14–C32). The peak of the carbon distribution in the hydrocarbon molecules of the petro-diesel is between 14 and 20 carbon, i.e. C14–C20 (Fig. 19.35).

It is observed that there is a gradual increase in the mass percentage occurrence of carbon from C8 to C18 in petro-diesel which reduced beyond C20. In the case of pongam biodiesel the carbon distribution is restricted to 16–24 (C16–C24) with the highest peak observed at C18 as seen in Fig. 19.36. The higher chain hydrocarbon gradually reduced from C20–C24 (Fig. 19.36).

The distribution of carbon number among the molecules of pongam biodiesel is a reflection of the composition of various fatty acids present in the pongam oil as could be seen in Fig. 19.37.

It indicates that the unsaturated fatty acid oleic acid C18:1 is in rich quantity and the rest of the fatty acids are far lower in concentration.

The oxygen content of the hydrocarbon molecule of the pongam biodiesel is 11% which is higher (2.9%) than that of the regular diesel. As a result, the proportion of

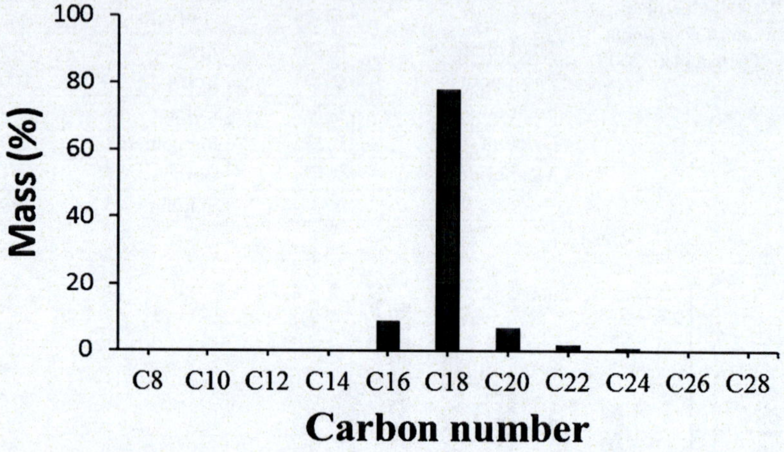

Fig. 19.36 Carbon number distribution of *P. pinnata*-biodiesel (Satyanarayana and Rao 2009)

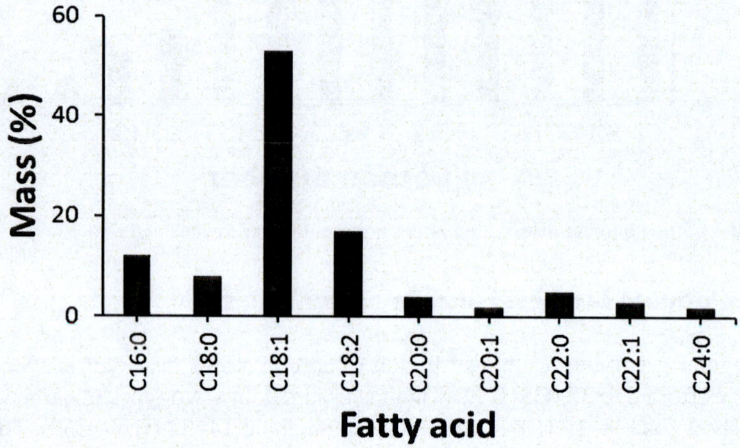

Fig. 19.37 Fatty acid content of *P. pinnata* oil

other components such as carbon and hydrogen which are the sources of energy naturally reduced resulting in the lowering of the heating value (38,450 J g^{-1} against 42,450 J g^{-1} of the regular diesel). Thus pongam biodiesel has less number of carbon when compared to the regular diesel. Besides the biodiesel has the esters of fatty acids with various degree of saturation. The reduction in heating value is by 9.4%

$$\frac{42,450 - 38,450}{42,450} \times 100$$

It indicates that around 1094 mL of pongam biodiesel may be required to match with 1000 mL of ordinary diesel for covering an equal distance of travel. Because of

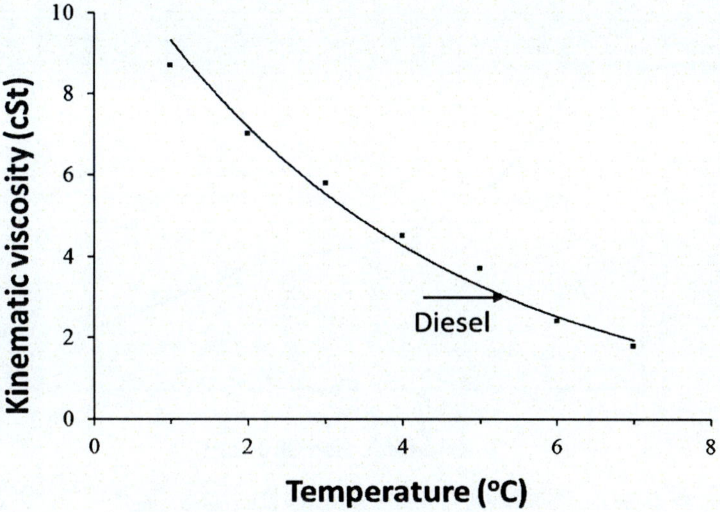

Fig. 19.38 The kinematic viscosity of pongam biodiesel as a function of the change in temperature. The arrow indicates the viscosity of diesel at 30 °C

the presence of the rich quantity of oxygen in biodiesel, the stoichiometric air fuel ratio is low. Moreover, the combustion efficiency of the pongam biodiesel is considered better than regular diesel.

In an automobile engine, the ignition of a fluid fuel is governed by two important factors (1) cetane number (CN) and (2) viscosity. The CN is indicative of the smooth combustion with low knocking of the engine. The CN of the pongam biodiesel is slightly higher than that of the petroleum diesel. However such benefit is offset by the higher viscosity of the pongam biodiesel which is 60% higher than that of the regular diesel though it is within the limit (ASTM Standard D 6751–02). The benefit thus being caused by the CN and acceptable viscosity of the biodiesel is largely dependent on the status of the automobile engine such as its design, size, speed and load. If the CN is low, the exhaust from the engine may contain excessive level of carbon monoxide, hydrocarbon and particulate matter. The CN of the biodiesel depends on the fatty acids. The kinematic viscosity if exceeds the optimum level, the ignition is adversely affected as it indirectly influences the spraying intensity and the fuel movement in the engine. To reduce the viscosity of pongam biodiesel further in certain applications, it is heated up marginally. The kinematic viscosity and the temperature of the pongam biodiesel are inversely proportional (Fig. 19.38).

The iodine value which indicates the quantity of iodine in gram which can be dissolved in 100 g of pongam biodiesel is higher than that of the petroleum diesel. The iodine value of pongam biodiesel is 96.5 (965 mg I_2 g^{-1}) against 38 in petroleum diesel. Higher iodine value indicates the degree of unsaturation of the fuel. Fuel with high iodine value has low melting point which makes a better biodiesel especially with reference to cold weather. At the same time such oil has the tendency to oxidize and polymerize in to a solid. Therefore pongam biodiesel has

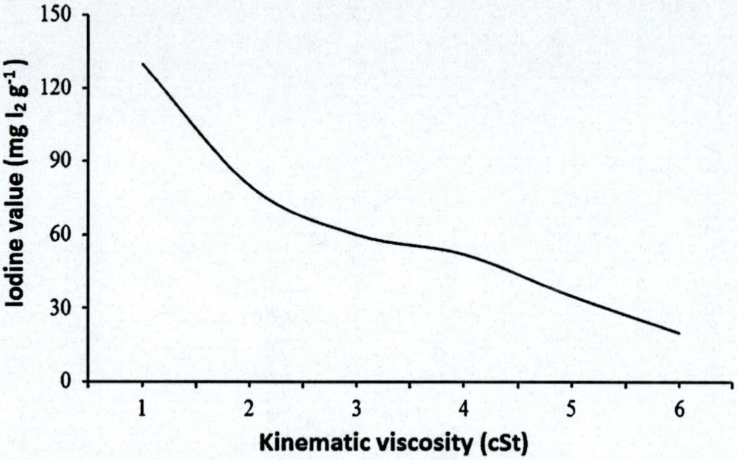

Fig. 19.39 The relationship between kinematic viscosity and iodine value

Table 19.22 Correlation coefficient existing between various properties of biodiesel and percentage of unsaturation (Gopinath et al. 2010)

Density at 40 °C	0.933
Kinematic viscosity at 40 °C	(−) 0.979
Cetane number	(−) 0.796
Heating value	(−) 0.796
Iodine value	0.927
Saponification value	(−) 0.496

to be utilized at the earliest without storing it long. Biodiesel with a high iodine value normally has an increased NO_x in the exhaust. Iodine value and kinematic viscosity normally have an inverse relationship (Fig. 19.39).

In pongam biodiesel, though the iodine value is fairly high the viscosity is not low. The correlative relationship between unsaturation and other parameters of the biodiesel is given in Table 19.22.

The density of biodiesel is higher than that of the petro-diesel but lower than that of water. As a result the bulk modulus (β) also is higher as β is a function of the density of the fluid. The Brake Specific Energy being a variable independent of the fuel source indicating the energy input required to produce unit power output shows that the Brake Specific Energy of pongam biodiesel is higher than that of the petroleum diesel. It may be due to low heating value and high viscosity of the biodiesel. Similarly the Brake Thermal Efficiency of pongam biodiesel is around 10% lower than that of the petroleum diesel. During the first stage of combustion which is generally referred as ignition delay the fuel droplets evaporate and mix with air at a high temperature. The ignition delay is influenced by cetane number, viscosity of the biodiesel and the temperature of the air. In the second stage which is referred as rapid combustion or premixed combustion the fuel mixture undergoes rapid combustion causing swift raise in pressure and heat. It is followed by slow diffusion which is often cited as third phase.

As for as the pongam oil is concerned during the first phase the mixing is incomplete because of the high viscosity of the fuel. In the premixed phase the burning efficiency is fairly good. But in the third phase the diffusion is limited. . The engine performance of blended biodiesel (B10, B20 and B30) is evaluated in terms of Brake Specific Fuel Consumption (BSFC), Brake Specific Energy Consumption (BSEC) and Brake Thermal Efficiency (BTE) at different loading conditions. B30 shows high BSFC (0.438 N m^2), whereas B20, B10 and conventional diesel has 0.403, 0.379 and 0.342 N m^2. Such higher BSFC in blended biodiesel is due to the combined effect of density, viscosity and low calorific value. The conventional diesel has higher BSEC (4.32 N m^2) than B30 (3.57 N m^2), B20 (4.08 N m^2) and B10 (4.1 N m^2). The BTE decreased by increasing the blending ratios. B30 has the lower BTE (20.01%) than conventional diesel fuel (28.06%) (Anjana et al. 2016).

The temperature of the exhaust gas of the engine running on pongamia biodiesel is relatively low due to the low heating value and subdued combustion caused by low evaporation and low spray characteristics. High share of oxygen present in the biodiesel molecule leads to the formation of more NO_x in the exhaust gas. The CO in the exhaust gas is observed to be low due to improved combustion with the additional utility of O_2 present in the biodiesel. Therefore the quantity of smoke being discharged from the pongam biodiesel is low.

19.4 African Oak (*Afzelia africana*)

The oil extracted from the seeds of *Afzelia africana* is now being considered as a raw material in the production of biodiesel in African continent specifically in Nigeria though they are not exploited economically. This tree grows to a height of 20 m and is being considered as a good timber for making essential furniture. This species is now enlisted in IUCN red list due to excessive felling for timber. It grows in humid and dry forest of Senegal, Uganda and Congo (Fig. 19.40) at an elevation ranging

Fig. 19.40 Geographical distribution of *Afzelia Africana* (indicated as dark shade)

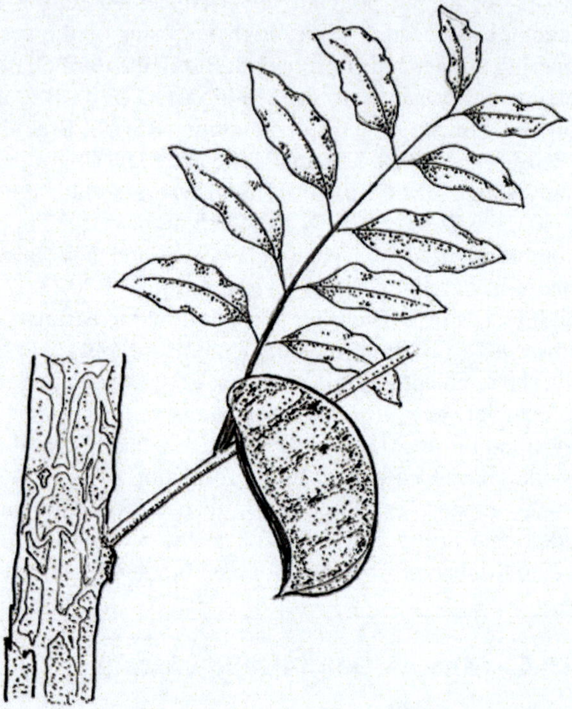

Fig. 19.41 The African oak, *Afzelia africana*

from 0 to 900 m with an annual day temp. from 20 to 32 °C and at a rainfall of 1000–2500 mm. They prefer slightly acidic soil (pH 5.5–6.5).

The leaves are alternate, petiolate and paripinnate. Each leaflet is 5–15 cm long and 3–8.5 cm broad. The inflorescence is 3–13 cm long. Flowers are white to yellow and discharge sweet cent. The fruit takes 6 months to ripe: The morphology of this fruit may please be seen in Fig. 19.41.

The fruits are hard, blackish with a red cap. On getting dry, the hard shell bursts and the seeds are discharged.

This species is referred by many location-specific vernacular names. They are African oak, bilinga, kawu, akparata, African mahogany, lenke, lengue, apa and doussi.

19.4.1 Oil Extraction

The seeds are unshelled, dried and crushed. The oil is extracted in petroleum ether (60–80 °C) and the solvent then distilled off at 80 °C (Ajiwe et al. 1995). The percentage yield of the oil is 25.8 ± 2.0. This oil is light yellow in colour and does not solidify at room temp. It is degummed to remove the phospholipids. To every lit. of oil 300 mL of water is added and stirred vigorously for 5 min. It is then loaded in a

Table 19.23 Physicochemical properties of the oil from *A. africana*

Parameters	Ogbu and Ajiwe (2016)	Otori et al. (2018)
Kinematic viscosity at 25 °C ($mm^2\ s^{-1}$)	–	30.37
Specific gravity	0.897	0.806
Acid value (g KOH kg^{-1})	–	4.4
Free fatty acid (%)	18.29	2.2
Iodine value (g I_2/100 g)	120.09	121.2

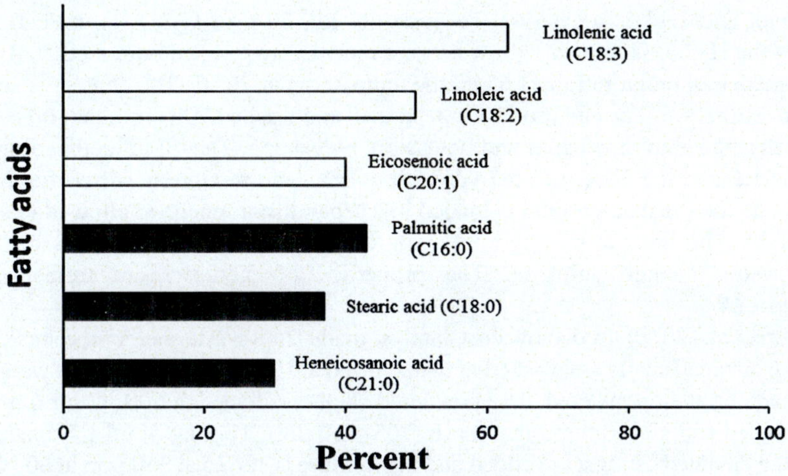

Fig. 19.42 Fatty acid composition of the oil from *A. africana*

separatory funnel and the aqueous layer containing the waste is discarded. This process is repeated thrice to ensure complete removal of the gum (Otori et al. 2018). This oil is neutralized by adding sodium hydroxide. The oil is first heated to 150 °C and sufficient sodium hydroxide (0.1 M) is added to neutralize the free fatty acid. The soap formed in it is precipitated by adding sodium chloride and the precipitate is separated and washed out. The properties of this oil are presented in Table 19.23.

The acid value is 4.4 g KOH kg^{-1}. This indicates the total acidity levels of the constituent fatty acids which make the glyceride molecule. It is also an index of the degradation of the triglyceride in the oil which would have decomposed into free fatty acids. It also refers to the level of rancidity and a measure of the freshness of the oil. The iodine value is similar to that of the dehydrated castor oil. The density and specific gravity are important parameters to be considered as the injection of fuel is measured in volume and not by weight. The viscosity also is high. The fatty acid composition of the oil is presented in Fig. 19.42. The unsaturated fatty acids comprise linolenic acid, linoleic acid and eicosenoic acid and the saturated fatty acid is formed of palmitic acid, stearic acid and heneicosanoic acid. The oil shows tendency to resist the formation of hydroperoxides compared to that of the methyl esters. This may be due to the natural antioxidants present in the vegetable oils which would have been lost during the transesterification process. (Ogbu and Ajiwe 2016).

19.4.2 Biodiesel Production

As the oil contains high free fatty acid (>1%), the transesterification is needed to be preceded with an esterification step wherein the free fatty acids are esterified using methanol and acid as a catalyst. To every lit. oil, 30 g of H_2SO_4 is used. Before adding the H_2SO_4 is diluted with water at a ratio 1:2 and mixed with methanol. The oil to methanol molar ratio is 1:6 and the temp. is set at 70 °C. The content is stirred for 3 h at 200 rpm. The end product is separated and washed. This resultant oil is then transesterified using methanol and potassium hydroxide. The oil to methanol molar ratio is 1:6 and the conc. of pot.hydroxide is 1%. The mixture is stirred for 2 h at 60 °C and the finished product is loaded in a separatory funnel and allowed to settle overnight. The layer at the top containing the biodiesel is separated, washed and dried using sodium sulphate. The properties of biodiesel are furnished in Table 19.24.

Otori et al. (2018) used a low cost catalyst made from waste shell of the snail. The shells from dead snails are washed in water, dried at 120 °C until a constant weight is obtained and then powdered. The powdered shell is calcined at 600 °C for 6 h. It is then cooled and sieved through a mesh (105–110 μm). To 1 kg of oil 1 L methanol and 30 g calcined catalyst are added and the mixture is stirred at 350 rpm at 60 °C for 60 min. Consequently the mixture is taken in a separatory funnel and the top layer containing the biodiesel is separated, washed and dried. The yield is around 85%. The viscosity (at 40 °C) and acid values of the biodiesels are higher than the standards specified in ASTM D 6751 and D 7467. Blends below B20 are recommended to meet the D 7467 specification (Ogbu and Ajiwe 2016).

The kinematic viscosity and acid value of biodiesel produced by one-step process using CaO as catalyst meet the ASTM D6751 standard (1.9–6.0 $mm^2 s^{-1}$ and 0.03

Table 19.24 The physicochemical properties of *A. africana* biodiesel

Parameters	Ogbu and Ajiwe (2016)	Otori et al. (2018)
Kinematic viscosity at 40 °C ($mm^2 s^{-1}$)	6.39	4.49
Relative density at 28 °C	0.92	–
Acid value (g KOH kg^{-1})	0.56	0.03
Flash point (°C)	108	165.17
Cetane number	–	51.2
Pour point (°C)	10	−7.8
Palmitic acid (C16:0) (%)	12.24	–
Stearic acid (C18:0) (%)	8.76	–
Heneicosanoic acid (C21:0) (%)	9.86	–
Linoleic acid (C18:2) (%)	26.23	–
Linolenic acid (C18:3) (%)	36.32	–

KOH kg^{-1}), whereas two stage process (esterification and transesterification) produces biodiesel of high viscosity (6.39 $mm^2\ s^{-1}$) and acid value (0.56 g KOH kg^{-1}). It has low flash point (108 °C) which does not meet the ASTM standard D6751 (Min. 130 °C), but it is advantageous as compared to high flash point 165.17 °C obtained in CaO and the flash point of conventional diesel is 70.5 °C (Thangaraj and Solomon 2020). Thus the catalyst influences the properties of biodiesel (Ogbu and Ajiwe 2016; Otori et al. 2018).

19.5 Babul (*Acacia nilotica*)

The babul *Acacia nilotica* is a medium sized plant which occurs in waste lands and grows profusely if the conditions are supportive or may remain shunted if the agro-climatic factors are unfavourable. It grows well at an annual rainfall of 600–2300 mm and a temperature range of 15–28 °C. The leaves are 5–15 cm long with compound and alternate leaflets. Each leaflet is 1.5–7 mm long and 0.5–2 mm broad. The flowers are scented and golden yellow in colour. The fruit is a pod of 4–22 cm long and 1–2 cm broad. The pod holds 5–15 flattened bean shaped seeds. The pod is necklace shaped (Fig. 19.43).

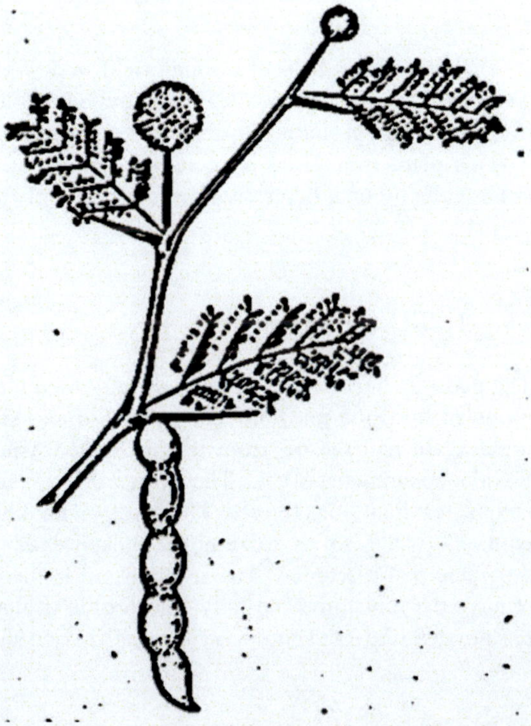

Fig. 19.43 The babul *A. nilotica*

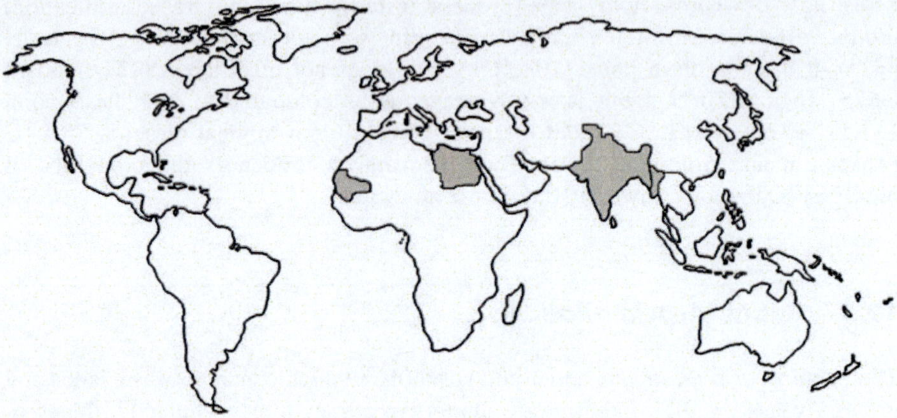

Fig. 19.44 The geographical distribution of *A. nilotica* (indicated by the dark shade)

Table 19.25 Fatty acid profile of the oil from *A. nilotica* (Adhikesavan et al. 2015)

Fatty acids	(%)
Palmitic acid (C16:0)	15.7
Stearic acid (C18:0)	9.0
Arachidic acid (C20:0)	1.2
Vaccenic acid (C18:1)	29.0
Linoleic acid (C18:2)	44.5

The tree begins to bear normal yield in 5–7 years and on an average produces 18 kg pods a year. This species is distributed within 30°N and 20°S and richly occurs in Senegal, Egypt, India, Burma, Sri Lanka and Pakistan (Fig. 19.44).

This plant is referred by certain common names: Babul, babool, gum arabic, kikar, nalla tumma, babli, thorn mimosa, prickly acacia, black piquant and karuvel.

19.5.1 Oil Extraction

The pods are harvested from the tree and dried repeatedly in hot sun. The seeds are removed from the pod, broken, powdered and sieved. The sieved seed powder is dried again and the oil from it is extracted using organic solvent *n*-hexane in a Soxhlet apparatus at 50 °C. The extract thus obtained is heated to expel the solvent and the resultant oil is filtered. The crude oil obtained is then degummed by stirring it with 2% water so as to remove phospholipids. The waste generated is phase separated and discarded. The resultant oil is then treated with sodium sulphate to remove the moisture. The fatty acid profile of the oil is given is Table 19.25. The unsaturated fatty acid component forms the rich share among the fatty acid contents.

19.5.2 Biodiesel Production

The oil has a high level of free fatty acid (Adhikesavan et al. 2015) and therefore two-step process is being followed. Sulphuric acid at the rate of 1% is mixed with adequate quantity of methanol to enable a molar ratio of 9:1 with oil. This mixture is then added to the preheated oil and stirred well for 2 h. Then it is transferred to a separatory funnel for phase separation from which the esterified oil is removed, washed and dried. Consequently transesterification reaction is carried out using 1% potassium hydroxide at a methanol oil ratio of 6:1, at 60 °C with a stirring rpm of 600 for an hour. The content is then allowed to settle.

The top layer having the biodiesel is separated washed and dried. Various concentrations of potassium hydroxide as catalyst (0.5, 1.0, 1.5 and 2.0% w/w) were attempted. The optimum concentration of 1% by weight gave a yield above 85%, whereas high concentration adversely affected the yield due to saponification (Adhikesavan et al. 2015).

The physicochemical properties and the ester profile of the biodiesel are given in Table 19.26.

The biodiesel produced out of the two stage process has its viscosity, density, acid value, cetane number and flash point agree well with that of the ASTM D6751 and EN 14214 standards.

Garba et al. (2018) employed extracellular enzyme to transesterify *A. nilotica* oil. At the outset the enzyme is prepared as per Shukla et al. (2007). The culture Pseudomonas aeruginosa is inoculated in a medium containing peptone, yeast extract, sodium chloride, magnesium sulphate and olive oil and incubated for 60 h at 37° C. At the end, the broth is centrifuged at 6000 rpm for 10 min. The solid which settled at the bottom is resuspended in 50 mM phosphate buffer (pH 8). The above material is then precipitated in 85% conc. ammonium sulphate solution. The precipitate obtained is dissolved in 0.05 molar tris-HCl buffer (pH 8) and dialyzed against the same buffer. The above material is then loaded in diethylaminoethyl cellulose column 50 × 25 cm pre-equilibrated with tris-HCl and eluted with a linear gradient of 0.1–1 mM sodium hydrochloride.

The lipase thus obtained is immobilized as per Bhushan et al. (2008). Five grams of sodium alginate is dissolved in 30 mL distilled water and autoclaved at 15 pounds

Table 19.26 The physicochemical properties and the ester profile of the biodiesel from *A. nilotica* oil prepared by a two stage process (Adhikesavan et al. 2015)

Parameters	Value
Kinematic viscosity ($mm^2\ s^{-1}$)	4.2
Density ($kg\ m^{-3}$)	890
Acid value ($g\ KOH\ kg^{-1}$)	0.19
Flash point (°C)	160
Cetane number	51
Palmitic acid methyl ester (%)	13.8
Stearic acid methyl ester (%)	4.88
Elaidic acid methyl ester (%)	11.2
Linoleic acid methyl ester (%)	68.3

Table 19.27 The properties of the biodiesel prepared due to enzyme catalysis (Garba et al. 2018)

Parameters	Value
Kinematic viscosity at 40 °C (cSt)	2.73
Refractive index	1.34
Flash point (°C)	113
Cloud point (°C)	1.06
Pour point (°C)	4.20
Iodine value (g I_2/100 g)	167.6
Acid value (g KOH kg^{-1})	0.61
Peroxide value (m eq kg^{-1})	6.6
Cetane number	32.8
Heating value (MJ kg^{-1})	29.0

per inch pressure for 10–15 min. It is then cooled and mixed with the prepared enzyme at the rate of 5 mL of the lipase in 20 mL of sodium alginate liquid. This mixture is then made to fall drop wise in 0.4 M calcium chloride solution and kept immersed for 1 h for hardening. The beads thus formed is rinsed in distilled water and stored in 0.05 M tris HCl buffer (pH 8.0) at 4 °C until used. The above enzyme is mixed with *A. nilotica* oil at the rate of 64.0 μg mL^{-1} and the transesterification is carried out under mild stirring (150 rpm) for 195 min at 45 °C. Various parameters such as alginate and $CaCl_2$ concentration, lipase units loading and bead size were investigated to obtain maximum immobilization yield. The activity of entrapped lipase was maximum at 1.5% alginate concentration with 2% $CaCl_2$ and decreased with increase in the amount of alginate. The size of the alginate beads affected the enzyme immobilized. The two different sizes of beads 1.2 and 2.1 mm were tried among them, 1.2 mm beads have 40% catalytic activity as small beads have high surface to volume ratio. The immobilized enzyme was observed to be stable under ideal operational conditions such as thermal stability, pH and storage. The immobilized enzyme has better catalytic activity than free enzyme. Free enzyme loses its activity completely beyond 40 °C, whereas the immobilized enzyme retains the activity even at 50–80 °C.

The properties of the biodiesel obtained due to enzymatic transesterification are given in Table 19.27.

The viscosity of the fuel is paradoxically low which indicates that there is substantial reduction in the lubricity which is likely to bring in friction and wear scar. Low viscosity may also cause leak in the fuel system Garba et al. (2018).

The low temperature properties of the fuel such as cloud point and pour point pose limitations. The iodine number is the crude reference to the degree of unsaturation. The value on refractive index is decided by the combination of molecular weight, fatty acid chain length and the degree of unsaturation (Sadrolhosseini et al. 2011). Low refractive index value in this case is due to the low viscosity. High iodine number may create deposit in the injectors (Samniang et al. 2014). The cetane number is lower than that of the standard which implies that there will be abnormality in startup, noise generation, knocking effect and smoke in the exhaust. The calorific value also is low and thus infers that this biodiesel may not offer a good fuel and in such event blending is one of the solutions.

19.6 Diesel Tree (*Copaifera langsdorffii*)

The diesel tree *Copaifera langsdorffii* is a medium sized tree of evergreen forest of the tropical and subtropical region. *C. nitida* and *C. sellowii* are the synonyms of this species. Though there are 35 species listed in the genus *Copaifera*, 4 are considered native to Africa. *C. langsdorffii* does not grow outside the tropics. This tree is called by different local names as shown hereunder (Table 19.28).

This tree grows to a height of around 15 m. The trunk is around 1 m diameter. The leaves are compound and paripinnate. The leaf is 5–13 cm long. Small whitish flowers are arranged in auxiliary panicle of 3–11 cm length. The fruit is egg shaped and 2.5–3.5 cm long and 2.5 cm wide. Each fruit carries a single large black seed of 0.45–0.7 g (Fig. 19.45). This species grows well in Venezuela, Guyana, Brazil, Bolivia, Paraguay and Argentina. It is also distributed in Congo, Cameroon, Guinea and Angola (Fig. 19.46).

Table 19.28 Local names of *C. langsdorffii* in different languages

Languages	Names
English	Diesel tree
Paraguay	Amacey, Copaiba, Kerosene tree
Portuguese	Caobi, capaiba, capaiuba, copai
Spanish	Cabismo

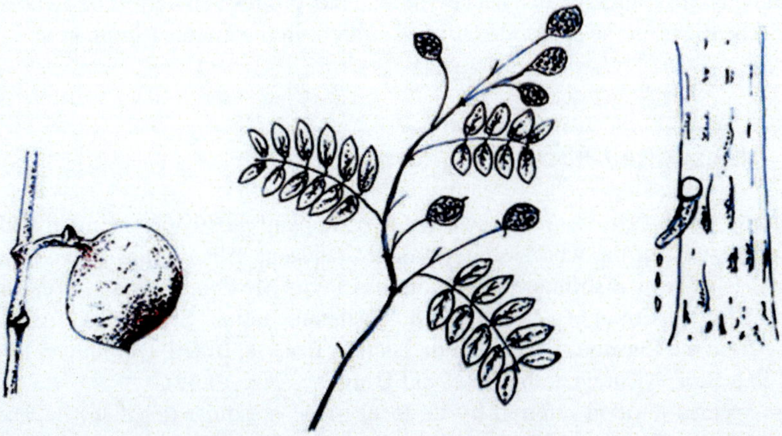

Fig. 19.45 The diesel tree *Copaifera langsdorffii*

Fig. 19.46 Geographical distribution of *Copaifera langsdorffii* (indicated by dark shade)

19.6.1 Biodiesel

The honey combed porous wood of the trunk holds a specialized oil in its capillary space. This oil is produced by the resiniferous channels of the tree (de Almeida et al. 2016). The secretion of the tree is oleoresin in nature. Its quality is on par with turpentine which breaks down to methanol and other simple compounds in due course. This oleoresin is highly transparent and yellowish to light brown in colour. Normally a hole of 4–5 cm diameter is made at the bark and a tunnel is made to touch the central core (inner heartwood) of the trunk through which 15–20 L of the resin is collected once in 6 months. Larger and smaller trees yield less oleoresin. In younger trees, since the heart wood is not developed in full the resin secretion is limited. It is demonstrated that this oleoresin can be directly fed into the fuel tank of the vehicle after it is freed from any suspended particles using a suitable filter (Duke 1983). It is reported that the plantation in a hectare of land can produce 25 barrels of oleoresin in a year. The oil from the seeds also is used straight in the engine (Stupp et al. 2008).

19.7 Mesquite (*Prosopis juliflora*)

The thorny shrub *Prosopis juliflora* is a universal plant mostly occurring in arid and semiarid regions of the world. Being drought resistant, it is used as a live fence to gardens. It is seen distributed in Colombia, Peru, Mexico, Ecuador, Venezuela, Sudan, Eritrea, Ethiopia, Kenya, Iraq, Pakistan, India, Sri Lanka, Australia, South Africa the Caribbean, the Atlantic Islands, Bolivia, Brazil, Dominican Republic, El Salvador, Nicaragua, the USA and Uruguay (Fig. 19.47).

This species is often referred by its local name mesquite (English) bayahonde blanca, Algarroba (Spanish), bayarone Francais (French), bayawonn (Creole), vilayati babul, vilayati khejra, gando baval, vilayati kikar, babul, kubli kikar, jaali,

Fig. 19.47 The geographical distribution of *Prosopis juliflora* (indicated by dark shade)

seemai karuvel, velikathan, mulla tumma, sarkar tumma (India) and uanni andara (Sri Lanka).

This plant grows up to 12 m in height and 1.2 m in dia. It is deciduous and the leaf has 12–20 leaflets. The flowers are straw yellow in colour and appear as a string of 4–8 cm long. The pod is 20–30 cm long, 0.5–1.6 cm wide and 0.4–0.9 cm thick each with 10–30 seeds. The seeds are protected by hard shell so that they remain viable even for 10 years. The seed is ovoid, brown and embedded in pulpy mesocarp. The seed is up to 0.65 cm dia and weigh 250–300 mg each. (250–300 seeds per kg) (Fig. 19.48).

19.7.1 Oil Extraction

The pods are dried repeatedly and powdered in a mill. The oil is extracted using organic solvents. Many polar and non-polar organic solvents (petroleum ether, *n*-pentane, ethyl acetate, isopropanol, *n*-hexane, methanol, ethanol) are tried and the best yield is reported in methanol (Rajeshwaran et al. 2016). The oil recovery ranges from 10% to 37%. Chosen properties of this oil are given in Table 19.29.

It is to be noted that the acid value is very high. In such high levels the transesterification process gets complicated due to the formation of soap if alkali is used as a catalyst. The soap thus formed creates problem in the separation of ester from glycerin. Therefore, at the outset the raw oil is treated with acid at the rate of 10 g H_2SO_4 for every litre of oil so as to reduce the acid value. To the preheated (110 °C) crude oil, acid and methanol (methanol to oil volumetric ratio 9:1) are added and stirred for 2 h at 600 rpm. The reacted contents are then poured in a separatory funnel. Excess methanol and acid move to the top and the ester stays as a bottom layer which is then separated. In this process the acid value may be reduced to anywhere around 10 g KOH kg^{-1}. Therefore it is again treated with acid so as to

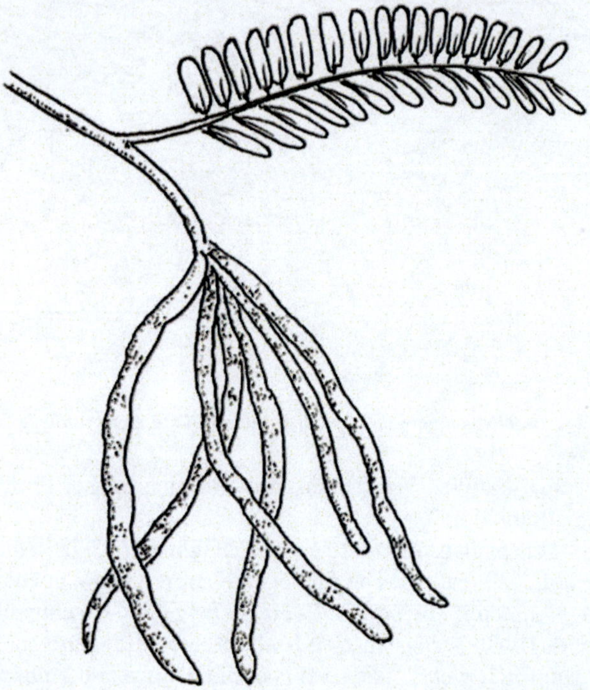

Fig. 19.48 The mesquite *Prosopis juliflora*

Table 19.29 The physicochemical properties of the oil from *Prosopis juliflora* (Rajeshwaran et al. 2016)

Parameters	Value
Kinematic viscosity at 40 °C (mm^2 s^{-1})	38–41.2
Density (kg m^{-3})	967–971
Acid value (g KOH kg^{-1})	39–43.7
Flash point (°C)	202–212
Calorific value (MJ kg^{-1})	38–41
Iodine value (I_2 g/100 g)	102–112
Palmitic acid (C16:0) (%)	10.6
Stearic acid (C18:0) (%)	5.2
Oleic acid (C18:1) (%)	34.7
Linoleic acid (C18:2) (%)	43.4

get the acid value of the oil around 1–3 g KOH kg^{-1}. The oil thus obtained is then used for alkali transesterification.

19.7.2 Biodiesel Production

The oil being the reaction product of the acid treatment, having a low acid value is used in the biodiesel production through alkali transesterification. To the above,

Table 19.30 Physicochemical properties of biodiesel prepared from the oil of *Prosopis juliflora* (Rajeshwaran et al. 2016)

Parameters	Value
Kinematic viscosity at 40 °C (mm^2 s^{-1})	4.9
Density (kg m^{-3})	893
Acid value (g KOH kg^{-1})	2.7
Flash point (°C)	120
Cloud point (°C)	4
Heating value (MJ kg^{-1})	39
Iodine value (I_2 g/100 g)	87

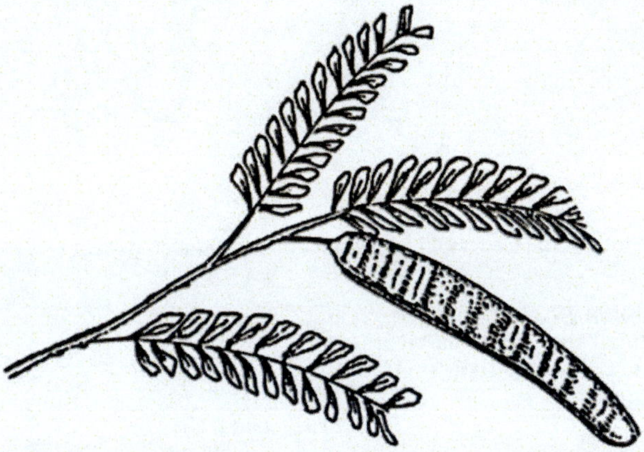

Fig. 19.49 The shikakai, *Acacia concinna*

NaOH at the rate of 10 g for every litre and methanol so as to form a molar ratio of 9:1 with oil are added, heated to 60 °C and stirred at the rate of 600 rpm for 2 h. In this process around 90% conversion is possible. The properties of the biodiesel obtained are given in Table 19.30.

Table 19.30 indicates that the oil from *P. juliflora* can be best utilized in biodiesel. It is reported that around 6.3 million litre of *P. juliflora* oil can be obtained from 1 million ha of plantation which is equivalent to the use of 4.4 million litres of biodiesel.

19.8 Shikakai (*Acacia concinna*)

The shikakai, *Acacia concinna* is a climbing shrub, and its seed oil is considered as a source of raw material for biodiesel. Very limited works are available on biodiesel from this species. The leaf stalk is 1–1.5 cm long and is double pinnate with 5–7 pairs of pinnae. Each pinna has 12–18 pairs of leaflets. The leaves of this plant are with caducous stipules (Fig. 19.49).

This plant is popularly called as shikakai, seege, soap pod, sheegae, reetha and sheekay. It grows in hot dry shrub jungles of Central Asia and the Far East (Fig. 19.50).

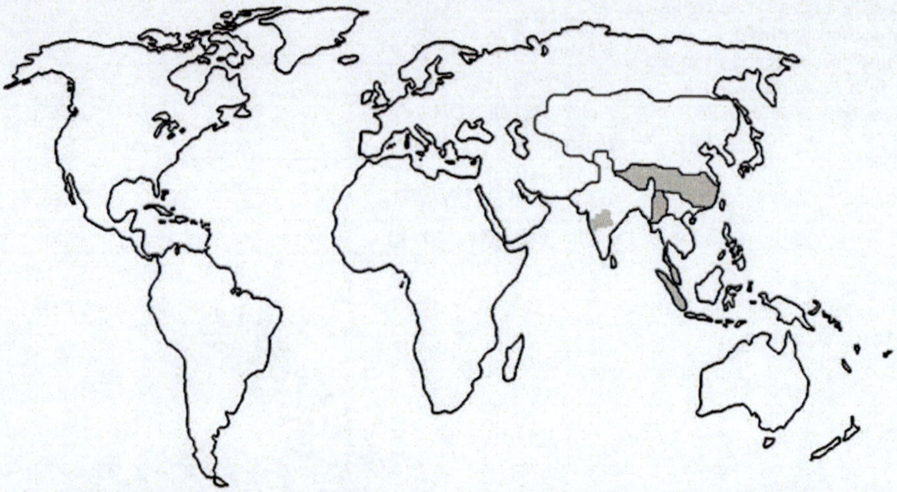

Fig. 19.50 Geographical distribution of shikakai, *Acacia concinna* (indicated by dark shade)

Table 19.31 Physicochemical properties of the oil of *A. concinna* (Saxena et al. 2018)

Parameters	
Kinematic viscosity at 40 °C ($mm^2\ s^{-1}$)	35.4
Density ($kg\ m^{-3}$)	924
Flash point (°C)	186
Free fatty acid (%)	8.9
Cloud point (°C)	15.5
Pour point (°C)	11.4
Heating value ($MJ\ kg^{-1}$)	32.8
Palmitic acid (C16:0) (%)	12.8
Stearic acid (C18:0) (%)	7.3
Oleic acid (C18:1) (%)	21.0
Linoleic acid (C18:2) (%)	41.6
Linolenic acid (C18:3) (%)	15.5
Average molecular weight ($g\ mol^{-1}$)	919

The flowers are bisexual, cream or white in colour. The fruit is a pod of 15 cm long and 2.5 cm wide. Each pad has 10–14 seeds. The pod yield is around 80–120 kg per shrub per year, and the oil extracted is being considered as a raw material for biodiesel. The crude oil comprises around 20% saturated fatty acids and 78% unsaturated fatty acids.

19.8.1 Oil Extraction

Oil is extracted by pressing the seed. The properties of the crude oil are given in Table 19.31.

The viscosity, density, flash point, free fatty acid and cloud point are higher than the standard.

19.8.2 Biodiesel Production

As the free fatty acid of the oil is high, double stage process (esterification and transesterification) is recommended. Normally the esterification is done using H_2SO_4 as a catalyst. But in the present case physical separation process is done (Saxena et al. 2018). The oil is mixed with hot water at 1:3 ratio by volume and stirred vigorously for 15 min. It is then allowed to settle in a separatory funnel for 4 h (Saxena et al. 2018). The FFA comes down from the original value and the yield of esterified oil is 85–90%. The oil thus purified is then separated and subsequently used for alkali based transesterification. The potassium hydroxide at a conc. of 1% of the whole mixture is first mixed with methanol so as to form a final molar ratio of 1:8 (oil to methanol). It is then poured into hot (60 °C) oil and stirred for 1 h. The content is then transferred to a separatory funnel and waited till the phase separation completes. The biodiesel moves to the top and is removed from the funnel, washed repeatedly to free it from alkali and then heated to expel methanol and dried using sodium sulphate. The properties of the biodiesel are given in Table 19.32.

The transesterification process has brought down the viscosity, density, flash point, free fatty acid, pour point, cloud point and the average molecular weight by 84, 3.5, 26, 96, 46, 30, 67% from that of the crude oil. The heating value of the biodiesel increased by 5% from that of the oil.

Table 19.32 Physicochemical properties of the biodiesel from the oil of *A. concinna* (Saxena et al. 2018)

Parameters	Value
Kinematic viscosity at 40 °C (mm^2 s^{-1})	5.76
Density (kg m^{-3})	892
Flash point (°C)	138
Acid value (g KOH kg^{-1})	0.7
Free fatty acid (%)	0.35
Cloud point (°C)	10.8
Pour point (°C)	6.2
Cetane number	54.98
Heating value (MJ kg^{-1})	37.95
Average molecular weight (g mol^{-1})	307.7
Palmitic acid (C16:0) (%)	8.6
Stearic acid (C18:0) (%)	4.9
Oleic acid (C18:1) (%)	17.2
Linoleic acid (C18:2) (%)	52.4
Linolenic acid (C18:3) (%)	10.8

19.9 Shittim (*Acacia raddiana*)

The shittim tree *Acacia raddiana* is one among the 300-odd species of the genus *Acacia*. The shittim wood was used to make the *Ark of Covenant* of the Tabernacle—the earthly abode of the Lord (Exodus 25: 10–22 of the Holy Bible) and used to worship by the Israelites. Its common names are tahi, talah, talh alluki, aluki and twisted acacia. It grows as a solitary tree widely spaced in deserts of the elevation between 0 and 2100 m above MSL. It is geographically represented in Senegal, Somalia, Africa, Middle East and Arabia (Fig. 19.51). The tree crown looks like an umbrella. The leaf is alternate, bipinnate and the fruit is long and curly. The tree has huge sharp thorns all over the body (Fig. 19.52).

19.9.1 Oil Extraction

The seeds are separated from the pods, dried repeatedly in hot sun, powdered and the oil extracted from it using a non-polar organic solvent *n*-hexane for more than 6 h. The extract is then vacuum distilled to separate the oil from the hexane. Chosen characteristics of this oil are given in Table 19.33.

The oil yield from *A. raddiana* seeds is around 11% only which is considered to be higher than that of *A. nilotica*, *A. longifolia*, *A. dealbata*, *A. melanoxylon* and *A. retinodes*. The share of unsaturated fatty acids (55%) is higher than saturated fatty acids.

Fig. 19.51 Geographical distribution of *Acacia raddiana* (indicated by dark shade)

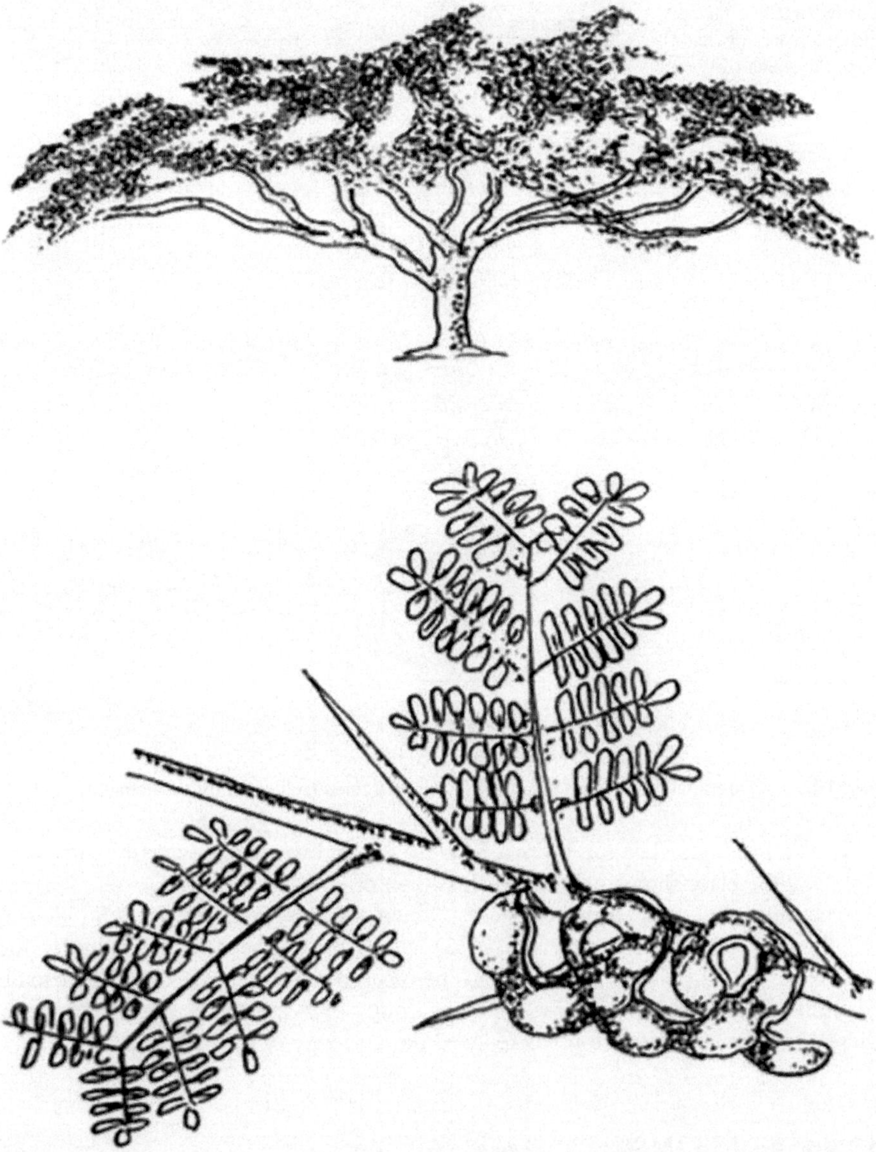

Fig. 19.52 *Acacia raddiana* tree

19.9.2 Biodiesel Preparation

The oil is allowed to react with methanol at oil to molar ratio of 1:5. The reaction is catalysed by sodium hydroxide at 0.1% of the oil. The mixture is stored for 4 h at a temperature of 55 °C; by then the ester is formed which on settlement separates and

Table 19.33 Chosen characteristics of the oil from the seeds of *A. raddiana* (Cheriti et al. 2011)

Parameters	Value
Kinematic viscosity at 40 °C (cSt)	37.4
Density (kg m^{-3})	912
Acid value (g KOH kg^{-1})	0.4
Iodine value (g I_2/100 g)	63.3
Palmitic acid (C16:0) (%)	38.3
Stearic acid (C18:0) (%)	6
Oleic acid (C18:1) (%)	34
Linoleic acid (C18:2) (%)	21

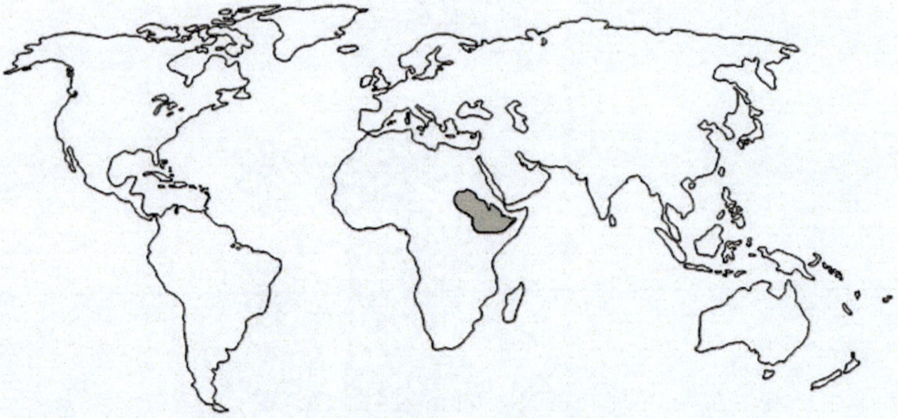

Fig. 19.53 Geographical distribution of *Millettia ferruginea* (indicated by dark shade)

stays as a top layer. The top layer is brought out and washed repeatedly with warm water till the ester shows neutral pH. It is then dried using $MgSO_4$.

The biodiesel thus prepared has a kinematic viscosity of 4.1 cSt, density 876 kg m^{-3} and an iodine value of 63 g I_2/100 g. The calorific value is low (32.78 MJ kg^{-1}) (Cheriti et al. 2011). The oxygen present in the ester has actually brought down the calorific value which is around 29% lower than that of the petro-diesel. The transesterification has reduced the viscosity of oil by 89%.

19.10 Brebra (*Millettia ferruginea*)

Millettia ferruginea is a species well established in Ethiopia (Fig. 19.53) and its adjacent area and grows resisting heat, drought, salinity, grazing and repeated cutting. It prefers to establish at an elevation of 1000–2600 m above mean sea level. The leaves are compound with many leaflets.

Each leaflet is around 9 cm long and 3 cm wide. The flowers are fairly large (2–3 cm dia.) and violet in colour. The pod, which carries 5–10 oil bearing seeds is 25 cm long and 3 cm wide (Fig. 19.54).

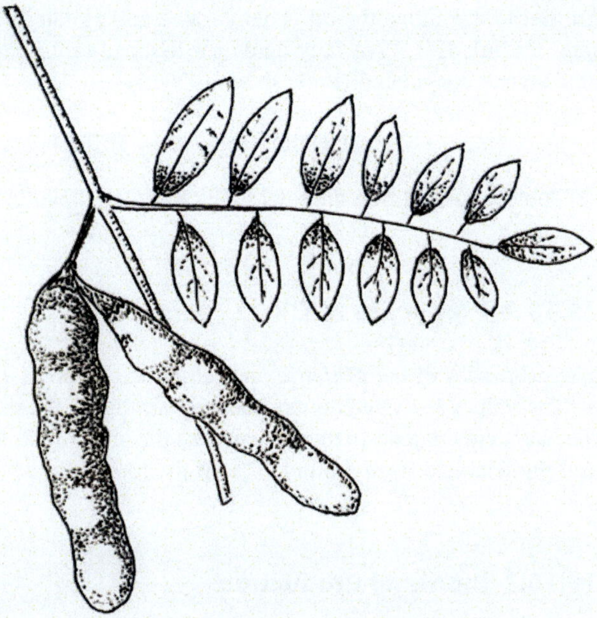

Fig. 19.54 The brebra *Millettia ferruginea*

This species is referred popularly as asra aladu, brebra, birbra, birbiraso, bibero, dedatu, ingidicho, sotellu, zagie and zagiya. The seed is reported to have as high as 48.5% oil (Andualem and Gessesse 2012a). It is reported that *M. ferruginea* produces 1350 kg of seed per hectare. Each tree yields 150 kg of pod and 37.5 kg of seed. The tree density is 36 ha^{-1}.

19.10.1 Oil Extraction

The oil from the seeds of *M. ferruginea* is extracted either by the conventional pressing or by *n*-hexane extraction. The oil recovery is reported to be 12.2% and 15.3% in pressing and in solvent (*n*-hexane) extraction processes, respectively (Andualem and Gessesse 2012a). The crude oil normally contains phospholipids (hydratable phospholipid-HPL and non-hydratable phospholipid-NHPL) and they are removed by degumming. To every 100 mL of oil 2 mL of hot sulphuric acid (0.1%) is added, stirred for 60 min at 45 °C and centrifuged to separate the phospholipid. The resultant oil is again treated with 2 mL of 0.1% hot phosphoric acid, stirred and centrifuged to separate the residual phospholipids. The degummed oil thus obtained is then refined. In order to refine the above material the hexane-oil mixture is mixed with equal volume of 90% ethanol and stirred. It is then transferred to a separatory funnel for phase separation. The hexane-oil mixture sinks down to the bottom which is then separated. The supernatant layer is discarded. The hexane-oil mixture is distilled and the oil recovered. The oil thus obtained is yellow in colour.

Further to the above the oil is dehydrated and agitated with hydrochloric acid at the rate of 4 mL L^{-1}. Free fatty acids get reduced to a minimum level as the free fatty acid stands coagulated (Prakash et al. 2016).

$$RCOOH + NaOH \rightarrow RCOONa + H_2O$$

The coagulated free fatty acid which settles down is filtered and discarded.

19.10.2 Properties of Oil

The properties of the *Millettia ferruginea* are given in Table 19.34.

The values for viscosity and density are high. These properties are governed by the fatty acid content. In the present case the unsaturated fatty acids such as oleic acid and linoleic acid form around 77% of the total fatty acids.

19.10.3 Biodiesel Production

To 100 g oil, methanol–KOH solution (22.6 g methanol and 1 g KOH) is added. The addition is made in full or in two equal split doses. The content is stirred for 2 h and at the end it is transferred to a separatory funnel and allowed to stand overnight for the phase separation. In the event of considering the transesterification with the split doses of methanol–KOH solution the phase separation is done at each stage. The glycerol is discarded and the methyl ester phase is separated and neutralized using 6 N HCl. The methanol if retained in the ester is distilled out and the ester is washed with warm (50 °C) water (Andualem and Gessesse 2012b).

19.10.4 Properties of Biodiesel

The physicochemical properties of *M. ferruginea* biodiesel are presented in Table 19.35.

Table 19.34 Physicochemical properties of the oil of *Millettia ferruginea* (Prakash et al. 2016)

Parameters	Value
Kinematic viscosity at 31 °C (cSt)	40.4
Density (kg m^{-3})	920
Acid value (g KOH kg^{-1})	38.2
Iodine value (I_2 g/100 g)	101.7
Palmitic acid (C16:0) (%)	4.2
Stearic acid (C18:0) (%)	6.9
Oleic acid (C18:1) (%)	43.1
Linoleic acid (C18:2) (%)	34.3

Table 19.35 The physicochemical properties of *M. ferruginea* biodiesel (Prakash et al. 2016)

Parameters	Value
Kinematic viscosity at 40 °C (cSt)	5.06
Density at 15 °C (kg m^{-3})	890
Acid value (g KOH kg^{-1})	0.69
Flash point (°C)	145.8
Cloud point (°C)	26.0
Pour point (°C)	21.0
Cetane number	52
Iodine value (I_2 g/100 g)	104.5
Heating value (MJ kg^{-1})	37.6
Copper strip corrosion	1b

Though the viscosity, density, cetane number and iodine vale are satisfactory the cold temp. properties are beyond the limit.

References

Adhikesavan C, Rajagopal K, Selwinrajadurai J (2015) Production and characterization of biodiesel from *Acacia nilotica* seeds. Int J Chem Tech Res 8(2):854–859

Ahmad M, Rashid S, Khan MA, Zafar M, Sultana S, Gulzar S (2009) Optimization of basic catalyzed transesterification of peanut oil biodiesel. Afr J Biotechnol 8(3):441–446

Ajiwe VIE, Okeke CA, Agbo HU (1995) Extraction and utilization of *Afzelia africana* seed oil. Bioresour Technol 53:89–90

Ali Y, Hanna MA, Cuppett SL (1995) Fuel properties of tallow and soybean oil esters. J Am Oil Chem Soc 72:1557–1564

Al-Widyan MI, Tashtoush G, Abu-Qudais M (2002) Utilization of ethyl ester of waste vegetable oils as fuel in diesel engines. Fuel Process Technol 76(2):91–103

Andualem B, Gessesse A (2012a) Methods of refining of Brebra (*Millettia ferruginea*) oil for the production of biodiesel. World Appl Sci J 17(3):407–413

Andualem B, Gessesse A (2012b) Production and characterization of biodiesel from Brebra (*M. Ferruginea*) seed non-edible oil. Biotechnology 11(4):217–224

Anggraini AA, Wiederwertung VG (1999) Speiseolen/-fetten im Energetisch – Technischen Bereicg: Ein Verfahren und Dessen Bewertung. Department of AgrarTechnik, Universität Hohenheim, Witzenhausen, Germany, p 193

Anitha A, Dawn SS (2010) Spent groundnut oil for biodiesel production using supported heteropolyacids. In: Proc. IEEE, 2nd international conference on chemical, biological and environmental engineering (ICBEE) - Cairo, Egypt (11.2–2010.11.4), pp 317–321. https://doi.org/10.1109/ICBEE.2010.5651565

Anjana PA, Niju S, Meera Sheriffa Begum KM, Anantharaman N, Anand R, Babu D (2016) Studies on biodiesel production from Pongamia oil using heterogeneous catalyst and its effect on diesel engine performance and emission characteristics. Biofuels 7(4):377–387

Anonymous (2004) World geography of the peanut. University of Georgia. http://www.lanra.uga.edu/peanut/Knowledgebase/

Azad AK, Rasul MG, Giannangelo B, Islam R (2015) Comparative study of diesel engine performance and emission with soybean and waste oil biodiesel fuels. Int J Autom Mech Eng 12(1):2866–2881

Bello EI, Agge M (2012) Biodiesel production from groundnut oil. J Eng Trends Eng Appl Sci 3 (2):276–280

Bhushan I, Parshad R, Nabi-Qazi G, Gupta V (2008) Immobilisation of lipase by entrapment in ca-alginate beads. J Bioactive Compat Polym 23(6):552–562

Canakci M, Gerpen JHV (2003) Comparison of engine performance and emissions for petroleum diesel fuel, yellow grease, bio-diesel and soybean oil bio-diesel. Am Soc Agric Eng 46 (4):937–944

Chand P, Chintareddy VR, Verkade JG, Grewell D (2010) Enhancing bio-diesel production from soybean oil using ultrasounds. Energy Fuels 24(3):2010–2015

Cheriti A, Talhi MF, Belboukhari N, Belhadjadji Y, Ghezali S (2011) Biodiesel production by transesterification of *Acacia raddiana* seed oils. Chem Technol 6(1):13–17

Cravotto G, Boffa L, Mantegna S, Perego P, Avogadro M, Cintas P (2008) Improved extraction of vegetable oils under high intensity ultrasound and/or microwaves. Ultrason Sonochem 15 (5):898–902

Davis JP, Dean LO, Faircloth WH, Sanders TH (2008) Physical and chemical characterization of normal and high oleic oils from nine commercial cultivars of peanut. J Am Oil Chem Soc 85:235–243

Davis JP, Geller D, Faircloth WH, Sanders TH (2009) Comparisons of biodiesel produced from unrefined oils of different peanut cultivars. J Am Oil Chem Soc 86:353–361

de Almeida LFR, Portilla RDO, Bufalo J, Marques MOM, Facanali R, Frei F (2016) Non oxygenated sesquiterpenes in the essential oil of *Copaifera langsdorffii* Desf. Increase during the day in dry season. PLoS One 11(2):0149332. https://doi.org/10.1371/journal.pone.0149332

de Alvarado D (1995) Mechanical properties of vegetable oils and fats. Grasas Aceites 46:264–269

Dorado MP, Ballesteros E, de Almeida JA, Schellert C, Loehrlein HP, Krause R (2002) An alkali-catalyzed transesterification process for high free fatty acid waste oils. Trans ASAE 45 (3):525–529

Duke JA (1983) *Copaifera langsdorffii* Desf. Handbook of energy crops. https://hort.purdue.edu/newcrop/duke_energy/Copaifera_langsdorfii.html

Dwivedi G, Mahendra PS (2015) Investigation and improvement in cold flow properties of pongamia biodiesel. Waste Biomass Valorization 6:73–79

Encinar JM, Gonzalez JF, Rodriguez JJ, Tejedor A (2002) Biodiesel fuels from vegetable oils: transesterification of *Cynara cardunculus* L. oils with ethanol. Energy Fuel 16(2):443–450

Fabbri D, Bevoni V, Notari M, Rivetti F (2007) Properties of a potential bio-fuel obtained from soybean oil by transmethylation with dimethyl carbonate. Fuel 86(5):690–697

Faria AE, Marques JS, Dias IM, Andrae RDA, Suarez PAZ, Prado AGS (2009) Nanosized and reusable SiO_2/ZrO_2 catalyst for highly efficient biodiesel production by soybean transesterification. J Braz Chem Soc 20(9):1732–1737

Filippis PD, Giavarini C, Scarsella M, Sorrentino M (1995) Transesterification processes for vegetable oils: a simple control method of methyl ester content. J Am Oil Chem Soc 72 (11):1399–1404

Freedman B, Butterfield RO, Pryde EH (1986) Transesterification kinetics of soybean oil. J Am Oil Chem Soc 63:1375–1380

Fukuda H, Kondo A, Noda H (2001) Biodiesel fuel production by transesterification of oils. J Biosci Bioeng 92(5):405–416

Garba A, Sallan AB, Ibrahim S, Abarshi MM, Muhammad A, Galadima MS, Babangida S (2018) Biodiesel production by lipase mediated transesterification of *Acacia nilotica* seed oil. Nigerian J Basic Appl Sci 26(1):23–30

Gelbard G, Bres O, Vargas RM, Vielfaure F, Schuchardt UF (1995) ^{1}H nuclear magnetic resonance determination of the yield of the transesterification of rapeseed oil with methanol. J Am Oil Chem Soc 72(10):1239–1241

Georgogianni KG, Kontaminas MG, Pomonis PJ, Avlonitis D, Gergis V (2008) Conventional and insitu transesterification of sunflower seed oil for production of biodiesel. Fuel Process Technol 89(5):503–509

Goembira F, Saka S (2015) Advanced supercritical methyl acetate method for biodiesel production from *Pongamia pinnata* oil. Renew Energy 83:1245–1249

Gopinath A, Sukumar P, Nagarajan G (2010) Effect of unsaturated fatty acid esters of biodiesel fuels on combustion, performance and emission characteristics of a D1 diesel engine. Int J Energy Environ 1(3):411–430

Goswami AK, Usmani GA (2014) Characterization of biodiesel obtained from pure soybean oil and its various blends with petro diesel. Int J Innov Res Sci Eng Technol 3(9) ISSN:2319–8753

Graef GL, Miller LA, Fehr WR, Hammond EG (1985) Fatty acid development in a soybean mutant with high stearic acid. J Am Oil Chem Soc 62:773–775

Gunstone FD, Hamilton RJ (eds) (2001) Oleochemical manufacture and applications. Sheffield Academic Press, Liverpool, pp 107–163

Gupta A, Chandra R, Subbarao PMV, Vijay VK (2009) Kinetics of batch biomethanation process of jatropha and pongamia oil cakes and their co-digested substrates. J Sci Ind Res 68:624–629

Hammond EG, Johnson LA, Su C, Wang T, White PJ (2005) Soybean oil. In: Industrial oil and fat products. Wiley Online Library, Hoboken, NJ. https://doi.org/10.1002/047167849xbio041

Ibeto CN, Okoye COB, Ofoefule AU (2012) Comparative study of the physic chemical characterization of some oils as potential feedstock for biodiesel production. ISRN Renew Energy 2012:621518. https://doi.org/10.5402/2012/621518

Joshi H, Moser BR, Toler J, Walker T (2010) Preparation and fuel properties of mixtures of soybean oil methyl and ethyl esters. Biomass Bioenergy 34:14–20

Kao TH, Chien JT, Chen BH (2008) Extraction yield of isoflavones from soybean cake as affected by solvent and supercritical carbon dioxide. Food Chem 107(4):1728–1736

Karaosmanoglu F, Akdag A, Cigizoglu KB (1996) Biodiesel from rapeseed oil of Turkish origin as an alternative fuel. Appl Biochem Biotechnol 61(3):251–265

Kashyap MC, Agrawal YC, Ghosh PK, Jayas DS, Sarkar BC, Singh BPN (2007) Oil extraction rates of enzymatically hydrolyzed soybeans. J Food Eng 81(3):611–617

Kaya C, Hamamci C, Baysal A, Akba O, Erdogan S, Saydut A (2009) Methyl ester of peanut (*Arachis hypogaea* L.) seed oil as a potential feedstock for biodiesel production. Renew Energy 34(5):1257–1260

Knothe G (2005) Dependence of biodiesel fuel properties on the structure of fatty acid alkyl esters. Fuel Process Technol 86:1059–1070

Koc AB, McKenzie EH (2010) Effects of ultrasonification on glycerin separation during transesterification of soybean oil. Fuel Process Technol 91(7):743–748

Koc AB, Abdullah M, Fereidouni M (2011) Soybeans processing for biodiesel production. In: Ng T (ed) Soybean applications and technology. https://www.intechopen.com/books/soybean-applications-and-technology/soybeans-processing-for-biodiesel production

Koseki H, Natsume Y, Iwata Y (2001) Evaluation of the burning characteristics of vegetable oils in comparison with fuels and lubricating oils. J Fire Sci 19:31–44

Leung DYC, Guo Y (2006) Transesterification of neat and used frying oil: optimization for biodiesel production. Fuel Process Technol 87(10):883–890

Li H, Pordesimo L, Weiss J (2004) High intensity ultrasound – assisted extraction of oil from soybeans. Food Res Int 37(7):731–738

Liang YC, May CY, Foon CS, Ngan MA, Hock CC, Basiron Y (2006) The effect of natural and synthetic antioxidants on the oxidative stability of palm diesel. Fuel 85(5–6):867–870

Lin BF, Huang JH, Huang DY (2009) Experimental study of the effects of vegetable oil methyl ester on DI diesel engine performance characteristics and pollutant emissions. Fuel 88 (9):1779–1785

Liu CC, May WC, Liu TJ (2012) Transesterification of soybean oil using CsF/CaO catalysts. Energy Fuel 26(9):5400–5407

Luthria DL, Biswas R, Natarajan S (2007) Comparison of extraction solvents and techniques used for the assay of isoflavones from soybean. Food Chem 105(1):325–333

Manneganti V, Rachapudi BNP, Bethala LAPD (2014) SO_3H – carbon derived from glycerol. An efficient and recyclable catalyst for smooth and regioselective azidolysis of oxiranes in water. Eur J Chem 5(1):167–170

Martins MI, Pires RF, Alves MJ, Hori CE, Reis MHM, Cardoso VL (2013) Transesterification of soybean oil for biodiesel production using hydrotalcite as basic catalyst. Chem Eng Trans 32:817–822

Meher LC, Naik SN, Das LM (2004) Methanolysis of *Pongamia pinnata* oil for the production of biodiesel. J Sci Ind Res 63:913–918

Mendes MF, Pessova FLP, Uller AMC (2002) An economic evaluation based on an experimental study of the vitamin E concentration present in deodorizer distillate of soybean oil using supercritical CO_2. J Supercrit Fluids 23(3):257–265

Miller KS, Farkas BE, Singh RP (1994) Viscosity and heat transfer coefficient for canola, corn, palm and soybean oil. J Food Process Preserv 18:461–472

Mittelbach M (1990) Lipase catalysed alcoholysis of sunflower oil. J Am Oil Chem Soc 67:168–175

Munoz M, Moreno F, Morea J (2004) Emissions of an automobile diesel engine fueled with sunflower methyl ester. Trans Am Soc Agric Eng 47(1):5–11

Noureddini H, Zhu D (1997) Kinetics of transesterification of soybean oil. J Am Oil Chem Soc 74 (11):1457–1463

Ogbu IM, Ajiwe VIE (2016) FTIR studies of thermal stability of the oils and methyl esters from *Afzelia africana* and *Hura crepitans* seed. Renew Energy 96:203–208

Oghome PL (2012) Transesterification and optimization of groundnut oil using magnetic silver. Open Sci Reposit Eng 2012:e70081906. https://doi.org/10.7392/Engineering.70081906

Oniya OO, Bamgboye AI (2014) Production of biodiesel from groundnut (*Arachis hypogeae*, L.) oil. Agric Eng Int CIGR J 16(1):143–150

Osman M, Wani SP, Balloli SS, Sreedevi TK, Srinivasarao CH, D'Silva E (2009) Pongamia seed cake as a valuable source of plant nutrients for sustainable agriculture. Indian J Fert 5(2):25–26, 29–32

Otori AA, Mann A, Suleiman MAT, Egwim EC (2018) Synthesis of heterogeneous catalyst from waste snail shells for biodiesel production using *Afzelia africana* seed oil. Niger J Chem Res 23 (1):35–51

Park JY, Kim DK, Lee JP, Park SC, Kim YJ, Lee JS (2008) Blending effects of biodiesels on oxidation stability and low temperature flow properties. Bioresour Technol 99:1196–1203

Perez A, Casas A, Fernandez CM, Ramos MJ, Rodriguez L (2010) Winterization of peanut biodiesel to improve the cold flow properties. Bioresour Technol 101(19):7375–7381

Prakash SR, Aravindhkumar R, Madhav R, Vinoth B (2016) Biodiesel production from *Millettia pinnata* oil and its characterization. Int Res J Eng Technol 3(9):512–516

Pulkrabek WW (2004) Engineering fundamentals of the internal combustion engine, 2nd edn. Pearson Prentice Hall, Upper Saddle River, NJ

Rajeshwaran M, Raja K, Marimuthu M, Duraimurugan MD (2016) Biodiesel production and optimization from *Prosopis juliflora* oil - a three step method. J Adv Eng Technol 7(2):214–224

Ramadhas AS, Muraleedharan C, Jayaraj S (2005) Performance and emission evaluation of a diesel engine fueled with methyl esters of rubber seed oil. Renew Energy 30(12):1789–1800

Sadrolhosseini AR, Moksin MM, Nang HLL, Norozi M, Yunus MMW, Zakaria A (2011) Physical properties of normal grade biodiesel and winter grade biodiesel. Int J Mol Sci 12(4):2100–2111

SAEJ 1349: SAE standards (2008) Engine power test code – spark ignition and compression ignition – net power rating. SAE, Troy

Salgin U (2007) Extraction of jojoba seed oil using supercritical CO_2 + ethanol mixture in green and high-tech separation process. J Supercrit Fluids 39(3):330–337

Samniang A, Tipachan C, Kajorncheappunngam S (2014) Comparison of biodiesel production from crude jatropha oil and krating oil by supercritical methanol transesterification. Renew Energy 68:351–355

Santos BS, Capareda SC, Capunitan JA (2013) Engine performance and exhaust emissions of peanut oil biodiesel. J Sustain Bioenergy Syst 3(4):272–286

Satyanarayana CH, Rao PV (2009) Influence of key properties of pongamia biodiesel on performance, combustion and emission characteristics of a DI diesel engine. WSEAS Trans Heat Mass Transf 2(4):34–44

Saxena V, Kumar N, Saxena VK (2018) Biodiesel synthesis from *Acacia concinna* seed oil: a comprehensive study. Energy Sources A Recov Utiliz Environ Effects. https://doi.org/10.1080/15567036.2018,1486912

Schuchardt U, Sercheli R, Matheus VR (1998) Transesterification of vegetable oils: a review. J Braz Chem Soc 9(3):199–210

Schumacher LG, Borgelt SC, Fosseen D, Goetz W, Hires WG (1996) Heavy duty engine exhaust emission tests using methyl ester soybean oil/diesel fuel blends. Bioresour Technol 57(1):31–36

Seth S, Agrawal JC, Ghosh PK, Jayas DS, Singh BPN (2007) Oil extraction rates of soya bean using isopropyl alcohol as solvent. Biosyst Eng 97(2):209–217

Shukla P, Gupta K, Kango N (2007) Production of lipase by hyper-lipolytic *Rhizopus oryzae* KG 10 on low value oil emulsions. Resour J Microbiol 2:671–677

Silva GF, Camargo FL, Ferreira ALO (2010) Application of response surface methodology for optimization of biodiesel production by transesterification of soybean oil with ethanol. Fuel Process Technol. https://doi.org/10.1016/j.fuproc.2010.10.002

Stupp T, deFreitas RA, Sierakowski MR, Deschamps FC, Wisniewski A Jr, Biavatti MF (2008) Characterization and potential uses of *Copaifera langsdorffii* seeds and seed oil. Bioresour Technol 99:2659–2663

Thangaraj B, Solomon PR (2020) Scope of biodiesel from oils of woody plants: a review. Clean Energy 4(2):2–23

Uquiche E, Jerez M, Ortiz J (2008) Effect of pretreatment with microwaves on mechanical extraction yield and quality of vegetable oil from Chilean hazelnuts (*Gevuina avellana* Mol.). Innov Food Sci Emerg Technol 9(4):495–500

USDA (2021) Foreign Agricultural Service, Feb 2021

Wang PS, Thompson J, Clemente TE, Gerpen JHV (2010) Improving the fuel properties of soy biodiesel. Am Soc Agric Biol Eng 53(6):1853–1858

Wang JX, Chen KT, Huang ST, Chen KT, Chen CC (2012) Biodiesel production from soybean oil catalysed by Li_2CO_3. J Am Oil Chem Soc 89(9):1619–1625

Wesolowski WM (1993) Quality control of soybean oils by thermogravimetry. Food Sci Technol 95:377–383

Xue J, Grift TE, Hansen AC (2011) Effect of biodiesel on engine performances and emissions. Renew Sust Energ Rev 15(2):1098–1116

Yu D, Tian L, Wu H, Wang S, Wang Y, Ma D, Fang X (2010) Ultrasonic irradiation with vibration for biodiesel production from soybean oil by Novozym 435. Process Biochem 45(4):519–525

Yusuf N, Sirajo M (2009) An experimental study of biodiesel synthesis from groundnut oil. Part I. Synthesis of biodiesel from groundnut oil under varying operating conditions. Australian J Basic Appl Sci 3(3):1623–1629

Irvingiaceae

20

Family Irvingiaceae comprises ten species of tropical trees found in Africa and from Southeast Asia to western Malaysia. It grows at altitudes from 200 to 500 m with annual rainfall from 1200 to 1500 mm and at temperature ranging from 20 to 38 °C. The fruit is a large drupe with fibrous flesh. Fruit pulp is palatable and be used for beverages and for jam production. The fruits form part of the staple diet of Nigerian and Cameroonian tribes. The pounded seed is added to meat and various vegetable dishes as a sauce. The kernel is a source of vegetable oil. The oil from wild mango (*Irvingia gabonensis*) is tried for biodiesel production and hence dealt in this chapter.

20.1 Wild Mango (*Irvingia gabonensis*)

The kernel of the wild mango, *Irvingia gabonensis* is now being considered for oil extraction as this oil is used as raw material for biodiesel. *I. gabonensis* is a tree indigenous to many forests, growing to a maximum height of 40 m with a wide canopy. This tree freely grows at an altitude of 200–500 m above sea level, with an average rainfall of 1300 mm, temperature 20–38 °C in fertile soil with a pH ranging from 4.5 to 7.5. It belongs to the family Irvingiaceae. The leaf of this tree is elliptical with a length of 5–15 cm and yellowish or greenish white and occurs in small panicles (Fig. 20.1).

The fruit is ellipsoidal and looks like a mango in all respects. The fruits are 4–6.5 cm long, 4–6.5 cm wide and 3–6 cm thick. An average sized tree yields around 1000 fruits weighing around 180–200 kg. Total kernel yield will be around 100 kg per tree per year. The oil content of the kernel is around 50–65%. This species is distributed in Angola, Congo, Nigeria, Cote d'ivoire, Uganda, Cameron, Ghana, Togo, Benin and Senegal (Fig. 20.2).

It is called by many vernacular names as shown hereunder:

S. P. Raj et al., *Biodiesel from Flowering Plants*,
https://doi.org/10.1007/978-981-16-4775-8_20

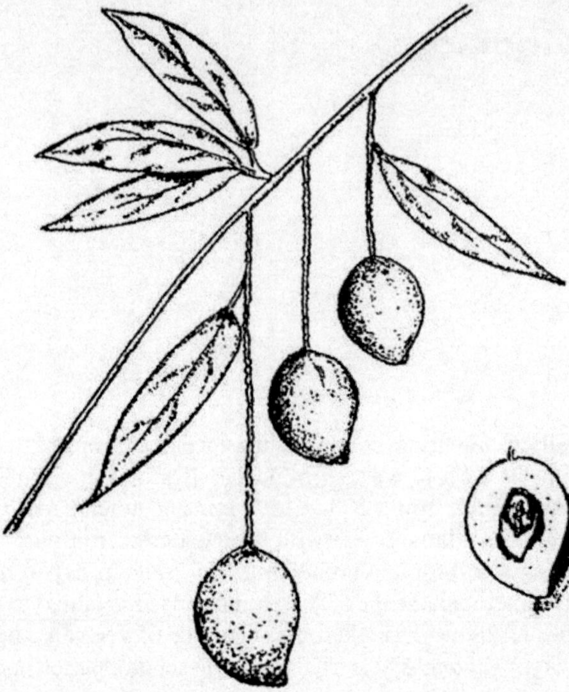

Fig. 20.1 The wild mango, *Irvingia gabonensis*

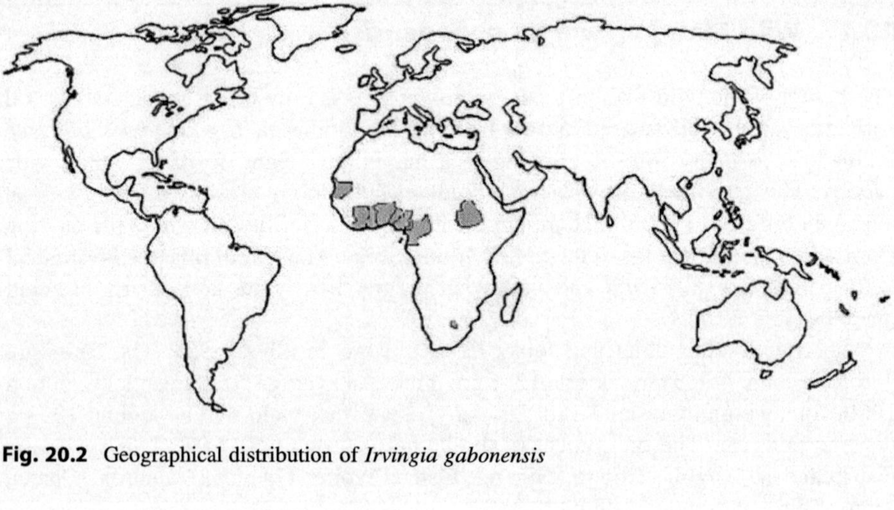

Fig. 20.2 Geographical distribution of *Irvingia gabonensis*

Countries	Name
United Kingdom	African mango, African wild mango, bush mango, dika du Gabon, manguier sauvage
Spain	Arbol chocolate, irvingia
Japan	Afurika mango no ki
Cameron	Andok
Congo Democratic Republic	Meba, Mueba
Nigeria	Oba, oro, kuwing
Côte d'lvoire	Boboru, wanini
Siera leone	Bobo
Germany	Wilder mango baun

20.1.1 Oil Production

The fruits normally fall down when ripe which are then collected, pooled and allowed to stay in a place till its external flesh is decayed completely. The seeds from the decayed fruits are separated, spilt open and the kernels removed. The kernels are dried, powdered and extracted in *n*-hexane at 60 °C (Bello et al. 2011) and the solvent is then expelled to get the oil. The oil obtained is of waxy consistency and needs to be degummed to remove phospholipids, calcium and magnesium salt of phosphatidic and lysophosphatidic acids. These impurities are known to interfere with the separation of glycerol consequent to the transesterification (Igbum et al. 2012). For degumming, the waxy oil of irvingia is mixed with warm water at the rate of 30 mL L^{-1} and agitated for 30 min at 70 °C. This will make the phospholipids and gums separated from oil (Nwufo et al. 2016). They are separated by settling and then discarded. The physicochemical properties of the oil thus obtained are given in Table 20.1.

The iodine value of the oil is paradoxically low indicating the very poor level of unsaturated fatty acid content. More than 90% of the fatty acids are saturated indicating a high oxidation stability. Hence often this oil is referred as myristic-lauric oil.

20.1.2 Biodiesel Preparation

In view of the waxy consistency, this oil is heated to 120 °C for 1 h during which time the oil liquifies and water if present is expelled. Transesterification is proceeded then. The required quantity of the catalyst (NaOH) is weighed (20 g) and dissolved in methyl alcohol whose volume corresponds to a molar ratio of 6:1 to the oil. The above alkali alcohol mixture is stirred at the rate of 500 rpm until the NaOH dissolves. The sodium methoxide thus prepared is emptied in the oil (1 L) which is kept heated and stabilized at 60 °C. The oil methoxide mixture is smoothly stirred

Table 20.1 Physicochemical properties of irvingia oil

Parameters	Bello et al. (2011)	Nwufo et al. (2016)	Ekpe et al. (2018)
Kinematic viscosity at 40 °C (cSt)	45[a]	45	–
Density (kg m^{-3})	930	930	850
Acid number (g KOH kg^{-1})	1.2	–	11.35
Flash point (°C)	300	300	120
Cloud point (°C)	23	23	–
Pour point (°C)	28	28	–
Heating value (°C)	28	28	–
Iodine value (g I_2/100 g)	–	–	0.42
Cetane number	–	44	–
Capric acid (C10:0) (%)	–	–	1.6
Lauric acid (C12:0) (%)	–	–	40.5
Myristic acid (C14:0) (%)	–	–	42.8
Palmitic acid (C16:0) (%)	–	–	4.7
Stearic acid (C18:0) (%)	–	–	0.9
Oleic acid (C18:1) (%)	–	–	4.5

[a]At 60 °C

Table 20.2 Physicochemical properties of irvingia biodiesel

Parameters	Bello et al. (2011) and Nwufo et al. (2016)
Kinematic viscosity at 40 °C (cSt)	3.2
Density (kg m^{-3})	910
Cloud point (°C)	−14
Flash point (°C)	140
Pour point (°C)	−6
Acid value (g KOH kg^{-1})	0.01
Heating value (MJ kg^{-1})	39
Iodine value (g I_2/100 g)	90
Cetane number	52

at 500 rpm for 2 h and the same is then transferred to a separating funnel and kept undisturbed overnight. The dark glycerin layer at the bottom is collected and discarded and the overlying pale yellow layer containing the biodiesel is taken to a container and washed repeatedly using 200 mL of water every time (Nwufo et al. 2016). This will effectively remove residual alcohol, glycerine and soap. The washed biodiesel is finally heated to expel the moisture. The physicochemical characteristics of the biodiesel thus obtained are given in Table 20.2.

The kinematic viscosity of this biodiesel is within the limit and therefore it can be best used with or without blending. Absence of unreacted triglyceride brings down the viscosity of the biodiesel. The cetane number of the oil (44) has increased (52) in the biodiesel due to the transesterification process. This biodiesel has an ideal cloud and pour points making it splendidly suitable in colder climates.

20.1.3 Engine Performance

Considering certain favourable points in the characteristics of irvingia-biodiesel, Bello et al. (2011) chose to use it in engines without actually blending it with diesel. It is observed that the clean biodiesel is found suitable in an engine working at a speed between 1200 and 1800 rpm. Thus this biodiesel proves better in slow speed engines. One of the striking information is that the specific fuel consumption of the irvingia biodiesel is lower than that of the regular diesel by 8% though the heating value is relatively low (Bello et al. 2011). Similarly it is noted that viscosity unusually decreases when the blend concentration is increased. It is one of the favourable factors (Nwufo et al. 2016). The low heating value of the biodiesel is due to the high level of oxygen in the molecule which ultimately reduces the share of other elements (C and H) in the molecule. Since the C and H present in the oil being the energy sources are reduced, the heating value reduces proportionally.

References

Bello EI, Fade-Aluko AO, Anjorin SA, Mogaji TS (2011) Characterization and evaluation of African bush mango nut (Dika nut) *Irvingia gabonensis* oil biodiesel as alternative fuel for diesel engines. J Petrol Technol Altern Fuels 2(9):176–180

Ekpe OO, Bassey SO, Udefa AL, Essien NM (2018) Physicochemical properties and fatty acid profile of *Irvingia gabonensis* (Kuwing) seed oil. Int J Food Sci Nutr 3(4):153–156

Igbum OG, Eloka-Eboka AC, Nwadinigwe CA (2012) Effects of transesterification variable on yield and properties of biodiesel fuels produced from four virgin tropical seed oils. Int J Environ Bioenergy 1(2):119–130

Nwufo OC, Nwaiwu CF, Ezeji KMD, Okoronkwo CA, Ogueke NV, Anyanwu EE (2016) Synthesis and characterisation of biodiesel from Nigerian physic nut, castor bean, dika nut and sandbox seed oils. Int J Ambient Energy 37(3):247–255

Linaceae

21

Members of the family Linaceae are herbaceous annuals or perennials. They are often with slender stems although some species are woody shrubs, sub-shrubs and small trees. This family is alternatively referred as flax family and are cosmopolitan in distribution with approximately 250 species in 14 genera. Flax is one of the oldest fibre crops in the world. These crops of late are cultivated for oil production. This chapter deals with linseed (*Linum usitatissimum*) and its role in biodiesel production.

21.1 Linseed (*Linum usitatissimum*)

Linum usitatissimum popularly referred as linseed belongs to the family Linaceae. It is being distributed and cultivated in many countries for its oil which has the characteristic of fast drying and hence richly needed in paint industries. Every year around 3 million tonnes of linseed are being produced all over the world from 2.8 million ha. It is grown in Russia, Canada, Kazakhstan, China, USA, India, Ukraine, Ethiopia, UK, France, Sweden, Argentina, Brazil, Belgium, Poland, Nepal, Germany, Australia, Uruguay, Bangladesh, Egypt, Romania, Tunisia, New Zealand, Pakistan and The Netherlands (Fig. 21.1).

It is an annual herb having erect, lean stem of around a metre height and richly bears blue coloured flowers. The leaves are long and alternate. The fruits are in the form of capsules which are spherical in shape. The seed is 4.6 mm long and oval in shape (Fig. 21.2). Linseed is cultivated in prime agricultural lands with adequate rainfall, moisture and low temperature. The oil content ranges from 33% to 47% (Dixit et al. 2012). The variations in the oil content of the seeds are governed by the agroclimate and geographical location in which the plants are grown and the variety being chosen for cultivation. Linseed is referred by different names (Table 21.1).

S. P. Raj et al., *Biodiesel from Flowering Plants*,
https://doi.org/10.1007/978-981-16-4775-8_21

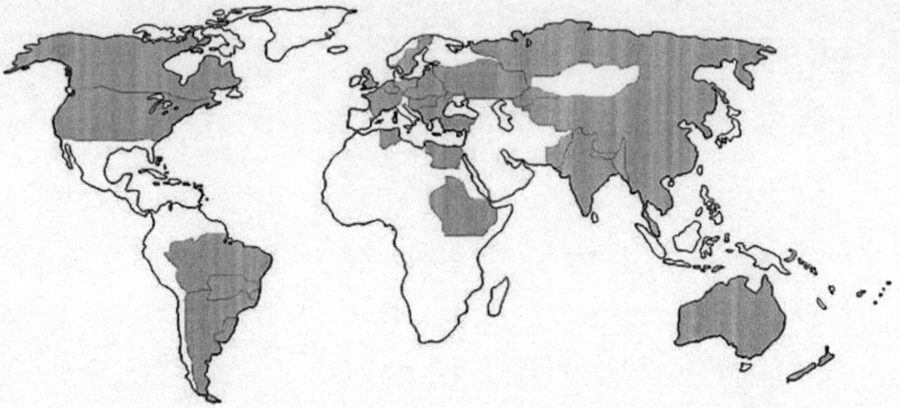

Fig. 21.1 Geographical distribution of *Linum usitatissimum*

Fig. 21.2 Linseed plant
Linum usitatissimum

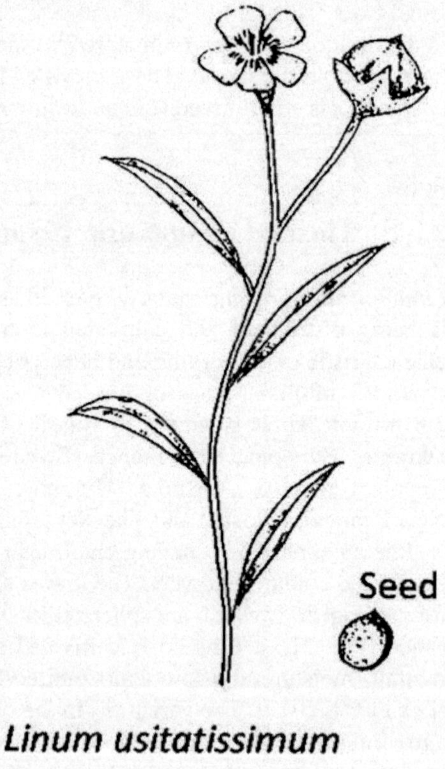

21.1.1 Production of Linseed Oil

Riped seeds are cleaned, sun dried and are passed through screw press or expeller to get the first stage oil. The cake is then treated with *n*-hexane so as to extract the

Table 21.1 Popular names of *L. usitatissimum* in different languages

Language	Popular names
English	Flax, lin, ilion, liner, linum linen, lein and lan
French	Graines de lin
Spanish	Linaza, semilias de lino
Portuguese	Linhaca
Dutch	Vlaszaad
German	Leinsamen
Italian	Semi di lino
Polish	Siemie lniane
Turkish	Keten tohumlari

Table 21.2 Properties of linseed oil (Demirbas 2009; Dixit et al. 2012; Kumar et al. 2013)

Parameter	Value
Viscosity at 40 °C (cSt)	27.2, 26.2
Density (kg m^{-3})	926.6, 893
Free fatty acid (%)	1.94
Iodine value (g I_2/100 g)	156.7
Saponific value (mg KOH g^{-1})	187.7
Flash point (°C)	242, 265
Cetane number	28
Heating value (MJ kg^{-1})	39.84
Pour point (°C)	−14.65
Cloud point (°C)	−3.5
Oleic acid (C18:1) (%)	18.9
Linoleic acid (C18:2) (%)	18.1
Palmitic acid (C16:0) (%)	5.1
Stearic acid (C18:0) (%)	2.5
Linolenic acid (C18:3) (%)	55.1

residual oil. This oil is available in market in different forms as cold pressed, alkali refined, sun bleached and sun dried oil.

21.1.2 Properties of Linseed Oil

The physicochemical properties of linseed oil are given in Table 21.2. Linseed oil contains trace level of glycosides, namely linolein, linolenin and olein. Linolenic acid, linoleic acid and oleic acid are the major fatty acids present. It has a low level of (1.94%) free fatty acid (Dixit et al. 2012) and cetane number. The oil is characterized by high viscosity and low volatility due to the long chain.

21.1.3 Preparation of Biodiesel

The oil is converted to biodiesel through different processes. They are pyrolysis, micro-emulsion, dilution or blending and transesterification. Out of the above, transesterification is considered as a popular method in which hydroxide of sodium or potassium is used as a catalyst. Methanol or ethanol is used as a reactant. Normally a 6:1 molar ratio of methanol to oil gives a good yield. Around 0.5% of catalyst (NaOH or KOH) over the total mixture is used . The optimum temperature is 65 °C.

The alcohol and alkali are allowed to react with oil at 65 °C for 3 h (Ullah et al. 2013). Consequent to the completion of the reaction the whole content is allowed to settle for 8 h. The supernatant solution is the biodiesel which is separated and heated to expel residual methanol if any and it is then washed 2–3 times with warm water to remove the alkali. The resultant biodiesel is dried by heating or passing through anhydrous calcium chloride. Often ethanol, propanol and butanol are employed in the place of methanol.

21.1.4 Properties of Biodiesel

Fuel properties of linseed biodiesel (methyl ester) are given in Table 21.3. The kinematic viscosity and flash point reduced significantly consequent to the transesterification. The viscosity is the measure of internal friction or resistance of a fluid to flow through an orifice. Viscosity refers to the duration in seconds, a fluid of 50 mL volume flow out of a standard Redwood viscometer. The viscosity and flashpoint are low in methyl ester and ethyl ester (Singh and Singh 2010). The internal friction caused by the ester is several fold lower than that of the oil. When the ester is heated the friction further reduced due to the reduction in viscosity. High viscosity leads to poor atomization of the fuel spray. The heating value of the biodiesel is lower by 10% of that of diesel. Considerable portion of the linseed oil biodiesel is formed of unsaturated fatty acid esters which is having more than two double bonds. According to the standard, no limit is prescribed for cloud and pour point since the readings on the above parameters are likely to vary in accordance with the climatic conditions of the area where the plant grows.

Table 21.3 Properties of linseed oil biodiesel (Demirbas 2009; Dixit et al. 2012; Kumar et al. 2013; Ullah et al. 2013; Karthikeyan et al. 2017; Tunio et al. 2018)

Parameter	Value
Viscosity at 40 °C (cSt)	3.36, 5.05, 3.95, 3.32, 3.75
Density (kg m^{-3})	888.2, 885, 852, 887
Flash point (°C)	176, 188, 151, 177
Cetane number	55
Heating value (MJ kg^{-1})	40
Pour point (°C)	−1, −6.25, −9, −7
Cloud point (°C)	2, 3.17, 0, 0
Acid value (mg KOH kg^{-1})	0.338

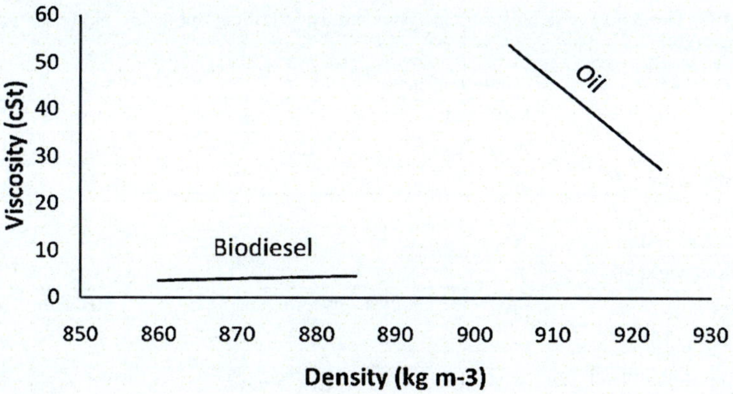

Fig. 21.3 Relationship between density and viscosity

During the transesterification, the molecular weight of the triglyceride reduced to one-third, whereas the viscosity gets reduced by eight times (Demirbas 2009). Density of the biodiesel also is higher (~880 kg m^{-3}). Since injection of biodiesel takes place on a volume basis (litre and not on kg) more quantity by weight of it enters the cylinder which causes high specific fuel consumption on mass basis.

Oil and biodiesel have contradictory behaviours especially on the relationship between density and viscosity. If the density of the oil increases, the viscosity decreases and thus they are negatively correlated. On the contrary, when the density of biodiesel increases there is a tendency of viscosity to increase. Thus there is positive correlation (Fig. 21.3).

The cetane number is based on two major components, namely hexadecane with a cetane of 100 and heptamethylnonane with a cetane of 15. Straight chain saturated hydrocarbon has higher cetane number. Branched chain aromatic compounds show low CN number. Fuel with high CN number has short ignition delay.

Supercritical methanol transesterification has also been developed in which the yield is reported to be 98% with a processing time of 15 min (Demirbas 2002, 2009). In the supercritical process high temperature (239.2 °C), pressure (8.1 MPa) and high molar ratio of methanol to oil (41:1) are used. The reaction parameters are given in Table 21.4. In the supercritical process ethanol, 1-propanol and 1 butanol are often used instead of methanol. In such contingency the critical temperature and pressure differ (Table 21.5).

The viscosity and flash point values of linseed oil methyl, ethyl, 1-propyl and butyl esters are given in Table 21.6.

The viscosity level is very low in methyl and ethyl esters. Heating value is high in methyl and butyl esters. Oxidation stability is an important parameter. Autoxidation takes place during long storage and in this process peroxides and hydro-peroxides are formed first and aldehydes and ketones are formed next. High degree of unsaturation in linseed oil biodiesel is one of the major reasons for the high reactivity and consequent low stability.

Table 21.4 The process details of catalytic and supercritical mode of biodiesel production (Demirbas 2009)

Process parameters	Catalytic process	Supercritical process
Alcohol	Methanol	Methanol
Catalyst	Acid/alkali	Nil
Max. temperature (°C)	65	300
Max. pressure (MPa)	0.1	25
Max. duration (min.)	360	15
Methyl ester yield (%)	97	98
By-products	Methanol, catalyst, glycerol, soap etc.	Methanol

Table 21.5 The critical temp and pressure of alcohols (Demirbas 2009)

Alcohol	Crit. temp. (°C)	Crit. press. (MPa)
Methanol	239.2	8.1
Ethanol	243.2	6.4
1-Proponal	264.2	5.1
1-Butanol	287.2	4.9

Table 21.6 Properties of linseed oil alkyl esters (Lang et al. 2001; Demirbas 2009)

Parameters	Methyl ester	Ethyl ester	1-Propyl ester	1-Butyl ester
Viscosity at 40 °C (cSt)	3.36	3.64	4.88	4.06
Density (kg m^{-3})	887	884	888	877
Cloud point (°C)	0	−2	3	−10
Pour point (°C)	−9	−6	−12	−13
Heating value (MJ kg^{-1})	40	39.65	39.56	40.38

21.1.5 Engine Performance

In order to improve the thermal stability of the linseed biodiesel, it is blended with diesel at a definite ratio. Such a blend normally shares the benefit of the biodiesel and diesel. The blends are reported to show improved thermal efficiency (Dixit et al. 2012). The blends also show increased brake thermal efficiency which is the conversion of heat energy into mechanical energy. The heat energy loss due to the coolant is observed to be less in blends. It is observed that the lubrication effect also is high in internal combustion engines (Tunio et al. 2018). Due to the low calorific value the brake specific fuel consumption is higher in biodiesel than that of diesel.

CO level in the exhaust of biodiesel fed engine is low. Due to the presence of oxygen in the biodiesel molecule C present in the fuel is oxidized to CO_2 to the maximum, leaving less scope to generate CO. There is a feeble chance that the CO_2 dissociates to form CO and O_2 at a high temperature (Nabi and Hoque 2008). Even at this stage additional oxygen pushes the reaction forward to form CO_2 thereby the level of CO is consistently low.

References

Demirbas A (2002) Biodiesel from vegetable oils via transesterification in supercritical methanol. Energy Convers Manag 43:2349–2356

Demirbas A (2009) Production of biodiesel fuel from linseed oil using methanol and ethanol in non-catalytic SCF conditions. Biomass Bioenergy 33:113–118

Dixit S, Kanakraj S, Rehman A (2012) Linseed oil as a potential resource for biodiesel: a review. Renew Sust Energ Rev 16:4415–4421

Karthikeyan P, Lokesh P, Suneel P (2017) Performance and emission characteristics of direct injection diesel engine using linseed oil as biodiesel by varying injection timing. Int J Ambient Energy. https://doi.org/10.1080/01430750.2017.1360201

Kumar R, Tiwari P, Garg S (2013) Alkali transesterification of linseed oil for biodiesel production. Fuel 104:553–560

Lang X, Dalai AK, Bakhshi NN, Reaney MJ, Hertz PB (2001) Preparation and characterization of biodiesel from various bio-oils. Bioresour Technol 80:53–62

Nabi MN, Hoque SM (2008) Biodiesel production from linseed oil and performance study of diesel engine. J Mech Eng 39(1):40–44

Singh SP, Singh D (2010) Biodiesel production through the use of different sources and characterization of oils and their esters as the substitute of diesel a review. Renew Sust Energ Rev 14:200–216

Tunio MM, Luhur MR, Ali ZM, Daher U (2018) Performance and emission analysis of a diesel engine using linseed biodiesel blends. Eng Technol Appl Sci Res 8(3):2958–2962

Ullah F, Bano A, Ali S (2013) Optimization of protocol for biodiesel production of linseed (*Linum usitatissimum* L.) oil. Pol J Chem Technol 15(1):74–77

22 Magnoliaceae

The family Magnoliaceae or the Magnolia family has limited species. The members of this family belong to the temperate regions of northern hemisphere, with centres of distribution in Eastern Asia, Malaysia, Eastern North America, West Indies, Brazil and North-East and South East India. The bark, dried roots, flowers and fruits are having abundant medicinal values. The timber is very valuable and is being used for making musical instruments and toys. The seeds are large with oily endosperm. Couple of research works were made on one of its species, namely Champaca (*Michelia champaca*) for its use in biodiesel.

22.1 Champaca (*Michelia champaca*)

Michelia champaca (syn: *Magnolia* champaca) named after the Italian botanist Peter A. Michel (1679–1737) belongs to the family Magnoliaceae. It is an evergreen tree growing to a maximum height of 50 m. The leaves are simple and arranged spirally. The tree gives forth flowers and fruits throughout the year (Fig. 22.1). It is distributed at an altitude of 600–2000 m, temperature 7–38 °C and grows well in moist deep and fertile soil. The tree is propagated through seeds as well as vegetatively. Trees developed from seeds come to flowering in 8–10 years and it is 2–3 years when developed vegetatively. The fruit size ranges from 7 to 15 cm. Each carpel carries 2–4 seeds.

This tree is the native of India, but distributed in Sri Lanka, Maldives, Bangladesh, China, Indonesia, Malaysia, Myanmar, Nepal, Thailand, Philippines and Vietnam (Fig. 22.2). It has different local names according to the languages (Table 22.1).

S. P. Raj et al., *Biodiesel from Flowering Plants*,
https://doi.org/10.1007/978-981-16-4775-8_22

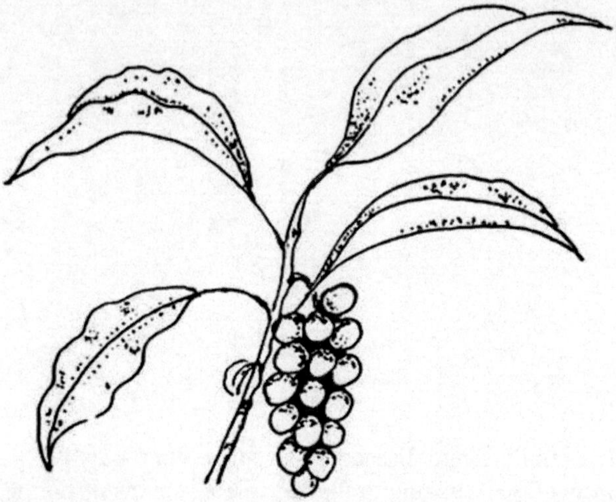

Fig. 22.1 The Champaca
Michelia champaca

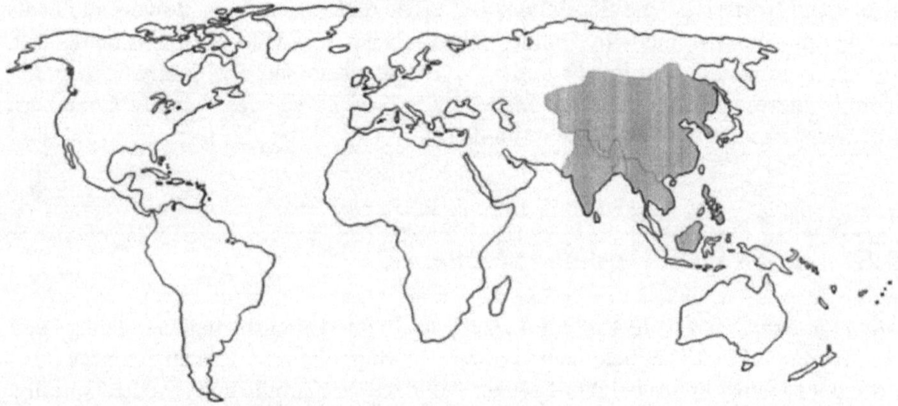

Fig. 22.2 Geographical distribution of champaca *M. champaca*

22.1.1 Oil Production

Though *M. champaca* oil is known to be potential, its commercial exploitation is not yet in sight. For oil extraction, the pods are first sun dried and the seeds are separated. The seeds are then passed through an expeller. Normally all the oil contained in the seeds are not released in the above mode of extraction. Therefore, the expeller process is followed by chemical extraction by treating it with petroleum ether. The crude oil thus obtained is allowed to settle for 48 h.

Table 22.1 Local names of *M. champaca* in different languages

Language	Names
English	Golden champa, yellow champa, fragrant champa
French	Ilang-ilang
Indonesian	Chempaka, kuning, capaka, cempak
Malay	Chempaka merah, champaka
Sinhala	Sapu
Spanish	Champaka
Thai	Champa pa, champa khao
Hindi	Chempaka
Gujarati	Rae-champo
Bengali	Champa
Burmese	Mawk-sam-lung
Cantonese	Sampige
Filipino	Champaca, tsampaka, sampaga

Table 22.2 Physicochemical properties of *M. champaca* oil (Hotti and Hebbal 2015)

Parameters	Value
Kinematic viscosity at 40 °C (cSt)	47.94
Density at 15 °C (kg m^{-3})	920
Free fatty acid (%)	5.3
Flash point (°C)	232
Saponification value (mg KOH g^{-1})	209.2
Iodine value (g I_2/100 g)	122.7
Cetane number	44.79
Heating value (MJ kg^{-1})	36.28
Palmitic acid (C16:0)	32.52
Stearic acid (C18:0)	8.88
Palmitoleic acid (C16:1)	13.39
Oleic acid (C18:1)	6.03
Linoleic acid (C18:2)	30.72
Ecosenoic acid (C20:1)	0.71
Average molecular weight (g mol^{-1})	840.08

22.1.2 Properties of Oil

The analytical properties of *M. champaca* oil are given in Table 22.2. The oil extracted by mechanical means is 18% by weight of the seed and it is 33% when the above is followed by solvent extraction. The oil, on extraction looks deep brown in colour.

The viscosity, density and the flash point are high. Similarly the free fatty acid also is high which makes the oil to be transesterified through two-step process. The free fatty acid value is 5.3 mg KOH g^{-1} (Hotti and Hebbal 2015). Around 50% of the fatty acids are unsaturated in nature.

22.1.3 Biodiesel Production

As the free fatty acid content of the oil is high, the usual base catalysed transesterification cannot be employed directly. In such situations enzymatic or acid catalysis is tried. Often acid and alkali treatments are made in series. In such events, the oil is first treated with the methanol sulphuric acid mixture so as to reduce the free fatty acid. The product is allowed to settle for 2–3 h and washed with warm water and dried. The above material is then taken to the base catalytic transesterification. Thus this sequential process is called the two step transesterification and the details of the reaction conditions are given in Table 22.3.

At the end of the base catalytic transesterification (second step) the product is allowed to settle for 8–12 h. During the settling period the biodiesel moves to the top and glycerin settles down. The bottom layer is then removed. The remaining material containing the biodiesel is washed repeatedly with warm distilled water till the residual biodiesel becomes neutral in pH. The finished biodiesel is dried by heating.

22.1.4 Properties of Biodiesel

The physicochemical characteristics of the biodiesel are given in Table 22.4. High iodine value indicates the unsaturated nature. The density also is high probably due to the high molecular weight. The density is a function between mass and volume. As fuel flow is measured in volume, the mass is high due to the high density. This results in high specific fuel consumption in terms of weight. The viscosity also is high marginally. High viscosity and density cause difficulty in atomization. The heat value of biodiesel is around 8% lower than that of the regular diesel understandably due to the 10% additional oxygen occupied in the molecular mass of biodiesel. It has a satisfactory cetane number.

Table 22.3 The reaction conditions for the two-step process of biodiesel production (Hotti and Hebbal 2015)

Step	Alcohol	Alcohol to oil molar ratio	Catalyst	Catalyst con. (%)	Temperature (°C)	Duration (min)
First	Methanol	24:1	H_2SO_4	1.5	60	75
Second	Methanol	6:1	NaOH	0.65	65	75

Table 22.4 Physicochemical properties of *M. champaca* oil biodiesel (Hotti and Hebbal 2015)

Parameters	Value
Kinematic viscosity at 40 °C (cSt)	7.9, 5.11
Density at 15 °C (kg m^{-3})	895, 870
Iodine value (g I_2/100 g)	101
Heating value (MJ kg^{-1})	39.7, 39.5
Flash point (°C)	134, 158
Cetane number	50.70
Cloud point (°C)	4

Reference

Hotti RS, Hebbal OD (2015) Biodiesel production and fuel properties from non-edible champaca (Michelia champaca) seed oil for use in diesel engine. J Thermal Eng 1(1):330–336

23 Malvaceae

The cotton family Malvaceae (mallows) has more than 4000 species in 244 genera. Members of this family are cosmopolitan in distribution, but they occur more abundantly in tropical and subtropical regions. They are mostly annual or perennial herbs, sometimes shrubs (*Hibiscus rosa-sinensis*) or trees. Many species are considered edible and also for use in medicine. Published information indicates that the oil obtained from two of its members, namely kapok (*Ceiba pentandra*) and cotton (*Gossypium hirsutum*) are used as raw material for biodiesel production.

23.1 Kapok (*Ceiba pentandra*)

Ceiba pentandra is a tropical tree and a native of Mexico, Central America, Caribbean, and coastal Africa. This tree is popularly referred as kapok and is also known as java cotton, java kapok, silk cotton, samauma and ceiba. It grows to a height of 75 m with a trunk diameter of 3 m and the buttress diameter goes up to 6 m. The crown is often as broad as 16 m. Thus, it is one of the largest trees of the world. It grows at elevations up to 1200 m and prefers an annual day time temperature of 17–38 °C and a rainfall of 1500–2500 mm. This tree comes to yield in 4–5 years and has a life span of 60 years. The fruits (pods) ripen in 80–100 days after flowering. A single tree can give forth around 1000 pods a year. Each pod has around 200 seeds. The fruit (pod) looks like a cucumber and hangs like a pendulum (Fig. 23.1).

This tree is distributed in Angola, Antigua, Barbados, Berlin, Bolivia, Brazil, Cameroon, Colombia, Congo, Costa Rica, Cuba, Ecuador, Ghana, Guatemala, Honduras, India, Jamaica, Mali, Mexico, Mozambique, Nicaragua, Nigeria, Panama, Peru, Philippines, South America, Sudan, Tanzania, Uganda, Vietnam and Zambia (Fig. 23.2).

This tree is referred by different location specific names (Table 23.1).

S. P. Raj et al., *Biodiesel from Flowering Plants*,
https://doi.org/10.1007/978-981-16-4775-8_23

Fig. 23.1 The kapok *Ceiba pentandra*

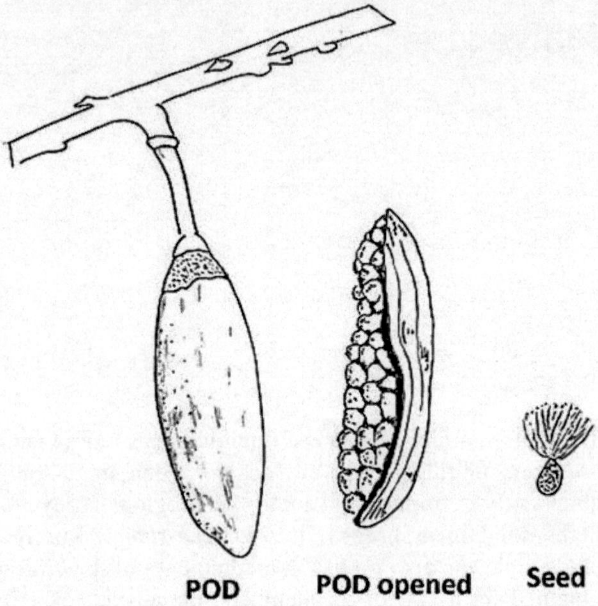

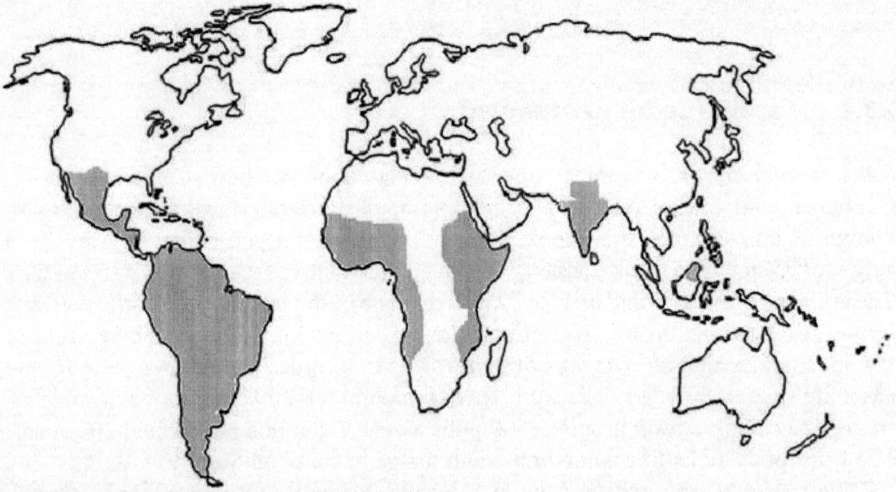

Fig. 23.2 Geographical distribution of *C. pentandra*

23.1.1 Oil Extraction

The cucumber shaped pod is split open longitudinally. From the 5 internal chambers around 200 seeds are released. These seeds are round and black in colour. Fifty eight percent by weight of the seed is formed of kernel. The oil content of the whole seed is around 28%. The trees standing in a hectare yield around 1300 kg of oil. The oils are

Table 23.1 Popular names of *C. pentandra* being in vogue in different countries

Country	Popular names	Country	Popular names
Benin	Hounti	Mexico	Pochota
Bolivia	Hoja de yuca	Nicaragua	Panya
Cambodia	Koo	Nigeria	Araba
Congo	M'Fuma	Peru	Lupuna blanca
East Africa	Msufi	Philippines	Balios
Ecuador	Uchiputu	Sri lanka	Kotta
Finland	Amazoninkapokkipuu	Sierra Leone	Banda
France	Fromager	Thailand	Nun
Germany	Kapok baum	The Netherlands	Fuma
Ghana	Enia, onyina	Toga	Hounti
Indonesia	Kapu	United Kingdom	Cotton wood
India	Ilavam, dudi, simolu	USA	Ceiba
Italy	Capoc	Venezuela	Ceiba yuca
Jamaica	Panya	Vietnam	Gao
Laos	Ngiuz baanz	West Africa	Odouma
Malaysia	Kakabu		

Table 23.2 Physicochemical properties of *C. pentandra* oil (Sivakumar et al. 2013; Vedharaj et al. 2013; Kathirvelu et al. 2014; Parthiban and Perumalsamy 2015)

Parameters	Value
Kinematic viscosity at 40 °C (cSt)	28.5, 30.4, 31.2, 29.3
Density (kg m^{-3})	925, 916, 923, 921
Iodine value (g I_2/100 g)	98, 100.6, 102
Saponification value	193, 195
Free fatty acid (wt%)	13.4, 14.7, 14.35
Refractive index	1.46
Flash point (°C)	297, 170
Cetane number	49, 38
Heating value (MJ kg^{-1})	41.6, 39.1
Pour point (°C)	−10
Average mol. wt (g mol^{-1})	848

extracted in a screw press. The resultant oil is kept undisturbed for 5 days so as to enable the settlement of suspended particles. The clear oil is separated and then heated to 120 °C to expel moisture if any.

Alternatively, the seeds are dried, dehulled, powdered and extracted in *n*-hexane at 60–70 °C.

23.1.2 Characterization of Oil

The raw oil has high viscosity and high molecular weight (Table 23.2) which does not support its use in engine. The oil is pale yellow in colour and it contains medium chain fatty acids. The iodine value also is high. High free fatty acid content prevents

Table 23.3 Fatty acid composition of *Ceiba pentandra* oil (Sivakumar et al. 2013; Kathirvelu et al. 2014)

Fatty acids	Percent
Myristic acid (C14:0)	0.1, 0.11
Palmitic acid (C16:0)	22.6, 23.2
Stearic acid (C18:0)	5.2, 5.68
Arachidic acid (C20:0)	1.7, 1.89
Behenic acid (C22:0)	0.3, 0.25
Lignoceric acid (C24:0)	1.5, 1.5
Oleic acid (C18:1)	30.14, 29.69
Linoleic acid (C18:2)	37.45, 35.11

the straight application of alkali catalysed transesterification. The unsaturated fatty acids, oleic and linoleic form the major portion of the fatty acid profile (Table 23.3). The saponification value indicates that the oil comprises medium chain fatty acids (C16 and C18). The high value for unsaturated fatty acids suggests the liquid stability at room temperature.

23.1.3 Biodiesel Preparation

Since the free fatty acid content is high, double-step processes are being followed. In the first step, acid catalysed transesterification is carried out using sulphuric acid. To every litre of the oil 10–12 mL of sulphuric acid is needed. The required molar ratio of methanol with oil is 8:1. The mixture is stirred continuously at 1000 rpm for 2 h at a constant temperature of 60 °C. After the esterification, the content is allowed to settle for 6 h. The esterified oil moves to the top and is then separated. It is repeatedly washed to remove acid trace if any. During the acid catalysed transesterification, the free fatty acid will be reduced to a satisfactory level. Care is exercised to add the exact quantity of the sulphuric acid. Excess addition darkens the colour of the product and low amount negatively affects the process. The acid catalysis is fast during the first 45 min and the speed of the reaction reduces after 45 min. The best process is the one which brings the FFA to less than 2%.

To the above esterified oil, alkali (NaOH) 1% by weight and methanol to have 8:1 methanol oil ratio is mixed and stirred at 1000 rpm and at 50 °C. The end product is allowed to settle for 6 h. The lower layer comprising impurities is removed and the overlying material, namely biodiesel is repeatedly washed to free it from alcohol and alkali. The ultimate material is dried by exposing the product over $CaCl_2$ for 1 day and then by Na_2SO_4 for 3 h.

Heterogeneous solid catalysts are in use, which includes alkaline earth metal hydroxides, alumina loaded with compounds such as zeolites with high basicity and pore size. Though, these types of catalysts are successful, they become sick when the free fatty acid content is high. In such context, solid acid catalysts though low in activity express high stability. The solid acid catalysts normally do not cause corrosion. It differs in acidity, surface area and thermal stability. They function well in higher temperature and are insoluble both in oil and in methanol. The solid acid catalysts amberlyst-15, SO_4^{2-}/SnO_2, sulphated zirconia (SO_4^{2-}/ZrO_2) and sulphated titanium oxide (SO_4^{2-}/TiO_2) are being utilized in biodiesel preparation (Chavan et al. 1996, 2001; Furuta et al. 2004a, b). Incomplete carbonization of D-glucose so as to form a carbon material and sulphonation makes high density active sites. This material does not leach. Asphalt from vegetable oil and petroleum, corn straw and canola cake are being tried as robust catalyst (Shu et al. 2010; Rao et al. 2011; Liu et al. 2013). *C. pentandra* stalks are also used as catalyst. The stalks are washed with distilled water and dried in hot air oven at 85 °C for 48 h. The dried material is broken in to small pieces and powdered. It is then burned in muffle furnace at 250 °C for 4 h by increasing the temperature at the rate of 50 °C at every 30 min. The material is then removed, cooled and powdered. To this 20% sulphuric acid is added and kept for a day. It is then repeatedly rinsed with excess water and filtered till it is neutral. The material is dried at 65 °C for 3 h. It is then transferred to a muffle furnace maintained at 200 °C for 2 h. This is used as a sulphonated nanocarbon catalyst. This particle is porous and has a surface area of 714 $m^2\ g^{-1}$ (Parthiban and Perumalsamy 2015).

23.1.4 Properties of Biodiesel

Properties of *C. pentandra* biodiesel are given in Table 23.4 and the fatty acid methyl ester profile is furnished in Table 23.5. One of the many factors which needs to the mentioned is cetane number that indicates the ignition character of the fuel. The cetane number decreases with decreasing chain length and increasing degree of unsaturation and branching. High cetane number indicates the shorter duration

Table 23.4 Properties of *C. pentandra* biodiesel (Norazahar et al. 2012; Silitonga et al. 2013; Sivakumar et al. 2013; Vedharaj et al. 2013; Kathirvelu et al. 2014; Parthiban and Perumalsamy 2015)

Parameters	Value
Kinematic viscosity at 40 °C (cSt)	4.3, 4.6, 4.36, 1.8, 5.4, 4.17
Density (kg m^{-3})	867, 877, 885, 860, 875, 876
Flash point (°C)	164, 156.5, 158, 148, 156, 169
Pour point (°C)	−2, 2.8, 1.0, −8
Cloud point (°C)	3, 3, 4.4, 1.0
Heating value (MJ kg^{-1})	38.1, 40.5, 41.8, 39.4, 36.2
Oxidative stability (h)	4.1, 4.42, 12.3
Cetane number	52, 57, 49, 57.5, 54, 47
Carbon (wt%)	78
Hydrogen (wt%)	12.5
Oxygen (wt%)	11.68
Cold filter plugging point (°C)	1.0, 4.0
Copper strip corrosion at 50 °C, 3 h	1a
Iodine value (g I_2/100 g)	107

Table 23.5 Fatty acid methyl ester profile of *C. pentandra*, biodiesel

Fatty acid methyl esters	Percent
Methyl palmitate	23.17
Methyl linoleate	30.0
Methyl oleate	22.9
Methyl caproate	9.42
Methyl sterculate	8.7
Methyl stearate	4.7
Methyl arachidate	1.2

between the initiation of fuel injection and ignition. The transesterification process reduces the density and viscosity. The low temperature properties such as pour point, cloud point and cold filter plugging point are not satisfactory but can be improved by incorporating additives. The oxidation of the biodiesel poses a major challenge during long storage. Rapid oxidation is a function of unsaturated bonds, inherent oxygen, storage temperature, light and chosen metals.

The peak of oxidation is often referred as induction period. The induction time of the *Ceiba* biodiesel is quite satisfactory. The heating value is normally lower than the diesel.

23.1.5 Engine Performance

The brake specific fuel consumption depends largely on the calorific value and the quantity of the fuel being injected. Since the calorific value is low in biodiesel, it demands more consumption to pull the load. The biodiesel expresses poor transmission due to low calorific value and high viscosity. The brake thermal efficiency of the engine depends on the extent to which the fuel is burnt inside the combustion chamber. The increased viscosity affects the fuel atomization causing the efficiency to drop. Since the cetane number is higher than that of the diesel the startup combustion is little bit earlier in biodiesel. The peak heat release rate is around 30% of that of diesel. This is due to the decreased calorific value and increased viscosity. In many occasions the CO level in the exhaust increases though the biodiesel contains more oxygen. This is due to the fact that more biodiesel enter in to the chamber to compensate the low calorific value thereby to generate the required energy as equal to that of diesel. This phenomenon automatically increases the fuel air ratio causing incomplete combustion. In addition to the above, poor atomization as a function of the viscosity makes the CO in the exhaust to increase. The NO_x emission in biodiesel increases since the oxygen present in the biodiesel molecule joins with nitrogen and releases the NO_x. The smoke value also is higher than that of the diesel. Under normal condition soot is generated during the premixed combustion phase which is relatively lengthy and in such case significant time gap is available for the carbon particle to combine with oxygen, but in the case of biodiesel the premixed combustion phase is short, which in effect increases the smoke.

23.2 Cotton (*Gossypium hirsutum*)

The genus *Gossypium* popularly referred as cotton comprises around 50 species, in which only 4 are considered useful in a commercial sense. They are *Gossypium hirsutum*, *G. barbadense*, *G. herbaceum* and *G. arboreum*. Around 90% of the cotton grown all over the world is formed of *G. hirsutum* although all the four species are being cultivated in many parts of the world for reasons not clearly known (http://www.gktoday.in/blog/important-species-of-cotton/). Other three species are grown to a relatively low extent (*G. barbadense* 3–4%, *G. arboreum* <2% and *G. herbaceum* <2%). The oil characteristics, biodiesel productions and engine performance analysis are mostly available for the species *G. hirsutum*. Other species are not yet considered for oil extraction or biodiesel production.

23.2.1 Characteristics of *Gossypium hirsutum* L.

The cotton, *G. hirsutum* is considered as a cross between *G. herbaceum* L. and *G. raimondii* (Ulbrich) or *G. gossypoides* (Ulbrich). This species is referred by its synonyms *G. religiosum* L., *G. latifolium* Murray, *H. punctatum* Schumach., *G. taitense* Park., *G. mexicanum* Tod. and *G. marie*-galante Watt. *G. hirsutum* is a shrub which grows to a maximum height of 5 m with a stem girth of 16 cm. The bark is grey and smooth. The young leaves are without lobes but the old leaves have not less than 3 lobes which are deeply incised (Fig. 23.3). The flowers have cream to pale yellow coloured petals with red patches at the base of the petals. At the end of summer, vegetative growth stops and following that the plant puts forth flowers. After a week the flower dries and falls down from the plant and consequently the

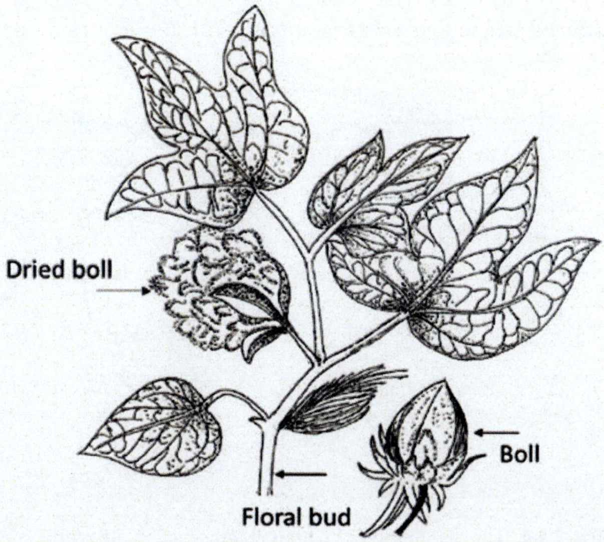

Fig. 23.3 *Gossypium hirsutum* L.

fruit or boll is exposed. The boll continues to increase in its size for the next 3 weeks from the time of its formation. During the above period the boll appears to be pale green, spherical or ovoid. In about 6 weeks the boll dries up. Each dried boll has 3–5 locks of cotton arranged in cluster each containing several seeds. Each seed is oval in shape and is around 8 mm in length consisting of a hull and kernel. Approximately 25–40 seeds are extracted from each boll. On an average each seed weigh 0.085 g and around 11,000–12,000 seeds may present in a kilogram.

23.2.2 Distribution

G. hirsutum is distributed in India, Southern Africa, Namibia, Angola, Botswana, Zimbabwe, Mozambique, Limpopo, Mpumalanga and Kwazulu-Natal, Central America, northern part of South America, Florida, Bahamas, Cuba, Jamaica, Puerto Rico, Mexico, Galapagos, Ecuador, Peru, Bolivia, Paraguay, Brazil, China, Egypt, Pakistan, Baluchistan, Tanzania, Uzbekistan, Japan and Korea (Fig. 23.4).

23.2.3 Vernacular Names

The species *G. hirsutum* referred by different vernacular names in different countries are given in Table 23.6.

23.2.4 Habitat

G. hirsutum grows spontaneously in coastal and lower coastal plains. It is found along road sides, river banks and in disturbed soils. It prefers to grow in a well-drained soil with a pH range of 5.5–7.0. It tolerates moderate level of salt in soil and

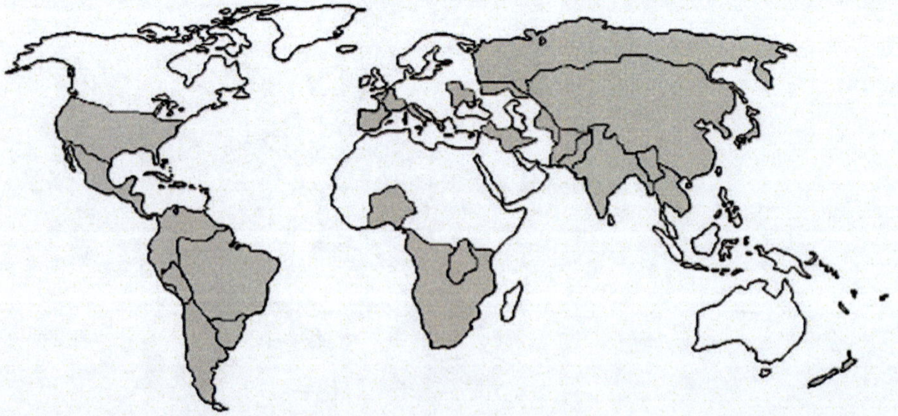

Fig. 23.4 The distribution of cotton *G. hirsutum* in the world

Table 23.6 The vernacular names of *G. hirsutum* being in force in different countries

Country	Vernacular names
England	Cotton, upland cotton, American cotton, American upland cotton, green seed cotton
France	Coton, cotonnier, cotton velu
Poland	Algodoeiro, algodoeiro
Sweden	Pamba
India	Panchu, panch, cotton, paruthi

irrigation water. However it does not tolerate shade. The preferred rainfall is 800–1700 mm. It grows well in a moderate elevation of 300 m above sea level.

23.2.5 Cultivation

The seeds are sown in a well prepared soil. The preferred temperature during sowing is 18–20 °C. It germinates depending upon the agro-climatic conditions in 15–60 days and grows to a height of 1.5 m during the first year. Approximately 15 or more nodes are required to be formed to enable the plant to bloom. Normally the plant comes to harvest during the second year. Ideal temperature helps to maximize the yield. They are likely to be killed by continuous frost and normally 180–200 frost-free days are considered suitable. One of the special features is that all the plants in the field put forth leaves, stem, flowers, bolls and seeds all at the same time. Each plant has two types of branches. The vegetative branch grows incessantly as a long and straight forward growth and the fruiting branches have multiple meristems each starting slightly after the position of the previous fruiting bud. This type of branches has zig-zag growth.

23.2.6 Oil Extraction

Oil extraction is often cumbersome due to the difficultly in dehulling. Mechanical dehullers are developed in the recent past to separate the kernels. These kernels are then partially cooked to destroy the cell walls and then passed through expellers. The expelled oil is separated from the cake and purified. This procedure suffers mostly due to long extraction time. As days passed on alternate methods came in to being such as extracting the smashed kernel in *n*-hexane at 50 °C for not less than 8 h. The hexane is then evaporated so as to relieve the oil.

23.2.7 Microwave Assisted Extraction

Cotton seeds are dehulled and the kernels are powdered using a mill. The moisture content is then assessed. Normally the moisture content of the powdered seeds is around 7%. Additional water is sprayed over the mass of kernel so as to make the final moisture content as 14%. In order to have a uniform moisture content of this material, the moisturized powder is stored in a refrigerator (4 °C) for 24 h. For every 1 kg of mass 3 kg of *n*-hexane solvent is added and placed in the microwave oven for 5–6 min and the contents are continuously stirred during the irradiation. After the due irradiation the resultant material is passed through a filter and then centrifuged. The clear fluid thus obtained is evaporated in a rotary evaporator connected to a suction device and the oil is then separated.

23.2.8 Subcritical Water Extraction

Of late subcritical water technology has taken the stage in the extraction of oil from cottonseed. The advantages of subcritical fluids in altering the solubility and diffusivity in relation to temperature and pressure were understood as early as 1970. In 1980s reactor designs concerned with extraction of materials at high temperature and pressure were developed. The whole kernel mass is taken in a stainless steel and a known quantity of water is added along with it. The tube is then sealed. The sealed tube shall have free space to accommodate the expanding liquid and vapour pressure. This tube is immersed in oil–paraffin bath maintained at a high temperature. After the reaction the tube is removed from the bath and cooled in cold water. The tube is then opened and the solid matter is filtered and discarded. The liquid is centrifuged to separate the oil. The efficiency of this procedure is largely governed by the temperature, duration, size of the whole kernel, seed to water ratio and pretreatment of the seed. The tube loaded with uncrushed (whole) seed and water at a water–seed ratio of 2:1 by wt. at a reaction duration of 35 min is tried in different temperatures of the bath. The extractable oil content of the liquid phase is observed to increase steadily as the (Fig. 23.5) temperature of the bath is increased from 150 to 280 °C and the highest percentage of oil during the extraction is reported when the bath is maintained at an ambient temperature of 270–275 °C.

As the temperature of the bath is increased beyond 280 °C the oil gets denatured and charred. As a result the recovery level lowered. It is also known from Fig. 23.5 that in low temperature, without changing the reaction time of 35 min, the recovery is low (e.g. in 180 °C the recovery is 12%). At the optimum temperature of 275 °C and at the reaction duration of 35 min the recovery of oil is around 22%. The seed water ratio 1:1 to 2:1 is observed to be ideal. It is also observed that the size of the whole seed has a negative influence on the extractability.

23.2.9 Properties of Cottonseed Oil

The properties of the cottonseed oil are given in Table 23.7 and the fatty acid profile is given in Table 23.8.

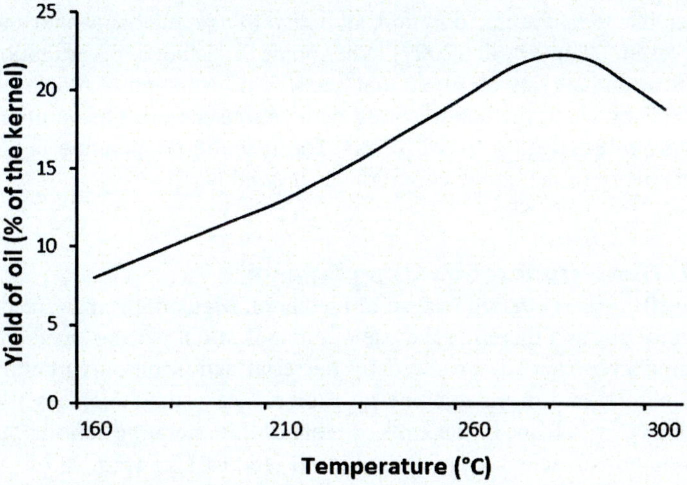

Fig. 23.5 Effect of bath temperature on the oil recovery in subcritical water extraction at a reaction duration of 35 min and at a seed water ratio of 2:1 by weight

Table 23.7 Properties of cottonseed oil (Shahid and Jamal 2011; Suresh et al. 2014)

Properties	Value
Density (kg m^{-3})	900
Calorific value (kJ kg^{-1})	38,049
Viscosity at 40 °C (cSt)	50
Flash point (°C)	210
Cetane number	48
Carbon residue (%)	0.66
Acid value (mg KOH g^{-1})	0.30
Saponification number (mg KOH g^{-1})	187
Molecular mass	852

Table 23.8 Fatty acid content of the oil of *G. hirsutum* (Pandey 1998)

Fatty acid	Percent
Palmitic acid	23.1–28.0
Stearic acid	02.4–03.4
Oleic acid	14.7–20.9
Linoleic acid	47.6–55.4

23.2.10 Preparation of Biodiesel

23.2.10.1 Transesterification Using Methanol

Methanol of 99.9% purity is mixed with sodium hydroxide to prepare sodium methoxide. It is then added to hot (40–50 °C) cottonseed oil and stirred gently without forming foam at an ambient temperature of 55–60 °C for 8 h. The above is then allowed to stand undisturbed for 24 h. A clear liquid (biodiesel) moves to the surface leaving a mass of glycerin at the bottom. The efficiency of the whole process depends on the temperature, duration, quantity of the methanol and the sodium hydroxide used. Nabi et al. (2009) who made a detailed investigation on the transesterification pathway observed that maximum formation of biodiesel is possible when 20% by wt. of methanol is used with cottonseed oil. The optimum level of sodium hydroxide is 1.25% by wt. of oil. The reaction temperature is 55 °C. The biodiesel recovery in the above conditions is around 80%.

23.2.10.2 Transesterification Using Ethanol

Jinsi et al. (2011) used ethanol instead of methanol. Methanol is manufactured from coal or natural gas and therefore the use of ethanol in the place of methanol gained momentum since ethanol is prepared by bacterial action on agricultural products which is a natural and recyclable energy source. The usual stoichiometric ratio of alcohol with oil is 3:1 but in the case of ethanol the operating ratio is fairly high. High conversion is reached at an ethanol to oil ratio of 12:1 (Fig. 23.6).

It may be good to increase the quantity of ethanol involved in the transesterification of cottonseed oil but it may pose problem in the isolation of excess ethanol from the reaction product. Besides, ethanol is likely to develop stable emulsion when there is excess quantity of free fatty acid present in the oil. The optimum level of ethanol to oil ratio thus studied is 12:1 at a transesterification temperature of 60 °C with a potassium hydroxide level of 1.25%. As a matter of fact further increase in ethanol normally do not give a desired result since addition of more ethanol results in the dilution of potassium hydroxide in the ethoxide mixture causing low recovery of biodiesel. Similarly the concentration of potassium hydroxide also has a massive influence over the recovery of the product. As the level of potassium hydroxide is increased from 0.25% to 1.25% in a batch reaction, the maximum effect is obtained at 1.25% (Fig. 23.7).

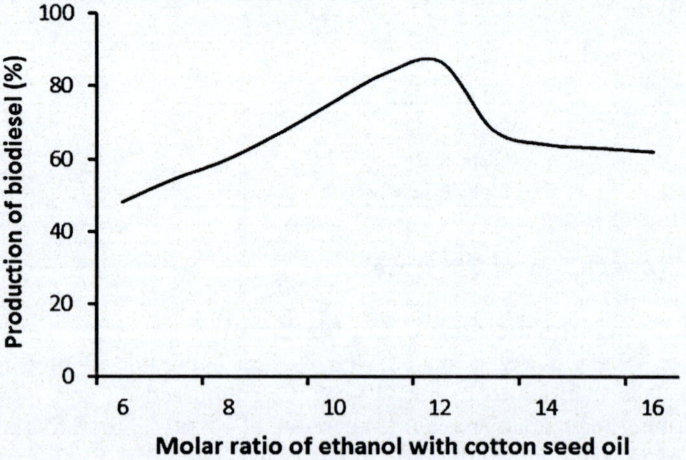

Fig. 23.6 The impact of ethanol to oil ratio in the transesterification of cottonseed oil

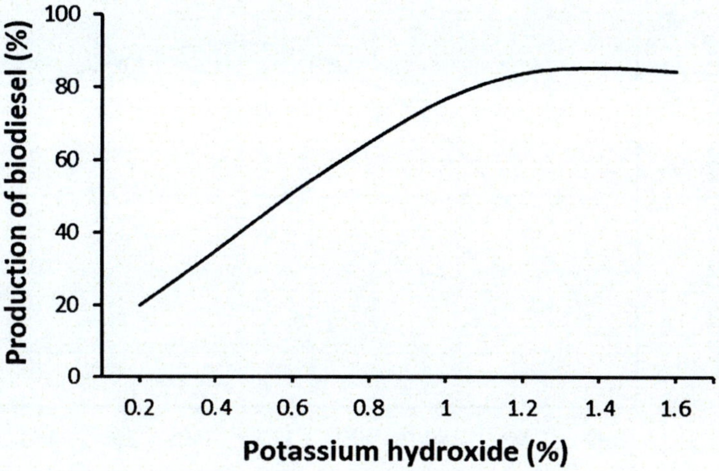

Fig. 23.7 Effect of potassium hydroxide concentration on the transesterification of cottonseed oil employing ethanol

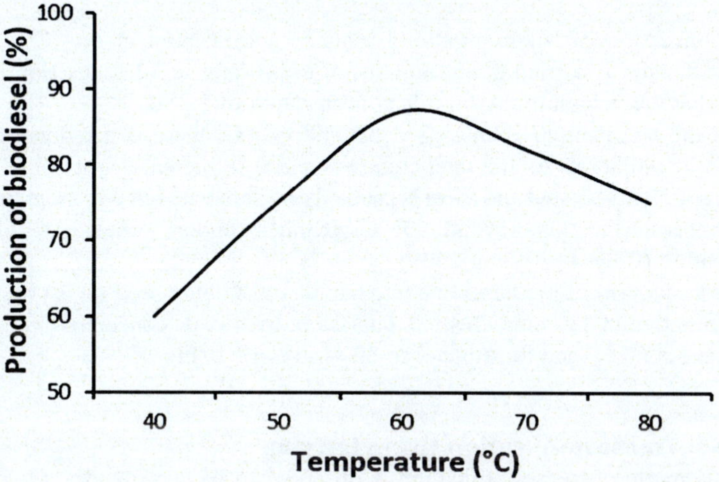

Fig. 23.8 Effect of reaction temperature on the transesterification of cottonseed oil using ethanol

Any further increase in potassium hydroxide beyond 1.25% promotes emulsion of the whole content and there is gel formation. Besides, there is large scale shift in the pH of the reaction medium (Guoying et al. 2003). Other parameters are 12:1 ethanol to oil ratio, temperature 60 °C and transesterification duration of 2 h. The effect of different reaction temperature ranging from 40 to 80 °C on the transesterification is shown in Fig. 23.8.

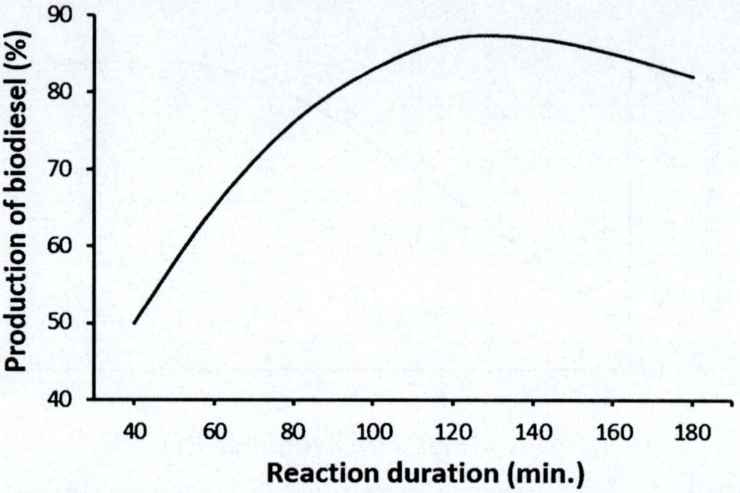

Fig. 23.9 Effect of reaction duration in the transesterification of cottonseed oil using ethanol and potassium hydroxide

The conversion of oil to biodiesel reaches a maximum at 60 °C when the transesterification is carried out at an ethanol to oil ratio of 12:1, reaction duration of 2 h and at the potassium hydroxide concentration of 1.25%.

When the temperature is increased beyond 60 °C, there is a reduction in the formation of biodiesel. As the temperature is made to increase slowly the level of reactants slowly exhausted and there is probably no substrate left any more to take up further transesterification beyond 60 °C. Besides, higher temperature promotes saponification (Chen and Wang 2005).

Highest recovery of biodiesel due to ethanolysis of cottonseed oil is possible at a reaction duration of 130 min when the potassium hydroxide concentration is 1.25%, temperature is 60 °C and the ethanol to oil ratio is 12:1 (Fig. 23.9).

23.2.10.3 Transesterification Using Enzyme

Enzymatic mode of transesterification is observed to be eco-friendly. However, if enzymes are used straight in the reaction, reusability is not possible and therefore the waste of enzyme cannot be averted. Hence, enzymes are initially incorporated in porous support materials so that consequent to the transesterification the enzymes already loaded in the porous material can be easily separated. Normally, Novozym 435 a branded enzyme obtained from *Candida antarctica* is used. This enzyme is known to be utilized in several batches repeatedly if loaded in a solid substrate (Watanabe et al. 2000).

Methanol is normally used in the enzyme mode of transesterification. When the level of methanol is increased beyond a molar ratio of 6 (Methanol:oil) the reaction invariably slows down due to the inhibitory effect of methanol. Besides, the glycerin formed during the reaction makes a coating over the enzyme and thus blocks the

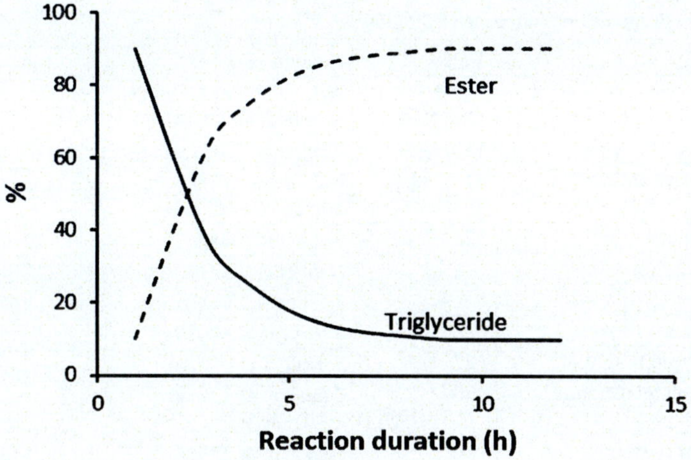

Fig. 23.10 The effect of duration on the transesterification of cottonseed oil using Novozym 435 enzyme and the consequent level of triglyceride

action-sites of the enzyme. To avoid this contingency the required methanol is added in 3 to 4 split doses (Shimada et al. 1999). In many of such cases certain quantity of *t*-Butanol is used as a solvent. This solvent enhances the yield. *t*-Butanol in the absence of methanol cannot promote alcoholysis (Royon et al. 2007). *t*-Butanol has the ability to dissolve both methanol and glycerol and therefore also prevents the coating of glycerol over the lipase besides it abates the inhibitory effect of methanol (Dossat et al. 1999). The reaction takes around 10 h when the ratio of methanol to oil is 6:1, the addition of *t*-Butanol is around 30% and the level of enzyme is 1.7% (Fig. 23.10).

As high as 90% efficiency is reached when the working temperature is around 60 °C. Around 500 cycles are expected to be run using the same enzyme if it is loaded on a solid substrate.

23.2.11 Properties of Cottonseed Biodiesel

The properties of cottonseed oil biodiesel are given in Table 23.9.

The viscosity is one of the important parameters. By increasing the temperature of the biodiesel the viscosity of it may be marginally reduced. When the temperature is increased, the intermolecular attraction among different layers of the biodiesel weakens causing the decrease in viscosity. For the normal function of the internal combustion engine there should be optimum viscosity. If the viscosity is higher than the standard, the mass of the fluid may not break in to tiny particles as the fuel is sprayed inside the combustion cylinder. If the fuel continues to remain as large particles even after the spray, the burning capacity of the fuel is diminished resulting in to poor performance of the engine. On the contrary if the viscosity is too low there

Table 23.9 Properties of cottonseed oil biodiesel

Parameters	Value	References
Kinematic viscosity at 40 °C ($mm^2\ s^{-1}$)	6, 4.2 (25 °C), 5	Nabi et al. (2009), Shahid and Jamal (2011), Suresh et al. (2014)
Density at 15 °C ($kg\ m^{-3}$)	850, 885	Nabi et al. (2009), Suresh et al. (2014)
Calorific value ($MJ\ kg^{-1}$)	41.7, 38.5, 39.6	Nabi et al. (2009), Shahid and Jamal (2011), Suresh et al. (2014)
Cetane number	52, 45	Nabi et al. (2009), Suresh et al. (2014)
Cloud point (°C)	−2, 7	Nabi et al. (2009), Suresh et al. (2014)
Pour point (°C)	−4, 4	Nabi et al. (2009), Suresh et al. (2014)
Carbon residue (%)	0.01	Suresh et al. (2014)
Molecular weight ($g\ mol^{-1}$)	227	Suresh et al. (2014)
Auto ignition temperature (°C)	470	Suresh et al. (2014)
Ash content (wt%)	0.01	Suresh et al. (2014)
Flash point (°C)	220, 167	Shahid and Jamal (2011), Suresh et al. (2014)
Carbon mass (wt%)	77	Nabi et al. (2009)
Sulphur (wt%)	<0.005	Nabi et al. (2009)
Total glycerin (%)	0.028	Nabi et al. (2009)
Hydrogen (wt%)	12.5	Nabi et al. (2009)
Oxygen (wt%)	10.5	Nabi et al. (2009)
C/H ratio	6.16	Nabi et al. (2009)
pH	6.90	Shahid and Jamal (2011)

will be low lubrication effect. As a result the moving parts will wear quickly. It is felt that cottonseed oil biodiesel and regular diesel are mixed at 1:1 ratio by volume and used in engines there may not be any problem (Nabi et al. 2009). The engine life is reported to increase if the fuel maintains adequate lubricity (Goodrum and Geller 2004). As a matter of fact, there is a recommendation from He and Bao (2005) that the oil of the cottonseed may be best utilized after mixing it with 7 portion of diesel since the cottonseed oil has an appreciable calorific value when compared to other oils. The highest efficiency be obtained in an injection angle of 20° (a shift from that of the pure diesel by 3° to 5°).

23.2.12 Engine Performance

The fuel consumption expressed as kg h^{-1} against a range of load (3–33 kW) and at a fixed engine speed of 1500 rpm is given in Fig. 23.11.

The biodiesel consumption pattern due to different load (kW) is always higher than that of the diesel. In the middle range of load, i.e. 9–23 kW the consumption of both the fuels is observed to be stable. The increase in consumption above 25 kW load is apparently due to the incomplete combustion as a consequence of lack of time to burn the fuel. The consumption of biodiesel for the load 5–28 kW ranges from 5.3 to 7.0 kg h^{-1}, whereas it is 5–5.5 kg h^{-1} for diesel.

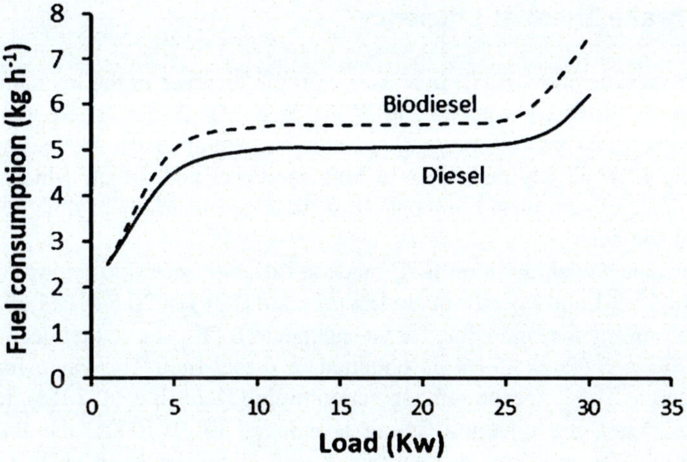

Fig. 23.11 Fuel consumption pattern of biodiesel made from cottonseed oil and diesel for different load at a constant speed (1500 rpm) of the engine

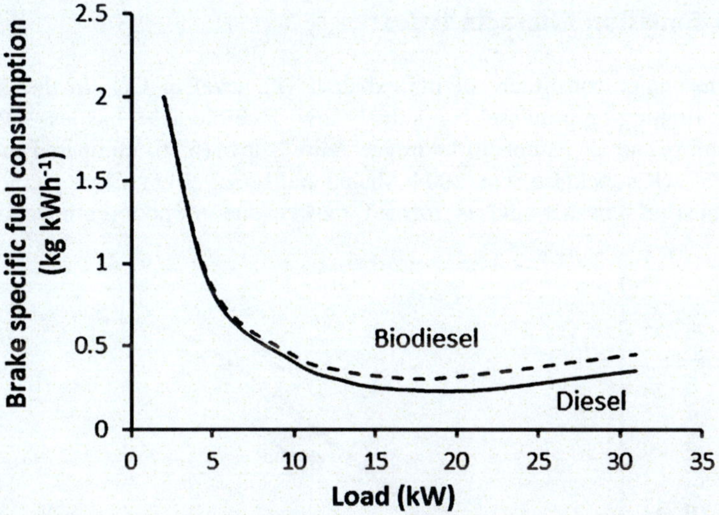

Fig. 23.12 Rate of brake specific fuel consumption against different load in an internal combustion engine with a speed of 1500 rpm

The Brake Specific Fuel Consumption (BSFC) showed rapid reduction in the first half of the load (Fig. 23.12). From 30% to 75% of the load, i.e. 16–25 kW the BSFC stabilized. But a feeble increase is noticed above 80% load (26–33 kW).

The BSFC values are always higher than that of the diesel since the biodiesel has low calorific value compared to diesel and therefore more biodiesel is needed to be spent for maintaining a constant output.

23.2.13 Brake Thermal Efficiency

Brake thermal efficiency (BTE) increases with the increase in the load and its peak value is reported to be between 24 and 27 kW load (Fig. 23.13). There will be better combustion and low loss of energy at the above range of loads. As load marginally exceeds, the BTE slowly decreases in both biodiesel and diesel, which is understandably due to incomplete combustion of fuels and insufficient time available in transferring the heat.

A performance evaluation on BTE made at different injection timings, compression ratio of 17.5:1 and injector operating pressure (IOP) of 20.5 MPa indicated that the injection timing is responsible for the change in BTE. At a static injection timing of 23° bTDC a BTE of 25.6% is obtained in diesel fuel. The value for BTE in biodiesel fuel is 21.5% for the same injection timing (Suresh et al. 2014). In the case of biodiesel when the injection timing is reduced to 19 bTDC the BTE of the biodiesel increases to 23%. A retarding injection timing by a change in crank angle by 4° helps to improve the BTE.

23.2.14 Emission Characteristics

CO_2 is the major constituent of the exhaust. The level of CO_2 in the emission of engine running on biodiesel is relatively low since biodiesel has less number of carbon atoms and its carbon to hydrogen ratio is low (6.16) compared to that of diesel (6.35) (Karabektasa et al. 2008; Shahid and Jamal 2011). CO is an intermediate combustion product and is formed mostly due to poor combustion as a

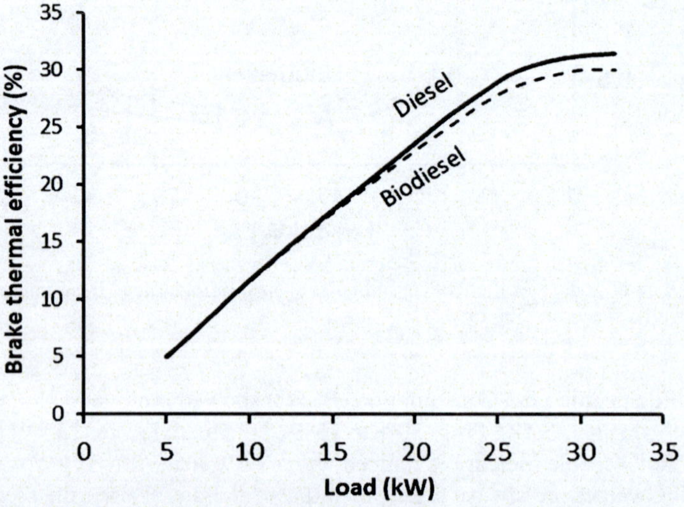

Fig. 23.13 Brake thermal efficiency against the load in an internal combustion engine operating with different load and with a speed of 1500 rpm

consequence of shortage of oxygen and low combustion temperature. As extra oxygen is present in the biodiesel molecule the chances of forming CO are rare since the presence of oxygen helps in complete combustion of the fuel so as to make a shift of CO to CO_2. During the burning of fuel each molecule of cottonseed oil biodiesel generates two atoms of oxygen which helps to reduce the generation of CO (Shahid and Jamal 2011).

The level of NO_x is high in the emission of the engine running on biodiesel and the NO_x concentration increases as the torque is increased. The inert nitrogen though it stays inactive till 1100 °C enters in to a reactive phase as the temperature of the combustion cylinder starts reaching 1500 °C. On such events nitrogen in the fuel is converted to its oxides. Once such oxide is formed at high temperature it does not revert to its elemental stage even it is cooled. The level of NO_x is higher for biodiesel since significant oxygen is available for easy conversion from nitrogen to its oxides at a conveniently high temperature. As such this extra oxygen is responsible for the high level of NO_x emission. Even if the biodiesel is mixed with diesel at 30% level the NO_x is reported to increase marginally (Nabi et al. 2009). The raised levels can be managed by a process of catalytic conversion or through exhaust gas recirculation (Lopez et al. 2009).

Often the total hydrocarbon level is reported to be present sparingly in the emission when the engine is operated with biodiesel at increased load. Efficient combustion due to the presence of rich oxygen helps to minimize the level of total hydrocarbon in the emission.

If the combustion is not proper or the combustion is made on heavy lubricating fuel, particulate matter is discharged in large quantities along the exhaust which is often referred as exhaust particle. These particles are highly heterogeneous and comprise soot, ash, sulphate or silicate. These particles may remain as individual fractions or they may stay as aggregates. They may be in the particle size range of respirable matter. Concentration of these particles is low in the case of biodiesel-diesel blend.

Few recommendations are made for the successful use of cottonseed oil biodiesel (Suresh et al. 2014). They are modification of injection timings, change in injector opening pressure and nozzle geometry reducing the injection timing, increasing the injection pressure and increasing the number of nozzle holes from three to four in the fuel injector assembly.

References

Chavan SP, Zubaidha PK, Dantale SW, Keshavaraja A, Ramaswamy AV, Ravindranathan T (1996) Use of solid super acid (sulphated SnO_2) as efficient catalyst in tacile transesterification of ketoesters. Tetrahedron Lett 37(2):233–236. https://doi.org/10.1016/0040-4039(95)02136-1

Chavan SP, Subbarao YT, Dantale SW, Sivappa (2001) Transesterification of ketoesters using amberlyst-15. Synth Commun 31(2):289–294. https://doi.org/10.1081/scc-100000212

Chen H, Wang J (2005) Kinetics of KOH catalysed transesterification of cottonseed oil for biodiesel production. J Chem Ind Eng 56(10):1971–1974

Dossat V, Combes D, Marty A (1999) Continuous enzymatic transesterification of high oleic sunflower oil in a packed bed reactor: influence of the glycerol production. Enzym Microb Technol 25:194–200

Furuta S, Matsuhashi HH, Arata K (2004a) Catalytic action of sulfated tin oxide for etherification and esterification in comparison with sulfated zirconia. Appl Catal A Gen 269(1):187–191. https://doi.org/10.1016/j.apcata.2004.04.017

Furuta S, Matsuhashi H, Arata K (2004b) Biodiesel fuel production with solid superacid catalysis in fixed bed reactor under atmospheric pressure. Catal Commun 5(12):721–723. https://doi.org/10.1016/j.catcom.2004.09.001

Goodrum W, Geller PD (2004) Effect of specific fatty acid methyl esters on diesel fuel lubricity. Fuel 83:2351–2356

Guoying W, Xiping L, Miaoxin W, Lin W, Yajun Z (2003) Kinetics of cottonseed oil transesterification in a batch reactor. J Chem Eng Chin Univ 17(3):314–318

He Y, Bao YD (2005) Study on cottonseed oil as a practical substitute for diesel oil in fuel for single cylinder diesel engine. Renew Energy 30:805–813

Jinsi C, Xiangyang W, Enzhu H, Yufu X, Xianguo H, Lijun P, Shaotong J (2011) Biodiesel production by transesterification of cottonseed oil with ethanol using tetrahydrofuran as co-solvent. Int Conf Mater Renew Energy Environ (ICMREE) 1:288–293

Karabektasa M, Ergena G, Hosoz M (2008) The effects of preheated cottonseed oil methyl ester on the performance and exhaust emissions of a diesel engine. Appl Therm Eng 28 (17,18):2136–2143

Kathirvelu S, Moorthi SVN, Krishnan SN, Mayilsamy K, Krishnaswamy T (2014) Production of biodiesel from non-edible *Ceiba pentandra* seed oil having high FFA content. ARPN J Eng Appl Sci 9(12):2625–2634

Liu T, Li Z, Li W, Shi C, Wang Y (2013) Preparation and characterization of biomass carbon-based solid acid catalyst for the esterification of oleic acid with methanol. Bioresour Technol 133:618–621. https://doi.org/10.1016/j.biortech.2013.01.163

Lopez JM, Jimenez F, Aparico F, Flores N (2009) On-road emissions from urban buses with SCR + urea and EGR + DPF systems using diesel and biodiesel. Transp Res Part D Transp Environ 14 (1):1–5

Nabi MN, Rahman MM, Akhter MS (2009) Biodiesel from cottonseed oil and its effect on engine performance and exhaust emissions. Appl Therm Eng 29(11–12):2265–2270

Norazahar N, Yusup S, Ahmad MM, Bakar SA, Ahmad J (2012) Parametric optimization of kapok (*Ceiba pentandra*) oil methyl ester production using Taguchi approach. Int J Energy Environ 6 (6):541–548

Pandey SN (1998) Cottonseed and its utilization. Indian Council of Agricultural Research, New Delhi, p 212

Parthiban KS, Perumalsamy M (2015) Nano sized heterogeneous acid catalyst from Ceiba pentandra stalks for production of biodiesel using extracted oil from *Ceiba pentandra* seeds. RSC Adv 5(15):11180–11187. https://doi.org/10.1039/c4ra13328e

Rao BVSK, Chandramouli K, Rambabu N, Dalai AK, Prasad RBN (2011) Carbon-based solid acid catalyst from de-oiled canola meal for biodiesel production. Catal Commun 14(1):20–26. https://doi.org/10.1016/j.catcom.2011.07.011

Royon D, Daz M, Ellenrieder G, Locatelli S (2007) Enzymatic production of biodiesel from cotton seed oil using t-butanol as a solvent. Bioresour Technol 98(3):648–653

Shahid EM, Jamal Y (2011) Performance evaluation of a diesel engine using biodiesel. Pak J Eng Appl Sci 9:68–75

Shimada Y, Watanabe Y, Samukawa T (1999) Conversion of vegetable oil to biodiesel using immobilized *Candida antarctica* lipase. J Am Oil Chem Soc 76:789–793

Shu Q, Gao J, Nawaz Z, Liao Y, Wang D, Wang J (2010) Synthesis of biodiesel from waste vegetable oil with large amounts of free fatty acids using a carbon-based solid acid catalyst. Appl Energy 87(8):2589–2596. https://doi.org/10.1016/j.apenergy.2010.03.024

Silitonga AS, Masjuki HH, Mahlia TMI, Ong HC, Chong WT (2013) Experimental study on performance and exhaust emissions of a diesel engine fuelled with *Ceiba pentandra* biodiesel blends. Energy Convers Manag 76:828–836. https://doi.org/10.1016/j.enconman.2013.08.032

Sivakumar P, Sindhanaiselvan S, Gandhi NN, Devi SS, Renganathan S (2013) Optimization and kinetic studies on biodiesel production from underutilized *Ceiba pentandra* oil. Fuel 103:693–698. https://doi.org/10.1016/j.fuel.2012.06.029

Suresh G, Kamath HC, Banapurmath NR (2014) Effects of injection timing, injector opening pressure and nozzle geometry on the performance of cottonseed oil methyl ester-fuelled diesel engine. Int J Sustain Eng 7(1):82–92

Vedharaj S, Vallinayagam R, Yang WM, Chou SK, Chua KJE, Lee PS (2013) Experimental investigation of kapok (Ceibapentandra) oil biodiesel as an alternate fuel for diesel engine. Energy Convers Manag 75:773–779. https://doi.org/10.1016/j.enconman.2013.08.042

Watanabe Y, Shimada Y, Sugihara A (2000) Continuous production of biodiesel fuel from vegetable oil using immobilized *Candida antarctica* lipase. J Am Oil Chem Soc 77:355–360

Meliaceae

24

The family Meliaceae has more than 1000 species in 50 genera and is popularly referred as Mahogany family. Many of them are evergreen, but few are deciduous either during dry season or in winter. Most of them are used as high density timbers of substantial economic value. Barring few species, they are not known for oil yielding. However, the neem (*Azadirachta indica*), andiroba (*Carapa guianensis*) and syringa (*Melia azedarach*) are reported to produce oil and are utilized as raw material for biodiesel.

24.1 Neem (*Azadirachta indica*)

The neem tree *Azadirachta indica* is also known by the names *Antelaea azadirachta* (L.) and *Melia indica*. The word azadirachta is derived from the Persian word azaddhirakt which means *a noble tree*. The species *indica* is one of the two species recorded under the genus *Azadirachta*. It is a fast growing tree reaching a maximum height of 35–40 m. It is mostly bushy and green but during severe drought it may shed its entire leaves as a mechanism to tide over. The crown is mostly oval or round. The trunk is relatively short and straight which may reach a maximum diameter of 1.2 m. It is a drought resistant tree and thrives at sub-arid or sub-humid conditions with an annual rainfall ranging from 400 to 1200 mm. It prefers a well-drained deep and sandy soil with a pH 6.2–7.0 and at a temperature range of 21–35 °C but unable to tolerate any temperature below 5 °C. In India this tree is referred by many names such as sacred tree, heal all, nature's drug store and village pharmacy as it is a panacea for all diseases. Every part of this tree has pharmaceutical value. Concoction of the bark is used as a spermicide. The seed oil is used in the manufacture of soaps, shampoos, balms, creams and mosquito repellents.

S. P. Raj et al., *Biodiesel from Flowering Plants*,
https://doi.org/10.1007/978-981-16-4775-8_24

24.1.1 Vernacular Names

The neem tree is called by different names all over the world (Table 24.1).

As this tree has wide popularity as a medicinal plant, people living in different States of India personally know this tree and therefore it is called by different names prevalent in the Indian states according to the respective language (Table 24.2).

Table 24.1 The vernacular names of neem in different countries

Countries	Names
England	Indian lilac, Indian cedar, cornucopia, margosa
Nigeria	Dogon Yaro
Arabia	Neeb, Kinin, neem, nim
Sri Lanka	Kohomba
Burma	Tamar bowtamaka, tamabin, tamakha, tamarkha
Vietnam	Xoan An Do, saafu daau, sau-dau
Spain	Paraiso
Africa (East)	Muarubaini
France	Azadiracdel' Inde, margosier, margousier, neem, im
Indonesia	Imba, intaran, membha, mempheuh, mamba, mind
Persia	Azad Dirakht
Java	Mimba
Khmer	Sdau
Tibet	Dao, kadau
Malaysia	Baypay, mambu, sadu, veppam
Nepal	Neem
Thailand	Cha-tang, kadao, kwinin, sadao

Table 24.2 The vernacular names of neem in different languages in different states of India

Language	Names
Hindi	Neem, im, nind, balnimb
Urdu	Neem
Bengali	Nim
Punjabi	Nimm, bakam
Malayalam	Ariya veppu
Sanskrit	Nimba
Oriya	Nimba
Marathi	Kadu-limba, kadu nimb
Telugu	Nimtree, vepu, vembu, vepa
Kannada	Bevu
Tamil	Vembu, vepa, veppam, veppu
Gujarati	Limbado, limado

24.1.2 Geographical Distribution

A. indica is a native of India, Burma, Bangladesh, Sri Lanka, Malaysia and Pakistan. No other tree in the world has an extensive distribution as that of *A. indica*. It is successfully grown in Saudi Arabia, Seychelles, Sierra Leone, Singapore, Somalia, South Africa, Trinidad and Tobago, Uganda, Tunisia, United States of America, Uruguay, Ghana, Guatemala, Haiti, Honduras, Jamaica, Venezuela, Virgin Islands, Zambia, Zimbabwe, Mauritania, Mexico, Morocco, Mozambique, Namibia, Nepal, Nicaragua, Chad, Chile, China, Colombia, Congo, Cote d'Ivoire, Cuba, Egypt, Guinea, Eritrea, Ethiopia, Guyana, Gabon, The Gambia, Kenya, Lesotho, Liberia, Madagascar, Malawi, Mali, Algeria, Angola, Antigua and Barbuda, Argentina, Australia, Barbados, Nigeria, Panama, Paraguay, Peru, Philippines, Puerto Rico, Rwanda, Benin, Botswana, Brazil, Burkina Faso, Burundi, Cameroon, Cape Verde, Djibouti, Dominica, Ecuador, Saint Vincent and the Grenada, Sudan, Suriname, Swaziland, Tanzania, Togo, Laos, Thailand, Africa, Fiji, Mauritius, Indonesia, Cambodia and Latin America (Fig. 24.1).

24.1.3 Botanical Features

It has opposite pinnate leaves with 20–31 medium to dark green leaflets of about 3–7 cm length. Tender leaves are mostly reddish purple in colour which slowly changes first to pale green and then to dark green. Such mature leaflet is more or less asymmetric and the margins are dentate. The flowers are white, fragrant and arranged as drooping panicle of 25 cm length. Each inflorescence bears 150–250 flowers. Such inflorescence is 5–6 cm long and 8–11 cm broad. Flowers are bisexual. Fruit is drupe ellipsoidal, 1–2 cm long, greenish or greenish-yellow or yellow or purple when ripe with a single seed. The exocarp is thin, mesocarp pulpy, endocarp

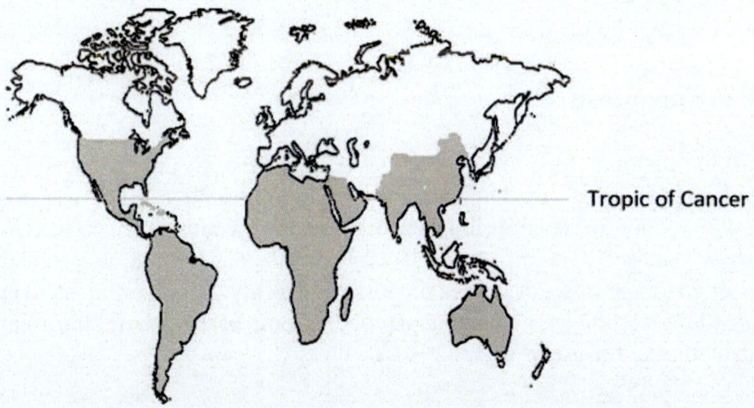

Fig. 24.1 Distribution pattern of Neem in the world as seen in the hatched area

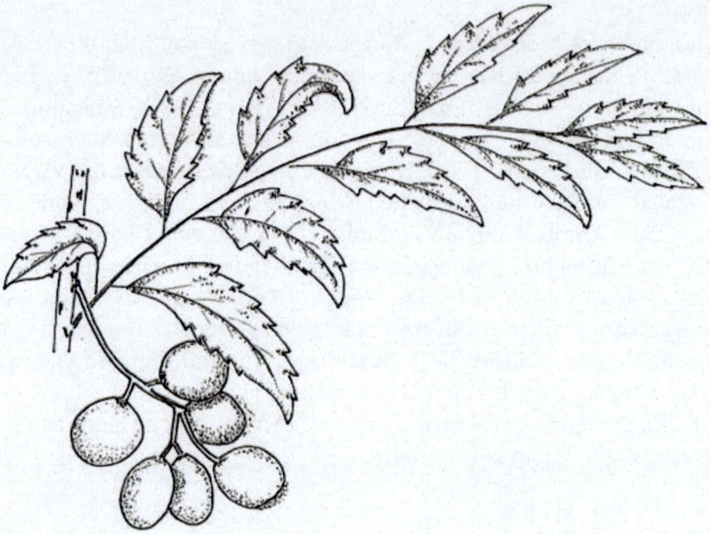

Fig. 24.2 The neem *Azadirachta indica*

Table 24.3 Parameters considered for raising neem plantation (Azam et al. 2005)

Spacing of trees (m × m)	5 × 5
Number of trees ha^{-1}	400
Average yield of seed $tree^{-1}$ $year^{-1}$ (kg)	60.0
Average yield of oil $tree^{-1}$ $year^{-1}$ (kg)	25.0
Total yield of oil ha^{-1} $year^{-1}$ (kg)	10,000
Biodiesel production ha^{-1} $year^{-1}$ (kg)	8000

cartilaginous, seed ovoid or spherical with the apex pointed, testa thin and composed of an outer shell and a kernel (Fig. 24.2).

24.1.4 Propagation

The fruits are first depulped to separate the seeds. The seeds are then washed in tap water and dried in open sun. They are sown straight into 30 × 30 × 30 cm pit at a spacing of 5 × 5 m. In extreme climates nursery raised seedlings are planted. Mature and current year seeds germinate in 10 days. It has a germination percentage of 80–90. Long storage of seeds causes the loss of viability. Air layering, stem cutting, grafting and marcotting are also being practiced. Some of the general information of neem plantation is furnished in Table 24.3.

24.1.5 Pests and Diseases

Aonidiella orientalis, a scale insect infests young branches and feeds on its sap. *Pulvinaria maxima* feeds on tender shoots. The nymphs of *Helopeltis antonii* also feed on the sap. The insects *Microtermes* sp. and *Loranthus* sp. also damage the tender leaves. Mistletoes *Dendrophthoe falcata* and *Tapinanthus* sp. also affect this tree. The fungus *Pseudocercospora subsessilis* attacks the leaves and causes shot hole effect. The bacteria *Pseudomonas azadirachtae* damages the leaves. However no specialized treatment is suggested as this tree recovers fast from any ailment. No permanent and extensive damage is reported so far.

24.1.6 Fruits and Seeds

The neem tree starts yielding fruits when it attains the age of 3. A fully grown mature tree yields around 60 kg of fruits in a year. The fruit is smooth glabrous olive like drupe. It is elongatedly oval or round in shape (1.5–2.7 cm long and 1.0–1.5 cm broad). The skin (exocarp) is thin which encloses a gelatinous 0.3–0.5 cm thick mesocarp (pulp). The pulp is bitter. The seed is embedded in the pulp. It is fairly hard and encloses the kernel. Around 4000–5000 seeds weigh a kg. Annually around 24 t of seeds are produced from a hectare which may yield around 8 t of biodiesel.

24.1.7 Oil Extraction

The potential of neem oil production in India is around 25,000 t year^{-1}. The quality of the oil largely depends on the mode of its extraction. The oil is extracted by pressing the seeds between two rollers or extracting the powdered kernels in *n*-hexane as solvent. Oil obtained through solvent extraction is generally of inferior quality. The oil recovery from seeds ranges from 25% to 45%. Hexane gives a recovery of around 40%, whereas ethyl alcohol gives a slightly higher recovery. In solvent extraction the dried seeds are first decorticated and the kernels are separated. The kernels are then dried in oven at 100 °C for a period of 2 h. The dried kernels are crushed, screened in a sieve of 0.25 mm hole and then subjected to solvent extraction.

24.1.8 Properties of Oil

The oil is light to dark brown in colour. The intensity of its colour depends on the harvest conditions. The oil has a repulsive odour due to the presence of a large amount of triterpenoid compounds and steroids (campesterol, beta-sitosterol and stigmasterol). This oil is commercially used as an insecticide, lubricant and as medicine (for external application) for many diseases (Johnson et al. 1996; Ragasa

Table 24.4 The physico-chemical characteristics of the neem oil

Parameter	Value
Colour	Brown
Odour	Garlic like
pH	5.5
Viscosity at 30 °C (cSt)	36.2
Density at 25 °C (g cc^{-1})	0.927
Free fatty acid (%)	8.98
Saponification value (mg KOH g^{-1})	210.8
Boiling point (°C)	>258
Iodine value (mg I_2 g^{-1})	885
Acid value (mg KOH g^{-1})	47
Heating value (MJ kg^{-1})	44.06

Table 24.5 Fatty acid profile of the neem oil

Fatty acid	%	Total
Saturated		
Palmitic acid C16:0	12.01	
Stearic acid C18:0	21.50	
Lignoceric acid C24:0	0.30	33.81
Unsaturated		
Oleic acid C18:1	34.80	
Linoleic acid C18:2	29.70	64.50

et al. 1997; Puri 1999; Barnwal and Sharma 2005). Chosen properties of the oil are given in Table 24.4.

It contains four major fatty acids. They are palmitic acid, stearic acid, oleic acid and linoleic acid. Very low quantity of lignoceric acid also is present (Table 24.5). The quantum of unsaturated fatty acid is high. Fatty acid with low carbon number is observed to be absent indicating the difficulty of using neem oil based biodiesel in automobiles.

24.1.9 Biodiesel Preparation

The usual problem in the processing of neem oil for biodiesel is its rich quantity of free fatty acid (>8%). Therefore, this oil cannot be converted to biodiesel so easily using alkali as catalyst due to the risk of foaming because of soap formation. Soap if formed in the process will hamper the separation of the biodiesel from its co-product, namely glycerin. Alternatively, acid catalysts are considered which are tolerant to free fatty acids (Aksoy et al. 1988). The esters (biodiesel) formed in the process relates to the type of alcohol used. The esters of methanol, ethanol, 1 propanol, 2-propanol, 2-butanol and 2-ethoxyethanol are formed depending on the alcohol used. The ester prepared through acid catalysed reactions has low viscosity when compared to alkali catalysed esters. Departing from such contention methanol based ester (methyl ester) has the lowest viscosity. High viscosity of a fuel or biodiesel or ester is often characterized by the presence of unreacted or partially reacted oils.

As the free fatty acid content of the neem oil is high, two stage process is recommended. During the first stage the free fatty acid content of the oil is reduced which is called as pre-esterification. In the second stage the usual transesterification is carried out. The pre-esterification (first stage) is carried out in the presence of catalysts such as sulphuric acid, phosphoric acid or sulphonic acid. These catalysts are referred as homogeneous catalysts as they have the same phase (i.e. liquid state to act on liquid substrate). Generally, the acid catalysts require more alcohol in the reaction to achieve a satisfactory level of conversion. Safe removal of the acid catalyst after the reaction is not possible and therefore the acids added in the process have to be washed away. The resultant effluent ultimately reaches the environment which is a cause for concern. To solve this problem solid catalysts are being considered. Such solid catalysts are termed as heterogeneous catalysts due to the physical difference of the phase (i.e. solid state of the catalyst to act on liquid substrate). The solid catalyst can easily be removed from the reaction mixture. They also have the potentiality to be regenerated and therefore they are environmentally benign and can be used repeatedly without the generation of any waste materials. Several solid catalysts are being employed. They are zeolites, ion exchange resins, mixed metal oxides and sulphated zirconia. Among the above, sulphated zirconia has received considerable attention during the last 20 years due to its strong acid properties (Muthu et al. 2010). The particle size of sulphated zirconia is less than 50 μm and is in an agglomerated form. This catalyst is used at the rate of 1% at 65 °C with a methanol to oil molar ratio of 9:1. At the above temperature alcohol is likely to evaporate. Therefore it is condensed back using a reflex arrangement. The contents are gently stirred. The reaction time is 2 h. Consequently, the whole contents are centrifuged at 8000 rpm for 10 min. The excess methanol which moves to the upper region is removed. The catalyst which remains as solid form also is separated. The remaining material is used for the second stage process, namely alkali catalysed transesterification. The recovery in the first stage will be more than 95%. Phosphoric acid modified mordenite also is used as an acid catalyst to reduce the free fatty acid present in the neem oil (Sathyaselvabala et al. 2010). The modified mordenite is able to reduce even 93% of the free fatty acid. A methanol to oil molar ratio of 6:1 is observed to be optimum. The catalyst loading at the rate of 1% is observed to maximize the conversion. The optimum reaction temperature is 60 °C.

Amberlyst BD 20 is a polymer based commercial catalyst being used to reduce the free fatty acid of the neem oil. The neem oil is heated to 80 °C and mixed with 85% phosphoric acid at the rate of 0.09% by weight of the oil. This is blended at 2000 rpm for 2 min to ensure complete mixing. It is then neutralized using 0.5 N sodium hydroxide solution. After 10 min the above content is centrifuged at 1800 rpm for 10 min at 85 °C so that water and gums are separated. The oil layer which remains at the top of the centrifuge tube is separated and treated with silica and diatomaceous earth so as to bleach the oil. The oil is then drained and dried at vacuum. The dried and bleached oil is then filtered through 2 μm filter in Buchner funnel. To 1600 g of the above oil 350 g of methanol is added and mixed. To this

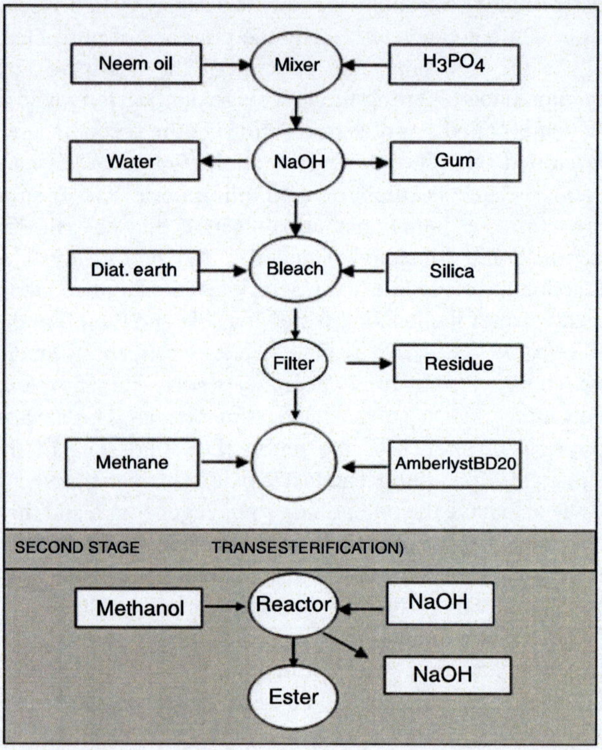

Fig. 24.3 Flow chart of Amberlyst BD20 treatment on neem oil

mixture (1600 + 350 = 1950 g) 300 g of activated Amberlyst BD 20 is added and the whole content is heated and refluxed in a condenser. After 1 h the methanol is distilled out at 80–85 °C under vacuum to ensure the complete removal of methanol. The free fatty acid content is then analysed. If the free fatty acid level is more than 0.5%, another dose of 350 mL of methanol is added. It is refluxed for one more hour and distilled out subsequently. The Amberlyst BD 20 is then removed, repeatedly washed in methanol and stored in methanol for further use. The finished product, namely the oil thus obtained with the free fatty acid reduced to a greater extent is further processed using alkali catalyst and methanol to remove the glycerin so as to obtain the biodiesel (Fig. 24.3).

During the second stage also the above material is heated to reach 60 °C and then treated with 1% KOH in methanol. The molar ratio of methanol to oil is 6:1. The contents are stirred gently for 2 h and allowed to stand undisturbed thereafter. The lower layer containing glycerin and other impurities is drained out. The remaining content is stirred with double of its quantity of hot (60 °C) water. The wash water which moves to the bottom is drained. Such washing process is repeated thrice. The biodiesel which remains at the upper layer is separated and heated to expel the moisture. A recovery of around 95% may be obtained. Thus in the two stage process the total recovery of biodiesel from 1 L of the neem oil is 902.5 mL (95 × 9.5).

24.1.10 Characteristics of the Biodiesel

The characteristics of the biodiesel obtained from neem oil indicate that this biodiesel is unable to reach the required viscosity and heating value. Otherwise certain other parameters such as density and cetane number nearly matched with that of the petro-diesel (Table 24.6).

The chances of using neem biodiesel in its pure form are very meek in view of its high viscosity and low heating value. Hence it is suggested to blend it with regular diesel at a maximum level of 15%. Nevertheless, this proposal has a tremendous scope due to the fact that neem is being grown universally and the quantity of oil available is bountiful so as to mix it with a large quantity of the petro-diesel all over the world.

24.1.11 Engine Performance

The performance of neem based biodiesel in a naturally aspirated direct injection diesel engine was studied. In a blend of 15% with petro-diesel the emission characteristics were studied at an engine speed of 1000 rpm with a fuel injection timing of 13 ATDC (after top dead centre). The dead centre refers to the position of the piston with reference to the crank shaft in a reciprocating engine where the applied force is straight along its axis meaning that no turning force is applied to it. It is observed that the level of CO and smoke discharge in the exhaust are lower by 5% when 15% blend is used. Such reduction is due to the fact that the biodiesel contains higher quantity of inherent oxygen. Excess oxygen causes the additional generation of NO_x to the tune of around 6%.

In order to reduce the NO_x emission, the technique of Exhaust Gas Circulation Rate (EGR) is adopted in diesel injection four stroke engine at an engine speed of 1000 rpm and at a brake mean effective pressure (BMEP) of 0.37 MPa (Nabi et al. 2006). In EGR a portion of the exhaust gas of the engine is sent back to the cylinders of the engine. This inert exhaust gas displaces the amount of combustible matter in the cylinder (Fig. 24.4).

Table 24.6 Characteristics of the neem biodiesel (Nabi et al. 2006)

Property	Value
Viscosity at 30 °C (cSt)	8.8
Density at 25 °C (g cc^{-1})	0.82
Cetane number	51.0
Carbon (%)	76.7
Hydrogen (%)	12.1
Oxygen (%)	11.15
C/H ratio	6.33
Sulphur (%)	<0.004
Total glycerin (%)	0.027
Heating value (MJ kg^{-1})	40.10

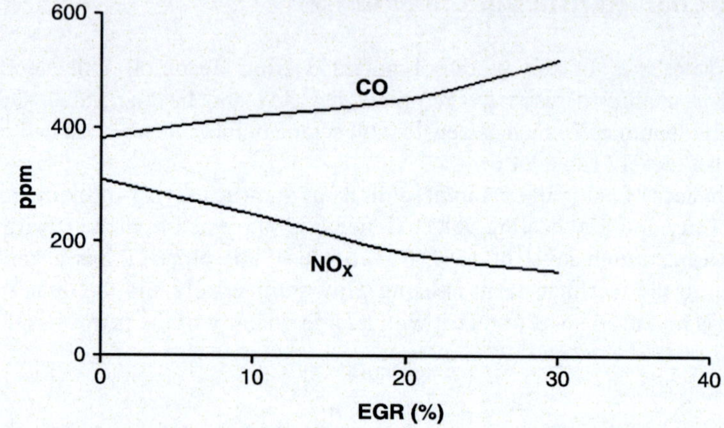

Fig. 24.4 Impact of EGR on NO_x and CO emission in 15% neem biodiesel and petro-diesel blend. The engine speed is 1000 rpm and the BMEP is 0.37 MPa

It is observed that when the EGR is increased the NO_x emission quantity decreases. At the same time the concentration of CO_2 and H_2O increases. As a result, the heat capacity of CO_2 and H_2O becomes higher. Such increase in the heat capacity with reference to CO_2 and H_2O results in the lowering of NO_x production. When the EGR is further increased CO and smoke emission increases. Figure 24.4 delineates the gradual reduction in NO_x when EGR is increased in 15% neem biodiesel blend. The NO_x and CO level shows inverse relationship. Almost a mirror image is established. However, the brake thermal efficiency decreases along with the EGR.

The effect of adiabatic flame temperature (AFT refers to the theoretically maximum temperature obtained by the combustion products if no energy is dissipated to the outside environment) and NO_x emission is compared against equivalence ratio (it refers to the ratio of air and fuel by mass in an internal combustion engine). The AFT increases when the equivalence ratio is increased up to 1.10. When the ratio is increased further, the AFT reduces. This decline in AFT is due to the higher proportion of fuel contained in the mixture at the higher equivalence ratio. Similarly, the NO_x emission gradually increases till the ratio of 0.97 and beyond the above value NO_x level reduces.

24.2 Andiroba (*Carapa guianensis*)

Andiroba *Carapa guianensis* (also called as crab wood) is a massive tree growing to a maximum height of 45 m and a diameter of 2 m (McHargue and Hartshorn 1983). It prefers to grow in swampy, inundated land, marshy edges and riverbanks. They grow at an altitude of 0–350 m above MSL (Pennington 1981). The flowers are arranged in large inflorescences of 20–80 cm long. The fruit is large with a four

Fig. 24.5 The andiroba, *Carapa guianensis*

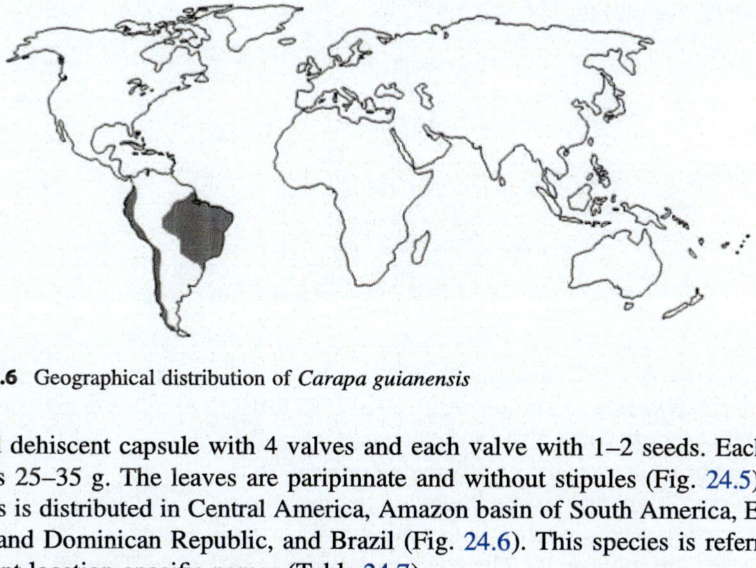

Fig. 24.6 Geographical distribution of *Carapa guianensis*

angled dehiscent capsule with 4 valves and each valve with 1–2 seeds. Each seed weighs 25–35 g. The leaves are paripinnate and without stipules (Fig. 24.5). This species is distributed in Central America, Amazon basin of South America, Eastern Cuba and Dominican Republic, and Brazil (Fig. 24.6). This species is referred by different location-specific names (Table 24.7).

24.2.1 Oil and Biodiesel Preparation

The seeds removed from the capsules are cut into small pieces, sun dried, powdered and the oil extracted from it using *n*-hexane for 4 h. The extracted material is distilled to expel the hexane and to recover the oil. The extraction yield is around 49%. The properties of the oil are given in Table 24.8.

Table 24.7 Location-specific names of *C. guianensis* in different countries

Country	Names
England US	Carapa, Crab wood
Belize	Bastard mahogany, Warawere
Brazil	Andiroba
Colombia	Masábolo, Tangarillo, Tangaré
Costa Rica	Cedro bateo, Cedro Macho
Cuba	Najesi
Dominica	Acajou
Dominican Republic	Cabirma de Guinea
French Guiana	Carapa blanc, Carapa rouge
Grenada	Crappo, Crappowood
Guyana	Carapa, Crabwood
Surinam	Krappa, Krappaboom
Venezuela	Carapa

Table 24.8 Properties of *C. guianensis* oil (Cabral et al. 2013; Iha et al. 2014)

Properties	Value
Kinematic viscosity at 40 °C (cSt)	38.4
Density at 20 °C (kg m^{-3})	915
Oxidative stability (h)	2.8
Acidity (mg KOH g^{-1})	36.1
Heat of combustion (MJ kg^{-1})	37.6
Myristic acid (C14:0) (%)	0.5
Palmitic acid (C16:0) (%)	25.3, 25.8
Stearic acid (C18:0) (%)	10.5, 9.1
Oleic acid (C18: 1) (%)	57.8, 51.8
Linoleic acid (C18:2) (%)	5.9, 8.3

It has around 3.5% unsaponifiable matter comprising triterpenes, steroids, coumarins, flavonoids and limonoids (Cabral et al. 2013). This oil is in high demand in cosmetics industry and also used as an insect repellent.

The biodiesel is prepared by a two-step process. In the first step the oil is treated with methanol-sulphuric acid mixture at a temperature of 55 °C for 2 h. Then the contents are repeatedly washed with 5% sodium bicarbonate solution. This is dried over magnesium sulphate for 2 h and filtered. It is then treated for 2 h with potassium hydroxide-methanol mixture for the transesterification. On completion of the reaction the whole content is allowed to stand undisturbed overnight. The lower layer containing the glycerol is removed. The upper layer containing the biodiesel is heated to 70 °C to expel methanol. The methyl ester is then washed with 5% (by volume) phosphoric acid and then with water. The water is removed by heating. Table 24.9 provides the properties of biodiesel as given by Iha et al. 2014.

Very few works are available on the conversion of this oil into biodiesel and consequently there is no work on the engine performance. Many of the parameters indicated in Table 24.3 match well with the standard specification.

Table 24.9 Properties of *C. guianensis* biodiesel

Properties	Value
Viscosity (cSt)	4.6
Density (g m^{-3})	0.875
Heat of combustion (MJ kg^{-1})	39.1
Oxidative stability (h)	6.7
Copper strip corrosion	1 a

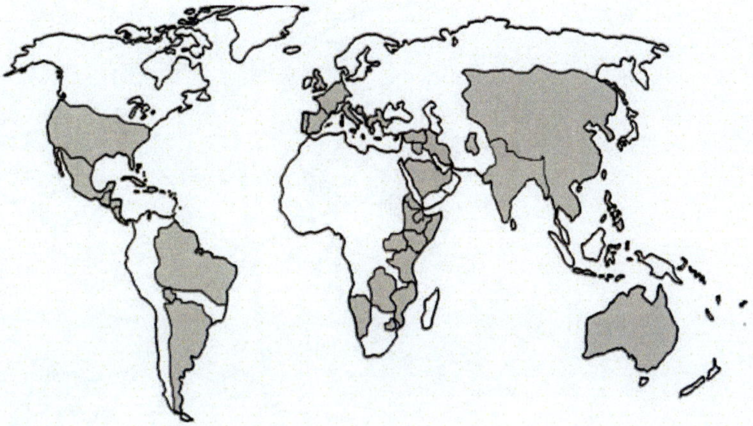

Fig. 24.7 The geographical distribution of *Melia azedarach*

24.3 Syringa (*Melia azedarach*)

The syringa, *Melia azedarach* is a tree of the subtropical zone of the seasonal forest. It has its best growth in the altitude 0–1800 m, annual temp. 23–27 °C and mean rainfall 350–2000 mm. It is a native of India, Bangladesh, Indonesia, Laos, Malaysia, Sri Lanka, Vietnam, Thailand, Myanmar, Nepal, Pakistan and Papua New Guinea. It is now distributed in Afghanistan, Albania, Argentina, Australia, Botswana, Brazil, Brunei, China, Croatia, Cuba, Djibouti, Dominican Republic, Eritrea, Ethiopia, France, Greece, Guatemala, Honduras, Iran, Iraq, Italy, India, Jamaica, Kenya, Korea, Malta, Mexico, Mozambique, Namibia, Nicaragua, Panama, Paraguay, Philippines, Portugal, Puerto Rico, Saudi Arabia, Singapore, Solomon Islands, Somalia, South Africa, Spain, Swaziland, Tanzania, Tonga, Turkey Uganda, United Kingdom, USA and Zanzibar (Fig. 24.7).

This tree grows to a height of 12 m and has a round crown. Leaf has a length of 50 cm with dark green serrated leaflets. The flowers are in inflorescence of 20 cm long. Flowers occur in numerous numbers and are white in colour with lobed petals. The fruit is tiny (15 mm dia.) and slightly fleshy (Fig. 24.8). The seed is oval in shape (2.5 cm long and 1.8 cm wide). Every part of this tree has reputed therapeutic values.

This tree is called by many local names (Table 24.10).

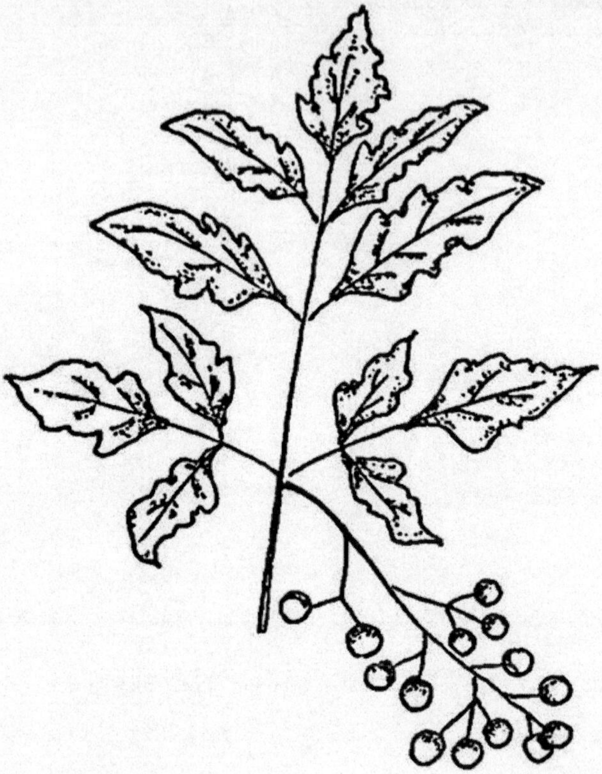

Fig. 24.8 The syringa *Melia azedarach*

Table 24.10 Local names of *M. azedarach* in many languages

Languages	Names
English	China berry, China tree, Persian lilae, pride of China, pride of India, syringa, azedarach, white cedar, bead tree
Indian	Bengali: mahanim, mahnim, bakarjan, ghora nim Gujarati: bakam limdo Hindi: betain, deikna, dek, drek, bakain, mallan, bakarja Sanskrit: mahanimba Tamil: puvempu, mallay vembu, malai vembu
Filipino	Paraiso, pagalunga, balagango
French	Lilas des Antilles, lilas des Indes, fleurs lias, piment d' eau
German	Zedarach Baum, paterroster Bawn
Hausa	Kurnam nasara
Indonesian	Marambung, mindi, gringging
Italian	Alberto dis paternostri
Malay	Mindi kecil
Spanish	Arbal enano, Alilila, violeta, kohomba

24.3.1 Extraction of Oil

The seeds are first cleaned, dried and crushed. During this process the kernels are separated which is further dried in hot sun, powdered and extracted using *n*-hexane in a Soxhlet apparatus for 10 h. The extract obtained is filtered and distilled to free it from hexane.

24.3.2 Properties of Oil

The physicochemical properties of the *M. azedarach* oil are given in Table 24.11.

The oil content of the *M. azedarach* seed is 32% (Akhtar et al. 2016). The unsaturated fatty acid forms around 86% of the total fatty acids which implies that this oil is susceptible to quick denaturation due to auto oxidation.

24.3.3 Preparation of Biodiesel

In 420 mL of methanol, 10 g of NaOH is dissolved to get methoxide which is then added to 1000 mL of freshly prepared oil. This content is sonicated in an ultrasonic irradiator at 36 °C for 20 min. The reacted product is taken in a separatory funnel, and is allowed to stand undisturbed until the phase separation is completed. The ester moves to the top and the glycerol settles down at the bottom. The catalyst and the unreacted methanol stand in between both the layers as a third phase. This middle layer slightly diffuses in the upper phase and liberally dissolves in the lower phase as the alcohol present in the middle layer has a tendency to dissolve in the glycerol. The top layer having the ester is isolated and washed with hot water repeatedly and dried by passing it over anhydrous calcium chloride. The biodiesel production in this process is around 93%. The use of ultrasonic irradiation is preferred over alkali catalysed transesterification due to the presence of free fatty acid at high level. Instead of methanol, ethanol is considered in view of the fact that ethanol is a renewable material and is eco-friendly. The ethanol has an advantage of high miscibility with oil as compared to that of methanol. The ethanol driven reaction

Table 24.11 Physicochemical properties of *M. azedarach* oil

Parameters	Stavarache et al. (2008)	Akhtar et al. (2016)
Kinematic viscosity at 40 °C (cSt)	15.46	–
Density at 25 °C (kg m^{-3})	902	–
Acid value (g KOH kg^{-1})	5.75	–
Iodine value (g I_2/100 g)	126.6	–
Palmitic acid (C16:0) (%)	–	9.54
Stearic acid (C18:0) (%)	–	4.35
Oleic acid (C18:1) (%)	–	69.55
Linoleic acid (C18:2) (%)	–	16.53

Table 24.12 Physicochemical properties of the fatty acid methyl ester and ethyl ester (Stavarache et al. 2008)

Parameters	Fatty acid methyl ester	Fatty acid ethyl ester
Kinematic viscosity at 40 °C ($mm^2\ s^{-1}$)	4.32	4.6
Density ($kg\ m^{-3}$)	892	888
Iodine value (g I_2/100 g)	131	124
Cetane number	48	53
Neutralization number (g KOH kg^{-1})	0.5	0.72
Palmitic acid (C16:0) (%)	10.1	10.1
Stearic acid (C18:0) (%)	3.67	3.64
Oleic acid (C18:1) (%)	22.3	22.97
Linoleic acid (C18:2) (%)	64.89	66.43

also is faster than that of methanol. However, the separation of ethyl ester from glycerol is cumbersome to a certain extent and hence ethanol is often not favoured. Besides, the phospholipids present in the oil as impurity are not acted upon by ethanol. Thus the phospholipids continue to remain as impurity even after the formation of ester. During the washing of the esters with warm water, the phospholipid impurities are hydrated and move into the water. Thus there is low recovery of esters in the event of using ethyl alcohol. Besides, the ethyl ester mixture has a high neutralization number in comparison with methyl esters. During the neutralization, soap will be formed which disintegrates into salt and free fatty acid. This free fatty acid thus formed has an affinity with the ester. Thus the neutralization is further increased. Moreover the ester derived from the use of ethanol normally has a relatively high viscosity. The physicochemical properties of the methyl ester and ethyl ester are given in Table 24.12.

The kinematic viscosity is within the limit. The combustion efficiency of the fuel is normally governed by the cetane number. The cetane number is marginally high in this biodiesel. The neutralization number reflects the presence of free fatty acid. Iodine value is an index of stability. This biodiesel has a high unsaturation and hence, it is susceptible to thermal and oxidative polymerization. The polymerized component ultimately forms insoluble fractions which interfere with the smooth flow of the fuel along the injection system.

References

Akhtar T, Tariq MJ, Rana SI (2016) Production of biodiesel from *Melia azedarach* seed oil a non-edible feedstock for biodiesel. Sci Int (Lahore) 28(4):3805–3807

Aksoy HA, Kahraman I, Karaosmanoglu F, Civelekoglu H (1988) Evaluation of Turkish sulfur olive oil as an alternative diesel fuel. JAOCS 65(6):936–938

Azam MM, Waris A, Nahar NM (2005) Prospects and potential of fatty acid methyl esters of some non-traditional seed oils for use as biodiesel in India. Biomass Bioenergy 29:293–302

Barnwal BK, Sharma MP (2005) Prospects of biodiesel production from vegetable oils in India. Renew Sustain Energy Rev 9:363–378

Cabral EC, de Cruz GF, Simas RC, Sanvido GB, Goncalves LV, Real RVP, da Silva RCF, da Silva JCT, de Sa GF, Eberlin MN (2013) Typication and quality control of the andiroba (*Carapa guianensis*) oil via mass spectrometry finger printing. Anal Methods 5:1385–1391

Iha OK, Alves FCSC, Suarez PAZ, Silv CRP, Meneghetti MR, Meneghetti SMP (2014) Potential application of *Terminalia catappa L.* and *Carapa guianensis* oils for biofuel production: physical-chemical properties of neat vegetable oils, their methyl-esters and bio-oils (hydrocarbons). Ind Crops Prod 52:95–98

Johnson S, Morgan ED, Peiris CN (1996) Development of the major triterpenoids and oil in the fruit and seeds of neem *Azadirachta indica*. J Ann Bot 78:383–388

McHargue LA, Hartshorn GS (1983) Seed and seedling ecology of *Carapa guianensis*. Turrialba 33:399–404

Muthu H, Sathyaselvabala V, Varathachary TK, Selvaraj DK, Nandagopal J, Subramanian S (2010) Synthesis of biodiesel from neem oil using sulfated zirconia via transesterification. Braz J Chem Eng 27(4):601–608

Nabi MN, Akhter MS, Shahadat MMZ (2006) Improvement of engine emissions with conventional diesel fuel and diesel-biodiesel blends. Bioresour Technol 97:372–378

Pennington TD (1981) Meliaceae (Flora Neotropica, monograph No: 28). New York Botanical garden Pr. Department, New York, p 470

Puri HS (1999) Neem-the devine tree. Harwood Academic Publishers, Amsterdam

Ragasa CY, Nacpil ZD, Natividad GM, Tada M, Coll JC, Rideout JA (1997) Tetranortriterpenoids from *Azadirachta indica*. J Phytochem 46:555–558

Sathyaselvabala V, Varathachary TK, Selvaraj DK, Ponnusamy V, Subramanian S (2010) Removal of free fatty acid in *Azadirachta indica* (neem) seed oil using phosphoric acid modified mordenite for biodiesel production. Bioresour Technol 101:5897–5902

Stavarache CE, Morris J, Maeda Y, Oyane I, Vinatoru M (2008) Syringa (*Melia azedarach*): a potential source for biodiesel fuel. http://www.researchgate.net/Publication/268176979

25 Nyssaceae

The family Nyssaceae composed of around 40 species. They are found in river swamps, floodplains, lake margins, bay heads and bottomlands. These plants are either shrubs or trees of deciduous nature. Most of the members of this family are reported to have been distributed in North America, Mexico, Central America (Costa Rica, Panama) and Asia. The seeds of one of the members, namely happy tree (*Camptotheca acuminat*) are used to extract oil for biodiesel.

25.1 Happy Tree (*Camptotheca acuminata*)

The happy tree, *Camptotheca acuminata*, a deciduous tree growing to the height of 20 m is a native of Southern China (Fig. 25.1). It is also distributed in Tibet. It belongs to the family Nyssaceae. It grows well when there is sufficient water in a nutrient rich soil. It is mostly located on the banks of streams and rivers at an elevation of 100 m above MSL. The leaves are simple, alternate, pinnate and ovate and 15 cm long. This tree is often called as tree of life, wild anise, wild banana and cancer tree. The bark of this tree contains an alkaloid camptothecin which is used in the treatment of cancer. The oil from the seeds of this tree has a high acid value. These seeds were discarded for a long time as it has no tangible use. Now the seed oil is actively considered as a raw material for biodiesel. Figure 25.2 shows a twig of the tree with fruits. Each fruit weighs 40–70 mg and seed weighs 30–38 mg.

25.1.1 Oil and Biodiesel

The tiny seeds in bulk are dried, ground and extracted for oil using *n*-hexane. The extract is then distilled to remove the solvent. The property of the oil thus obtained is given in Table 25.1.

S. P. Raj et al., *Biodiesel from Flowering Plants*,
https://doi.org/10.1007/978-981-16-4775-8_25

Fig. 25.1 Geographical distribution of *Camptotheca acuminata*

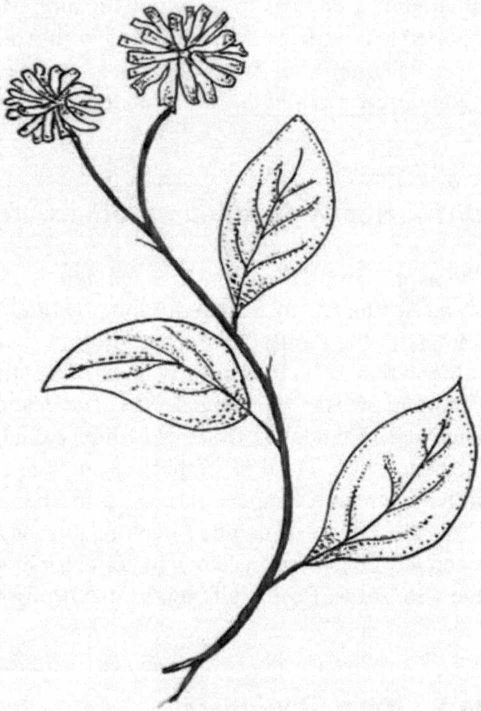

Fig. 25.2 The happy tree, *Camptotheca cuminata*

Biodiesel of *C. acuminata* is prepared through transesterification process using Brønsted acidic ionic liquid with metal sulphate (BSO_3HMIM) HSO_4 and $Fe_2(SO_4)_3$. The $Fe_2(SO_4)_3$/(BSO_3 HMIM) HSO_4 is prepared by mixing the $Fe_2(SO_4)_3$ in the ionic liquid at a molar ratio of 3:1. The ionic liquid metal sulphate

Table 25.1 The properties of *C. acuminata* seed oil (Li et al. 2014)

Parameters	Value
Density at 20 °C (kg m^{-3})	920
Acid value (mg KOH g^{-1})	23.7
Saponification value (mg KOH g^{-1})	205

Table 25.2 Properties of the *C. acuminata* biodiesel (Li et al. 2014)

Parameters	Value
Density at 20 °C (kg m^{-3})	880
Viscosity at 40 °C (cSt)	4.2
Acid value (mg KOH g^{-1})	0.37
Flash point (°C)	109
Cetane number	57

catalyst is calcinated at 350 °C for 2 h (Tanabe and Takeshita 1967; Takeshita et al. 1973).

The methanol to oil ratio is 5:1. The catalyst concentration is 4% of the oil. The reaction temperature is 60 °C and the duration is 2 h. The oil, methanol and the catalyst are mixed and stirred. The content after the reaction is poured into a separating funnel. The upper layer containing the biodiesel is removed and freed from methanol by distillation. The properties of the biodiesel thus obtained are given in Table 25.2.

The biodiesel from *C. acuminata* did not receive any commercial status since the resource is limited by volume. Therefore, works on engine performance are not yet taken off.

References

Li J, Peng X, Luo M, Zhao C, Gu C, Zhu Y, Fu Y (2014) Biodiesel production from *Camptotheca acuminata* seed oil catalyzed by novel Brønsted-Lewis acidic ionic liquid. Appl Energy 115:438–444

Takeshita T, Ohnishi R, Tanabe K (1973) Recent survey of catalysis by solid metal sulfates. Catal Rev 8:29–63

Tanabe K, Takeshita T (1967) Catalytic activity and acidic property of solid metal sulfates. Adv Catal 17:315–349

Oleaceae

26

The members of the family Oleaceae are distinctive in being woody trees or shrubs (rarely lianas), usually with peltate secretory trichomes. The shrubs are often scandent, climbing by scrambling into other vegetation. They are cosmopolitan in distribution but occur in great diversity in temperate and tropical Asia. In certain plants of this family the fruit is a drupe with a fleshy outer layer and a hard inner layer containing a single seed. In few species the fruit is a berry and fleshy throughout with one or more seeds. Olive oil is obtained from the fruit pulp and seeds of *Olea europaea*.

26.1 Olive (*Olea europaea*)

Olea europaea of the family Oleaceae is an evergreen tree native to the Mediterranean region, and is recorded to have distributed in Afghanistan, Albania, Algeria, Angola, Baleares, Botswana, Burundi, Canary Is., Cape Provinces, China South-Central, Cyprus, Djibouti, East Aegean Is., Eritrea, Ethiopia, Free State, Greece, Italy, Kenya, Kriti, KwaZulu-Natal, Lesotho, Libya, Madeira, Malawi, Mauritius, Morocco, Mozambique, Namibia, Niger, Northern Provinces, Oman, Pakistan, Palestine, Portugal, Rwanda, Réunion, Sardegna, Saudi Arabia, Sicilia, Somalia, Spain, Sudan, Swaziland, Tanzania, Tunisia, Turkey, Turkey-in-Europe, Uganda, West Himalaya, Yemen, Yugoslavia, Zambia, Zimbabwe (www.plantsoftheworldonline.org).

They grow on rocky hillsides having a low quality soil with a pH 5.5–7.5. They occur at a latitude ranging from 30° to 45° both in the Northern and Southern hemisphere. They are reported to establish at a mean sea level of 1200–2000 m with a rainfall of 500 mm per year having a temperature range of 22–28 °C (Fig. 26.1).

It grows to a height of 8–15 m. The leaves are silvery, oblong, 4–10 cm long and 1–3 cm broad. The flowers are small, white and feathery. The fruit is a small drupe of

S. P. Raj et al., *Biodiesel from Flowering Plants*,
https://doi.org/10.1007/978-981-16-4775-8_26

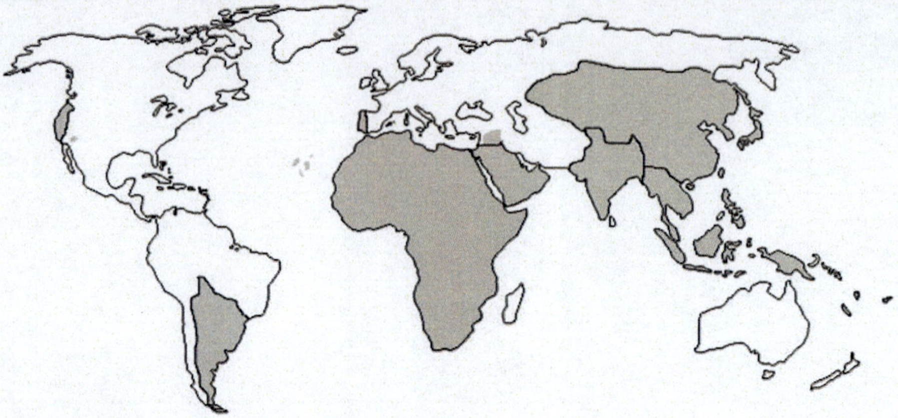

Fig. 26.1 Geographical distribution of olive, *Olea europeaea*

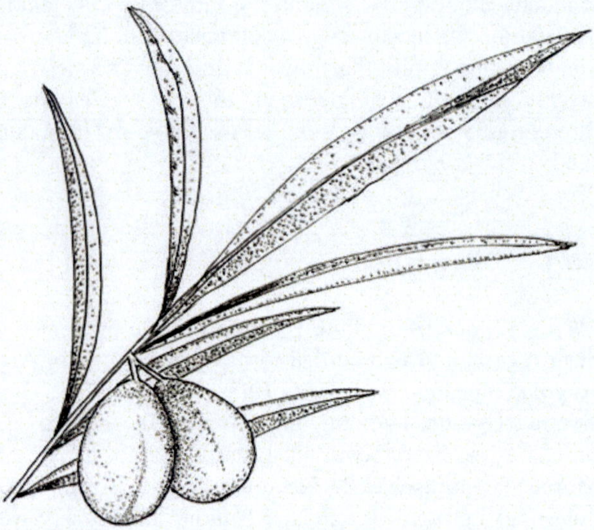

Fig. 26.2 The olive *Olea europaea* plant

1–2.5 cm long (Fig. 26.2) and each drupe carries a seed. These drupes are harvested at green or purple stage. During the year 2016 the world production of drupe was 1.9 million tonnes from a cultivated area of 1.06 million ha (1.8 t ha^{-1}). This tree has a life span of around 50 years.

It is popularly referred as olive tree (English), aceituna, oliva, olivo (Spanish), Olivier (French), az-zeitoun (Arabic), mu xi lan (Chinese), oliveira (Portuguese), drive (French Polynesia), oelbaum, olivenbaum (Germany), zayit (Israel), oliboom (The Netherlands), olivtraed (Sweden), oliwa, oliwa haole (USA/Hawaii), elia (Greek), olive (Italian) and zeytin (Turkish).

26.1.1 Extraction of Oil

The drupe has around 35% oil. At the first instance the drupes are washed to remove any contaminants and then made as a paste by grinding. The paste is then pressed so as to release a liquid which predominantly contains oil. This liquid is allowed to stand undisturbed in a container (Fig. 26.3). Often water is added and the contents diluted to enhance the phase separation.

The top layer contains oil, under beneath the watery content stands which is often called as vegetation water. The bottom layer is formed of slurry popularly referred as pomace. The pomace is resuspended in water and centrifuged to separate a low quantity of oil. The separated oil is rerun in a centrifuge at 6000 rpm that enables the maximum separation of the oil. If the oil has peroxide value higher than 10 meq kg^{-1} it is less stable.

The pomace is often used as a raw material for biodiesel. The annual production of olive pomace is 40,000 t in Greece. As the pomace is not good for human consumption, attempts are being made to make use of it for biodiesel production. The fatty acid composition of the olive pomace is given in Table 26.1.

26.1.2 Biodiesel Production

The raw oil (pomace) normally has an acid value not less than 20 g KOH kg^{-1}, which if treated through alkali catalysis is likely to generate considerable amount of soap. This soap may prevent the phase separation of the biodiesel. Therefore, this oil

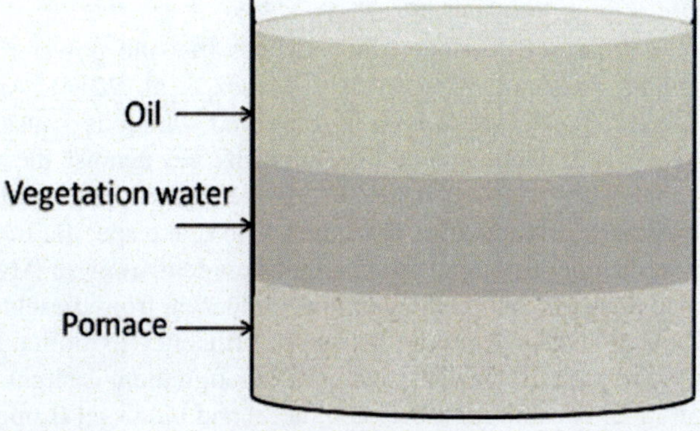

Fig. 26.3 The extract of olive fruit

Table 26.1 Fatty acid composition of the olive pomace (Che et al. 2012)

Parameters	Value	
Palmitic acid	(C16:0) (%)	9.37
Stearic acid	(C18:0) (%)	2.68
Arachidic acid	(C20:0) (%)	0.58
Palmitoleic acid	(C16:1) (%)	0.68
Oleic acid	(C18:1) (%)	75.21
Linoleic acid	(C18:2) (%)	7.84
Linolenic acid	(C18:3) (%)	0.87

Table 26.2 Physicochemical properties of the treated olive pomace (Lopez et al. 2014)

Parameters	Value
Kinematic viscosity at 40 °C ($mm^2\ s^{-1}$)	46.27
Density at 15 °C ($kg\ m^{-3}$)	908
Acid value ($g\ KOH\ kg^{-1}$)	0.6
Iodine value ($g\ I_2/100\ g$)	99.8
Heating value ($MJ\ kg^{-1}$)	40.49

is first treated with acid once or twice till the pomace has acid value to the tune of around 1 g KOH kg^{-1}. The raw oil having an acid value of around 20 g KOH kg^{-1} is stirred with methanol 0.45% by volume of oil and 1% H_2SO_4 for a total duration of 60 min. It is then settled to have the oil separated. This oil is analysed for acid value. If the reduction in acid value is not satisfactory, it is once again treated with methanol and acid. The treated oil is then processed for biodiesel using alkali as catalyst (Che et al. 2012). The physicochemical properties of the aid treated pomace are given in Table 26.2.

To a litre of the above acid treated oil, (pomace) 12.0 g KOH and 300 g of methanol are added. The molar ratio of methanol to oil is 8.6:1. The content is stirred for 40 min. It is then settled and the top layer having the biodiesel (Fig. 26.4) is separated, and is washed with warm distilled water (Lopez et al. 2014).

The properties of the biodiesel thus obtained are given in Table 26.3.

The properties of the biodiesel given in Table 26.3 qualify the standard. The viscosity, density, cold weather properties, oxidation stability, cetane number all show ideal characteristics. High oleic acid appears to balance the above parameters (Pinzi et al. 2009) (Table 26.4).

Engine performance trial indicates that the power release in the biodiesel run engine does not differ grossly (Lopez et al. 2014). An increase (15%) in brake specific fuel consumption is observed which is caused by the marginally high viscosity reflecting on the volume of fuel against the mass and the low heating value. The brake specific fuel equivalence was 327.4 and 386.7 g kWh in diesel and biodiesel, respectively. The increase in brake specific fuel consumption is inversely proportional to the heating value as evident from an ANOVA test carried out. It is also reported that the power value reduction from diesel to biodiesel is by a factor of 1.69. It is also seen that the energy efficiency exhibited by the internal combustion engine does not deviate much even though there is slight difference in engine power and fuel consumption between diesel and biodiesel (Lopez et al. 2014). The pomace

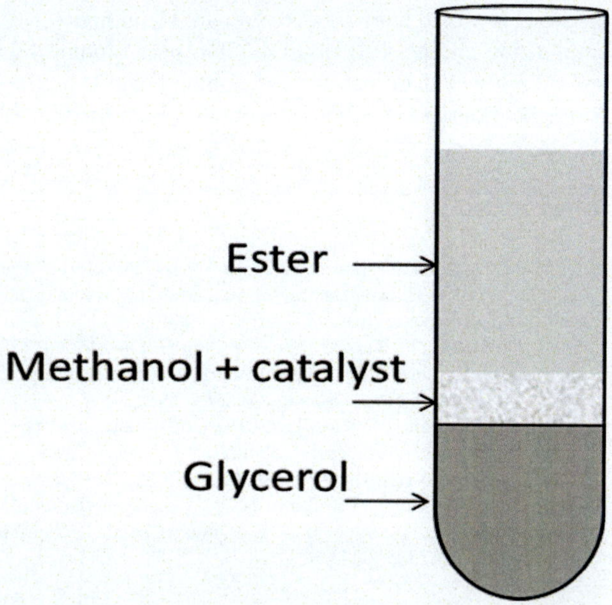

Fig. 26.4 Phase separation of transesterified pomace oil

Table 26.3 Physicochemical properties of the biodiesel from olive pomace (Lopez et al. 2014)

Parameters	Value
Kinematic viscosity at 40 °C ($mm^2\ s^{-1}$)	4.53
Density ($kg\ m^{-3}$)	876
Acid value ($g\ KOH\ kg^{-1}$)	0.1
Flash point (°C)	138
Cetane number	58.7
Iodine number ($g\ I_2/100\ g$)	78
Cloud point (°C)	−2
Cold filter plugging point (°C)	−9
Pour point (°C)	−6
Heating value ($MJ\ kg^{-1}$)	39.96
Copper strip corrosion	1a
Oxidation stability (110 °C) (h)	15.5

Table 26.4 Fatty acid ester composition of olive pomace biodiesel (Lopez et al. 2014)

Fatty acid ester	Value
Palmitic (C16:0)	11
Stearic (C18:0)	3
Palmitoleic (C16:1)	0.8
Oleic (C18:1)	75.2
Linoleic (C18:2)	7.2
Linolenic (C18:3)	0.5

biodiesel is mixed with diesel to prepare blend of 5%, 7%, 10% and tested in internal combustion engine. It is observed that pure biodiesel gave power and torque 23.5 and 22.5 lower than that of the 5% blend (5 mL biodiesel and 95 mL diesel). The corresponding specific fuel consumption increased by 64.5% (Özçelik et al. 2019).

References

Che F, Sarantopoulos I, Tsoutsos T, Gekas V (2012) Exploring a promising feedstock for biodiesel production in Mediterranean countries: a study on free fatty acid esterification of olive pomace oil. Biomass Bioenergy 36:427–431

Lopez I, Quintana CE, Ruiz JJ, Cruz-Peragon F, Dorado MP (2014) Effect of the use of olive-pomace oil biodiesel/diesel fuel blends in a compression ignition engine: preliminary energy analysis. Energy Convers Manag 85:227–233

Özçelik AE, Acaroğlu M, Köse H (2019) Determination of combustion characteristics of olive pomace biodiesel–eurodiesel fuel mixtures. Energy Sour A Recov Utiliz Environ Effects. https://doi.org/10.1080/15567036.2019.1604862

Pinzi S, Garcia IL, Lopez FJ, Castro MOL, Dorado G, Dorado MP (2009) The ideal vegetable oil-based biodiesel composition: a review of social, economical and technical implications. Energy Fuel 23:2325–2341

Papaveraceae

27

Papaveraceae is an economically important family comprising around 800 species. This family is also referred as poppy family. The members are cosmopolitan in their occurrence but they are richly distributed in temperate and subtropical climates (mostly in the northern hemisphere), and are not much known in the tropics. Most of them are annual or biennial or perennial herbs. The fruits are non-fleshy and break open on maturity to release the seeds. The Mexican poppy *Argemone mexicana* an oil yielding plant for biodiesel is dealt in this chapter.

27.1 Mexican Poppy (*Argemone mexicana*)

The Mexican poppy, *Argemone mexicana*, an annual herb growing to a maximum height of 1.5 m is hitherto not been used in any commercial purposes. It is believed to be a native of Florida. It is a hardy plant growing in roadside, derelict community dry-tank-beds, flood plains, fallow lands and in riverbanks. It remains unaffected by extreme drought conditions.

This species is often grown as an ornamental plant. Its leaves are with fissured margins having sharp teeth and thistles. The whole body is covered with prickles. The leaves are bluish green with grey-white vein and have no petioles. The flowers are located at the top of the branches. Each flower has six petals, 2–5 cm diameter and bright yellow (Fig. 27.1). The fruit is an oblong capsule insulated with spines carrying 300–400 seeds, each 1 mm diameter. A single plant has 60–90 capsules holding 30,000 seeds on an average. The seeds are dispersed by water, wind or through birds.

This plant is distributed in Florida, Mexico, Central America, the Caribbean, Spain, France, Germany, Haiti, India, Pakistan, Paraguay, Bangladesh, Puerto Rico and Zimbabwe (Fig. 27.2).

The vernacular names by which it is called are Mexican poppy, Mexican prickly poppy, medicinal weed, satyanashi, prickly poppy, ekijembajembe, ankole,

S. P. Raj et al., *Biodiesel from Flowering Plants*,
https://doi.org/10.1007/978-981-16-4775-8_27

Fig. 27.1 The mexican poppy, *Argemone mexicana*

mkumajalaga, yao, brachma thandu, agara, chardon beni, umjelemani, texas poppy, devil's fig, geelblombloudissel, ugudhithukela, yellow flowered mexican poppy and yellow mexican poppy.

27.1.1 Oil Extraction

The fruit of *A. mexicana* is harvested, dried and dehulled to obtain the seed. The seeds are then pressed in an expeller to separate the oil. The residue retained is extracted next with petroleum ether. The oils thus obtained from both the processes are then pooled. To each litre of the oil 250 mL of methanolic orthophosphoric acid (12% v/v) is added, mixed and allowed to stand overnight. The oil phase separated from the methanol layer is filtered through 60–120 mesh silica gel using a suction device. The resultant oil is then heated with dilute (0.1%) sodium hydroxide solution. The aqueous layer formed after a day is discarded and the oil is washed repeatedly to remove any residual hydroxide. It is then heated to expel moisture and passed through anhydrous sodium sulphate (Pramanik et al. 2012). The physico-chemical properties of the resultant oil are given in Table 27.1.

The *A. mexicana* seed yields around 30% oil of its weight (Sharma et al. 2012). The acid value is considerably high, warranting a two stage process in biodiesel

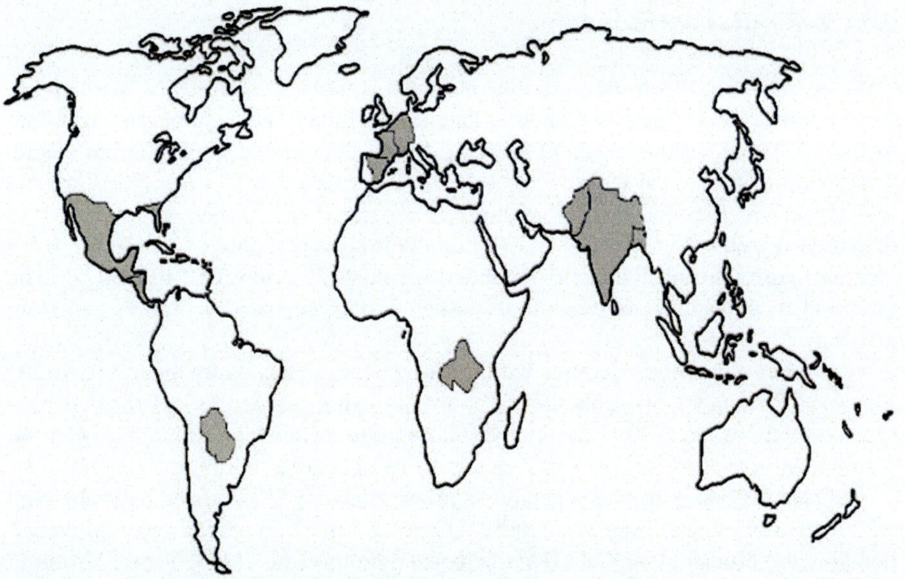

Fig. 27.2 Geographical distribution of *Argemone mexicana*

Table 27.1 Physicochemical characteristics of the *A. mexicana* oil

Parameters	Rao et al. (2012)	Sharma et al. (2012)	Sagar et al. (2016)
Kinematic viscosity at 40 °C (cSt)	29.6	31.1	6.5
Density at 15 °C (kg m^{-3})	910	904[a]	865[a]
Acid value (g KOH kg^{-1})	76.2	–	–
Flash point (°C)	235	–	193
Cloud point (°C)	12	−3	–
Pour point (°C)	−12	−6	–
Iodine value	–	111.4	–
Oxidation stability 110 °C (h)	–	1.8	–
Heating value (MJ kg^{-1})	–	–	37.5
Palmitic acid (C16: 0) (%)	14.7	11.4	–
Stearic acid (C18:0) (%)	6.8	4.0	–
Oleic acid (C18:1) (%)	40.0	28.7	–
Linoleic acid (C18:2) (%)	36.6	55.9	–

[a]At 15 °C

preparation. The viscosity has a positive correlation with monounsaturated fatty acid and negatively correlated with polyunsaturated fatty acid.

27.1.2 Biodiesel Production

The degummed moisture-free oil, thus obtained is transesterified using methanol in the presence of a suitable catalyst. The oil is mixed with sulphuric acid and methanol. The combination is 50 mL oil, 10 mL alcohol and 1 mL sulphuric acid. This mixture is stirred at 200 rpm at 60 °C for 3 h. During the course of this process two layers are formed. The upper layer is separated for further processing. It is thereafter mixed with methanol and sodium hydroxide at a ratio of 25 mL oil, 0.2 g sodium hydroxide and 5 mL methanol and stirred at 250 rpm for 4 h at 50 °C. The biodiesel thus obtained is separated, washed and heated to dry (Parida and Rout 2017).

A number of novel procedures with heterogeneous catalysts are emerging for the transesterification of oil. Calcium carbonate is being utilized as a Bronsted base catalyst (Suppes et al. 2001). Alkali metal compounds such as KNO_3 if loaded on alumina and calcined at 773–873 K act as an ideal catalyst (Cruz et al. 2007).

CaO also is being used as a catalyst and its efficiency is enhanced by preheating with methanol (Kawashima et al. 2009). Often Ca-Zn mixed oxides are employed in boosting the efficiency of CaO (Ngamcharussrivichai et al. 2008). Tetrahydrofuran (Gryglewicz 1999) is used as a co-solvent to increase the transesterification turnover by CaO as a catalyst.

Sharma et al. (2012) used CaO in combination with $Ca_{12}Al_{14}O_{33}$ as heterogeneous catalyst in the transesterification of *A. mexicana* oil. Accordingly CaO-$Ca_{12}Al_{14}O_{33}$ complex catalyst is prepared by adding calcined calcium hydroxide and aluminium nitrate monohydrate $Al(NO_3)_3{\cdot}9H_2O$ which is then mixed with 2-propanol so as to have the ratio of CaO with the newly formed $Ca_{12}Al_{14}O_{33}$ as 75:25. The content is stirred for 1 h at 75 °C till it forms a slurry. After 2 h, the slurry is transferred to a shallow dish and dried at 120 °C for 8 h. The white material, thus formed is calcined first at 500 °C for 4 h and then at 900 °C for 4 h. The resultant product is made as tiny pellets and is ready for transesterification. It has a surface area of 16.6 $m^2\ g^{-1}$ at a particle size of 14.4 μm (Sharma et al. 2012). To 100 mL of methanol, catalyst (to form 5% by wt. of oil) is added (for an oil to alcohol ratio 1:10) and the reaction proceeded by refluxing under controlled temperature (65 °C). After 4 h the content is transferred to a separating funnel and waited till clear phases are emerging. The upper phase containing the ester is separated. The catalyst used in the above process is cleaned and employed for many cycles to follow. It maintains its activity even after 5 cycles.

Crystalline manganese carbonate also is used as a catalyst in the transesterification of oil (Rao et al. 2012). The manganese carbonate 1%, molar ratio of methanol to oil 5:1 and the reaction duration of 45 min give a yield of 90% ester. The physicochemical properties of the biodiesel from *A. mexicana* are given in Table 27.2.

Table 27.2 Physicochemical properties of the *A. mexicana* biodiesel

Parameters	Rao et al. (2012)	Sharma et al. (2012)	Parida and Rout (2017)
Kinematic viscosity at 20 °C (cSt)	11.1	4.6	5.1
Density at 40 °C (kg m^{-3})	860	–	868[a]
Acid value (g KOH kg^{-1})	0.95	–	–
Flash point (°C)	235	135	130
Cloud point (°C)	12	−1	–
Pour point (°C)	−12	–	–
Cetane number	–	47.1	–
Oxidation stability at 110 °C (h)	–	1.4	–
Heating value (MJ kg^{-1})	–	–	41.5
Methyl palmitate (C16:0) (%)	3.9	–	–
Methyl stearate (C18:0) (%)	32.5	–	–
Methyl oleate (C18:1) (%)	15.0	–	–
Methyl linoleate (C18:2) (%)	47.5	–	–

[a]At 20 °C

27.1.3 Engine Performance

Marginally high viscosity and density compel the use of biodiesel in engines as blends by mixing the biodiesel with regular diesel in certain proportion. In doing so the brake thermal efficiency (BTE) normally increases as the load increases. With the increase in rpm the brake thermal efficiency decreases at a low load. In higher load the BTE increases along with the increase in rpm. But such increase is feeble. The low heating value of the blend increases the brake specific fuel consumption (BSFC). An increase of around 13% of BSFC in the blend (B40) is reported at 25% load at 400 rpm, whereas it is slightly lower (7.6%) in 75% load at 2500 rpm for B30 when compared to regular diesel (Sagar et al. 2016) .The NO_x concentration is high in biodiesel blend as compared to that of diesel. An adjustment in engine geometry, compression ratio, reduction in reaction time and reduced temperature may bring out NO_x reduction in the exhaust (Qi et al. 2009). With reference to hydrocarbon emission the level increases in the exhaust as the rpm is increased due to limited time available for combustion. At low blend (<B30) the level of hydrocarbon in the exhaust is low and it starts increasing when the blend level (>B30) is increased. The level of CO in the exhaust largely depends on the combination of air-fuel mixture. When the duration for combustion is less due to the increase in rpm the CO level in the exhaust increases. Such increase is reportedly caused by the increase in viscosity and density, which may reflect in poor combustion as a consequence of low atomization. The increased level of CO_2 in the exhaust indicates that the combustion level is appreciable. In addition, high level of CO_2 in the exhaust is the reason for the high density of biodiesel translating into high fuel mass (Ng et al.

2011). In high load there is a possibility of increased CO_2 as the air-fuel mixture becomes rich and the overall fuel mass brought into the cylinder is more.

References

Cruz AD, Kulkarni MG, Meher LC, Dalai AK (2007) Synthesis of biodiesel from canola oil using heterogeneous base catalyst. J Am Oil Chem Soc 84:937–943

Gryglewicz S (1999) Rapseed oil methyl ester preparation using heterogeneous catalysts. Bioresour Technol 70:249–253

Kawashima A, Matsubara K, Honda K (2009) Acceleration of catalytic activity of calcium oxide for biodiesel production. Bioresour Technol 100(2):696–700

Ng JH, Ng HK, Gan S (2011) Engine-out characterization using speed–load mapping and reduced test cycle of a light duty diesel engine fuelled with biodiesel blends. Fuel 90:2700–2709

Ngamcharussrivichai C, Totarat P, Bunyakiat K (2008) Ca and Zn mixed oxide as heterogeneous base catalyst for transesterification of palm kernel oil. Appl Catal A Gen 341(1):77–85

Parida MK, Rout AK (2017) Combustion analysis of Argemone mexicana biodiesel blends. Energy Sour A Recov Utiliz Environ Effects 39(7):698–705

Pramanik P, Das P, Kim PJ (2012) Preparation of biofuel from argemone seed oil by an alternative cost-effective technique. Fuel 91:81–86

Qi DH, Geng LM, Chen H, Bian YZ, Liu J, Ren XC (2009) Combustion and performance evaluation of a diesel engine fuelled with biodiesel produced from soybean crude oil. Renew Energy 34(12):2706–2713

Rao YR, Zubaidha PK, Kondhare DD, Reddy NJ, Deshmukh SS (2012) Biodiesel production from *Argemone mexicana* seed oil using crystalline manganese carbonate. Pol J Chem Technol 14 (1):65–70

Sagar A, Uppal N, Singh B (2016) Utilization of argemone oil biodiesel in commercial IDI-CI engine. Eng Technol India 7(1):33–44

Sharma M, Khan AA, Dohhen KC, Christopher J, Puri SK, Tuli DK, Sarin R (2012) A heterogeneous catalyst for transesterification of *Argemone mexicana* oil. J Am OilChem Soc 89:1545–1555

Suppes GT, Bockwinkel K, Lucas S, Botts JB, Mason MH, Heppert JA (2001) Calcium carbonate catalyzed alcoholysis of fats and oils. J Am Oil Chem Soc 78(2):139–145

Pedaliaceae

28

Family Pedaliaceae comprises around 70 annual or perennial herbs and shrubs. The fruiting body of these plants is a loculicidal capsule or nut, often spiny or with wings, hooks or thorns, beaked or barbed. The seeds are smooth with thin fleshy endosperm. They mostly grow in deserts or cultivated as a semi-arid crop. In this chapter the role of sesame the gingelly (*Sesamum indicum*) as an oil yielding crop for biodiesel production is considered.

28.1 Sesame (*Sesamum indicum*)

The sesame (*Sesamum indicum*) is a valuable oilseed crop extensively cultivated in arid and semi-arid tropics. It is a member of the family Pedaliaceae. *Sesamum indicum* (syn: *Sesamum mulayanum* and *Sesamum orientale*) is referred by different location-specific popular names (Table 28.1). Sesame grows well in neutral (pH 6.8–7.2) well-drained fertile soil. It requires 90–120 frost free days. It has a deep root system by which it gets water during drought season. It grows up to a height of 200 cm. It takes around 125–135 days from sowing to harvesting. The leaves are hairy, long and ovate (3–15 cm long and 1–7 cm broad). The fruit is a deeply grooved capsule of 2.5–3.5 cm long having 8 rows of 100–150 seeds in total (Fig. 28.1). Since flowering, it takes 4–7 weeks to come to maturity. Around 30,000–35,000 seeds weigh 1 kg. Total word production of sesame is around 5.5 million metric tonnes a year in which around 65% (3.6 million metric tonnes) is diverted exclusively for oil production for domestic purposes, and the rest is directly utilized for human consumption in confectionary and for household food items. The seeds contain around 60% oil (Saydut et al. 2008). It is cultivated in many countries, more so in Sudan, Burma, Tunisia, Egypt, Thailand, Mexico, Guatemala, El Salvador, Afghanistan, Pakistan, Bangladesh, Indonesia, Sri Lanka, Saudi Arabia, Turkey, Myanmar, Ethiopia, Uganda, Nigeria, Tanzania, Mozambique (Fig. 28.2)

S. P. Raj et al., *Biodiesel from Flowering Plants*,
https://doi.org/10.1007/978-981-16-4775-8_28

Table 28.1 Popular names of *S. indicum* in different languages

Language	Popular names
English	Benne, Beni, Beni seed, gingelly
French	Sesame
Spanish	Ajonjoli, Sesamo
Portuguese	Gergelim, sesame
Afrikaans	Sesam
Croatian	Sezan
Indonesian	Wijen
Serbian	Indiskisezam
Turkish	Susam
Tamil	Ellu

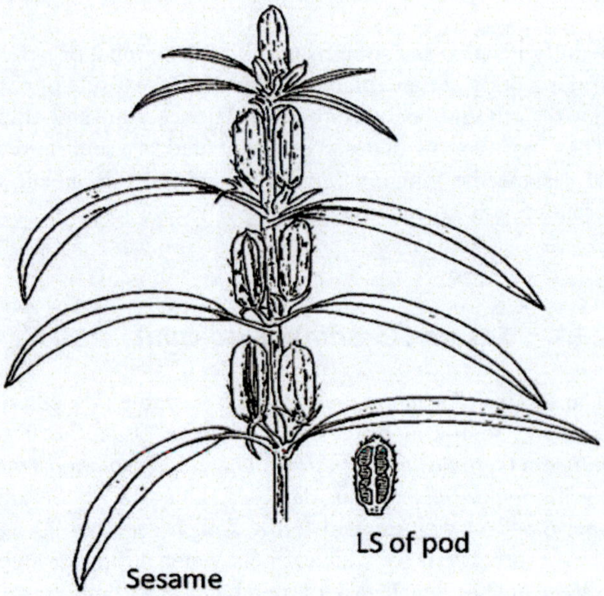

Fig. 28.1 Sesame plant and longitudinal section of the pod exposing the array of seeds

(Elleuch et al. 2007; Koca et al. 2007; Uzun et al. 2007; Wu 2007; Saydut et al. 2008; Dawodu et al. 2014).

28.1.1 Production of Oil

The seeds separated from the dry pod are pressed either manually or in an expeller to squeeze out the oil. This oil is often referred as virgin oil or good oil. The resultant cake is heated and passed through the pressing machine and the oil released from it is coloured and considered as second quality. The press cake is again treated with *n*-hexane and the oil obtained is mostly of non-edible quality.

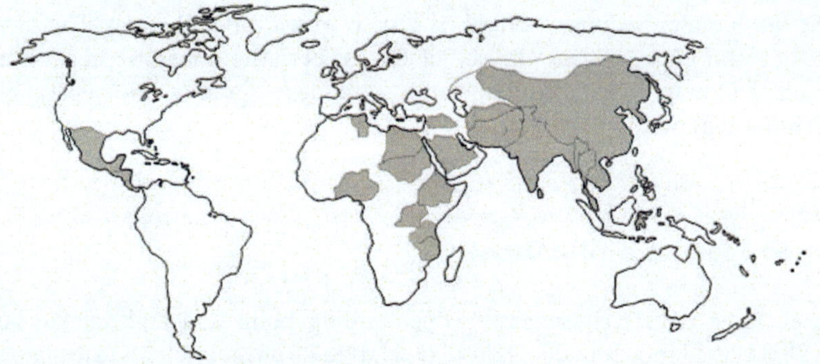

Fig. 28.2 Geographical distribution of *Sesamum indicum*

Table 28.2 Properties of sesame oil (Saydut et al. 2008; Bhave et al. 2013; Gadhavel and Ragit 2016)

Parameter	Value
Viscosity at 40 °C (cSt)	25.78
Density (kg m^{-3})	913, 899, 826
Iodine value (g I_2/100 g)	82.45, 105
Flash point (°C)	142, 245, 260
Cetane number	42
Heating value (MJ kg^{-1})	38.48, 39.5, 39.3
Pour point (°C)	6, −10, −3.9
Cloud point (°C)	18, 1.0, −9.4
Oleic acid (%)	43
Linoleic acid (%)	35
Palmitic acid (%)	11
Stearic acid (%)	7

28.1.2 Properties of Oil

The sesame seeds contain 55–65% oil. This oil has sesamin, sesamolin and sesamol which are known for its medicinal properties. Rich unsaponifiable matter present in the oil keeps the oil stable. The oil is characteristic of high viscosity, low volatility and poor cold flow properties which make it not possible to be used directly in the engines. Detailed analysis of oil is given in Table 28.2.

28.1.3 Biodiesel Production

Biodiesel from sesame oil is prepared by alkali transesterification. To 1 kg of oil, 200 g of methyl alcohol and 5 g NaOH are added. The content is placed in a heating mantle set at 60 °C and then refluxed for 2 h at constant stirring. On completion the product is allowed to stand for 8 h. By then the glycerol and water as a by-product will settle down and the biodiesel and the unreacted methanol float at the top. The top layer is decanted and distilled to expel the methanol and washed with warm water until the wash water becomes neutral. If time is given for settlement. The biodiesel becomes crystal clear. During the settlement process the transesterification also is taking place slowly. Therefore longer the settlement duration, the separation and conversion are good.

28.1.4 Properties of Biodiesel

Biodiesel has a reduced viscosity, proper boiling point and high cetane number (Table 28.3). It is biodegradable, non-toxic and free from sulphur and aromatics. The transesterification reduces the viscosity more than five times. The density also is reduced marginally. The flash point of the biodiesel is higher than that of the diesel and thus it is safe to transport the biodiesel. The cloud point and pour point of the biodiesel are almost double than that of the diesel and thus this biodiesel has a limitation at areas having low-temperature climate. Cetane number is a reference of ignition characteristics.

More the cetane number higher is the smoothness in combustion. The ignition properties are satisfactory when the cetane number is high. Cetane number also influences stability, drivability, occurrence of white smoke, noise and emission of CO_2 and HC. The cetane number of the biodiesel depends on the type of processing and the agro-climatic conditions in which the oil seed is collected. Iodine value is a measure of the unsaturation of the fatty acid which regulates the fuel oxidation.

28.1.5 Engine Performance

At a peak load, the brake specific fuel consumption of biodiesel is feebly higher due to the better combustion facilitated by the oxygen in the fuel molecule. Similarly, the brake specific fuel consumption is high in biodiesel. Poor fuel atomization and vaporization of the biodiesel are as a consequence of increased viscosity results in the late burning of the injected fuel reflecting in the exhaust gas temperature. The CO and CO_2 emissions are relatively low. Higher fuel entry in the combustion chamber may change the mixing pattern.

Table 28.3 Properties of sesame oil biodiesel

Parameter	Value
Viscosity at 40 °C (cSt)	4.2
Density (kg m^{-3})	867
Iodine value (g I_2/100 g)	80.3
Flash point (°C)	170
Cetane number	50.5
Heating value (MJ kg^{-1})	40.4
Pour point (°C)	−14
Cloud point (°C)	−6

References

Bhave NA, Ugare VM, Andhare AM (2013) Performance and emission studies of sesame oil methyl ester in compression ignition engine. Int J Emerg Technol Adv Eng 3(5):344–351

Dawodu FA, Ayodele OO, Bolanle-Ojo T (2014) Biodiesel production from *Sesamum indicum* L. seed oil: an optimization study. Egypt J Pet 23(2):191–199. https://doi.org/10.1016/j.ejpe.2014.05.006

Elleuch M, Besbes S, Roiseux O, Blecker C, Attia H (2007) Quality characteristics of sesame seeds and by-products. Food Chem 103(2):641–650. https://doi.org/10.1016/j.foodchem.2006.09.008

Gadhavel SL, Ragit SS (2016) Biodiesel development from sesame oil: transesterification and fuel characterization. Int J Sci Res Dev 4(5):14–17

Koca H, Bor M, Özdemir F, Türkan İ (2007) The effect of salt stress on lipid peroxidation, antioxidative enzymes and proline content of sesame cultivars. Environ Exp Bot 60 (3):344–351. https://doi.org/10.1016/j.envexpbot.2006.12.005

Saydut A, Duz MZ, Kaya C, Kafadar A, Hamamci C (2008) Transesterified sesame (*Sesamum indicum* L.) seed oil as a biodiesel fuel. Bioresour Technol 99(14):6656–6660. https://doi.org/10.1016/j.biortech.2007.11.063

Uzun B, Arslan C, Karhan M, Toker C (2007) Fat and fatty acids of white lupin (*Lupinus albus* L.) in comparison to sesame (*Sesamum indicum* L.). Food Chem 102(1):45–49. https://doi.org/10.1016/j.foodchem.2006.03.059

Wu WH (2007) The contents of lignans in commercial sesame oils of Taiwan and their changes during heating. Food Chem 104(1):341–344. https://doi.org/10.1016/j.foodchem.2006.11.055

Poaceae

29

The members of the family Poaceae are nearly ubiquitous and occupy around 40% of the land area excluding Greenland and Antarctica. They are often referred as grass family and are annual or perennial herbs and rarely tree. Poaceae ranks among the top five families of the flowering plants in terms of the number of species with more than 10,000 species in 600 genera. Chosen plants of this family are used as fodder to cattles, feeds to poultry, cereals and millets for human consumption and are raw materials for alcohol synthesis and for paper industry. Oils extracted from two species, namely rice *Oryza sativa* and corn *Zea mays* are reported to be raw materials for biodiesel.

29.1 Rice (*Oryza sativa*)

The rice, *Oryza sativa* is being cultivated for its grain. It is one of the staple foods for many nations. The straw (hay) is used as cattle feed. It is a monoculture wet land crop richly cultivated in China, India, Indonesia, Bangladesh, Vietnam, Thailand, Myanmar, Philippines, Brazil, Japan, and Pakistan (Fig. 29.1). Other countries which produce rice are Cambodia, USA, Egypt, South Korea, Nepal, Nigeria, Laos, Madagascar and Sri Lanka. The local names of rice in different languages are given in Table 29.1. The rice is a water-demanding wetland crop which comes to yield in 110–120 days. The riped ears (Fig. 29.2) are thrashed to separate the grains. The grains have an outer coat called husk, which is dehulled to extract polished grain for human consumption. The local names of this species are given in Table 29.1.

The materials of the outer coat and the inner coat of the rice seed are the by-products which otherwise go waste are used for extracting the oil. The above waste is the dehulled reject which is often called as bran which contains 15–23% oil (Sinha et al. 2008). Around 600 million tonnes of paddy are being cultivated all over the world and this has the potential of 7.25 million metric tonnes of oil which is equal to 2.5% of the total diesel consumption in India. This is likely to replace 1.2% of the

S. P. Raj et al., *Biodiesel from Flowering Plants*,
https://doi.org/10.1007/978-981-16-4775-8_29

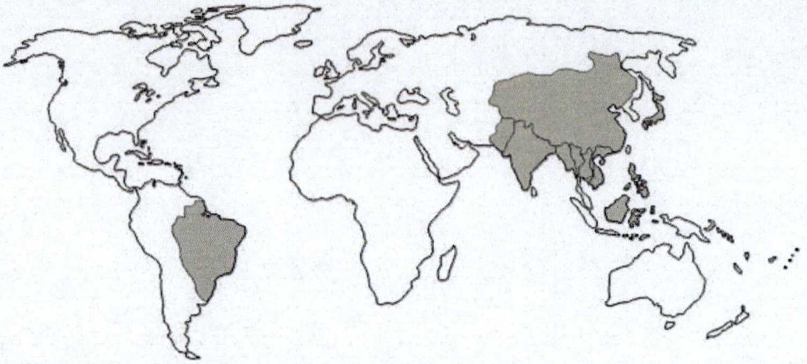

Fig. 29.1 Geographical distribution of the rice, *Oryza sativa*

Table 29.1 Local names of rice in different languages

Language	Name	Language	Name
English	Asian rice	Malay	Beras
Arabic	Orz	Malayalam	Nellu
Finnish	Riisi	Portuguese	Arroz
French	Riz	Russian	Ris
German	Reis	Spanish	Arroz
Italian	Riso	Sundanese	Pare
Japanese	Ine	Vietnamese	Lua

diesel consumption in road transport. Many attempts were made to utilize the rice bran oil directly in the internal combustion engines (Rao et al. 2005; Sinha and Agarwal 2005).

29.1.1 Preparation of Rice Bran Oil

Rice comprises 65% grain, 8% husk and 27% oil (Kumar and Manimaran 2014). The oil from the bran is extracted using hexane as a solvent. The oil thus extracted contains wax and gum which are separated at the first instance (Kaimal et al. 2002). The oil contains lipase which cleaves the 1, 3 sites of the triacylglycerol. The lipase along with oxidase deteriorates the oil present in the bran immediately after production. As the spoilage continues the free fatty acid content increases. This may create off-flavour and cause a change in functional properties (Takano 1993; Ramezanzadeh et al. 1999). Heat treatment to the raw bran immediately after milling is likely to inactivate the lipase and arrest the chain reaction leading to the formation of FFA. This is not a practical approach especially when the economic utility of the oil is not in sight. Due to rapid splitting of the oil the FFA content may go even up to 40–50%.

Fig. 29.2 The rice plant, *Oryza sativa*

Table 29.2 Physicochemical properties of rice bran oil (Sinha et al. 2008; Kumar and Manimaran 2014)

Parameters	Value
Kinematic viscosity at 40 °C (cSt)	43.5
Density at 15 °C (kg m^{-3})	922
Cetane number	50.1
Heating value (MJ kg^{-1})	41.1
Flash point (°C)	316
Fire point (°C)	337
Cloud point (°C)	13
Pour point (°C)	01
Free fatty acid (%)	17–21.5

29.1.2 Properties of Rice Bran Oil

The high viscosity of oil (43.5 cSt) as compared to diesel interfaces with the pumping and spray characteristics. Reduced atomization of the oil causes inefficient mixing of air resulting in incomplete combustion. High flash point (Table 29.2) as a consequence of low volatility promotes carbon deposit in piston rings, causes injector coking and denatures the lubricating oil. Often it is considered that heating of the oil reduces the viscosity, but it is reported that hot oil often causes increased deposits.

The free fatty acids of this oil are relatively high (Table 29.3). Besides, long storage causes an increase in FFA content. The FFA level also greatly depends on

Table 29.3 Fatty acid composition of rice bran oil (Zullaikah et al. 2005; Sinha et al. 2008)

Fatty acid	%
Lauric acid (C12:0)	0.2
Myristic acid (C14:0)	0.8
Palmitic acid (C16:0)	17.7, 18.8
Palmitoleic acid (C16:1)	0.23
Stearic acid (C18:0)	2.2, 2.4
Oleic acid (C18:1)	40.6, 43.1
Linoleic acid (C18:2)	35.6, 33.2
Linolenic acid (C18:3)	1.8, 0.6
Arachidic acid (C20:0)	0.2, 0.7
Behenic acid (C22:0)	0.3
Lignoceric acid (C24:0)	0.6

the moisture content and temperature during the storage of bran. Rice bran oil contains around 3% wax and 2% phospholipids. Oil with high FFA needs special pre-treatment for easy conversion into biodiesel. It has a high viscosity which leads to low atomization on the fuel in the injection cylinder. High acidity coupled with high viscosity brings problem in cold engine startup, creates misfire and causes ignition delay. The calorific value of this oil is around 15% lesser than that of the diesel.

29.1.3 Preparation of Biodiesel

29.1.3.1 Modified In-Situ Esterification

Crude rice bran is refluxed in methanol-sulphuric acid mixture for 5 h. Each kg of rice bran needs 15 L of methanol-sulphuric acid mixture. The methanol-sulphuric acid mixture is formed of 0.5–1.0 mL of sulphuric acid per litre of methanol. After 5 h, the refluxing is stopped and the liquid content is distilled to remove the methanol. The remnant is extracted with hexane to secure the biodiesel. This is then washed with distilled water repeatedly till it comes to neutral pH. The organic layer remains consequent to the final wash is the fatty acid methyl ester which is heated to 80 °C for 2 h to expel the moisture if any. The biodiesel yield was reported to be 17.99% by weight of the bran (Zullaikah et al. 2014).

29.1.3.2 Two-Step Methanolysis

Two-step methanolysis is resorted especially when the FFA content of the oil is high. In the first step, acid catalysed (H_2SO_4) treatment is made. This reaction is carried out at 60 °C at a methanol to FFA molar ratio of 5:1. Sulphuric acid 2% by weight of methanol is mixed with methanol and this mixture is allowed to react with the oil for 2 h. The contents are then washed with copious amount of warm water till the washed material becomes neutral. It is then analysed for FFA and if the value is less than 0.5%, alkali catalysed methanolysis is carried out on the above material. If the

FFA is more than 0.5%, once again it is subjected to acid catalysed methanolysis before taken up for alkali catalysed methanolysis.

29.1.4 Properties of Biodiesel

The properties of biodiesel thus obtained are given in Table 29.4. The parameters such as viscosity, density, cetane number show a tendency closer to the diesel.

29.1.5 Engine Performance

The specific energy consumption with respect to biodiesel being the energy utilized by the engine to generate 1 kWh of energy is high. The leaner combustion of biodiesel and the consequent ignition causes the burning of more fuel. Brake specific fuel combustion of biodiesel is higher than that of the diesel at all loading conditions. The power developed by the engine at varying load suggests that the maximum power is low in biodiesel because of its low heating value. The fuel flow by volume is slightly high. The high density of biodiesel leads to more quantity of fuel entering through the injection system to match the energy requirement thereby the consumption increases. The CO and HC in the exhaust of the engine running with the biodiesel are low due to the increased combustion assisted by the rich oxygen. The oxygen content of the exhaust with biodiesel also is reduced as high as 20% in many cases. The smoke capacity of the exhaust of the biodiesel is high (Subbaiah and Gopal 2013). NO_x of the exhaust formed by the combination of oxygen with nitrogen at elevated temperature is high. This specific issue is a matter of concern in the context of its nucleus of activity pertaining to acid rain.

Table 29.4 Physicochemical properties of rice bran oil biodiesel (Subbaiah and Gopal 2013; Mohanty 2013; Sinha et al. 2008)

Parameters	Value
Kinematic viscosity at 40 °C ($mm^2\ s^{-1}$)	4.63, 4.81, 5.3
Density at 20 °C ($g\ m^{-3}$)	0.874, 0.872, 0.877
Iodine value (g I_2/100 g)	102
Acid value (mg KOH g^{-1})	0.25
Heating value (MJ kg^{-1})	38.7, 41.4, 42.2
Flash point (°C)	165, 157, 183
Oxygen content (wt%)	11.25
Cetane number	56.2, 51.6, 63.8
Pour point (°C)	3, −2

29.2 Corn (*Zea mays*)

The corn *Zea mays* is a subtropical plant, known to grow in latitude from 58°N to 40°S and at an altitude below 4000 m. It prefers frost free zone with an annual rainfall of about 700 mm or more. It grows in a variety of soil types. The preferred soil pH is 4.5–8.5 and is propagated through seed.

Corn which is also called as maize (Spanish) belongs to the class: Liliopsida, order: Poidae and family: Poaceae. It is known to be domesticated in Mexico 1000 years ago. There are six varieties of *Zea mays*. They are flint corn *Z. mays* var. *indurata* with hard kernel, dent corn *Z. mays* var. *indentata* with its kernel showing dent when dried, sweet corn *Z. mays* var. *saccharata* with soft kernel, flour corn *Z. mays* var. *amylacea* with rich flour, popcorn *Z. mays* var. *everta* and waxy corn *Z. mays* var. *ceratina*.

This species is distributed in USA, China, Brazil, Mexico, India, Romania, Philippines, Indonesia, Canada, Russia, New Zealand and the South American continent (Fig. 29.3).

Global production of maize surpasses the yield of wheat or rice. The maize being used in human consumption is also used in the manufacture of alcohol, animal feed, food products such as starch and syrup and biodiesel.

The maize plant grows to a maximum height of 3 m. The stem comprises 20 internodes of 18 cm in length. The leaf which emanates from each node is approximately 9 cm width and 120 cm length. Ears are often located in the mid-section of the stem positioning themselves between the stem and sheath of the leaf. These fruiting bodies emerge as female inflorescence which is covered by several layers of specialized leaves often referred as husks. The apex of the plant ends as male inflorescence, which is often referred as tassel. Under suitable conditions the male inflorescence dehisces and releases pollen and fertilizes the female inflorescence. The fertilized female inflorescence emerges as cob which has around 600 kernels (Fig. 29.4).

There are other names for *Zea mays* in different languages (Table 29.5).

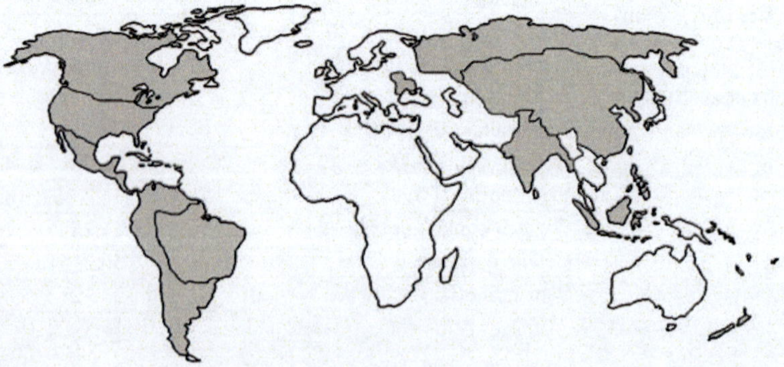

Fig. 29.3 Geographical distribution of *Zea mays*

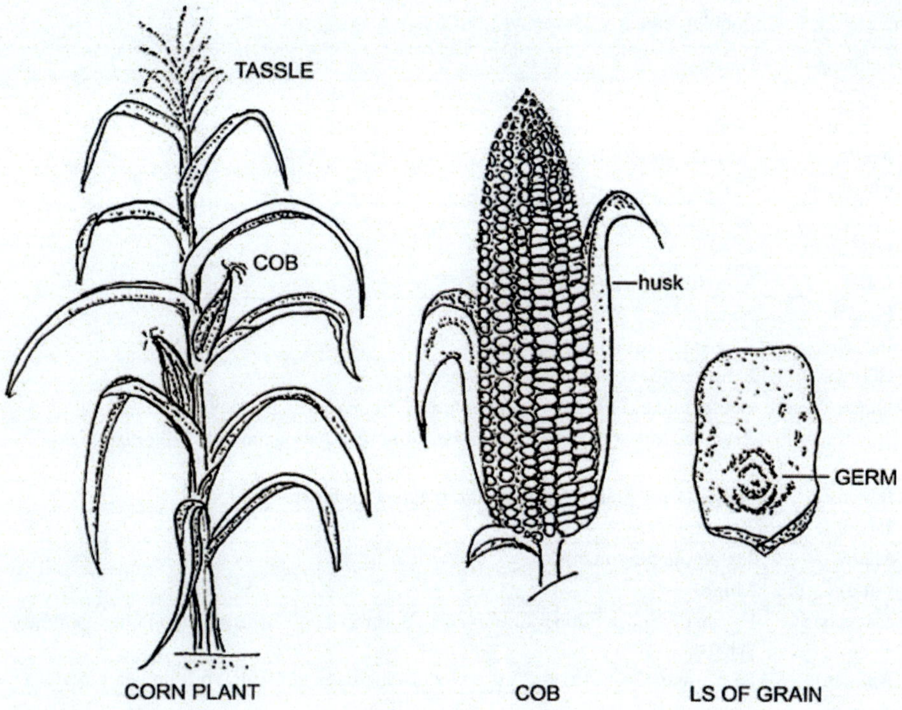

Fig. 29.4 The corn *Zea mays*

29.2.1 Cultivation

As maize does not tolerate cold it has to be sown in warm climates. The seeds are sown in furrows at a depth of 2 cm at 60–120 cm distance. The density of planting is 25,000 to 75,000 ha^{-1}. The seed rate is 20 kg ha^{-1}. It comes to yield in 60–70 days. The total yield will be 5–8 t ha^{-1}. The matured kernels are removed and dried in open fields for 7–8 days and the grains are separated by beating the cobs by sticks or by using maize shellers.

29.2.2 Oil Extraction

The corn being a coarse grain is rich in flour and its oil content is low (around 6%). But the maize germ which is an important constituent of the corn forms around 5–14% of the total mass of the kernel and its oil content is 18–41% (Johnston et al. 2005; MPOC 2008).

During the past years, oil extraction is made by cold pressing method in which considerable quantity is retained in the resultant cake and thus it is not found to be efficient. Solvent extraction process is an improvement on the cold pressing method and the solvent normally being used is *n*-hexane (Abdulkadir and Abubakar 2011).

Table 29.5 Colloquial names of *Zea mays* in different languages

Language	Names
English	Corn, Indian corn, maize, teasinthe, Turkey wheat
African	Mielie
Arabic	Dhurah, dhurah shâmîyah
Bulgarian	Carevica
Croatian	Kokavac, kukuruz, mekunac, zubanac, secerac
Danish	Majs
Dutch	Korrelmais, mais
Finnish	Maissi
French	Blîd indie, maize, mais commun
German	Echter mais, körnermais, türkisches korn
Indian	Bhuta, makkai makka cholam, chujak, mokka javanalu makkejola, mako
Hungarian	Ezemege kukorica, kukorica, lisztes kukorica, pattogatni valo kukorica, viasz-kukorica
Italian	Formentone, grano di turchia, granoturco, granturco
Japanese	Tomorokoshi
Malay	Jagong, jagung manis
Portuguese	Milho
Romanian	Diente de cal, porumb, porumb ceros porumb de cocosei, porumb fainos, porumb zaharat
Russian	Krachmalistaja kukuruza, kukuraza, sacharnaja kukuruza, voskoidnaja kukuruza
Serbian	Kokavac, kukuruz, mekunac, zuban, zubanac
Spanish	Diente de caballo, maiz, trigo de indias
Swedish	Majs
Turkish	Kokoroz
Vietnamese	Ngo
Welsh	Corn melys

Other solvent used is ethanol and its efficiency is governed by temperature, duration, solvent solid ratio and ethanol concentration. It is good if the temperature of the extraction is maintained slightly above the boiling point of the alcohol. If the extraction is carried out in higher atmospheric pressure, the efficiency is bound to increase. Supercritical fluid extraction is a new method wherein CO_2 under pressure is passed over flaked germ at 50 °C (Reverchan 1994) and at 5000 psi pressure. In another method the tissue of the grain is disrupted by heat at moist state to allow oil separation. The material is then crushed, cooked and pressed. This method in due course of time underwent continued improvement such as malaxation treatment (McWilliams 2001). Malaxation helped to improve the recovery of oil. Recently enzyme assisted aqueous extraction came in to being, in which the materials are treated enzymatically. This process eliminates the involvement of solvent (Fig. 29.5). Besides this method does not need degumming of the finished product. The enzymes hydrolyse the structural polysaccharides of the cell wall and release the oil locked inside the oil sac bound by the membrane. The enzymes being used are cellulase, *a*-amylase and pectinase (Rosenthal et al. 1996). Oil also is formed as a

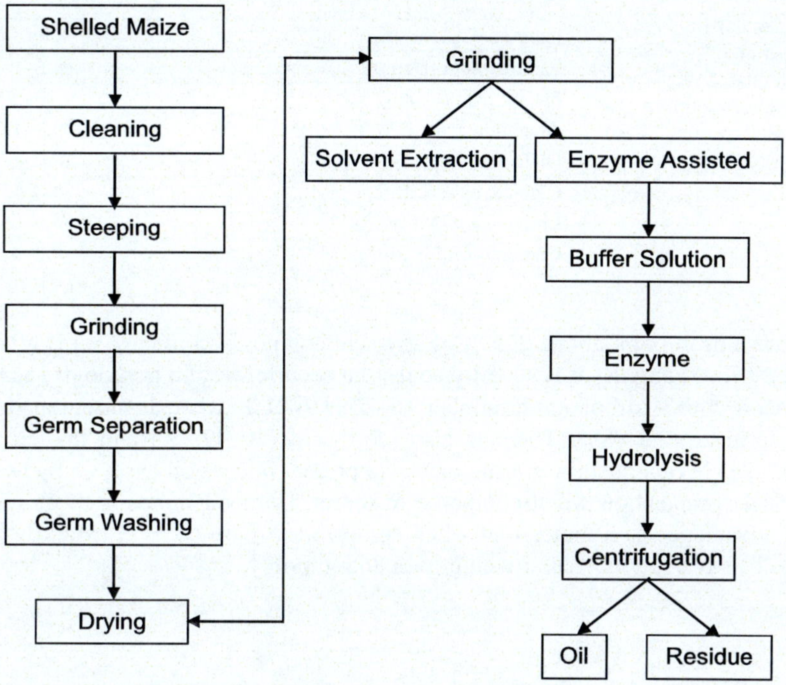

Fig. 29.5 Process flow chart for aqueous enzymatic oil extraction

Table 29.6 Properties of corn oil (Boulifi et al. 2010; Mata et al. 2012)

Parameters	Value
Density at 20 °C (kg m^{-3})	919 ± 0
Kinematic viscosity at 40 °C (mm^2 s^{-1})	36.27 ± 0.01
Iodine value (g I_2/100 g)	127 ± 1
Acid value (mg KOH g^{-1})	0.26 ± 0.02
Heat value (MJ kg^{-1})	39.8 ± 1.8
Peroxide value (meq kg^{-1})	2.26 ± 0.2
Average molecular weight (g mol^{-1})	865.4 ± 8

by-product during the production of ethanol from corn. Ethanol is formed when starch is hydrolysed and the glucose is fermented. During this process corn oil is formed as a by-product along with gluten.

29.2.3 Properties of Corn Oil

Properties of corn oil are given in Table 29.6. Corn oil is a light-yellow liquid with a freezing point ranging from −10 to −20 °C with a density ranging from 914 to 926 kg m^{-3}. It has a good oxidative stability in spite of its high unsaturated fatty acid content (Rasimoglu and Temur 2014).

Table 29.7 Fatty acid composition of corn oil (Boulifi et al. 2010; Rasimoglu and Temur 2014; Markov et al. 2016)

Fatty acid	% by weight
Palmitic acid (C16:0)	8.6–16.5
Stearic acid (C18:0)	0–3.3
Arachidic acid (C20:0)	0.3–1.0
Oleic acid (C18:1)	20.0–42.2
Linoleic acid (C18:2)	34.0–65.6
Linolenic acid (C18:3)	34.0–65.6
Gadoleic acid (C20:1)	0.2–0.6

It has a flash point above 225 °C. Fatty acid composition of corn oil is given in Table 29.7. Among the various fatty acids, the occurrence of a portion of saturated fatty acids with a carbon atom number 16–20 is 9–21%, whereas unsaturated fatty acids form around 88%. There is also oleic acid (20–42.2%) and linoleic acid (34–65.6%) in rich quantities. Linolenic acid present in corn oil has a vastly heavier fractional composition when compared to diesel. Corn oil mixed with diesel has many advantages as a motor fuel when compared to pure corn oil (Markov et al. 2016). The average molecular weight was 865 g mol^{-1}.

29.2.4 Biodiesel Production

29.2.4.1 Transesterification

Alkali catalysed transesterification is popularly considered in the preparation of biodiesel. Initially the methanol and potassium hydroxide are mixed to form methoxide. The methoxide is then added with a calculated amount of oil so as to have the alcohol oil ratio of 6:1 and stirred at a temp. of 50 °C for 60 min. At this temperature the solubility of methanol in oil increases resulting in the active formation of ester. At low temperature the methanol does not dissolve in oil. Stirring helps to initiate the reaction. When the ester formation is initiated the alcohol solubilizes progressively since ester is a solvent for both alcohol and oil. During the course of the 60 min the biodiesel is formed and slowly stratification commences. The mixture is then kept undisturbed overnight and the top layer separates itself as biodiesel and glycerol move to the bottom. In this transesterification or alcoholysis process, 1 mol of triglyceride reacts with 3 mol of alcohol (Stoichiometric molar ratio of methanol to vegetable oil is 3:1) to form 1 mol of glycerol and 3 mol of ester. The process is a sequence of three reversible reactions, in which the triglyceride molecule is converted step by step in to diglyceride, monoglyceride and glycerol. This process is often called as methanolysis or alcoholysis (Ozturk et al. 2010). Once the transesterification reaction is seemingly completed the mixture is allowed to stand undisturbed for 10–20 h. As a matter of fact phase separation commences just in 10 min. The ester standing as upper layer for a long time will be opaque indicating the incompleteness of separation. It is known that during the long duration of settling there will be mild transesterification reaction still going on making use of the

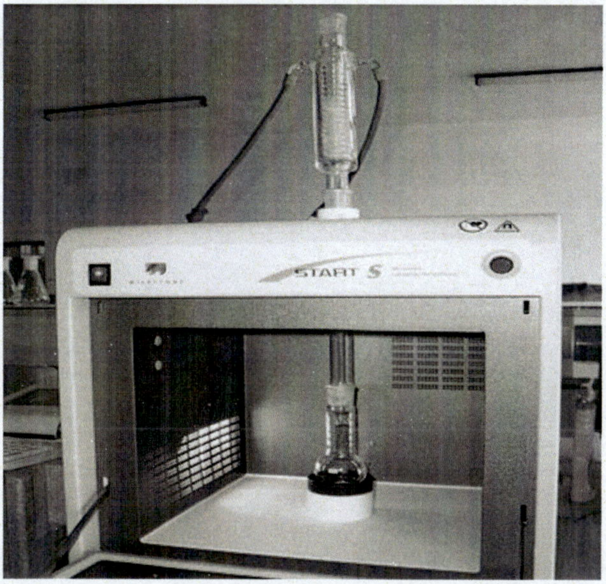

Fig. 29.6 Transesterification of corn oil using a microwave (Ozturk et al. 2010)

residual substrate. Thus longer the settling duration better the conversion and separation.

The top layer containing the clear biodiesel is isolated and washed with hot distilled water repeatedly to get rid of the excess alkali, glycerol and soap. By direct heating of the biodiesel thus obtained, the water present if any is removed to the maximum so as to maintain a moisture level not exceeding 0.01% as presence of water beyond that level promotes microbial growth, tank corrosion, emulsion and hydrolytic oxidation. Sodium methoxide $NaOCH_3$ or sodium ethoxide $NaOCH_2CH_3$ or potassium methoxide $KOCH_3$ or potassium ethoxide $KOCH_2CH_3$ were tried as catalyst. It was observed that sodium methoxides and sodium ethoxide expressed higher performance compared to that of the other catalyst. Heterogeneous solid catalysts were attempted in the transesterification of oil which include MgO, CaO, SrO and ZnO.

Transesterification aided by microwave is being carried out on lab-scale (Ozturk et al. 2010). The advantage is that the reaction can be completed in 5 min against the 2 h of reaction time in the conventional method. The oil is initially heated to 60 °C in the microwave. Without removing the flask, methanol and sodium hydroxide are introduced through the condenser so as to maintain an alcohol oil ratio of 10:1 and alkali conc. of 1.5% (Figs. 29.6 and 29.7).

In microwave the smallest polar molecules and ions are subjected to dynamic magnetic field. The electrical field created in the microwave vibrates the molecules and ions so as to activate the transesterification process as a consequence of molecular friction. Once the reaction is completed the mixture is allowed to stand overnight and the glycerol is then separated by gravity using a separating funnel. The

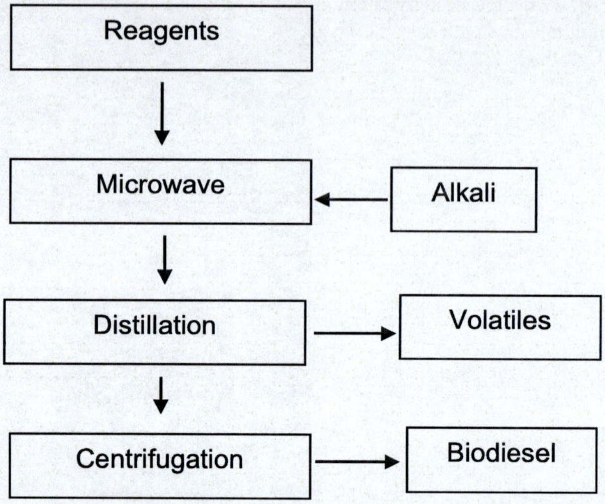

Fig. 29.7 Process details of transesterification by microwave

ester is repeatedly washed with hot water till the phenolphthalein indicator gives a negative inference. Residual moisture and unreacted methanol are distilled out.

Lipase enzymes are effectively used in the transesterification process especially in the ethanolysis of triglyceride. The immobilized lipases are reutilized for five successive cycles without losing its potency. The composition of corn oil biodiesel is given in Table 29.8 and the fatty acid profile is furnished in Table 29.9.

29.2.5 Fuel Characteristics

One of the major issues which hinders the liberal use of biodiesel is its poor cold flow properties. The saturated fatty acyl esters are responsible for the high CP (Cloud point), CFPP (Cold filter plugging point) and PP (Pour point). Cloud point is the cut-off temperature at which the biodiesel begins to make a cloudy appearance when it is slowly cooled. The fuel then slowly solidifies and starts crystalizing. The flow of the crystallized fuel slows down in the fuel conduits. Further decrease in temperature causes the fuel to block the pipes and filters of the fuel system of the vehicle. The temperature at which the fuel flow stops completely is referred as pour point. CFPP is also a reference point which is defined as the highest temperature at which a given amount of fuel sample cooled under certain experimental conditions cannot pass through a standard filter in a specific time. Saturated monoacylglycerols or free steryl glucosides are mainly responsible for the clogging of the fuel pipes and filters. One important problem is that it has not been possible to obtain a biodiesel having favourable cold flow properties without causing a negative effect in another property. Certain cold flow properties are said to make a favourable change when unsaturated fatty acid is high. At the same time cetane number and oxidative stability improve when there is high saturated fatty acid content. Efforts are in progress to

Table 29.8 Composition of corn oil biodiesel (Ozturk et al. 2010; Aydin et al. 2011; Rashid et al. 2012; Senthilkumar et al. 2014)

Fuel property	Value
Density at 25 °C (kg m^{-3})	886, 864, 850 (15 °C), 875 (20 °C)
Kinematic viscosity at 40 °C (mm^2 s^{-1})	3.83, 4.07, 3.5, 4.6 (30 °C)
Cetane number	56.00, 52.5, 52
Oxidative stability (h)	2.03
Cloud point (°C)	−2.00, −15, −14
Pour point (°C)	−4.00, −30, −31
Cold filter plugging point (°C)	−1.00
Flash point (°C)	164, 168, 167, 143
Sulphur content (%)	0.012
Ash content (%)	0.016
Acid value (mg KOH g^{-1})	0.18
Iodine value (g I_2/100 g)	73.7, 93.5
Copper strip corrosion (50 °C, 3 h)	1a
Higher heating value (MJ kg^{-1})	45.35, 39.9, 42.56
Ester content (%)	97.02
Methanol content (%)	0.180
Free glycerine (%)	0.013
Total glycerine (%)	0.235

Table 29.9 Fatty acid composition of corn oil biodiesel (Rashid et al. 2012)

Fatty acid	% by weight
Palmitic acid (C16:0)	9.98 ± 0.51
Stearic acid (C18:0)	1.8 ± 0.02
Arachidic acid (C20:0)	0.30 ± 0.02
Oleic acid (C18:1)	36.00 ± 1.73
Linoleic acid (C18:2)	50.89 ± 2.06
Linolenic acid (C18:3)	0.98 ± 0.01
Total saturated fatty acid	12.08
Total unsaturated fatty acid	87.87

modify the processing procedures to moderate the cold flow properties of corn oil biodiesel. Alterations made on the reaction time, temperature and stirring speed have not made any change in cold flow properties. However marginal reduction in the amount of catalyst and methanol oil ratio without grossly affecting the transesterification process favourably changes the cold flow properties (Rasimoglu and Temur 2014). Usually, cetane number, heat of combustion, melting point, and viscosity increase with the increasing chain length and they decrease with the increasing unsaturation.

The corn oil biodiesel being a mixture of long chain fatty acid methyl esters is prone to be oxidized. Such oxidation is largely governed by the number and location of methylene interrupted double bond present in the biodiesel. Besides the carbons which are allylic to the olefinic groups are susceptible to peroxidation. The oxidation

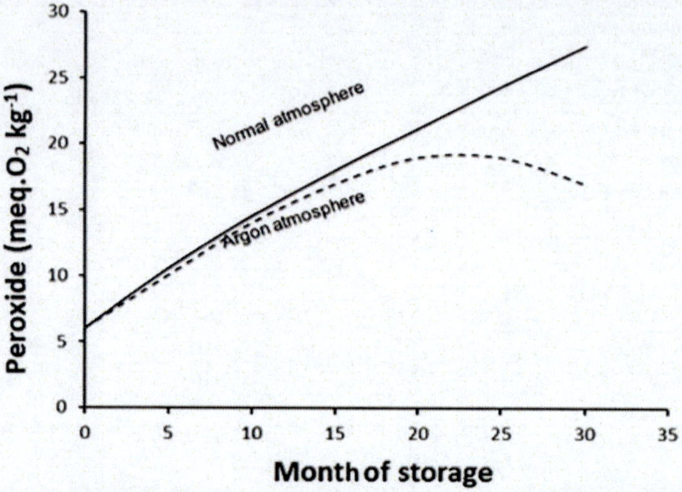

Fig. 29.8 Change in peroxide in ester due to long storage in normal and argon atmosphere

also depends on the raw material and the extraction methodology involved (Chahine and MacNeill 1974). The degradation in the quality of the biodiesel due to long storage is an important aspect needs consideration. The biodiesel which has rich oxygen tends to oxidize the ester link. Under normal storage condition the peroxide value will increase continuously as a function of the length in storage. When the biodiesel is stored in argon atmosphere the peroxide content increases during the initial stage of storage and the rate of increase slows down subsequently when the storage period extends for a long time (Fig. 29.8).

The peroxide thus formed degrades to aldehyde at first and then to acid. Acids and alcohols are also likely to be formed by the hydrolysis of ester in the presence of a trace of water if present consequent to the washing process (Boulifi et al. 2010). Acid content may increase when biodiesel is continued to be stored in normal atmosphere. But, when it is stored in argon atmosphere the acid value increases during the initial period of storage (Fig. 29.9).

It reaches a static level thereafter and the final value normally may not cross the standard (0.5 mg KOH g^{-1}). The increase in acid content also is caused by the rich dissolved oxygen present in the ester which generates short chain fatty acid. Storage also induces polymer formation which parallely increases the viscosity (Fig. 29.10). In such event high molecular weight species will develop which also contributes to the increases in viscosity (Moser et al. 1966).

To arrest such increase in viscosity it is required to store the corn oil biodiesel in argon atmosphere. Such increase in viscosity is a function of oxidation process which forms fatty acid and double bond isomerization products. The fatty acid moieties are known to join by C–O–C and C–C linkages. Under inert atmosphere ROOH decay takes place and C–C linkage alone will be formed. Iodine value is an index to understand the level of unsaturation of the oil or ester. Normally when

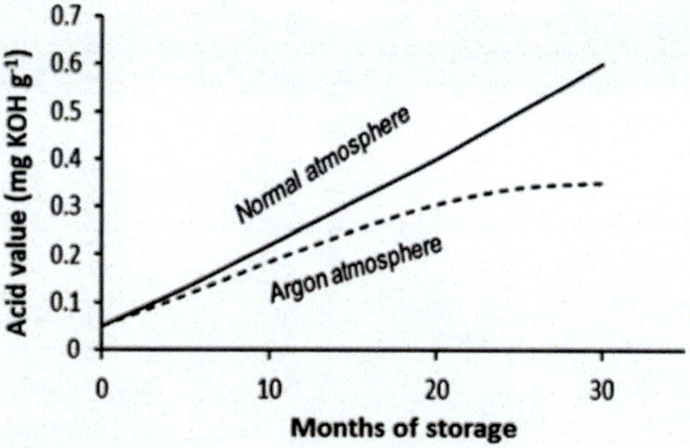

Fig. 29.9 Change in acid value in ester due to long storage in normal and argon atmosphere

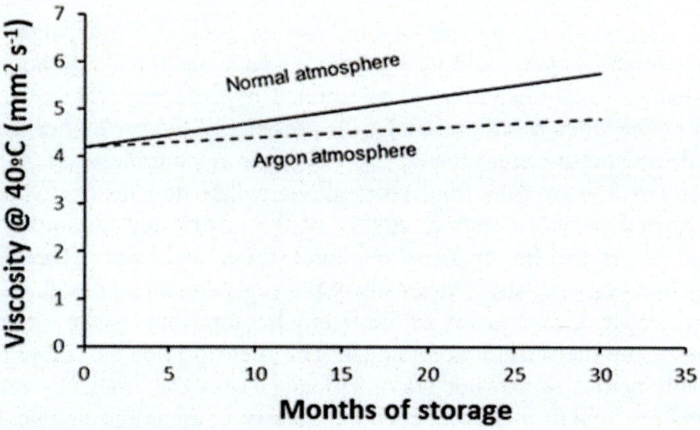

Fig. 29.10 Change in viscosity of ester due to long storage in normal and argon atmosphere

oxidative reaction takes place there will be reduction in the content of unsaturated fractions. It is mostly due to the elimination of hydrogen located adjacent to a double bond resulting in the formation of free radicals. The biodiesel stored in a normal atmosphere will show rapid reduction in iodine value indicating the conversion of polyunsaturated fatty acid into hydroperoxides. Corn oil ester is known to have a high iodine value indicating the presence of rich polyunsaturated fatty acid posing susceptibility to oxidation.

29.2.6 Preheating and Blending with Petroleum Diesel

Effect of temp. on the fuel viscosity without blending also was studied (Senthilkumar et al. 2014) by preheating the fuel. It was known that the kinematic viscosity reduced as the heating progressed from 30 to 70 °C. It was also observed that there was mild variation in viscosity as the fuel was preheated from 70 to 90 °C

Fig. 29.11 Viscosity reduction due to preheating of the biodiesel and diesel

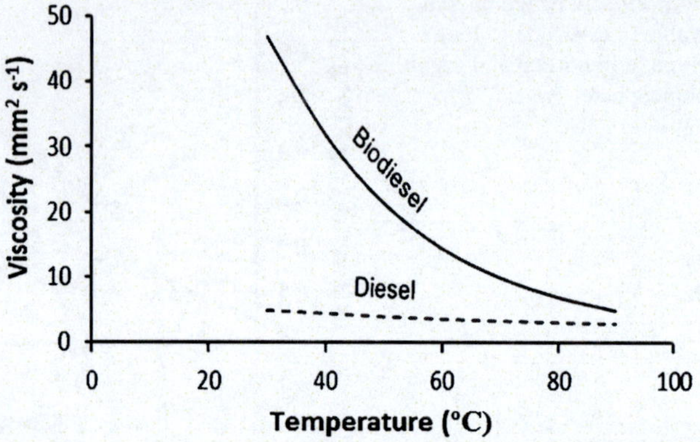

(Fig. 29.11). Major disadvantages of biodiesel in general are its high viscosity, density, low volatility, poor cold flow properties with functional consequences due to the presence of rich quantities of unsaturated hydrocarbon. One of the factors differentiating the biodiesel from diesel is the presence of oxygen higher than 10% in its chemical structure, whereas there is no oxygen in regular diesel. High viscosity, low volatility and poor cold flow properties are key fuel factors which causes unusually high deposits in engine, coking of the nostril and gumming in piston rings. These factors prevent the use of biodiesel straight in engines. There is a direct relationship between viscosity and density of the vegetable oil biodiesel. Density and viscosity are negative parameters in the internal combustion as they influence the atomization of the fuel inside the cylinder. The fuel injection assembly assess the fuel flow volumetrically and not gravimetrically. Thus the efficiency reference is based on volume and as a consequence the density is important in calculating the mass of fuel entering the combustion process. The density plays a role in the start of injection thus interfering the spray pressure, particle size of the droplet, engine function, combustion process and exhaust quality. The density of the biodiesel ranging from 0.86 to 0.9 kg L^{-1}, whereas it is around 0.864 kg L^{-1} for regular diesel. Besides, the fuel pump demands more energy and consequently there will be high wear and tear. High viscosity is a problem in cold weather as viscosity increases when the temperature decreases.

The density, heating value and cetane number are interrelated. To manage this problem, the practice of mixing the biodiesel with regular diesel has come in to vogue. These two fuels are compatible for blending without any stratification. Such mixture is referred as blend. The nature of blend is alphanumerically referred as Bx; B refers the blend, x refers the percent of biodiesel contained in the mixture (e.g. B10 means a mixture of diesel and biodiesel in which biodiesel forms 10% by volume). When the content of biodiesel in the blend increases the density and viscosity tend to increase. It is generally agreed that the content of biodiesel in the mixture can go up to 20% (B20) without grossly affecting the performance of the engine (Alptekin and Canakci 2008). The effect of blending on the viscosity is given in Fig. 29.12. The change in density due to blending is given in Fig. 29.13.

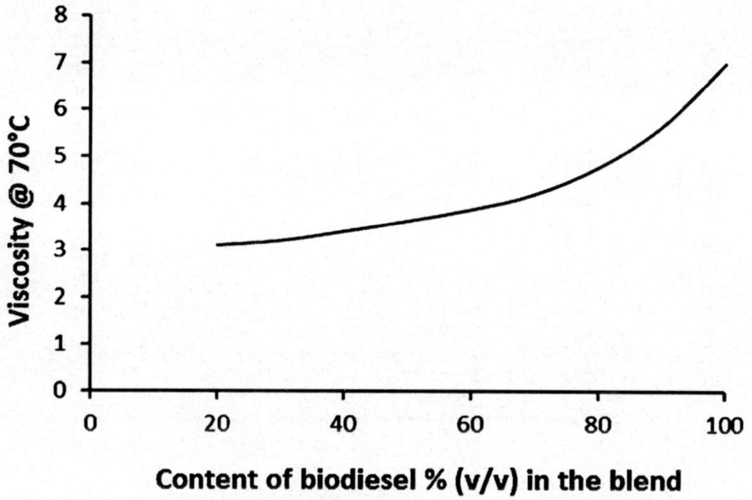

Fig. 29.12 Effect of blending of biodiesel on the viscosity at 70 °C

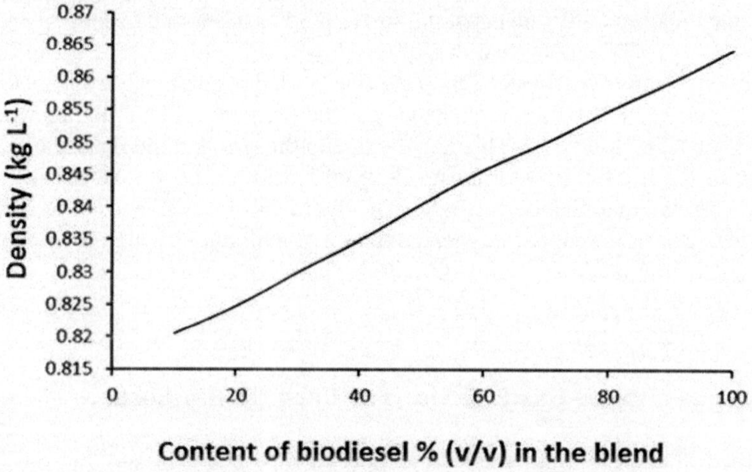

Fig. 29.13 The change in density of the biodiesel as a function of blend

29.2.7 Engine Performance

In order to combat the issue of viscosity the fuel is fed in the engine in a hot condition and its performance shows marked improvement. The brake specific fuel consumption (BSFC) decreases in B20 fuel at an optimum preheated temperature of 70 °C due to better fuel spray. However even under hot condition the consumption of biodiesel is slightly higher than that of the diesel but when the temperature increased beyond the optimum limit the fuel starts vaporizing and locks the fuel line. In

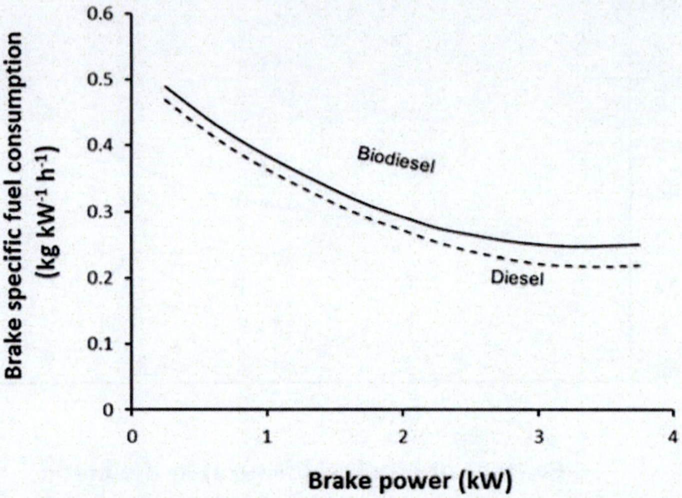

Fig. 29.14 Brake specific fuel consumption in biodiesel and diesel at 70 °C

general, the brake specific fuel consumption of biodiesel is found to show marginally higher value (15–20% more than that of the diesel) (Fig. 29.14). The brake thermal efficiency also improves in hot biodiesel due to better combustion characteristics (Senthilkumar et al. 2014). The exhaust gas temperature will be higher when the engine is run using biodiesel. This may be due to the fact that the marginally higher viscosity in the hot biodiesel imposes delayed burning. The smoke density of the biodiesel is lower than that of the diesel. The hydrocarbon level tends to be low even in high load due to the higher oxygen content in the biodiesel which favoured clean burning. The NO_x level is around 10–15% higher than that of the diesel.

29.2.8 Greenhouse Gas Emissions of Corn Oil Biodiesel

The United States Environmental Protection Agency has assessed that 2.6 billion litres of corn oil biodiesel will be available in 2022. It is possible that corn oil biodiesel may qualify Federal Renewable Identification Number (RIN) credits as per the renewable fuel standard. The greenhouse gas (GHG) emission intensity credits will also be available under the California's low carbon fuel standard if corn oil biodiesel achieves more than 50% reduction in life-cycle greenhouse gas (GHG) emissions when compared to that of conventional diesel. The California's Low Carbon Fuel standard(LCFS) targets a 10% reduction in the average life-cycle GHG intensity of the ground transportation fuel pool by 2020. Eligible fuels may receive LCFS credits in accordance with the carbon intensity (CI) value.

References

Abdulkadir M, Abubakar GI (2011) Production and refining of corn oil from hominy feed a byproduct of dehulling operation. ARPN J Eng Appl Sci 6(4):22–28
Alptekin E, Canakci M (2008) Determination of the density and the viscosities of biodiesel and diesel fuel blends. Renew Energy 33:2623–2630
Aydin F, Kafadar AB, Erdogan S, Saydut A, Kaya C, Hamamci C (2011) The basic properties of transesterified corn oil and biodiesel – diesel blends. Energy Sour A 33:745–751
Boulifi NEI, Bouaid A, Martinez M, Aracil J (2010) Process optimization for biodiesel production from corn oil and its oxidative stability. Int J Chem Eng 2010:518070, 9 pages
Chahine MH, MacNeill RF (1974) Effect of stabilisation of crude whale oil with tertiary butyl-hidroquinine and other antioxidants upon keeping quality of resultant deodorized oil. A feasibility study. J Am Oil Chem Soc 51(3):37–41
Johnston DB, McAloon AJ, Moreau RA, Hicks KB, Singh V (2005) Composition and economic comparison of germ fractions derived from modified corn processing techniques. J Am Oil Chem Soc 82:603–608
Kaimal TNB, Vali SR, Rao BVSK, Chakrabarti PP, Vijayalakshmi P, Kale V, Rao TC (2002) Origin of problems encountered in rice bran oil processing. Eur J Lipid Sci Technol 104 (4):203–211
Kumar RS, Manimaran R (2014) Evaluation of a diesel engine using rice bran oil as a bio-diesel. Int J Sci Eng Technol 3(6):1779–1789
Markov VA, Kamaltdinov VG, Loboda SS (2016) Optimization of diesel fuel and corn oil mixtures composition. Int Conf Ind Eng Proc Eng 150(2016):225–234
Mata TM, Sousa IRBG, Vieira SS, Cetano NSL (2012) Biodiesel production from corn oil via enzymatic catalysis with ethanol. Energy Fuel 26(5):3034–3041
McWilliams M (2001) Foods: experimental perspectives, 4th edn. Prentice Hall, Hoboken, NJ
Mohanty SK (2013) A production of biodiesel from rice bran oil and experimenting on small capacity diesel engine. Int J Modern Eng Res 3(2):920–923
Moser HA, Cooney PC, Evans CD, Cowan JC (1966) The stability of soybean oil: effect of time and temperature on deodorization. J Am Oil Chem Soc 43(11):632–634
MPOC (Malaysian Palm Oil Council) (2008) Glob Oils Fast Bus Mag 5(1):33–34
Ozturk G, Kafadar AB, Duz MZ, Saydut A, Hamamci C (2010) Microwave assisted transesterification of maize (*Zea maize* L.) oil as a biodiesel fuel. Energy Explor Exploit 28 (1):47–57
Ramezanzadeh FM, Rao RM, Windhauser M, Prinyawiwatkul W, Marshall WE (1999) Prevention of oxidative rancidity in rice bran during storage. J Agric Food Chem 47(8):2997–3000
Rao GLN, Saravanan S, Sampath S, Rajagopal K (2005) Combustion characteristics of a direct injection diesel engine fuelled with rice bran oil and its blends. In: Proc. of the nineteenth national conference on I.C. engines and combustion. Allied Publishers Private Ltd, New Delhi, India, pp 415–418
Rashid U, Ibrahim M, Ali S, Adil M, Hina S, Bukhari IH, Yunus R (2012) Comparative study of the methanolysis and ethanolysis of maize oils using alkaline catalysts. Grasasy Aceites 63 (1):35–41
Rasimoglu N, Temur H (2014) Cold flow properties of biodiesel obtained from corn oil. Energy 68 (2014):57–60
Reverchan E (1994) Comparison of processes for the supercritical carbon dioxide extraction of oil from soybean seeds. J Am Oil Chem Soc 71:1007–1012
Rosenthal A, Pyle DL, Niranchan K (1996) Aqueous and enzymatic processes for edible oil extraction. Enzym Microbial Technol 19:402–420
Senthilkumar R, Loganathan M, Tamilarasan P (2014) Performance and emission characteristic of DI diesel engine with preheating corn oil methyl ester. Int J Mech Eng 5(1):79–89
Sinha S, Agarwal AK (2005) Combustion characteristics of rice bran oil derived biodiesel in a transportation diesel engine. SAE Paper No. 2005-26-354

Sinha S, Agarwal AK, Garg S (2008) Biodiesel development from rice bran oil: transesterification process optimization and fuel characterization. Energy Convers Manag 49(5):1248–1257

Subbaiah GV, Gopal KR (2013) An experimental investigation on the performance of emission characteristics of a diesel engine fuelled with rice bran biodiesel and ethanol blends. Int J Green Energy 8(2):197–208. https://doi.org/10.1080/15435075.2010.548539

Takano K (1993) Mechanism of lipid hydrolysis in rice bran. Cereal Foods World 38:695–698

Zullaikah S, Lai CC, Vali SR, Ju YH (2005) A two-step acid-catalyzed process for the production of biodiesel from rice bran oil. Bioresour Technol 96(17):1889–1896

Zullaikah S, Rakhadima YT, Rachimoellah M, Widjaja T, Sumarno S (2014) An efficient method for the production of biodiesel from rice bran. IPTEK J Proc Ser 1(1):351–354

30 Putranjivaceae

Putranjivaceae has around 200 species of evergreen tropical trees or shrubs in two genera and they predominantly occur in Africa. The fruits are indehiscent, drupaceous, and globose to ovoid. The exocarp and mesocarp are fleshy but the endocarp is crustaceous with 1–2 seeds. The seeds are ellipsoid to globose, smooth, anatropous carunculate with copious endosperm.

30.1 Lucky Bean (*Putranjiva roxburghii*)

The lucky bean, *Putranjiva roxburghii*, is a medium sized tree growing to a height of 18 m with a girth of 2 m. This tree occurs in arid or in cultivated lands of India. They are sparingly distributed in South East Asia, Japan, India, Southern China Pakistan, Bangladesh, Nepal, Sri Lanka, Indo China and New Guinea (Fig. 30.1). They grow at an elevation of 800–1000 m above MSL.

It is an evergreen tree with pendant branches bearing elongated, simple, alternate, dark green, shiny elliptic leaves. It has unisexual flowers in separate trees. Fruits are ellipsoid or round in shape. The surface of the fruit is velvety in nature. Each fruit has a single stony seed of 2.5 cm long (Fig. 30.2). This tree is locally referred in India as jiaputta, putranjiva, putajan, kudrajivi, kuduru, irukolli, amani, kurupallai, parupala, pongalam and poothilanji.

30.1.1 Oil Production

The fruits are dried in hot sun till it has a constant weight. The seeds are then removed from the dried mass and passed through rollers to break the hard shell. The oil from the powdered material is extracted using *n*-hexane or petroleum ether (Nag et al. 1995). The extract thus obtained is filtered to remove any suspended materials and dried over anhydrous Na_2SO_4. It is then subjected to vacuum distillation to expel

S. P. Raj et al., *Biodiesel from Flowering Plants*,
https://doi.org/10.1007/978-981-16-4775-8_30

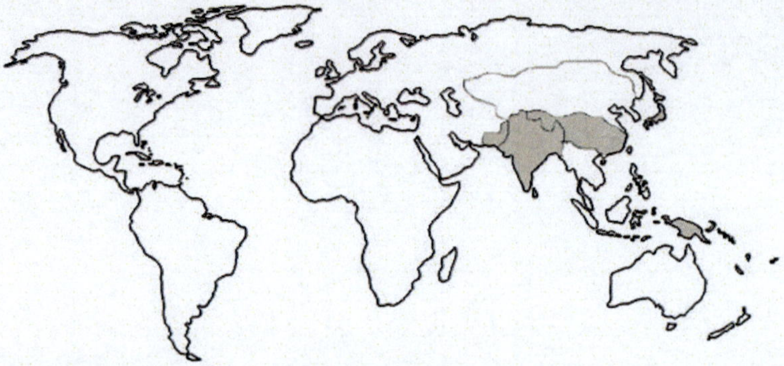

Fig. 30.1 Geographical distribution of *Putranjiva roxburghii*

Fig. 30.2 The lucky bean, *P. roxburghii*

the solvent. The oil recovery is around 35% by weight of the seed. The oil is yellow, pungent and volatile. The properties of *P. roxburghii* seed oil are given in Table 30.1.

The flash point is paradoxically low and the pour point is favourable in colder climate. The oil without transesterification is tried in engine as a fuel. It is observed that the raw oil when blended with diesel at 30% level gives satisfactory performance with respect to brake specific fuel consumption and brake thermal efficiency. The blend shows higher ignition temperature than that of diesel and accordingly there is less exhaust emission. However, NO_x emission of the blend is slightly higher than that of diesel at low load which is apparently due to the relatively low cetane

Table 30.1 Properties of *P. roxburghii* seed oil (Nag et al. 1995; Halder et al. 2009; Katagi et al. 2014; Subramanian et al. 2015)

Parameters	Value
Viscosity at 32 °C (cSt)	37.6, 34.7
Density (kg m^{-3})	917, 918, 923
Flash point (°C) Max.	55, 48, 51
Cetane number	55, 58.5, 31
Acid value (g KOH kg^{-1})	0.6
Calorific value (MJ kg^{-1})	36.6, 38.5, 39.5, 40.0
Iodine value (g I_2/100 g)	69.5
Pour point (°C)	−3
Aniline point (°C)	57
Palmitic acid (C16:0) (%)	10.2
Stearic acid (C18:0) (%)	10.6
Oleic acid (C18:1) (%)	48.65
Linoleic acid (C18:2) (%)	27.5
Arachidic acid (C20:0) (%)	1.05

Table 30.2 Properties of *P. roxburghii* biodiesel (Ghosh et al. 2008)

Parameters	Value
Viscosity at 40 °C (cSt)	5.8
Density at 25 °C	883
Cetane number	40.2
Calorific value (MJ kg^{-1})	39.3

number (Halder et al. 2009). At high loads NO_x emission in biodiesel is lower than that of diesel. The ignition timing of the oil is slightly higher (45°CA bTDC) than that of the diesel (40°CA bTDC). The volatile nature of the oil increases hydrocarbon emission when compared to that of regular diesel.

30.1.2 Biodiesel Production

Biodiesel (ester) from this oil is prepared by alkali catalyst transesterification. The biodiesel characteristic are given in Table 30.2.

It is known that the brake specific fuel consumption is higher, whereas the brake thermal efficiency is lower in biodiesel when compared to that of diesel (Ghosh et al. 2008). The emission of NO_x, hydrocarbon, CO, particulate matter is lower than that of the diesel under standard conditions. The basic characteristics of the biodiesel indicate that the diesel engine can be run with 50% blend and for a short duration even 100% biodiesel as fuel is considered (Ghosh et al. 2008).

References

Ghosh BB, Haldar SK, Nag A (2008) Synthesis of biodiesel from oils of Jatropha, Karanja and Putranjiva to utilize in ricardo engine and its performance and emission measurement. In: Proc.

Fourth BSME-ASME international conference on thermal engineering, 27–29 Dec, Dhaka, Bangladesh

Halder SK, Ghosh BB, Nag A (2009) Utilization of unattended *Putranjiva roxburghii* non-edible oil as fuel in diesel engine. Renew Energy 34:343–347

Katagi SK, Naikawadi SB, Koli AB, Kulkarni SB (2014) Biodiesel properties of *Putranjiva roxburghii* and *Plumieria rubra* seed oils: elevation based on fatty acid chemistry. J Applicable Chem 3(2):663–671

Nag A, Bhattacharya S, De KB (1995) New utilization of vegetable oils. JAOCS 72(12):1591–1593

Subramanian N, Mahendradas DK, Kasirajan R, Sahadevan R (2015) Bio-oil separation from potential non-edible urban waste source *Putranjiva roxburghii*. Sep Sci Technol. https://doi.org/10.1080/01496395.2015.1018439

31 Rosaceae

The family Rosaceae is known for edible fruits (apple, apricot, cherry, loquat, peach, pear, plum, quince, raspberry and strawberry), nuts (almond) and ornamental flowers (rose). Thus its members carry economic importance. They are primarily found in the north temperate zone and occur in a wide variety of habitats and form the third economically important group of species after Poaceae. The seeds of the wild apricot *Prunus armeniaca*, siberian apricot (*Prunus sibirica*), sweet almond (*Prunus amygdalus*) are reported to have been extracted for oil and the oil then used in biodiesel production.

31.1 Wild Apricot (*Prunus armeniaca*)

The wild apricot *Prunus armeniaca* (Armenian plum) is a cultivable species and is often referred as ansu apricot and Tibetan apricot. The exact nativity of this species is not known, but it is grown in many countries. It is often thought to have been originated from Armenia (Morikian 1983) and is recognized as a celebrated species (Fig. 31.1).

It is known to have been domesticated in China. It is also believed by some Scientists that this species is first cultivated in India (Huxley 1992). It is a tree of moderate size (8–12 m height) with a trunk dia. of 40 cm. The leaves are ovate and approximately 7–9 cm long. The fruits are pinkish red (Fig. 31.2). It thrives well at an elevation of 100–1600 m above sea level. Around 100 cm rainfall is sufficient. The propagation is through seeds and also through grafting. The tree starts bearing at the age of 5 and maintains the yield till 30 years. Each tree bears fruit to the tune of 80–120 kg (Papitha and Paulraj 2007). Being a deciduous tree, it is grown in Central Asia, Russia, USA, Iran, Iraq, India, Afghanistan, Pakistan, Syria and Turkey (Raj et al. 2012) (Fig. 31.3).

The seeds are removed from the fruits and the kernels isolated. The kernels are dried, powdered and extracted with petroleum ether. Chosen properties of oil are

S. P. Raj et al., *Biodiesel from Flowering Plants*,
https://doi.org/10.1007/978-981-16-4775-8_31

Fig. 31.1 An Armenian stamp on *Prunus armeniaca*

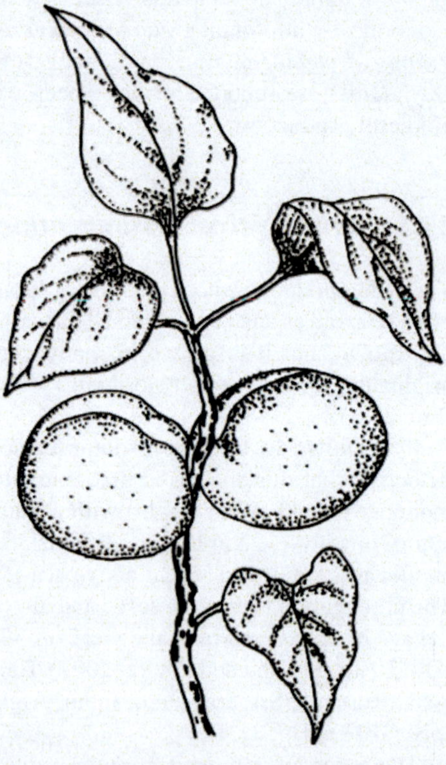

Fig. 31.2 The wild apricot, *P. armeniaca*

Fig. 31.3 Geographical distribution of the wild apricot, *P. armeniaca*

Table 31.1 Chosen physicochemical properties of *P. armeniaca* oil (Ullah et al. 2009)

Parameters	Value
Specific gravity at 25 °C (kg m^{-3})	910
Free fatty acid (%)	0.84
Iodine value (g I_2/100 g)	103
Saponification number (mg KOH g^{-1})	173
Average molecular weight (g mol^{-1})	867

Table 31.2 Properties of biodiesel from *P. armeniaca* oil (Papitha and Paulraj 2007; Ullah et al. 2009)

Parameters	Value
Kinematic viscosity at 25 °C (cSt)	4.26
Density at 15 °C (kg m^{-3})	855
Flash point (°C)	105
Cloud point (°C)	−4
Pour point (°C)	−15
Palmitic acid (C16:0) (%)	3.5
Stearic acid (C18:0) (%)	2.0
Oleic acid (C18:1) (%)	73.4
Linoleic acid (C18:2) (%)	20.0

given in Table 31.1. The oil is treated with sodium methoxide. Five grams of sodium hydroxide pellets are dissolved in 200 g of methanol to prepare sodium methoxide which is then mixed with 1 L of oil and stirred for 1 h. The content is allowed to settle for 8–12 h. The top layer (biodiesel) is isolated, distilled to expel methanol if any and washed subsequently to remove the alkali and dried by heating.

Normally around 90% of the oil is converted to biodiesel. If molecular sieve is used to separate the water in the preparation of sodium methoxide the yield will rise to 93%. If molecular sieve is used the chances of soap formation are limited (Wang et al. 2006). The quality of biodiesel is given in Table 31.2.

The density is 855 kg m^{-3} which is equivalent to a specific gravity of 0.877. The ASTM standard for the specific gravity of the biodiesel is 0.877. The viscosity also meets the standard (1.9–6.0) of the ASTM D6751. The flash point is higher than the regular diesel by 93 °C.

31.2 Siberian Apricot (*Prunus sibirica*)

The Siberian apricot *Prunus sibirica* is a deciduous shrub and grows in temperate climate of rich solar radiation. It bears flowers and fruits during the fourth year of its planting (Fig. 31.4).

It is grown mainly to prevent soil erosion, water conservation and as a wind shelter (Zhang et al. 2006). It grows in eastern China, Japan, Korea, Mongolia and Eastern Siberia (Fig. 31.5).

31.2.1 Preparation of Oil and Biodiesel

The seeds are removed from the fruits, dried and crushed. It is then extracted with *n*-hexane at 45–50 °C for 6–8 h. The characteristics of the oil thus obtained are given in Table 31.3.

Ten gram of potassium hydroxide and 200 g of anhydrous methanol are mixed. To this 1 L of oil is added and the entire content is stirred at 600 rpm at an ambient temperature of 60 °C for 1 h. It is then allowed to stay undisturbed overnight. The lower layer is discarded and the upper layer containing the biodiesel is heated to

Fig. 31.4 The Siberian apricot, *Prunus sibirica*

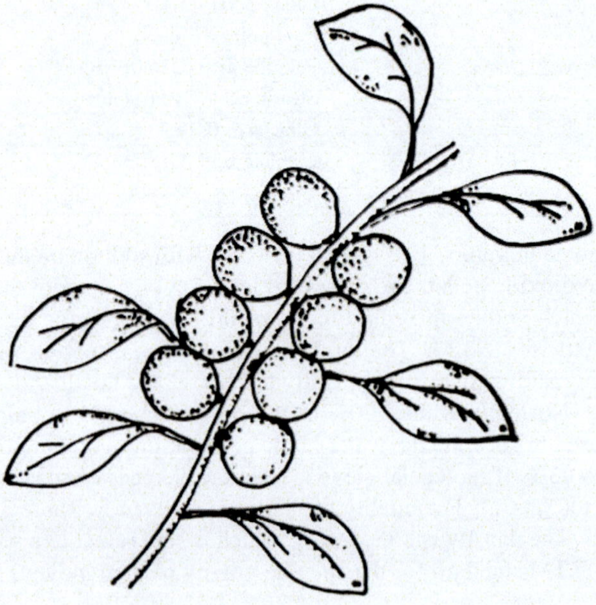

Fig. 31.5 Geographical distribution of the Siberian apricot, *P. sibirica*

Table 31.3 The properties of *P. sibirica* seed oil (Wang and Yu 2012)

Parameters	Value
Oil content of the kernel (% wt)	50.18
Myristic acid (C14:0) (%)	0.03
Palmitic acid (C16:0) (%)	3.79
Stearic acid (C18:0) (%)	1.01
Arachidic acid (C20:0) (%)	0.09
Palmitoleic acid (C16:1) (%)	0.67
Oleic acid (C18:1) (%)	65.23
Linoleic acid (C18:2) (%)	28.92
Linolenic acid (C18:3) (%)	0.14

remove the residual methanol and then washed repeatedly with warm distilled water to remove the alkali. The properties of the biodiesel are given in Table 31.4.

Since the oil contained unsaturated fatty acid to the extent of 95%, the biodiesel prepared out of it shows extremely satisfactory cold filter plugging point. The CN number also is good, ably assisted by the rich presence of linoleic acid in the oil (Knothe et al. 2003). The oxidation stability is paradoxically low apparently due to the high methyl linoleate content which has a tendency to oxidize rapidly. The oxidation stability of methyl linoleate is 0.94 h (Knothe 2008). In order to improve the oxidation stability synthetic antioxidant *tert*-butylhydroquinone is added so as to meet the standard. The flash point, viscosity and density fairly satisfy the limit. There is not much work carried out on its performance in engines.

Table 31.4 The properties of the *P. sibirica* oil biodiesel (Wang and Yu 2012)

Parameter	Value
Kinematic viscosity at 40 °C (cSt)	4.3
Density at 15 °C (kg m^{-3})	878
Flash point (°C)	173
Cetane number	48.8
Oxidative stability at 110 °C (h)	2.7
Cold filter plugging point (°C)	−14

31.3 Sweet Almond (*Prunus amygdalus* var. *dulcis*)

Two varieties of the almond *Prunus amygdalus* known to yield oil are the bitter almond *P. amygdalus* var. *amara* and sweet almond *P. amygdalus* var. *dulcis*. The sweet almond seeds carry not less than 51% oil by weight. The *P. amygdalus* var. *dulcis* is a deciduous tree which grows to 4–10 m in height with a trunk dia. of 30 cm. The leaves are 7–12 cm long with an undulated margin. They prefer to grow well at an ambient temperature of 15–30 °C. It comes to economic yield in 3 years and full bearing in 5–6 years. Throughout the year the tree bears flowers. The flower gives the fruit in 8 months of flowering. The fruit measures 3.5–6 cm. The fruit is often referred as drupe (Fig. 31.6).

This species is distributed in Middle East, India, North Africa, USA and Australia (Fig. 31.7). Around 3 million tonnes of fruits are produced in a year. This tree is called by different names as shown in Table 31.5.

31.3.1 Oil Extraction and Biodiesel Production

The seeds containing the kernel are removed from the fruits. The seeds are then cracked and the kernel isolated. The kernel is dried, powdered and extracted in *n*-hexane at 45–50 °C for 6–8 h until all the oil is extracted. The properties of oil are given in Table 31.6.

The free fatty acid is high (4.9%) which warrants a pretreatment before the alkali catalytic mode of transesterification is commenced. The FFA level is reduced through pre-treatment with potassium hydroxide (Giwa and Ogunbona 2014). The saturated and unsaturated fatty acids form a ratio of 11:89. To every kg of preheated

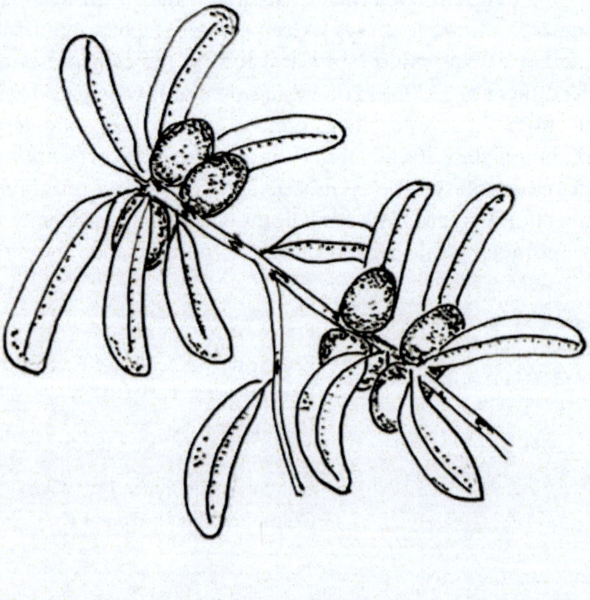

Fig. 31.6 The sweet almond plant, *Prunus amygdalus* var. *dulcis*

Fig. 31.7 Geographical distribution of the sweet almond, *P. amygdalus* var. *dulcis*

Table 31.5 The local names of *Prunus amygdalus* var. *dulcis* in different languages

Language	Names
English	Almond
German	Mandel, knack mandel
Italian	Mandorlo
Portuguese	Amêndova
Spanish	Almendra

Table 31.6 Physicochemical properties of *P. amygdalus* var. *dulcis* oil (Giwa and Ogunbona 2014)

Parameters	Value
Oil content (wt %)	51.45
Free fatty acid (%)	4.9
Palmitic acid (C16:0) (%)	9.3
Stearic acid (C18:0) (%)	1.8
Palmitoleic acid (C16:1) (%)	0.3
Oleic acid (C18:1) (%)	69.7
Linoleic acid (C18:2) (%)	18.2
Linolenic acid (C18:3) (%)	0.7
Average molecular weight (g mol^{-1})	874.5

oil sodium methoxide mixture containing 220 g of methanol and 5 g of NaOH is added and stirred for 1 h at 600 rpm in 60 °C. Consequently, the contents are allowed to settle for 8–12 h. The lower layer containing glycerol is discarded and the upper layer containing the biodiesel is heated at 80 °C so as to remove the residual methanol. It is then washed repeatedly to free it from any alkali or soap and dried by passing it over anhydrous sodium sulphate. The analytical details of biodiesel thus obtained are presented in Table 31.7.

The viscosity is fairly high, but it is within the range specified by ASTM D6751 and EN 14214. Oxidative stability is an important parameter. A minimum value of 3–6 h is prescribed as limit. The relatively satisfactory level (3.1 h) of oxidation stability is due to the poor quantity of methyl linoleate present in the biodiesel

Table 31.7 The properties of *P. amygdalus* var. *dulcis* oil biodiesel (Giwa and Ogunbona 2014)

Parameters	Value
Kinematic viscosity at 40 °C (cSt)	4.23
Iodine value (g I_2/100 g)	92.3
Oxidative stability (h)	3.1
Flash point (°C)	169
Cloud point (°C)	−3
Pour point (°C)	−9
Cetane number	58

(Knothe 2008). The flash point is above the minimum limits as per standard. The CN of the biodiesel is reasonably good. The iodine value is within the limit of 120 g I_2/100 g as recommended by EN 14214.

References

Giwa S, Ogunbona C (2014) Sweet almond (*Prunus amygdalus* dulcis) seeds as a potential feedstock for Nigerian Biodiesel Automotive Project. Ambiente Água Interdiscip J Appl Sci 9:37–45. ISSN: 1980-993X. https://doi.org/10.4136/1980-993X

Huxley A (ed) (1992) New RHS dictionary of gardening, vol 1. Macmillan, London, pp 203–205. ISBN: 0-333-47494-5

Knothe G (2008) Designer biodiesel: optimizing fatty ester composition to improve fuel properties. Energy Fuel 22:1358–1364

Knothe G, Matheans A, Ryan T (2003) Cetane numbers of branched and straight chain fatty esters determined in an ignition quality tester. Fuel 82(8):971–975

Morikian ES (1983) Apricots of Armenia: origin and classification of varieties. Acta Hortic 121:271–274. https://doi.org/10.17660/ActaHortic.1983.121.36

Papitha P, Paulraj S (2007) Tree borne oil seeds. In: Proc. natl. workshop on biodiesel, 17 and 18 Oct 2007. Madurai Kamaraj University, Madurai, pp 55–78

Raj V, Jain A, Chaudhary J (2012) *Prunus armeniaca* (apricot): an overview. J Pharm Res 5 (8):3964–3966

Ullah F, Noshen A, Hussain I, Bano A (2009) Base catalysed transesterification of wild apricot kernel oil for biodiesel production. Afr J Biotechnol 8(14):3289–3293

Wang L, Yu H (2012) Biodiesel from Siberian apricot (*Prunus sibirica* L.) seed kernel oil. Bioresour Technol 112:355–358

Wang L, Du W, Liu DH, Li LL, Dai MM (2006) Lipase catalysed biodiesel production from soybean oil deodorizer distillate with absorbent present in *tert.* butanol system. J Mol Catal B Enzym 43(1–4):29–32

Zhang J, Fu D, Wei Z, Zhao H, Zhang T (2006) Determination of ability of several arbor and shrub species to endure and survive extreme aridity with limited-areas methods under field conditions in Horqin Sandy Land. Acta Ecol Sin 26(2):467–474

32 Rutaceae

Members of the family Rutaceae have worldwide distribution, but appear richly in tropical regions. They are known for many important fruits (oranges, grapefruits, lemons, limes). Rutaceae is commonly known as citrus family. Most of them are trees or shrubs and a few are herbs. Family Rutaceae comprises more than 2000 species across 160 genera. A number of these plants, notably members of the genus *Citrus*, are important food crops, and some are grown as garden ornamentals. The seeds are endospermic containing oil or exalbuminous. In this chapter two species, namely wood apple *Aegle marmelos* and Chinese red pepper *Zanthoxylum bungeanum* known for their role in biodiesel production are dealt.

32.1 Wood Apple (*Aegle marmelos*)

The wood apple, *Aegle marmelos*, the only member of the genus *Aegle* is a native of Indian subcontinent. It is a deciduous shrub or medium sized tree growing to a maximum height of 13 m. The canopy is irregular with drooping branches. The leaf is trifoliate alternate, shallow with rounded teeth. Young leaf is pale green which turns into dark green as it matures. It prefers an altitude of 0–1200 m and a rainfall of 550–2000 mm. The flowers are 1.5–2.0 cm dia., pale green or yellow, sweetly scented and bisexual. The fruits are of the size 5–12 cm and take 11 months to mature fully. The seeds are embedded in a mucilaginous mass. This species is commonly referred as bael, bile, bhel, bengal quince, golden apple, Japanese bitter orange, stone apple and wood apple (Fig. 32.1).

This plant is distributed at parts of Asia, India, Sri Lanka, Thailand and Malaysia (Fig. 32.2).

S. P. Raj et al., *Biodiesel from Flowering Plants*,
https://doi.org/10.1007/978-981-16-4775-8_32

Fig. 32.1 The wood apple
Aegle marmelos

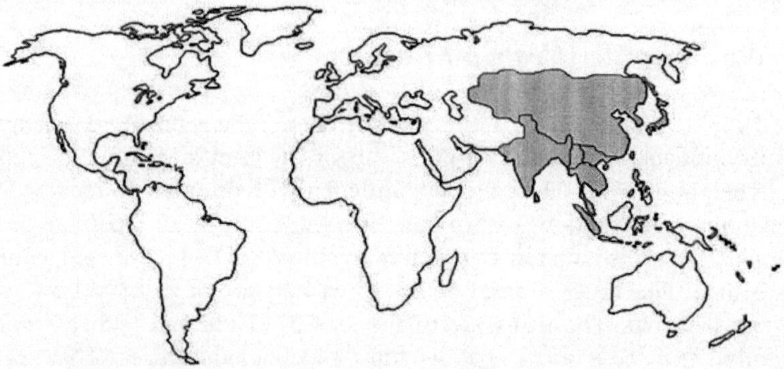

Fig. 32.2 Geographical distribution of *A. marmelos*

32.1.1 Oil Extraction

The hard, woody outer shell of the fruit is hammered out and the seeds are isolated from the mucilaginous pulp. The cleaned seeds are dried well. It is then crushed to have a particle size of 0.75 mm. This powder is extracted with *n*-hexane. The solvent-oil mixture is subsequently heated to expel the hexane.

Table 32.1 Physicochemical properties of the oil from *A. marmelos*

Parameters	Dhanamurugan and Subramanian (2015)	Thangarasu and Anand (2019)
Kinematic viscosity at 40 °C ($mm^2\ s^{-1}$)	36.22	27.92
Density at 15 °C ($kg\ m^{-3}$)	870	910
Flash point (°C)	–	325
Cloud point (°C)	–	4
Pour point (°C)	–	0
Acid value ($g\ KOH\ kg^{-1}$)	3.7	5.6
Iodine value ($g\ I_2/100\ g$)	–	85
Heating value ($MJ\ kg^{-1}$)	43.26	39.46

Ultrasonic sonicator with a maximum power output of 900 W also is used in oil extraction (Thangarasu and Anand 2019). The seed powder is taken in a beaker and mixed with acetone-heptane mixture (50:50 by volume). This is treated at 35% of ultrasonic power output at 40 °C. After 20 min of extraction the solvent-oil is separated from the solid by gravity filtration. The solvent is then distilled out in a rotary evaporator. The microwave process also is undertaken to separate the oil (Ramanathan and Thangarasu 2019). The seed powder (0.75 mm) is mixed 8 times of its vol. of acetone: heptane mixture (50:50 by volume) at 340 W power for 20 min. On completion of the extraction the mixture is centrifuged for 15 min at 3000 rpm. The analytical properties of the oil are given in Table 32.1.

The viscosity and the acid value are high. High acid value necessitates esterification of the oil before transesterification commences.

32.1.2 Biodiesel Preparation

The oil is taken in a two necked flat bottom glass reactor and is heated to 60 °C. The methanol acid mixture (methanol to oil molar ratio of 6:1 and acid (H_2SO_4) at a conc. of 0.5% of oil) is then poured on the above hot oil and stirred at 650 rpm. The reaction product is taken in a separatory funnel and waited till the phase separation completes. The esterified oil which settles as a bottom layer is drained and washed repeatedly with hot distiller water and dried after the last wash in a rotary evaporator.

The dried oil is then transesterified using methanol and sodium hydroxide. At the outset the methanol (6:1 molar ratio of methanol to oil) and sodium hydroxide (1% of oil) are mixed to get methoxide. This mixture (methoxide) is then poured on to the hot oil and stirred till the reaction completes. The finished product is then loaded in a separatory funnel. When kept waiting, two layers are formed in the separatory funnel. The upper layer comprises the biodiesel and the unreacted methanol. The bottom layer comprises glycerin which is drained. The biodiesel cum methanol layer is repeatedly washed to remove traces of glycerol, catalyst and other impurities if any. It is finally heated to expel the moisture.

Table 32.2 Physicochemical properties of the biodiesel from *A. marmelos*

Parameters	Dhanamurugan and Subramanian (2015)	Thangarasu and Anand (2019)
Kinematic viscosity at 40 °C ($mm^2\ s^{-1}$)	4.81	3.75
Density at 15 °C (kg m^{-3})	876	880
Flash point (°C)	128	164
Cloud point (°C)	–	−1
Pour point (°C)	–	−4
Acid value(g KOH kg^{-1})	0.28	0.2
Iodine value (g I_2/100 g)	–	56
Heating value (MJ kg^{-1})	40.52	37.5
Cetane value	52.3	–

Table 32.3 Specification of the diesel engine used in the trial of *A. marmelos* biodiesel (Dhanamurugan and Subramanian 2015)

Details	Specification
Number of cylinder	Single, vertical
Bore	87.5 × 110 mm
Compression ratio	17.5:1
Rated power (kW)	5.2
Rated speed (rpm)	1500
Injection timing	23° bTDC
Injection pressure	220 bar

The ultrasonic processor (900 W) or microwave oven (850 W) also is used in the transesterification process. The stirring speed is 600 rpm. The methanol to oil molar ratio is 6:1, the alkali (NaOH) conc. is 1% of the oil, the duration is 60 m at the maximum and the temp. is 60 °C. The biodiesel is separated using a separatory funnel and the product is water-washed and dried by heating (Ramanathan and Thangarasu 2019). The properties of the biodiesel obtained are given in Table 32.2.

The kinematic viscosity, flash point, cloud point, pour point, acid value, iodine value are seen reduced in biodiesel as compared to that of its parent oil due to the esterification and transesterification effects. The viscosity plays a crucial role in the combustion process inside the engine cylinder. Ideal viscosity of the fuel ensures effective spray pattern. If the viscosity is too low, there will be internal pump leakage which is likely to bring down the pressure considerably which ultimately affects the fuel spray. Normally, if the chain length of the fatty acid increases in the parent oil, the viscosity automatically increases.

32.1.3 Engine Performance

A series of trials were made by Dhanamurugan and Subramanian (2015) employing *A. marmelos* biodiesel in a single cylinder, four stroke water cooled engine with the details as shown in Table 32.3.

All the experiments were made at a constant speed (1500 rpm). The brake specific fuel consumption (BSFC) of biodiesel was more by 11.34% compared to that of diesel. The low heating value and the increased viscosity cause the decrease in brake thermal efficiency as there is limited atomization. The percentage reduction in brake thermal efficiency of this biodiesel is 6.89 when compared to that of diesel. The NO_x in the exhaust increases by 14.38%, whereas the CO level reduces by 31.25% as compared to that of the diesel. The maximum heat release in diesel is 109.565 kJ m^{-3}, whereas it is 98.206 in this biodiesel (Dhanamurugan and Subramanian 2015).

32.2 Chinese Red Pepper (*Zanthoxylum bungeanum*)

The Chinese red pepper *Zanthoxylum bungeanum* is a deciduous shrub and grows to a maximum height of 6 m. Its stem and branches are covered with prickles. The leaves are with 5–13 foliolates. The blades of the leaflets are sessile and opposite. The fruits are purple red and 4–5 mm dia. The seeds are 3.5–4.5 mm dia. Each seed weighs 15–16 mg (Fig. 32.3).

This species is reported to occur at an elevation lower than 3200 m in China, Japan, India, Korea and Taiwan (Fig. 32.4). In China its seed potential is around 1 million metric tonne (Jian et al. 2015). A single tree yields 2–3 kg of seeds.

Fig. 32.3 The Chinese red pepper *Zanthoxylum bungeanum*

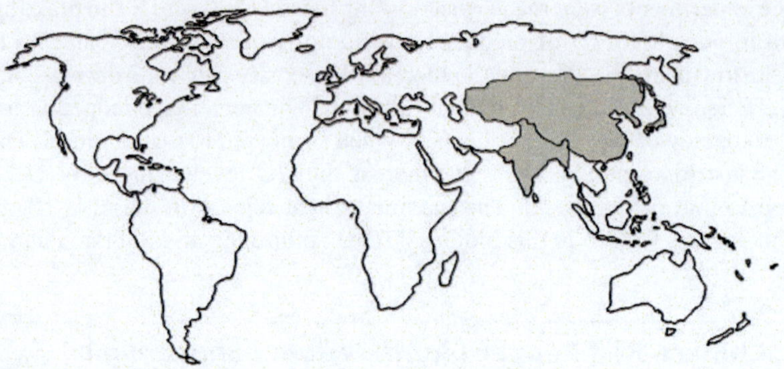

Fig. 32.4 The geographical distribution of *Z. bungeanum*

This species is popularly called as Chinese red pepper, Chinese pepper, flatspine prickly ash, Da-Hong-Pao, Red huajiao, prickly ash and red shell.

32.2.1 Oil Production

This oil has a great potential of being utilized as a raw material for biodiesel (Yang et al. 2008; Zhang et al. 2010, 2013, 2014). The seeds are dried repeatedly and powdered. The oil is extracted from this powder in a Soxhlet apparatus. Various solvents tried for extraction are methanol, ethanol, petroleum ether and *n*-hexane. Out of the four solvents, petroleum ether and *n*-hexane are proved to be efficient. The desired solvent and seed powder ratio is 4:1–7:1 by weight (Zhang et al. 2015). The extract is then heated to expel the solvent. The highest recovery of oil is obtained in *n*-hexane (24.7%). In methanol and ethanol the oil yield is 19.2% and 18.9%, respectively. It is reported that the oil extracted using methanol has more free fatty acid than that extracted by ethanol. It is due to the high solubility of free fatty acid in methanol. More palmitic acid and less linoleic acid are present in oil when the seed is extracted using methanol as compared to that of ethanol, petroleum ether and *n*-hexane (Table 32.1). The acid value of the oil obtained due to the extraction by methanol and ethanol is 68.9 and 44.6 g KOH kg^{-1} at a solvent-to-seed ratio of 4. The fatty acid composition of *Z. bungeanum* oil extracted in different solvent is given in Table 32.4.

32.2.2 Biodiesel Production

The free fatty acid of the oil is relatively high (>25%). Therefore, it is esterified twice. Acid catalysed esterification carried out at the first instance reduces the free fatty acid content to an acceptable level. In case the free fatty acid does not come down to an acceptable level the esterification process is repeated. The values for the

Table 32.4 Fatty acid composition of *Z. bungeanum* oil extracted using different solvents (Zhang et al. 2015)

	Composition (%)			
Fatty acid	Methanol	Ethanol	Pet. ether	*n*-Hexane
Palmitic acid (C16:0)	16.54	13.42	13.69	13.50
Stearic acid (C18:0)	1.27	1.29	1.36	1.32
Palmitoleic acid (C16:1)	7.39	6.20	6.23	6.23
Oleic acid (C18:1)	31.54	32.26	32.89	32.25
Linoleic acid (C18:2)	19.44	22.58	22.42	22.54
Linolenic acid (C18:3)	23.80	24.24	23.41	24.16

Table 32.5 The free fatty acid content of the oil due to esterification process (as adopted from Zhang and Jiang 2008)

Number of steps	Molar ratio of methanol to oil	Free fatty acid (%)
First esterification	9:1	2.90
Second esterification	9:1	0.94

Table 32.6 The free fatty acid level of *Z. bungeanum* oil after esterification with certain Lewis acids from the original value of 28.5% (as adopted from Zhang et al. 2012a, b)

Catalyst	Final free fatty acid (%)	Reduction from original (%)
$SnCl_2 \cdot 2H_2O$	12.7	55.4
$SnCl_4 \cdot 5H_2O$	2.4	91.6
$Fe_2(SO_4)_3 \cdot XH_2O$	2.6	90.9
$CuSO_4 \cdot 5H_2O$	19.0	33.3
$Al_2(SO_4)_3 \cdot 18H_2O$	6.9	75.8

free fatty acid content in the repeated acid treatment as adopted from Zhang and Jiang (2008) are presented in Table 32.5.

The value for free fatty acid in the oil before the first esterification is 22.75% (Zhang and Jiang 2008) which is reduced to 0.94% at the end of second repetitive esterification. The sulphuric acid used in the above process is one% and the methanol to oil ratio is 9:1. The stirring speed is 600 rpm and the ambient temp. is 95 °C. The process is carried out in a reflux unit to avoid any wastage of methanol. After the esterification process the content is transferred to a separatory funnel and allowed to settle overnight. The esterified oil stays at the bottom which is then removed, washed repeatedly and dried by anhydrous sodium sulphate.

Many other Lewis acids also are being used in the esterification process. They are $SnCl_2 \cdot 2H_2O$, $SnCl_4 \cdot 5H_2O$, $Fe_2(SO_4)_3 \cdot XH_2O$, $CuSO_4 \cdot 5H_2O$ and $Al_2(SO_4)_3 \cdot 18H_2O$. Esterification of *Z. bungeanum* oil using the above catalysts is carried out at a conc. of 3%, with methanol to oil molar ratio of 20:1 in a reaction duration of 2 h. The reduction of free fatty acid from the original free fatty acid value (28.5%) is given in Table 32.6.

$SnCl_4 \cdot 5H_2O$ gave the highest reduction (91.6%). These types of catalysts are less corrosive. $SnCl_4 \cdot 5H_2$ is known to generate hydrogen chloride gas during the

Table 32.7 The physicochemical properties of *Z. bungeanum* biodiesel

Parameters	Yang et al. (2008)	Jian et al. (2015)
Kinematic viscosity at 40 °C ($mm^2\ s^{-1}$)	4.8	4.7
Density at 25 °C (kg m^{-3})	892	878.8
Flash point (°C)	>174	–
Cold filter plugging point (°C)	−3	–
Cloud point (°C)	−6	–
Cetane number	45	54
Acid value (g KOH kg^{-1})	0.3	–
Copper corrosion	1	–
Heating value (MJ kg^{-1})	–	39.3
Palmitic acid methyl ester (%)	–	10.7
Methyl palmitoleate (%)	–	7.0
Stearic acid methyl ester (%)	–	1.6
Oleic acid methyl ester (%)	–	27.6
Linoleic acid methyl ester (%)	–	24.6
Linolenic acid methyl ester (%)	–	21.6

esterification process. However, $SnCl_4{\cdot}5H_2$ is considered less dangerous compared to sulphuric acid.

Ferric sulphate as a catalyst at a conc. of 2% in an oil methanol ratio of 1:24 was also reported (Zhang et al. 2012a, b) which gave a reduction in free fatty acid to the tune of 90.9%. The ferric sulphate during the process of esterification converts itself to $[Fe(H_2O)_6]^{3+}$ which then releases protons. If the time duration extends beyond the optimum, the water generated during the reaction inhibits the esterification.

The esterified oil thus obtained is transesterified using potassium hydroxide as a catalyst. This catalyst is mixed with methanol so as to have a final conc. of 1% with the total mixture at a methanol to oil molar ratio of 6.5:1. This methanol-alkali mixture is added to the oil at hot condition (60 °C) and stirred incessantly (600 rpm) for 2 h. This is then loaded in a separatory funnel and waited overnight for phase separation. The biodiesel standing as a top layer is drained, washed repeatedly with hot distilled water to remove residual alkali and alcohol. The properties of the biodiesel are presented in Table 32.7.

The viscosity of this oil is within the standard specification and therefore there may not be any problem in the atomization and fuel distribution. Similarly the density, cetane number and flash point are all within the limit. The mass fraction of C, H and O are 77%, 12% and 11%, respectively (Jian et al. 2015). According to Knothe (2005) the viscosity, cold flow and cetane number are governed by the chain length, and unsaturation of the fatty acids. Excess of linolenic acid methyl ester is reported to be unfavourable to the biodiesel.

It is known that the catalyst concentration is one of the essential factors in the transesterification. The factors governing the conversion are in the order: concentration of the catalyst > duration > temperature > methanol to oil molar ratio (Zhang et al. 2012a, b).

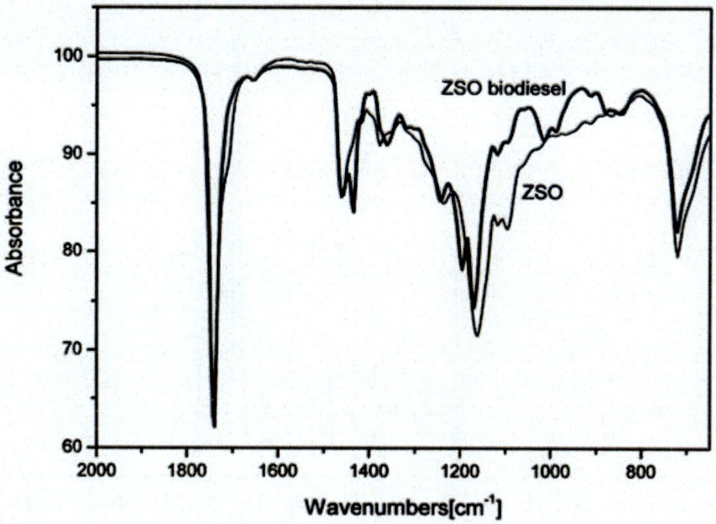

Fig. 32.5 FTIR spectra of oil (ZSO) and biodiesel (ZSO biodiesel) from *Z. bungeanum* (having high free fatty acids). (Reprinted with permission from Zhang et al. 2012a, b American Chemical Society)

Fourier transformed infrared spectroscopy analysis (Fig. 32.5) at the wave number between 2000 and 600 cm^{-1} indicates that the *Z. bungeanum* oil and its biodiesel more or less superimposes at 1740 cm^{-1} which is related to the carbonyl group. The band at 1169 cm^{-1} is mainly on C–CH_2–O, and its conc. is observed to be reduced on account of the transesterification. The availability of $(CH_2)_n$ is known by the band at 725 cm^{-1}. The bands at 1196 and 1435 cm^{-1} indicate the presence of O–CH_3 and CH_3 in the biodiesel which confirms the transformation of ester from the oil.

In a ^{1}H NMR study made by the same author (Zhang et al. 2012a, b) no signal related to glycerol was recorded in 4.1–4.4 ppm which indicated that the oil had already been converted into methyl ester. The spectrum at 3.66 ppm explicitly indicates this contention (Fig. 32.6).

In the transesterification process heterogeneous solid acid catalysts with –SO_3H are considered. Carbon based solid acid catalyst made by carbonization-sulphonation using glycerol and sulphuric acid is employed (Devi et al. 2009). This catalyst is prepared by mixing the *Z. bungeanum* oil with concentrated sulphuric acid at 1:4 ratio by weight. This mixture is slowly heated until it reaches 220 °C (in 20 min). The heating is continued gently for 30–40 min, by then the foaming ceases. The black material formed eventually is repeatedly washed using dist. water till it becomes neutral pH. The catalyst thus developed is observed to be an irregular mass with a high dispersion of elemental sulphur indicating the presence of –SO_3H group. The pore size is 20 nm, volume is 0.029 cm^3 g^{-1}. The BET surface area is 3.54 m^2 g^{-1}. The acid density is 4.39 mmol g^{-1}. It is then filtered and dried at 120 °C for 3 h. The oil, methanol and the above catalyst are mixed in a reactor. It is

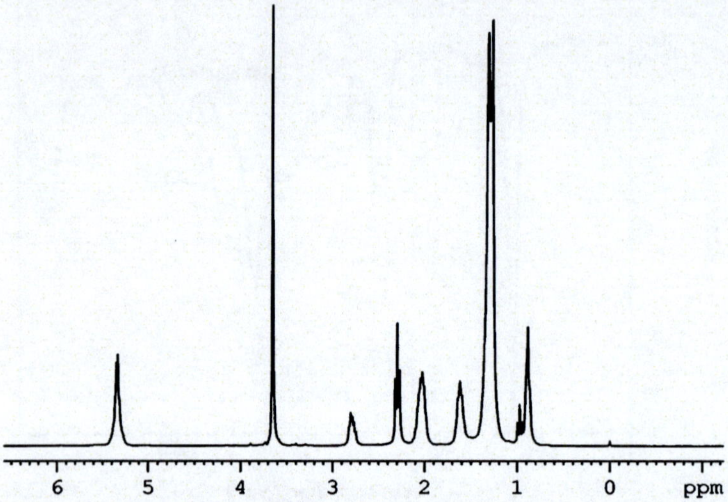

Fig. 32.6 ^{1}H NMR spectrum of *Z. bungeanum* biodiesel. (Reprinted with permission from Zhang et al. 2012a, b American Chemical Society)

then heated while stirring for 2 h. The recommended experimental parameters are catalyst 10%, temp. 140 °C, duration 3 h and the methanol to oil ratio is 30:1. On completion of the reaction the content is cooled and the catalyst is filtered out. The filtrate is loaded in a separatory funnel and is kept undisturbed until the phase separation completes. The biodiesel formed is then separated, washed repeatedly using warm dist. water and dried using anhydrous sodium sulphate. This process is found to be favourable as the esterification and transesterification processes are carried out simultaneously (Wang et al. 2017). During the above process, the carbonyl group present in the free fatty acid is activated by the hydrogen present in the catalyst (Yaroshevich et al. 2015) and form the protonated carbonyl group. Parallelly the methanol is absorbed by the hydrogen bond. The absorbed methanol gradually attacks the protonated free fatty acid and yields ester and water as by-products (Miao and Shanks 2011). In the above reaction, as the methanol is protonated, the process of attacking the protonated carbonyl group gets inhibited and thus more methanol is demanded, necessitating a high molar to oil ratio.

The catalyst involved in the above reaction can be reused after separating it from the mixture. It is reported that the recovered catalyst, when used at the second time there is a reduction of 11% in the activity. The reduction in activity is reported to be 31%, 38% and 40% in the third, fourth and fifth successive cycles. Such reduction in the activity is due to the leaching of C, O and S from the surface of the catalyst complex. The acid density also is reduced. It is proved that the extraction of the oil from the seed powder, esterification and transesterification can be performed jointly in the Soxhlet apparatus as a single step using methanol or ethanol along with sulphuric acid as a catalyst. The reaction is extended up to 16 h with a reaction temp. of 95 °C so that the biodiesel is obtained as a final product. This avoids phase separation and solvent removal (Zhang et al. 2015).

32.2.3 Engine Performance

The biodiesel produced is blended with diesel and tested in a four stroke, water cooled direct injection diesel engine (Jian et al. 2015). It is known that the cylinder pressure decreases as the share of biodiesel is increased in the blend. The ignition delay also decreases with the increase in biodiesel blend. There is a marked reduction in the pressure lifting ratio. Such reduction is high, especially when the engine is made to run in low rpm. Low ignition delay causes low instantaneous heat release, especially at low rpm. According to Jian et al. (2015) NO_x emission reduced to a maximum of 25%, 26% and 29% in B10, B20 and B30 blends, respectively, when the engine runs at low rpm. The NO_x conc. in the exhaust and cylinder pressure is observed to be directly proportional. The size of the particulate matter present in the exhaust ranged between 0.01 and 0.1 μm. The concentration of this particle is observed to be reduced in exhaust by 35%, 46% and 56% in B10, B20 and B30 as compared to that of petro-diesel when the engine is made to run in 1500 rpm. The reduction values are 38%, 50% and 56%, respectively at 2000 rpm. Thus high rotation reduces the particulate exhaust implying that there is a near complete combustion. The mass of particle increases when the size of the particle increases. The biodiesel consumption increases by 5.7%, 3.13% and 24% (Jian et al. 2015) in B10, B20 and B30 blends. One of the major reasons is the low heating value of the biodiesel. Another reason is the change in combustion pattern due to the high cetane number and injection timing.

References

Devi BL, Gangadhar KN, Prasad PS, Jagannadh B, Prasad RB (2009) A glycerol based carbon catalyst for the preparation of biodiesel. ChemSusChem 2(7):617–620

Dhanamurugan A, Subramanian R (2015) Emission and performance characteristics of a diesel engine operating on diesel-bael (*Aegle marmelos*) biodiesel blends. Nat Environ Pollut Technol 14(2):331–336

Jian Z, Xuanjun W, Pilong H, Mingjun H, Shuyan L (2015) Physicochemical properties, combustion and emission performance of a novel *Zanthoxylum bungeanum* seed oil methyl ester biodiesel. Int J Green Energy 12(12):1255–1262

Knothe G (2005) Dependence of biodiesel fuel properties on the structure of fatty acid alkyl esters. Fuel Process Technol 86:1059–1070

Miao SJ, Shanks BH (2011) Mechanism of acetic acid esterification over sulfonic acid-functionalised mesoporous silica. J Catal 279:136–143

Ramanathan A, Thangarasu V (2019) Effect of high-frequency microwave irradiation on *Aegle marmelos* correa oil extraction: kinetic and thermodynamic study. Energy Procedia 158:1046–1051

Thangarasu V, Anand R (2019) Physicochemical fuel properties and tribological behaviour of *Aegle marmelos* correa biodiesel. In: Advances in ecofuels for a sustainable environment. https://doi.org/10.1016/B978-0-08-102728-8.00011-5

Wang W, Lu P, Tang H, Ma Y, Yang X (2017) *Zanthoxylum bungeanum* seed oil based carbon solid acid catalyst for the production of biodiesel. New J Chem 41(17):9256–9261

Yang F, Su Y, Li X, Zhang Q, Sun R (2008) Studies on the preparation of biodiesel from *Zanthoxylum bungeanum* Maxim seed oil. J Agric Food Chem 56:7891–7896

Yaroshevich IA, Krasilnikov PM, Rubin AB (2015) Functional interpretation of the role of cyclic carotenoids in photosynthetic antennas via quantum chemical calculations. Comput Theor Chem 1070:27–32

Zhang J, Jiang L (2008) Acid catalysed esterification of *Zanthoxylum bungeanum* seed oil with high free fatty acids for biodiesel production. Bioresour Technol 99:8995–8998

Zhang J, Chen S, Yang R, Yan YY (2010) Biodiesel production from vegetable oil using heterogeneous acid and alkali catalyst. Fuel 89:2939–2944

Zhang J, Xuanjun W, Hejun G, Shenghua L, Jing M (2012a) Synthesis and properties of a novel ethylene glycol monobutyl ether palm oil monoester biofuel. Int J Green Energy 9:573–583

Zhang J, Zhang L, Jia L (2012b) Variables affecting biodiesel production from *Zanthoxylum bungeanum* seed oil with high free fatty acids. Ind Eng Chem Res 51(9):3525–3530

Zhang J, Wang X, Huang Z, Han QL (2013) Preparation and reaction dynamics on biodiesel made from *Zanthoxylum bungeanum* seed oil and methanol. J Renew Sustain Energy 5:043123

Zhang J, Cui C, Chem H, Liu J (2014) The completion of esterification of free fatty acids in *Zanthoxylum bungeanum* seed oil with ethanol. Int J Green Energy 11:822–832

Zhang J, Wu H, Yang F, Zhang J (2015) Evaluation of Soxhlet extractor for one-step biodiesel production from *Zanthoxylum bungeanum* seeds. Fuel Process Technol 131:452–457

Salicaceae

33

Salicaceae is a willow family and the members include willows, poplar, aspen, and cottonwoods. They grow along any stream, lake or mountain meadow. This family comprises bushes and trees with simple and alternate leaves. Plants of this family have medicinal importance as analgesic, anti-inflammatory, astringent and diuretic. By and large they contain varying amounts of phenolic glycosides: populin, salicin and methyl salicylate from which the common aspirin was originally prepared. These properties are rich in the inner bark, but are also present in the leaves. Willow family also is used to treat fevers, headaches, arthritis and other inflammations, particularly in the urinary tract. One of the members, namely wonder tree (*Idesia polycarpa*) has some importance as a resource for biodiesel.

33.1 Wonder Tree (*Idesia polycarpa*)

The wonder tree *Idesia polycarpa* vernacularly referred as Chinese wonder tree or igiri tree or shantongzit is usually a tall tree of a maximum height of 15 m. A fully grown tree has a spread of 12 m. It is a deciduous, dioecious tree growing liberally at hills and mountainous forests. For an ideal plantation both male and female trees are necessary. They prefer to grow in sandy, loamy and clay soil of far-reaching diversified soil-pH. The leaves are heart shaped with pointed tip. The leaf is glossy and pure green. It is 8–20 cm long and 7–20 cm broad. The flowers are yellow or green. Male flowers are large in size (12–16 mm dia.) and the female flowers are relatively small (5–9 mm dia.). The fruits are red or orange in colour and are a berry of 5–10 mm dia. Figure 33.1 indicates the cluster of fruits hanging from the branch of the tree. The berry contains seed of 2–3 mm dia. Each. Both the pulp of the berry and the seed contain oil.

This tree has a longevity of 50–150 years and is richly grown in East Asia, China, Japan, Korea and Taiwan (Fig. 33.2).

S. P. Raj et al., *Biodiesel from Flowering Plants*,
https://doi.org/10.1007/978-981-16-4775-8_33

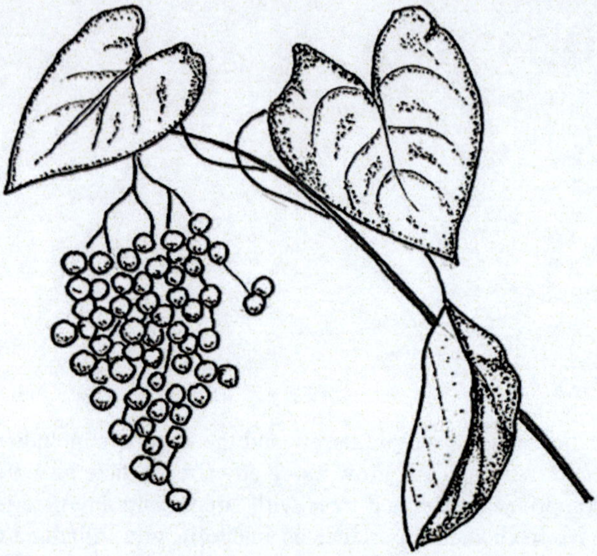

Fig. 33.1 The wonder tree *Idesia polycarpa*

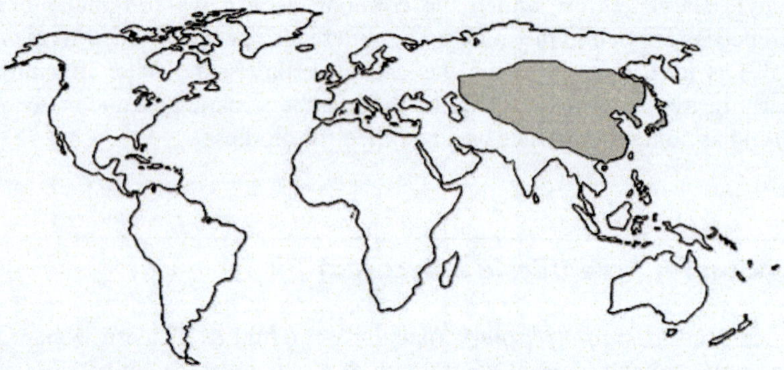

Fig. 33.2 Geographical distribution of *I. polycarpa*

The oil content of the pulp is reported to be 26.15%, whereas it is 26.26 in the seed kernel. The total yield of oil per tree is 1.5–2.5 kg and the overall yield is 2.25–3.5 tonnes ha^{-1} $year^{-1}$ (Yang et al. 2009). The tree comes to yield once a year.

33.1.1 Oil Extraction

The dried berry is passed through a mechanical expeller and the oil thus obtained is neutralized by treating with sodium carbonate at the rate of 150 g kg^{-1} of oil. The neutralized oil is normally stored at low temperature (4.8 °C) (Yang et al. 2009). Alternatively, the fruits are dried at 60 °C and made as a powder. Oil is extracted

Table 33.1 The fatty acid profile of *I. polycarpa* fruit oil (Yang et al. 2009)

Fatty acid	(%)
Palmitic acid (C16:0)	15.06
Stearic acid (C18:0)	1.18
Palmitoleic acid (C16:1)	6.5
Oleic acid (C18:1)	5.5
Linoleic acid (C18:2)	70.6
Linolenic acid (C18:3)	1.1

Table 33.2 Physicochemical properties of *I. polycarpa* biodiesel (Yang et al. 2009)

Parameters	Value
Kinematic viscosity at 40 °C ($mm^2\ s^{-1}$)	4.12
Density at 15 °C ($kg\ m^{-3}$)	886
Flash point (°C)	>174
Acid value ($g\ KOH\ kg^{-1}$)	0.27
Cloud point (°C)	−4
Cold filter plugging point (°C)	−2
Cetane number	47
Copper corrosion at 50 °C 3 h	1a

from it using a Soxhlet apparatus with petroleum ether as solvent (Li et al. 2016). The fatty acid composition of the oil is given in Table 33.1.

Among the various fatty acids, linoleic acid is the most prevalent unsaturated fatty acid in this oil.

33.1.2 Biodiesel Preparation

Alkali catalysed transesterification procedure is being followed for biodiesel preparation from the seed oil of this species. Potassium hydroxide (1% of oil) is dissolved in methanol (molar to oil ratio 6:1) and kept aside. Then the oil is heated to 30 °C and the potassium hydroxide methanol mixture is poured in it and stirred at 600 rpm for 40 min. Then the reacted product is taken to a separatory funnel and kept undisturbed overnight to enable phase separation. At the end the biodiesel is seen as a top layer and glycerin moves down to the bottom which is drained and discarded. The biodiesel layer is repeatedly washed with water and the washed biodiesel is gently heated to remove the moisture. The properties of the biodiesel thus obtained are given in Table 33.2.

Among the few characteristics presented in Table 33.2 the cetane number is slightly lower than the standard. The properties such as viscosity, cetane number and cold flow characteristics of the biodiesel are largely governed by the fatty acid profiles. The viscosity increases with the increase in chain length and decreases in accordance with the increase in unsaturation.

References

Li R, Giao X, Li L, Liu X, Wang Z, Lu S (2016) De novo assembly and characterization of the fruit transcriptome of *Idesia polycarpa* reveals candidate genes for lipid biosynthesis. Front Plant Sci 71:801. https://doi.org/10.3389/fpls.2016.00801

Yang G, Su Y, Li X, Zhang Q, Sun R (2009) Preparation of biodiesel from *Idesia polycarpa* var. *vestitia* fruit oil. Ind Crops Prod 29:622–628

Salvadoraceae

34

Salvadoraceae is a small family of plants of the order Brassicales (formerly Celastrales) in 3 genera with a less than 60 plants. They occur in hot dry areas of Africa (including Madagascar), Southeast Asia, and Java. They are either shrubs or small trees often covered with spines. Few works are available for biodiesel on the plants jhal (*Salvadora oleoides*) and miswak (*Salvadora persica*).

34.1 Jhal (*Salvadora oleoides*)

The jhal (*Salvadora oleoides*) tree is often called as desert grapes and grows to a maximum height of 9 m. The trunk is short and twisted. The leaf is glaucous, linear and ovate and is 3–10 cm long and around 1 cm wide. The flowers are greenish white and very small (around 2.5 mm across) (Fig. 34.1). The fruit is a drupe of 5 mm dia. The seeds are greenish yellow and are about 3 mm diameter. The seeds contain 40–45% oil (Arora et al. 2014).

This species is commonly referred as jhal, badapilu, pilu, vridhpilu and khakan. It grows at an altitude of 1000 m above MSL and is distributed in tropical Africa, Asia, Egypt, China, Mascarene Islands and India (Fig. 34.2). It occupies lands of extreme environmental conditions and is able to tolerate low rainfall (less than 200 mm).

34.1.1 Oil and Biodiesel

The seed is isolated from the drupe,dried, powdered and extracted with *n*-hexane. The oil thus obtained is transesterified with methanol in the presence of alkali. The resultant mixture is allowed to settle overnight. The glycerol formed during the reaction is dense, which moves to the bottom and is removed. The overlying biodiesel is heated to expel the methanol. It is then repeatedly washed with water

S. P. Raj et al., *Biodiesel from Flowering Plants*,
https://doi.org/10.1007/978-981-16-4775-8_34

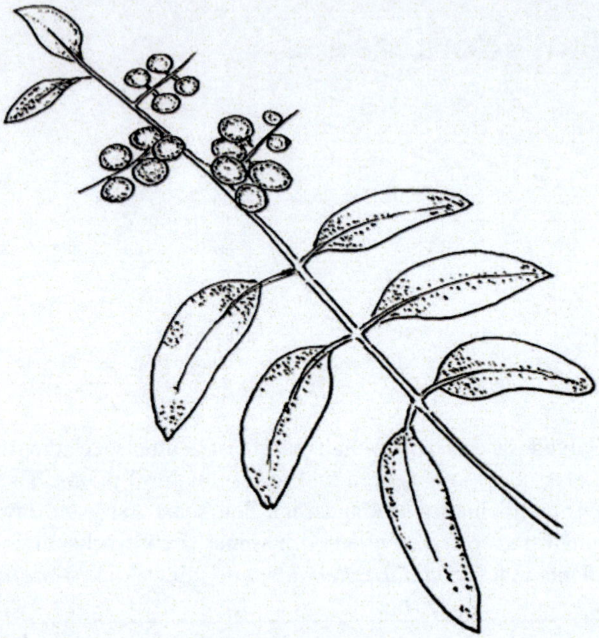

Fig. 34.1 The jhal, *Salvadora oleoides*

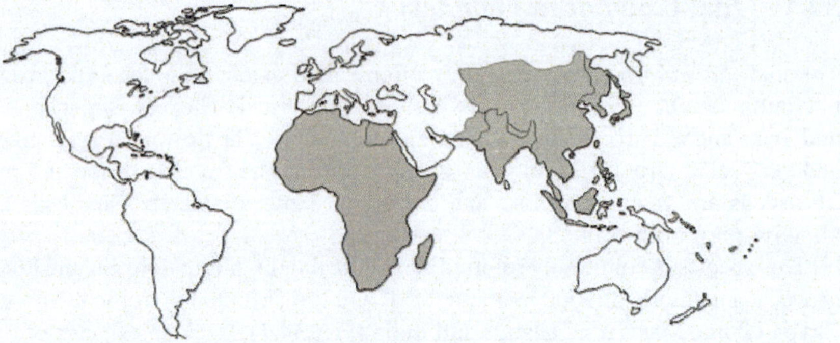

Fig. 34.2 Geographical distribution of *Salvadora oleoides*

to remove any trace of alkali. The residual moisture is removed by heating the oil or dried through molecular sieves.

The chosen characteristics of *S. oleoides* biodiesel are given in Table 34.1.

The kinematic viscosity of this biodiesel is paradoxically low and the cetane number is appreciably high. It has a high oxidative stability. The saturated fatty acid ester forms around 90% of the total. In order to qualify Euro III and Euro IV of any fuel, the sulphur content of it has to be removed which ultimately results in the loss of lubricity. This naturally increases the wear and tear of the moving parts of the engine. Therefore it is recommended to blend the diesel with a portion of biodiesel (not exceeding 20%). The fatty acid esters act as a lubricating agent (Bhatnagar et al.

Table 34.1 Properties of *S. oleoides* biodiesel (Bhatnagar et al. 2006)

Properties	Value
Kinematic viscosity at 40 °C (cSt)	3.25
Density at 15 °C (kg m^{-3})	876
Oxidative stability (h)	14.6
Flash point (°C)	130
Total sulphur (mg L^{-1})	1200
Pour point (°C)	3
Molecular weight (g mol^{-1})	258
Acid value (mg KOH g^{-1})	0.45
Cetane number	59.9

Table 34.2 HFRR friction coefficient of diesel (HSD) and biodiesel (*S. oleoides*) blend (Bhatnagar et al. 2006)

Properties	Friction coefficient (μ)
HSD + 5% biodiesel	0.27
HSD + 10% biodiesel	0.28
HSD + 15% biodiesel	0.28
HSD + 20% biodiesel	0.28
HSD + 50% biodiesel	0.28
Biodiesel 100%	0.28

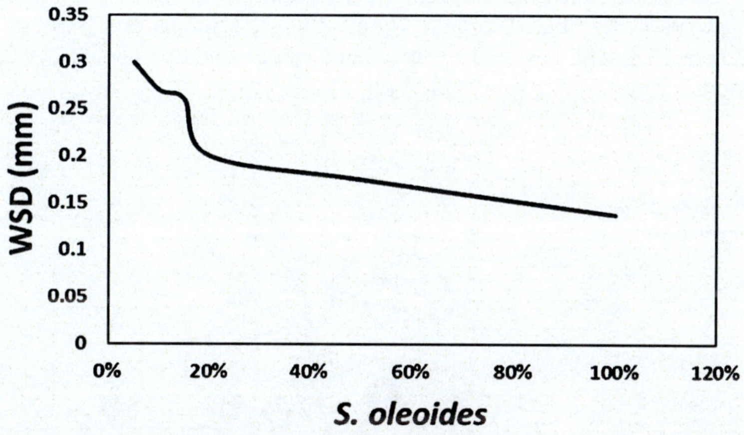

Fig. 34.3 Effect of *S. oleoides* biodiesel blend with diesel on the wear scar diameter (WSD)

2006). High frequency reciprocating rig (HFRR) friction coefficient with respect to the high speed diesel (HSD) blended with different percent of biodiesel is given in Table 34.2 and the corresponding Wear Scar Diameters (WSD) are shown in Fig. 34.3.

It is to be noted that when the portion of the biodiesel of *S. oleoides* in the blend is increased the WSD reduced. The WSD values for neat diesel (HSD) and biodiesel from *Jatropha curcas*, *Pongamia glabra*, *Madhuca indica* and *S. oleoides* are 0.372, 0.137, 0.200, 0.171 and 0.138 mm, respectively. This indicates the superiority of

S. oleoides as a lubricant. High lubricity of the oil of *S. oleoides* is due to the high sulphur content present as thioglycoside, glucotropaeolin dibenzylthioureas and dibenzylurea (Bhatnagar et al. 2006). The engine performance study is not yet made in this biodiesel.

34.2 Miswak (*Salvadora persica*)

The shrub, *Salvadora persica* popularly known as miswak (tooth brush) is often found in the coastal area and grows resisting saline conditions (Reddy et al. 2008). It is an evergreen plant growing to a height of 3 m. It flourishes well at an elevation up to 1800 m above mean sea level. The leaves are fleshy, glaucous, 4–6 cm long and 2–3 cm wide. These leaves are elliptic, lanceolate and obtuse. The flowers are greenish yellow. The fruit is a drupe of 3 mm dia., globose and smooth (Fig. 34.4). It becomes red when ripe. The seeds yield a semi-solid fat which richly contain lauric acid (40–45%). The seeds are harvested after 3 months of seed setting. Around 3400 seeds form a kilogram. As these seeds are rich in oil they cannot be stored for long (Kumar et al. 2012). A 5 year old plantation yields 1800 kg ha^{-1} of oil. This plant is reported to live around 100 years.

This species is distributed in Iran, Egypt, Sudan, Ethiopia and parts of India, Pakistan and Sri Lanka (Fig. 34.5). It is commonly called as peelu, kharijal and miswak.

Fig. 34.4 The miswak *Salvadora persica*

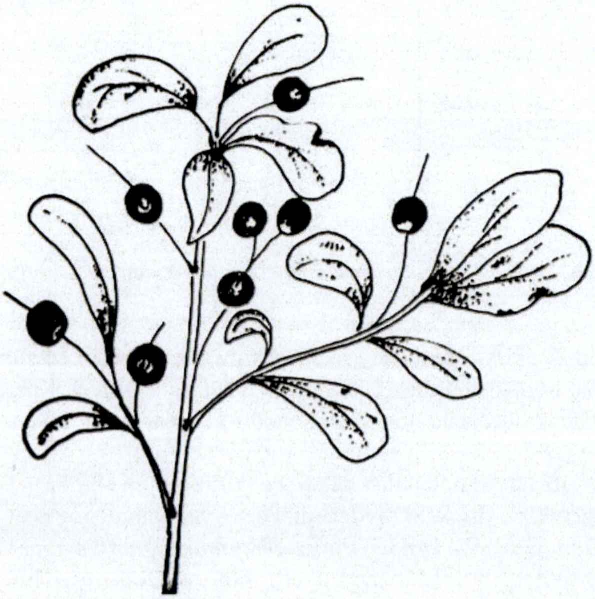

Fig. 34.5 The geographical distribution of *S. persica*

Table 34.3 The physico-chemical properties of *S. persica* oil (Ali et al. 2018)

Parameters	Value
Kinematic viscosity at 40 °C (mm^2 s^{-1})	30.7
Density at 28 °C (kg m^{-3})	920
Flash point (°C)	210.5
Acid value (g KOH kg^{-1})	3.35
Free fatty acid (%)	1.68
Iodine value (g I_2/100 g)	36.09
Heating value (MJ kg^{-1})	39.48
Lauric acid (C12:0) (%)	32.76
Myristic acid (C14:0) (%)	18.82
Palmitic acid (C16:0) (%)	11.28
Oleic acid (C18:1) (%)	20.12
Linoleic acid (C12:0) (%)	9.22

34.2.1 Oil Extraction

The seeds are separated from the fruits and dried in sunlight for many days. These seeds are then powdered, sieved and extracted using ethyl alcohol in a reflux unit. The alcohol present in the extract is then distilled out and the oil is then recovered. The properties of the oil are given in Table 34.3.

The kinematic viscosity, density and flash point are high, though within the upper limit. Among the total fatty acids, the share of saturated fatty acids is high. The free fatty acid is low enough to transesterify the oil using alkali as catalyst.

34.2.2 Biodiesel Preparation

The conventional alkali catalysed transesterification process is followed in the biodiesel production. The end product containing ester and glyceride is loaded in a separatory funnel and kept undisturbed for phase separation. The upper phase

Table 34.4 Physicochemical properties of *S. persica* biodiesel (Ali et al. 2018)

Parameters	Value
Kinematic viscosity at 40 °C ($mm^2\ s^{-1}$)	5.51
Density at 28 °C ($kg\ m^{-3}$)	894
Flash point (°C)	178.5
Heating value ($MJ\ kg^{-1}$)	35.26
Cetane number	61
Sulphur content ($mg\ kg^{-1}$)	844.17

formed in the funnel is biodiesel which is yellow in colour. The dark brown bottom layer is formed of glycerol which is discarded. The biodiesel is separated and washed repeatedly with hot water and dried by heating it at 80 °C so as to dispel the moisture. The properties of the biodiesel thus obtained are analysed and given in Table 34.4.

The viscosity and density are within the limit. The cetane number is good which indicates the ignition quality and the engine is prone to start quickly due to the short ignition delay. The iodine value is an indirect reference to the unsaturated fatty acids with double bond (c–c) originally present in the crude oil. Normally any biodiesel when ignited releases low SO_2. But in the present case the biodiesel contains high total sulphur (Kaul et al. 2007) as compared to that of many other biodiesel and petro-diesel (500 ppm). This is one of the major impediments in popularizing *S. persica* biodiesel. *S. persica* biodiesel is reported to create corrosion in the engine ten fold higher than that of the petro-diesel (Kaul et al. 2007). It is reported that the sulphur quantity in the biodiesel can be minimized if the oil extraction is made through mechanical means instead of extracting it by solvent (He et al. 2009). Vacuum distillation of the methyl ester is known to remove the sulphur content by flashing out H_2S and SO_2 present in it. According to Privitera and Borgese (2005), if hydrogen is allowed to react with biodiesel through a suitable catalytic column (Co–Mo or Ni–Mo) at high temperature and pressure, the sulphur will be converted to H_2S and then may be expelled.

References

Ali M, Naqvi B, Watson IA (2018) Possibility of converting indigenous *Salvadora persica* seed oil into biodiesel in Pakistan. Int J Green Energy. https://doi.org/10.1080/15435075.2018.1472603

Arora M, Siddiqui AA, Palival S, Sood P (2014) A phyto-pharmacological overview on *Salvadora oleoides* Decne. Indian J Nat Prod Resour 5(3):209–214

Bhatnagar AK, Kaul S, Chhibber VK, Gupta AK (2006) HFRR studies on methyl esters of non edible vegetable oils. Energy Fuels 20:1341–1344

He BB, Gerpen JHV, Thompson JC (2009) Sulphur content in selected oils and fats and their corresponding methyl esters. Am Soc Agric Biol Eng 25(2):223–226

Kaul S, Saxena RC, Kumar A, Negi MS, Bhatnagar AK, Goya HB, Gupta AK (2007) Corrosion behavior of biodiesel from seed oils of Indian origin on diesel engine parts. Fuel Process Technology 88(3):303–307

Kumar S, Rani C, Mangal MA (2012) Critical review on *Salvadora persica*: an important medicinal plant of arid zone. Int J Phytomed 4:292–303

Privitera M, Borgese C (2005) Sulphur removal strategies for biodiesel, 15 Jan 2005. Preprocess Inc. www.preprocessinc.com

Reddy MP, Shah MT, Patolia JS (2008) *Salvadora persica* a potential species for industrial oil production in semi arid saline and alkali soils. Ind Crop Prod 28(3):273–278

35 Sapindaceae

The members of the family Sapindaceae are trees, shrubs or climbers with spring like tendrils. Sapindaceae is often referred as litchi family and has more than 1800 species. The family is mainly restricted to tropical and subtropical regions. They richly occur in North Eastern and North Western Himalayas. The species connected with biodiesel are triangle top (*Blighia unijugata*), soap nut (*Sapindus mukorossi*), kusum (*Schleichera oleosa*) and yellow horn (*Xanthoceras sorbifolia*).

35.1 Triangle Tops (*Blighia unijugata*)

The triangle top *Blighia unijugata* is a stout tree with thick canopy. This tree grows to a maximum height of 35 m. The trunk is short and it is up to 180 cm in dia. This tree is often grown for shade and occurs in wooded grass lands. It is also found at the vicinity of termite mound and flourish well at an elevation ranging up to 1900 m. This tree is also cultivated in public gardens for ornamental purposes and it responds to coppicing and pollarding. Its branches are hairy. The leaves are shiny green at the upper surface and dull coloured at the reverse side. The leaves are 8–30 cm long and they occur in pairs with the uppermost pair usually being the largest. Young leaves usually appear red. Flowers are unisexual, yellow, small and pleasantly scented. Fruits are 3 cm long and pear shaped. Each fruit has 3 seeds (Fig. 35.1).

This tree is geographically distributed in Guinea, Ethiopia, Zambia and Mozambique. Thus has a limited area of occurrence (Fig. 35.2). This species is locally called as triangle tops, musadema, umdiaguva, sundo and mkivule.

35.1.1 Oil Extraction

The seeds are isolated from the fruit, dried and powdered. This powder is extracted using *n*-hexane in a Soxhlet apparatus. The extract is then freed from the solvent in

S. P. Raj et al., *Biodiesel from Flowering Plants*,
https://doi.org/10.1007/978-981-16-4775-8_35

Fig. 35.1 The triangle tops *B. unijugata*

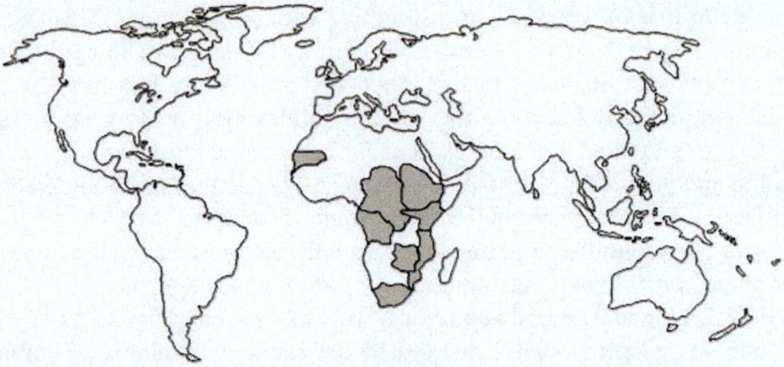

Fig. 35.2 The geographical distribution of *B. unijugata*

vacuum distillation. The oil thus obtained is degummed by mixing with distilled water forming around 2% of the total oil. The contents are vigorously stirred and subsequently centrifuged at 1800 rpm for 45 min (Adewuyi et al. 2012). The analytical details of the oil of *B. unijugata* are given in Table 35.1.

The oil yield of *B. unijugata* is 50.8. The fatty acid profile of the *B. unijugata* oil is given in Table 35.2.

Table 35.1 The physico-chemical properties of *B. unijugata* oil (Adewuyi et al. 2012)

Parameters	Value
Oil yield (recovery from the seed) (%)	50.82
Free fatty acid (%)	7.00
Iodine value (g I_2/100 g)	47.80
Saponification value (g KOH kg^{-1})	239.20

Table 35.2 The fatty acid profile of *B. unijugata* oil (Adewuyi et al. 2012)

Fatty acids	Value
Palmitic acid (C16:0) (%)	34.5
Stearic acid (C18:0) (%)	14.1
Oleic acid (C18:1) (%)	48.1
Linoleic acid (C18:2) (%)	1.8

Table 35.3 The physico-chemical characteristics of *B. unijugata* biodiesel (Adewuyi et al. 2012)

Parameters	Value
Kinematic viscosity at 40 °C (mm^2 s^{-1})	4.52
Density at 15 °C (kg m^{-3})	880
Flash point (°C)	185
Acid value (g KOH kg^{-1})	0.04
Pour point (°C)	12.0
Iodine value (g I_2/100 g)	47.6
Copper corrosion test	1a
Oxidative stability 110 °C (h)	44.3

The saturated and unsaturated fatty acid contents are more or less on par. The saturated fatty acid is led by palmitic acid, whereas the oleic acid dominated the unsaturated fatty acids. The degummed oil is treated with 2% sulphuric acid in methanol (a combination of 98 mL methanol and 2 mL sulphuric acid). The methanol sulphuric acid content for mixing is calculated so as to make an oil alcohol molar ratio of 1:6. The content is stirred for 3 h at 600 rpm. The resultant product is extracted with ethyl acetate (Adewuyi et al. 2012), washed repeatedly with water and dried by passing it over sodium sulphate. The esterified oil is then transesterified using 1% KOH in methanol at 70 °C. The product is again extracted with ethyl acetate, washed with distilled water and dried using sodium sulphate. The properties of the biodiesel obtained from the oil of *B. unijugata* are given in Table 35.3.

The kinematic viscosity and density are within the standard limit. The oxidative stability is extraordinarily high against a standard value of 6 (EN14214).

35.2 Soap Nut (*Sapindus mukorossi*)

The soap nut *Sapindus mukorossi* is a common tree of the Indo Gangetic plains and its seeds are considered as a raw material for biodiesel. They occur at an elevation ranging from 200 to 1500 m with an annual rainfall of 1500–2000 m. It is a deciduous tree growing up to 20 m tall and 1.8 m girth. The leaf is pinnate and

Fig. 35.3 The soap nut *Sapindus mukorossi*

alternate with 15–25 leaflets each 20–40 cm long. The flowers occur in large panicles with small creamy white flowers. The fruit is a shiny drupe of 1.5–2.5 cm dia. and black in colour containing one to three seeds (Fig. 35.3).

The drupe is golden yellow when harvested and becomes dark when long stored. The pulp of the soap nut is used as a substitute for soap and it contains saponin a natural surfactant and is on use as a detergent. A single tree produces around 125 kg of fruits in a year. The seed contained in the nut does not carry any material value. Therefore its use as a raw material for biodiesel production is a boon for the soap nut growers. This species is exclusively distributed in Shivaliks, outer Himalayas, Uttaranchal, Himachal Pradesh, Haryana, Jammu and Kashmir and Fujian province of China (Fig. 35.4). This species is referred by its vernacular names: Indian soap berry, wash nut, ritha, reetha and areetha.

35.2.1 Oil Extraction

The freshly collected fruits are dried at 80 °C for 4 h and shelled to separate the kernel. The kernels form around 15% of the fruit mass (Chakraborty and Baruah 2013). The kernels are further dried till the moisture content is around 5%. The

Fig. 35.4 The geographical distribution of *Sapindus mukorossi*

Table 35.4 Physicochemical properties of *S. mukorossi* oil

Parameters	Chakraborty and Baruah (2013)	Chen et al. (2013)
Kinematic viscosity at 40 °C (cSt)	32.1	39.6
Density at 20 °C	923	917
Flash point (°C)	159	–
Pour point (°C)	6	–
Iodine number (g I_2/100 g)	–	85
Oxidation stability at 110 °C (h)	–	>50
Acid value (g KOH kg^{-1})	15.6	13.4
Calorific value (MJ kg^{-1})	38	38.2
Palmitic acid (C16:0) (%)	5.5	–
Stearic acid (C18:0) (%)	2.3	–
Oleic acid (C18:1) (%)	58.4	–
Linoleic acid (C18:2) (%)	5.4	–
Arachidic acid (C20:2)	7.5	–
Linolenic acid (C18:3)	17.1	–
Behenic acid (C22:0) (%)	1	–

kernel is powdered and extracted in petroleum ether (bp 40–60 °C) for 4 h with a minimum of 8 cycles per hour in a Soxhlet apparatus. The extract is then distilled in vacuum to remove the solvent. The oil forms around 39% by weight of the kernel. The physicochemical properties of the oil thus obtained are given in Table 35.4.

Around 80% of the oil is formed of unsaturated fatty acid. The pour point appears to be unfavourable. The oxidation stability is high due to the presence of antioxidants (Chen et al. 2013).

35.2.2 Biodiesel Production

Due to the high acid value, single alkali transesterification may create gel formation. Hence double stage processes are recommended. As a first step, acid esterification is carried out. Consequently, the alkali catalysed transesterification is made. The former is often referred to as pretreatment.

Methanol and the raw oil are mixed at 6:1 molar ratio and heated to an ambient temperature of 55 °C. It is then added drop by drop with sulphuric acid at the rate of 10 g L^{-1}. The content is kept on stirring for 60 min at the same temperature. On completion of the said period, it is transferred to a separating funnel and waited for the phase separation to complete.

The top layer comprising a part of methanol and free fatty acid is separated and discarded. The lower layer containing oil is collected, heated and dried. It is then passed through anhydrous sodium sulphate to remove any residual moisture.

The above acid treated dried material is transesterified with alkali catalyst. Sodium hydroxide is dissolved in methanol to prepare sodium methoxide. The quantity of sodium hydroxide is 1% and the methanol volume used is equivalent to form a molar ratio of 8:1 (methanol to oil). The oil is heated to 60 °C and to that the methoxide is added and stirred at 600 rpm for 120 min. On completion of the above period the content is transferred to a separating funnel and allowed to settle for 6 h by then the contents divide into upper and lower zones. The upper zone containing the biodiesel is removed and washed repeatedly with warm water so as to get rid of glycerol and alkali if present. The washed biodiesel is then heated to dry and then passed over sodium sulphate to make it completely moisture free. The physicochemical characteristics of the biodiesel thus prepared are given in Table 35.5 and the ester profile is given in Table 35.6.

The alkaline transesterification gives more than 90% conversion (Chakraborty and Baruah 2013). The viscosity and density are well within the norms. The viscosity value is between 4.5 and 5 cSt. The viscosity is governed by the constituent

Table 35.5 Physicochemical properties of the biodiesel prepared from *S. mukorossi*

Parameters	Chakraborty and Baruah (2013)	Chen et al. (2013)
Kinematic viscosity at 40 °C (cSt)	4.63	4.88
Density (kg m^{-3})	876	878
Flash point (°C)	140	177
Pour point (°C)	−4	–
Cloud point (°C)	−1	–
Copper strip corrosion	1a	–
Cetane number	56	58
Iodine number (g I_2/100 g)	–	83.6
Oxidation stability at 110 °C (h)	1.2	–
Acid value (g KOH kg^{-1})	0.14	0.12
Cold filter plugging point (°C)	–	6
Calorific value (MJ kg^{-1})	40	–

Table 35.6 The ester profile of the biodiesel from *S. mukorossi*

Name of fatty acid methyl esters	Chen et al. (2013)	Shaw et al. (2014)
Methyl palmitate (C16:0) (%)	3.8	4.8
Methyl arachidate (C20:0) (%)	5.3	8.6
Methyl oleate (C18:1) (%)	55.2	58.9
Methyl linoleate (C18:2) (%)	8.3	–
Methyl linolenate (C18:3) (%)	2.4	–
Methyl eicosenoate (C20:1) (%)	23.9	25.9

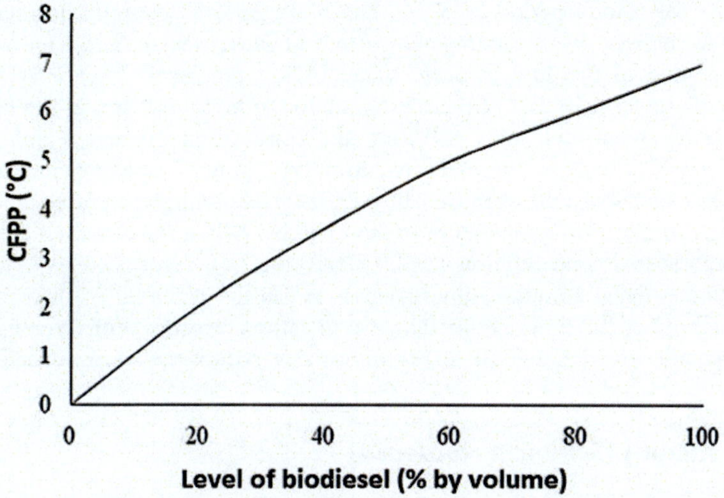

Fig. 35.5 The CFPP in accordance with the blend level

esters, namely oleic, linolenic and palmitic whose viscosities at 40 °C are 4.5, 3.2 and 4.4 cSt. Often it is reported that the *S. mukorossi* oil has a considerable amount of sulphur (Chakraborty and Baruah 2013). Such high content of sulphur is apparently due to the use of sulphuric acid in the esterification process and due to the sodium sulphate in the drying process. In this context, it is suggested that *p*-toluenesulfonic acid in the place of sulphuric acid and molecular sieve in the place of sodium sulphate be tried.

Lubricity is another factor worth consideration. Wear scar diameter is an index of the lubricity. It is reported that *S. mukorossi* biodiesel has a wear scar diameter of 0.181 mm against 0.45 mm for diesel (Chakraborty and Baruah 2013). The oxidation stability also is needed to be improved through the addition of suitable stabilizers.

The cold filter plugging point (CFPP) is not satisfactory for which the blending of the biodiesel with neat diesel is considered (Chen et al. 2013). The CFPP shows its decreasing trend when the blending level is decreased. In B20 (i.e. 20 mL of biodiesel with 80 mL of diesel) the CFPP reduces rapidly. (Fig. 35.5).

The CFPP value is high when the level of biodiesel is high in the blend (R^2 = 0.987). Similarly, the density also decreases when the level of biodiesel is low in the blend ($R^2 = 0.999$). If blending is resorted at a level of 20% (B20) the oxidation stability is known to improve automatically, obviating the use of antioxidants.

35.2.3 Engine Performance

A short term engine performance is carried out using small sized water cooled, single cylinder four stroke diesel engine. The performance evaluation is made by running the engine for a total period of 50 h. The brake specific energy consumption is reported to increase by 8% compared to that of diesel (Misra and Murthy 2011) apparently due to the low calorific value of the biodiesel. The brake thermal efficiency is lower than that of the diesel due to the high viscosity of the biodiesel which creates poor fuel spray affecting the combustion process which in turn increases the CO level in the emission. Similarly, poor combustion increases the hydrocarbon content in the exhaust. The NO_x increases with the increase in load due to the high combustion temperature in the cylinder. When the biodiesel is blended with diesel, at low concentration (<B20 blend) the NO_x level decreases. The NO_x level in the exhaust assumes importance as it has the potential of increasing the greenhouse effect 292 times higher than that of carbon dioxide. With reference to the smoke opacity, low blend (B10–20) is preferred to reduce the smoke opacity.

35.3 Kusum (*Schleichera oleosa*)

The kusum *Schleichera oleosa* (Oken) is an oil seed bearing tree and is also identified as *Cussambium oleosum* (Kuntze), *Pistacia oleosa* (Lour) and *Schleichera trijuga* (Willd.). It grows up to 40 m height with its bole rarely grow up to 2 m in diameter. The trunk is crooked and buttressed. It has a potential of 66,000 tonnes of oil production per year in India alone (Acharya et al. 2009) and has not been fully exploited for the purpose of biodiesel.

35.3.1 Habitat

These trees are naturally growing at a height of 900–1200 m from the main sea level and at an annual rainfall ranging from 750 to 2500 mm. They prefer a temperature range of 25–35 °C but are also seen distributed in extreme temperatures (2.5–47.5 °C). They occur in dry deciduous forests and savanna as a solo tree at scattered positions or they may grow gregariously. They are also existing in companion with teak tree. Often they grow at dry areas or occasionally at swampy habitats having rocks or gravels. *S. oleosa* prefers acidic soils. These trees mostly serve as a host for lac insects.

S. oleosa is called by different names according to the country in which they are found (Table 35.7).

Table 35.7 Names of kusum in different countries

Countries	Names
Cambodia	Pongro
The Philippines	Gum-lac tree
France	Pongro, gum-lac-tree
Indonesia	Kasambi
Tibet	Phen
Malaysia	Kusambi
Thailand	Takhro
Vietnam	Cay van rao

Table 35.8 Different vernacular names of kusum in India

Languages	Names
English	Ceylon oak, akota, Kusum, gum-lac
Kannada	Chakota
Malayalam	Dhoothalam, Poovanam, Puvam
Tamil	Puvam, Puvatthi, Pumaratha, Karanachi, Pulichi

Fig. 35.6 Kusum *Schleichera oleosa*

In India they are being called by different vernacular names (Table 35.8).

35.3.2 Botanical Features

The bark of the tree is smooth and grey. The branches are terete, striate, with sericeous hairs and sessile glands. Leaves are parinnate. The topmost leaflet is often situated like a terminal leaflet, axial parts usually glabrescent, deltoid, about 1.5 mm long, obtuse with thin hairs on both sides, margin ciliate and sometimes glandular. The tree loses its leaf during the month of December.

It comes to flowering in India at the onset of dry season (February to April). The flowers do not have petals. There are 5–9 stamens. In female flowers the ovary is ovoid and is 1.3 mm long. The fruit is edible and broadly ovoid with 1–2 seed

(Fig. 35.6). The seed is subglobular, testa brown, smooth and glabrous. The seed coat is brown and brittle which breaks at a slight pressure bringing out a U shaped kernel.

35.3.3 Distribution

In India the tree occurs mainly at sub-Himalayan tracts (Orissa, Jharkhand, Western Deccan and South Karnataka). It is distributed in Sri Lanka, Indo-China, Malaysia, Indonesia, Java, Myanmar, Timor, Sunda Islands, Sulawesi, Molucas, Seram and Kai Islands (Fig. 35.7).

35.3.4 Propagation

The seeds maintain the viability for a maximum of 1 year. This tree can be propagated by seeds as well as by suckers. The seeds are sown immediately after collection in nursery bed 7.5 cm apart from each seed. The stem of the seedling reaches a dia of 1 cm in 1 year. The plant is cut at a height of 6 cm and the stem with root is then transferred to the fields and planted in pits of the size 30 × 30 × 30 cm at a spacing of 8 × 8 m.

35.3.5 Pest Management

These trees are often afflicted by stem blight (*Rosellinia* sp. and *Hexagonia apiaria*), yellow cork rot (*Polyporus weberianus*), white fibrous rot (*Irpex flavus*) and white spongy rot (*Daedalea flavida* and *Hexagonia apiaria*). The seeds are also often infested with bug (*Serinetha augur*). This bug feed gregariously on the fallen seeds in the field.

Fig. 35.7 Distribution (darkened area) of kusum in Asia

35.3.6 Seed

The fruits are normally stored as heaps for 3–4 days and then squeezed to release the seeds. The depulped seeds are washed and air dried. A fully grown mature tree yields 20–25 kg of seed. Each seed weighs 0.5–0.7 g. The oil is extracted by boiling the kernel or by pressing the kernel between rollers.

35.3.7 Oil

Though the kernels contain 55–70% oil, only around 30% is extractable by boiling or pressing. The oil is yellowish brown in colour and has a semi-solid consistency. This oil contains cyanogenic compounds and hence not suitable for human consumption. The characteristics of the oil are given in Table 35.9.

The viscosity and iodine values are very high. More than 50% of the fatty acids present in it is formed of saturated fatty acids. The fatty acid profile is given in Table 35.10.

The oil of kusum is most often used in engines without transesterification. One of the problematic issues of this oil is its high viscosity. High viscosity is likely to cause carbon deposit in the fuel conduits besides there may be incomplete combustion due

Table 35.9 The characteristics of the kusum oil (Acharya et al. 2009; Sharma and Singh 2010)

Parameters	Value
Viscosity at 40 °C (cSt)	42.08
Density at 30 °C (g cc^{-1})	0.893
Iodine value (mg I_2 g^{-1})	215
Free fatty acid (%)	8.0
Total fatty acid (%)	91.6
Total glycerine (%)	3.6
Heating value (MJ kg^{-1})	35.0

Table 35.10 Fatty acid composition of kusum oil (Sharma and Singh 2010)

Fatty acid	%	Total
Saturated		
Lauric C12:0	0.31	
Myristic C14:0	15.54	
Palmitic C16:0	10.35	
Stearic C18:0	11.11	
Arachidic C20:0	15.79	
Behenic C22:0	0.01	53.11
Unsaturated		
Oleic C18:1	27.08	
Linoleic C18:2	6.14	
Gondoic C20:1	6.17	
Eicosadienoic C20:2	0.08	39.47
Unidentified		7.42

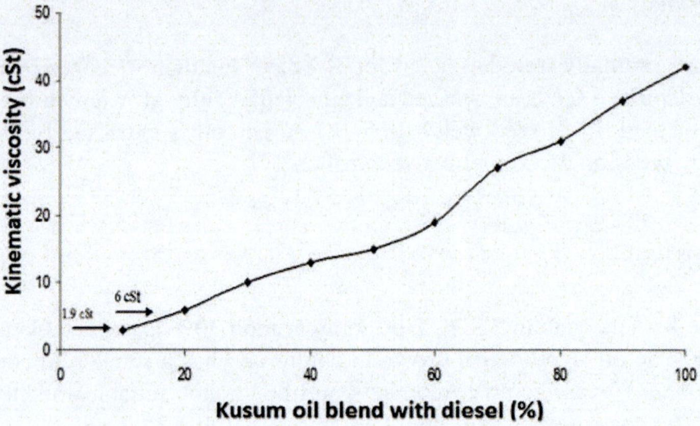

Fig. 35.8 The relationship between kusum oil blend (with petro-diesel) and the viscosity. The standard limit for petro-diesel 1.9–6 cSt is also indicated

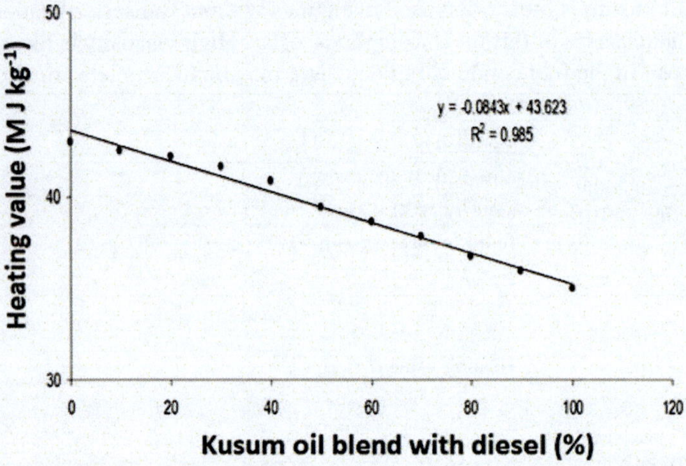

Fig. 35.9 The relationship between kusum oil blend (with petro-diesel) and the heating value

to reduced flow characteristics. Therefore attempts are being made to blend this oil with petro-diesel so as to feed it in the automobile engine. The changes in the kinematic viscosity in relation to the degree of blending are indicated in Fig. 35.8. Figure 35.8 also indicates the range prescribed by ASTM. It is known that a maximum of 15% blend is possible for routine operation. The changes in heating values according to the blending percent are given in Fig. 35.9. The regression indicates that the heating value (y) established a negative relationship with the percent blend (x) as y: $-84.273x + 43.623$. Accordingly a 15% blend (i.e. 150 mL oil + 900 mL petro-diesel) makes a very marginal reduction in the heating value (i.e. less than 3% reduction over that of the petro-diesel). The change in the specific

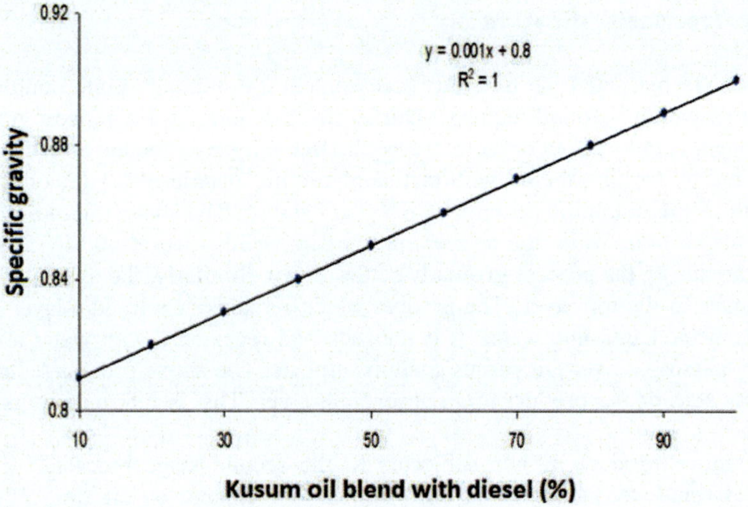

Fig. 35.10 The relationship between kusum oil blend (with petro-diesel) and the specific gravity

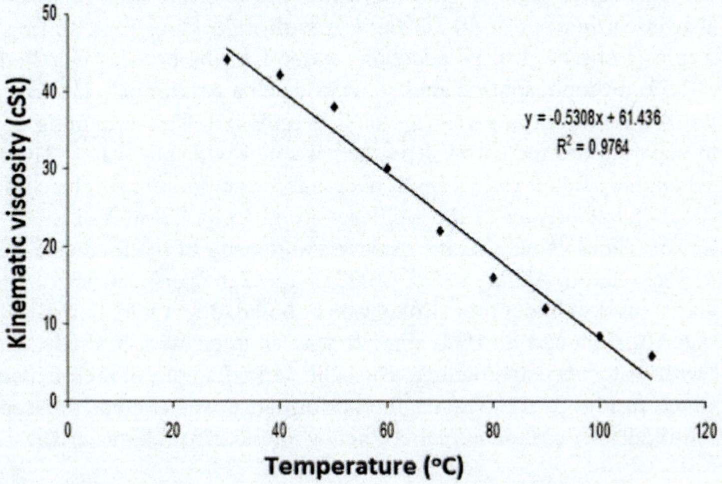

Fig. 35.11 The pattern of reduction in viscosity of the kusum oil due to heating

gravity in relation to the blending percent is given in Fig. 35.10. A 15% blend would make an increase of 2% in specific gravity over the regular petro-diesel based on the regression equation $y = 0.001x + 0.8$ where y is the specific gravity and x is the percent content of kusum oil in the blend. It is known that the viscosity actively reduces when the oil is heated. The relationship between temperature and viscosity of the oil is indicated in Fig. 35.11. It is seen that the oil reaches the standard viscosity of 3 cSt at 114 °C. The relationship between kinematic viscosity (y) and the temperature (x) is $y = -0.4988x + 59.624$.

35.3.8 Transesterification

Since the viscosity and the free fatty acid content are mutually high, double stage transesterification is found required. During the first stage of the process free fatty acid content is reduced. In order to reduce the free fatty acid content esterification is carried out by treating the oil with methanol at a molar ratio of 6:1 (alcohol to oil) and with 15 mL sulphuric acid per lit. of oil as catalyst. The above mixture is mildly stirred for an hour while the temperature is kept maintained at 50 °C. The water generated during the process gradually settles down. Similarly, the unreacted methanol moves to the top layer. The product which remains as a middle layer is then freed from methanol and water. It is then washed repeatedly with water and then dried by passing it over anhydrous sodium sulphate. The above process reduces the free fatty acid of the product to an acceptable level. This is referred as first stage process.

The above product is then subjected to the second stage process, i.e. alkali catalysed transesterification. For this reaction a methanol to oil ratio of 8:1 is considered as optimum. Potassium hydroxide as a catalyst at the rate of 0.7% is observed to be adequate. If the KOH conc. is increased beyond the above limit, saponification product is likely to develop which will interfere with the process. This reaction also is carried out at 50 °C for 1 h with mild stirring. A stirring rate of 800–1000 rpm is observed to be adequate. After 1 h, the process is halted by the addition of 75% orthophosphoric acid so as to reach a neutral pH. The layer which sinks down to the bottom is glycerin which is removed. The remaining content is distilled to separate the methanol. It is then washed with hot water. The finished product is then dehydrated by passing it over anhydrous sodium sulphate (Na_2SO_4). The analytical characteristics of the biodiesel are given in Table 35.11.

The viscosity, acid value, specific gravity and density of the biodiesel are nearly within the specification. Acharya et al. (2009) studied the performance of kusum oil in a stationary single cylinder four stroke, water cooled, direct injection diesel engine with its speed maintained at 1500 rpm. It was coupled with a single phase AC alternator with water rheostat loading. The schematic diagram of their experimental set up is given in Fig. 35.12. The engine performance was evaluated based on the fuel consumption rate, brake thermal efficiency and exhaust characteristics.

Table 35.11 The physicochemical parameters of kusum biodiesel (Sharma and Singh 2010)

Parameters	Values
Viscosity at 40 °C (cSt)	4.9
Acid value (mg KOH g^{-1})	0.45
Specific gravity	0.874
Density at 30 °C (g cc^{-1})	0.871
Elemental composition	
Carbon (%)	75.08
Hydrogen (%)	14.22
Oxygen (%)	10.70
Nitrogen (%)	0
Sulphur (%)	(<0.005)

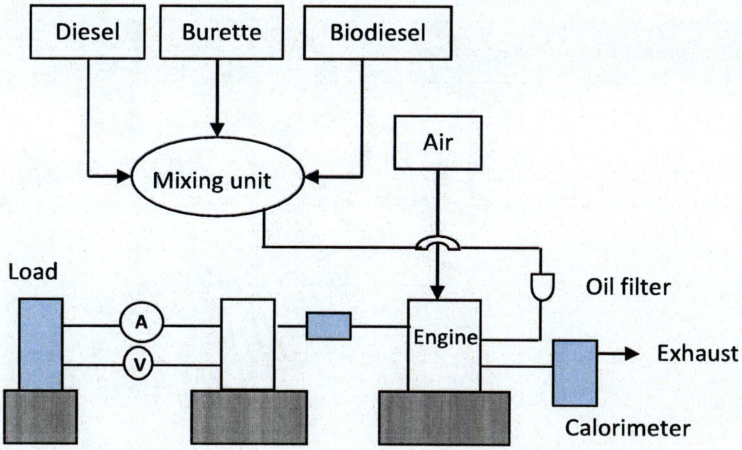

Fig. 35.12 Schematic diagram of the unit for engine performance study using kusum oil. (Adopted from Acharya et al. 2009)

The consumption of kusum oil was observed to be higher than that of the petro-diesel at different load. It is inferred that the brake specific fuel consumption increased due to its higher density and viscosity and also due to the low heating value. It is also observed that the brake thermal efficiency of the kusum oil is lower due to the poor combustion. The temperature of the exhaust also is lower in kusum oil when compared to that of the petro-diesel. The CO, CO_2, HC and NO_x emissions were observed to be higher even when the kusum oil was used at 10–20 blend level. This strongly suggests kusum oil cannot be used as such in engines and it is required to be transesterified.

35.4 Yellow Horn (*Xanthoceras sorbifolia*)

The yellow horn, *Xanthoceras sorbifolia* is an oilseed tree of China. It has the capacity to flourish, resisting environmental extremes. It can survive in saline and alkaline tracts and at a temp. regime of even −40 °C for a short period. The oil from the seeds of this tree is considered as a raw material for biodiesel (Fu et al. 2008; Li et al. 2012; Zhang et al. 2016). Being a deciduous tree it grows to a maximum height of 10 m with a trunk dia. of 90 cm at breast height. The leaves are alternate with 9–10 leaflets. The leaflets are lanceolate, sessile and elliptic, 2–6 cm long, 1–1.5 cm broad with serrated margins. The flowers occur in inflorescence of 15–25 cm long racemose with around 50 flowers. The fruit is a leathery capsule (Fig. 35.13) of 5–6 cm dia. with 4–6 black spherical (1.5 cm dia.) seeds. This tree grows at an altitude of 300–2000 m in China and Russia (Fig. 35.14).

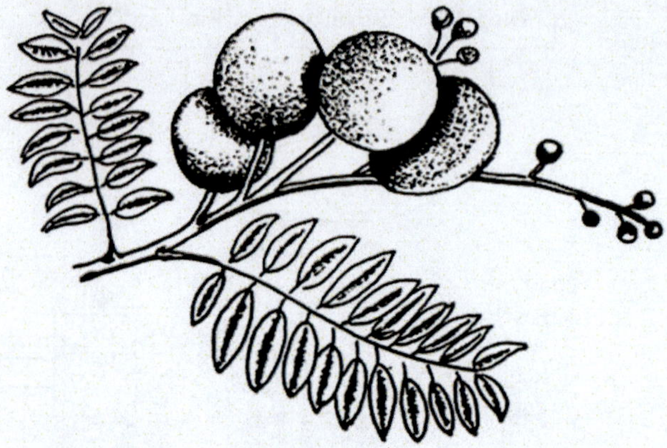

Fig. 35.13 The yellow horn, *Xanthoceras sorbifolia*

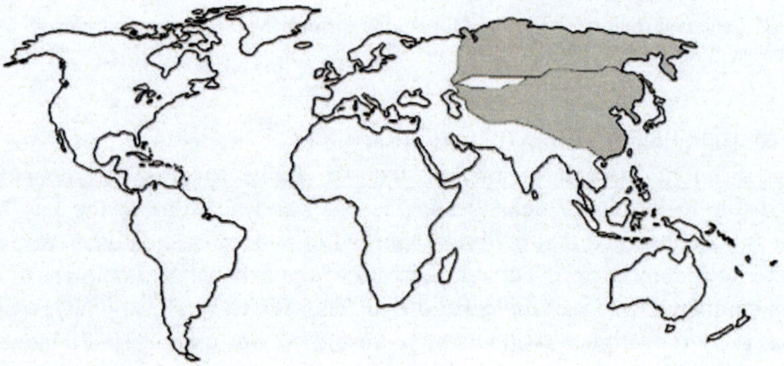

Fig. 35.14 Geographical distribution of *X. sorbifolia*

35.4.1 Oil Extraction

The fruit of *X. sorbifolia* is formed of 53% husk and 47% seed by weight. The seeds contain 51% kernel and 49% seed coat. The fruits are dried in the hot sun. The seeds are collected, cracked and the kernels removed. The kernel contains 65% oil (Yao et al. 2013) (Fig. 35.15). The dried kernels are then milled and extracted with *n*-hexane at 85 °C for 6–8 h. The oil is then separated from the mixture and dried (Yu et al. 2017).

The properties of the seed oil are given in Table 35.12.

Table 35.12 indicates that the unsaturated fatty acids have a share of more than 90% among the total fatty acids. The dominant fatty acids are oleic and linoleic acids. Favourable tribology of the oil is characterized by a low value of the ratio between C20–24 and C16–18 fatty acids. In the present case the value is just 0.25. Similarly the ratio between PUFA and MUFA is expected to be low. In the oil of *X. sorbifolia* the value is 0.9. Both the values imply that this oil resists thermo-oxidation and solidification at low temperature (Yu et al. 2017).

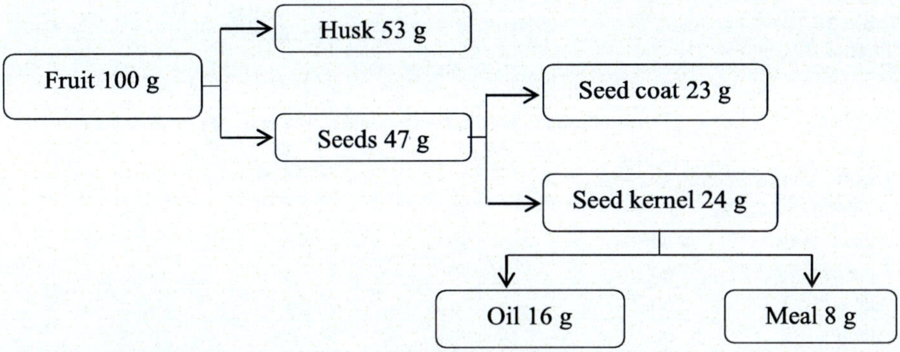

Fig. 35.15 Fruit fraction of *X. sorbifolia* (Yao et al. 2013)

Table 35.12 The physicochemical characteristics of *X. sorbifolia* oil (Yao et al. 2013)

Parameters	Value
Kinematic viscosity at 40 °C (cSt)	38.1
Density at 20 °C (kg m^{-3})	914
Free fatty acid (%)	0.3
Iodine value (g I_2/100 g)	113
Heating value (MJ kg^{-1})	39.7
Palmitic acid (C16:0) (%)	5.2
Stearic acid (C18:0) (%)	2.2
Oleic acid (C18:1) (%)	28.6
Linoleic acid (C18:2) (%)	43.3
Linolenic acid (C18:3) (%)	0.5
Arachidic acid (C20:0) (%)	0.4
Gadoleic acid (C20:1) (%)	6.8
Heneicosanoic acid (C21:0)	0.4
Behenic acid (C22:0) (%)	0.6
Erucic acid (C22:1) (%)	8.7
Lignoceric acid (C24:0) (%)	0.3
Nervonic acid (C24:1) (%)	3.0

35.4.2 Biodiesel Production

Under the conventional mode of alkali catalysed transesterification, potassium hydroxide at the 5% level of oil to methanol ratio of 1:8 is tried. The process is exposed to 60 °C under constant stirring for 3 h. On completion of the reaction, the product is allowed to stand overnight in a separating funnel and the supernatant phase with biodiesel is removed, washed and dried.

Recently the heating process is carried out in the microwave which gives as high as 97.5% of conversion for the alkali catalysed transesterification (Table 35.13). In the microwave aided transesterification, there is electromagnetic agitation of the

Table 35.13 Comparison of the transesterification processes using different ion exchange resins (Li et al. 2012)

Catalyst	Transesterification Temperature (°C)	Duration (min)	Conc. (%)	Conversion (%)
Heating source: microwave				
Amberlite IRA-900	60	90	5	96.3
Amberlite IRA-72	60	90	5	55.7
Amberlite IRC-93	60	90	5	57.4
Amberlyst-15	60	90	5	83.5
Alkali KOH	60	60	5	97.5
Heating source: water bath				
Amberlite IRA-900	60	180	5	25.8
Amberlyst-15	60	180	5	23.6

Table 35.14 The variation in the content of chosen methyl esters in the biodiesel prepared using amberlite IRA-900 and KOH (Li et al. 2012)

Fatty acid methyl ester	Amberlite IRA-900	KOH
Palmitic acid	5.27	5.94
Linoleic acid	42.12	41.27
Oleic acid	30.40	29.04
Stearic acid	2.03	3.12
Eicosenoic acid	10.09	9.78
Docosenoic acid	3.49	3.67

participating molecules. There is volumetrically distributed heat, causing impact between oil and methanol (Oliveira and Franca 2002).

Li et al. (2012) used macroporous ion exchange resins as catalyst in the transesterification process. They are amberlite IRA-900 (anion-strongly alkaline), amberlite IRC-72 (cation-weakly acidic), amberlite IRC-93 (anion-weakly alkaline) and amberlyst-15 (cation-strongly acidic). These resins are used along with methanol. The dose, duration and temperature during the transesterification and the conversion efficiency of the above catalysts are given in Table 35.13.

Among the various catalysts used in the microwave heating process KOH shows the highest performance. The anionic alkaline catalyst amberlite IRA-900 also performs well (96.3%), whereas amberlite IRC-93 does not show good result, though it is also an alkaline catalyst as it is weakly alkaline. Among the cation resin the performance of amberlyst-15 is satisfactory in terms of its ability to convert the oil (83.5%). The catalysts amberlite IRA-900 and amberlyst-15 do not perform well in the water bath heating source.

One of the major advantages in using the ion exchange catalysts is the recovery of them and putting them back in the next process by cleaning them using methanol. It is also known that the composition with respect to the amino acid (methyl ester) also varies in the biodiesel prepared using an ion exchange catalyst and KOH catalyst (Table 35.14).

Table 35.15 The physicochemical properties of the biodiesel from the oil of *X. sorbifolia* (Li et al. 2012; Yao et al. 2013)

Parameters	Value
Kinematic viscosity at 40 °C (cSt)	4.4, 4.2
Acid value (g KOH kg^{-1})	0.06, 0.36
Flash point (°C)	165
Cetane number	56.1, 58
Density at 20 °C (kg m^{-3})	872
Copper strip corrosion	1

The level of the various fatty acid esters present in the biodiesel prepared using amberlite IRA-900 differs by 2–35% of that of the biodiesel prepared using KOH as catalyst.

Chosen properties of the biodiesel prepared are shown in Table 35.15.

The kinematic viscosity, acid value, flash point and cetane number are considered to be satisfactory. However, there is no extensive work done on the engine performance of this biodiesel.

References

Acharya SK, Mohanty MK, Swain RK (2009) Kusum oil as a fuel for small horse power diesel engine. Int J Eng Technol 1(3):219–223

Adewuyi A, Oderinde RA, Rao BVSK, Prasad RBN, Anjaneyulu B (2012) Blighia unijugata ad Luffa cylindrica seed oils: renewable sources of energy for sustainable development in rural Africa. Bioenergy Res 5:713–718

Chakraborty M, Baruah DC (2013) Production and characterization of biodiesel obtained from *Sapindus mukorossi* kernel oil. Energy 60:159–167

Chen Y, Chiang T, Chen J (2013) Properties of soap nut (*Sapindus mukorossi*) oil biodiesel and its blends with diesel. Biomass Bioenergy 52:15–21

Fu YJ, Zu YG, Wang LL, Zhang NJ, Liu W, Li SM, Zhang S (2008) Determination of fatty acid methyl esters in biodiesel produced from yellow horn oil by L.C. Chromatographia 67:9–14

Li J, Fu Y, Qu X, Wang W, Luo M, Zhao C, Zu Y (2012) Biodiesel production from yellow horn (Xanthoceras sorbifolia Bunge) seed oil using ion-exchange resin as heterogeneous catalyst. Bioresour Technol 108:112–118

Misra RD, Murthy MS (2011) Performance, emission and combustion evaluation of soap nut oil-diesel blends in a compression ignition engine. Fuel 90:2514–2518

Oliveira MEC, Franca AS (2002) Microwave heating of foodstuffs. J Food Eng 53:347–348

Sharma YC, Singh B (2010) An ideal feed stock, kusum (*Schleichera trijuga*) for preparation of biodiesel: optimization of parameters. Fuel 89:1470–1474

Shaw MAH, Dutta K, Deka DC (2014) Fatty acid composition of *Sapindus mukorossi* seed oil. Adv Appl Sci Res 5(4):43–50

Yao Z, Qi J, Yin L (2013) Biodiesel production from Xanthoceras sorbifolia in China: opportunities and challenges. Renew Sust Energ Rev 24:57–65

Yu H, Fan S, Bi Q, Wang S, Hu X, Chen M, Wang L (2017) Seed morphology, oil content and fatty acid composition variability assessment in yellow horn (Xanthoceras sorbifolium Bunge) germplasm for optimum biodiesel production. Ind Crop Prod 97:425–430

Zhang DY, Yao XH, Luo M, Zhao CJ, Fu YJ (2016) Optimization of negative pressure cavitation-microwave assisted extraction of yellow horn seed oil and its application on the biodiesel production. Fuel 166:67–72

36 Sapotaceae

Sapotaceae is one of the latex yielding family comprising 800 species of evergreen trees whose gum is used for gutta percha and chicle gum. Seeds are hard, shiny, with oily endosperm. The seeds of certain plants (e.g. *Madhuca butyracea*) produce vegetable butter (phulwa) used as cold cream, lip salve leminant and vegetable ghee. Biodiesel is prepared from the seed oil of mahua (*Madhuca indica*).

36.1 Mahua (*Madhuca indica*)

Madhuca indica, popularly known as mahua is a medium to large slow growing tree. It has a wide and round canopy. It attains a maximum height of 20 m in its life time. It grows along road side, canal banks and in public places. It prefers an altitude up to 1200 m and an annual rainfall of 1000–1800 mm. It gives forth flowers in February to April and the fruit ripens in August and September. The tree comes to flowering in 10–15 years. The oil potential of this species in India is around 180 million tonnes a year (Kulkarni et al. 2013).

This tree grows richly in peninsular India, Nepal, Sri Lanka and Myanmar and does not occur in other parts of the world. Hence the biodiesel works made on this oil are restricted to the above countries. In India this tree occurs in the state of Bihar, Chhattisgarh, Gujarat, Jharkhand, Kerala, Maharashtra, Madhya Pradesh, Odisha, Tamil Nadu, Telangana, Uttar Pradesh and West Bengal (Fig. 36.1).

The leaves of this tree are clustered at the tip of the branches and are pubescent or tomentose, coriaceous and elliptic with short petiole (Fig. 36.2). The fruits are ovoid, fleshy and green with 1–4 dark brown seeds. Each seed contains two kernels which form around 70% by weight of the seed. The kernels have 52% oil. *Madhuca indica* is known as butter tree (English), mahua (Hindi) and iluppai (Tamil).

S. P. Raj et al., *Biodiesel from Flowering Plants*,
https://doi.org/10.1007/978-981-16-4775-8_36

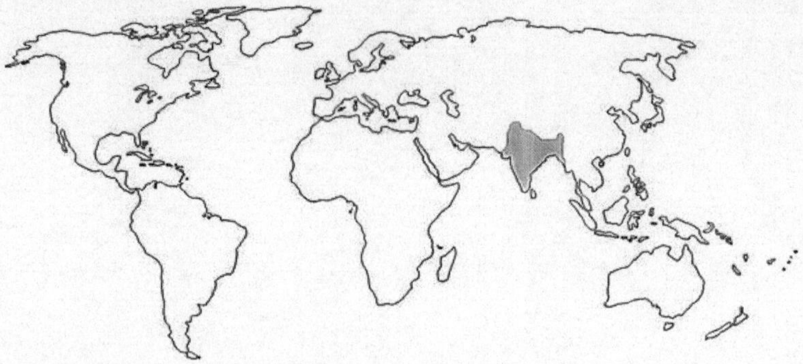

Fig. 36.1 Occurrence of *Madhuca indica* in Asia

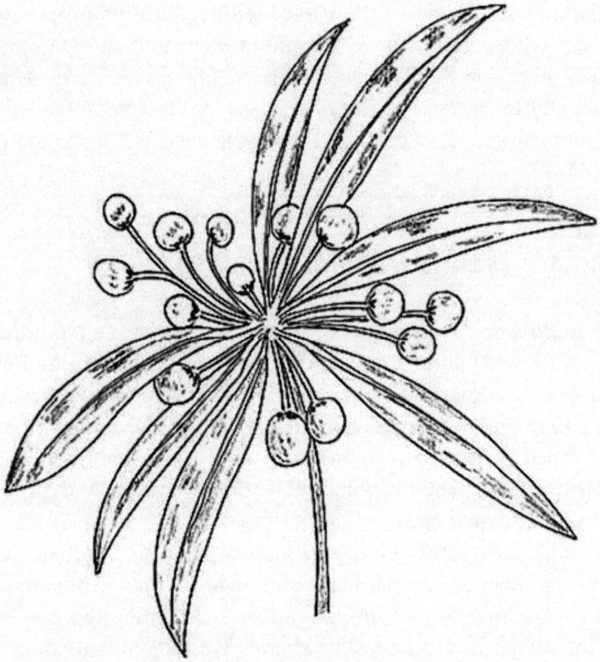

Fig. 36.2 Mahua, *Madhuca indica* tree branch with fruits

36.1.1 Mahua Oil

The seeds are collected, dried by exposing in hot sun, powdered in mill and the oil extracted. On pressing mode, the oil recovery will be around 26%. If *n*-hexane of boiling point 65–70 °C is used, the recovery will be around 50–52%. In industrial process the seeds are initially broken, flaked and steam cooked. The cooked flakes are extracted using solvent grade hexane at 63 °C.

36.1.2 Physicochemical Characteristics of Oil

The unrefined crude oil is normally greenish yellow in colour. It has a high free fatty acid which makes the oil unamenable to the normal process of transesterification. It the FFA is more than 1% by weight, soap may be formed during transesterification which greatly interferes with the production of biodiesel if alkali catalysis is followed. The soap thus formed is likely to hamper the separation of biodiesel from glycerin. Such situation demands a pretreatment process to reduce the FFA to an acceptable level (Table 36.1).

The oil is characterized by high density, viscosity, flash and fire points but the calorific value is low. The cetane number also is lesser than that of the diesel. The cetane number is needed to be sufficient to cause self-ignition when the fuel is sprayed inside the cylinder. High viscosity is likely to increase the system pressure there by the injection will be affected. Incomplete combustion causes nozzle coking, engine deposits, lube oil dilution and unreliable ignition. The molecular weight of the oil is roughly three times higher than that of the biodiesel and hence the viscosity of the oil is high. The saturated fatty acids comprise palmitic, stearic and arachidic acids. The unsaturated fatty acids comprise oleic and linoleic acids. The unsaturated fatty acids are more than 50% of the total fatty acids and as a result the oil is likely to oxidize rapidly. The methylene group being located at the proximity of the double bonds is amenable to the attack of free radical.

Table 36.1 The physicochemical characteristics of mahua oil (Ghadge and Raheman 2005; Kapilan et al. 2009; Kulkarni et al. 2013; Acharya et al. 2018)

Parameters	Value
Viscosity at 40 °C ($mm^2\ s^{-1}$)	39, 27.6, 24.6, 38.9
Density at 15 °C ($kg\ m^{-3}$)	924, 915, 960, 904
Iodine value (g I_2/100 g)	59, 60
Acid value (mg KOH g^{-1})	16, 38
Free fatty acid (%)	19
Flash point (°C)	230, 212, 232, 220
Cloud point (°C)	18
Pour point (°C)	14, 15
Ash content (%)	0.9
Carbon residue (%)	3.7
Calorific value (MJ kg^{-1})	32, 35.6, 36, 37.7
Palmitic acid (16:0) (%)	22
Stearic acid (18:0) (%)	22.5
Arachidic acid (20:0) (%)	1.6
Oleic acid (18:1) (%)	46
Linoleic acid (18:2) (%)	11.3

36.1.3 Biodiesel Preparation

Since the oil is characterized by high free fatty acid, two step transesterification is carried out. High level of FFA present in the oil is converted to ester through an acid aided catalysis using methanol. Consequent to the reaction the contents are allowed to settle for 1 h. The supernatant layer containing methanol and water is removed and the acid value of the lower layer is analysed. If the FFA value is lower than 2%, it is then subjected to alkali catalysis.

Methanol at 0.25% in the total volume (i.e. methanol to oil at a molar ratio of 6:1) and potassium hydroxide by 0.7% of the total volume (or any other quantity to neutralize the FFA plus 0.5% KOH) are added and the contents are stirred continuously for an hour at 60 °C. The reacted contents are allowed to settle overnight. The clear ester moves to the top and the glycerol moves down to enable the separation of the two phases. The overlying biodiesel is washed repeatedly with warm water till the entire KOH residue is removed. The washed materiel is heated to remove the moisture.

Methanol is mostly used in the transesterification process due to its low cost and also due to its tendency to dissolve KOH in it. Methanol has a low molecular weight (32.04), boiling point (64.7 °C) and specific gravity (0.792). Since its molecular weight is low, less quantity of methanol is required to complete the reaction. As the boiling point also is relatively low, the transesterification reaction can be accomplished by expending low energy. Sodium hydroxide, potassium hydroxide, ethoxide, propoxide, butoxide, sulphuric acid and sulphonic acid are used as catalysts. Among them alkali is the fastest and ideal catalyst and hence used in commercial processes. In the presence of alkali catalyst the transesterification is around 4000 times faster than that of the acid catalyst.

The rate of reaction in the acid treatment is largely governed by the quaintly of methanol and duration of the transesterification. In the initial period there is rapid reduction in FFA and in the later part of transesterification the rate of reduction is low. This might be due to the generation of water at the initial stage of reaction which interferes with the continued process of esterification and thus necessitates more methanol. During the next phase, namely the alkali treatment, relatively less quantity of methanol is needed since major quantity of FFA has already been esterified.

36.1.4 Physicochemical Characteristics of Biodiesel

The viscosity when reduced to an acceptable level in the biodiesel helps to have its easy mobility. The density also reduced which favours the mass flow of the fuel per every stroke of the plunger of the pump and thus helps to have efficient power output of the engine. The cetane number also satisfies the standard. During the long storage of biodiesel, hydroperoxides increases which subsequently decomposes to form short chain carboxylic acid, ketone, epoxide and hydroxy compounds. Soluble polymers are ultimately formed which tends to increase the viscosity. The acid

Table 36.2 Physicochemical characterization of biodiesel (Ghadge and Raheman 2005; Kapilan et al. 2009; Acharya et al. 2018)

Parameters	Value
Viscosity at 40 °C ($mm^2 s^{-1}$)	4.0, 4.85, 4.2
Density at 15 °C ($kg m^{-3}$)	880, 883, 882
Flash point (°C)	208, 129, 170
Cloud point (°C)	9
Pour point (°C)	6, 5
Acid value ($mg KOH g^{-1}$)	0.4
Ash content (%)	0.01
Carbon residue (%)	0.20
Cetane number	51
Calorific value ($MJ kg^{-1}$)	37, 36.9, 38.5

value also increases during long storage which later forms hexanals, heptanals, propanals, pentane and heptadienal (Table 36.2).

Along with the above, density also increases during the storage due to insoluble high molecular weight sediments. The iodine value is the index of the total number of double bonds. The iodine value normally decreases during storage. The degradation initiates normally at the site between olefinic carbons and as a result conjugated diene is likely to be formed. Consequently there will be aldehydes, alcohols, carboxylic acid and high molecular weight oligomers as final products and thus the iodine value decreases.

36.1.5 Engine Performance

The brake thermal efficiency (BTE) which is an index of a portion of the fuel energy being converted to useful energy normally increases as the load is enhanced, but the BTE of mahua biodiesel is normally around 5% lower than that of the regular diesel in normal injection timings.

The brake specific fuel consumption (BSFC) of the biodiesel usually decreases with the increases in load. But when the BSFC of diesel and biodiesel is compared, the BSFC of biodiesel is relatively high. The high BSFC in biodiesel is due to the low calorific value and high viscosity.

The total consumption also is high in biodiesel. The biodiesel has better combustion quality due to the presence of oxygen (around 10%). However heat release does not depend only on oxygen but richly rely on the heating value which is relatively low in biodiesel. Normally the air fuel ratio decreases in biodiesel when the load is increased. The load and fuel consumption is directly proportional and hence the air fuel ratio reduces at high load. Characteristically the thermal efficiency of biodiesel decreases with load when compared to diesel at normal compression ratio (Magadum and Sridhara 2017).

The exhaust gas temperature (EGT) is an index of the burning efficiency. Since the combustion process is better, the thermal efficiency is good. Therefore, the EGT is relatively high.

36.1.6 Emission Characteristics

The emission of CO in biodiesel is low in all loads when compared to diesel. The oxygen present in the biodiesel aids the combustion thereby CO_2 is generated leaving less scope for the discharge of CO. Thus the level of CO_2 is relatively high. The hydrocarbon content of the exhaust of the biodiesel fed engine is lower than that of the diesel. The NO_x release in the exhaust depends on the oxygen content of the fuel, temperature of the cylinder, peak pressure inside the cylinder and the rate of combustion. In biodiesel the NO_x emission is higher. The smoke intensity of biodiesel is relatively lower than that of diesel (Puhan et al. 2005).

References

Acharya N, Nanda P, Panda S (2018) A study of engine performance and emission characteristics of mahua biodiesel and it's blends. Int J Sci Eng Res 9(4):36–41

Ghadge SV, Raheman H (2005) Biodiesel production from mahua (*Madhuca indica*) oil having high free fatty acids. Biomass Bioenergy 28(6):601–605

Kapilan NT, Babu PA, Reddy RP (2009) Evaluation of properties and storage stability of *Madhuca indica* biodiesel. J Oleo Sci 58(12):617–622

Kulkarni PS, Sharanappa G, Ramesh MR (2013) Mahua (*Madhuca indica*) as a source of biodiesel in India. Int J Sci Eng Res 4(7):2319–2328

Magadum A, Sridhara SN (2017) Performance of IC engine by using Mahua oil (*Madhuca indica*) biodiesel. Int J Sci Res Publ 7(9):345–349

Puhan S, Vedaraman N, Rambrahamam BV, Nagarajan G (2005) Mahua (*Madhuca indica*) seed oil: a source of renewable energy in India. J Sci Ind Res 64:890–896

Schisandraceae

37

Schisandraceae family comprises around 90 species of tropical and subtropical woody plants in three genera. They are evergreen and rarely deciduous. Fruit is a berry either in single follicle or in star-like aggregate with 1–5 seeds each per follicle. The fruit on maturity dehisces explosively to disperse seeds. Seeds are flat in appearance and the endosperm is rich with oil. Very few works are available on magnolia berry (*Fructus schisandrae*) for yielding oil for biodiesel production.

37.1 Magnolia Berry (*Fructus schisandra chinensis*)

The magnolia berry *Fructus schisandrae* grows in north-eastern China (Fig. 37.1).

The *F. schisandra chinensis* is referred popularly as magnolia vine, Chinese magnolia vine, schisandra and five flavour fruit. This plant is a creeper with a maximum length of 8 m. It prefers shade and diffused light and grows in a well-drained soil. It is a dioecious plant and therefore the male and female plants are to be grown together. The leaves are oval shaped (5–11 cm long and 2–5 cm wide), flowers unisexual, fruits yellowish white to pinkish red with a dia. of 5–8 mm. Each fruit carries 1–2 seeds of 4–5 mm long and 3–5 mm dia. (Fig. 37.2).

37.1.1 Oil Extraction

The berries along with the seeds are dried in hot sun repeatedly and grinded to powder. It is then extracted with petroleum ether. The extract obtained is centrifuged (Fan et al. 2017) and then the solvent distilled out. In this process as high as 41.92% oil is obtained. The fatty acid content of the seed oil is given in Table 37.1.

Bulk of the fatty acids are unsaturated in nature. The fatty acids of the oil contain 16–20 carbon atoms in straight chain and therefore this oil can be effectively used in

S. P. Raj et al., *Biodiesel from Flowering Plants*,
https://doi.org/10.1007/978-981-16-4775-8_37

Fig. 37.1 The geographical distribution of *Fructus schisandra chinensis*

Fig. 37.2 Magnolia berry *Fructus schisandra chinensis*

Table 37.1 Fatty acid content of the oil of *F. schisandra chinensis* (Fan et al. 2017)

Fatty acids	Value
Palmitic acid (C16:0) (%)	2.63
Oleic acid (C18:1) (%)	16.32
Linoleic acid (C18:2) (%)	75.21
Linolenic acid (C18:3) (%)	0.70

the biodiesel production. The acid value is 2.14 g KOH kg^{-1} and the mol. weight is 996.86.

37.1.2 Biodiesel Production

Fan et al. (2017) employed sulphuric acid as a catalyst at the rate of 2.2 g/100 g of oil. Oil to methanol ratio is 1:6, the reaction temperature is 66 °C and the reaction duration is 2.5 h. It is to be noted that the acid catalysed reaction is around 4000 times slower than that of alkali catalysed reaction. However, the option is in favour of acid catalyst since acid easily manages high level of fatty acids (Lotero et al. 2005). In this process as high as 95% of ester yield is reported.

Immobilized 5-amino-1-naphthol catalysed transesterification is also attempted using coconut shell charcoal as carrier. Coconut shell charcoal is activated at 400 °C. It is then treated with hydrochloric acid (pH 1) for a while and then with phosphate buffer (pH 5) for 3 h. Consequently the extra phosphate buffer if remain in it is washed out repeatedly with warm distilled water. It is then dried at 200 °C for 6 h. Separately 5-amino-1-naphthol is dissolved in phosphate buffer of pH 6 and centrifuged at 800 rpm for 10 min. The supernatant is then mixed with the processed activated charcoal (mass ratio of 1 part of 5-amino-1-naphthol with 35 parts of charcoal). The content is stirred for 5 h in a shaker (180 rpm) at 40 °C. It is then freeze dried and stored in a closed container. To this the oil and methanol are added so as to have an oil to methanol molar ratio of 1:2.3. The 5-amino-1-naphthol and charcoal processed as explained above is incorporated at the rate of 24 g/100 g of oil. The reaction temperature is 37.5 °C and the duration is 16 h. The recovery of ester is 61.6 only (Fan et al. 2017).

The treatment of charcoal with 5-amino-1-naphthol at the rate of 19.6 g/100 mL of oil, at an oil to methanol ratio of 1:3.5 along with tertiary butanol at a molar ratio of 1:0.8 (oil to tertiary butanol) is also attempted. In this process the yield is reported to be 89.14% at a treatment duration of 11.0 h at 37.5 °C. It is to be noted that the addition of tertiary butanol has helped to increase the yield. The amino naphthol charcoal mixture can be used for 20 batches consecutively and the efficiency after 20 batches is around 50%.

The biodiesel produced is analysed and the results are given in Table 37.2.

The parameters referred above conform to the standard. The 5-amino-1-naphthol immobilized on activated charcoal produced biodiesel which can be easily separated

Table 37.2 Physicochemical properties of *F. schisandrae chinensis* biodiesel (Fan et al. 2017)

Parameters	Value
Kinematic viscosity at 20 °C (mm^2 s^{-1})	4.2
Density at 20 °C (kg m^{-3})	868
Flash point (°C)	137
Acid value (g KOH kg^{-1})	0.154
Copper corrosion test	1a
Cetane number	51

from the by-products, besides the catalyst also remains stable. The tertiary butanol added to it acted as a buffer to reduce the toxic effect of methanol on the amino naphthol.

References

Fan R, Zhao J, Du Y, Zhao W, Guo W, Yang J, Chen X (2017) Biodiesel production from *Fructus schisandrae* seed oil. Indian J Biotechnol 16:114–118

Lotero E, Liu Y, Lopez DE, Suwannakarn K, Bruce DA, Godwin JG Jr (2005) Synthesis of biodiesel via acid catalysis. Ind Eng Chem Res 44:5353–5363

Simaroubaceae

38

The Simaroubaceae family includes 32 genera and more than 170 species of trees and bushes of pantropical distribution. Their population is localized at tropical America, extending to Africa, Madagascar and also at Australia. This family is characterized by the presence of quassinoids a group of compounds responsible for antitumor, antimalarial, antiviral, insecticidal, amoebicidal and herbicidal activities. The seeds of one of the plants, namely paradise tree (*Simarouba glauca*) yield oil for biodiesel production.

38.1 Paradise Tree (*Simarouba glauca*)

The seed oil of the paradise tree, *Simarouba glauca* is a raw material for biodiesel. It is one of the promising biodiesel trees with a potential of 1000–2000 kg oil ha^{-1} $year^{-1}$ with a life span of 70 years. It grows at an elevation of 1000 m above main sea level with a rainfall ranging from 250 to 2500 mm and at a soil pH of 5.5–8.0. This tree is characterized by a circular crown with a maximum height of 15 m and a trunk dia. of 60 cm. The leaves are prominently compound with a maximum of 21 leaflets arranged alternatively. The leaflets are oval in shape with a sparkling green waxy consistency. The flower has five petals arranged in an overlapping pattern. The fruit is a drupe and looks like grapes both in size and in colour. The drupe is around 2 cm long and oval in shape. The kernel forms around 40% of the drupe and the kernel has 50–65% oil (Mishra et al. 2012) (Fig. 38.1).

This tree is referred by certain location specific vernacular names, namely: Simarouba, gavilan, negrito, maruba, marupa, dysentery bark, bitter wood, paradise tree, robieceillo, caixeta, palo blanco, daguilla, cedro blanco malacacheta, palo amargo, pitomba, bois amer, bois blanc, boisfrene, boisnegresse, simaba, lakshmi taru, lakshmi tharu and pasac.

S. P. Raj et al., *Biodiesel from Flowering Plants*,
https://doi.org/10.1007/978-981-16-4775-8_38

Fig. 38.1 The paradise tree *Simarouba glauca*

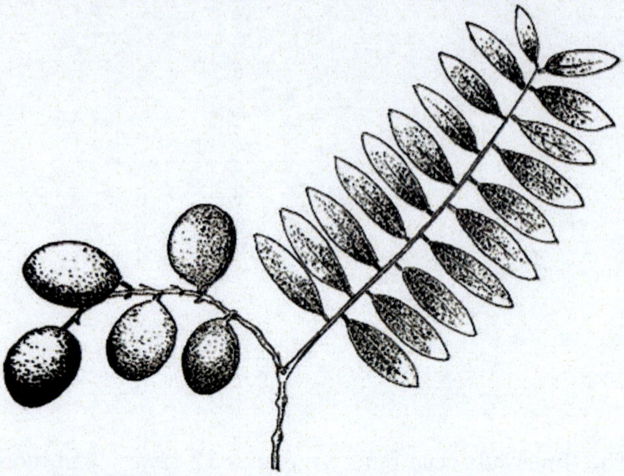

Fig. 38.2 The geographical distribution of *Simarouba glauca*

S. galuca is distributed in El-Salvador, Florida, Costa Rica, Caribbean, Puerto Rico, India, Sri Lanka, Nicaragua, Honduras, Guatemala, Mexico, Cuba, Jamaica, West Indies, USA, Philippines, Myanmar, Kenya and Burundi (Fig. 38.2).

The oil is extracted from the seeds by mechanical expeller or by solvent extraction using n-hexane in a Soxhlet apparatus. In mechanical expeller, the oil recovery is around 15%, whereas it is 55–60% through Soxhlet extraction (Mishra et al. 2012).

The physicochemical properties of the oil are given in Table 38.1.

The viscosity, flashpoint and unsaturated fatty acid values are fairly high. The iodine value is low.

Table 38.1 Physicochemical characteristics of *S. glauca* oil (Mishra et al. 2012, 2014)

Parameters	Mishra et al. (2012)	Mishra et al. (2014)
Kinematic viscosity at 40 °C (cSt)	–	17.3
Flash point (°C)	–	230
Iodine value (g I_2/100 g)	83.4	–
Acid value (g KOH kg^{-1})	5.34	10
Palmitic acid (C16:0) (%)	12.3	–
Stearic acid (C18:0) (%)	27.3	–
Oleic acid (C18:1) (%)	54.6	–
Linoleic acid (C18:2) (%)	2.3	–
Arachidic acid (C20:2) (%)	1.2	–

Table 38.2 Physicochemical properties of *S. glauca* biodiesel (Mishra et al. 2012)

Parameters	Value
Kinematic viscosity at 40 °C	4.68
Density at 15 °C (kg m^{-3})	865
Flash point (°C)	165
Cloud point (°C)	19
Pour point (°C)	14.2

38.1.1 Biodiesel Production

The free fatty acid content is high and it is more than 50% of the total acid value as shown in Table 38.1. Therefore a two-step conversion process is recommended. In the first step, acid esterification is carried out. On completion of the esterification, the finished product is subjected to alkaline catalysed transesterification process so that mono-ester and glycerol are produced.

In 1 L of raw oil, 250 mL of methanol and 10 mL (1%) of sulphuric acid are added and stirred for an hour at a 50 °C. Then the whole material is transferred to a separating funnel. Excess alcohol and acid moves to the top layer and the oil phase moves to the bottom. The bottom layer is collected and used for the next treatment. To 1 L of the above, 6 g of KOH and 250 mL methanol are added and stirred for 2 h at 600 rpm and at 60 °C. The content is then transferred to a separating funnel and is allowed to stay undisturbed for 24 h. The supernatant clear biodiesel phase is separated, washed repeatedly in dist. water and dried by heating. The properties of the resultant biodiesel are given in Table 38.2.

During the transesterification of oil to biodiesel, there is normalization of viscosity, density and flash point. The values for the cloud point and pour point are not satisfactory. Therefore, it is recommended to mix it with diesel and ethanol so as to moderate its properties (Vijayaragavan and Ramachandran 2019).

It is known that blending the biodiesel with diesel has not actively reduced the brake specific fuel consumption. But the addition of ethanol at 5% and 10% level in the B50 blend reduces the brake specific fuel consumption and increases the brake thermal efficiency. The ethanol incorporation has reduced the CO and HC emission at full load condition (Vijayaragavan and Ramachandran 2019). As the ethanol increases the latent heat of vaporization, the NO_x content in the emission also reduces. Parallelly the addition of ethanol reduces the smoke emission due to excess oxygen which is known to improve the combustion.

References

Mishra SR, Mohanty MK, Das SP, Pattanaik AK (2012) Production of biodiesel (methyl ester) from *Simarouba glauca* oil. Res J Chem Sci 2(5):66–71

Mishra SR, Mohanty MK, Das SP, Pattanaik AK (2014) Optimisation of base-catalyzed transesterification of *Simarouba glauca* oil for biodiesel production. Int J Sustain Energy 33 (6):1033–1040

Vijayaragavan MR, Ramachandran GSL (2019) An experimental study on the performance, combustion and emission characteristics on diesel: *Simarouba glauca* biodiesel blended fuel for diesel engine with ethanol as an additive. Wiley Online Library. wileyonlinelibrary.com. https://doi.org/10.1002/ep.13206

Simmondsiaceae 39

Simmondsiaceae or the jojoba family is represented by a single genus and a single species *Simmondsia chinensis* which is native to the Southwestern United States. It is cultivated as a commercial crop. Fruit is a triangular capsule mostly with single seed (rarely 2–3-seeds). Seeds are brownish-black with thick and fleshy cotyledons containing liquid wax.

39.1 Jojoba (*Simmondsia chinensis*)

Simmondsia chinensis is an exclusive species of the genus Simmondsia. Vernacularly it is referred as goat nut, deer nut, pignut, wild hazel, coffee berry, grey box bush and quinine plant. It is a small shrub growing not more than 2 m height. The crown is, with dense, succulent oval shaped leaves arranged opposite to each other showing all characteristics of an xerophytic plant. These leaves are 30–40 mm long and 15–25 mm broad and looks grey, which are vertically oriented to meet the extremes of environmental exigency. The flowers are greenish yellow in colour with 5–6 sepals. The fruits are oval shaped and 10–20 mm long (Fig. 39.1). The tree is dioecious and the fruits are dehiscent with a single seed. This plant has a life span of 100–200 years.

This shrub is domained in the Sonoran and Colorado desert, California, Arizona and Mexico. It has also spread in Sudan and Ghana. It is endemic to North America and occupies an area of 260,000 km^2 between 25° and 31° North latitudes and between 109° and 117° West the longitudes. It prefers to grow at an elevation ranging from 600 to 1200 m above MSL with a rainfall of around 100 mm (Fig. 39.2).

The oil production is at the rate of 1800 kg ha^{-1} (Yang et al. 2014). The estimated jojoba oil demand of the world is around 0.2 million tonnes a year, which is far below the current production (Sánchez et al. 2016).

S. P. Raj et al., *Biodiesel from Flowering Plants*,
https://doi.org/10.1007/978-981-16-4775-8_39

Fig. 39.1 The jojoba *Simmondsia chinensis*

Fig. 39.2 Geographical distribution of *S. chinensis*

Table 39.1 Jojoba oil extraction using different solvents (Sánchez et al. 2016)

Solvents	Duration (h)	Recovery (%)
Hexane	18	52
Chloroform	18	33
Isopropanol	18	45
Toluene	18	45

39.1.1 Oil Extraction

The seeds of jojoba are dried, crushed and pressed in roller in which the oil recovery is hardly 8% (Sánchez et al. 2016). To increase the recovery organic solvents are

Table 39.2 Supercritical process in the extraction of jojoba oil using CO_2 as co-solvent (Salgin 2007)

Co-solvent	Temperature (K)	Pressure (bar)	Recovery (%)
CO_2	343	450	44
CO_2	600	363	51
30% CO_2 + propane	313	70	49

These values are based on 98% of the extractable quantity

Table 39.3 Physicochemical properties of the jojoba oil (Sánchez et al. 2016; Al-Hamamre and Al-Salaymeh 2014)

Parameters	Value
Kinematic viscosity at 40 °C (cSt)	27, 24
Density (kg m^{-3})	862, 865
Acid value (g KOH kg^{-1})	0.36, 0.7
Iodine value (g I_2/100 g)	83, 56
Flash point (°C)	225, 295
Cold filter plugging point	10
Oxidation stability 110 °C (h)	41
Wax ester C38 (%)	6.6
Wax ester C40 (%)	30
Wax ester C42 (%)	51
Wax ester C44 (%)	9
Heating value (MJ kg^{-1})	49.3

used (Table 39.1). The seed powder (<1 mm size) is used in the soxhlet extraction with 8–10 cycles per hour.

Out of the above hexane is considered as an effective solvent. But the quantity needed and the time involved are certain of the negative factors limiting its use. Supercritical CO_2 is also considered in the extraction due to its low cost and environmental inertness. However, the involvement of high pressure is an impediment. To improve the supercritical process, CO_2 is added as a co-solvent (Salgin 2007). There are possibilities that this co-solvent may remain in the oil as residues unless they are carefully removed. The percentage of extractable oil due to CO_2 as co-solvent is seed in Table 39.2.

Chosen properties of the jojoba oil are given in Table 39.3.

Unlike any other vegetable oils, the oil of jojoba is not a triglyceride but predominantly formed of long chain esters (around 98%) of fatty acids and fatty alcohols and therefore it has become pertinent to refer the oil as wax or wax ester or wax oil. Nevertheless the wax contains trace of free fatty acids, alcohols and triacylglycerol. The flash point is very high (225–295 °C). The kinematic viscosity is also relatively low. The molecular makeup of this oil is of monounsaturated esters and therefore the product of alcoholysis does not give glycerol and therefore it obviates the phase separation step (Sanchez et al. 2015). Jojoba oil is a source of 11-eicosenol ($C_{20}H_{38}O_2$), 13-docosenol ($C_{22}H_{42}O_2$) and 15-tetracosenol ($C_{24}H_{46}O_2$) which are value added straight monounsaturated alcohols. Jojoba oil has extraordinarily high oxidation stability.

39.1.2 Biodiesel Production

In the transesterification process, fatty acid methyl esters and monounsaturated alcohols are produced. The alkali catalysed transesterification is considered better than that of heterogeneous and enzyme catalyst as the acidity of the oil is lower than 1%. Consequent to the transesterification, the product contains monounsaturated alcohols, alkyl esters, unreacted oil, short chain alcohol and the catalyst, all in a single phase. The washing is needed to remove the alkali and any unreacted alcohol. In the catalytic transesterification different alcohols, methanol, ethanol, propanol and butanol are tried in which high efficiency (94%) is reported in methanol with a catalyst (NaOH) concentration of 1.5% at an operational duration of 5 h. NaOH, KOH and dibutyltin diacetate at a concentration of 1% at a methanol to oil molar ratio of 23:1, at 60 °C temperature and in an experimental duration of 33 h are reported (Shah et al. 2014). Methanolysis of this oil is successfully carried out at a low temperature (30 °C) at a catalyst concentration of 1.25%, methanol to oil molar ratio of 6:1, 600 rpm agitation in an experimental duration of 1 h. The yield obtained is 83% only (Bouaid et al. 2007). When the above experiment is carried out in the microwave, the yield is around 90% in a reaction duration of 3 min (Hussein et al. 2014). As heterogeneous catalysts CaO from calcinated mussel shells (Sanchez et al. 2015) dibutyltin oxide, diocotyltin oxide, diphenyltin oxide, monobutyltin chloride dihydroxide and monobutyltin hydroxide (Shah et al. 2014). Organotin carboxylates (Sirajuddin et al. 2015) are tried with a limited success. The properties of the biodiesel developed from jojoba oil using different alcohols are given in Table 39.4. Viscosity, density, cloud point, cold filter plugging point, oxidation stability are fairly influenced by the type of oil used. The increase in the oxidation stability is due to the increase in the molecular weight. Consequently, the kinematic viscosity is high.

Table 39.4 The physicochemical properties of the jojoba oil biodiesel prepared using different alcohols (Al-Awad et al. 2014; El-Boulifi et al. 2015)

	Alcohols in the transesterification process			
Parameters	Methanol	Ethanol	1-Propanol	1-Butanol
Viscosity at 40 °C (cSt)	19	14	17	18
Density (kg m^{-3})	866	882	891	907
Flash point (°C)	75	–	–	–
Cloud point (°C)	6	5	4	1
Pour point (°C)	6	8	0	-3
Iodine value (g I_2/100 g)	80	–	–	69
Cold filter plugging point (°C)	7	6	3	-1
Oxidation stability 110 °C (h)	2	3	4.4	6
Calorific value (MJ kg^{-1})	47.4	47.4	–	–

39.1.3 Fuel Blends

Three types of blends are tried. They are jojoba oil + petro-diesel, jojoba oil + biodiesel, biodiesel + petro-diesel (Al-Hamamre and Al-Salaymeh 2014). When raw jojoba oil is mixed with petro-diesel, the density is reported to increase as a function of jojoba oil content at a regression of $y = 0.8393 + 0.0003x$, when jojoba oil is mixed with the biodiesel the density reduces in accordance with the concentration of jojoba oil ($y = 0.8963 - 0.0003x$).When jojoba biodiesel is mixed with petro-diesel the density increases linearly ($y = 0.8377 + 0.0006x$). In all the three, y is the density and x is the level of jojoba oil or jojoba oil biodiesel. When all the three types of mixtures are subjected to heating the density decreases more or less with the same pattern. The viscosity increased in all the three set of blends as the level of jojaba oil or jojoba oil biodiesel is increased. In all the three combinations the flash point increases almost as a sigmoidal pattern.

Regarding the heating value, there is a systematic increase in accordance with the concentration of jojoba oil with biodiesel, whereas there is a steady decrease in the heating value as a function of jojoba oil with petro-diesel. Similarly the heating value shows a negative slope when biodiesel content is increased in the combination of biodiesel and petro-diesel. Saleh (2009) observed that the NO_x emission level can be contained through exhaust gas recirculation (EGR) method. Though it reduces NO_x, there is increase in CO and HC, besides, there is an increase in brake specific fuel consumption. In this context, alumina nanoparticles are supplemented so as to improve the combustion quality, thereby reducing both NO_x and CO in the emission (Attia and Hassaneen 2016).

References

Al-Awad AS, Selim MYE, Zeibak AF, Moussa R (2014) Jojoba ethyl ester production and properties of ethanol blends. Fuel 124:73–75

Al-Hamamre Z, Al-Salaymeh A (2014) Physical properties of (Jojoba oil + biodiesel), (Jojoba oil + diesel) and (biodiesel + diesel) blends. Fuel 123:175–188

Attia AMA, Hassaneen AE (2016) Influence of diesel fuel blended with petrodiesel produced from waste cooking oil on diesel engine performance. Fuel 167:316–328

Bouaid A, Bajo L, Martinez M, Aracil J (2007) Optimization of biodiesel production from jojoba oil. Proc Saf Environ Prot 85(B5):378–382

El-Boulifi N, Sãnchez M, Martinez M, Aracil J (2015) Fatty acid alkyl esters and monounsaturated alcohols production from jojoba oil using short chain alcohols for biorefinery concepts. Ind Crops Prod 69:244–250

Hussein LI, Elsabee MZ, Ismail EA, Naguib HF, Aziz HA, Elsawy MA (2014) Transesterification of jojoba oil-wax using microwave techniques. World Acad Sci Eng Technol Int J Environ Chem Ecol Geol Geophysics Eng 27:314–322

Saleh HE (2009) Experimental study on diesel engine nitrogen oxide reduction running with jojoba methyl ester by exhaust gas recirculation. Fuel 88:1357–1364

Salgin U (2007) Extraction of jojoba seed oil using a supercritical CO2+ethanol mixture in green and high tech separation process. J Supercrit Fluids 39:330–337

Sanchez M, Marchetti J, El-Boulifi N, Aracil J, Martinez M (2015) Jojoba oil biorefinery using a green catalyst, part II: feasibility study and economical assessment. Biofuels Bioprod Biorefin 9 (2):139–146

Sánchez M, Avhad MR, Marchetti JM, Martínez M, Aracil J (2016) Jojoba oil: a state of the art review and future prospects. Energy Convers Manag 129:293–304

Shah M, Tariq M, Ali S, Guo Q, Fu Y (2014) Transesterification of jojoba oil, sunflower oil, neem oil, rocket seed oil and linseed oil by tin catalysts. Biomass Bioenergy 70:225–229

Sirajuddin M, Tariq M, Ali S (2015) Organotin (IV) carboxylates as an effective catalyst for the conversion of corn oil into biodiesel. J Organomet Chem 779:30–38

Yang I, Takase M, Zhang M, Zhao T, Wu X (2014) Potential non-edible oil feedstock for biodiesel production in Africa: a survey. Renew Sust Energ Rev 38:461–477

40 Solanaceae

The family Solanaceae with 3000 species often referred as night shade family or potato family comprises annual herbs, shrubs, trees of rainforest or lianas. They occur world-wide and include many agro-crops, medicinal plants and spices. Few members of the family produce alkaloids in which some are highly toxic. Many are being cultivated as olericulture crops (tomatoes, potatoes, egg plants and bell and chilli peppers). Fruit is either capsule or berry. Seeds are endospermic. The tobacco plant *Nicotiana tabacum* is dealt in this chapter as its seeds are oil bearing.

40.1 Tobacco (*Nicotiana tabacum*)

The seeds of tobacco (*Nicotiana tabacum*) was not considered for biodiesel production for a long time till it was comprehensively tested and reported in 2005 (Usta 2005). Tobacco is a popular herbaceous plant extensively cultivated for its leaves which are being used for cigarettes, cigar and for snuff. This species belongs to the family Solanaceae. It grows to a maximum height of 2.5 m. The leaves are large and magnificent with a maximum length of 60 cm and the flower is long and trumpet shaped with white pinkish colour (Fig. 40.1).

Its seeds are now considered as raw material for biodiesel production. The seeds are very small and tiny (0.1 mg) but the quantity by weight is high which ranges from 0.6 to 1.2 tonnes ha^{-1} (Usta 2005). The seeds can be stored for a long time if they are dry. It has a strong shell and its endosperm has rich oil. The tobacco plant is cultivated in more than 100 countries. Top ten countries in the order of cultivation are China, Brazil, India, USA, Argentina, Indonesia, Turkey, Greece, Italy and Pakistan (Barla and Kumar 2016) (Fig. 40.2).

S. P. Raj et al., *Biodiesel from Flowering Plants*,
https://doi.org/10.1007/978-981-16-4775-8_40

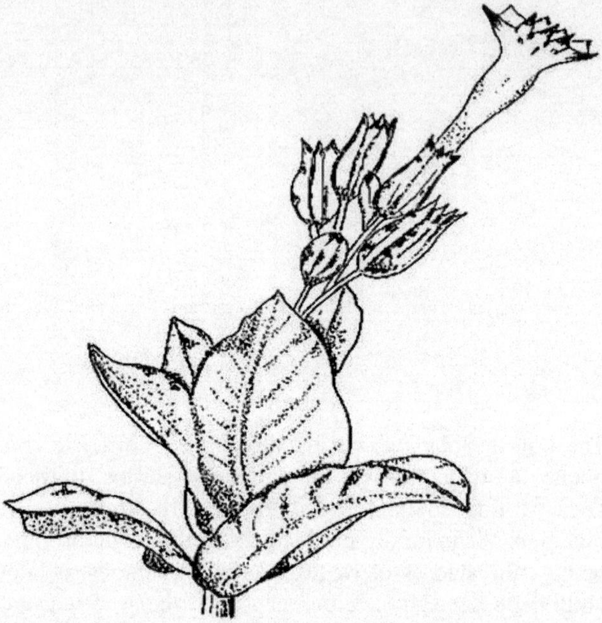

Fig. 40.1 The tobacco *Nicotiana tabacum* plant

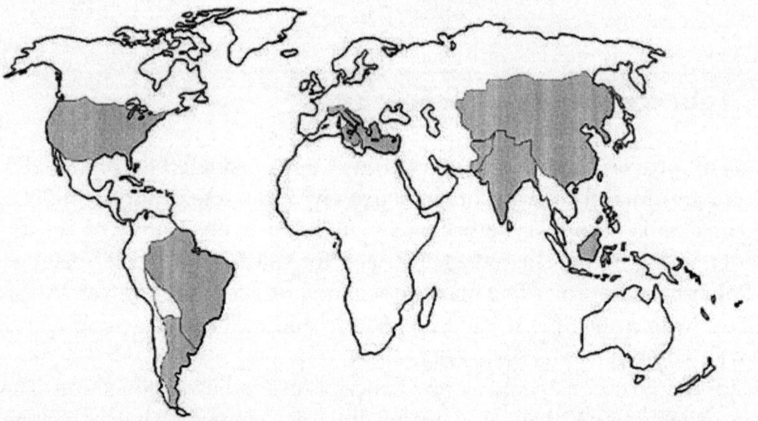

Fig. 40.2 Geographical distribution of *N. tabacum*

40.1.1 Oil Production

The seeds are extracted for its oil using organic solvents (*n*-hexane or diethyl ether). The grinded seeds along with the solvent are loaded in Soxhlet apparatus. The ratio between the solvent and seed powder is 10:1. The grinded seeds are left in the Soxhlet for 90 min and then the extraction proceeded with a minimum of 8 cycles per

Table 40.1 Fatty acid content of *N. tabacum* oil extracted by *n*-hexane and petroleum ether (Stanisavljević et al. 2009)

Fatty acid	*n*-Hexane	Petroleum ether
Palmitic acid (C16:0) (%)	8.42	8.69
Stearic acid (C18:0) (%)	3.78	2.90
Oleic acid (C18:1) (%)	11.78	12.54
Linoleic acid (C18:2) (%)	72.22	72.92
Linolenic acid (C18:3) (%)	0.84	0.89
Saturated fatty acids (%)	12.2	11.59
Unsaturated fatty acid (%)	84.8	86.4

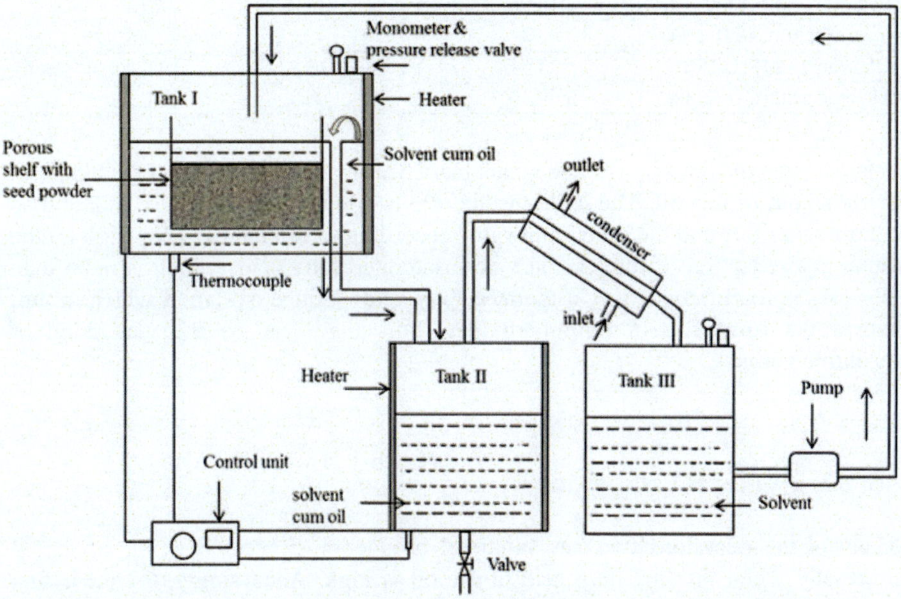

Fig. 40.3 Schematic diagram of commercial type oil extractor

hour for a total of 8 h. The oil extracted either by *n*-hexane or by petroleum ether makes not much difference on the fatty acid composition as shown in Table 40.1.

Usually, a commercial type extractor is used with a simple architecture. The details may be seen in Fig. 40.3.

The seed powder is loaded in a porous shelf of Tank 1 and kept immersed in the solvent. When the level of incoming solvent raises in Tank 1 the oil-solvent mixture flows into the next tank (Tank II). From this tank, the solvent is distilled out and cooled in a condenser and the solvent enters into the third tank (Tank III) from which solvent is pumped back into the porous shelf. The tank having the porous shelf and the tank receiving the oil-solvent mixture are thermostated. The oil retained in the Tank III is drained out. The oil characteristics are given in Table 40.2.

The yield of oil due to hexane extraction ranges from 36% to 41.5% (Giannelos et al. 2002). The kinematic viscosity and density are phenomenally high. The density

Table 40.2 The physicochemical characteristics of the oil of *N. tabacum*

Parameters	Giannelos et al. (2002)	Murthy (2010)	Usta et al. (2011)
Kinematic viscosity at 40 °C (cSt)	–	64.9	–
Density (kg m^{-3})	–	954.8	–
Acid value (g KOH kg^{-1})	–	3.2	–
Flash point (°C)	–	290	–
Iodine number (g I_2/100 g)	135	135	–
Palmitic acid (C16:0) (%)	10.96	15.2	8.16
Stearic acid (C18:0) (%)	3.34	4.8	3.56
Oleic acid (C18:1) (%)	14.54	13.2	12.14
Linoleic acid (C18:2) (%)	69.49	66.7	72.98

value crosses the limit prescribed which has a bearing on the density of the biodiesel prepared out of this oil. The iodine value of this oil (135 g I_2/100 g) is greater than certain other oils and the maximum value according to standard is 120. High iodine value is a reflection of the presence of unsaturated fatty acids which is more than 80% of the total fatty acid. Unsaturated fatty acids induce oxidation which in turn causes the formation of hydroperoxides. This may lead to the generation of insoluble gums.

40.1.2 Biodiesel Production

Many of the occasions two-step biodiesel production process is being attempted especially when the free fatty acid of the oil is high. Accordingly, the oil is first esterified by acid treatment. Sulphuric acid 1–2% of the oil depending upon the free fatty acid content is used. The acid is first mixed with methanol. The quantity of methanol so taken has to form 10:1 molar ratio with oil. Both the oil and the sulphuric acid-methanol mixture are heated to 60 °C separately and mixed and stirred for an hour. Subsequently, the above mixture is taken to a separating funnel. The lower phase containing the oil is taken for further processing and the upper phase is discarded (Veljković et al. 2006). The oil is then washed repeatedly to rid of any acid present in it. The resultant oil has low level of free fatty acid.

In the next stage, the oil thus processed is to be treated with alkali-methanol solution. A known quantity of KOH (around 1% of the oil) is dissolved in methanol. The quantity of methanol needed is to be decided so as to form an oil to methanol ratio of 1:6. This solution is heated to 60 °C and then mixed with the preprocessed hot oil (60 °C) and stirred (600 rpm) for 2 h. The resultant material is loaded in a separating funnel and allowed to stand overnight. The supernatant biodiesel is separated, washed repeatedly with distilled water and heated to dry.

In the acid treatment (esterification) mentioned above the reaction is governed by the amount of catalyst (H_2SO_4) and the correct molar ratio of oil to methanol. The

Table 40.3 The physicochemical characteristics of the biodiesel from the oil of *N. tabacum*

Parameters	Usta (2005)	Veljković et al. (2006)	Usta et al. (2011)
Kinematic viscosity at 40 °C (cSt)	3.5	5.2	4.23
Density at 15 °C (kg m^{-3})	887	882	888.5
Flash point (°C)	–	–	165.4
Copper strip corrosion (3 h at 50 °C)	–	–	1A
Acid value (g KOH kg^{-1})	0.45	0.66	0.3
Cetane number	51	–	51.6
Iodine number (g I_2/100 g)	112	109.8	136
Oxidation stability 110 °C (h)	–	–	0.8
Cold filter plugging point (°C)	–	–	−5
Heating value (MJ kg^{-1})	39.8	–	–

initial phase of the acid esterification reaction is relatively faster. As a result the free fatty acid is reduced to less than 1%. During the above process the triglyceride content also reduces feebly and along with it a minimum quantity (around 20%) of ester also is formed. This indicates that the free fatty acid is esterified along with partial methanolysis of the triglyceride (Veljković et al. 2006).

Similarly the base catalysed transesterification takes a rapid speed within 10 min of the startup and subsequently the process slows down due to the decrease in the quantity of methanol and glyceride as reactants. The physicochemical properties of the *N. tabacum* biodiesel thus prepared are presented in Table 40.3.

Iodine value and oxidation stability are the two parameters grossly deviating. Since the biodiesel has high quantity of unsaturated fatty esters, the iodine value is high and consequently the oxidation stability is low. In order to improve the oxidation stability certain additives as antioxidants are added. These are tertiary butyl hydroquinone, butylated hydroxytoluene, propyl gallate, pyrogallol, α-tocopherol and butylated hydroxyl anisole. The efficiency of these additives in improving the oxidation stability of the biodiesel thus made from the oil of *N. tabacum* is delineated in Fig. 40.4.

Among the various antioxidants tried, pyrogallol offers the best stability. It is also known that the addition of pyrogallol does not interfere with the cold filter plugging point, viscosity, iodine value and cetane number (Tang et al. 2008).

The cold filter plugging point, cloud point and pour point are collectively referred to as cold flow properties. When the cold filter plugging point is at a disadvantageous level, incorporation of cold flow improvers such as ethylene-vinyl acetate copolymer, and octadecene-1-maleic anhydride copolymer are suggested. The corrosion effect of this biodiesel is low since the copper strip corrosion is reported to be 1 A. The cetane number is marginally higher than the minimum limit prescribed. The issue of cetane number is considered important in the field of automobile engineering as the cetane number reflects the ignition delay which is the time taken between the start of injection and start of combustion. If the cetane number is low, the ignition

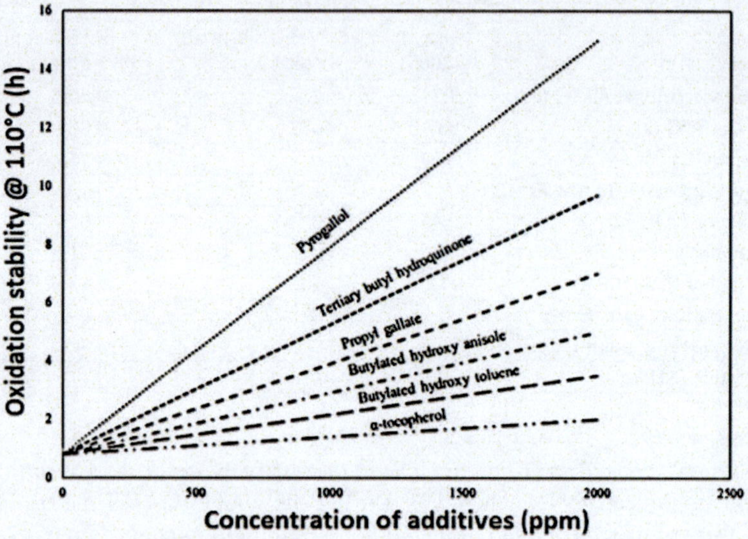

Fig. 40.4 Effect of various additives on the oxidation stability of *N. tabacum* biodiesel

delay is high which causes delay in the combustion resulting in fuel knock. This needs the incorporation of anti-knocking agent.

Murthy (2010) who studied the performance of the blends of tobacco biodiesel with petro- (neat) diesel at the level of 2% and 5% (B2 and B5) in a four-stroke direct injection single cylinder diesel engine, indicates that the B5 gives a maximum thermal efficiency of 29% against 27% in petro-diesel. Interestingly, Murthy (2010) also noted that the brake specific fuel consumption of B2 and B5 blends is lower than that of neat diesel presumably due to the better combustion because of rich oxygen in the biodiesel. The engine torque, power and specific fuel consumption are in accordance with the standard limit in force. The CO and NO level in the emission decreases while hydrocarbon marginally increases in the blend. Usta (2005) who carried out combustion experiments in a four cycle, four cylinder turbo charged indirect injection diesel engine at full and partial loads reports that the engine performance does not show any adverse impact for B10 and B25 fuel blends.

References

Barla FG, Kumar S (2016) Tobacco biomass as a source of advanced biofuels. Biofuels. https://doi.org/10.1080/17597269.2016.1242684

Giannelos PN, Zannikos F, Stournas S, Lois E, Anastopoulos G (2002) Tobacco seed oil as alternative diesel fuel: physical and chemical properties. Ind Crop Prod 16:1–9

Murthy SYVV (2010) Performance of tobacco oil-based biodiesel fuel in a single cylinder direct injection engine. Int J Phys Sci 5(13):2066–2074

Stanisavljević IT, Lakicevic S, Velickovic D et al (2009) The extraction of oil from tobacco (*Nicotiana tabacum* L.) seeds. Chem Ind Chem Eng Q 13(1):41–50

Tang HY, Wang AF, Salley SO, Ng KYS (2008) The effect of natural and synthetic antioxidants on the oxidative stability of biodiesel. J Am Oil Chem Soc 85(4):373–382

Usta N (2005) Use of tobacco seed oil methyl ester in a turbocharged indirect injection diesel engine. Biomass Bioenergy 28:77–86

Usta N, Aydoğan B, Con AH, Uğuzdoğan E, Ozkal SG (2011) Properties and quality verification of biodiesel produced from tobacco seed oil. Energy Convers Manag 52:2031–2039

Veljković VB, Lakićević SH, Stamenković OS, Todorović ZB, Lazić ML (2006) Biodiesel production from tobacco (*Nicotiana tabacum* L.) seed oil with a high content of free fatty acids. Fuel 85:2671–2675

41 Sterculiaceae

The members of the family Sterculiaceae also called as cacao family are trees, shrubs or herbs comprising 1000 species. They occur in tropical and subtropical regions. These plants are characterized by the presence of star-shaped hairs with mucilaginous secretions. Chocolate, cacao powder, cacao butter, caffeine and theobromine are produced from the seeds of chosen plants of this family.

41.1 Java Olive (*Sterculia foetida*)

The java olive tree, *Sterculia foetida* is a tropical and subtropical tree of the family Sterculiaceae. This tree mostly occurs in an area lying between 30°N and 35°S-latitudes. Vernacular names of this tree are bastard poon tree, java olive oil tree, hazel sterculia and wild almond tree. It is a large deciduous tree with a straight trunk growing up to a height of 40 m and a girth of 3 m. The leaves (12–15 cm long) are positioned at the tip of the branch (Fig. 41.1).

The flowers are purple in colour and are arranged in panicles of 10–15 cm long. Fruit formation takes around two months from the flowering. Each fruit has 4–5 follicles and each follicle has 10–15 seeds. The fruit is around 10 cm long. Each tree bears 200–350 kg of seed per year and the oil is around 50–60% (Sivakumar et al. 2012). This species is distributed in Taiwan, Indonesia, India, Philippines, Ghana, Australia, Egypt, Pakistan, Myanmar, Sri Lanka, Puerto Rico, Thailand, Uganda, Mozambique, Togo, Burma and Malaysia (Fig. 41.2).

41.1.1 Preparation of Seed Oil

The seeds are sundried and dehulled to obtain the kernels. The dry kernels are powdered and the oils from them are extracted using Soxhlet apparatus. The kernel to solvent ratio is 1:10 by volume which however depends on the type of solvent.

S. P. Raj et al., *Biodiesel from Flowering Plants*,
https://doi.org/10.1007/978-981-16-4775-8_41

Fig. 41.1 The plant *Sterculia foetida*

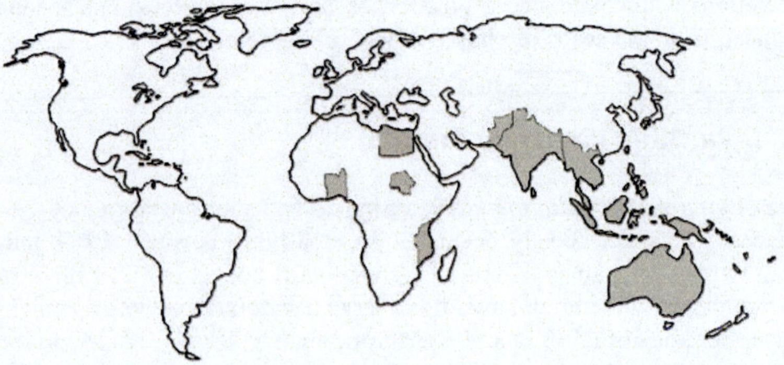

Fig. 41.2 Geographical distribution of *S. foetida*

Table 41.1 Effect of different solvents on the extraction yield of *S. foetida* oil from the kernel (Sivakumar et al. 2012)

Solvent	Yield (wt%)
n-Hexane	57
Petroleum ether	59
Tetrahydrofuran	62
Methanol	45
Chloroform	47

The solvents normally used are *n*-hexane, petroleum ether, tetrahydrofuran, methanol and chloroform. The temperature of the extraction is adjusted in such a way that there are 8–10 cycles per hour for a total extraction time of 10 h. The effect of different solvents in the extraction is given in Table 41.1.

High yield was obtained in tetrahydrofuran extraction. The extraction of oil using methanol is low apparently due to its polar nature which destroys intracellular

Table 41.2 Physicochemical characteristics of *S. foetida* oil (Bindhu et al. 2012; Ong et al. 2013; Subramanian and Muthiah 2016)

Parameters	Bindhu et al. (2012)	Ong et al. (2013)	Subramanian and Muthiah (2016)
Kinematic viscosity at 40 °C (cSt)	54	63.9	52
Density (kg m^{-3})	920	937	925
Free fatty acid (wt%)	2.1	–	11.2
Iodine value (g I_2/100 g)	76	88.7	84
Phosphorous content (mg L^{-1})	217	–	199
Flash point (°C)	–	198.5	–
Heating value (MJ kg^{-1})	–	39.8	–
Palmitic acid (C16:0) (%)	–	23	–
Stearic acid (C18:0) (%)	–	8.2	–
Oleic acid (C18:1) (%)	–	1.0	–
Linoleic acid (C18:2) (%)	–	11.0	–
Malvaloyl (C18:CE)	–	10.2	–
Sterculoyl (C19:CE)	–	44.3	–

compartment in the kernel mass (Sivakumar et al. 2012). Normally *n*-hexane is preferred since its extract has low saponifiable matter. The analytical features of *S. foetida* oil are given in Table 41.2.

Higher viscosity of the oil is due to the presence of (>50% by wt) cyclopropane fatty acids (Malvaloyl and Sterculoyl). The phosphorous content also is high. Before subjecting the oil to transesterification the oil is degummed. Dilute (20%) phosphoric acid is added to the oil at the rate of 0.5% by volume of the oil at hot condition (60 °C) and stirred for 30 min. The mixture is then transferred to a separating funnel and the gum compounds settle at the bottom. The oil which remains at the upper phase is collected and washed with distilled water so as to make it free from acid and soluble matter. It is then heated to expel the moisture.

41.1.2 Biodiesel Preparation

Often the level of free fatty acid in the oil is high which necessitates an initial esterification process (by acid) and is followed by transesterification by alkali. If alkali transesterification alone is considered, there may be excessive soap and gel formation due to the high level of free fatty acid which may interfere with the phase separation.

In the double stage process the crude oil is first mixed with methanol at a molar ratio of 1:10 (oil to methanol) and the content is stirred at a speed of 1200 rpm for 15 min. Sulphuric acid is then added to the above dropwise at the rate of 10 mL L^{-1} of oil and the stirring continued for 3 h. It is then transferred to a separating funnel

Table 41.3 Physicochemical characteristics of *S. foetida* oil biodiesel (Bindhu et al. 2012; Ong et al. 2013; Subramanian and Muthiah 2016)

Parameters	Bindhu et al. (2012)	Ong et al. (2013)	Subramanian and Muthiah (2016)
Kinematic viscosity at 40 °C (cSt)	4.7	4.9	4.7
Density (kg m^{-3})	850	879	850
Flash point (°C)	179	160.5	159
Pour point (°C)	3	−3	−3
Cloud point (°C)	3	1.2	2
Cold filter plugging point (°C)	–	−2.5	–
Iodine value (g I_2/100 g)	72.6	103	–
Acid value (g KOH kg^{-1})	–	0.14	–
Copper strip corrosion	–	1a	–
Oxidation stability 110 °C (h)	3.4	–	4.9
Cetane number	–	56.5	–
Heating value (MJ kg^{-1})	–	40.2	–

where it is allowed to settle. The lower layer containing the esterified oil is separated (Ong et al. 2013).

In the next step sodium hydroxide at the rate of 1% of the treated oil is dissolved in methanol. The quantity of methanol is decided so as to form 1:10 molar ratio of oil to methanol. This mixture is added to the acid treated oil obtained through esterification process as explained above. This mixture is then stirred at 1200 rpm at 50 °C for 2 h. Upon completion of the reaction, it is transferred to a separatory funnel and kept undisturbed overnight so as to have the completion of the phase separation. The upper phase is the biodiesel which is removed and washed repeatedly with warm distilled water until the biodiesel shows neutral pH. It is then dried using anhydrous $CaCl_2$ for a day and consequently by Na_2SO_4 for 3 h. Still if there is any moisture, the biodiesel is heated in vacuum to expel the moisture. The properties of biodiesel are given in Table 41.3.

The kinematic viscosity, density, cloud point and flash point are within the standard. The density normally increases with the increase in the number of double bonds. The higher pour point is due to the poor unsaturation. It is reported that the cyclopropane ring present in this biodiesel limiting the stability. The level of phosphorous is either zero or tending towards zero. The absence of phosphorous is a good factor since phosphorous at any level can damage the catalytic converters employed in the popular emission control systems.

41.1.3 Engine Performance

Brake thermal performance is fairly lower in biodiesel than that of diesel. If the biodiesel is blended with diesel even at 5% (B5), the brake thermal efficiency shows a difference of 2.8% from that of diesel (Rao et al. 2015) in a four stroke single cylinder engine. The break specific fuel consumption also is relatively high. With respect to the mechanical efficiency, the values are high since there is low friction.

References

Bindhu CJ, Reddy JRC, Rao BVSK, Ravinder T, Chakrabarti PP, Karuna MSL, Prasad RBN (2012) Preparation and evaluation of biodiesel form *Sterculia foetida* seed oil. J Am Oil Chem Soc 89:891–896

Ong HC, Silitonga AS, Masjuki HH, Mahalia TMI, Chong WT, Boosroh MH (2013) Production and comparative fuel properties of biodiesel from non-edible oils: *Jatropha curcas*, *Sterculia foetida* and *Ceiba pentandra*. Energy Convers Manag 73:245–255

Rao VP, Ramesh S, Kumar AS (2015) Study of C.I.Engine performance with diesel-biodiesel (*Sterculia foetida*) blend as fuel. Int J Trend Res Dev 2(4):76–79

Sivakumar P, Parthiban KS, Sivakumar P, Vinoba M, Renganathan S (2012) Optimization of extraction process and kinetics of *Sterculia foetida* seed oil and its process augmentation for biodiesel production. Ind Eng Chem Res 51(26):8992–8998

Subramanian PK, Muthiah P (2016) Development of functionalized heterogeneous catalyst from *Ceiba pentandra* stalk for biodiesel production using *Sterculia foetida* seed oil. Environ Prog Sustain Energy 35(1):308–314

Theaceae

42

Family Theaceae is referred as tea family which comprises around 600 species of tree or shrubs or scrambling climbers. They are usually distributed in tropics and subtropics. Most of them are ornamental plants due to their elegant foliage and flowers. The fruit is usually a loculicidal capsule or berry or achene. *Camellia* sp. is dealt in this chapter due to its oil rich seeds used in biodiesel production.

42.1 Tea (*Camellia* Species)

The genus *Camellia* (Family: Theaceae) is a tea plant formed of many species which includes *C. reticulata*, *C. sinensis*, *C. oleifera* and *C. japonica*. This genus is named after Rev. George Kamel, a jesuit missionary. *Camellia* grows at an elevation of 1800–2700 m above MSL. The *Camellia* seeds contain oil ranging from 24% to 50% (Wang et al. 2011). A picture of the *C. reticulata* is shown in Fig. 42.1. *Camellia* sp. is a shrub of 10–15 m height. The leaves are elliptic to oblong (10 cm long and 5 cm wide). The flower is 10 cm dia. The specific details of *C. reticulata* fruits are given in Table 42.1. The occurrence of all the four species *C. reticulata*, *C. sinensis*, *C. oleifera* and *C. japonica* are presented in Fig. 42.2.

42.1.1 Oil and Biodiesel

As for as *C. reticulata* is concerned its oil content is inversely proportional to the terrain elevation. It is known (Fig. 42.3) that the oil content decreases at the rate of 1% for every 45 m ($y = 84.488 - 0.022x$, elevation in metre 6.849 as per Huang et al. 2013).

The seeds are dried and the kernel removed. The kernels are then powdered and the oil extracted out of it using *n*-hexane as solvent in a Soxhlet apparatus. Often the

S. P. Raj et al., *Biodiesel from Flowering Plants*,
https://doi.org/10.1007/978-981-16-4775-8_42

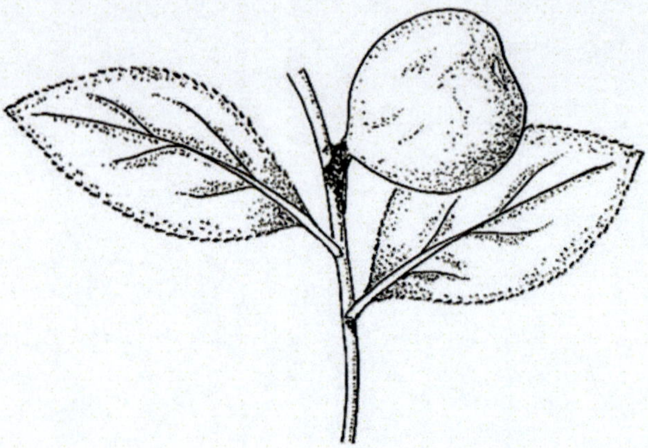

Fig. 42.1 The plant *Camellia reticulata*

Table 42.1 Details of the fruits and kernels of *C. reticulata* (Huang et al. 2013)

Parameters	Value
Fruit length (mm)	45.36 ± 0.2
Fruit diameter (mm)	56.34 ± 0.3
Fruit weight (g)	75.3 ± 0.9
Number of seeds per fruit	8.4
Oil (%)	34.13 ± 0.3

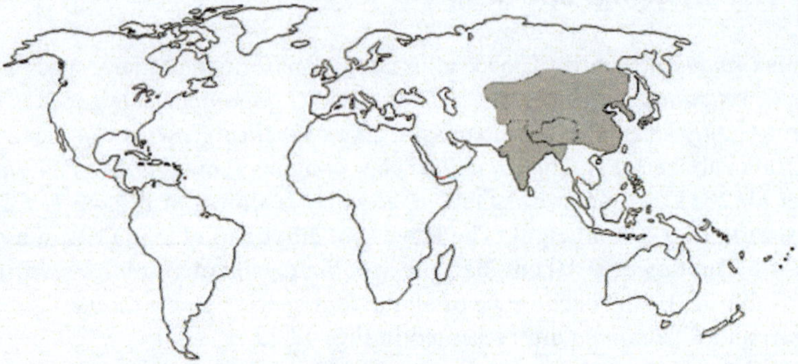

Fig. 42.2 Geographical distribution of *Camellia reticulata, C. sinensis, C. oleifera* and *C. japonica*

oil is extracted with acetone through supercritical means (Demirbas 2010). The properties of the oil from *C. sinensis* and *C. japonica* are given in Table 42.2.

The conversion of the oil into biodiesel is carried out through alkali catalysed transesterification. The catalyst employed is KOH at the rate of 1% of the oil used. The KOH is dissolved in methanol. The actual quantity of the methanol used in the process is of 6 molar quantity with 1 mol of oil. Once the alkali is dissolved in methanol (methoxide) it is slowly added to the oil at an ambient temperature of 65 °C and then the content is stirred at the rate of 600 rpm for 3 h. Consequently, the biodiesel separates and stay at the top when the entire content is allowed to stand

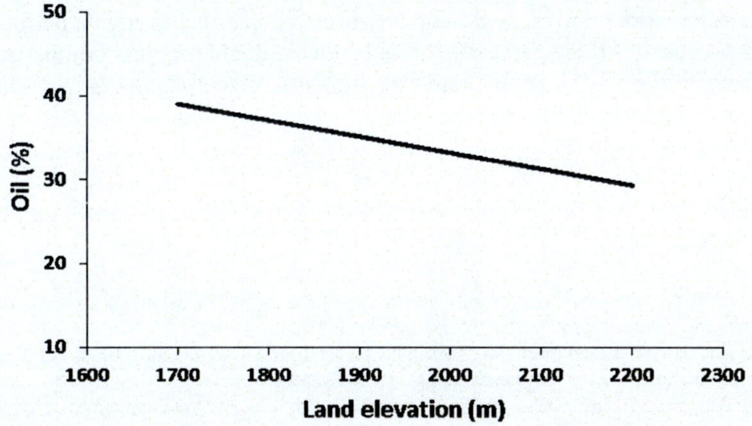

Fig. 42.3 The inverse relationship between the seed oil content and the elevation at which the plant *Camellia reticulate* grows

Table 42.2 Physicochemical properties of the oil from the seeds of *C. sinensis* and *C. japonica* (Chung 2010; Demirbas 2010)

Parameters	*C. sinensis*	*C. japonica*
Kinematic viscosity at 40 °C (cSt)	–	43.7
Density (kg m^{-3})	–	905
Saponification value (mg KOH g^{-1})	195	–
Acid value (mg KOH g^{-1})	0.21	0.98
Iodine value (g I_2/100 g)	107	–
Palmitic acid (C16:0) (%)	10.4	14.5
Stearic acid (C18:0) (%)	1.8	12.1
Oleic acid (C18:1) (%)	62.5	40.3
Linoleic acid (C18:2) (%)	18.1	33.1
Linolenic acid (C18:3) (%)	2.4	–
Gadoleic acid (C20:1) (%)	0.8	–

Table 42.3 Properties of *C. japonica* biodiesel (Chung 2010)

Parameters	Value
FAME content (%)	97.7
Acid value (mg KOH g^{-1})	0.16
Density (kg m^{-3})	877
Kinematic viscosity (cSt)	4.7
Flash point (°C)	193
Cetane number	54

undisturbed in a separating funnel. The biodiesel thus obtained is washed repeatedly in salt-free water so as to remove the residual alkali and then heated to expel the methanol and moisture. The properties of *C. japonica* oil biodiesel are given in Table 42.3.

Systematic work on the oil extraction, transesterification and engine performance for all the 4 species of the genus *Camellia* is not available for close comparison and for understanding the scope of biodiesel from the genus *Camellia*. Some of the studies made hitherto on the characteristics of the oil and biodiesel on chosen species indicate that the biodiesel if prepared on a large scale can support a mini industry.

References

Chung KH (2010) Transesterification of *Camellia japonica* and *Vernicia fordii* seed oils on alkali catalysts for biodiesel production. J Ind Eng Chem 16:506–509

Demirbas A (2010) Tea seed upgrading facilities and economic assessment of biodiesel production from tea seed oil. Energy Convers Manag 51:2595–2599

Huang J, Ahrends A, He J, Gui H, Xu J, Mortimer PE (2013) An evaluation of the factors influencing seed oil production in *Camellia reticulata* L. plants. Ind Crop Prod 50:797–802

Wang Y, Sun D, Chen H, Qian L, Xu P (2011) Fatty acid composition and antioxidant activity of tea (*Camellia sinensis* L.) seed oil extracted by optimized supercritical carbon dioxide. Int J Mol Sci 12:7708–7719

43 Zygophyllaceae

Zygophyllaceae is a family of around 300 flowering plants. They may be trees, shrubs or herbs. They are often found in dry habitats. Some of them are caltrops and live in halophytic salt deserts. The stems often have swollen nodes. Few plants are succulent and perennial. One of the members the desert date *Balanites aegyptiaca* and its role in biodiesel are dealt in this chapter.

43.1 Desert Date (*Balanites aegyptiaca*)

The desert date (*Balanites aegyptiaca*) is a candidate species hitherto not explored as raw material for preparing biodiesel. It is a multibranched, spiny, evergreen, Afro-Asiatic tree adopted to Sahelian climate characterized by low to very low rainfall (<40 mm for at least 3 summer months). It grows to a maximum height of 10 m. Its growth and existence are indifferent to the type of soil (sand to clay). It normally grows at an altitude of 100–2000 m above MSL and with a rainfall of 150–1300 mm a year. This species comprises three varieties, namely *Balanites aegyptiaca* var. *aegyptiaca*, *B. aegyptiaca* var. *tomentosa* and *B. aegyptiaca* var. *quarrei* in which *B. aegyptiaca* var. *aegyptiaca* is popular and its distribution is wide. It belongs to the family Zygophyllaceae. The leaves are leathery around 6 cm long and 4 cm broad (Fig. 43.1). The tree comes to yield in 6 years of planting and the yield sustains up to 25 years. The flower is greenish yellow, 10 mm dia. and blossoms during the dry season. The fruit appears in 45 days of flowering and takes a year to mature. The fruit is ellipsoid, brown when ripe, with a brittle outer coat enclosing a sticky scented pulp. At the centre of the drupe, there is a single hard stony seed. The fruit (drupe) is around 5 cm long and 2.5 cm dia. The seed is 2 cm long and weighs around 7 g at wet condition. Each tree yields around 10,000 fruits which contain 75 kg seeds. This species is distributed in arid and subhumid countries: Algeria, Angola, Benin, Burundi, Cameroon, Chad, Cote d'Ivoire, Congo, Djibouti, Egypt, Eritrea, Ethiopia, Gambia, Ghana, Guinea, India, Israel, Kenya, Morocco, Myanmar, Nigeria, Saudi

S. P. Raj et al., *Biodiesel from Flowering Plants*,
https://doi.org/10.1007/978-981-16-4775-8_43

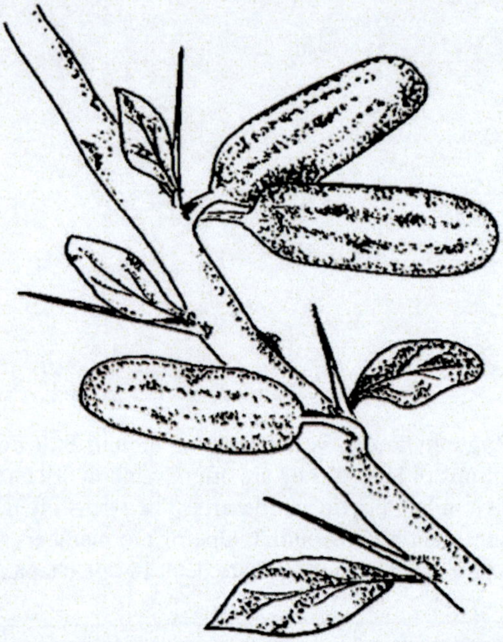

Fig. 43.1 *Balanites aegyptiaca* plant

Fig. 43.2 Geographical distribution of *Balanites aegyptiaca*

Arabia, Senegal, Somalia, Sudan, Tanzania, Uganda, Zambia and Zimbabwe (Chapagain et al. 2009) (Fig. 43.2).

B. aegyptiaca is called by popular names in different languages (Table 43.1).

43.1.1 Oil Production

The ripened fruits cannot be harvested through manual means due to the sharp spines present all through the branches. The arboreal animals chew the fruits and discard the

Table 43.1 Popular names of *B. aegyptiaca* in different languages

Languages	Names
English	Torch wood, Jericho balsam desert date, soap berry
French	Myrobalan d' Egypte
Spanish	Corona di Jesus
Arabic	Heglig
Amharic	Jemo

Table 43.2 The physico-chemical characteristics of the *B. aegyptiaca* oil (Chapagain et al. 2009; Deshmukh and Bhuyar 2009; Dass et al. 2018)

Parameters	Value
Viscosity at 40 °C (cSt)	38.6
Density (kg m^{-3})	886, 927
Acid value (g KOH kg^{-1})	1.96
Flash point (°C)	230
Pour point (°C)	−3.5
Calorific value (MJ kg^{-1})	39.8
Iodine value (g I_2/100 g)	78.7
Palmitic acid (C16:0)	17.0, 16.7
Stearic acid (C18:0)	7.8, 11.7
Palmitoleic acid (C16:1)	4.3
Oleic acid (C18:1)	32.4, 22.8
Linoleic acid (C18:2)	31.3, 47.9
Linolenic acid (C18:3)	7.2

stony seeds which can be collected from the ground in large numbers. Alternatively, the entire tree is shaken manually so as to make the ripened fruits to fall down on the ground. The fruit being a drupe contains rich quantities of pulp. Therefore at the outset, the fruits are soaked in warm water overnight and then squeezed well so as to get rid of the pulp. The seeds are then washed and dried. The seeds thus obtained along with the batch of seeds collected from the regurgitations of tree-borne animals are metal hammered to break the hard shell. It is then crushed in an electrical crusher and the oil is extracted using *n*-hexane at 65 °C (Dass et al. 2018). The solvent is then distilled off and the oil recovered. The physicochemical characteristics of the oil thus obtained are given in Table 43.2.

The oil is greaseless, non-sticky and non-drying (Dass et al. 2018). Since the unsaturated fatty acid fraction of the oil is high the fuel properties are satisfactory.

43.1.2 Biodiesel Production

Two types of transesterification procedures are being followed (1) in the first method the oil is transesterified using an alkali catalyst. Potassium hydroxide at the rate of 1.7 g/100 g of oil is taken and dissolved in methanol. The methanol quantity is decided to form an oil to methanol molar ratio of 1:6. The potassium hydroxide as it dissolves in methanol forms potassium methoxide. It is then added to the hot oil

Table 43.3 Properties of the *B. aegyptiaca* oil (Chapagain et al. 2009; Deshmukh and Bhuyar 2009)

Parameters	Value
Kinematic viscosity at 40 °C (cSt)	3.7, 3.98
Density (kg m^{-3})	870, 860
Flash point (°C)	126, 75
Cloud point (°C)	3–7
Cold filter plugging point (°C)	1–3
Copper strip corrosion	1
Calorific value (MJ kg^{-1})	39.65
Iodine value (g I_2/100 g)	100
Cetane number	53.6
Palmitic acid (C16:0) (%)	15.0
Stearic acid (C18:0) (%)	13.3
Oleic acid (C18:1) (%)	26.3
Linoleic acid (C18:2) (%)	44.4

(60 °C) and stirred for 1 h at an rpm of 600. After 1 h the heating and stirring stopped and the contents are allowed to settle overnight. The upper phase containing the biodiesel is separated and washed with dilute phosphoric acid (pH 4.0) so as to have the pH neutralized. The methanol content present in it is expelled by heating. (2) Second method is referred to as in-situ transesterification. The dried seeds are powdered, added with ethanol at two times the quantity of the seed powder and is homogenized for one hour at 1100 rpm. To this potassium hydroxide is added at the rate of 2 g/100 mL of the homogenate. The whole content is loaded in a flask and put on a rotary shaker for 6 h at 60 °C. On completion of the above, it is added with petroleum ether (bp 60–80 °C) at 1:1 ratio. The etheric layer is separated and treated with dilute phosphoric acid-sodium chloride solution so as to neutralize it. The resultant ether layer is dried over sodium sulphate and filtered. The solvent is removed by evaporating the ether so as to leave behind the ester. The properties of the *B. aegyptiaca* biodiesel are given in Table 43.3. The analytical data given in Table 43.3 indicate that the major parameters are in agreement with the standard.

43.1.3 Engine Performance

The brake specific fuel consumption of biodiesel is more than that of the diesel as biodiesel with low calorific value has to be pumped in large quantity to discharge constant energy required to meet the load. The brake thermal efficiency of biodiesel is on par with that of diesel at low load. But at high load, the brake thermal efficiency lowers due to the low (around 7%) calorific value. The exhaust temperature with biodiesel is observed to be higher than that of the diesel since dense biodiesel in high quantity is burned in the cylinder.

The smoke density is relatively low in biodiesel on the fact that biodiesel ensures efficient burning due to the presence of oxygen in the biodiesel molecule. The CO and HC emissions are low due to the reason cited above. Unlike other fuels the NO_x

level also is low. It is reasoned that the available oxygen in the ester is first utilized to convert the carbon to carbon dioxide leaving less oxygen for the conversion of nitrogen to its oxides (Deshmukh and Bhuyar 2009).

References

Chapagain BP, Yehoshua Y, Wiesman Z (2009) Desert date (*Balanites aegyptiaca*) as an arid lands sustainable bioresource for biodiesel. Bioresour Technol 100:1221–1226

Dass PM, Louis H, Alheri A, Amos PI, Bifam M, Ago MA (2018) Production of biodiesel oil from desert dates *Balanites aegyptiaca* seeds oil using a heterogeneous catalyst produced from Mahogany (*Khaya senegalensis*) fruit shells. Anal Chem Ind J 18(1):129

Deshmukh SJ, Bhuyar LB (2009) Transesterified hingan oil (*Balanites*) oil as a fuel for compression ignition engines. Biomass Bioenergy 33:108–112

Catalysts for Transesterification 44

Catalyst is a substance in small amount which involves greatly in modifying the speed of a reaction, without actually being consumed in the process. It provides an active mechanism to break the bond between elements (or other electronic charges in the reactants). Catalyst also involves in the formation of intermediates affecting the rate of reaction, i.e. kinetics while maintaining the thermodynamics. Catalysts offer new pathways and reduce the energy necessary to proceed along the reaction (Fig. 44.1).

44.1 Types of Catalytic Reactions

Catalysts may be divided in to two main types. They are heterogeneous and homogeneous. A heterogeneous catalyst is in a phase (often solid state) other than that of the reactants (often gas or liquid or solution) (Fig. 44.2). The reaction takes place on the surface of the catalyst which is normally a transition metal or one of its compounds. The reactants are adsorbed at the active sites located at the surface of the catalyst and are physically adsorbed or weakly bonded. However, such active sites are poorly defined. On either way, it has the effect of concentrating the reactants close to each other and weakening the original intra-molecular bonds within the reactant molecules and thus allowing a greater chance of collision. Heterogeneous catalysts are easily removed from the products by physical separation or easily recycled. Such catalysts are effective only on the surface and the reaction mechanisms are poorly understood.

Homogeneous catalyst is in the same phase as that of the reactants (Fig. 44.3). The reaction mechanisms of such catalysts are well understood as it has well defined process of diffusion and heat transfer. Such catalysts have well defined active sites. These catalysts cannot be removed easily from the products as they cannot be differentiated physically from the reactants.

S. P. Raj et al., *Biodiesel from Flowering Plants*,
https://doi.org/10.1007/978-981-16-4775-8_44

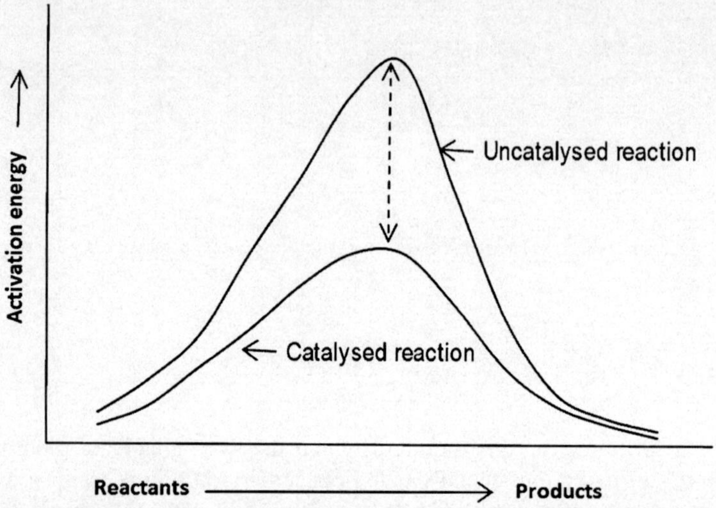

Fig. 44.1 The activation energy involved is higher in an uncatalysed reaction. The two-way arrow indicates the difference in the activation energy between catalysed and uncatalysed path way of common reactants and products

Ordinary strong acids and bases are used as homogeneous catalysts. Heterogeneous types of catalysts are formed of solid acids and solid bases. Various types of catalysts with examples are given below.

44.1.1 Homogeneous Catalysts

When the catalyst is in the physical state of the reactants it is known as homogeneous catalyst. The alkali catalysts sodium hydroxide and calcium hydroxide and the acid catalysts sulphuric acid, sulphonic acid and hydrochloric acids are examples of homogeneous catalysts for biodiesel production. The alkali catalysts sodium hydroxide and potassium hydroxide though in solid form go in solution with alcohol so as to form alkoxide which is in liquid form. The alkoxide is referred by the metal associated with it (e.g. sodium methoxide).

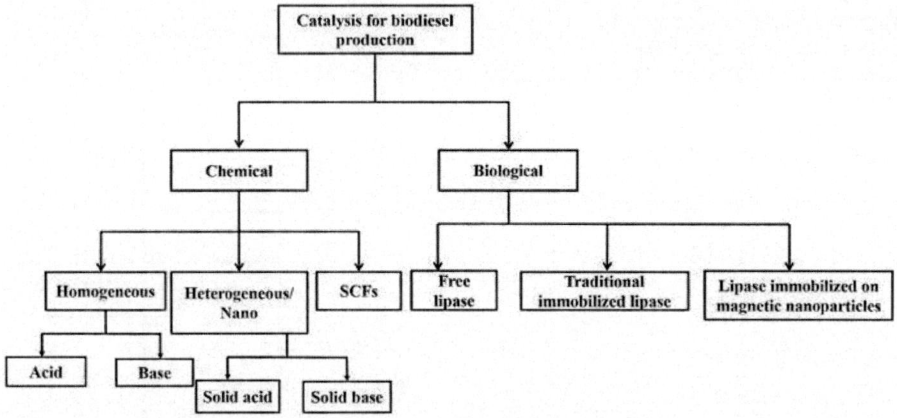

Classification of the catalysts which are being used in transesterification

$$\underset{\text{(solid)}}{NaOH} + \underset{\text{(liquid)}}{CH_3OH} \rightarrow H_2O + \underset{\text{(liquid)}}{CH_3O^-Na^+}$$

The alkoxide liquid reacts with oil during transesterification. This process is relatively less expensive. The water generated during the above reaction is not sufficient to cause saponification. Contrary to the above, water if any present in the oil would promote saponification. Therefore, the oil is required to be without moisture and hence the oil is preheated to expel water if any. Alcohol used in the process also is to be free from water. Similarly, if the oils contain free fatty acids in considerable quantities saponification may take place resulting into the loss of catalysts. Saponification reaction is irreversible besides it inhibits transesterification. The most desirable level of free fatty acid in oil as raw material for biodiesel is less than 0.5%. If the level ranges from 0.5% to 2.5%, the transesterification is manageable. In case the free fatty acid level is more than 2.5%, additional treatment course is recommended to minimize the level of free fatty acid before transesterification process is commenced.

When the metallic alkoxide is added to the oil (triglyceride) the alkyl group is replaced in a series of reactions. The carbon atom present in the ester unit of the triglyceride is feebly positive in its charge. The oxygen present in the alkyl group is negative in its charge. Thus there is attraction between them and the ^-OR is virtually brought to the reaction site. Consequently, there develops a transition junction which has a pair of electrons due to the C=O bond newly formed. These electrons then fall back to the carbon and push off the glycerol.

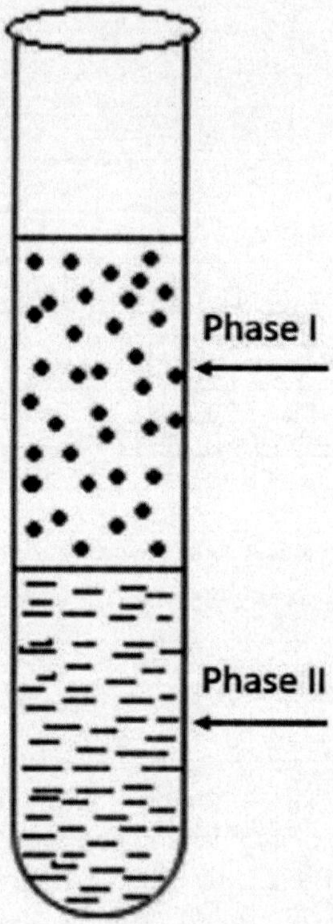

Fig. 44.2 Illustration of a process medium in which the catalyst and reactants exist in two different phases

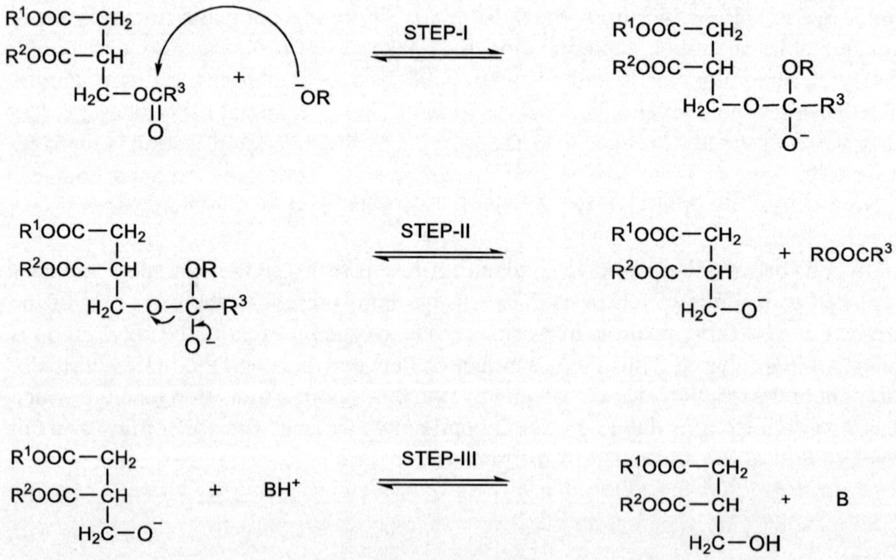

Reaction mechanism of transesterification of triglycerides with base catalysts

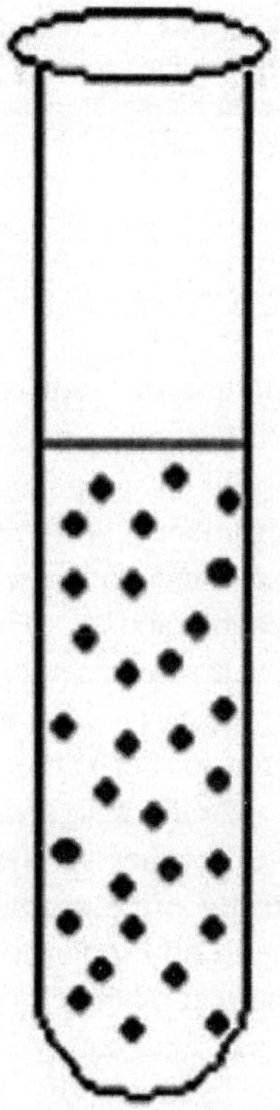

Fig. 44.3 Illustration of a process medium in which the catalyst exists along with reactants as a single phase

Similarly, two more $^{-}$OR groups react in the similar pattern and release the glycerol in total. However, the above reaction is very specific since technically the $^{-}$OR has to fit in a narrow socket whose charge is very feeble. As the R of the $^{-}$OR gets bigger the rate of reaction retards and this phenomenon is called steric hindrance. The reaction mechanism is given below. In the first step a catalytically active alkoxide ion ($^{-}$OR) is formed and this ion attacks the carbonyl carbon of the triglyceride so as to form a tetrahedral intermediate. In the second step a fatty acid ester and a degradation product, namely diglyceride are formed due to the decomposition of the tetrahedral intermediate. In the third step proton transfer from alcohol to the diglyceride takes place so as to regenerate the nucleophile $^{-}$OR. The above sequence of reaction is repeated twice to yield monoglyceride and the glycerol separates along with the formation of ester.

The alkali catalyst is normally preferred by many biodiesel industries because of its low cost and efficiency in short reaction time at low temperature. However, the process is considered energy intensive. The removal of residual glycerol is often cumbersome due to the incipient formation of soap when water is used to wash away the alkali. There will be environmental problem in disposing the wash water having high pH. Sodium hydroxide has a low molecular weight compared to that of

Table 44.1 Comparison of alkaline catalysts required for efficient transesterification of oil

Catalyst	%	Mol. wt.	Moles base for kg of oil
$KOCH_3$	0.25	70	0.003
$NaOCH_3$	0.25	54	0.004
KOH	1.00	56	0.016
NaOH	1.00	40	0.023

Adopted from Reaney et al. (2005)

potassium hydroxide. Thus it has more molecules per unit weight. Though potassium hydroxide has a larger molecular weight, its solubility is relatively more. On mole basis potassium hydroxide is relatively efficient but on weight basis the efficiency is on par. On account of the inactivation of the catalyst by free fatty acid and water the effect is equal when one mole of fatty acid reacts with one mole of each catalyst. Thus sodium hydroxide can face more free fatty acid as more molecules of sodium hydroxide is accommodated in unit weight. Therefore, when more fatty acid is available it is better to use sodium hydroxide than potassium hydroxide. Table 44.1 indicates that the moles base for kg of oil is higher in NaOH.

Acid catalysed transesterification has a poor turnover. It proceeds approximately 4000 times slower than that of the alkali catalyst. Substantial amount of unreacted triglyceride would remain in the finished product even after prolonged treatment with high quantity of alcohol especially when the oil has more than 20% free fatty acid and hence not recommended in the usual operation. However, as a preliminary step, acid catalyst is employed to reduce the content of free fatty acid as acid is not affected by the free fatty acid and water content. Acids actually suppress the saponification. The acids catalyse the esterification of the free fatty acid to produce fatty acid alkyl ester. Thus the free fatty acid instead of transforming in to soap gets converted to useful product, namely ester or biodiesel. Thus, if the oil or fat as raw material has rich quantity of free fatty acid, employing acid as catalyst is considered to be appropriate. When the raw material has high quantity of free fatty acid two stage processes is recommended. Accordingly, acid catalysed treatment is made in the first stage so as to reduce the free fatty acid. Once the free fatty acid is reduced alkali catalysed treatment is taken up on the resultant material in the second stage.

The first stage process, namely the acid catalysed reaction has two steps. In the first step there is protonation of the carbonyl group which results in the formation of carbon cation. The second step involves the nucleophilic attack of alcohol producing a tetrahedral intermediate which rearranges to cause the release of alkyl ester and proton catalyst. This reaction proceeds in high temperature and excess methanol.

The acid catalysts are thus richly capable of converting the free fatty acid to ester but are not effective in the transesterification of triglyceride. Acid catalysed esterification is performed using sulphuric acid at a concentration ranging from 1% to 5% at an oil methanol ratio of around 1:5 taking a reaction time of around 2 h. Normally more than 98% of the free fatty acid is converted to ester. The resultant material is washed profusely with warm water to remove the acid so as to take it to the next stage, i.e. the alkali treatment.

In general catalysts are long lasting, hasten the speed of the reaction and lower the energy inputs. But in the case of alkali, it physically mixes in the medium and therefore cannot be recovered. Hence it is unavailable for further reaction. Similarly, acid homogeneous catalysts also react with oil having high acid number which tend to dissolve in the glycerol phase and thus may not be available for further use.

In an alternate procedure the free fatty acid present in the oil is washed away using methanol since the free fatty acid is selectively miscible in methanol. Consequently, the oil free from free fatty acid is used in the transesterification process. The resultant methanol carrying the free fatty acid is treated with acid catalyst so as to get ester.

In order to enhance the function of homogeneous catalyst an ideal alcohol source is necessary. Some of the alcohols tried are methanol, ethanol, isopropanol, *n*-propanol, butanol and pentanol. The actual quantity of alcohol required in the reaction depends on the molecular weight of the alcohol and the oil. When the molecular weight of the oil is low the methanol required also is low.

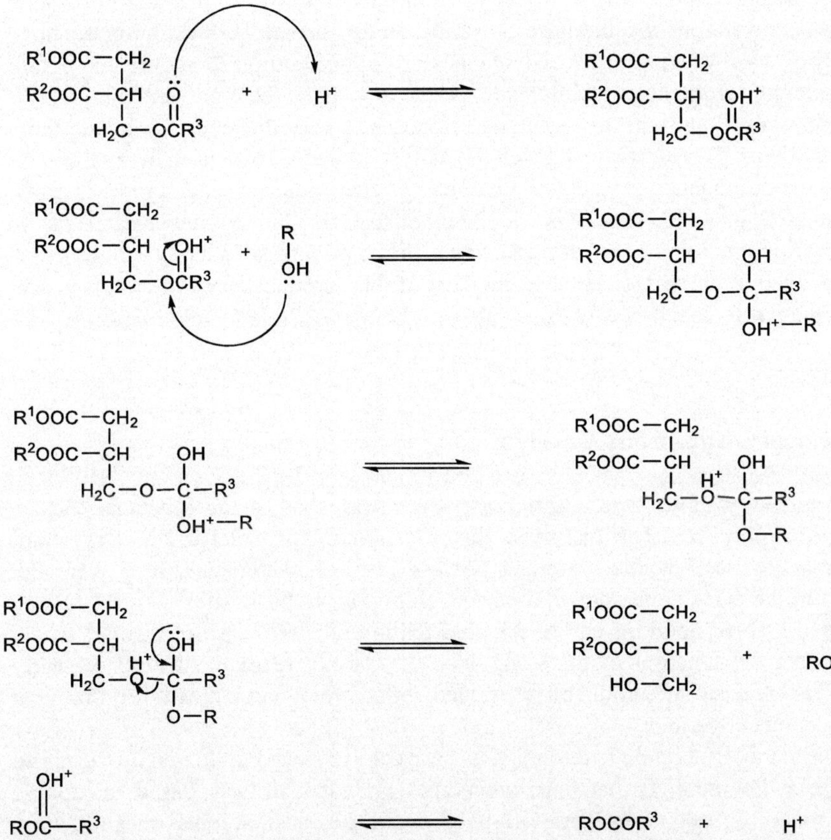

Reaction mechanism of transesterification of triglycerides with acid catalysts

Table 44.2 Methanol requirement of different oils (Reaney et al. 2005)

Oil type	Oil moles kg^{-1}	Methanol kg kg^{-1} oil
Palm oil	1.17	0.112
Canola oil	1.13	0.108
Rapeseed oil	1.04	0.100

The changes in the chain length of the fatty acid greatly influence the ester recovery but the saturation level of the oil has no effect on ester production. If the oil has more palmitic acid, more alcohol is required. If the oil has more oleic acid, less alcohol is required. If the oil contains more erucic acid, the alcohol requirement is still lower than that of the oleic acid. The methanol requirement for palm oil, canola oil and rapeseed oil is given in Table 44.2.

The homogeneous catalysts though found to be simple and effective, they carry with themselves series of problems of product purification and catalyst recovery. Even a trace of acid or alkali if present in the biodiesel would cause engine problems. The alkali may induce high levels of incombustible ash. The acid may also corrode the metallic parts of the engine. Therefore, complete removal of the alkali or acid is to be ensured before the biodiesel is made ready for use. While doing so huge quantity of wash water is generated which needs to be neutralized and treated before it is discharged from the production site. As already stated the homogeneous catalyst fully identifies along with the reactants. Hence it is very difficult to separate them and reuse them. Therefore, fresh batch of catalyst is needed to be used every time. As most of the homogeneous catalysts are corrosive the reactor is needed to be made of corrosion free material. Thus, it is widely understood that homogeneous catalysts are not eco-friendly. Even if all the associated problems are adequately addressed to make the whole process acceptable, the cost of final product may not be lower than that of the petro-diesel.

44.1.2 Heterogeneous Catalyst

Heterogeneous catalyst has a different phase from that of the reactants (*hetero implies a difference*). The history of heterogeneous base catalysis is very short. Although few solid acidic materials possess ideal characteristics, they were not utilized in the early developmental period. Similarly, in basic sites CO_2, H_2O, and O_2 and the corresponding carbonate, hydroxide and peroxide are formed which incapacitate the function of the basic sites. It was not recognized until the early 1970s that the basic sites could be resurfaced with thermal pretreatment for removing the poisonous coverings.

In 1955, initial study on heterogeneous base catalysis was conducted by Pines and Eschinazi who reported that heterogeneous base catalysis was based on double bonds migration and dehydrogenation of cyclic alkenes. Studies on solid base catalysts rapidly progressed in 1970s and most dealt with basic single metal oxides such as alkaline oxides and rare earth metal oxides. The studies were then expanded to alkali metal exchanged zeolites, alkali metal ion-supported catalysts and clay minerals such as hydrotalcites.

In general, heterogeneous catalysts are not dispensed either in methanol or in any reaction mixture of oil. It generates a separate layer in the reaction mixture. Heterogeneous catalysts are widely categorized as solid acid and solid base. Solid base catalysts include a wide group of compounds in the category of alkaline earth metal hydroxides, hydrotalcites layered double hydroxides, alumina loaded with various compounds, zeolites, and various other compounds. Solid base catalysts have been quite successful with high conversion and yield of biodiesel. Solid catalysts have high mass transfer limitations as most of the solid catalysts are prepared on porous supports, and the reaction is mostly three phase (solid–liquid–liquid, i.e. solid catalyst, polar methanol and non-polar triglyceride).

The metal oxides are made up of (positive) metal ions (cations) which possess Lewis acidity, i.e. they behave as electron acceptors and (the negative) oxygen ions (anions) which behave as proton acceptors and are thus having Bronsted bases. Catalysts with proper configuration and ideal operating conditions are essential in minimizing the mass transfer limitation and for achieving efficient production of biodiesel. Most of the solid catalysts are prepared by impregnation of active compounds into the surface of porous support materials. Therefore, the catalytic activity of the coated compounds and the density of the active compounds coated are the most important factors determining the activity of solid catalysts. To maximize the catalytic activity of solid catalysts, it is necessary to identify highly active compounds for coating.

One major disadvantage of the solid catalyst is the formation of three phases together with oil and alcohol, which leads to diffusion limitations thereby decreasing the rate of the reaction (Mbaraka and Shanks 2006). Such mass transfer difficulty is overcome by using co-solvent materials such as tetrahydrofuran (THF), dimethyl sulphoxide (DMSO), *n*-hexane and ethanol, which assist in the miscibility of oil and methanol leading to the increase in the rate of reaction. Catalyst supports provide more specific surface area and pore space for anchoring the enzyme thereby to make it react actively with large triglyceride molecules. This would easily encounter the poor mass transfer (Zabeti et al. 2009).

44.2 Preparation of Catalysts

Several methods are available for the preparation of solid catalysts being used in the transesterification process. These methods include:

- Thermal treatment
- Hydrothermal synthesis
- Physical mixing
- Impregnation
- Precipitation and co-precipitation

44.2.1 Thermal Treatment (Calcination)

The surfaces of metal oxides are covered with carbon dioxide, water and in some cases with oxygen as they are exposed in air. Often heat treatment is given to such surface sites. The improvement of the basic sites due to heat treatment varies according to the pretreatment conditions. Besides, removal of water and carbon dioxide, rearrangement of the bulk atoms present on the surface occurs during such treatment. High temperature alters the number and nature of the basic sites. However, treatment temperature varies in accordance with the type of reaction involved.

44.2.2 Hydrothermal Synthesis

In hydrothermal synthesis high temperature aqueous solution with high vapour pressure is involved which is known to alter the surface characteristics of the solid support.

44.2.3 Physical Mixing

Physical mixing improves the efficiency of the transesterification process (Wang et al. 2006b). Physical mixing often results in low crystallinity of the catalyst which causes negative effect on the efficiency. Furthermore, a high loading catalyst often results in a low specific surface area, and thus offering low surface concentration of catalytic sites for transesterification.

44.2.4 Impregnation

Impregnation is one of the procedures of catalyst preparation which is achieved by filling the pores of the support with a solution of the metal salt from which the solvent is subsequently evaporated. The catalyst is prepared either by spraying the support with a solution of the metal compound or by adding the support material to a solution of a suitable metal salt. In such processes the required weight of the active component is incorporated in to the support without the involvement of excess solution. It is then dried at high temperature, either by thermal decomposition or through reduction. In such preparation it is essential to have a clear understanding of both chemical and physical properties of the support and impregnating solution in order to control the physical properties of the finished catalyst. Often more than one metal is used. In such events, care has to be taken to confirm that a component in an impregnating solution of metal salts is not selectively adsorbed resulting in an unexpectedly different and undesirable combination of metals in a mixed metal catalyst.

44.2.5 Precipitation and Co-precipitation

The preparation of supported catalysts by the co-precipitation of metal ions along with the support ions usually brings in an intimate mixing of catalysts and support. A typical example is the co-precipitation of metal ions with aluminium ions to produce an alumina gel containing the metal hydroxide. This product when calcined produces a refractory support with active component dispersed throughout the bulk as well as at the surface. However, in the preparation of multi-component catalysts, heterogeneous product may be formed because of the different solubility products of the constituents.

44.2.6 Alumina

Aluminium oxide which exists as γ-alumina or η-alumina or α-alumina has been extensively used as a support material in catalysis processes due to its extremely favourable thermal and mechanical stability, high specific surface area, large pore size and pore volume.

Kaita et al. (2002) patented aluminium phosphate solid acid catalyst for transesterification of palm kernel oil with methanol. The catalytic efficiency of potassium supported by alumina as a solid base catalyst was evaluated for the production of biodiesel from soybean oil (Xie et al. 2006). The catalyst was prepared by impregnating aqueous solution of potassium nitrate on alumina. $Al_2O_3/ZrO_2/WO_3$ solid acid catalyst was prepared by co-precipitation method and applied in methanolysis of soybean oil (Furuta et al. 2006). The catalyst was compatible for both esterification and transesterification reaction. Titanium oxide–zinc oxide supported by alumina (ATZ) and titanium oxide–bismuth oxide supported by alumina (ATB) were employed in the transesterification of colza oil with methanol. These catalysts were prepared by co-mixing followed by calcination at 600 °C for 3 h. The specific surface areas of ATZ and ATB were 62 and 54 $m^2\ g^{-1}$, respectively.

Alumina supported coconut chaff (ASCC) is prepared by impregnation method and is used as a catalyst for biodiesel production from waste frying oil. It is incorporated with SiO_2 or K_2O or Al_2O_3 or CuO. The catalyst thus prepared has achieved a biodiesel yield of 91.05 wt% in optimum condition. Besides, the catalyst is able to maintain its activity up to 5 cycles (Yusuff and Owolabi 2019). Alumina supported zinc oxide catalyst is synthesized by the precipitation and gel method. It has exhibited a high surface area (71.56 $m^2\ g^{-1}$), total pore volume (0.137 $cm^3\ g^{-1}$) and mean pore diameter (8.149 nm). Using this catalyst high biodiesel yield is achieved (92.29%) from *Schleichera oleosa* oil by 4% by weight of catalyst, 12:1 methanol to oil molar ratio in 6 h at 65 °C (Asri et al. 2018).

Mesoporous γ-alumina is synthesized from aluminium waste by sol-gel method using potassium hydroxide and sulphuric acid. The above mixture is stirred at room temperature until there is gel formation. Then, it is subjected to ageing for 6 and 24 h which are denoted as AL-6-30 and AL-24-30. Above alumina gels are recovered by centrifugation and dried at 80 °C in oven. The dried alumina particles are calcined in a furnace at 500 °C. It is then impregnated with KNO_3 by stirring (250 rpm) and heating at 85 °C for 4 h. The product obtained is further dried at 120 °C for 16 h and calcined at 500 °C for 5 h. These two alumina catalysts are used for biodiesel production on waste cooking oil. The synthesized biodiesel by KNO_3/AL-24-30 and KNO_3/Al-commercial has been analysed by GC-MS to identify methyl esters according to their retention time (Fig. 44.4). Biodiesel produced by both the catalysts exhibits identical mass spectra (Table 44.3). The biodiesel peaks are indicated by circles (Mokaizh et al. 2019).

Different heterogeneous catalysts used in the transesterification of vegetable oils are given in Table 44.4.

The catalytic activity of Al_2O_3 loaded with different potassium compounds on the transesterification as indicated by the conversion to methyl ester is in the order $KI/Al_2O_3 > KF/Al_2O_3 > KOH/Al_2O_3 > KNO_3/Al_2O_3 > K_2CO_3/Al_2O_3 > KBr/Al_2O_3$ (Table 44.5). Potassium iodide supported by alumina as a solid base catalyst was obtained by loading 35% by weight of KI on the support and calcined at 500 °C for 3 h and the basicity recorded was 1.57 mmol g^{-1}. These results were directly correlated with the conversion of oil. The reaction was performed in soybean oil to methanol molar ratio of 1:15, with 2% by weight of catalyst in 6 h of reaction duration.

Kim et al. (2004) utilized alumina loaded with sodium and sodium hydroxide as a solid base catalyst in the transesterification of soybean oil along with hexane and methanol. The specific surface area of γ-alumina decreased from 143.1 to 83.2 $m^2\ g^{-1}$ when Na and NaOH were loaded onto the support. However, the biodiesel yield reached 83% using Al_2O_3/Na/NaOH while it was only 1% when γ-alumina alone was used without any modification. The activity was closely correlated to the basic active sites which was determined by temperature programmed desorption of carbon dioxide (TPD-CO_2).

Bournay et al. (2005) developed a new process for obtaining continuous yield of biodiesel in which a solid catalyst containing zinc oxide and alumina was used. The results showed that the process does not require any post-treatment to remove the catalyst from biodiesel and the yield of methyl ester at high pressure and temperature was close to the theoretical value. Moreover, glycerol obtained through this process had a purity of approximately 98%.

Transesterification of canola oil with methanol was studied in a heterogeneous system, using γ-Al_2O_3 supported with various alkaline catalysts. γ-Al_2O_3 catalyst loaded with KOH having the highest basicity showed the best catalytic activity in the transesterification of canola oil. When the reaction was carried out at 60 °C with 12:1 molar ratio of methanol to canola oil with 3% catalyst gave the highest FAME yield (89.40%) after 9 h of reaction duration (Ilgen and Akin 2009). Catalytic activity of γ-Al_2O_3 support loaded with different alkali metal compounds is given in Table 44.6.

Liu et al. (2010) employed calcined catalyst K_2CO_3/γ-Al_2O_3 for the transesterification of rapeseed oil with methanol. Alumina was first pretreated at 823 K for 4 h for the removal of unwanted chemical species present on the surface.

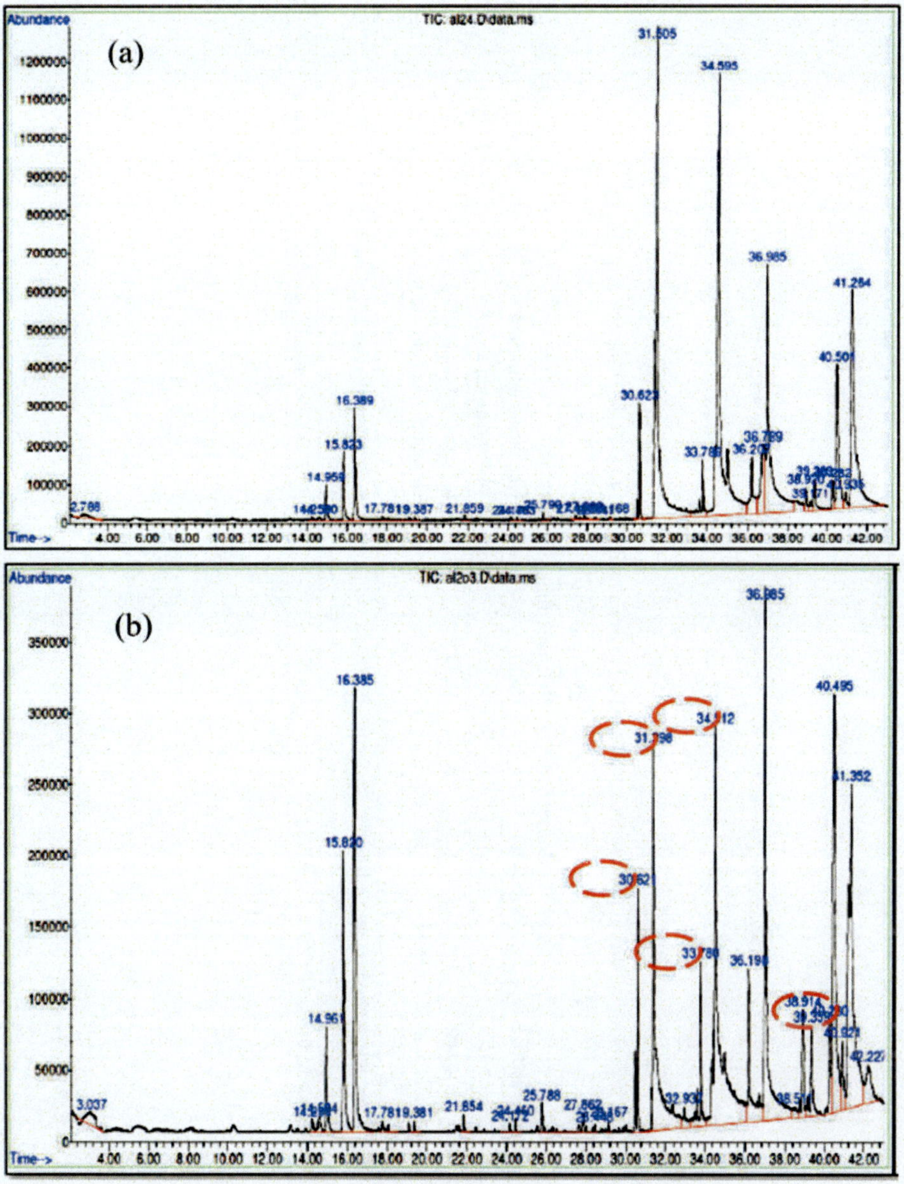

Fig. 44.4 GC-MS analysis on biodiesel by using catalyst: (**a**) KNO_3/AL-24-30. (**b**) KNO_3/AL-commercial. (Source: https://doi.org/10.1088/1757-899X/702/1/012038)

Potassium carbonate was then loaded onto the alumina surface by an impregnation method. The prepared catalysts were characterized by SEM, IR and BET and their catalytic activity was assessed based on the conversion of rapeseed oil to methyl ester. SEM analysis of the particles of the catalyst showed irregular and polydispersed size ranging from 0.2 to 10 μm. If it is calcined at high temperature

Table 44.3 The conversion of waste cooking oil due to two catalysts KNO_3/Al-commercial and KNO_3/AL-24-30 (Mokaizh et al. 2019)

KNO_3/Al-commercial			KNO_3/AL-24-30		
Retention time (min)	Chemical compound	Conversion (%)	Retention time (min)	Chemical compound	Conversion (%)
37.88	9-Octadecenoic acid	24.93	34.51	14-Pentadecenoic acid	17.96
31.50	*n*-Hexadecanoic acid	22.67	31.39	*n*-Hexadecanoic acid	11.35
33.78	9-Octadecenoic acid (*Z*) methyl ester	1.96	30.62	Hexadecanoic acid methyl ester	2.78
30.62	Hexadecanoic acid methyl ester	1.78	33.78	9-Octadecenoic acid (*E*), methyl ester	2.58
40.28	9,12-Octadecadienoic acid (Z,Z)	0.73	40.28	9,12-Octadecadienoic acid (Z,Z)	1.73

Table 44.4 Different heterogeneous catalysts used in the transesterification of vegetable oils

Catalyst	Oil	Yield (%)	Reference
Calcined Mg–Al hydrotalcites	Soybean oil	67	Xie et al. (2005)
Na/NaOH/-Al_2O_3	Vegetable oil		Kim et al. (2004)
K/KOH/c-Al_2O_3	Rapeseed oil	85	Ma et al. (2008)
Zn/I_2	Soybean oil	96	Li et al. (2006)
Ba–ZnO	Soybean oil	95	Xie and Yang (2007)
KF/ZnO	Soybean oil	87	Xie and Huang (2006)
KF/Al_2O_3	Vegetable oil		Boz et al. (2009)
Mg/Zr	Vegetable oil		Sree et al. (2009)
KNO_3/Al_2O_3	Jatropha oil	84	Vyas et al. (2009)
Ca and Zn mixed oxide	Palm kernel oil		Ngamcharussrivichai et al. (2008)
KI/Al_2O_3	Soybean oil	96	Xie and Li (2006)
K_2CO_3/MgO	Soybean oil		Liang et al. (2009a)
KF/MgO	Vegetable oil		Liang et al. (2009b)
Mesoporous silica loaded with MgO	Vegetable oil	96	Li and Rudolph (2008)
WO_3/ZrO_2, zirconia–alumina and sulphated tin oxide	Soybean oil	90	Furuta et al. (2006)
Calcined LDH (Li–Al)	Soybean oil	72	Shumaker et al. (2008)
Mg–Al–CO_3 (hydrotalcite)	Palm oil	87	Trakarnpruk and Porntangjitlikit (2008)
La/zeolite beta	Soybean oil	49	Shu et al. (2007)
$MgAl_2O_4$	Soybean oil	57	Wang et al. (2008)
NaOH/alumina	Sunflower oil	99	Arzamendi et al. (2007)
MgO, ZnO, Al_2O_3	Soybean oil	82	Antunes et al. (2008)
CaO–ZrO_2	Soybean oil	70	Wang et al. (2006b)
CaO/SBA-15	Sunflower oil	95	Albuquerque et al. (2008)
CaO	Jatropha oil	93	Huaping et al. (2006)
WO_3/ZrO_2	Vegetable oil	70	Park et al. (2008)
Mg–Al HT	Rapeseed oil	91	Zeng et al. (2008)
SrO, CaO	Soybean oil	95	Liu et al. (2007, 2008)
ETS-10	Soybean oil	95	Suppes et al. (2004)
Mg–Al–CO_3 (hydrotalcite)	Cottonseed oil	87	Barakos et al. (2008)

Table 44.5 Catalytic activities and base strengths of alumina loaded with different potassium compounds on soybean oil (Xie and Li 2006)

Catalysts	Basic strength	Conversion (%)
Al_2O_3	<7.2	No reaction
KF/Al_2O_3	$15.0 < H- < 18.4$	85.8
KCl/Al_2O_3	<7.2	No reaction
KBr/Al_2O_3	$9.3 < H- < 15.0$	16.7
KI/Al_2O_3	$15.0 < H- < 18.4$	87.4
K_2CO_3/Al_2O_3	$9.3 < H- < 15.0$	48.0
KNO_3/Al_2O_3	$15.0 < H- < 18.4$	67.4
KOH/Al_2O_3	$15.0 < H- < 18.4$	80.2

Reaction conditions: Methanol to oil molar ratio 15:1; catalyst amount (wt) 2%; reaction duration 6 h; the loading amount of potassium compounds 2.1 mmol g^{-1} on alumina

Table 44.6 Catalytic activity of γ-Al_2O_3 support loaded with different alkali metal compounds (Ilgen and Akin 2009)

Catalyst	BET surface ($m^2\ g^{-1}$)	FAME yield (wt%)
γ-Al_2O_3	120–190	0.7
$NaOH/\gamma$-Al_2O_3	49.64	73.44
$LiOH/\gamma$-Al_2O_3	90.6	8.99
$Na_2\ CO_3/\gamma$-Al_2O_3	89.57	74.57
K_2CO_3/γ-Al_2O_3	89.12	76.56
KOH/γ-Al_2O_3	80.98	82.29

Reaction conditions: 6:1 methanol/oil molar ratio, catalyst 3% by weight, temp. 333 K and reaction duration 9 h

Table 44.7 Effect of calcination temperature and BET surface area on the conversion of rapeseed oil to methyl esters Liu et al. (2010)

Catalyst	BET surface area ($m^2\ g^{-1}$)	FAME yield (wt%)
γ-Al_2O_3 calcined at 823 K	128.9	0
K_2CO_3/γ-Al_2O_3 (not calcined)	48.8	–
K_2CO_3/γ-Al_2O_3 calcined at 623 K	49.3	52.7
K_2CO_3/γ-Al_2O_3 calcined at 823 K	31.1	98.2
K_2CO_3/γ-Al_2O_3 calcined at 1023 K	36.3	2.16

(1023 K), the crystal structure collapses and forms new substance of different structure expressing poor yield of FAME (Table 44.7).

The surface area of K_2CO_3/γ-Al_2O_3 samples reduced during the calcinations at 823 K. The loss in surface area is due to the decomposition of CO_3^{2-} present on the catalyst surface. The specific surface area of γ-Al_2O_3 calcined at 823 K was very large and may be due to the absence of K_2CO_3. It did not show any catalytic activity in transesterification of rapeseed oil with methanol. However, reduction in specific surface area of K_2CO_3/γ-Al_2O_3 calcined at 823 K showed higher catalytic activity expressing a yield of 98%.

44.2.7 Calcium Oxide

Among the oxides of alkali earth metal, calcium oxide has attracted attention of many workers since it holds relatively high basic strength with less environmental impacts due to its low solubility in methanol. Besides, that can be synthesized from limestone and calcium hydroxide. The catalytic activity of CaO can be accelerated with an initial pretreatment with methanol at room temperature. Generally activated CaO can be obtained by calcining commercial CaO or its precursor. A solid super base of CaO prepared by the treatment with an ammonium carbonate solution and calcinations at a high temperature was used for the transesterification (Zhu et al. 2006). The sequence of catalytic activity of calcium oxides for transesterification was reported and its efficiency was in the order CaO > $Ca(OH)_2$ > $CaCO_3$. The rapid deactivation of CaO catalysts by water and CO_2 is also a matter of concern.

CaO was rapidly hydrated and carbonated in air. Active surface sites of CaO were bound to be poisoned with CO_2 and covered with H_2O. In order to avoid the loss of catalytic activity, CaO was thermally treated at 700 °C so as to desorb CO_2 before being used in the reaction. Leaching of catalyst was often observed in the reaction media when the catalyst was activated at high temperature. However, the amount of leaching did not result in massive reduction in the catalytic activity even it was used for many cycles. It was also found that small amounts of water could improve catalytic activity of CaO and biodiesel yields because of the presence of water. O_2 present on the surface of the catalyst joined with H^+ present in the water molecules and formed OH^- which subsequently reacted with methanol to form methoxide. This acted as a powerful catalyst in the transesterification reaction.

KCl-doped CaO catalyst synthesized by a wet impregnation method under microwave irradiation was employed in the transesterification of refined soybean oil. KCl/CaO catalyst showed increase in biodiesel yield up to 93.5%. Alkaline earth metal compounds were used in the methanolysis of sunflower oil (Arzamendi et al. 2007). Recently alkali-doped metal oxide catalysts, $Ca(NO_3)_2$/CaO, $LiNO_3$/CaO, $NaNO_3$/CaO, KNO_3/CaO and $LiNO_3$/MgO were examined for employing them in the transesterification of vegetable oils (Granados et al. 2007). Mixed metal oxides of BaO/CaO/ZnO were used as catalyst in the production of biodiesel from *Spirulina platensis* which showed maximum (98.94%) conversion of fatty acid methyl ester at 2.5 wt% catalyst, 1:18 oil to methanol molar ratio, 600 rpm stirring rate, 65 °C temperature for 120 min. The catalyst activity was observed to decrease from 98.94% to 69.56% after 6 cycles (Fig. 44.5) due to blockage of active sites of the surface of the catalyst. The products biodiesel and glycerol have layered the catalyst surface which automatically reduced the contact existing between catalyst and methanol (Singh et al. 2019).

44.2.8 Magnesium Oxide

Magnesium oxide produced by direct heating of magnesium carbonate or magnesium hydroxide is rarely used in the production of biodiesel since it is a weak base.

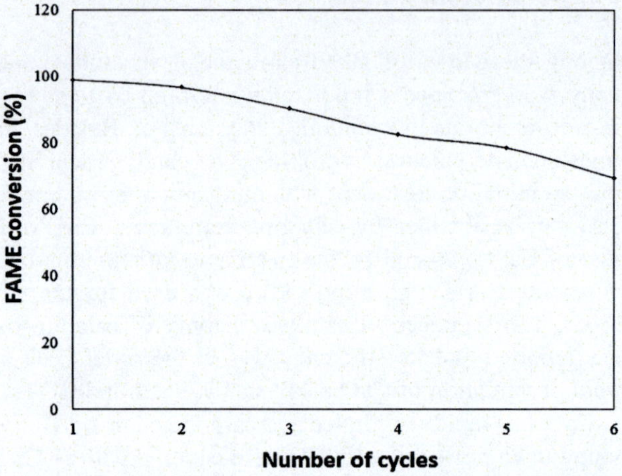

Fig. 44.5 Reusability test of synthesized catalyst (Ca, Ba and Zn mixed oxides) on *Spirulina platensis* oil (molar ratio 1:18, catalyst weight 2.5 wt%, time 120 min, temperature 65 °C, and stirring speed 600 rpm). (Adopted from Singh et al. 2019)

However, nano magnesium oxide was used in the transesterification of soybean oil which gave a yield of 99% in 10 min at a supercritical temperature of 523 °C and high pressure (24 MPa). The results showed that this catalyst has more activity at higher pressures and temperatures. The preparation of a Li-doped MgO for biodiesel synthesis was investigated by optimizing the catalyst composition and calcination temperatures. The results showed that the formation of strong base sites was promoted by the addition of Li, thus resulting in an increase in the biodiesel synthesis. The catalyst Li/Mg at a molar ratio of 0.08 and calcination temperature of 823 K gave the best performance. The biodiesel conversion decreased when Li/Mg molar ratio was increased beyond 0.08. This could be due to the formation of lithium hydroxide due to excess Li ions and concomitant decrease of BET value (Wen et al. 2010).

A similar approach was made by Wan et al. (2008) in which biodiesel was synthesized from rapeseed oil by transesterification over magnesium oxide loaded with KF. It was also shown that the catalytic activity strongly depended on the loading amount of KF and calcined temperature. The authors found that the reaction reached a yield of 79.37% when the loading of KF was at 35% and calcined at 500 °C. Incorporation of KF at 30% level in KF/MgO at 80 °C was found to give equally good results from the catalyst calcined at 500 °C. Biogenic mediated synthesis of the nanocomposite Cs2-MgO/mesoporous carbon used as catalyst for biodiesel from olive oil yielded 95.18% of biodiesel under optimum reaction time 6 h, catalyst 5.2%, reaction temperature 65 °C and oil to methanol molar ratio 1:15. In addition, the catalyst maintained a good catalytic activity up to 4 cycles without a major variation in conversion up to 3% (Hassan et al. 2020).

$Ca(OH)_2$ was impregnated in diatomite through conventional impregnation method which was then chemically precipitated with MgO to develop a composite catalyst of diatomite@CaO/MgO. The prepared catalyst was used in waste cooking oil to produce biodiesel. Effect of various reaction temperatures ranging from 40 to 120 °C using catalyst 6 wt%, oil to methanol molar ratio 1:15 for 180 min was evaluated. Maximum conversion was 96.2% at 90 °C through active collision between methanol and triglycerides, whereas in other high reaction temperatures 100, 110 and 120 °C the conversion reduced slightly (96.1%, 95.88% and 95.67%). Thus, the biodiesel conversion is not conspicuously high in the temperature range of 90–120 °C. On energy consumption point of view, the optimum reaction temperature was considered to be 90 °C (Rabie et al. 2019).

44.2.9 Strontium Oxide

Not much work was carried out on strontium oxide as catalyst. Strontium oxide is an active metal oxide soluble in the reaction medium. Transesterification of soybean oil with SrO as a solid base catalyst showed that the specific surface area of the catalyst was as low as 1.05 $m^2\ g^{-1}$ giving 90% yields of methyl ester in 30 min at 65 °C with a methanol to oil molar ratio of 12:1 at 3% weight of catalyst. The catalyst was stable even after 10 cycles.

Strontium metal doped ZnO was prepared by impregnation followed by calcination at a higher temperature (Yang and Xie 2007). It was used as a catalyst in the transesterification of soybean oil with methanol. Loading 2.5 mmol $Sr(NO_3)_2$ on ZnO and calcination at 873 K for 5 h was found to be the most appropriate conditions for preparing the optimum catalyst which exhibited the highest basicity with best catalytic activity. A maximum conversion of 94.7% was achieved in 1:12 molar ratio of soybean oil to methanol with 5% by weight of catalyst in a reaction duration of 5 h. However, the used catalyst was significantly deactivated and could not be directly reused for transesterification. Such deactivation was caused by the deposition of reactants and products on the active sites.

Solid base catalyst SrO/MgO with differential molar ratios of Sr and Mg (2:8, 3:7, 4:6 and 5:5) was synthesized by co-precipitation method and was followed by calcination at 850 °C for 5 h. The SrO/MgO at 3:7 molar ratio showed higher catalytic activity than other ratios as it had high basic strength $22.5 < H- \leq 27$. Effect of calcination temperature was also investigated on a temperature range of 750–1050 °C (Fig. 44.6).

Biodiesel yield was observed to be poor (25%) at 750 °C, whereas it was 79% at 850 °C. The yield was observed to be a function of temperature between 750 and 850 °C. But the yield gradually decreased as the temperature was raised from 850 to 1050 °C. At 850 °C, the catalyst commenced decomposition of carbonate compounds and consequent reduction of catalytic active sites of the crystalline phases of SrO, MgO, $Sr(OH)_2$ and residual $SrCO_3$. The $Sr(OH)_2$ and $SrCO_3$ completed decomposition above 850 °C and converted into the crystalline phases (SrO and MgO). They are agglomerated due to the sintering effect (Shahbazi et al. 2020).

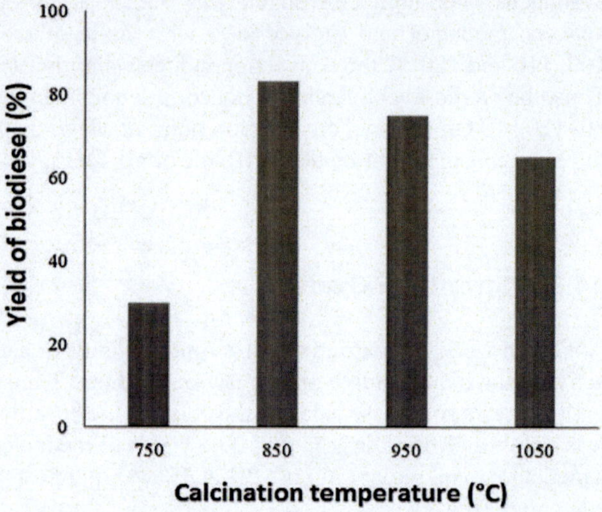

Fig. 44.6 Effect of calcination temperatures on the yield (%) of biodiesel in the transesterification of waste cooking oil with methanol using SrO/MgO (3:7) as catalyst. Reaction condition: 0.1 g SrO/MgO (3:7) catalyst with the grain size of 250–280 mesh, methanol to oil molar ratio 9:1. Reflux temperature 60 °C; reaction duration 2 h, the stirring rate 850 cycle/min (Shahbazi et al. 2020)

44.2.10 Zeolite

Zeolite as a catalyst is prepared with extensive variation in acidic and textural properties. It is synthesized to overcome the diffusional limitations so that optimum biodiesel production can be achieved. Zeolites can also be modulated to exhibit hydrophobic characteristics without compromising its functional ability of the acidic sites. Catalytic efficiency is sustained by incorporating hetero poly acids.

Zeolites are generally recognized as a versatile catalyst known by its chemical composition, pore size distribution and ion exchange abilities. The acid-base properties of zeolites are controlled by the kinds and quantities of ion exchanged cations, and also by the Si/Al ratio of the zeolite framework. To control the basicity of zeolite, two approaches are generally made. They are (1) ion exchange with alkali metal ions and (2) the impregnation of basic components on the inner surface of the zeolite pores. It is reported that the former product has relatively weak basic sites, and the latter material has relatively strong basic sites (Hattori 1995).

Among the zeolite family, Zeolite-X, titanosilicates and mesoporous zeolites have attracted the attention of researchers working in biodiesel production. Zeolite X is generally accepted as one of the most basic zeolites, and the ion exchange of Na–X with larger monovalent cations such as K remarkably increases the basicity of the catalyst (Barthomeuf 1996).

Engelhard titanosilicate structure (ETS) has gained attention recently due to its strong basic characteristics, high cation exchange and also due to large pore structures, which give advantage over other zeolite for liquid phase reactions. Application of titanosilicate for biodiesel synthesis came in the recent past. The soluble titanates are quite active in the homogeneous base-catalysed transesterification, but the problem is on the equilibrium existing between the monomeric and dimeric titanium species, the latter being prevalent but catalytically inactive at low temperature (Nasr et al. 1995).

44.2.11 Zirconium Oxide

Zirconia are being used in a variety of reactions, such as alkylation of isobutane (Xiao et al. 1999) and isomerization of *n*-butane and showed high activity correlated with acidity. There have been several reports on the usage of zirconia as a solid acid catalyst for transesterification of different oils due to its strong acidity. It was observed that the acidity increased when the surface of these metal oxides held anions such as sulphate and tungstate.

Jitputti et al. (2006) assessed the catalytic performance of zinc oxide and zirconium oxide as solid acid catalysts in the process of palm kernel oil in supercritical methanol. Zinc oxide catalyst at 3% by weight with methanol to oil at a molar ratio of 6:1, 86% methyl ester was formed. The corresponding value for zirconium oxide was 64.5%. However, with sulphated zirconia the yield of methyl ester increased to 90.3% which might be due to the high acidic strength of sulphate anions on the surface of the zirconia.

The catalytic activity of ZrO_2/SO_4^{2-} and ZrO_2/WO_3^{2-} was compared. Sulphated zirconia expressed high activity in the conversion of triacetin than that of tungstated zirconia owing to its higher specific surface area coupled with active phase concentration under identical conditions of reaction. Various concentrations (0, 1, 2, 3 and 5 wt%) of La_2O_3 were used to prepare La_2O_3-ZrO_2 mixed metal oxides by sol-gel method. They were calcined at 600 °C for 6 h. This catalyst at 3% by weight achieved 56% conversion of canola oil in 4 h due to its high basicity (Salinas et al. 2018).

44.2.12 Zinc Oxide

Yang and Xie (2007) compared the performance of alkali earth metals loaded on different catalyst supports in the conversion of soybean oil to biodiesel. The performance of such catalysts largely depended on the concentration of basic sites on the surface of the catalyst. Aqueous solution of an alkali earth metal nitrate was impregnated and calcined at 600 °C for 5 h. Among them $ZnO/Sr(NO_3)_2$ showed the best catalytic activity and the maximum conversion achieved was 93.7%. Xie and Huang (2006) loaded potassium on zinc oxide as a solid base catalyst. An aqueous solution of potassium fluoride was used and then the loaded support was calcined. Concentration of basic sites reached 1.47 mmol g^{-1} when 15% by weight of KF was loaded on ZnO. Zn/I_2 also was used as a heterogeneous catalyst (Li et al. 2006).

Calcium-doped zinc oxide was prepared by co-precipitation with different ratios of $Ca(NO_3)_2{\cdot}4H_2O$ (0.01, 0.03, 0.05 M) and loaded on zinc nitrate ($Zn(NO_3)_2{\cdot}6H_2O$) with UV and without UV(NUV) source. Low concentration of Ca exhibited lengthy

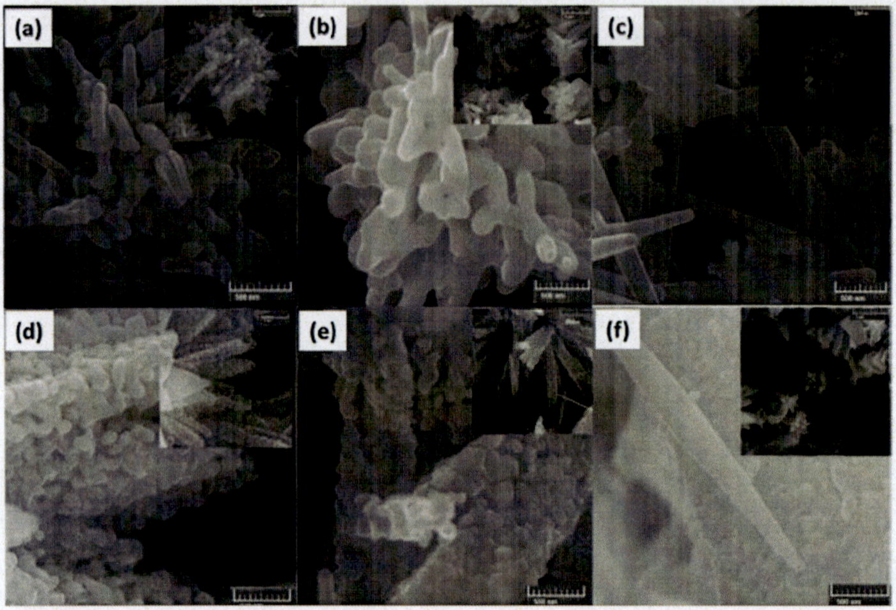

Fig. 44.7 SEM images of (**a**) $0.01CZO_{UV}$, (**b**) $0.01CZON_{UV}$ calcined 700 °C, (**c**) $0.03CZO_{UV}$, (**d**) $0.03CZO_{NUV}$, (**e**) $0.05CZO_{UV}$, and (**f**) $0.05CZO_{NUV}$. (Reprinted with permission from Abdala et al. 2020 American Chemical Society)

nanoparticles with agglomeration, whereas high concentration showed better defined shapes. UV light assisted the formation of nanorods and nanospherical particles at low concentration of Ca (Fig. 44.7). Calcination exhibited yellow colour at 600 °C, light yellow at 700 °C and white colour at 800 °C. High concentration of Ca (0.05 M) loaded on ZnO at a calcination temperature of 700 °C achieved 99% biodiesel yield (Abdala et al. 2020).

Esterification of waste cooking oil containing high FFA was made by photocatalyst $La^{3+}/ZnO\text{-}TiO_2$ and the reaction mechanism was studied as per Langmuir Hinshelwood Mechanism (Fig. 44.8). The catalyst was irradiated by ultraviolet rays to create electrons (e^-) in the valence band which ultimately stimulated and transferred into the conduction band where equal number of holes (h^+) developed. In the transesterification reaction, ethanol and FFA migrated to the surface of photocatalyst and created a large number of active sites. Ethanol adsorbed on the surface of the catalyst reacted with photogenerated hole (h^+) to form H^+ and $CH_3CH_2O^-$. RCOOH adsorbed on the surface of the catalyst reduced to form RCOOH· free radical by accepting photogenerated electron (e^-) on the surface. Further, the reaction took place between the free radicals RCOO·, H^+ and CH_3CH_2O· to produce biodiesel and water (Guo et al. 2021).

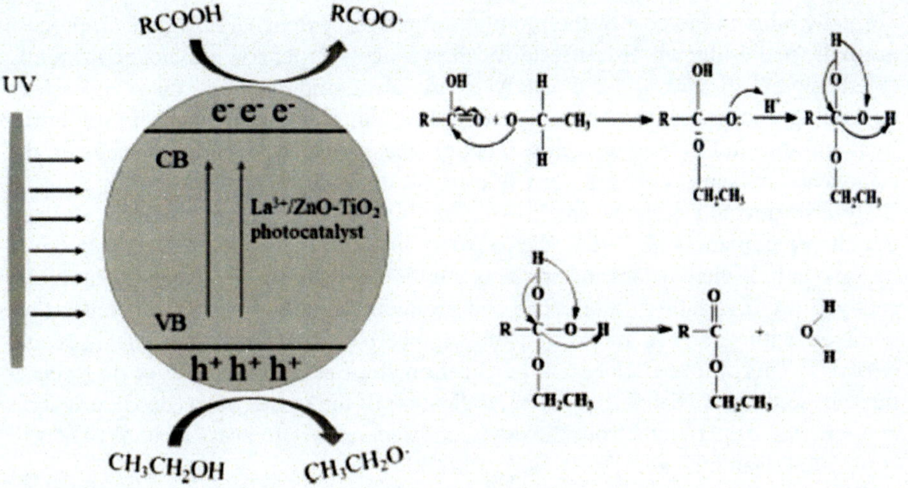

Fig. 44.8 The mechanism of FFA esterification catalysed by $La^{3+}/ZnO\text{-}TiO_2$. (Reprinted with permission from Guo et al. 2021 Elsevier)

44.3 Enzyme Catalysed Transesterification

Alkali catalysed transesterification of oil is richly undertaken in the commercial sector due to high efficiency. Sodium hydroxide and potassium hydroxide (Schuchardt et al. 1998; Marchetti et al. 2008; Robles-Medina et al. 2009) are popularly used as alkali catalysts. Other alkalies used for this purpose include carbonates, methoxide, ethoxide, propoxide and butoxide of sodium (Fukuda et al. 2001). The above alkali catalysts mostly act under conditions of low temperature, pressure and in short duration (Bacovsky et al. 2007; Leung et al. 2010). These alkalis are often interfered by the presence of free fatty acid. As a result, unwanted soap develops. The soap cannot be separated easily form the product. Therefore, it has become necessary to reduce the level of free fatty acid by any one of the following pretreatments. A maximum of 2.5% of free fatty acid in the oil is manageable.

(a) Acid esterification
(b) Ion exchange process
(c) Extraction with chosen alcohol

The above process is to be resorted before undertaking alkali catalysis and is considered as additional works. Besides, in such eventualities multistage purification is needed in isolating the biodiesel with satisfactory purity. The water needed for such purification process is around 200 mL L^{-1} of biodiesel. The wash water ultimately enters in to the environment which is needed to be treated. Such treatment normally adds to the cost of the product. Moreover, extensive downstream process normally makes the whole transesterification process expensive and not eco-friendly (Fjerbaek et al. 2009). Acid catalysts are also employed on many occasions depending on the quality of raw oil used. Most commonly used acids are sulphuric, hydrochloric and sulphonic acids. Though the use of acid catalysts negates the occurrence of soap, they are corrosive in nature and damage the reactor vessels. They also develop pungent smell. Thus the use of acid as catalyst has restricted option (Freedman et al. 1984; Bacovsky et al. 2007). By nature, catalysts never become part of the reactants or products but they only assist the reaction. Therefore, catalysts are invariably wasted. The end products, namely the ester generally stays admixed with glycerol, mono and diacylglycerols, pigments, extra alcohols and catalysts. This warrants elaborate purification steps such as separation of glycerol through settling or centrifugation, neutralization of the waste water, decolourization and removal of pigments from the esters and separation of excess alcohol (Antczak et al. 2009; Banerjee and Chakraborty 2009).

Enzymatic conversion of vegetable oils in to biodiesel offers an eco-friendly benign alternative to the chemical process. Enzymes do not form soap even high level of free fatty acid is present in the oil (Harding et al. 2007; Fjerbaek et al. 2009). Trace quantity of water if present in the oil or develops during the course of the transesterification will not disturb the process (Dizge and Keskinler 2008). Enzyme can be effectively used in the transesterification of low cost oils and lard. There is no need for a washing as free fatty acid and triglycerides are totally catalysed at a single step. The whole process is very mild and there will be low energy consumption (Narasimharao et al. 2007; Fjerbaek et al. 2009). The comparison of enzyme process and the alkali process is presented in Table 44.8.

The lipases act on the ester bonds of carboxylic acids.

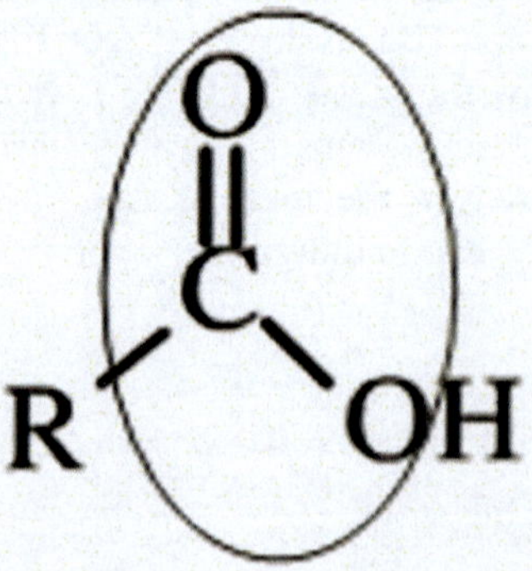

The efficiency of the lipases largely depends on the length of the chain of fatty acid, nature of the fat and regioselectivity. In general lipases convert triglycerides, diglycerides, monoglycerides and free fatty acids to esters. Lipase has the ability to work in biphasic (hydrophilic) and monophasic (hydrophobic) systems. Besides, lipases can be manufactured in bulk quantities through bacterial sources. Many

Table 44.8 Comparison of lipase catalysed and alkali catalysed transesterifications

	Transesterification	
Conditions	Lipase catalysed	Alkali catalysed
Reaction temperature	30–40 °C	60–70 °C
Rich free fatty acids in oil	Methyl esters	Saponified products
Yield of methyl esters	Higher	Normal
Recovery of glycerol	Easy	Difficult
Quality of glycerol	Purification easy	Purification process results in soap formation
Purification of methyl esters	Not needed	Washing needed
Catalyst recovery	Enzymes may be removed along with regenerated glycerol	Cannot be reused
Wash water	No wash water	Wash water pollutes the environment
Specificity[a]	Highly specific method	Nonspecific reaction leading to wastage of chemicals and poor quality biodiesel

[a]Since there is no fool proof mechanism to confirm the completeness of esterification, the exact quantity of reactants participating in the chemical reaction cannot be arrived. If the quantity of the reactants is increased, there is a chance of wasting the excess chemicals. On the contrary, if the reactants quantity is reduced, the esterification may be incomplete. In such case, biodiesel will be of poor quality as highly unsaturated fatty acids will be produced

lipases effectively carry out the transesterification in the presence of long and branched chain alcohols. These alcohols normally will not affect transesterification in their normal mode. The by-products of transesterification, namely biodiesel and glycerin can be easily separated if enzymes are used as a catalyst. Lipase has a reasonably good thermal stability and tolerates short chain alcohol (Bacovsky et al. 2007; Kato et al. 2007; Robles-Medina et al. 2009).

44.3.1 Microbial Lipases

The lipases produced by various microorganisms are used in the transesterification processes. Based on the production mode, they are classified as extracellular and intracellular. Extracellular enzyme refers to the production and release of the same by the microorganism which is extracted and purified for ready use. Any enzyme which stays intact in the cell of the microbes is referred as intracellular.

44.3.1.1 Extracellular Lipase

Microbial lipases are mostly extracellular, produced by a microbial culture in suitable reactors either through submerged culture or through solid state culture. The product obtained is purified before being used. The degree of purification depends on the origin and its structure (Palekar et al. 2000; Saxena et al. 2003). The cost of the enzyme directly rests on its potency and on the yield at short duration. Extracellular enzymes are relatively costlier. Very popular extracellular enzymes in the transesterification process are Novozym 435 from *Candida antarctica*, Lipozyme RM 1M from *Rhizomucor miehei* and Lipozyme TL 1M for *Thermomyces lanuginosus* (Robles-Medina et al. 2009). Due to the cumbersome processes involved in the purification of extracellular enzymes it is considered cost prohibitive in the application of biodiesel production.

44.3.1.2 Intracellular Lipase

The enzyme which is harboured inside the cell is referred to as intracellular enzyme. Use of such microbial cells holding lipase in its body directly in the transesterification process eliminates the drudgery of the enzyme purification processes. These microbes are often spontaneously incorporated in solid support and thus immobilized. The efficiency of the intracellular enzyme is relatively lower when compared to that of the extracellular enzyme (Robles-Medina et al. 2009). But the intracellular enzyme is relatively stable (Klibanov 1983; Ranganathan et al. 2008). *Candida antarctica*, *Rhizopus chinensis*, *R. oryzae* and *Saccharomyces cerevisiae* are used as intracellular lipases (Fukuda et al. 2008; Qin et al. 2008). The direct use of cells in the biocatalytic process is called as whole-cell biocatalyst.

44.3.2 Regiospecific Enzymes

Enzymes which choose to act on a specific bond of the triglyceride molecule are referred as regiospecific enzymes (Robles-Medina et al. 2009). Regiospecific enzyme known for its selective action on the specific acyl position on the glycerol back bone (Chandler 2001) is grouped in to three types: (1) 1,3 specific, (2) 2 specific and (3) nonspecific (Koskinen and Klibanov 1996; Rahman et al. 2005).

1,3 Specific lipases act mostly on the ester bonds at the extreme position of the triglyceride molecule and rarely attack the middle ester bond. *Rhizopus oryzae*, *R. delemar*, *Rhizomucor miehei*, *Thermomyces lanuginosus*, *Aspergillus niger* (Shimada et al. 1997; Fukuda et al. 2001; Lanser et al. 2002; Robles-Medina et al. 2009) are the example for the 1,3 specific lipases. 2 Specific lipases are known to attack the middle ester bond on the triglyceride molecule. *Geotrichum candidum* is an example for this type (Macrae 1983).

The third type, namely the nonspecific lipase expresses no preference to its action on any specific bonds. *Candida antarctica*, *C. cylindracea*, *C. rugosa*, *Pseudomonas cepacia* and *P. fluorescens* (Fukuda et al. 2001) are examples for this type. Thus this type of enzymes is best preferred.

Regioselective enzyme though does not act on all bonds is still considered efficient predominantly due to spontaneous acyl migration. The acyl moieties migrate from the second position to the first or third position of the triglyceride in sparingly aqueous environment (Fukuda et al. 2009). To accomplish acyl migration, polar immobilization support is needed. Often silica gel is added to the reaction mixture (Akoh et al. 2007; Robles-Medina et al. 2009) for this purpose.

44.3.3 Enzyme Stability

Maintaining the stability of the enzyme is a challenge especially in the transesterification process. Enzymes being used in the biodiesel reactor have to face extremes of conditions such as far-reaching temperatures, inactivating impurities, reactants such as alcohols, by-product such as glycerol and water content (Malcata et al. 1990; Marchetti et al. 2007; Torres et al. 2008; Robles-Medina et al. 2009). Some of the suggestions made to improve the stability of the enzyme are genetic improvement of the strains, molecular configurations, physical and chemical treatments, immobilization improvements and reactor modifications (Malcata et al. 1990; Reetz 2002; Mateo et al. 2007).

44.3.4 Microorganisms

The microorganisms considered for lipase production and for use in the transesterification process are given in Table 44.9.

The lipases produced from the cultures of *Candida antarctica*, *C. rugosa*, *Pseudomonas cepacia*, *P. fluorescens*, *Rhizomucor miehei*, *Rhizopus chinensis*, *R. oryzae and Thermomyces lanuginosus* were observed to be most effective in the transesterification (Vasudevan and Briggs 2008). *C. antarctica* lipase expressed fairly good activity during methanolysis and ethanolysis but found not to be

Table 44.9 List of microbes whose lipases are used in the transesterification of oil

Microbes	Microbes
Aspergillus niger	*P. fluorescens*
Bacillus thermoleovorans	*P. putida*
Burkholderia cepacia	*Rhizomucor miehei*
Candida antarctica	*Rhizopus arrhizus*
C. cylindracea	*R. chinensis*
C. rugosa	*R. circinans*
Chromobacterium viscosum	*R. delemar*
Enterobacter aerogenes	*R. fusiformis*
Fusarium heterosporum	*R. japonicus* NR 400
Geotrichum candidum	*R. oryzae*
Humicola lanuginosa	*R. stolonifer* NRRL 1478
Oospora lactis	*Rhodotorula rubra*
Penicillium cyclopium	*Saccharomyces cerevisiae*
P. roqueforti	*Staphylococcus hyicus*
Pseudomonas aeruginosa	*Thermomyces lanuginosus*
P. cepacia	

matching with certain other alcohols. Methanolysis using *C. antarctica* lipase in a solvent free medium gave good conversion (>90%). Similarly, as high as 82% conversion could be obtained during ethanolysis using the above enzyme (Mittelbach 1990). Thus *C. antarctica* enzyme manages the inhibitory effect of methanol and ethanol. The yield of ester is inversely proportional to the carbon length of the alcohol (Rodrigues et al. 2008). *Pseudomonas cepacia* gave a conversion more than 65% in methanol and ethanol (Noureddini et al. 2005). It gave more than 98% yield (Salis et al. 2005) in butanol. *Rhizomucor miehei* also gave a highest yield as that of *P. cepacia* when butanol was used.

44.3.5 Enzyme Tuning

The catalytic efficiency of an enzyme can be increased by tuning the same. Tuning is all the more necessary especially when the enzyme is put to use in non-aqueous conditions or under low water conditions. Under such state the enzyme expresses low molecular flexibility due to rigidity of the protein molecule which is H-bonded to many residues. To avoid the rigidity, the enzyme is dissolved in a buffer and dehydrated by lyophilization. In the catalytic activity during the transesterification of the oil the tuned enzymes may remember the pH of the buffer in which they were tuned already. This is referred as the enzymes pH memory. The lyophilization process of the enzyme after exposing it in a buffer solution induces perturbation on the secondary structure of the enzyme. Such enzymes are stable during the transesterification.

44.3.6 Enzyme Lid

In aqueous medium many lipases act in the presence of a water–lipid interface. Such activity of the enzyme at the interface involves the displacement of a surface structure called lid. Enzymes known to act on the interface normally in alternative conformational states endowed with different activity. In a closed conformation, the lid prevents the active site of the enzymes from exposure to the molecules of the substrate. In the open conformation the lid opens the entrance of the catalytic pathway. Not all lipases are activated by the interface. The quality and quantity of the interface play a conspicuous role in the catalytic function of the enzyme. The serine residue present in the active site of the lipase is occluded by a polypeptide flap which is referred as lid whose job is to make the enzyme inaccessible to substrate in aqueous media which makes the enzyme not responsive to the reaction. One of the underlying facts is that the reaction has to take place in a medium which is required to create conducive conditions for action by both enzyme and substrate. Most lipases experience active conformational changes in the interfacial zone. The lid hitherto blocked the active site would move from closed to an open conformation. During this interim period lipase may stand as solution waiting for the reaction to start. Some of the organic solvents also help to open the enzyme lid.

44.3.7 Limitations of Enzymatic Transesterification

Enzymatic transesterification requires long reaction duration, higher quantity of catalyst entailing high cost of production. It also loses its activity during repeated use. The duration of the transesterification process is almost double when compared to that of the chemically catalysed transesterification. Normally it takes 16–20 h to complete the process. During the initial 4 h the reaction is rapid. As high as 70–80% turnover takes place within 4 h expressing a rate of 20% h^{-1}. As the reaction proceeds the rate of conversion slows down (Fig. 44.9). During 4–14 h the total conversion will be around 95%. The rate of reaction during that period will be around 15% h^{-1}. From 14 to 20 h the reaction is dead slow and it will be less than 5% h^{-1}. There are occasions that the completion of transesterification reaction takes around 36 h (Watanabe et al. 2002).

The quantity of enzyme to be involved is very high which often limits the popularization of this process. As high as 30% of lipase is found to be optimum in the case of Novozym 435 (*Candida antarctica*) on cotton seed oil, rape seed oil, lea seed oil, soy bean oil and lard especially when used with methyl acetate as acyl acceptor (Huang et al. 2010). Shu et al. (2007) observed that maximum of 15% of Novozym 435 is sufficient to carry out the enzymatic transesterification of soybean oil, rapeseed oil, corn oil, pea nut oil, sunflower oil, sesame oil, castor oil, olive oil and cotton seed oil. In this case dimethyl carbonate was used as the acyl acceptor thus it is known that the acyl acceptor has a conspicuous role in deciding the enzyme concentration required. Shah et al. (2004) observed that a maximum of 15% of enzyme was doing well when *Chromobacterium viscosum* lipase was used in the

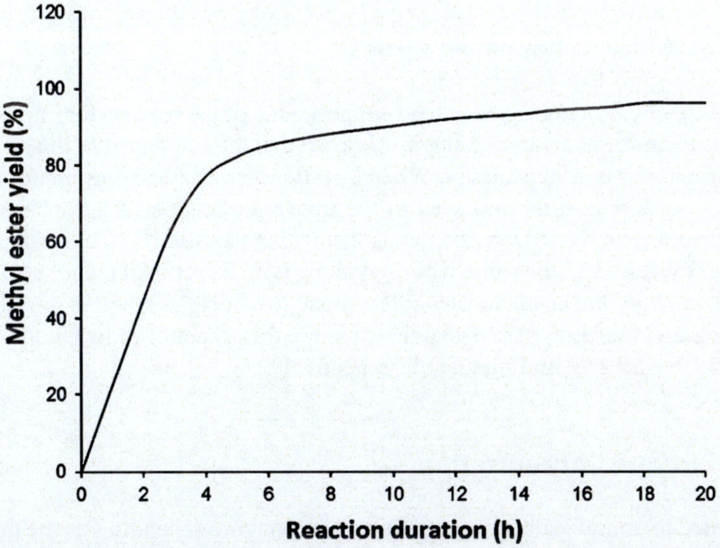

Fig. 44.9 The methyl ester yields due to enzymatic transesterification over a period of 20 h

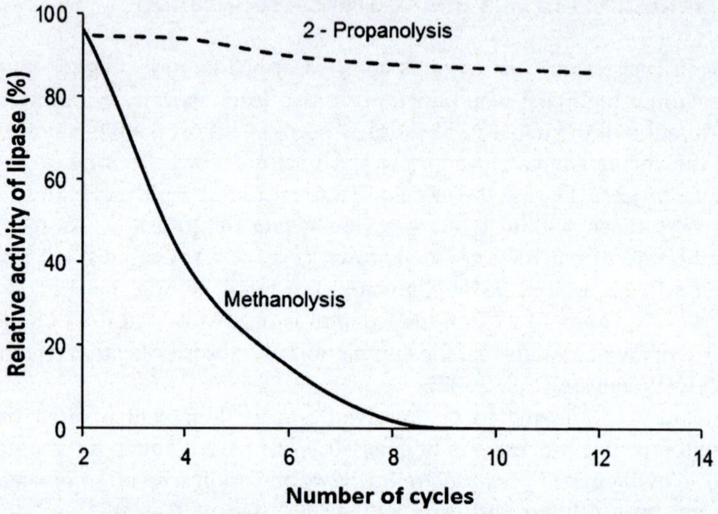

Fig. 44.10 Operational stability of immobilized lipase during repeated cycles of transesterification of jatropha oil

presence of ethanol. Modi et al. (2006) who worked on the lipase mediated transformation of jatropha oil (*Jatropha curcas*), Karanj oil (*Pongamia pinnata*) and sunflower oil (*Helianthus annuus*) suggested a maximum of 10% lipase (*Candida antarctica*). If the levels of enzymes are increased, the reaction mixture becomes viscous and as a result the recovery of ester reduced.

44.3.8 Stability in Repeated Cycles

Enzymes are often immobilized in hard support. One of the reasons for immobilizing is to reuse them for many cycles but it is known that the impregnated lipase loses its potency when it is used repeatedly. When *C. antarctica* enzyme was immobilized in macro porous acrylic resin and used in the transesterification of jatropha oil along with methanol (Fig. 44.10), its potency is lost in 7 cycles but its stability maintained beyond 12 cycles when the same was used along with 2 propanol (Modi et al. 2006). Contrary to short linear chain alcohols propan-2-ol displays better miscibility in triglycerols and are less polar. The polarity index for propan-2-ol is 3.9 only against 5.2 and 5.1 for ethanol and methanol, respectively.

44.3.9 Enzyme Concentration

It is required to optimize the concentration of the enzyme depending on the quality of the oil and temperature. Use of enzyme in higher concentration beyond what is required causes aggregation (Fig. 44.11).

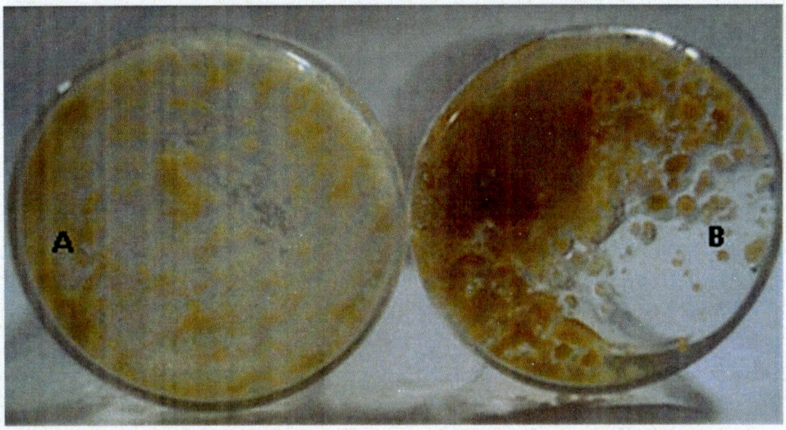

Fig. 44.11 Aggregation of enzymes in the reaction medium containing different enzyme concentrations (**a**—lower concentration, **b**—higher concentration)

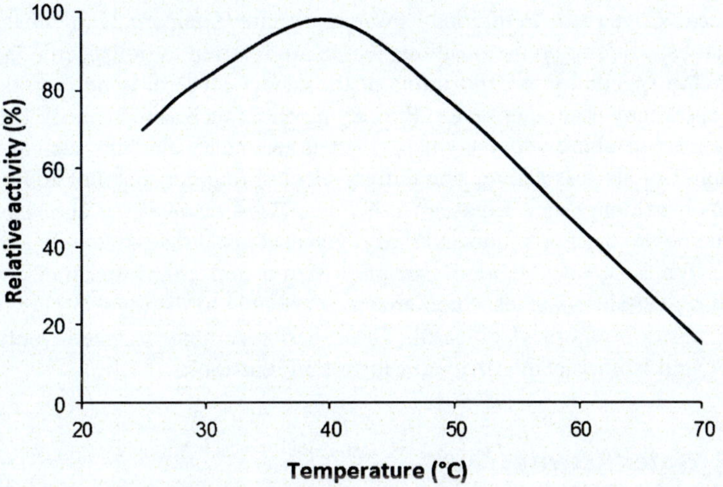

Fig. 44.12 Effect of temperature on the relative esterification activity of immobilized lipase

44.3.10 Thermal Impact

The heat applied during the transesterification process may retard the function of lipase therefore cursory test may be necessary before proceeding towards large scale production. The lipase shows its optimum function around 40 °C. At higher temperature the function of the lipase is affected (Fig. 44.12) as the enzymes suffer conformational transition so that catalytically inactive form of enzyme is produced. In many of the cases the change is uni-molecular resulting in to aggregation. The classical globular structure of the enzyme is likely to alter as a consequence of enzyme unfolding. Usually enzymes are stored in low temperature (−20 °C) to maintain its full activity. In such low temperature the original conformation tends to remain stable.

44.3.11 The Enzyme Specificity

The specificity of an enzyme is largely oriented to its relationship with the substrate on which it is acting. The substrate specificity refers to many parameters such as fatty acid, bond, alcohol, acylglycerol and stereo and chiral specificity. Fatty acid specificity is affected by the carbon length of fatty acids and the number and position of the double bond present in it. Most of the lipases act strongly on a carbon chain of C8–C24, whereas few lipases (*Candida rugosa*) act very feebly on a carbon chain C20 or greater than that.

In terms of region selectivity certain of the enzymes such as *Rhizomucor miehei*, *R. oryzae*, *Thermomyces lanuginosus* show specificity to 1,3 ester bonds (i.e. ester bonds at atom C1 and C3). Another set of lipases from organisms such as *Candida cylindracea*, *C. rugosa*, *Penicillium cyclopium* and *Humicola lanuginosa* do not express any specificity on the positions which are referred as nonspecific lipases. A lipase having specificity to 2 positions of the triacyl glycerol is not yet identified. Alcohol specificity is another factor. Primary alcohols such as methanol, ethanol and butanol are active along with enzymes, whereas secondary alcohols such as propan-2-ol, butan-2-ol are less active. The tertiary alcohol (e.g. tert-butanol and 2-methyl propan-2-ol) seldom react. Lipases from *Penicillium camemberti* and *Aspergillus flavus* are known to act on mono and diacyl glycerol faster than triacylglycerol. Thus certain enzymes are acylglycerol specific. Stereo and chiral specificity is often exhibited by certain enzymes. Such enzymes are able to distinguish between *sn-1* and *sn-3* position of triacyl glycerol. The stereo specificity is recently utilized to produce chiral isomer as intermediates in certain reactions.

44.3.12 Water Activity

Water activity which is a measure of water content was usually considered as an essential parameter in the non-aqueous enzymatic catalysis. Water is considered essential in order to maintain the specific three dimensional structure of the enzyme (Lu et al. 2009). Enzymatic transesterification requires the unmarking and restructuring of the active site of the enzyme which will take place only in the oil–water interface. Optimum water content minimizes the hydrolysis and maximizes the enzyme activity leading towards the successful transesterification (Noureddini et al. 2005; Jegannathan et al. 2008). If the system is water free, no reaction takes place with the enzyme from organisms such as *Candida rugosa* and *Pseudomonas cepacia*. The enzymes from the above species stay active once the water content at an optimum level is provided (Fig. 44.13). On the contrary any excess water is detrimental since the excess water flood the site of the reaction thereby decreasing

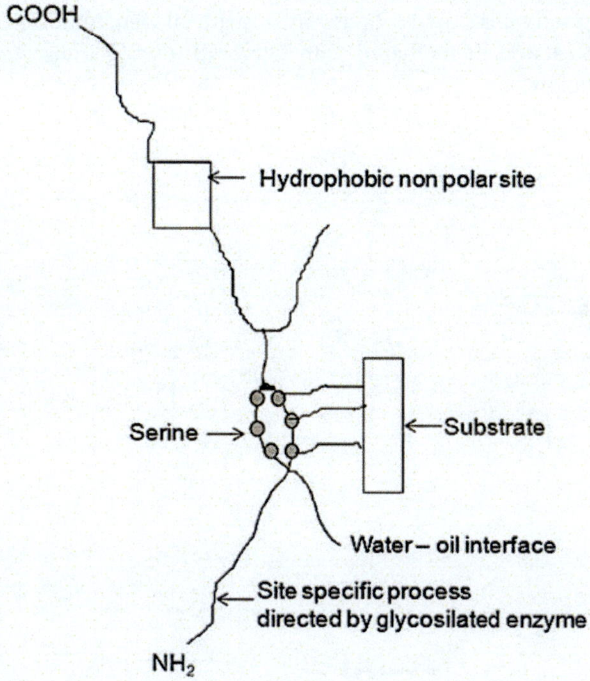

Fig. 44.13 The effect of water–oil interface in the presence of glycosylated enzyme

the exposure of the enzyme. The non-polar liquids tend to form a film or meniscus since the molecules of the non-polar liquids tend to arrange themselves in such a way that there will be minimum contact between the two. When polar liquid is added the changed sites try to make contact between the non-polar liquids. Some enzymes do not fold appropriately especially when they are not glycosylated.

44.3.13 Alcohol

Primary, secondary, straight chained and branched alcohols are being employed as acyl acceptors in the transesterification processes. Commonly used alcohols are methanol, ethanol, propanol, isopropanol, 2-propanol, *n*-butanol and isobutanol. The methyl group (R′) of the methanol is donated to the fatty acyl group (R-COO−) of the triglycerides so that a molecule of fatty acid methyl ester is formed. Thus all the three fatty acids of a triglyceride molecule are ultimately converted to methyl esters on reaction with methanol. As the reaction proceeds the reaction yield is also decreased due to the loss of catalytic activity of lipase as the essential water layer that cover the enzyme is solubilized and removed by methanol since the methanol is a hydrophilic solvent (Koskinen and Klibanov 1996; Kaieda et al. 2001). Water content in the reaction medium is required to be maintained at the optimal level because an insufficient amount of water may result in the irreversible inactivation of the lipase as a result of denaturation of the enzyme by methanol. Reaction of methanol with fatty acid during transesterification by lipase is given below.

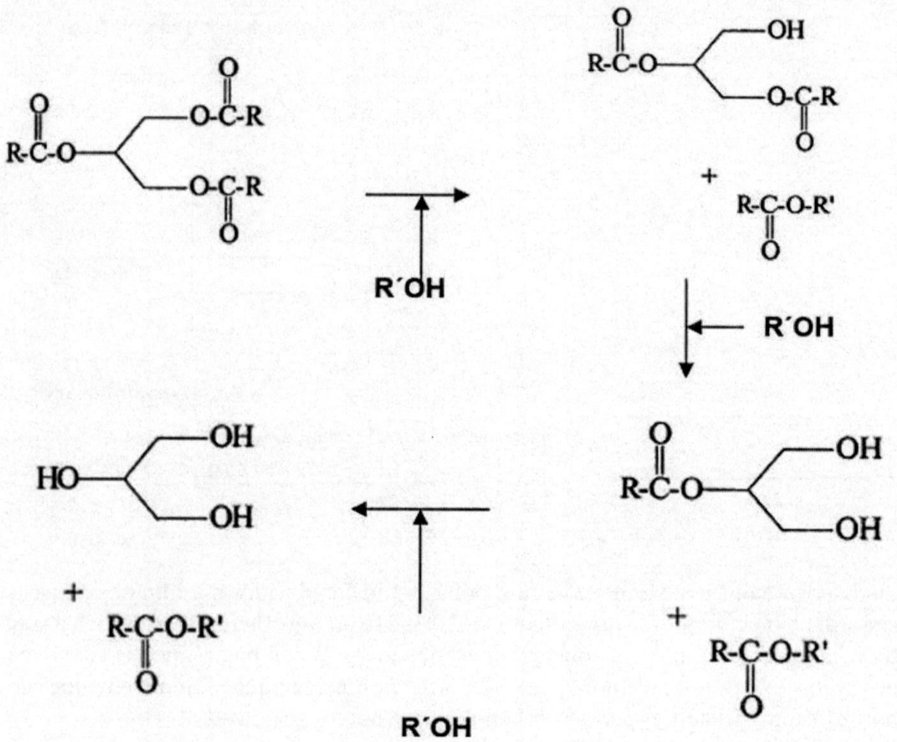

R = Fatty acids present in triglyceride; R′ = CH_3

Use of methanol and ethanol is economically feasible but they are known to deactivate the enzyme. The glycerol which is formed as by-product during the process adsorbs as a monolayer on the lipase and cause the inactivation of the enzyme. The extent of deactivation is inversely proportional to the number of carbon atoms present in the alcohol. In that way methanol is the most deactivating alcohol. Efficiency of the lipase increases when the length of the carbon chain of the alcohol is more. Thus the use of ethanol is more advantageous than methanol. Ethanol has an advantage over methanol as ethanol is manufactured from renewable sources. Methanol is prepared from fossil fuel sources. As the production of ethanol is bound to increase in the near future there is a chance that ethanol application will be preferred for enzymatic transesterification. Normally if the alcohol is added in excess, the transesterification proceeds at a faster rate to a successful end. In such event a portion of the alcohol may remain non-reactive and stay as droplet. The droplets coat the enzyme and deactivate it. If the alcohol has less than three carbon

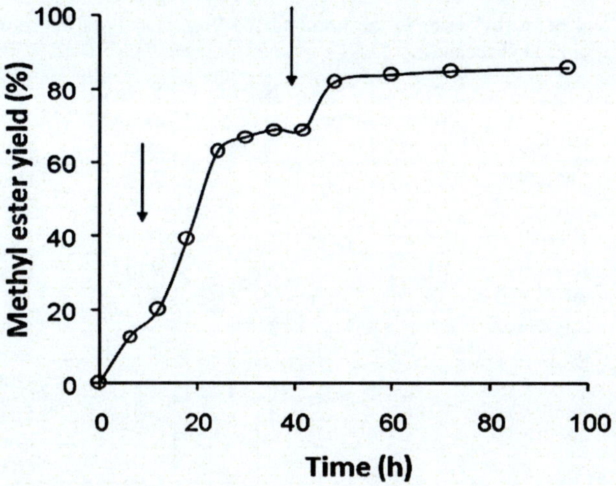

Fig. 44.14 Transesterification of neem oil with three stepwise addition of methanol for biodiesel production. Arrows indicate the addition of 1 M equiv. of methanol (Indumathi 2013)

atoms, it is likely to inhibit the enzyme as the concerned alcohol's solubility is less than the stoichiometry. Methanol and ethanol are soluble at 1/2 and 2/3 of their stoichiometric ratios.

Alcohol is known to separate the monolayer of water which is essential for enzyme functioning (Bernardes et al. 2007). The reaction yield is also reduced due to the loss of catalytic activity of lipase as the essential water layer that cover the enzyme is removed by methanol as methanol is a hydrophilic solvent (Koskinen and Klibanov 1996; Kaieda et al. 2001). Water content in the reaction medium should be maintained at the optimal level required for the reaction. Insufficient amount of water results in the irreversible inactivation of the lipase due to denaturation of the enzyme by methanol.

In order to overcome the deactivating effect, it is suggested lower chained alcohols are either added stepwise or sequentially. Stepwise addition methanol is considered favourable (Fig. 44.14 and Table 44.10). As ethanol has minimum deactivation effect, ethanol is normally not considered for stepwise addition. Methanol deactivation normally takes place if methanol is added at a methanol to oil ratio not less than 3.

When methanol is added at 3 mol equiv. the methyl ester yield dwindled slowly during the progress of the reaction. Instead, the whole process proceeds on a higher scale if the above quantity of methanol is added in split doses of 1 mol equiv. each.

44.3.14 Solvents

Insolubility of the lower chained alcohol is one of the reasons for the deactivation of the enzyme. Therefore, solvents are used to improve the alcohol stability thereby to

Table 44.10 Yield of methyl ester in the immobilized lipase catalysed reaction with stepwise addition of methanol as acyl acceptor

Enzyme	Feedstock	Molar ratio of methanol to oil	No. of split doses	Yield of methyl ester	Authors
Candida antarctica	Sunflower oil	3:1	4	93.0	Deng et al. (2005)
C. antarctica	Tallow	3:1	3	74.0	Lee et al. (2002)
Candida sp. 99-125	Rapeseed oil	3:1	3	83.0	Deng et al. (2005)
Pseudomonas fluorescens	Sunflower oil	4.5:1	3	>95	Soumanou and Bornscheuer (2003)
Rhizomucor miehei	Sunflower oil	3:1	3	>80	Soumanou and Bornscheuer (2003)
Thermomyces lanuginosus	Sunflower oil	3:1	3	>60	Soumanou and Bornscheuer (2003)

restrain the deactivation to a possible extent (Kumari et al. 2009). Often the by-product, namely the glycerol also coats the enzyme. Besides the lipase has to work on a medium which is more viscous. In such events solvents are considered to induce the solubility of the glycerol (Royon et al. 2007). Thus the reactant and the product require the support of solvents. One of the most important criteria in selecting a non-aqueous solvent is its compatibility with the maintenance of the catalytic activity of the reactants. Therefore, a common solvent would help to carry the job. Such solvent is required to maintain the homogeneity and viscosity of the mixture besides it would improve the stability of the enzyme. Homogeneous mixture limits the problems associated with multiple phase reaction mixture and the reduced viscosity helps to contain problems associated with the mass transfer. Addition of solvents helps to increase the reaction rate in comparison to that of the solvent free systems (Vasudevan and Briggs 2008). Some of the solvents used in the transesterification processes are hexane, isooctane, *n*-heptane, petroleum ether, cyclohexane, 2-butanol and tertiary butanol. Among the above solvents, tertiary butanol is often considered most useful since it is moderately polar and uninfluenced by the polarity of water or any other components of the mixture (Fjerbaek et al. 2009). Tertiary butanol and 2 butanol are used in the treatments for the regeneration of the deactivated lipase (Robles-Medina et al. 2009).

Use of solvents is not free from problems. Additional processing is required to separate the solvents from the biodiesel. The solvents are volatile in nature and thus bound to be inflammable. Addition of solvents tends to increase the capacity in terms of volume of the reactors. Thus extra investment may be required to combat the problems associated with the addition of solvents. Solvents most commonly used in biodiesel production are listed in Table 44.11.

Table 44.11 Application of chosen solvents in the transesterification of different oils using microbial enzymes

Solvents	Feedstock	Enzyme	Yield of methyl ester (%)	Reference
Petroleum ether	Sunflower oil Rapeseed oil Waste oil	*Candida antarctica* *Candida* sp. 99-125 *Candida* sp. 99-125	79 83 92	Mittelbach (1990) Deng et al. (2005), Nie et al. (2006), Tan et al. (2006) Deng et al. (2005), Nie et al. (2006), Tan et al. (2006)
n-Hexane	Salad oil Sunflower oil Soybean oil Sunflower oil	*Candida* sp. 99-125 *Rhizomucor miehei* *Rhizomucor miehei* *Thermomyces lanuginosus*	95 >80 92 >60	Deng et al. (2005), Nie et al. (2006), Tan et al. (2006) Soumanou and Bornscheuer (2003) Shieh et al. (2003) Soumanou and Bornscheuer (2003)
n-Heptane	Soybean oil	*Pseudomonas fluorescens*	92	Lou et al. (2006)
Propanol	Sunflower oil	*Candida antarctica*	93	Deng et al. (2005)
Tert-butanol	Rapeseed oil Cottonseed oil	*Rhizopus oryzae* *Candida antarctica*	95 97	Li et al. (2007) Royon et al. (2007)

44.3.15 Agitation Speed

In case of immobilized catalysts, the reactants need to move from the mixture to the external surface of the support system and subsequently to the interior of the pores. In such events the mass transfer limitations can be minimized by carrying out the reaction at an optimum speed of agitation. The agitation is largely governed by the nature of the substrate, temperature, molar ratio of the alcohol and also on the nature enzyme used. Chosen agitation speed is given in Table 44.12.

44.3.16 Lipase Treatment

Immobilized lipases are often soaked in suitable medium so as to activate the enzyme. If an inactive enzyme is immersed in a polar organic solvent, the closed active site is opened so as to enhance its activity (Jegannathan et al. 2008). Isopropanol, methyl oleate and tertiary butanol are often employed as a soaking medium to activate the enzyme.

Enzymes are also activated by immersing them in the feed stock for a long time before they are used in the subsequent process. Often glutaraldehyde treatment is

Table 44.12 Agitation speed (rpm) followed in various processes involving immobilized enzymes

rpm	Substrate	Enzyme	Yield of ester (%)	Reference
130	Sunflower oil	*Candida antarctica*	97	Belafi-Bako et al. (2002)
	Rapeseed oil	*Thermomyces lanuginosus*	95	Li et al. (2007)
150	Soybean oil	*C. antarctica*	92	Du et al. (2004)
	Jatropha oil	*C. antarctica*	93	Modi et al. (2006)
	Soybean oil	*Rhizopus oryzae*	71	Matsumoto et al. (2001)
180	Rapeseed oil	*Candida* sp. 99-125	83	Deng et al. (2005), Nie et al. (2006), Tan et al. (2006)
	Salad oil	*Candida* sp. 99-125	95	Deng et al. (2005), Nie et al. (2006), Tan et al. (2006)
	Waste oil	*Candida* sp. 99-125	92	Deng et al. (2005), Nie et al. (2006), Tan et al. (2006)
	Vegetable oil	*Candida* sp. 99-125	96	Deng et al. (2005), Nie et al. (2006)
200	Tallow	*C. antarctica*	74	Lee et al. (2002)
	Mahua oil	*P. cepacia*	96	Kumari et al. (2007)
	Jatropha oil	*P. cepacia*	98	Shah and Gupta (2007)
	Sunflower oil	*P. fluorescens*	>95	Soumanou and Bornscheuer (2003)
	Sunflower oil	*Rhizomucor miehei*	>80	Soumanou and Bornscheuer (2003)
	Sunflower oil	*T. lanuginosus*	>60	Soumanou and Bornscheuer (2003)
	Jatropha oil	*Chromobacterium viscosum*	92	Shah et al. (2004)

given to activate the function of the enzyme. When the immobilized enzyme is appropriately pretreated the activity increases multifold. The ability of the enzyme to resist deactivation by methanol is also high.

44.3.17 Enzyme Deactivation and Regeneration

The linear alcohols (methanol, ethanol, propanol and butanol) are toxic to the enzymes in varying degrees. The degree of deactivation is inversely proportional to the number of carbons in the linear alcohols. The degree of deactivation by the branched alcohols (isopropanol, 2-butanol and isobutanol) is relatively low. One of the main causes of deactivation of the enzyme is the poor miscibility between the feed material and the alcohol. Alcohols are often attracted by the materials used in the immobilization support.

Getting back the efficiency of the used enzyme is one of the challenges in the economy of the enzyme-based transesterification. To economize the whole operation, it is necessary to use the enzyme repeatedly. The used enzymes are often washed by certain solvents before it is put on use in the next batch. Washing the enzyme by tertiary butanol helps to maintain the potency even beyond 100 cycles (Huang et al. 2010; Li et al. 2007). Similarly, isopropanol gives fairly good result (Lee et al. 2008). However certain solvents such as hexane offer limited success (Salah et al. 2007).

44.3.18 Methyl Acetate as Acyl Acceptor

In order to combat the problems created by straight alcohols, it is suggested to have methyl acetate as an acyl acceptor. Methyl acetate will enhance the stability of the lipase. If methyl acetate is used as acyl acceptor the by-product is triacetylglycerol which has no adverse effect on the activity of lipase. The reactant and product may please be seen in the following equation where the transesterification pathway of triglycerides and methyl acetate is indicated.

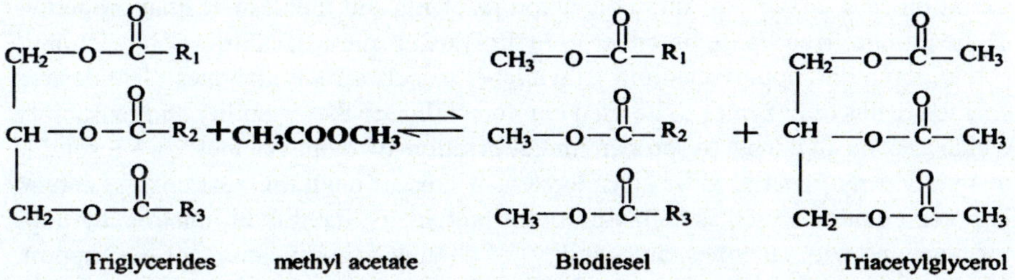

In the event of using methyl acetate as an acyl acceptor in a normal course, more lipases and more methyl acetate are required.

44.3.19 Dimethyl Carbonate as Acyl Acceptor

Dimethyl carbonate also is recommended to be used. Dimethyl carbonate (DMC) is neutral, cheap, odourless, non-corrosive and non-toxic. It has effective solvent properties. The reaction is irreversible. Dimethyl carbonate is eco-friendly. The conversion is higher than that of other acyl acceptors.

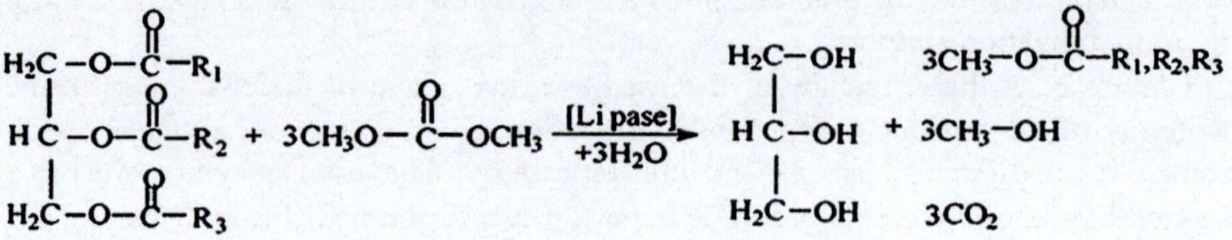

Transesterification of vegetable oil with dimethyl carbonate

44.3.20 Immobilized Lipase in Biodiesel Production

The success of any enzymatic process largely depends on the effective immobilization of the enzymes. Successful immobilization is necessary to recover the enzyme and reuse it (Noureddini et al. 2005). Immobilized lipase may also be employed in continuous process where there is adequate control of reactions and thus economically feasible.

The immobilized lipase normally has a good thermal stability, storage stability and reusability. Hence it is a potential candidate for large scale application (Wang et al. 2006a). Lipase is more reactive in silicone oils and silicone elastomers than in hydrocarbons. Selection of a suitable immobilization strategy is mainly based on the process specifications for overall enzymatic activity, effectiveness of its utilization, its deactivation and regeneration properties, overall cost involved in the immobilization process, toxicity of immobilization reactants and the desired final properties of the immobilized lipase on either in hydrolysis or in transesterification reactions.

No common support medium is available to suit all the enzymes. The desired characteristics of enzymes to be used commercially are their stability and reusability which can be obtained by proper immobilization of them on inert and insoluble materials. Any material to be considered as a support medium must possess certain important qualities such as high affinity for protein, availability of functional groups for reactions with enzymes, amenability for chemical modifications of the support, mechanical stability, rigidity, feasibility of repeated regeneration, high loading capacity, low biodegradability and non-toxicity (Rahman et al. 2005; Mateo et al. 2007).

Supports chosen for multipoint immobilization should have adequate internal surface so as to have a good geometrical congruence with enzyme surface, availability of large number of reactive sites on the surface, development of minimal steric hindrances between the reactive sites of the enzyme and the support and stable bonding with the enzyme. After immobilization, it should be convenient to block or destroy the remaining reactive groups of the support. Reaction duration, pH of the medium, temperature, buffers and inhibitors or other protein protectors are the major factors often govern the successful immobilization (Mateo et al. 2007).

A successful support material shall have high specific surface area and numerous active sites so as to fix the enzyme molecule. If such support particles are made, as small as nanoparticles there may be better distribution of enzymes since the surface area offered is relatively larger.

Lipase from *Pseudomonas cepacia* immobilized in a phyllosilicate sol-gel matrix with silicate polymers produced by controlled hydrolysis of tetramethyl orthosilicate was observed to be more stable and had higher activity compared to that of free lipase. Also, the phyllosilicate sol-gel-immobilized lipase was reusable for at least five transesterification cycles without major loss of activity.

Pseudomonas fluorescens lipase immobilized in porous kaolinite particle expressed higher activity in the transesterification reaction when compared to that of the free lipase. Immobilization of microbial enzymes through microencapsulation on chitosan beads was greatly influenced by the molecular weight of the enzyme and the microporosity of the support which in turn influence the diffusion of the substrate and products thereby varying the reaction kinetics. Due to the large surface area, quite disproportionate to the volume of the particulate size offered by nanostructured materials, the enzyme loading and the consequent catalytic efficiency of the immobilized enzyme greatly increased. However, such particles disperse in the reaction medium making their recovery and reuse difficult. In such context nanofibres can overcome this problem and can be applied in continuous operations.

Lipase from *P. cepacia* and *C. rugosa* entrapped in a sol-gel structure prepared by polycondensation of hydrolysed tetra methoxy silane and iso-butyl tri methoxy silane was employed in the transesterification of soybean oil. Such immobilized lipase was observed to be stable in repeated uses.

The concept of immobilization came in to force in 1953. Incorporating the enzymes in semisolid carrier is called enzyme immobilization. Such protected enzymes are named as immobilized enzymes. The immobilization of the enzyme depends on the specific reactions between the enzymes and the carrier. Immobilized enzymes have more stability than free enzymes as they are firmly attached to solid materials. Wastage of enzymes during extraction is avoided. It enhances reuse thereby enabling overall economy. Enzyme activity of lipase is high during continuous batch operation. Thus immobilization enables better control over the process with more benefits. In 1969 large scale application of the immobilized enzyme was attempted. The use of lipase is cost effective when it is employed in an immobilized form. The limitations of lipase catalysed transesterification such as loss of stability, wastage of enzyme and the cost prohibitiveness can be overcome by immobilizing the free enzyme.

Immobilization of the enzyme inside the porous structure of a solid causes the dispersion of the enzyme and the possibility of interacting with the enzyme by any other non-target component of the feedstock is effectively minimized. Besides stabilization, the immobilization prevents aggregation, autolysis of the enzyme. The immobilized enzyme molecules inside a porous support will not be in contact with any air bubbles generated during vigorous stirring thereby averting any inactivation due to air bubbles. Figure 44.15 indicates the impact of bubble on the enzyme. When the enzyme is safely in position inside the pore, bubble is unable to enter and thus the functioning of the enzyme is not affected.

At the same time any attachment of enzyme on nonporous solid and smooth support has a risk of being attacked by bubbles. Thus porous media are best preferred.

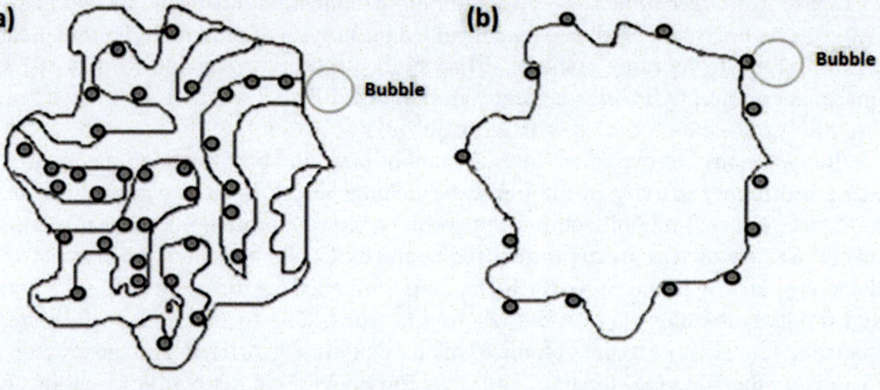

Fig. 44.15 The advantage of porous support and the stability of the enzyme in a medium having bubbles. (**a**) Bubble could not enter the pore to disturb the functioning of the enzyme. (**b**) The enzyme remains susceptible to bubble

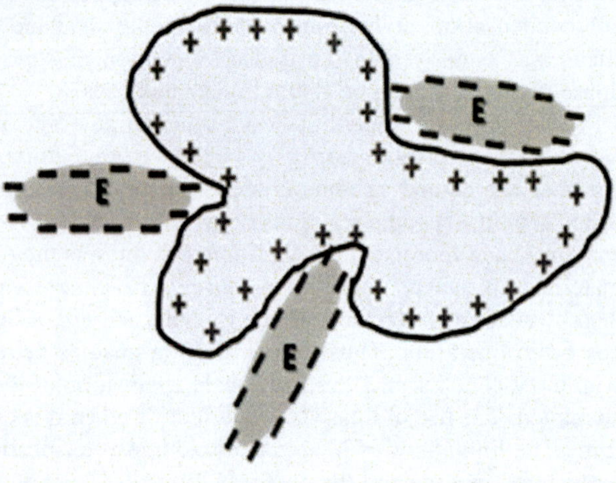

Fig. 44.16 The enzyme seen attached on the surface of the support through the mechanism of adsorption

Six major types of immobilization are available. They are:

1. Adsorption
2. Entrapment
3. Encapsulation in semi-permeable membrane
4. Covalent attachment
5. Cross linking
6. Ionic binding

44.3.20.1 Adsorption

Adsorption is the simplest method of immobilization. Enzyme is attached to the surface of the hydrophobic supports (Fig. 44.16) by the combination of hydrophobic, Van der Waals, or electrostatic forces (Yong and Al-Duri 1996; Fernandez-Lafuente

Table 44.13 Carriers used in the immobilization of lipase through adsorption

Carrier	Lipase	Reference
Polystyrene	*Pseudomonas cepacia*	Li and Yan (2010)
Carbon cloth	*P. cepacia*	Naranjo et al. (2010)
Polyacrylonitrile	*P. cepacia*	Sakai et al. (2010)
Ceramics	*P. cepacia*	Shah and Gupta (2007)
Hydrophilic resins	*Rhizomucor miehei*	DePaola et al. (2009)
Silica	*P. fluorescens*	Chen et al. (2009a, b), Salis et al. (2009)
Mg-Al hydrotalcite	*Saccharomyces cerevisiae*	Zeng et al. (2009)
Resin D4020	*Penicillium expansum*	Li et al. (2009)
Polymethacrylate	*Pseudomonas fluorescens*	Salis et al. (2009)
Organosilicate	*P. fluorescens*	Salis et al. (2009)
Hydrotalcite	*Thermomyces lanuginosus*	Yagiz et al. (2007)
Zeolites	*T. lanuginosus*	Yagiz et al. (2007)
Activated carbon	*Candida antarctica*	Orcaire et al. (2006)
	C. rugosa	Moreno-Parajàn and Giraldo (2011)
Celite	*C. rugosa*	Shah and Gupta (2007)
	P. cepacia	Shah and Gupta (2007)
	Pseudomonas aeruginosa	Ji et al. (2010)
	P. fluorescens	Shah and Gupta (2007)

et al. 1998). The most common carriers used in the adsorption are toyonite, celite, cellulose polypropylene, acrylic, silica gel, textile membranes spherosil, sepharose, sephadex and siliconized glass (Malcata et al. 1990; Jegannathan et al. 2008). Few carriers used in adsorption are porous glass and ceramics, sand, cellulose, synthetic polymers, activated carbon and metallic oxides (Klibanov 1983). This technique may have a higher commercial potential because it is (a) simpler, (b) less expensive, (c) without chemical additives (d) capable of carrying large mass transfer and (e) with high catalytic activity (Fukuda et al. 2001; Gao et al. 2006).

The adsorption of lipase on to porous support is one of the most widely employed methods in continuously operated packed beds and stirred tank reactors (Gao et al. 2006). The major limitations of adsorption are the risk of the enzyme being stripped off the support. The stability of the enzyme when adsorbed is very low, which makes the reuse of the enzyme difficult when compared to other immobilization methods (Jegannathan et al. 2008). There is a possibility of leakage of the enzyme from the support during mixing. Certain carriers used in the immobilization of lipase through adsorption are given in Table 44.13.

44.3.20.2 Entrapment

Entrapment entails the capture of lipase and arresting the movement within the lattice structure of a gel (Cheetham et al. 1979; Malcata et al. 1990; Shtelzer et al. 1992). Lipases that are immobilized by entrapment are more stable and display better activities than those immobilized by adsorption (Malcata et al. 1990). Gels often used are methylene bisacrylamide, calcium alginate and kappa carrageenan (Klibanov 1983).

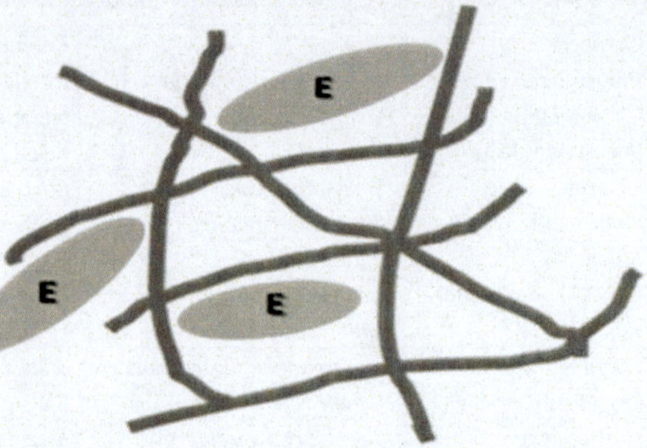

Fig. 44.17 Lipase seen trapped in a lattice structure

Table 44.14 Carrier used for lipase immobilization through entrapment

Carrier	Lipase	Reference
Carrageenan	*Candida antarctica*	Jegannathan et al. (2010)
	C. rugosa	Jegannathan et al. (2010)
	Burkholderia cepacia	Jegannathan et al. (2009, 2010)
	Pseudomonas fluorescens	Jegannathan et al. (2010)
	Aspergillus niger	Jegannathan et al. (2010)
Silica gel	*Thermomyces lanuginosus*	
	P. cepacia	Noureddini et al. (2005)
Celite supported sol-gel	*Candida antarctica*	Meunier and Legge (2010)
	Lipase NS44035	Meunier and Legge (2010)
Silica aerogel	*C. antarctica*	Nassreddine et al. (2008)
	C. antarctica	Orcaire et al. (2006)
	Burkholderia cepacia	Orcaire et al. (2006)

The procedure used to entrap the lipase is relatively simple, quite robust and easy to make continuous operation but the cost factor is not as favourable as adsorption (Meter et al. 2007). The biggest disadvantage to entrapment is the mass transfer. A diagrammatic representation of lipase trapped in a lattice structure is shown in Fig. 44.17.

Temperature induced gelatin, polymerization and ionotropic gelation of macromolecules with multivalent cations are some of the procedures followed in entrapment. Some of the carriers used in lipase entrapment process are given in Table 44.14.

44.3.20.3 Encapsulation

Encapsulation is almost like entrapment but encapsulation involves the confinement of the enzyme in small beads or capsules (Malcata et al. 1990; Serralheiro et al. 1990; Vicente et al. 1994; Khan and Vulfson 2001). Figure 44.18 indicates how the enzymes are confined to a capsule. The encapsulation helps to separate the enzyme from the reaction mixture with a semipermeable membrane. In such process the

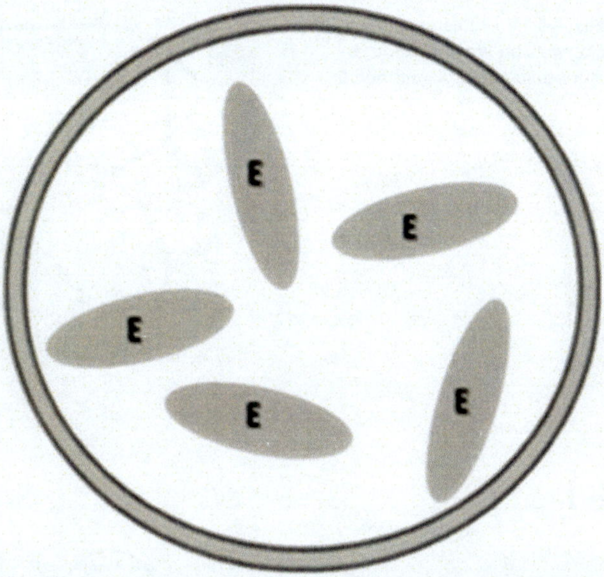

Fig. 44.18 Diagrammatic representation of confinement of enzyme inside a semipermeable membrane

conversion is low because of the limited permeability of the membrane which adversely affects the activity of lipase on large molecules such as triglycerides (Malcata et al. 1990). Large molecule cannot pass through the pores of the membrane. It is also known that the membrane may become clogged or a thin layer may be formed around the capsule either of which would severely limit the reaction (Antczak et al. 2009; Fjerbaek et al. 2009).

44.3.20.4 Covalent Attachment

The immobilization is often effected by establishing covalent bonds between solid supports and functional groups of the amino acid residues present at the surface of the enzymes. Normally the support is activated chemically so as to make its functional groups strongly electrophilic. These functional groups then react with the nucleophilic groups of the enzymes. The covalent bonds are usually strong and offer high stability. Such immobilization is relatively costlier and the yields are moderate. Figure 44.19 illustrates the mode of attachment of the enzyme with the support through covalent bond.

The covalent bonds are of two types. They are

1. single point covalent attachment and
2. multipoint covalent attachment

Normally the single point covalent bond immobilization has long spacer arms and therefore it is not rigid but in multipoint covalent attachment of enzymes the enzymes are attached with the support with many short spacer arms thereby causing the whole complex very rigid (Fig. 44.20). This phenomenon actively reduces the

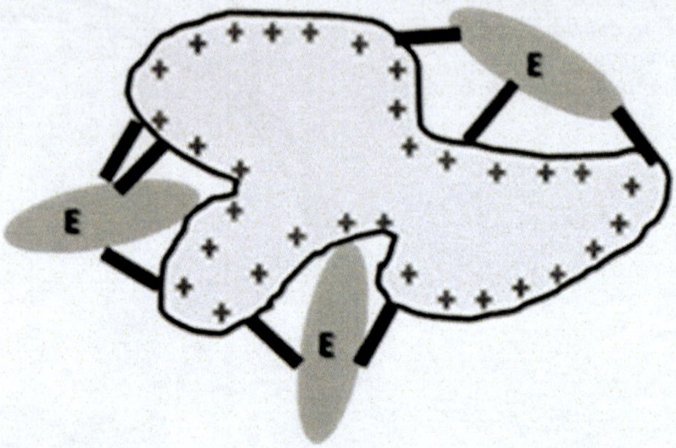

Fig. 44.19 Attachment of enzymes on support through more than one covalent bonds

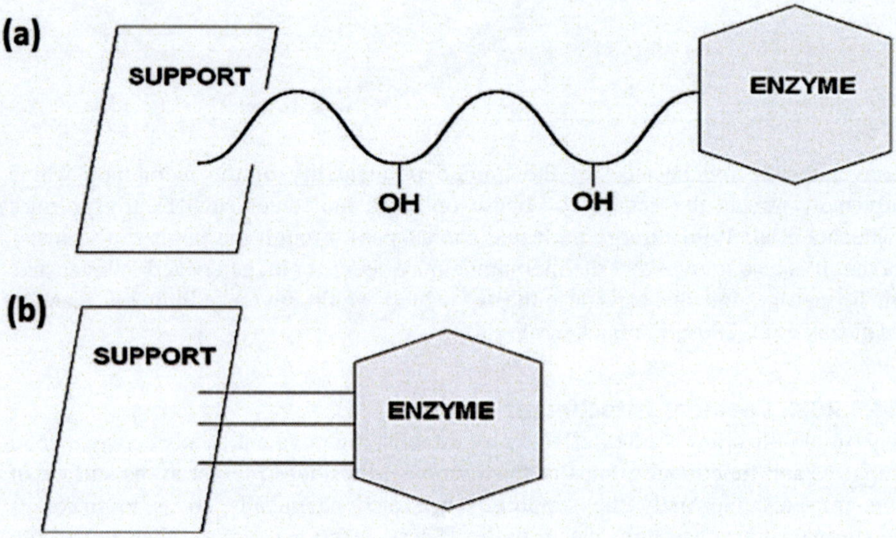

Fig. 44.20 Illustration of the type of covalent attachment between support and an enzyme (**a**) single point attachment and (**b**) multiple point attachment

conformational changes involved in enzyme inactivation which increases the stability of the enzyme. Various carriers employed in the lipase immobilization through covalent bonds are listed in Table 44.15.

44.3.20.5 Cross Linking

In cross linking the lipase molecules are attached one another through the use of chemicals (Fig. 44.21) so as to form a more robust structure. Such product has a high concentration of enzyme per unit volume (Malcata et al. 1990; Lopez-Serrano et al. 2002). The reagents used include glutaraldehyde, bisdiazobenxidine and hexamethylene diisocyanate, among which the most commonly used is glutaraldehyde

Table 44.15 Carrier used for lipase immobilization through covalent attachment

Carrier	Lipase	Reference
Olive pomace	*Thermomyces lanuginosus*	Yücel (2011)
Resins	*T. lanuginosus*	
	Pseudomonas fluorescens	Mendes et al. (2011)
Polymers	*T. lanuginosus*	Dizge et al. (2009a, b)
Polyurethane foam	*T. lanuginosus*	Dizge and Keskinler (2008)
Chitosan	*Candida rugosa*	Shao et al. (2008)
Resin	*T. lanuginosus*	Rodrigues et al. (2010)
Silica	*Rhizopus oryzae + Candida rugosa*	Lee et al. (2008)
	Enterobacter aerogenes	Kumari et al. (2009)
Magnetic nanostructures	*C. rugosa* *C. antarctica*	Dussán et al. (2007) Xie and Ma (2010)

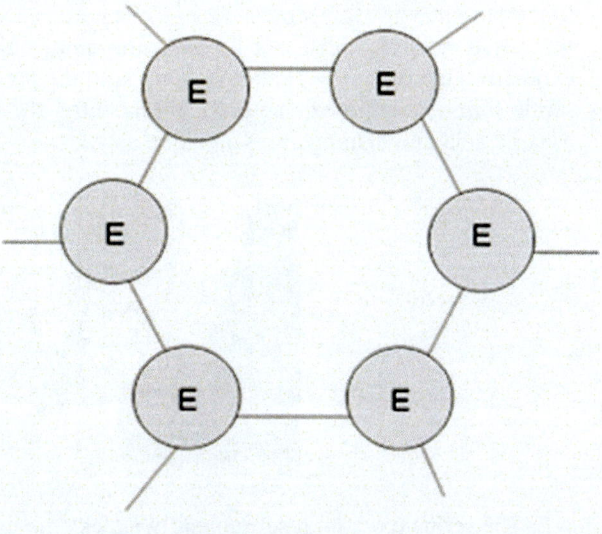

Fig. 44.21 The lipase molecules attached one another through the use of chemicals

(Jegannathan et al. 2008). Immobilization of this sort does not involve any matrices. Cross linking occurs both inter molecularly and intra molecularly (Klibanov 1983). The cross linked enzyme accelerates the rate of transesterification.

The glutaraldehyde treatment as a matter of cross linking is to be done with great care using appropriate concentration and temperature. If not there will not be uniform linking.

44.3.20.6 Ionic Binding

It is an important mode of adsorption in which the immobilization is through electrostatic forces of the differently charged ionic groups of the support and enzymes. The degree of ionic binding conformation is slightly higher than adsorption and lower than covalent binding. There is no universal support medium commonly available to all the enzymes. Any support material shall have high affinity for protein and the functional groups of it shall freely react with enzymes. Besides the support is to be amenable to chemical modifications and capable of withstanding mechanical stability and rigidity. There shall be scope for regeneration, high loading capacity, biodegradability and non-toxicity.

Supports for multipoint immobilization through covalent binding shall have large internal surface to have good geometrical congruence with the enzyme surface. If the carrier particles are made as small as nanoparticles, there will be scope for better distribution of enzymes since the surface area offered is disproportionately larger against unit volume. The action of the lipase is initiated by a charge relay system at the active site of the lipase (Al-Zuhair 2007). Such active site has a triad of amino acids, namely aspartate, histidine and serine. The carboxylate group of aspartic acid (COO^-), nitrogen of histidine and alcohol group of serine are involved in the reaction. The general mechanism of lipase is illustrated below in six steps.

Step 1:

In the presence of triglycerides the enzyme commences its activity and the lid at the active site is opened. As a result, the amino acids at the active sites are exposed to the reactants. Histidine along with the proton of the alcohol group of serine forms a serine oxy anion. Consequently, the oxy anion is stabilized by aspartic acid and histidine present at the active site.

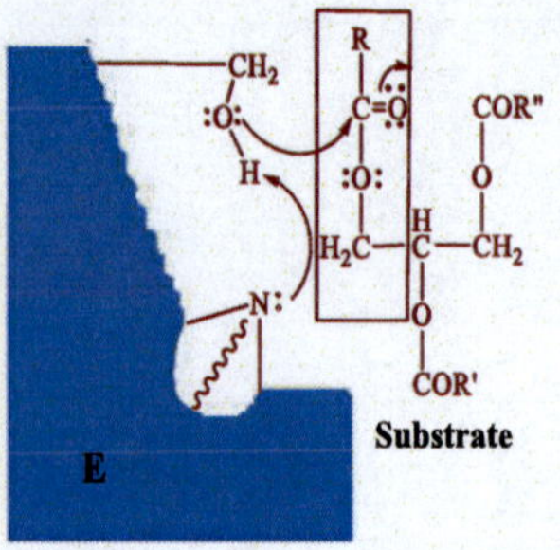

Step 2: The serine oxy anion subsequently attacks the carbon atom of the carbonyl group present in the substrate. Then the serine is linked to a fatty acid of triglyceride through ester linkage forming a tetrahedral intermediate (ES Complex).

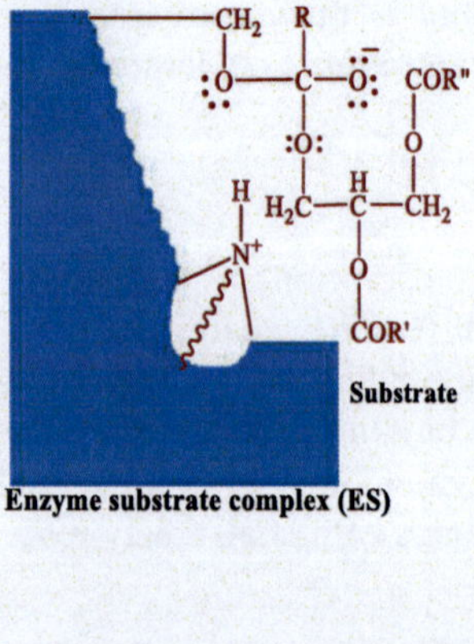

Enzyme substrate complex (ES)

Step 3: The electron on the oxy anion is then pushed back to the carbonyl group. The proton (H^+) present on the histidine is transferred to the diglyceride which is subsequently released for forming acylated enzyme (enzyme linked to acyl group of a fatty acid).

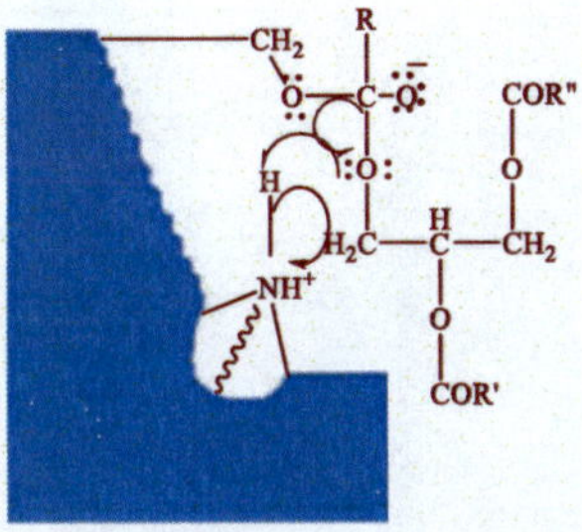

ES

Step 4: The serine ester thus formed reacts with the alcohol to complete the transesterification process. The nitrogen in histidine removes the hydrogen from the alcohol molecule so as to form an alkyl oxide anion. Tetrahedral intermediate 2 is formed.

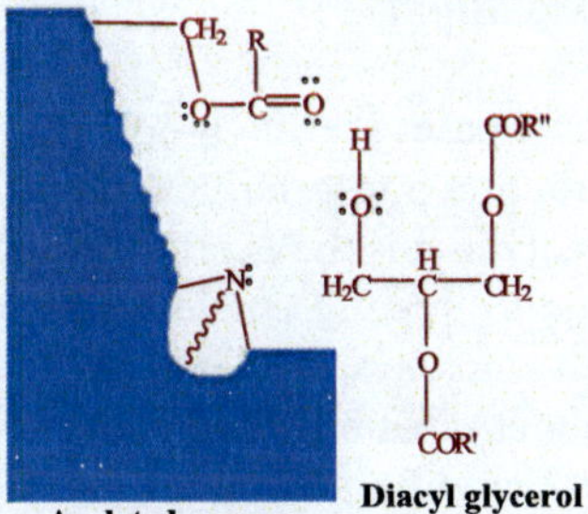

Acylated enzyme

Step 5: The electrons are pushed back to the carbonyl carbon atom. The histidine residue donates a proton to the oxygen atom of the active serine residue.

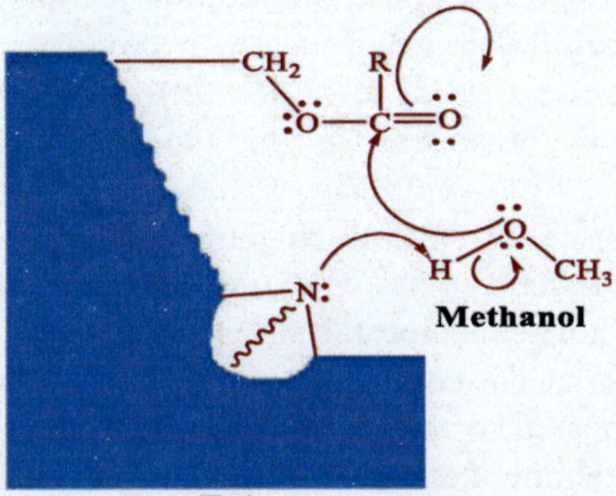

E-Ac

Step 6: At this stage ester bond existing between serine and acyl component is broken. The free fatty acid methyl ester is then released. The catalytic triad at the active site is ready for the next cycle.

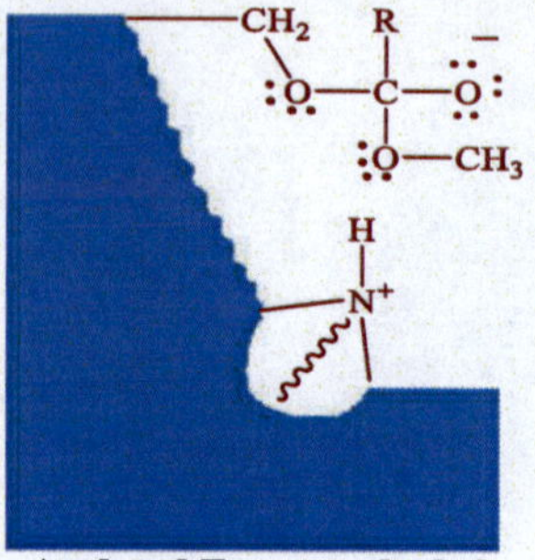

Acylated Enzyme alcohol complex (E-AcA)

44.3.21 Whole-Cell Biocatalyst

Application of extracellular lipase in the biodiesel production has a definite eco-friendly edge over the routine transesterification process involving alkaline or acid catalyst. The only adversity in it is the exorbitant cost of enzyme involved. As a result, efforts are being made to directly employ the microbes responsible for the production of enzyme. Such microbes are referred as whole-cell biocatalysts. In these organisms the lipase either remains inside the cell or localized in the cell wall or cell membrane. Among the different microbes, the filamentous fungi offer far-reaching scope (Ban et al. 2001; Fujita et al. 2002; Narita et al. 2006). Open demonstration of the application of the fungus *Rhizopus chinensis* was initially made by Ban et al. (2001). The use of another fungus *R. oryzae* also came to light (Oda et al. 2005; Hama et al. 2008; Li et al. 2008). The specialty of *R. oryzae* is that it produces a 1,3 position specific lipase. In addition to fungi, efforts are made on bacteria *Escherichia coli* (Kalscheuer et al. 2006) and yeast *Saccharomyces cerevisiae* (Matsumoto et al. 2002) to utilize them in biodiesel production. The employment of different organisms in the biodiesel production is given in Table 44.16.

It is known that exposure of the cells to glutaraldehyde increases the enzymatic ability of the cells over several cycles. One of the parameters observed to enhance the durability of the whole-cell biocatalysts is the type of culture method being followed (Oda et al. 2005). Cells cultivated in an air lift system gave a higher yield of methyl ester when compared to that of shake flask culture. Enzyme activity also increased when the mycelium of the fungus (*R. chinensis*) was treated with yatalase (chitinase) just prior to its use in biodiesel production. The yatalase partially dissolves the cell wall and allows the membrane bound lipase to have an easy

Table 44.16 The whole-cell biocatalysts immobilized in biomass support particles (BSPs) for biodiesel production

Lipase source	Oil source	Methyl ester (%)	Reference
Aspergillus niger	Waste cooking	86.4	Xiao et al. (2010)
A. niger	Palm	>90	Xiao et al. (2011)
A. niger	Palm	87	Xiao et al. (2010)
A. oryzae NS4	Soybean	98	Adachi et al. (2011)
A. oryzae carrying r-CALB	Soybean	85–90	Adachi et al. (2013)
A. oryzae expressing r-FHLc	Rapeseed	96, 94[e], 96[p], 97[b]	Koda et al. (2010)
Escherichia coli BL21	Rapeseed	97.7	Gao et al. (2009)
Rhizopus chinensis CCTCC M201021	Soybean	>86	He et al. (2008)
Rhizomucor miehei display *Pichia pastoris*	Soybean	83.1	Huang et al. (2012)
Rhizopus oryzae IFO 4697	Refined rapeseed Crude rapeseed Acidified rapeseed	~60 ~60 ~70	Li et al. (2007)
R. oryzae IFO 4697	Soybean	~90	Sun et al. (2010)
R. oryzae IFO 4697	Soybean	~85	Sun et al. (2011)
R. oryzae IFO 4697	Soybean	71	Matsumoto et al. (2001)
R. oryzae IFO 4697 and *A. oryzae* nia D300 (*combined*)	Soybean	~100	Arai et al. (2010)
R. oryzae ATCC 24563	Soybean	97	Lin et al. (2011)
R. oryzae IFO 4697	Soybean	72	Li et al. (2007)
R. oryzae	Soybean	90	Hama et al. (2007)
R. oryzae	Jatropha	80	Tamalampudi et al. (2008)
R. oryzae IFO 4697	Soybean	86	Zeng et al. (2006)
R. oryzae	Rapeseed	83 79[e], 93[p], 69[b]	Koda et al. (2010)
Serratia marcescens YXJ-1002	Grease	97	Li et al. (2012)

[e]Ethanol, [p]1-propanol, [b]1-butanol

exposure to the substrate. The efficiency of the cell also depends on the permeability of the plasma membrane. Cells which are exposed to unsaturated fatty acid such as oleic or linoleic acid expressed enhanced activity at the first phase of methanolysis.

At the same time the cells exposed to saturated fatty acids (palmitic acid) expresses stability in lipase action. The fatty acid ratio (R_f) between oleic acid and oleic acid plus palmitic acid at weight basis (o/o + p) if remained around 0.67 showed high enzyme activity and stability. (Hama et al. 2004). The *R. oryzae* is observed to produce two types of lipases having differences in molecular mass (i.e. 31 and 34 kDa). The 31 kDa lipase is localized in the cell membrane. From the membrane it easily moves into the substrate and as a result the level of enzyme at the membrane slowly reduces over a period. However, in an immobilized cell, it remains for a long duration. It is also known that immobilization inhibits the secretion and reduces the discharge of 31 kDa enzyme in the substrate. The amount of 31 kDa enzyme in the membrane greatly relied on the nature of the substrate. The concentration is high if the substrate contains more oleic acid. On the other hand, the fraction having 34 kDa is localized in the cell wall and its quantity is independent of the composition of the substrate.

One of the major advantages of the whole-cell biocatalyst is that it can effectively takes up the job even at 5–15% moisture in the substrate. However, it is known that the efficiency of the whole-cell is relatively low. This issue is suitably addressed by Hama et al. (2007) who used whole-cell biocatalyst in a packed column reactor which resulted high conversion of methyl ester in a limited reaction duration. The fungal cells capable of producing lipase are immobilized in porous biomass support particles to enhance the efficiency and also to maintain its potency in many repeated cycles.

The fungus is first cultured in a basal medium. For culturing *R. oryzae* following composition of the medium is suggested (Xiao et al. 2010). Oil 30 g, peptone 70 g, $NaNO_3$ 1.2 g, KH_2PO_4 1.2 g, $MgSO_4{\cdot}7H_2O$ 0.5 g and water 1 L. Sterilized medium is inoculated with *R. oryzae* spores from a slant and incubated in a reciprocal shaker at 35 °C for 40 h. The cell mass is then filtered, washed with tap water, stored in a refrigerator and dried in vacuum for 10–12 h. In order to enhance the microbe's withstanding capacity in the oil it is pre-incubated for a few h in the corresponding oil. The cells are then recovered from the oil and treated with methyl ester for few hours. This would greatly enhance the efficiency of the whole-cells during methanolysis.

In order to engage the same fungus repeatedly in many cycles, reticulated polyurethane foam solid blocks with a particulate voidage higher than 97% having not less than 50 pores in a linear inch are added to the medium before sterilization. If the fungus is grown in such medium, the mycelia get incorporated in the foam blocks. These foam blocks containing the microbes are easily separated from the medium, washed in sterile water and dried at low temperature for a maximum of 2 days. It is then suspended in a solution of 0.1% glutaraldehyde for 1 h with mild shaking. Subsequently it is transferred to cold (4 °C) phosphate buffer and retained in it for 5 min. Following that, it is removed and dried at 25 °C for a day and used for methanolysis. In such blocks the cell-aggregation of fungus also is observed to be high. Such aggregation is reported to enhance the level of intracellular lipase.

The methanolysis is carried out in a reciprocal shaker (130 rpm) at 37 °C. The composition of the substrate is as follows: oil 9.7 g, methanol 0.35 g and acetate buffer (pH 5.6) 0.5 mL (Zeng et al. 2006). Another two doses of methanol at the rate of 0.35 g each are added to it at 24 and 48 h of the reaction. The fungal cultures are then inoculated in it. After 60 h the substrate is removed by centrifugation and the biodiesel remaining at the upper layer also is separated.

44.3.22 Immobilized Lipase on Nanoparticles

Recently nanomaterials are used as carriers for enzyme immobilization due to their large surface area which allows high enzyme loading and improves mass transfer (Hwang and Gu 2012). There are many non-magnetic nanoparticles such as zirconia, silica, polystyrene, chitosan and polylactic acid. The reuse of immobilized enzymes on non-magnetic nanoparticles need advanced centrifugation. Magnetic nanoparticles if used are easily isolated from the products by applying external magnetic field (Marta et al. 2017).

Magnetic nanoparticles (MNPs) as carrier for lipase immobilization are known to offer high specific surface area which favours binding efficiency, minimizes mass transfer resistance and fouling, enhances reaction rate, increases turnover number (TON), prevents lipase contamination of the product and increases the cycle which favour economic consideration (Thangaraj et al. 2019). In addition, MNPs have a unique property of superparamagnetism (Vaghari et al. 2016).

TEM images of lipase covalently bonded on the surface of (3-aminopropyl) triethoxysilane (APTES) and (3-mercaptopropyl) trimethoxysilane (MPTMS) functionalized-Fe_3O_4 MNPs are shown in Fig. 44.22. Biodiesel yield of 89% and 81% was achieved by APTES-Fe_3O_4 and MPTMS-Fe_3O_4 in soybean oil under

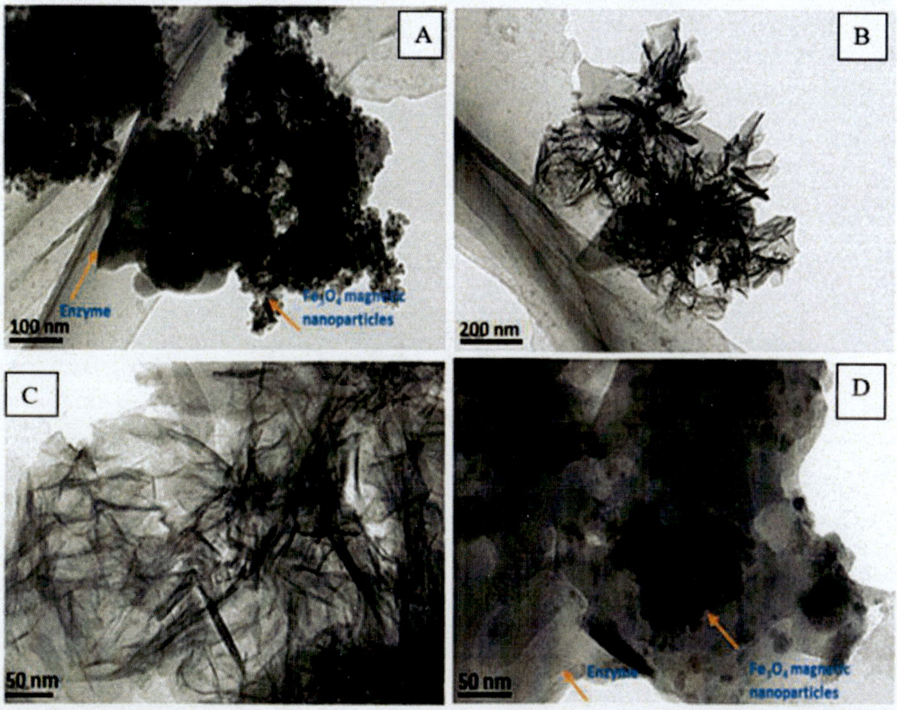

Fig. 44.22 TEM images of lipase immobilized on APTES-Fe_3O_4 (**a, b**) and MPTMS-Fe_3O_4 (**c, d**) magnetic nanoparticles. (Thangaraj et al. 2016; source: https://doi.org/10.1515/auoc-2016-0008)

optimum reaction conditions and at an oil to methanol molar ratio of 1:3 with three step addition of methanol, in a reaction temperature of 45 °C for 12 h. The immobilized lipase loses its activity gradually in every cycle due to mechanical blending which may cause the leakage of catalyst from the carrier (Thangaraj et al. 2016).

In order to improve the catalyst reusability, silica-coated MNPs were considered as carrier for lipase immobilization. Lipase-APTES-Fe_3O_4@SiO_2 and lipase-MPTMS-Fe_3O_4@SiO_2 maintained their catalytic activity up to fifth cycle which were higher than that of the lipase-APTES-Fe_3O_4 and lipase-MPTMS-Fe_3O_4 due to rich coating of silica on Fe_3O_4 (1:3). But, the biodiesel yield was moderate in Fe_3O_4@SiO_2 than that of Fe_3O_4 as the size of Fe_3O_4 increased by silica coating (Thangaraj et al. 2020). The experimental procedure of coating various ratios of silica on Fe_3O_4 magnetic nanoparticles is presented in Scheme 44.1 (Thangaraj et al. 2019).

MNPs of 3-glycidoxypropyl trimethoxysilane functionalized-Fe_3O4@SiO_2 were used as a carrier in *Candida antarctica* immobilization. The protein binding efficiency of the functionalized-Fe_3O4@SiO_2 was calculated as 84%. This preserved 97% of the specific activity of the free enzyme. The synthesized biocatalyst could be easily recovered from the reaction mixture and reused for 6 cycles without any appreciable decrease in the enzyme activity. The activity level remained unaffected until sixth cycle. In the tenth cycle the activity was 65% (Mehrasbi et al. 2017). Improvement which took place on the surface of MNPs is due to the increase in the ratio of silica coating (1:0.25, 1:0.5, 1:1, 1:2) (Fig. 44.23).

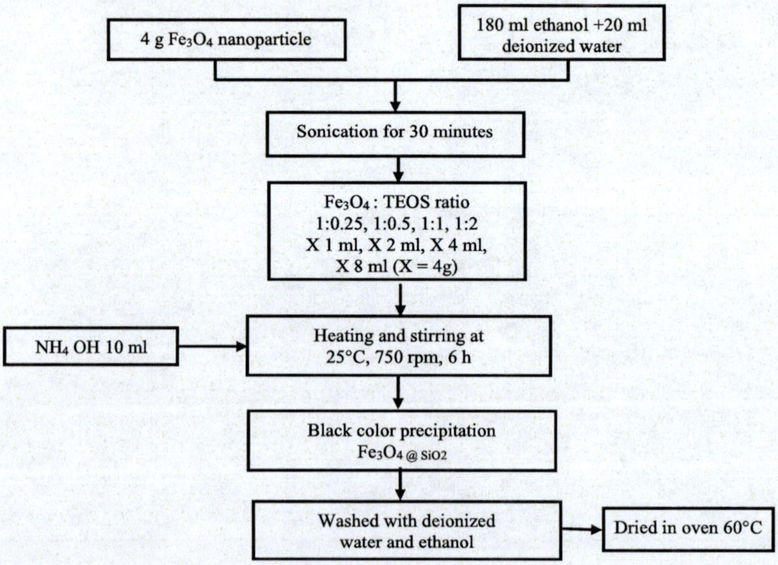

Scheme 44.1 Schematic diagram of preparing Fe_3O_4@SiO_2 magnetic nanoparticles with different quantities of TEOS (tetraethyl orthosilicate)

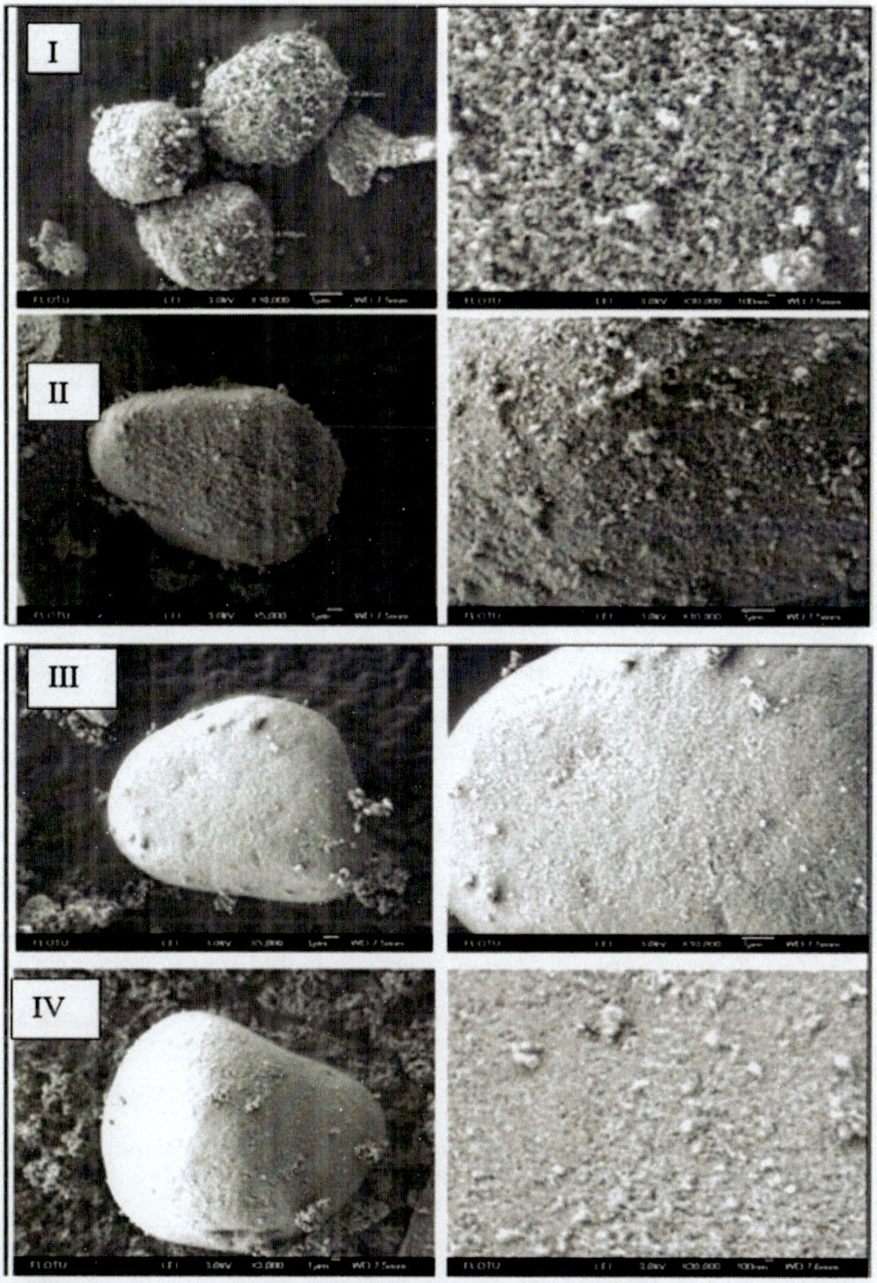

Fig. 44.23 SEM images of $Fe_3O_4@SiO_2$ developed by incorporating Fe_3O_4 with different ratios of tetra orthosilicate (I) 1:0.25, (II) 1:0.5, (III) 1:1 and (IV) 1:2. The left side images are the amplified portion of that of the right side. (Reprinted with permission from Thangaraj et al. 2019 Elsevier)

The images (Fig. 44.23) indicated that the particles were agglomerated due to magnetic dipole moment between the particles. The surface of the particles greatly changed while increasing the ratio of silica coating on the Fe_3O_4 MNPs nanoparticles.

References

Abdala E, Nur O, Mustafa MA (2020) Efficient biodiesel production from algae oil using Ca-doped ZnO nanocatalyst. Ind Eng Chem Res 59:19235–19243

Adachi D, Hama S, Numata T, Nakashima K, Ogino C, Fukuda H, Kondo A (2011) Development of an *Aspergillus oryzae* whole cell biocatalyst coexpressing triglyceride and partial glyceride lipases for biodiesel production. Bioresour Technol 102:6723–6729

Adachi D, Hama S, Nakashima K, Bogaki T, Ogino C, Kondo A (2013) Production of biodiesel from plant oil hydrolysates using an *Aspergillus oryzae* whole cell biocatalyst highly expressing *Candida antarctica lipase*. Bioresour Technol 135:410–416

Akoh CC, Chang S, Lee G, Shaw J (2007) Enzymatic approach to biodiesel production. J Agric Food Chem 55:8995–9005. https://doi.org/10.1021/jf071724y

Albuquerque MCG, Jimenez-Urbistondo I, Santamarıa-Gonzalez J, Merida-Robles JM, Moreno-Tostb R, Rodríguez-Castellónb E, Jiménez-Lópezb A, Azevedoa DCS, Cavalcante CL Jr, Torres PM (2008) CaO supported on mesoporous silicas as basic catalysts for transesterification reactions. Appl Catal A Gen 334(1–2):35–43

Al-Zuhair S (2007) Production of biodiesel: possibilities and challenges. Biofuels Bioprod Biorefin 1(1):57–66

Antczak MS, Kubiak A, Antczak T, Bielecki S (2009) Enzymatic biodiesel synthesis-key factors affecting efficiency of the process. Renew Energy 34(5):1185–1194. https://doi.org/10.1016/j.renene.2008.11.013

Antunes WM, Veloso CO, Henriques CA (2008) Transesterification of soybean oil with methanol catalyzed by basic solids. Catal Today 133–135:548–554

Arai S, Nakashima K, Tanino T, Ogino C, Kondo A, Fukuda H (2010) Production of biodiesel fuel from soybean oil catalyzed by fungus whole cell biocatalysts in ionic liquids. Enzyme Microb Technol 46:51–55

Arzamendi G, Campoa I, Arguinarena E, Sanchez M, Montes M, Gandia LM (2007) Synthesis of biodiesel with heterogeneous NaOH/alumina catalysts: comparison with homogeneous NaOH. Chem Eng J 134:123–130. https://doi.org/10.1016/j.cej.2007.03.049

Asri NP, Soe'eib S, Poedjojono B, Suprapto (2018) Alumina supported zinc oxide catalyst for production of biodiesel from kesambi oil and optimization to achieve highest yields of biodiesel. Euro-Mediterr J Environ Integr 3:1–10

Bacovsky D, Korbitz W, Mittelbach M, Worgetter M (2007) Biodiesel production: technologies and European providers. IEA, Task 39 report T39-B6, Graz, Austria, p 104

Ban K, Kaieda M, Matsumoto T, Kondo A, Fukuda H (2001) Whole cell biocatalyst for biodiesel fuel production utilizing *Rhizopus oryzae* cells immobilized within biomass support particles. Biochem Eng J 8:39–43

Banerjee A, Chakraborty R (2009) Parametric sensitivity in transesterification of waste cooking oil for biodiesel production—a review. Resour Conserv Recycl 53:490–497. https://doi.org/10.1016/j.rescomrec.2009.04.003

Barakos N, Pasias S, Papayannakos N (2008) Transesterification of triglycerides in high and low quality oil feeds over an HT2 hydrotalcite catalyst. Bioresour Technol 99:5037–5042

Barthomeuf D (1996) Basic zeolites: characterization and uses in adsorption and catalysis. Catal Rev Sci Eng 38:521–612

Belafi-Bako K, Kovacs F, Gubicza L, Hancsok J (2002) Enzymatic biodiesel production from sunflower oil by *Candida antarctica* lipase in a solvent-free system. Biocatal Biotransformation 20:437–439. https://doi.org/10.1080/10242420210.00040855

Bernardes OL, Bevilaqua JV, Leal MCM, Freire DMG, Langone MAP (2007) Biodiesel fuel production by the transesterification reaction of soybean oil using immobilized lipase. Appl Biochem Biotechnol 137:105–114. https://doi.org/10.1007/s12010=007-9043-5

Bournay L, Casanave D, Delfort B, Hillion G, Chodorge JA (2005) New heterogeneous process for biodiesel production, a way to improve the quality and the value of the crude glycerin produced by biodiesel plants. Catal Today 106:190–192

Boz N, Degirmenbasi N, Kalyon DM (2009) Conversion of biomass to fuel, transesterification of vegetable oil to biodiesel using KF loaded nano-γ-Al_2O_3 as catalyst. Appl Catal B Environ 89:590–596

Chandler IC (2001) Determining the regioselectivity of immobilized lipases in triacylglycerol acidolysis reactions. J Am Oil Chem Soc 78:737–742. https://doi.org/10.1007/s11746-001-0335-7

Cheetham PSJ, Blunt KW, Bocke C (1979) Physical studies on cell immobilization using calcium alginate gels. Biotechnol Bioeng 21:2155–2168. https://doi.org/10.1002/bit.260211202

Chen YZ, Ching CB, Xu R (2009a) Lipase immobilization on modified zirconia nanoparticles: studies on the effects of modifiers. Process Biochem 44:1245–1251

Chen Y, Xiao B, Chang J, Fu Y, Lv P, Wang X (2009b) Synthesis of biodiesel from waste cooking oil using immobilized lipase in fixed bed reactor. Energy Conver Manag 50:668–673

Deng L, Xu XB, Haraldsson GG, Tan TW, Wang F (2005) Enzymatic production of alkyl esters through alcoholysis: a critical evaluation of lipases and alcohols. J Am Oil Chem Soc 82:341–347. https://doi.org/10.1007/s11746-005-1076-3

DePaola MG, Ricca E, Calabrò V, Curcio S, Lorio G (2009) Factor analysis of transesterification reaction of waste oil for biodiesel production. Bioresour Technol 100:5126–5513

Dizge N, Keskinler B (2008) Enzymatic production of biodiesel from canola oil using immobilized lipase. Biomass Bioenergy 32:1274–1278. https://doi.org/10.1016/j.biombioe.2008.03.005

Dizge N, Keskinler B, Tanriseven A (2009a) Biodiesel production from canola oil by using lipase immobilized onto hydrophobic microporous styrene-divinyl benzene copolymer. Biochem Eng J 44:220–225. https://doi.org/10.1016/j.bej.2008.12.008

Dizge N, Aydiner C, Imer DY, Bayramoglu M, Tanriseven A et al (2009b) Biodiesel production from sunflower, soybean and waste cooking oils by transesterification using lipase immobilized onto a novel microporous polymer. Bioresour Technol 100:1983–1991. https://doi.org/10.1016/j.biortec.2008.10.008

Du W, Xu Y, Liu D, Zeng J (2004) Comparative study on lipase-catalyzed transformation of soybean oil for biodiesel production with different acyl acceptors. J Mol Catal B Enzym 30:125–129. https://doi.org/10.1016/j.molcatb.2004.04.004

Dussán K, Giraldo O, Cardona C (2007) Application of magnetic nanostructures in biotechnological processes: biodiesel production using lipase immobilized on magnetic carriers. In: Proceedings of European congress of chemical engineering (ECCE-6), Copenhagen, 16–20 Sept 2007

Fernandez-Lafuente R, Armisen P, Sabuquillo P, Fernandez-Lorente G, Guisan JM (1998) Immobilization of lipases by selective adsorption on hydrophobic supports. Chem Phys Lipids 93:185–197. https://doi.org/10.1016/S0009-3084(98)00042-5

Fjerbaek L, Christensen KV, Norddahl B (2009) A review of the current state of biodiesel production using enzymatic transesterification. Biotechnol Bioeng 102:1298–1315. https://doi.org/10.1002/bit.22256

Freedman B, Pryde EH, Mounts TL (1984) Variables affecting the yields of fatty esters from transesterified vegetable oils. J Am Oil Chem Soc 61:1638–1643. https://doi.org/10.1007/BF02541649

Fujita Y, Takahashi S, Ueda M, Tanaka A, Okada H, Morikawa Y, Kawaguchi T, Arai M, Fukuda H, Kondo A (2002) Direct and efficient production of ethanol from cellulosic material with a yeast strain displaying cellulolytic enzymes. Appl Environ Microbiol 68:5136–5141

Fukuda H, Kondo A, Noda H (2001) Biodiesel fuel production by transesterification of oils. J Biosci Bioeng 92:405–416

Fukuda H, Hama S, Tamalampudi S, Noda H (2008) Whole-cell biocatalysts for biodiesel fuel production. Trends Biotechnol 26:668–673

Fukuda H, Kondo A, Tamalampudi S (2009) Bioenergy: sustainable fuels from biomass by yeast and fungal whole-cell biocatalysts. Biochem Eng J 44:2–12. https://doi.org/10.1016/S1389-1723(01)80288-7

Furuta S, Matsuhashi H, Arata K (2006) Biodiesel fuel production with solid amorphous zirconia catalysis in fixed bed reactor. Biomass Bioenergy 30:870–873

Gao Y, Tan TW, Nie KL, Wang F (2006) Immobilization of lipase on macroporous resin and its application in synthesis of biodiesel. Chin J Biotechnol 22:114–118. https://doi.org/10.1016/S1872-207(06)60008-3

Gao B, Su E, Lin J, Jiang Z, Ma Y, Wei D (2009) Development of recombinant *Escherichia coli* whole cell biocatalyst expressing a novel alkaline lipase-coding gene from *Proteus* sp. for biodiesel production. J Biotechnol 13(9):169–175

Granados ML, Poves MDZ, Alonso DM, Mariscal R, Galisteo FC, Moreno-Tost R, Santamaría J, Fierro JLC (2007) Biodiesel from sunflower oil by using activated calcium oxide. Appl Catal B Environ 73:317–326

Guo M, Jiang W, Chen C, Qu S, Lu J, Yi W, Ding J (2021) Process optimization of biodiesel production from waste cooking oil by esterification of free fatty acids using $La^{3+}/ZnO\text{-}TiO_2$ photocatalyst. Energy Convers Manag 229:113745

Hama S, Yamaji H, Kaieda M, Oda M, Kondo A, Fukuda H (2004) Effect of fatty acid membrane composition on whole-cell biocatalysts for biodiesel-fuel production. Biochem Eng J 21:155–160

Hama S, Yamaji H, Fukumizu T, Numata T, Tamalampudi S, Kondo A, Noda H, Fukuda H (2007) Biodiesel fuel production in a packed-bed reactor using lipase-producing *Rhizopus oryzae* cells immobilized within biomass support particles. Biochem Eng J 34:273–278

Hama S, Tamalampudi S, Shindo N, Numata T, Yamaji H, Fukuda H, Kondo A (2008) Role of N-terminal 28-amino-acid region of Rhizopus oryzae lipase in directing proteins to secretory pathway of *Aspergillus oryzae*. Appl Microbiol Biotechnol 79:1009–1018

Harding CC, Chang S, Lee G, Shaw J (2007) Enzymatic approach to biodiesel production. J Agric Food Chem 55:8995–9005. https://doi.org/10.1016/j.Jclepro.2007.07.003

Hassan HMA, Alhumaimess MS, Alsohaimi IH, Essawy AA, Hussein MF, Alshammari HM, Aldosari OF (2020) Biogenic-mediated synthesis of the Cs_2O–MgO/MPC nanocomposite for biodiesel production from olive oil. ACS Omega 43:27811–27822

Hattori H (1995) Heterogeneous basic catalysis. Chem Rev 95:537–558

He Q, Xu Y, Teng Y, Wang D (2008) Biodiesel production catalyzed by wholecell lipase from *Rhizopus chinensis*. Chin J Catal 29(1):41–46

Huang Y, Zheng H, Yan Y (2010) Optimization of lipase-catalyzed transesterification of lard for biodiesel production using response surface methodology. Appl Biochem Biotechnol 160:504–515. https://doi.org/10.1007/s12010-008-8377-y

Huang D, Han S, Han Z, Lin Y (2012) Biodiesel production catalyzed by *Rhizomucor miehei* lipase displaying *Pichia pastoris* whole cells in an iso-octane system. Biochem Eng J 63:10–14

Huaping Z, Zongbin W, Yuanxiao C, Ping Z, Shije D, Xiaohua L, Zongqiang M (2006) Preparation of biodiesel catalyzed by solid super base of calcium oxide and its refining process. Chin J Catal 27:391–396

Hwang ET, Gu MB (2012) Enzyme stabilization by nano/microsized hybrid materials. Eng Life Sci 13:49–61

Ilgen O, Akin ASNU (2009) Development of alumina supported alkaline catalysts used for biodiesel production. Turk J Chem 33:281–287

Indumathi R (2013) Studies on lipase producing microbial whole cell biocatalyst and its application in biodiesel production. PhD thesis, Madurai Kamaraj University, India

Jegannathan KR, Abang S, Poncelet D, Chan ES, Ravindra P (2008) Production of biodiesel using immobilized lipase—a critical review. Crit Rev Biotechnol 28:253–264. https://doi.org/10.1080/07388550802428392

Jegannathan KR, Jun-Yee L, Chan E, Ravindra P (2009) Design an immobilized lipase enzyme for biodiesel production. J Renew Sustain Energy 1:063101-1–063101-8

Jegannathan KR, Jun-Yee L, Chan E, Ravindra P (2010) Production of biodiesel from palm oil using liquid core lipase encapsulated in κ-carrageenan. Fuel 89:2272–2277

Ji Q, Xiao S, He B, Liu X (2010) Purification and characterization of an organic solvent-tolerant lipase from *Pseudomonas aeruginosa* LX1 and its application for biodiesel production. J Mol Catal B Enzym 66:264–269

Jitputti J, Kitiyanan B, Rangsunvigit P, Bunyakiat K, Attanatho L, Jenvanitpanjakul P (2006) Transesterification of crude palm kernel oil and crude coconut oil by different solid catalysts. Chem Eng J 116:61–66

Kaieda M, Samukawa T, Kondo A, Fukuda H (2001) Effect of methanol and water contents on production of biodiesel fuel from plant oil catalyzed by various lipases in a solvent free system. J Biosci Bioeng 91:12–15. https://doi.org/10.1016/S1389-1723(01)80103-1

Kaita J, Mimura T, Fukuoka N, Hattori Y (2002) Catalyst for transesterification. United States Patent 6,407,269 B2

Kalscheuer R, Stölting T, Steinbüchel A (2006) Microdiesel: *Escherichia coli* engineered for fuel production. Microbiology 152:2529–2536

Kato M, Fuchimoto J, Tanino T, Kondo A, Fukuda H, Ueda M (2007) Preparation of a whole cell biocatalyst of mutated *Candida antarctica* lipase B (mCALB) by a yeast molecular display system and its practical properties. Appl Microbiol Biotechnol 75:549–555. https://doi.org/10.1007/s00253-006-0835-2

Khan JA, Vulfson EN (2001) Microencapsulation of enzymes and cells for nonaqueous biotransformations methods in biotechnology. In: Vulfson EN, Halling PJ, Holland HL (eds) Enzymes in nonaqueous solvents: methods and protocols, 2nd edn. Humana Press Inc., Totowa, NJ, pp 31–40. ISBN: 0896039293

Kim HJ, Kang BS, Kim MJ, Park YM, Kim DK, Lee JS, Lee KY (2004) Transesterification of vegetable oil to biodiesel using heterogeneous base catalyst. Catal Today 93–95:315–320

Klibanov AM (1983) Immobilized enzymes and cells as practical catalysts. Science 219:722–727. https://doi.org/10.1126/science.219.4585.722

Koda R, Numata T, Hama S, Tamalampudi S, Nakashima K, Tanaka T, Ogino C, Fukuda H, Kondo A (2010) Ethanolysis of rapeseed oil to produce biodiesel fuel catalyzed by *Fusarium heterosporum* lipase—expressing fungus immobilized whole-cell biocatalysts. J Mol Catal B Enzym 66:101–104

Koskinen AMP, Klibanov AM (1996) Enzymatic reactions in organic media, 1st edn. Chapman and Hall, London, p 272. ISBN: 0-7514-0258-1

Kumari V, Shah S, Gupta MN (2007) Preparation of biodiesel by lipase catalyzed transesterification of high free fatty acid containing oil from *Madhuca indica*. Energy Fuels 21:368–372. https://doi.org/10.1021/ef0602168

Kumari A, Mahapatra P, Garlapati VK, Banerjee R (2009) Enzymatic transesterification of Jatropha oil. Biotechnol Biofuels 2:1–7. https://doi.org/10.1186/1754-6834-2-1

Lanser AC, Manthey LK, Hou CT (2002) Regioselectivity of new bacterial lipases determined by hydrolysis of triolein. Curr Microbiol 44:336–340. https://doi.org/10.1007/s00284-001-0019-3

Lee K, Foglia T, Chang KS (2002) Production of alkyl ester as biodiesel from fractionated lard and restaurant grease. J Am Oil Chem Soc 79:191–195. https://doi.org/10.1007/s11746-002-0457-y

Lee JH, Lee DH, Lim JS, Um B, Park C, Kang SW, Kim SW (2008) Optimization of the process for biodiesel production using a mixture of immobilized *Rhizopus oryzae* and *Candida rugosa* lipases. J Microbiol Biotechnol 18:1927–1931. https://doi.org/10.4014/jmb.0800.054

Leung DYC, Wu X, Leung MKH (2010) A review on biodiesel production using catalyzed transesterification. Appl Energy 87:1083–1095. https://doi.org/10.1016/apennergy.2009.10.006

Li E, Rudolph V (2008) Transesterification of vegetable oil to biodiesel over MgO functionally mesoporous catalysts. Energy Fuels 22:143–149

Li Q, Yan Y (2010) Production of biodiesel catalyzed by immobilized *Pseudomonas cepacia* lipase from *Sapium sebiferum* oil in micro-aqueous phase. Appl Energy 87:3148–3154

Li L, Du W, Li D, Li H, Xie W (2006) Transesterification of soybean oil to biodiesel with Zn/I_2 catalyst. Catal Lett 107:25–30

Li W, Du W, Liu D (2007) *Rhizopus oryzae* IFO 4697 whole cell catalyzed methanolysis of crude and acidified rapeseed oils for biodiesel production in tert-butanol system. Process Biochem 42:1481–1485. https://doi.org/10.1016/j.procbio.2007.05.015

Li W, Du W, Liu D (2008) *Rhizopus oryzae* whole-cell-catalyzed biodiesel production from oleic acid in tert-butanol medium. Energy Fuels 22:155–158

Li N, Zong M, Wu H (2009) Highly efficient transformation of waste oil to biodiesel by immobilized lipase from *Penicillium expansum*. Process Biochem 44:685–688

Li A, Ngo TPN, Yan J, Tian K, Li Z (2012) Whole cell based solvent-free system for one pot production of biodiesel from waste grease. Bioresour Technol 114:725–729

Liang X, Gao S, Wu H, Yang J (2009a) Highly efficient procedure for the synthesis of biodiesel from soybean oil. Fuel Process Technol 90:701–704

Liang X, Gao S, Yang J, He M (2009b) Highly efficient procedure for the transesterification of vegetable oil. Renew Energy 34:2215–2217

Lin YH, Luo JJ, Hwang SCJ, Liau PR, Lu WJ, Lee HT (2011) The influence of free fatty acid intermediate on biodiesel production from soybean oil by whole cell biocatalyst. Biomass Bioenergy 35:2217–2223

Liu X, He H, Wang Y, Zhu S (2007) Transesterification of soybean oil to biodiesel using SrO as a solid base catalyst. Catal Commun 87:1107–1111

Liu X, He H, Wang Y, Zhu S (2008) Transesterification of soybean oil to biodiesel using CaO as a solid base catalyst. Fuel 87:216–221

Liu X, Xiong X, Liu C, Liu D, Wu A, Hu Q, Liu C (2010) Preparation of biodiesel by transesterification of rapeseed oil with methanol using solid base catalyst calcined K_2CO_3/γ-Al_2O_3. J Am Oil Chem Soc 87:817–823

Lopez-Serrano P, Cao L, Randwijk FV, Sheldon RA (2002) Cross-linked enzyme aggregates with enhanced activity: application to lipases. Biotechnol Lett 24:1379–1383. https://doi.org/10.1023/a:1019863314646

Lou Y, Zheng Y, Jiang Z, Ma Y, Wei D (2006) A novel psychrophilic lipase from *Pseudomonas fluorescens* with a unique property in chiral resolution and biodiesel production via transesterification. Appl Microbiol Biotechnol 73:349–355. https://doi.org/10.1007/s00253-006-0478-3

Lu J, Chen Y, Wang F, Tan T (2009) Effect of water on methanolysis of glycerol trioleate catalyzed by immobilized lipase *Candida* sp. 99-125 in organic solvent system. J Mol Catal B Enzym 56:122–125. https://doi.org/10.1016/j.molcath.2008.05.004

Ma H, Li S, Wang B, Wang R, Tian S (2008) Transesterification of rapeseed oil for synthesizing biodiesel by K/KOH/c-Al_2O_3 as heterogeneous base catalyst. J Am Oil Chem Soc 85:263–270

Macrae AR (1983) Lipase-catalyzed interesterification of oils and fats. J Am Oil Chem Soc 60:291–294. https://doi.org/10.1007/BF02543502

Malcata FX, Reyes HR, Garcia HS, Hill CG, Amundson CH (1990) Immobilized lipase reactors for modification of fats and oils—a review. J Am Oil Chem Soc 67:890–910. https://doi.org/10.1007/BF02541845

Marchetti JM, Miguel VU, Errazu AF (2007) Possible methods for biodiesel production. Renew Sust Energ Rev 11:1300–1311. https://doi.org/10.1016/j.eser.2005.08.006

Marchetti JM, Miguel VU, Errazu AF (2008) Techno-economic study of different alternatives for biodiesel production. Fuel Process Technol 89:740–748. https://doi.org/10.1016/j.fuproc-2008-01-007

Marta Z, Dorota C, Tomasz S, Adam S, Katarzyna W, Joanna S, Halina K, Marszałł MP (2017) Chitosan–collagen coated magnetic nanoparticles for lipase immobilization-new type of "enzyme friendly" polymer shell crosslinking with squaric acid. Catalysts 7:26

Mateo C, Palomo JM, Lorente GF, Guisan JM, Lafuente RF (2007) Improvement of enzyme activity, stability and selectivity via immobilization techniques. Enzyme Microb Technol 40:1451–1463. https://doi.org/10.1016/j.emictec.2007.01.018

Matsumoto T, Tkahashi S, Kaieda M, Ueda M, Tanaka A, Fukuda H, Kondo A (2001) Yeast whole-cell biocatalyst constructed by intracellular overproduction of *Rhizopus oryzae* lipase is applicable to biodiesel fuel production. Appl Microbiol Biotechnol 57:515–520. https://doi.org/10.1007/s002530100733

Matsumoto T, Fukuda H, Ueda M, Tanaka A, Kondo A (2002) Construction of yeast strains with high cell surface lipase activity by using novel display systems based on the Flo1p flocculation functional domain. Appl Environ Microbiol 68:4517–4522

Mbaraka IK, Shanks BH (2006) Conversion of oils and fats using advanced mesoporous heterogeneous catalysts. J Am Oil Chem Soc 83:79–91

Mehrasbi MR, Mohammadi J, Peyda M, Mohammadi M (2017) Covalent immobilization of Candida antarctica lipase on core-shell magnetic nanoparticles for production of biodiesel from waste cooking oil. Renew Energy 101:593–602

Mendes A, Giordano RC, Giordano RLC, Castro H (2011) Immobilization and stabilization of microbial lipases by multipoint covalent attachment on aldehyde-resin affinity: application of the biocatalysts in biodiesel synthesis. J Mol Catal B Enzym 68:109–115

Meter F, Zarcula C, Kiss C (2007) Enhancement of lipases enantioselectivity by entrapment in hydrophobic sol-gel materials: influence of silane precursors and immobilization parameters. J Biotechnol 131:S109. https://doi.org/10.1016/j.jbiotec.2007.07.187

Meunier S, Legge R (2010) Evaluation of diatomaceous earth as a support for sol–gel immobilized lipase for transesterification. J Mol Catal B Enzym 62:54–58

Mittelbach M (1990) Lipase catalyzed alcoholysis of sunflower oil. J Am Oil Chem Soc 67:168–170. https://doi.org/10.1007/BF02539619

Modi MK, Reddy JRC, Rao BVS, Prasad RBN (2006) Lipase-mediated transformation of vegetable oils into biodiesel using propan-2-ol as acyl acceptor. Biotechnol Lett 28:637–640. https://doi.org/10.1007/s10529-006-0027-2

Mokaizh AAB, Wirman N, Shariffuddin JH (2019) Synthesis of alumina from aluminium can waste to be applied as catalyst support for biodiesel production. IOP Conf Ser Mater Sci Eng 702:1–12

Moreno-Parajàn JC, Giraldo L (2011) Study of immobilized *Candida rugosa* lipase for biodiesel fuel production from palm oil by flow microcalorimetry. Arab J Chem 4:55–62

Naranjo J, Córdoba A, Giraldo L, García V, Moreno-Parajàn JC (2010) Lipase supported on granular activated carbon and activated carbon cloth as a catalyst in the synthesis of biodiesel fuel. J Mol Catal B Enzym 66:166–117

Narasimharao K, Lee A, Wilson K (2007) Catalysts in production of biodiesel: a review. J Biobased Mater Bioenergy 1:19–30. https://doi.org/10.1016/jbmb.2007.002

Narita J, Okano K, Tateno T, Tanino T, Sewaki T, Sung M, Fukuda H, Kondo A (2006) *Display* of active enzymes on the cell surface of *Escherichia coli* using PgsA anchor protein and their application to bioconversion. Appl Microbiol Biotechnol 70:564–572

Nasr E, Guirand R, Pellegatta JL, Blandy C (1995) Esterification of stearic acid using light alcohols— homogeneous catalysis with titanates. Can J Chem Eng 73:129–134

Nassreddine S, Karout A, Christ M, Pierre A (2008) Transesterification of a vegetable oil with methanol catalyzed by a silica fibre reinforced aerogel encapsulated lipase. Appl Catal A Gen 344:70–77

Ngamcharussrivichai C, Totarat P, Bunyakiat K (2008) Ca and Zn mixed oxide as a heterogeneous base catalyst for transesterification of palm kernel oil. Appl Catal A Gen 341:77–85

Nie K, Xie F, Wong F, Tan T (2006) Lipase catalyzed methanolysis to produce biodiesel: optimization of the biodiesel production. J Mol Catal B Enzym 43:142–147. https://doi.org/10.1016/j.molcatb.2006.07.016

Noureddini H, Gao X, Philkana RS (2005) Immobilized *Pseudomonas cepacia* lipase for biodiesel fuel production from soybean oil. Bioresour Technol 96:769–777. https://doi.org/10.1016//j.biotech.2004.05.029

Oda M, Kaieda M, Hama S (2005) Facilitatory effect of immobilized lipase-producing *Rhizopus oryzae* cells on acyl migration in biodiesel-fuel production. Biochem Eng J 23:45–51

Orcaire O, Buisson P, Pierre AC (2006) Application of silica aerogel encapsulated lipases in the synthesis of biodiesel by transesterification reactions. J Mol Catal B Enzym 42(3):106–113

Palekar AA, Vasudevan PT, Yan S (2000) Purification of lipase: a review. Biocatal Biotransform 18:177–200. https://doi.org/10.3109/10242420009015244

Park YM, Lee DW, Kim DK, Lee JS, Lee KY (2008) The heterogeneous catalyst system for the continuous conversion of free fatty acids in used vegetable oils for the production of biodiesel. Catal Today 131:238–243

Qin H, Yan X, Dong W (2008) Biodiesel production catalyzed by whole-cell lipase from *Rhizopus chinensis*. Chin J Catal 29:41–46. https://doi.org/10.1016/S1872-2067(08)600015-7

Rabie AM, Shaban MM, Abukhadra MR, Hosny R, Ahmed SA, Negem NA (2019) Diatomite supported by CaO/MgO nanocomposite as heterogeneous catalyst for biodiesel production from waste cooking oil. J Mol Liq 279:224–231

Rahman RNZRA, Baharum SN, Basri M, Salleh AB (2005) High-yield purification of an organic solvent-tolerant lipase from *Pseudomonas* sp. strain S5. Anal Biochem 34:267–274. https://doi.org/10.1016/j.ab.2005.03.006

Ranganathan SV, Narasimhan SL, Muthukumar K (2008) An overview of enzymatic production of biodiesel. Bioresour Technol 99:3975–3981. https://doi.org/10.1016/j.biortech.2007.04.060

Reaney MJT, Hertz PB, McCalley WW (2005) Vegetable oil as biodiesel. In: Shahidi F (ed) Industrial oil and fat products. John Wiley & Sons Inc., Hoboken, NJ

Reetz MT (2002) Lipases as practical biocatalysts. Curr Opin Chem Biol 6:145–150. https://doi.org/10.1016/S1368-5831(02)00297-1

Robles-Medina A, Gonzalez-Moreno PA, Esteban-Cerdán L, Molina-Grima E (2009) Biocatalysis: towards ever greener biodiesel production. Biotechnol Adv 27:398–408. https://doi.org/10.1016/j.biotechhadv.2008.10.008

Rodrigues RC, Volpato G, Wada K, Ayub MAZ (2008) Enzymatic synthesis of biodiesel from transesterification reactions of vegetable oils and short chain alcohols. J Am Oil Chem Soc 85:925–930. https://doi.org/10.1007/s11746-008-1284-0

Rodrigues RC, Pessela BCC, Volpato G, Fernandez-Lafuente R, Guisan JM, Ayub MAZ (2010) Two step ethanolysis: a simple and efficient way to improve the enzymatic biodiesel synthesis catalyzed by an immobilized–stabilized lipase from *Thermomyces lanuginosus*. Process Biochem 45:1268–1273

Royon D, Daz M, Ellenrieder G, Locatelli S (2007) Enzymatic production of biodiesel from cotton seed oil using t-butanol as a solvent. Bioresour Technol 98:648–653. https://doi.org/10.1016/j.boiretch.2006.02.021

Sakai S, Liu Y, Yamaguchi T, Watanabe R, Kawabe M, Kawakami K (2010) Production of butyl-biodiesel using lipase physically-adsorbed onto electrospun polyacrylonitrile fibers. Bioresour Technol 101:7344–7734

Salah RB, Ghamghui H, Miled N, Mejdoub H, Gargouri Y (2007) Production of butyl acetate ester by lipase from novel strain of *Rhizopus oryzae*. J Biosci Bioeng 103:368–373. https://doi.org/10.1263/jbb.103.368

Salinas D, Sepúlveda C, Escalona N, GFierro JL, Pecchi G (2018) Sol-gel La_2O_3–ZrO_2 mixed oxide catalysts for biodiesel production. J Energy Chem 27:565–572

Salis A, Pinna M, Monduzzi M, Solinas V (2005) Biodiesel production from triolein and short chain alcohols through biocatalysis. J Biotechnol 119:291–299. https://doi.org/10.1016/jbiotec.2005.04.009

Salis A, Bhattacharyya M, Monduzzi M, Solinas V (2009) Role of the support surface on the loading and the activity of *Pseudomonas fluorescens* lipase used for biodiesel synthesis. J Mol Catal B Enzym 57:262–269

Saxena RK, Sheoran A, Giri B, Davidson WS (2003) Review—Purification strategies for microbial lipases. J Microbiol Methods 52:1–18. https://doi.org/10.1016/S0167-7020(o2)00161-6

Schuchardt ULF, Sercheli R, Vargas RM (1998) Transesterification of vegetable oils: a review. J Biochem Soc 9:199–210. https://doi.org/10.1590/S0103-50531998000300002

Serralheiro ML, Empis JM, Carbal JMS (1990) Peptide synthesis by microencapsulated chymotrypsin. Ann N Y Acad Sci 613:638–642

Shah S, Gupta M (2007) Lipase catalyzed preparation of biodiesel from Jatropha oil in a solvent free system. Process Biochem 42:409–414

Shah S, Sharma S, Gupta MN (2004) Biodiesel preparation by lipase-catalyzed transesterification of jatropha oil. Energy Fuels 18:154–159. https://doi.org/10.1021/ef030075z

Shahbazi F, Mahdavi V, Zolgharnein J (2020) Preparation and characterization of SrO/MgO nanocomposite as a novel and efficient base catalyst for biodiesel production from waste cooking oil: a statistical approach for optimization. J Iran Chem Soc 17:333–349

Shao P, Meng X, He XJ, Sun P (2008) Analysis of immobilized *Candida rugosa* lipase catalyzed preparation of biodiesel from rapeseed soap stock. Food Bioprod Process 86:283–289

Shieh CJ, Lia HF, Lee CC (2003) Optimization of lipase catalyzed biodiesel by response surface methodology. Bioresour Technol 88:103–106. https://doi.org/10.1016/S0960-8524(02)00292-4

Shimada Y, Sugihara A, Nakano H, Nagao T, Suenaga A, Nakai S, Tominaga Y (1997) Fatty acid specificity of *Rhizopus delemar* lipase in acidolysis. J Ferment Bioeng 83:321–327. https://doi.org/10.1016/S0922-338X(97)8013605

Shtelzer S, Rappoport S, Avnir D, Ottolenghi M, Braun S (1992) Properties of trypsin and of acid phosphatase immobilized in sol-gel glass matrices. Appl Biochem Biotechnol 15:227–235. ISSN: 0885-4513

Shu Q, Yang B, Yuan H, Qing S, Zhu G (2007) Synthesis of biodiesel from soybean oil and methanol catalyzed by zeolite beta modified with La^{3+}. Catal Commun 8:2159–2165

Shumaker JL, Crofcheck C, Tackett SA, Jimenez SE, Morgan T, Ji Y, Crocker M, Toops TJ (2008) Biodiesel synthesis using calcined layered double hydroxide catalysts. Appl Catal B Environ 82:120–130

Singh R, Kumar A, Sharma YC (2019) Biodiesel production from microalgal oil using barium–calcium–zinc mixed oxide base catalyst: optimization and kinetic studies. Energy Fuel 33:1175–1184

Soumanou MM, Bornscheuer UT (2003) Improvement in lipase-catalyzed synthesis of fatty acid methyl esters from sunflower oil. Enzyme Microb Technol 33:97–103. https://doi.org/10.1016/S0141-0229(03)00090-5

Sree R, Babu NS, Saiprasad PS, Lingaiah N (2009) Transesterification of edible and non-edible oils over basic solid Mg/Zr catalysts. Fuel Process Technol 90:152–157

Sun T, Du W, Zeng J, Dai L, Liu D (2010) Exploring the effects of oil inducer on whole cell mediated methanolysis for biodiesel production. Process Biochem 45:514–518

Sun T, Du W, Zeng J, Liu D (2011) Comparative study on stability of whole cells during biodiesel production in solvent-free system. Process Biochem 46:661–664

Suppes GJ, Dasari MA, Doskocil EJ, Mankidy PJ, Goff MJ (2004) Transesterification of soybean oil with zeolite and metal catalysts. Appl Catal A Gen 257:213–223

Tamalampudi S, Talukder R, Hama S, Numata T, Kondo T, Fukada H (2008) Enzymatic production of biodiesel from jatropha oil: a comparative study of immobilized-whole cell and commercial lipases as a biocatalyst. Biochem Eng J 39:185–189

Tan T, Nie K, Wang F (2006) Production of biodiesel by immobilized *Candida* sp. lipase at high water content. Appl Biochem Biotechnol 128(2):109–116. https://doi.org/10.1385/ABAB/128:2:109

Thangaraj B, Jia Z, Dai L, Liu D, Du W (2016) Lipase NS81006 immobilized on Fe_3O_4 magnetic nanoparticles for biodiesel production. Ovidius Univ Annal Chem 27:13–21

Thangaraj B, Jia JZ, Dai L, Liu D, Du W (2019) Effect of silica coating on Fe_3O_4 magnetic nanoparticles for lipase immobilization and their application for biodiesel production. Arab J Chem 12:4694–4706

Thangaraj B, Jia Z, Dai L, Liu D, Du W (2020) Lipase NS81006 immobilized on functionalized ferric-silica magnetic nanoparticles for biodiesel production. Biofuels 11:811–819

Torres CF, Fornari T, Tenllado D, Senoráns FJ, Reglero G (2008) A predictive kinetic study of lipase-catalyzed ethanolysis reactions for the optimal reutilization of the biocatalyst. Biochem Eng J 42:105–110. https://doi.org/10.1016/j.bej.2008.06.004

Trakarnpruk W, Porntangjitlikit S (2008) Palm oil biodiesel synthesized with potassium loaded calcined hydrotalcite and effect of biodiesel blend on elastomers properties. Renew Energy 33:1558–1563

Vaghari H, Jafarizadeh-Malmiri H, Mohammadlou M, Berenjian A, Anarjan N, Jafari N, Nasiri S (2016) Application of magnetic nanoparticles in smart enzyme immobilization. Biotechnol Lett 38:223–233

Vasudevan PT, Briggs M (2008) Biodiesel production-current state of the art and challenges. J Ind Microbiol Biotechnol 35:421–430. https://doi.org/10.1007/s10295-008-0312-2

Vicente LC, Barros RA, Empis JMA (1994) Stability and proteolytic activity of papain in reverse micellar and aqueous media: a kinetic and spectroscopic study. J Chem Technol Biotechnol 60:291–297. https://doi.org/10.1002/jetb.2806003

Vyas AP, Subrahmanyam N, Patel PA (2009) Production of biodiesel through transesterification of Jatropha oil using KNO_3/Al_2O_3 solid catalyst. Fuel 88:625–628

Wan T, Yu P, Gong S, Li Q, Luo Y (2008) Application of KF/MgO as a heterogeneous catalyst in the production of biodiesel from rape seed oil. Korean J Chem Eng 25:998–1003

Wang YD, Shemmeri TA, Eames P, McMullan J, Hewitt N, Huang Y, Rezvani S (2006a) An experimental investigation of the performance and gaseous exhaust emissions of a diesel engine using blends of a vegetable oil. Appl Thermal Eng 26:1684–1691

Wang H, Wang M, Liu S, Zhao N, Wei W, Sun Y (2006b) Influence of preparation methods on the structure and performance of CaOZrO2 catalyst for the synthesis of dimethyl carbonate via transesterification. J Mol Catal A Chem 258(1–2):308–312

Wang Y, Zhang F, Yu S, Yang L, Li D, Evans DG, Duan X (2008) Preparation of macrospherical magnesia-rich magnesium aluminate spinel catalysts for methanolysis of soybean oil. Chem Eng Sci 63:4306–4312

Watanabe Y, Shimada Y, Sugihara A, Tominaga Y (2002) Conversion of degummed soybean oil to biodiesel fuel with immobilized *Candida antarctica* lipase. J Mol Catal B Enzym 17:151–155. https://doi.org/10.1016/S1381-1177(o2)00022-X

Wen Z, Yu X, Tu S, Yan J, Dahlquist E (2010) Synthesis of biodiesel from vegetable oil with methanol catalyzed by Li-doped magnesium oxide catalysts. Appl Energy 87:743–748

Xiao X, Tierney JW, Wender I (1999) Alkylation of isobutane with 2-butane over an ion modified zirconium oxide catalysts. Appl Catal A Gen 183:209–219

Xiao M, Mathew S, Obbard JP (2010) A newly isolated fungal strain used as whole-cell biocatalyst for biodiesel production from palm oil. GCB Bioenergy 2:45–51. https://doi.org/10.1111/j.1757-1707.2010.01038.x

Xıao M, Qi C, Obbard JP (2011) Biodiesel production using *Aspergillus niger* as a whole-cell biocatalyst in a packed-bed reactor. GCB Bioenergy 3:293–298

Xie W, Huang X (2006) Synthesis of biodiesel from soybean oil using heterogeneous KF/ZnO catalyst. Catal Lett 107:53–59

Xie W, Li H (2006) Alumina—supported potassium iodide as a heterogeneous catalyst for biodiesel production from soybean oil. J Mol Catal A Chem 255:1–9

Xie W, Ma N (2010) Enzymatic transesterification of soybean oil by using immobilized lipase on magnetic nano-particles. Biomass Bioenergy 34:890–896

Xie W, Yang Z (2007) Ba–ZnO catalysts for soybean oil transesterification. Catal Lett 117:159–165

Xie W, Peng H, Chen L (2005) Calcined Mg–Al hydrotalcites as solid base catalysts for methanolysis of soybean oil. J Mol Catal A Chem 246:24–32

Xie W, Peng H, Chen L (2006) Transesterification of soybean oil catalyzed by potassium loaded on alumina as a solid-base catalyst. Appl Catal A Gen 300:67–74

Yagiz F, Kazan D, Akin NA (2007) Biodiesel production from waste oils by using lipase immobilized on hydrotalcite and zeolites. Chem Eng J 134:262–267

Yang Z, Xie W (2007) Soybean oil transesterification over zinc oxide modified with alkali earth metals. Fuel Process Technol 88:631–638

Yong YP, Al-Duri B (1996) Kinetic studies on immobilized lipase esterification of oleic acid and octanol. J Am Oil Chem Soc 65:239–248. https://doi.org/10.1002/(SICI)1097-4660(199603)65:3

Yücel Y (2011) Biodiesel production from pomace oil by using lipase immobilized onto olive pomace. Bioresour Technol 102:3977–3980

Yusuff AS, Owolabi JO (2019) Synthesis and characterization of alumina supported coconut chaff catalyst for biodiesel production from waste frying oil. S Afr J Chem Eng 30:42–49

Zabeti M, Daud WMAW, Aroua MK (2009) Activity of solid catalysts for biodiesel production, a review. Fuel Process Technol 90:770–777

Zeng J, Du W, Liu X, Liu D, Dai L (2006) Study on the effect of cultivation parameters and pretreatment on *Rhizopus oryzae* cell-catalyzed transesterification of vegetable oils for biodiesel production. J Mol Catal B Enzym 43:15–18

Zeng HY, Feng Z, Deng X, Li YQ (2008) Activation of Mg–Al hydrotalcite catalysts for transesterification of rape oil. Fuel 87(13–14):3071–3076

Zeng H, Liao K, Deng X, Jiang H, Zhang F (2009) Characterization of the lipase immobilized on Mg–Al hydrotalcite for biodiesel. Process Biochem 44:791–798

Zhu H, Wu Z, Chen Y, Zhang P, Duan S, Liu X, Mao Z (2006) Preparation of biodiesel catalyzed by solid super base of calcium oxide and its refining process. Chin J Catal 27:391–396

Standards for Biodiesel

45

Prescribing standards for biodiesel is observed to be essential in the context of commercialization. Mandatory specifications are the need to evaluate the quality of biodiesel in the open market since its quality is likely to change according to the raw material from which the concerned biodiesel is produced. Besides, such standards provide the norms for safety, associated risk parameters and environmental pollution concerned with the use of that fuel (Prankl and Worgetter 1999; Anonymous 1995). As there is growing interest on biodiesel all over the world, standards have been established in various countries including the USA (ASTM D6751), European Union (EN 14214), Brazil, South Africa, Australia and China. Biodiesel also is being used as heating oil for which separate specification (EN 14213) is available.

Pure vegetable oils are not prescribed for biodiesel as they are not considered actively for use in engines due to their poor cold flow, high viscosity, low cetane number and low stability (Reaney et al. 2005). Therefore, these oils are scientifically processed to improve their quality so that the finished product, namely the ester is separated for use in the engine. This ester is devoid of sulphur and nitrogen and contains around 11% oxygen. This product is often called as *first generation* biodiesel and the process is called as transesterification. The transesterification involves the replacement of glycerol a tri-alcohol with a monoalcohol. Methanol is often used. Ethanol or iso-propanol may also be used but for their slow reaction and the need for more quantity for a unit operation. As water interferes with the transesterification reaction, the alcohol to be employed should be free from water.

Successful production of biodiesel from vegetable oil largely relies on three major factors: appropriate quantity of catalyst, extended reaction duration and excess alcohol. In an ideal reaction the expected yield of minimum ester content is 96.5% by weight. If that value is not reached, the process is considered incomplete and the finished product is considered as of poor quality. Thorough washing of the biodiesel is recommended to remove the impurities such as soap, catalyst and salts if chemical catalyst is used. The excess methanol also is to be expelled, if not the flash point of

S. P. Raj et al., *Biodiesel from Flowering Plants*,
https://doi.org/10.1007/978-981-16-4775-8_45

the final product will go down. Besides, the alcohol is likely to damage the elastomers and rubber caskets of the engine. The methanol is to be removed before the biodiesel is to be washed with water. If not the alcohol enters into the wash water and make the waste water treatment very difficult besides causing economic loss on account of the wastage of alcohol.

45.1 FFA and Acid Value

The acid value is one of the indices of monitoring fuel quality (ASTM D664 and EN 14104 standards). ASTM D664 is a potentiometric method which offers mediocre reproducibility, a problem mentioned in the method itself. The problem is likely due to the variability of electrodes. This method is based on the use of KOH in isopropanol with *p*-naphtholbenzein as indicator and is suitable even for coloured samples. Analytical results are more consistent with ASTM D974 than with ASTM D664. Therefore, ASTM D974 is considered as a more accurate method than ASTM D664. EN 14104 is a titration method and it uses dilute ethanolic KOH solution with phenolphthalein as indicator. In this method, 0.1–0.5 mL of biodiesel is taken in a conical flask. Solvent mixture (95% ethanol and diethyl ether in 1:1 ratio) 50 mL is added to the flask. The mixture is well shaken and titrated against 0.1 M KOH using phenolphthalein (1%) as indicator (Karmakar et al. 2018).

Neutralization number can be calculated from the acid value of the above two methods. Method for determining strong acids and FFA using potentiometry in a single measurement is developed. The potentiometry is more reliable than the use of indicators. The neutralization number calculated from the titration method shows 10–20% variation (Knothe 2006).

45.2 Density

Density of the biodiesel normally lies between 865 and 910 kg m^{-3} against the limit of 800–860 kg m^{-3} for diesel. The density variation depends on the type of raw material used. Presence of alcohol often reduces the density. Biodiesel with higher density is known for higher energy comparatively. High density for biodiesel is due to the high mass of the long chain fatty acid alkyl ester molecules. Although ASTM D6751 does not specify any limit, European standard EN 14214 prescribes a range of 860–900 kg m^{-3} at 15 °C. However, most of the biodiesels fall within the above range.

The relative density of biodiesel is measured by pycnometer. The pycnometer is kept inside a refrigerator after filling them up with biodiesel. They are taken out from the refrigerator when the temperature of biodiesel is reached to 15 °C. The mass and volume of the biodiesel are measured for the calculation of density of biodiesel. The formula for relative density is given below

$$\text{Density}\ \left(\text{kg m}^{-3}\right) = \frac{\text{Mass of the pycnometer containing the biodiesel} - \text{Mass of the empty pycnometer}}{\text{Volume of the biodiesel}}$$

45.3 Flash Point and Fire Point

Flash point is normally considered as the minimum temperature at which the vapour generated from the biodiesel catches fire when it comes into contact with open flame under specific condition. The flash point indicates the flammability of the liquid fuel. High level of flash point is considered safe in storage and in transport. It is necessary that the biodiesel be devoid of any residual alcohol. If alcohol is present, the flash point is considerably reduced and this would cause adverse combustion characteristics such as premature ignition of the fuel, inconsistent firing, irregular timing and excessive exhaust emissions. The flash point is often considered as an index of alcohol contamination. Presence of 0.2% methanol in biodiesel would pull down the flash point to less than 100 °C and 0.4% would cause the flash point to go around 50 °C.

The European standard EN 14214 indicates a minimum flash point of 120 °C, whereas it is 130 °C as per US ASTM D6751. However, the flash point of biodiesel ranges from 150 to 190 °C.

The biodiesel is filled inside the flash and fire point apparatus, and a cotton thread is placed in it. The biodiesel is heated with a gas stove. Another ignited cotton thread is dragged on the surface of the former thread. The temperature at which the spark came out of first thread is noted for flash point measurement, and the temperature at which the thread started burning is noted for measurement of fire point (Karmakar et al. 2018).

45.4 Calorific Value (CV)

Generally bomb calorimeter is used to measure the CV of biodiesel. The biodiesel (0.5 g) in a container is placed in the bomb. A 8 cm long cotton thread hanging from nichrome wire was dipped into the biodiesel. The bomb is filled with oxygen at 400 psi. Then, it is placed inside the well-insulated container containing distiller water and the fuse wires are placed in their position on the bomb. The electrical circuit was ensured. The initial temperature is noted and then it is reduced to 0 °C. The fire button is pressed to make a short circuit on the nichrome wire and ignite the biodiesel. The temperature kept on increasing for a certain time (Karmakar et al. 2018). Finally, the stable temperature is noted. The CV is calculated using the following formula.

$$\text{Calorific value }\left(\text{kJ kg}^{-1}\right) = \frac{(\text{Wt. of water} + \text{water equivalent}) \times \text{Temp. rise} \times \text{Spec. heat of water}}{\text{Wt. of the sample}}$$

45.5 Water Content

Water content of the oil induces the formation of soap during the transesterification. Besides, as the process reaction proceeds water is generated which in turn retards the rate of reaction. The water content in biodiesel is a menace which may occur due to incomplete drying or bad storage. This water may remain as dissolved or in suspension state. If biodiesel is stored in humid conditions, moisture may get accumulated in it. The solubility of water in biodiesel increases according to the ambient temperature. Biodiesel may absorb as high as 1 g L^{-1} of moisture. If the biodiesel has alcohol, it tends to absorb moisture more aggressively. If the biodiesel contains alcohol to the tune of 0.2% by mass, the natural absorption of moisture may go up to 1.5 g L^{-1}.

Free and settled water causing density gradient is yet another problem. Such water may become a cause of rusting the tank which in turn spoils the quality of the biodiesel. The rust will also transform into sediment and settle as deposits in engine parts which ultimately reduces the life of the engine (Knothe 2006).

The water if present in the biodiesel will cause

1. rust and corrosion in the engine components
2. hydrogen embrittlement of engine parts
3. water etching
4. erosion of tank wall
5. calcium scaling

Calcium scaling may cause blockages in fuel delivery pipes. Washing the ester with warm water often results in more water content in the biodiesel. However, ion exchange and absorbent methods have come into being as alternative processes to purify the biodiesel. High content of water is normally removed through vacuum driers and molecular sieves. ASTM specification indicates the use of centrifuge method which on operation proves to be not good in separating the water contained in the biodiesel. But EN specification is based on Karl Fisher (KF) method which covers the total water present in the biodiesel. Maximum water content agreed to be present in the biodiesel in chosen countries is given in Table 45.1.

Thus it is widely agreed that the water content in biodiesel should be below 500 mg kg^{-1}. The limit for the free water content as per EN 14214 and ASTM D6751 is $\leq$500 mg kg^{-1}.

Table 45.1 Maximum water content permitted in biodiesel in chosen countries

Country	Water (mg kg^{-1})	Country	Water (mg kg^{-1})
China	500	Japan	500
European Union	500	New Zealand	500
India	500	Thailand	500

45.6 Copper Strip Corrosion

Copper strip corrosion is a basic test to know the presence of acid and any other corrosive material which may be present. Certain components of the engine may be formed of bronze or other materials which may be affected by the presence of corrosive materials. Corrosive damages are likely on the seals and the rings of the fuel delivery system. Sulphur compounds and certain fatty acids are known to be corrosive to the fuel tank walls, engine system and few other components of the engine which may have contact with it. The corrosion level is measured by a copper strip immersed in it at 50 °C for 3 h. This indicates the probable chance and degree of corrosion caused by the biodiesel. Separate scale is set to indicate the degree of corrosion level. The European standard EN 14214 indicates a scale No. 1, whereas ASTM D6751 indicates scale No. 3 for the maximum corrosion.

45.7 Viscosity

Viscosity is a measure to indicate the ability of a fluid to flow. This parameter is very vital to a liquid fuel. If the fuel is thick, there will be great strain on the fuel pump resulting into loss of efficiency and ultimate failure of the pump. A thick fluid will not be atomized properly when it is forced to move through a tiny hole of the injector for spraying in to the combustion chamber (Allen and Watts 2000). In such events there will be incomplete combustion. Fuel dripping from the nozzle may cause hot spots on the piston head which eventually damages the whole piston assembly. The seeping fuel is likely to mix with the lubricating oil and consequently the lubricating oil may lose its quality. Besides such sunburn oil may be a cause for generating excessive exhaust. Since the spray pattern is altered viscous fluid is likely to cause excessive coking. Viscosity of biodiesel is generally double than that of the diesel. Vegetable oil in general is around 5–12 times more viscous than that of the diesel. The sole purpose of the transesterification of the oil is to reduce the viscosity to suit the engine characteristics. The fatty acid ethyl ester is slightly more viscous than the fatty acid methyl ester. Viscosity also affects the flow of fuels through pipes, injection nozzles and orifices thereby altering the temperature range for the proper operation of fuel in burners (Knothe and Steidley 2005). At the same time if the viscosity goes lower than the minimum limit, there is chance for wear and tear of the parts and loss of power. The European standard limit is 3.5–5.0 $mm^2\ s^{-1}$ at 40 °C. The US ASTM D6751 prescribes a limit of 1.9–6.0 $mm^2\ s^{-1}$. The upper limit of

6.0 $mm^2\ s^{-1}$ is higher than the regular diesel and therefore it should be considered during the blending. Often the viscosity of the fluid is referred in a unit cSt (Centi Stokes).

The surface tension and viscosity are two major factors which govern the atomization characteristics. The molecular weight also is another factor. The mean diameter of the droplet of coconut oil methyl ester comprising 12 and 14 carbon chains was almost on par with that of the diesel whereas the methyl ester having 18 carbon chains has the mean droplet diameter 20–30% higher than that of the diesel fuel. The rapeseed oil with a mean carbon chain of 20 has its droplet diameter around 40% higher than that of the diesel.

45.8 Wear

Moving parts of the automobile engines have to be protected from friction and therefore needs to be lubricated. The viscosity of the fuel is expected to lubricate the friction spots so as to increase the life of the engine. The reduction in friction is made by an effective combination of *hydrodynamic lubrication and boundary lubrication*. Hydrodynamic lubrication minimizes the frictional contact between the two surfaces which are opposing. On the contrary in boundary lubrication the viscous fluid has to offer protective layer to the parts especially when the engine takes strain due to high load and consequent loss in acceleration. If the boundary lubrication is not proper, a small portion of the opposing surface may develop frictional contact leading towards wear. One of the popular methods in assessing the lubricity is the *high frequency reciprocating rig* (HFRR). The limit for scar according to ASTM D975 is 520 μm at 60 °C, whereas in EN 590 limit it is 460 μm.

45.9 Cold Filter Plugging Point

Cold filter plugging point (CFPP) refers to a temperature at which the wax present in the liquid fuel crystallizes out and plug the filters. Above the cold filter plugging point temperature the fuel is expected to remain in liquid. In diesel the molecular configuration is very complex and each component has its own crystallization temperature. Therefore, the solidification pattern is not uniform. In such condition the solidification is gradual. But in the case of biodiesel the molecular configuration is relatively simple. Biodiesel has few components in which one or two major components tend to dominate. As a result, the low temperature solidification is rapid and it is very difficult to regulate it. The cold filter plugging point of diesel is normally 10 °C lower than cloud point, whereas in biodiesel both the values are almost closer.

45.10 Conductivity

The conductivity is the ability of a fuel to dissipate static electric charge. In a low conductivity fuel electric charge will accumulate which may result in a spark. The conductivity of the biodiesel is high (20–23 μSm^{-1}) which normally varies according to the FFA content of the oil used as the raw material.

45.11 Distillation Temperature

Distillation temperature is a fair index of the presence of specific components having characteristic boiling points. As the biodiesel comprises fractions having the carbon chain of narrow range C_{16}–C_{18} the distillation temperature of the fraction also is compact. Therefore, the different fractions boil almost at an identical temperature. Besides the boiling points of the biodiesels made from various vegetable oils also do not vary substantially since the composition of naturally occurring oils and fats are almost similar.

45.12 Thermal Stability

The degradation of the biodiesel takes place at an elevated temperature involving hyperperoxide and thus does not require oxygen. Near the injection point the temperature of the fuel normally remains around 150 °C. In the fuel conducts its temperature is around 60–70 °C. The blend is relatively stable when compared to pure biodiesel.

Neither the European Standard (EN 14214) nor the US Standard (ASTM D6751) prescribes any limit for the thermal stability of biodiesel.

45.13 Oxidation Stability

Test on oxidation stability is carried out to assess the duration the fuel takes to chemically react with oxygen molecule to form peroxide. Biodiesel molecule enriched with oxygen atom causes the biodiesel to degrade quickly. Methyl esters are normally more stable than ethyl esters. Containers play considerable role in maintaining the quality of the biodiesel. If the biodiesel is stored in a glass container especially in dark bottle the degradation is relatively slow as primary oxidative products remain accumulated. If the same material is stored in an iron container, the degradation is rapid. A naturally occurring enzyme lipoxidase in certain oils promotes polymerization. The oxidation stability is linked to the cetane number. It is widely believed that the biodiesel standard should include a measure of the oxidation tendency. Oxidation tendency is indicated by the iodine number and peroxide value (PV). PV beyond 70 in biodiesel is not appreciated. The product of oxidative instability may cause filter plugging, injector fouling and deposit formation in the

Table 45.2 Maximum oxidation stability of biodiesel at 110 °C admitted in certain countries

Country	Stability (h)
Australia	6
Brazil	6
China	6
European Union	6
India	6
Korea	6
Indonesia	1
New Zealand	6
The Philippines	1
Thailand	6
USA	3

Table 45.3 Maximum peroxide value permitted for biodiesel in certain countries

Country	Peroxide (mg KOH g^{-1})
Australia	0.8
Brazil	0.8
China	0.8
European Union	0.5
India	0.5
Indonesia	0.8
Japan	0.5
New Zealand	0.5
The Philippines	0.5
Korea	0.5
Thailand	0.5
USA	0.5

combustion chamber. The oxidation stability is closely related to viscosity and acid number. The viscosity decreases markedly, whereas the acid number decreases feebly during the period of instability. Similarly, Conradson carbon residue also changes. On oxidation biodiesel may produce gums and lacquers causing deposits. Biodiesel degrades around four times faster than diesel. It is known that when biodiesel is blended with diesel the degradation is activated. Twenty percent of biodiesel blend (B20) with conventional diesel is reported to degrade twice as fast as diesel alone. This necessitates the use of antioxidants. Rancimat method is used to assess the degradation at a specific temperature and specific time duration. A minimum oxidation stability of 6 h at 110 °C is suggested. Japan is recommending a minimum oxidation stability of 10 h. The specification prevailing in certain countries is given in Table 45.2.

In the initial stage of fuel oxidation peroxides are formed. The peroxide in turn generates acids and gums. Therefore, peroxide value is considered as an index of oxidation. Peroxide value being adopted in various countries is presented in Table 45.3.

Table 45.4 The Conradson Carbon Residue (CCR) standard in 10% distillation residue being permitted in certain countries

Country	CCR in 10% residue Max.%
Australia	0.3
Brazil	0.1
China	0.3
European Union	0.3
Indonesia	0.3
Japan	0.3
New Zealand	0.3
Thailand	0.3

The ASTM (USA) and Japanese Automotive Association have recommended to have stringent regulations with respect to the peroxide value as indicted in Table 45.3.

It is known that linolenic acid methyl ester if available at a maximum of 12% the stability of the diesel can be maintained. To control the deposits as a consequence of instability Conradson Carbon Residue (CCR) is considered as a reference value. It indicates gross impurities such as free fatty acid, glycerides, soaps, insolubles, polymers and high unsaturated fatty acids.

The EU standard for Conradson Carbon Residue is based on the 10% distillation residue. The CCR value at 10% distillation residue is given in Table 45.4.

45.14 Cetane Number

The cetane number refers to the readiness of a fuel to ignite and it is considered as an indicator of the smoothness of the fuel for combustion. This is also an index of the ignition time delay. The cetane number of the biodiesel depends on the distribution of the fatty acids in the raw material, namely the vegetable oil. If the fatty acid is longer and more saturated, the cetane number is higher. The cetane number is affected by the oxidation level of the biodiesel. Biodiesel contains 10–11% oxygen by weight which promotes complete combustion when compared to that of the regular diesel. If hydroxides are formed during the oxidation process, the cetane number increases. However, such effect is not pronounced beyond a cetane number of 70. Enhancement of oleic acid in the fatty acid profile is the common method to find a compromise between the issue of oxidative stability and cold flow properties without reducing the cetane number to an unacceptably low level (Knothe 2009). High cetane number ensures good starting of the engine at a low temperature and pressure and consequently there will be low knocking effect. At the same time fuels with high cetane number will burn before there is a proper mix between the fuel and air. White smoke tends to discharge if the cetane number is very high. Contrary to the above, biodiesel with a low cetane number is considered to have a poor ignition quality resulting in to misfiring, tarnish on pistons, engine deposits, rough operation and higher knocking causing increased noise level (Knothe et al. 2003, 2005).

The European Standard EN 14214 prescribes a standard of $\leq$51, whereas US standard ASTM D6751 indicates a cetane number of $\leq$47. The cetane number of biodiesel ranges from 45 to 70 while it is 40 to 52 for the usual diesel.

45.15 Diesel Index

The diesel index like cetane number refers to the ignition characteristics. The diesel index is calculated based on the fuel density and distillation range.

$$\text{Diesel Index} = \left[\text{aniline point } \left({}^{\circ}\text{F}\right) \times \text{API gravity } \left(60^{\circ}\text{F}\right)\right]/100$$

where API refers to American Petroleum Institute.

If the diesel index value is lesser than 50, there will be smoke in the exhaust indicating incomplete combustion.

45.16 Aniline Point

Aniline point refers to the lowest temperature at which equal volume of aniline and biodiesel is completely miscible and stands as a single phase. Lower the aniline point higher is the content of aromatic compounds. In such situations, the aromatic contents are higher than the alkane paraffinic compound. The aniline point of high speed diesel is 66 °C, whereas it is 25 °C for biodiesel.

45.17 Sulphated Ash Content

The residues of catalyst, metallic soaps and abrasive solids which remain in the fuel ultimately contribute to ash which is often specified as sulphated ash. Metals such as Ba, Ca, Mg, Na, K and Sn and S, P and Cl in ionic form are responsible for sulphated ash. There is active correlation between sulphated ash and phosphorus content. The ash thus formed may clog the fuel line and the filters. Moreover, such solids if pass through the filters may enhance the wear and tear of the injection, pump and piston. As the biodiesel is organic in nature sulphated ash content is normally low. However, if sodium or potassium hydroxides are used as catalyst there is chance for increased ash content. Both European and US Standards (EN 14214 and ASTM D6751) set a limit of $\leq$0.02% on weight basis.

The biodiesel sample (5 g) is taken in a pre-weighed quartz crucible and placed inside a muffle furnace (preheated at 450 °C). After half an hour, when the biodiesel burnt completely to ash, the crucible is taken out. The crucible is weighed again when its temperature dropped to at room temperature. The following formula is used to calculate ash content of biodiesel.

$$\text{Ash content of biodiesel (\%)} = \frac{\text{Initial wt. of the crucible} - \text{final wt. of the crucible}}{\text{Wt. of the biodiesel}} \times 100$$

45.18 Alcohol Content

Methanol and ethanol are commonly employed in the production of biodiesel. Iso propanol, *n*- ropanol, butanol and pentanol are also used (Lang et al. 2001). Residual alcohols are often found in the final product predominantly due to the use of extra alcohol in the transesterification process (Ma and Hanna 1999). Incidentally, ASTM does not specify any limit for methanol. As the alcohol has low flash point its residue tends to reduce the flash point of the biodiesel. Reduction in the flash point often threatens the safety of the engine. When alcohol content exceeds 5%, the cetane number of the biodiesel is altered. Besides, it is being associated with the lowering of the flash point. Methanol often induces the corrosion of the engine parts.

45.19 Oxygen Content

A typical biodiesel molecule contains 19 carbon atoms and 19 hydrogen molecules. Accordingly, the oxygen content is calculated as

$$\text{Oxygen content (\%)} = \frac{32}{32 + 228 + 38} \times 100$$

where

32 is the wt. of two oxygen atoms (2 × 16)
228 is the wt. of 19 carbon atoms (19 × 12)
38 is the wt. of 19 hydrogen molecules (19 × 2)
100 is to express in percent

which works out to 10.7%. But the ordinary diesel does not contain any oxygen. Biodiesel is formed predominantly of straight chain hydrocarbon esters, whereas diesel contains ring structures formed of aromatic molecules. Straight chain molecules (biodiesel) are normally characterized by higher cetane number and as a result they have easier and quicker ignition which results in smooth combustion as straight chains break more easily than ring structures.

45.20 Total Contamination

If the transesterification process is not carried out to the best of satisfaction, impurities are bound to stay. Residual catalyst and unsaponifiable neutral lipids which are insoluble in water after saponification contribute to total contamination. The neutral lipids are non-glyceridic in nature which include certain free fatty acids, fatty alcohols, hydrocarbons, triterpene alcohols, carotenoids and vitamins. Soap is formed when free fatty acids react with the alkali used in the transesterification. Such soaps if not washed away in full they may contaminate the fuel. The remains of unsaponifiable matter affect the storage properties of the biodiesel. The unsaponifiable matter normally has a higher boiling point and consequently the low temperature behaviour and the corresponding physical and chemical properties changes. The standard for total contamination is $\leq$24 mg kg^{-1} according to European standard but ASTM does not prescribe any standard.

45.21 Cloud and Pour Point

Biodiesel slowly becomes hazy or cloudy due to the development of wax crystals at certain low temperature. The point of temperature at which the biodiesel becomes cloudy is referred as cloud point and semi-solid is referred as pour point. The cloud point of ethyl esters is normally lower than that of the methyl esters. If the biodiesel maintains its clarity till 15 °C it is considered as a suitable biodiesel to be operated in colder climates. The cloud point of biodiesel varies with the type of oil being used as raw material to prepare the respective biodiesel. The cloud point and cold filter plug point usually have direct correlation. Normally the cloud point of biodiesel is higher than that of the regular diesel. Therefore, on account of the above, it is almost impossible for the biodiesel to meet the standard set for the diesel. Unless abundant care is taken, there is every likely the biodiesel plugs the filters and the fuel in the tank may also get solidified at low temperature. The fuel line is often warmed to solve this problem.

The cloud and pour point apparatus is filled up with ice. Biodiesel is taken in the glass vessel of this apparatus. It is then placed in the slot. The temperature at which the biodiesel started showing cloudiness indicates cloud point. The temp at which the biodiesel becomes semi-solid is referred as pour point (Karmakar et al. 2018).

45.22 Linolenic Acid Content

Linolenic acid content is specified as a parameter in European Standard EN 14214 but excluded in ASTM D6751. As per EN 14214 linolenic acid content shall not exceed 12% because of the propensity of methyl linolenate to oxidize. Such high limit is set so as to ensure that rapeseed oil is not rejected since rapeseed oil is the major oil source in Europe.

45.23 Iodine Number

Iodine number refers to the quantity of iodine in gram which reacts with a known quantity of oil in a specific condition. It is an indication of the number of double bond present in the oil or biodiesel and therefore it indirectly indicates the degree of unsaturation. The unsaturated components readily react with hydrogen, oxygen and iodine and thus iodine is an easy reference of unsaturation. The unsaturated fatty acid molecules do not contain adequate hydrogen in the molecule. Therefore, there will be double bond which decreases the fuel quality. The double bond normally enhances the cetane number of the oil. The unsaturated components readily react with hydrogen, oxygen and iodine and thus iodine is an easy reference of unsaturation. If unsaturated fatty acid content is high, there can be polymerization of glycerides which causes extra deposit in the tank. It is known that the iodine value above 115 g I_2/100 g (1150 mg I_2 g^{-1}) may increase the risk of polymerization. Oils from certain sources such as sunflower, soybean, cotton seed and maize have high proportion of unsaturated fatty acid with C18:2 (single:double C bonds) configuration. The biodiesel prepared from such oils has low melting point and cetane number. The iodine value also is high. A powerful oxidizing agent lipoxidase is present in certain oil. The lipoxidase increases the rate of oxidation as the activation energy required to commence the routine oxidation is much lower. In such events the polymerization increases rapidly causing negative effect to the engine. The lubricating oil of the engine also is likely to be thickened if the biodiesel passes through the crank case of the engine.

The limit mentioned by the European standard EN 14214 for polyunsaturated ($\leq$4 double bond) methyl ester is $\leq$1% (m m^{-1}). However, the ASTM D6751 does not prescribe any limit.

The member countries of the European Union recommend an iodine value of 115–125 g I_2/100 g.

The above value is drawn based on that of the canola. The ASTM (USA) and UNI (Italy) have not specified any standard for iodine. Iodine value is determined by adding iodine to the oil slowly till the whole titrate is consumed in 100 mL oil. High value for iodine indicates the tendency to solidify and thicken (polymerization) at high temperature of operation in the engine. Such solidification leads to the clogging of filters developing deposits inside fuel injection equipment, nozzle coking and seizure. The iodine value of few oils is indicated in Table 45.5.

Iodine value of the oil normally establishes a negative correlation with the melting point of it (Fig. 45.1.). The melting point of nine different oils was plotted against their corresponding values of iodine.

The iodine value reduces during the hydrogenation of the oil in which the double bond is broken so as to obtain a solid fat. Hydrogenation reduces the tendency of the oil to polymerize. Hydrogenation increases the pour point (melting point) of the oil. Thus, when iodine value is low there is great stability but such oil cannot be used in cold areas as oil is likely to be solidified. To solve such problem transesterification is recommended which breaks the complex triglyceride into small ester molecules.

Table 45.5 Iodine value of chosen oils

Oil	Iodine value (g I_2 /100 g)
Coconut	6–12
Palm	35–61
Castor	82–88
Peanut	80–106
Rapeseed	94–120
Cotton seed	90–119
Sunflower	119–138
Soybean	117–143

Ref.: Terry de Winne, Biofuels for sustainable transport (http://www.biofuels.fsnet.co.uk) and National Biodiesel Board

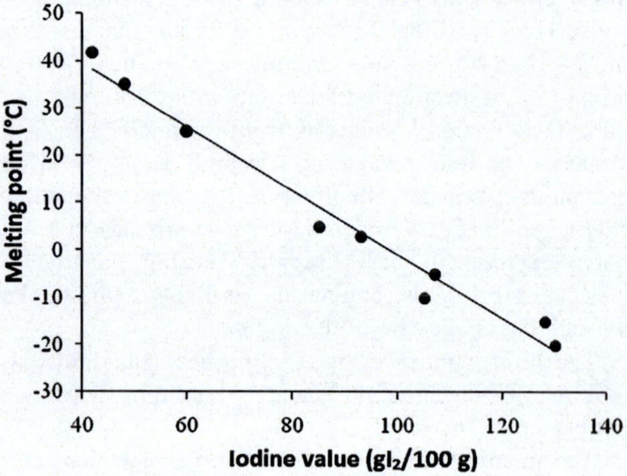

Fig. 45.1 Relationship between melting points of oil and the corresponding iodine values

45.24 Glycerides (Bound Glycerin)

The glycerin (mono, di and triglycerides) may present in the final products as impurity. Such glycerin is called bound glycerin. It causes carbon deposit in fuel injector tips and piston rings. Monoglycerides have a high melting point. They have a low solubility in methyl esters. Therefore, they require high temperature to maintain them from crystallization. Monoglyceride normally settles down in the storage tank if the ambient temperature is low. The ASTM is silent on the specification of mono, di and triglycerides separately as analysis of these parameters is cumbersome. Hence free and bound (total) glycerides alone are specified in standards since it is easy to analyse them by simple colorimetric iodimetry. The triglyceride content of biodiesel is an indicator of pure oil present (remaining unconverted) in the biodiesel.

45.25 MAG, DAG and TAG Content

During the process of biodiesel formation intermediate products such as mono acyl glycerols (MAG) and diacyl glycerols may often occur. The presence of MAG increases the viscosity and melting point of the biodiesel. In such events low quantity of alcohol is required to affect further transesterification. One mole of glycerol mono-oleate (356 g mol^{-1}) reacts with one mole of methanol to produce one mole of methyl ester (296 g mol^{-1}) and one mole of glycerol. The mass of alcohol utilized in the reaction is observed to be far lower than that of the mass of alcohol normally used when triacyl alcohol is used. Insufficiency of the quantity of catalyst also may govern the presence of glycerols. The quantity of alkali required to transesterify an oil differs in accordance with the type of oil. Therefore, alkali requirement is arrived at by titration. Normally a gram of fatty acid (as oleic acid) will act with 0.2 g of anhydrous potassium hydroxide or 0.14 g of sodium hydroxide.

45.26 Free Glycerin

During transesterification process the glycerin separates and settles down at the bottom of the container which is no longer bound with ester. This will be dark brown in colour and it is called free glycerin. This glycerin is needed to be removed. Glycerin forms around 16–18% of the total volume. Unremoved glycerin eventually settles at the bottom of the fuel tank and may create blockage in the fuel pump or strain the fuel pump. This may cause the failure of the pump. In most of the cases the glycerin may block the filter before it reaches the vital components. If the glycerin is present even in minimum quantities, it may create dirt in the injection system and consequently there will be poor atomization of the fuel resulting in to the generation of white smoke at the exhaust. The glycerin separated during the biodiesel formation is likely to contain methanol, alkali, salt, water and unreacted triglyceride, esters and soaps. Absorbents such as magnesium silicates or silica gel are used to selectively absorb the trace quantities glycerin if present in the biodiesel.

45.27 Unsaponifiable Matter

Unsaponifiable matter may also present in considerable quantities. The unsaponifiables are predominantly of alcohols and hydrocarbons which may not affect the transesterification process. Unsaponifiable matter present in the oil may not react with alcohol during transesterification and thus remains as impurity in the biodiesel. In certain oils the unsaponifiable matter comprises sterols whose molecular weight is higher than that of the other components.

45.28 Antifoaming

Pure biodiesel has antifoaming properties and therefore while transferring the biodiesel there will not be any foam leaks or overflows from the containers. Thus, there is a scope for fast filling of the fuel in vehicles.

45.29 Sulphur

Sulphur is one of the common contaminants in the petro-diesel causing the generation of sulphur dioxide in the exhaust gas which may harm the environment. They also develop acids in the combustion cylinder and corrode its wall. Besides, it affects the quality of the lubricating oil at the crank case resulting in to the replacement of the oil frequently. Biodiesel is properly known as a sulphur free fuel but there is chance for the presence of sulphur in trace quantities due to its availability in low quantities in certain oil used as raw material. Biodiesel prepared from certain vegetable oil contains sulphur to a maximum of 15 mg L^{-1}. Biodiesel normally has sulphur content ranging from 1 to 14 mg L^{-1}, whereas diesel has around 50 mg L^{-1} if it is ultra-low, or it may have up to 500 mg L^{-1}. If the biodiesel is made from used cooking oil, there is every chance for the sulphur content to cross 15 mg L^{-1}. Slaughter house wastes and cooked oil may contain sulphur to a certain extent. Combustion of fuels containing sulphur will cause the emission of sulphur dioxide and particulate matter. Low sulphur fuels are reported to have low lubricity.

The European standard for sulphur content in biodiesel is $\leq$10 mg L^{-1} (less than 0.001% by mass). The ASTM D6751 standard prescribes a limit of $\leq$50 mg L^{-1} (0.005% by mass).

As already indicated vegetable oil may contain sulphur to the tune of 5–50 mg L^{-1}. If the regulation prescribes $\leq$10 mg L^{-1} sulphur in the biodiesel, appropriate care should be exercised to select an oil of low sulphur. The permitted sulphur content of biodiesel in few countries is given in Table 45.6.

Table 45.6 Limits of sulphur in biodiesel being followed in certain countries

Country	Limit (mg L^{-1})
Australia	10
China	50
European Union	10
India	50
Indonesia	100
Japan	10
New Zealand	50
Korea	10
Thailand	10
USA	10

45.30 Carbon Residue

Carbon residue is an important reference to the quality of biodiesel. Carbon residue is normally a function of the presence of glycerides, free fatty acids, soaps, residual catalysts and other impurities. Carbon residue indicates the tendency of the fuel to form deposits in the engine parts. Such deposits on injection tips and other moving parts may cause wear and tear. Besides, these residues limit the fuel efficiency. The particulate matter present in the emission harbours carbon in the form of crystallites.

The European standard limit for carbon residue is $\leq 0.30\%$ by mass. As per ASTM D6751 it is $<0.05\%$ by mass.

The biodiesel (5 g) is taken in a pre-weighed heat proof glass bulb and placed inside the carbon residue apparatus (preheated at 450 °C) for 30 min. The weight of the bulb is taken after its heat comes down to room temperature. The following formula is used to calculate carbon residue content in biodiesel,

$$\text{Carbon residue } (\%) = \frac{\text{Initial wt. of the bulb} - \text{final wt. of the bulb}}{\text{Wt. of the biodiesel taken}}$$

45.31 Ester Content

The ester content of the biodiesel is a measure of the purity of the biodiesel. In case the ester content is low in biodiesel the reason may be an improper molar ratio of the oil to alcohol or a wrong selection of catalyst or incorrect reaction temperature or inadequate reaction duration or presence of high level of moisture or exceedingly high free fatty acid which promotes soap formation or high level of glycerol present in the oil or the selection of alcohol which is not appropriate. If the biodiesel is formed of pure esters, the engine performance will be very good. Ester formation if affected due to defective processing the biodiesel may contain considerable amount of mono, di and triglyceride. In such situation the viscosity will be very high. The partially reacted alcohol if remain along with the ester, it will get precipitated at a low temperature and cause negative effect to the fuel. The European standard indicates that the ester content shall be more than 96.5%. However, ASTM D6751 is silent on this parameter.

45.32 Ester with ≥4 Double Bond

This specification eliminates the use of fish oil as a source of biodiesel. As the fish oil has a high content of methylene interrupted double bonds, the fatty acids of fish oil are prone to oxidation. Therefore, ester with ≥4 double bond is restricted to 1% (mol mol^{-1}).

45.33 Group I Metals (Na and K)

The level for Na + K is 5 mg kg^{-1} both in ASTM D6751 and EN 14214.

45.34 Group II Metals (Ca and Mg)

Group II metals such as Ca and Mg are known to be contaminants to biodiesel. These metals are also associated with fuel instability even at trace concentrations (Table 45.7).

Softened water which is marginally acidic is used in the washing process, Ca and Mg content can be kept on check.

45.35 Raw Material Quality

Fatty acid alkyl ester (biodiesel) differs from regular diesel largely by its chemical and physical characteristics such as density, viscosity and cold flow properties. The presence of oxygen and polarity are the two qualifying characteristics of the biodiesel. The raw material, namely the type of vegetable oil largely governs the ultimate quality of the biodiesel.

The triglyceride content of the oil differs by the chain length, branching and degree of saturation. The fatty acid components of the triglyceride are of different types. Some of the characteristics of the biodiesel are inherited from the type of oil used as a raw material. They are density, viscosity, cetane number and iodine number. Such characteristics depend on the molecular weight and carbon configuration. Long and straight chains of the fatty acid imparts high cetane number. Cetane number lowers especially when the degree of saturation reduces. Iodine number is a measure of unsaturation. A high number indicates poor stability of the biodiesel. The oxidation stability of the biodiesel will be high if the oil used as a raw material contains more saturated fatty acids. If the number of double bond increases in the fatty acid profile, the instability also increases.

The rapeseed oil used in the biodiesel production has 89% of unsaturated fatty acid. The biodiesel prepared from rapeseed oil has a cold filter plugging point ranging from −10 to −12 °C. The soy oil has low oxidation stability. But it has

Table 45.7 The limit of Ca and Mg in the biodiesel in certain countries

Country	Limit (mg kg^{-1})
Australia	5
European Union	5
Japan	5
USA	5
New Zealand	5
Korea	5
Thailand	5

good cold flow properties. Palm oil has a high level of saturation offering high oxidation stability. But it has poor cold flow properties. Coconut oil has a high degree of saturation (45%) but with restricted C14 molecules and hence the cold filter plugging point is very low. The sunflower oil containing high level of unsaturated fatty acid develops deposits in the fuel line. Certain parameters such as free fatty acid, sulphur, phosphorus, water content, insoluble and contaminants change the treatment course.

45.36 Free Fatty Acid

If oil has high level of free fatty acid, it requires more catalyst for neutralization. Dilute ethanolic KOH solution with phenolphthalein as indicator is used to assess the neutralization number (NN) of biodiesel (Komers et al. 1997). High quantity of catalyst causes the ash content to increase in the product. The free fatty acid acts with alkali catalyst and forms soap. Excessive soap formation imposes difficulty in separating the glycerin from ester as the soap creates cloudiness imparting difficulties in phase separation. Besides, soap creates foam and emulsions requiring more water for cleaning the biodiesel. Consequently, waste water is needed to be treated. In order to solve this issue excessive free fatty acids if present in the oil are reduced to a level lesser than 0.5% by acid esterification before it is passed on through the regular transesterification.

45.37 Phosphorus

If phosphorus is present in the oil in trace quantities, it is likely to be transferred to the biodiesel. The phosphorus present in the oil creates emulsion which prevents the separation of biodiesel and glycerol. As a matter of fact, the emulsion formed due to the phosphorus stands as a third layer in between the glycerol and biodiesel. In large scale preparation of biodiesel, the unwanted emulsion layer is pumped out, which also contain a little quantity of ester (biodiesel) and glycerol. Therefore, high phosphorus content causes low recovery of ester causing economic loss. Phosphorus present in the fuel is likely to damage the catalytic converters being used in emission control process. Phosphorus content in the oil to the level of 3–4 mg kg^{-1} is acceptable and any concentration above may cause emulsion problem. The accepted level of phosphorus in biodiesel level is 10 mg kg^{-1} in many countries (Table 45.8).

45.38 Insoluble and Contaminants

If insoluble materials (dust and rust) are present in the oil, they are likely to pass through the process unnoticed which ultimately form a constituent of the biodiesel. Such materials may damage the injection system. Fuel tank hitherto used for filling regular diesel if needed to be used for biodiesel, it is necessary to ensure a complete

Table 45.8 Limit of phosphorus in biodiesel being accepted in certain countries

Country	Level of phosphorus (mg kg^{-1})
Australia	10
European Union	10
India	10
Indonesia	10
Japan	10
New Zealand	10
The Philippines	10
Korea	10
Thailand	10
USA	10

Table 45.9 Limit of contaminants in biodiesel being permitted in certain countries

Country	Limit of contaminants (mg kg^{-1})
Australia	24
European Union	24
India	24
Japan	24
New Zealand	24
Thailand	24

cleaning. Otherwise, the polar nature of the biodiesel is likely to dislocate the rust if any from the walls of the tank which may build up the sediment. The insoluble should not exceed 24 mg kg^{-1}.

The standard for the insoluble in biodiesel being followed in many countries is given in Table 45.9.

The alkaline earth or Group 2 metals (calcium and magnesium) are soft in nature and are known to be of low density. They react readily with water if present in the biodiesel and form alkaline hydroxides.

45.39 Life Cycle Assessment

The Life Cycle Assessment (LCA) is a measure of the net impact of a process or series of processes. The LCA values change according to the culture condition of the oil crop. Major components of the energy sink are tillage, synthetic fertilizer (N, P, K and S), oil recovery methodology and the quantity of methanol employed in transesterification. Pest resistant crops capable of growing in an area with low tillage and providing high yield substantially reduce the energy input. Nitrogen fixing oil crop has an enhanced LCA. The LCA can be greatly improved if the oil content of the seed is increased and the phosphorus content of the oil is decreased. Canola meal is rich in phytate inositol hexaphosphate. The phytate present in the meal is indigestible to non-ruminant animals. Complete or partial removal of phytate or conversion of phytate in to other forms may cause a net positive impact on LCA as the finished product is good as feed for animals. Popular corn varieties with low phytate mostly show poor seed yield. Therefore, genetic engineering is suggested in developing seed oil with low phosphorus and nitrogen. Similarly, research is called for to develop oil with low glycerol. Oil containing low level of glycerol demands less quantity of methanol for transesterification.

45.40 Toxicity

Acute oral toxicity of methyl and ethyl esters developed from canola oil for albino rats was observed to be higher than 5000 mg kg^{-1}. Dermal exposure up to 2000 mg kg^{-1} did not cause any visible effect except making mild redness and swelling on the applied area.

Biodiesel is observed to cause acute toxicity to many animals. However, biodiesel is known to degrade fast. It is also known that hydrocarbon, carbon monoxide and particulate matter occur in low concentration in the exhaust when biodiesel is used as a fuel. But, the concentration of the oxides of nitrogen in the exhaust is relatively higher. Though the occurrence of polyaromatic hydrocarbons is greatly reduced, formaldehyde and acrolein are observed to increase by 40%.

45.41 International Standards

As the biodiesel is being developed commercially from different oil sources available in different locations, a common quality is needed to be guaranteed for which acceptable standards are being developed by many institutions in different countries. Popular standards being followed in many countries are ASTM D6751 (USA), EN 14214 (European Union) and DIN E51606 (Germany). The limits prescribed by the above three standards are shown below (Table 45.10).

The ASTM (The American Society for Testing and Materials) was formed during the year 1898 by a few Chemists and Engineers of the Pennsylvania Railroad. During the period of its formation it was known as the American Section of the International Association for Testing and Materials. This society was formed by the initiation of Dr. Charles D. Dudley, a chemist of the Pennsylvania Railroad. In 2001 the society was renamed as ASTM International. Around 30,000 members of this society from 140 countries were reported to prepare and deliver various test methodologies so as to improve product quality, enhance safety and to facilitate market access and trade. As on date around 12,000 standards are in force. The origin of ASTM predated British Standards Institution (BSI), Deutsches Institut für Normung (DIN), Association francaise de Normalisation (AFNOR), Japanese Standard Association (JSA) and the International Organization for Standards (ISO). These organizations are dedicated to the need of the respective countries. The role of ASTM in USA is being taken care of by the American National Standards Institute (ANSI).

Table 45.10 Standards for biodiesel

Specification	USA	Europe	Germany
Density at 15 °C (kg m^{-3})	–	860–900	875–900
Viscosity at 40 °C (mm^2 S^{-1})	1.9–6.0	3.5–5.0	3.5–5.0
Distillation—max 90 (Recor) (°C)	360	–	–
Flash point—min (°C)	93	101	110
Sulphur (mg kg^{-1}—max)	15	10	0.01
Carbon residue in 10% dist—max (% w/w)	–	0.3	–
Carbon residue in 100% dist—residue max (% w/w)	0.05	–	0.05
Sulphated ash—max (% w/w)	0.02	0.02	0.03
Water and sediment (% vol)	0.05	0.05	–
Total contamination—max (mg kg^{-1})	–	24	20
Cu strip corrosion—max class 3 h at 50 °C	3	1	1
Oxidation stability—min h at 110 °C	3	6	–
Cetane number—min	47	51	49
Linolenic acid methyl ester—max (% w)	–	12	–
Polyunsaturated ≤4—max (mg kg^{-1})	–	1	–
Acid value—max (mg KOH kg^{-1})	0.5	0.5	0.5
Methanol—max (% w)	0.2	0.2	0.5
Ester content—min (% w)	–	96.5	–
Monoglyceride—max (% w)	–	0.8	0.8
Diglyceride—max (% w)	–	0.2	0.4
Triglyceride—max (% w)	–	0.2	0.4
Freeglyceride—max (% w)	0.02	0.02	0.02
Total glyceride—max (% w)	0.24	0.25	0.25
Iodine number—max (g I_2 g^{-100})	–	120	115
Phosphorous—max (mg kg^{-1})	10	4	10
Group I (Na + K)—max (mg kg^{-1})	5	5	5
Group II (Ca + Mg)—max (mg kg^{-1})	5	5	–
Cold Soak Filterability—max	360	–	–
Cloud point	–	–	–
CFPP	–	–	−10

The ASTM D6751 refers to the Standards and specifications for biodiesel. ASTM also suggests procedures for analysing the biodiesel which is normally being followed for petroleum products. The ASTM test methods are listed in Table 45.11.

The European Committee for Standardisation (CEN = Comite europeen de Normalisation) is a nonprofit body which developed standards and specifications. The CEN was established in 1961. CEN is officially recognized as European Standard (EN) body for the European Union. The European Standards are worked out by the Scientists from many nations (Austria, Belgium, Bulgaria, Cyprus, Czech Republic, Denmark, Estonia, Finland, France, Germany, Greece, Hungary, Ireland, Italy, Latvia, Lithuania, Luxembourg, Malta, The Netherlands, Poland, Portugal, Romania, Slovakia, Slovenia, Spain, Sweden, the United Kingdom, Iceland,

Table 45.11 ASTM Standard test methods

No.	Test method
D93	Flash point by Pensky-Martens closed cup tester
D130	Corrosiveness to copper from petroleum products by Copper Strip Test
D189	Conradson Carbon Residue of petroleum products
D445	Kinematic viscosity of transparent and opaque liquids (and calculation of dynamic viscosity)
D524	Ramsbottom Carbon Residue of petroleum products
D613	Cetane number of diesel fuel oil
D664	Acid number of petroleum products by potentiometric titration
D874	Sulphated Ash from Lubricating Oils and Additives
D974	Acid and Base Number by Colour-Indicator titration
D975	Specification for diesel fuel oils
D976	Calculated Cetane Index of distillate fuels
D1160	Distillation of petroleum products at reduced pressure
D1266	Sulphur in petroleum products (Lamp Method)
D1796	Water and sediment in fuel oils by the centrifuge method (Laboratory Procedure)
D2274	Oxidation stability of distillate fuel oil (Accelerated Method)
D2500	Cloud Point of petroleum products
D2622	Sulphur in petroleum products by Wavelength Dispersive X-ray Fluorescence Spectrometry
D2709	Water and sediment in middle distillate fuels by Centrifuge
D2880	Specification for gas turbine fuel oils
D3117	Wax Appearance Point of distillate fuels
D3120	Trace quantities of sulphur in light liquid petroleum hydrocarbons by oxidative microcoulometry
D3242	Acidity in aviation turbine fuel
D3828	Flash point by Small Scale Closed Cup Tester
D4057	Practice for manual sampling of petroleum and petroleum products
D4177	Practice for automatic sampling of petroleum and petroleum products
D4294	Sulphur in petroleum and petroleum products by energy Dispersive X-ray Fluorescence Spectrometry
D4530	Determination of carbon residue (Micro Method)
D4737	Calculated Cetane Index by four variable equation
D4865	Guide for generation and dissipation of static electricity in petroleum fuel systems
D4951	Determination of additive elements in lubricating oils by Inductively Coupled Plasma Atomic Emission Spectrometry
D5453	Determination of total sulphur in light hydrocarbons, spark ignition engine fuel, diesel engine fuel, and engine oil by ultraviolet fluorescence
D5771	Cloud point of petroleum products (Optical Detection Stepped Cooling Method)
D5772	Cloud point of petroleum products (Linear Cooling Rate Method)
D5773	Cloud point of petroleum products (Constant Cooling Rate Method)
D6217	Particulate contamination in middle distillate fuels by laboratory filtration
D6450	Flash point by continuously closed cup (CCCFP) Tester
D6469	Guide for microbial contamination in fuels and fuel systems

(continued)

Table 45.11 (continued)

No.	Test method
D6584	Determination of total monoglycerides, total diglycerides, total triglycerides, and free and total glycerin in B-100 biodiesel methyl esters by gas chromatography
D6890	Determination of Ignition Delay and Derived Cetane Number (DCN) of diesel fuel oils by combustion in a constant volume chamber
D7039	Sulphur in gasoline and diesel fuel by monochromatic wavelength dispersive X-ray Fluorescence Spectrometry
D7397	Cloud point of petroleum products (Miniaturized Optical Method)
D7501	Determination of fuel filter Blocking Potential of Biodiesel (B100) Blend Stock by Cold Soak Filtration Test (CSFT)

Norway, Switzerland, Croatia and Turkey). Currently Albania, Armenia, Azerbaijan, Belarus, Bosnia and Herzegovina, Egypt, Georgia, Israel, Jordan, Lebanon, Libya, Macedonia, Moldova, Montenegro, Morocco, Serbia, Tunisia, and Ukraine are affiliated to this body. The partner bodies are Australia and Mongolia. During July 2003 the European EN 14214 established the standards for biodiesel.

An agreement established between CEN and International Organization for Standardization (ISO) during 1991 came in to force in 2000 so as to avoid duplication between CEN and ISO.

EN 14214 is the European standard for biodiesel. This standard applies only to the biodiesel (methyl ester) produced using methanol and excludes ethyl ester. The EN 14214 exists in 3 different languages (English, French and Germany). EN prescribes parallel test methods which are shown in Table 45.12.

The standards being followed in certain European countries are given in Table 45.13.

The Austrian Standards Institute (ASI) [formerly Österreichisches Normungsinstitut (ONORM): German] is an organization dealing with standards which is a member of the ISO body of Austria. It is located at Vienna. In 1991 the standard for rape oil methyl ester ONORM C 1190 was published which was amended twice (1st January, 1995 and 1st November, 1996).

The Czech Republic published the biodiesel standard CSN 65507 in 1994.

In Italy the biodiesel quality is ensured by Ente Nazionate Italiano di Unificazione (ENIU). Often the above authority is referred as UNI and the standard is UNI 10635.

The Government of Sweden established the Swedish Standard Institute and the biodiesel standard is SS 155436.

In France the Institute Francais du Petrole published the fuel standard which later became the French law NF EN 590 on the 1st April 1995. The Journal officiel de la Republique Francaise is the official Gazette of the French Republic which publishes the major legal official information.

The German Institue for Standardisation Deutsches Institut für Normung e.V. (Berlin) published the standard DIN V 51606 in June 1994. The DIN standards for biodiesel are of three different types based on the sources of raw material (Table 45.14).

They are

Rapeseed methyl ester—DIN E 51606
Vegetable methyl ester (purely vegetable products)—DIN E 51606
Fat methyl ester, vegetable and animal products—DIN V 51606

In China on the 5th January, 2007 The General administration of Quality Supervision, Inspection and Quarantine and State Standardisation Supervision Committee jointly promulgated the National Standard of Biodiesel Blend stock for diesel engine fuel which took effect on 1st May, 2007. The biodiesel standard of China is GB/T 20828—2007 where GB (*Guobiao*) refers to the phonetic transcription of National Standards in Chinese language. T refers the abbreviation of the word *tuijian* which means recommended (Table 45.15).

Table 45.12 EN Standard test methods

No.	Test methods
EN 14103	Ester content
EN ISO 3675 and 12185	Density at 15 °C
EN ISO 3104	Viscosity
EN ISO 2719 and 3679	Flash point
EN ISO 20846 and 20884	Sulphur content
EN ISO 10370	Carbon residue remnant (at 10% distillation remnant)
EN ISO 5165	Cetane number
ISO 3987	Sulphated ash content
EN ISO 12937	Water content
EN ISO 12662	Total contamination
EN ISO 2160	Copper band corrosion (3 h at 50 °C)
EN 15751 and 14112	Oxidation stability (at 110 °C)
EN 14104	Acid value
EN 14111	Iodine value
EN 14103	Linolenic acid methyl ester, polyunsat (≥4 double bonds), methyl ester
EN 14110	Methanol content
EN 14105	Monoglyceride content, Diglyceride content, Triglyceride content
EN 14105 and 14106	Free glycerin
EN 14105	Total glycerine
EN 14108, 14109 and 14538	Group I metals (Na + K)
EN 14538	Group II metals (Ca + Mg)
EN 14107	Phosphorus content

Table 45.13 The standards being followed in certain European countries for biodiesel

Specification	Austria	Czech Republic	France	Italy	Sweden
Density at 15 °C (kg m^{-3})	850–890	855–885	870–900	860–900	870–900
Viscosity at 40 °C (mm^2 S^{-1})	3.5–5.0	3.5–5.0	3.5–5.0	3.5–5.0	3.5–5.0
Distillation—max 90 (Recor) (°C)	–	–	360	360	–
Flash point—min (°C)	100	110	100	100	100
Sulphur (mg kg^{-1}—max)	0.02	0.02	–	0.01	0.001
Carbon residue in 10% dist—max (% w/w)	–	–	0.3	0.5	–
Car Resin 100% dist—residue max (% w/w)	0.05	0.05	–	–	–
Sulphated ash—max (% w/w)	0.02	0.02	–	–	–
Water and sediment (% vol)	–	–	–	–	–
Total contamination—max (mg kg^{-1})	–	24	–	–	20
Cu strip corrosion—max class 3 h at 50 °C	–	1	–	–	–
Oxidation stability—min h at 110 °C	–	–	–	–	–
Cetane number—min	49	48	49	–	48
Linolenic acid methyl ester—max (% w)	–	–	–	–	–
Polyunsaturated ≤4—max (mg kg^{-1})	–	–	–	–	–
Acid value—max (mg KOH kg^{-1})	0.8	0.5	0.3	0.5	0.6
Methanol—max (% w)	0.2	0.3	0.1	0.2	0.2
Ester content—min (% w)	–	–	96.5	98	98
Monoglyceride—max (% w)	–	–	0.8	0.8	0.8
Diglyceride—max (% w)	–	–	0.2	0.2	0.1
Triglyceride—max (% w)	–	–	0.2	0.1	0.1
Freeglyceride—max (% w)	0.02	0.02	0.02	0.05	0.02
Total glyceride—max (% w)	0.24	0.24	0.25	–	–
Iodine number—max (g I_2/100 g)	120	–	115	–	125
Phosphorous—max (mg kg^{-1})	20	20	10	10	10
Group I (Na + K)—max (mg kg^{-1})	–	1.0	5	–	10
Group II (Ca + Mg)—max (mg kg^{-1})	–	–	–	–	–
Cold Soak Filterability—max (s)	–	–	–	–	–
Cloud point	–	–	–	–	–
CFPP	<01–15	−5	–	<01–15	<−5

Table 45.14 Standards being followed in various countries for biodiesel in automobiles

Countries	Standards	Year
Australia	Fuel Standard Determination	2003
Austria	ÖNC 1191	1997
Brazil	ANP 42	–
Czech Republic	CSN 656507	1998
Europe	EN 14214	2003
France	Journal officiel	1997
Germany	DIN E 51606	1997
Italy	UNI 10946	2001
India	IS 15607	–
Indonesia	SNI 04-7182	2006
New Zealand	NZS 7500	2005
Sweden	SS 155436	1996
The Philippines	PNS 2020	2003
USA	ASTM D6751	2002
RSA	SANS	–
Japan	JASO/JIS	–
China	GB/T 20828	2007

The Vietnam Standards Tiêu Chuân Viêt Nam abbreviated as TCVN is the national standards of Vietnam issued by the Vietnam Standard and Quality Institute. The standards are prefixed TCVN and is followed by the number, colon and the year of issue. The biodiesel standard in Vietnam is TCVN 7717:2007. The biodiesel standards followed in Australia, Brazil, New Zealand and South Africa are given in Table 45.16.

In Japan the Japanese Automotive Standards Organization is responsible for the regulation of its standard JASO M360.

The Indonesian National standard has prescribed a biodiesel standard SNI 04-7182-2006 through a decree No. 73/KEP/BSN/2006 on 22 February 2006 which partially follows ASTM and EN standards.

The Philippine National Standards (PNS) for biodiesel is PNS/DOE QS 002:2007 which is based on coconut oil methyl ester. In India Bureau of Indian Standards, 2005 recommends ISO 15607 for biodiesel.

The Brazilian Biodiesel Standard, Agência Nacional de Petróleo (ANP) Act No. 42 (2004) comprehensively deals with the quality of the process and the nature of the raw material. The fuel standard (Biodiesel) determination 2003 was promoted by David Alistair Kemp, Minister for Environment and Heritage, Govt. of Australia as a part of Section 21 of the Fuel Quality Standards Act 2000 on 18th September, 2003. This standard to a certain extent follows that of ASTM (USA). However, the Australian Federal Government has also incorporated few additional parameters present in the European standard.

In the New Zealand the biodiesel standard is NZS 7500:2005—Automotive Biodiesel specification for manufacture and blending.

SANS 1935:2011 is the standard for biodiesel developed by the South African National Standard of the Republic of South Africa.

Table 45.15 The standards being followed in certain Asian countries for biodiesel

Specification	China	India	Indonesia	Japan	Korea	The Philippines	Thailand
Density at 15 °C ($kg\ m^{-3}$)	820–900	860–900	850–890	860–900	860–900	–	860–900
Viscosity at 40 °C ($mm^2\ S^{-1}$)	1.9–6.0	3.5–5.0	2.3–6.0	3.5–5.0	1.9–5.0	2.0–4.5	3.5–5.0
Distillation—max 90 (Recor) (°C)	360	–	360	–	–	360	–
Flash point—min (°C)	130	101	100	120	120	100	120
Sulphur ($mg\ kg^{-1}$—max)	50	10	–	10	10	500	10
Carbon residue in 10% dist—max (% w/w)	0.3	0.3	0.3	0.3	–	–	0.3
Car. res in 100% dist—max (% w/w)	–	0.05	0.05	–	0.1	0.05	–
Sulphated ash—max (% w/w)	–	0.02	–	0.02	–	–	0.02
Water and sediment (% vol)	–	0.05	0.05	0.05	0.05	0.05	–
Total contamination—max ($mg\ kg^{-1}$)	–	24	–	24	–	–	24
Cu strip corrosion—max class 3 h at 50 °C	1	1	3	1	1	3	1
Oxidation stability—min h at 110 °C	6	6	–	4	6	–	6
Cetane number—min	49	51	51	51	–	42	51
Linolenic acid methyl ester—max (% w)	–	12	–	12	–	–	12
Polyunsaturated ≤4—max ($mg\ kg^{-1}$)	–	–	–	–	–	–	–
Acid value—max ($mg\ KOH\ kg^{-1}$)	0.8	0.5	0.8	0.5	0.5	0.5	0.5
Methanol—max (% w)	–	0.2	–	0.2	0.2	–	0.2
Ester content—min (% w)	–	96.5	96.5	96.5	96.5	–	96.5
Monoglyceride—max (% w)	–	0.8	–	0.8	–	–	0.8

(continued)

Table 45.15 (continued)

Specification	China	India	Indonesia	Japan	Korea	The Philippines	Thailand
Diglyceride—max (% w)	–	0.2	–	0.2	–	–	0.2
Triglyceride—max (% w)	–	0.2	–	0.2	–	–	0.2
Freeglyceride—max (% w)	0.02	0.02	0.02	0.02	–	0.02	0.02
Total glyceride—max (% w)	0.24	0.25	0.24	0.25	0.24	0.24	0.25
Iodine number—max (g I_2/100 g)	–	120	115	120	–	–	120
Phosphorous—max (mg kg^{-1})	10	10	10	10	10	10	10
Group I (Na + K)—max (mg kg^{-1})	–	5	18	5	5	–	5
Group II (Ca + Mg)—max (mg kg^{-1})	–	–	–	5	5	–	5
Cold Soak Filterability—max (s)	–	–	–	–	–	–	–
Cloud point	–	–	–	–	–	–	–
CFPP	–	–	–	–3	–	–	–

The Canadian Renewable Fuels Association and the National Biodiesel Board developed a quality standard BQ 9000.

The problem faced by automobiles due to the long usage of biodiesel is given in Tables 45.17 and 45.18.

The standards formulated by many institutions are highly ephemeral in nature as they face constant revision in accordance with the revelations due to continued research and development. As there is no definite raw material for biodiesel, uniform standard cannot be developed. Besides any standard has to be viewed in the context of standing environmental conditions, automobile profile of the concerned country and its geographical locations. The quality control, monitoring mechanism and the analytical precision also govern the implementation of the standard. The resolve of the enforcement authorities also plays a major role in maintaining the standard.

Table 45.16 The standards being followed in Australia, Brazil, New Zealand and South Africa for biodiesel

Specification	Australia	Brazil	New Zealand	South Africa
Density at 15 °C (kg m^{-3})	860–890	–	860–900	860–900
Viscosity at 40 °C (mm^2 S^{-1})	3.5–5.0	–	2.0–4.5	3.5–5.0
Distillation—max 90 (Recor) (°C)	360	360	–	–
Flash point—min (°C)	120	100	100	120
Sulphur (mg kg^{-1}—max)	10	–	10	10
Carbon residue in 10% dist—max (% w/w)	0.3	–	0.3	0.3
Carb res. in 100% dist—residue max (% w/w)	–	0.1	0.05	–
Sulphated ash—max (% w/w)	0.02	0.02	–	0.02
Water and sediment (% vol)	0.05	0.05	–	0.05
Total contamination—max (mg kg^{-1})	24	–	24	24
Cu strip corrosion—max class 3 h at 50 °C	1	1	1	1
Oxidation stability—min h at 110 °C	6	6	6	6
Cetane number—min	51	–	51	51
Linolenic acid methyl ester—max (% w)	–	–	12	12
Polyunsaturated $\leq$4—max (mg kg^{-1})	–	–	–	1
Acid value—max (mg KOH kg^{-1})	0.8	0.8	0.5	0.5
Methanol—max (% w)	0.2	–	0.2	0.2
Ester content—min (% w)	96.5	–	96.5	96.5
Monoglyceride—max (% w)	–	–	0.8	0.8
Diglyceride—max (% w)	–	–	–	0.2
Triglyceride—max (% w)	–	–	–	0.2
Freeglyceride—max (% w)	0.02	0.02	0.02	0.02
Total glyceride—max (% w)	0.25	0.38	0.24	0.25
Iodine number—max (g I_2/100 g)	–	120	120	140
Phosphorous—max (mg kg^{-1})	10	0.10	10	10
Group I (Na + K)—max (mg kg^{-1})	5	10	5	5
Group II (Ca + Mg)—max (mg kg^{-1})	5	–	5	5
Cold Soak Filterability—max (s)	–	–	–	–
Cloud point	–	–	–	–
CFPP	–	–	–	–

Table 45.17 Biodiesel parameters responsible for creating trouble in automobile engines

Troubles	Warning properties
Fuel line damage, corrosion and swelling of rubber	Acid value, methanol oxidation stability, ester, water
Filter clogging and impairment of fuel pump	Oxidation stability, polyunsaturated methyl ester, triglyceride, monoglyceride, diglyceride, glycerin, hard particles, water and metal
Emission problems	Triglyceride, metal and phosphorus
Starting trouble in cold condition	Low temperature performance

Table 45.18 Occurrence of certain problems in biodiesel applications in automobile

System	Problems	Occurrence (%)
Fuel supply system	Deterioration or swelling of elastomer	17
	Fuel leakage from elastomer parts	7
	Fuel filter clogging	44
	Fuel supply pump clogging	18
Engine	Deterioration of engine startup	20
	Instability of engine speed	16
	Engine seizing	3
	Other miscellaneous problems	11

References

Allen CAW, Watts KC (2000) Comparative analysis of the atomization characteristics of fifteen biodiesel fuel types. Trans Am Soc Agric Eng 43:207–211

Anonymous (1995) Proceedings of the international conference on standardization and analysis of biodiesel, Vienna, 6–7 Nov 1996. Research Institute for Chemistry and Technology of Petroleum Products (FICHTE), Vienna. ISBN: 3 9014 5701 1

Karmakar R, Krishnendu K, Rajor A (2018) Fuel properties and emission characteristics of biodiesel produced from unused algae grown in India. Pet Sci 15:385–395

Knothe G (2006) Analyzing biodiesel: standards and other methods. JAOCS 83(10):823–833

Knothe G (2009) Improving biodiesel fuel properties by modifying fatty ester composition. Energy Environ Sci 2:759–766

Knothe G, Steidley KR (2005) Kinematic viscosity of biodiesel fuel components and related compounds. Influence of compound structure and comparison to petrodiesel fuel components. Fuel 84:1059–1065

Knothe G, Matheaus A, Ryan TW III (2003) Cetane numbers of branched and straight chain fatty esters determined in an ignition quality tester. Fuel 82:971–975

Knothe G, Matheaus A, Krahl J (2005) The biodiesel handbook. AOCS Press, Champaign, IL

Komers K, Skopal F, Stloukal (1997) Determination of the neutralisation number for biodiesel fuel production. Fett-Lipid 99:52–54

Lang X, Dalai AK, Bakhshi NN, Reaney MJ, Hertz PB (2001) Preparation and characterization of biodiesels from various biooils. Bioresour Technol 80:53–62

Ma F, Hanna MA (1999) Biodiesel production: a review. Bioresour Technol 70:1–15

Prankl H, Worgetter M (1999) Standardisation of biodiesel on a European level. Third European Motor Biofuels Forum, 10–13 Oct 1999, Brussels

Reaney MJT, Hertz PB, McCalley WW (2005) Vegetable oils as biodiesel. In: Shahidi F (ed) Bailey's industrial oil and fat products, 6th edn. John Wiley & Sons Inc., Hoboken, NJ, pp 223–225

Index

S. P. Raj et al., *Biodiesel from Flowering Plants*,
https://doi.org/10.1007/978-981-16-4775-8